Analysis of Structures on Elastic Foundations

Edward Tsudik

Copyright © 2013 by J. Ross Publishing

ISBN-13: 978-1-60427-074-7

Printed and bound in the U.S.A. Printed on acid-free paper.

10 9 8 7 6 5 4 3 2 1

Library of Congress Cataloging-in-Publication Data

Tsudik, Edward, 1933–
    Analysis of structures on elastic foundations / by Edward Tsudik.
       pages cm
    Includes bibliographical references and index.
    ISBN 978-1-60427-074-7 (hardcover : alk. paper) 1. Soil-structure
interaction. 2. Foundations. 3. Elastic analysis (Engineering) I. Title.

    TA711.5.T78 2012
    624.1'5—dc23

                              2012019966

Phone: (954) 727-9333
Fax: (561) 892-0700
Web: www.jrosspub.com

Content

To Your Memory, to You, INNA

Preface

Analysis of beams and plates supported on elastic foundation, in most cases, is related to the design and analysis of various foundations such as individual footings, combined and mat foundations. Although methods of analysis of beams and plates on elastic foundation have been developed for a long time, up-to-date, practical application of these methods is a difficult problem. Analytical methods such as the method of initial parameters (Pouzirevsky, Dutov, Umansky, Kiselev, and others) and the method of superposition (Hetenyi) based on Winkler's soil model are complex and cannot be used by practicing engineers. Analysis based on the soil model with two coefficients of subgrade reactions (Filonenko-Borodich, Pasternak) produces results close to those obtained from analysis based on the Winkler foundation and are never used in practical applications. Analysis based on other soil models such as elastic half-space and elastic layer (Shechter, Gorbunov-Posadov, and Zhemochkin) is even more complex, although, in many cases, it reflects much better the actual behavior of the soil compared to Winkler foundation, and produces more accurate and reliable results. Most publications in the area of soil-foundation interaction are written by scientists and for scientists. These publications may be of a high scientific value, but they are not written for practicing engineers and cannot be used by practicing engineers. A practicing engineer is always in need of a method for hand or computer analysis that is easy to use and produces logical and reasonably accurate results. However, not too many publications and methods meet these requirements. Nevertheless, there are some good publications that allow analyzing various beams and plates on elastic foundation using tabulated solutions and simple equations. These methods are developed for Winkler's soil model as well as for elastic half-space and elastic layer. They are developed for simple, free supported beams, but, as shown by the author of this book, they can also be used for analysis of complex beams such as stepped beams, beams with various boundary conditions, and others. These methods are not simplified approximate methods. They are based on accurate analytical methods of analysis and can be used not only for preliminary calculations, but for final analysis as well. It is clear that large and much more complex structures such as frames, continuous and interconnected beams, mat foundations, and other large structures cannot be analyzed by hand. They require good, professional computer software based on analytical or numerical methods. However, such methods are not always available. Most of the available computer software uses Winkler's soil model. Practical methods of analysis of beams and plates on elastic half-space and elastic layer are not developed at all.

All structures such as buildings, bridges, dams, and many others are structures supported on soil and work as structures supported on elastic foundation. Loads applied to these structures and transferred to the soil produce soil reactions and soil settlements.

Settlements of the soil, in turn, affect the structural system producing additional deflections and stresses in the superstructure. So, the soil, foundation, and the superstructure work together as one whole system. However, up-to-date analyses of foundations and superstructure are performed separately. Analysis of the superstructure does not take into account deflections of the foundation, and analysis of the foundation does not take into account the rigidity of the superstructure. The primary purpose of this book is to provide the civil/structural engineer with simple methods for hand and computer analysis of various structures supported on elastic foundation including beams, plates, frames, and walls using the three most popular soil models: Winkler foundation, elastic half-space, and elastic layer.

This book includes tables and simple equations for hand analysis of various beams on elastic foundation. It also contains a detailed explanation of analytical and numerical methods for computer analysis of various beams and plates. It includes methods proposed by the author of this work for combined analysis of the total system Soil-Foundation-Superstructure as one structural system. This book contains ten chapters.

Chapter 1 provides a basic description of the soil models such as Winkler foundation, elastic half-space, and elastic layer. It includes practical recommendations for obtaining the soil properties and soil settlements. It shows advantages and disadvantages of each of the soil models and the use of these models in practical design. Basic equations for obtaining the modulus of subgrade reaction for elastic foundations as well as for totally rigid foundations are recommended.

Chapter 1 also includes equations for obtaining settlements and rotations of totally rigid foundations supported on elastic half-space and elastic layer, and also tables proposed by Giroud and Gorbunov-Posadov.

Chapter 2 is devoted to analysis of beams on Winkler foundation. It includes basic simple equations for the analysis of infinite and semi-infinite beams. Chapter 2 also includes a detailed explanation of two classical analytical methods of analysis of finite beams: method of superposition and method of initial parameters. Numerical examples illustrate applications of both methods to practical analysis. Using the method of initial parameters, the author obtained equations for beams with various boundary conditions and used these equations for analysis of statically indeterminate systems such as continuous beams on Winkler foundation and frames with continuous foundations. The author also obtained equations for beams with totally rigid elements at their ends and used them for analysis of frames on Winkler foundation taking into account the totally rigid elements located at points of intersection of the frame columns and continuous foundations. Numerical examples performed by the author show that when taking into account the totally rigid elements, results of analysis can be significantly specified. The same equations are applied to analysis of a system of interconnected beams taking into account the totally rigid elements at the area of beam intersection and the torsional rigidities of the beams. Numerical examples performed by the author proved that this exact specified analysis can reduce bending moments 2 to 2.5 times. Some of the material in Chapter 2 is devoted to analysis of piles loaded with lateral and vertical loads and moments applied to the heads of the piles.

Chapter 3 is devoted to hand analysis of beams supported on Winkler foundation. It includes equations and tables developed by Klepikov. Tables and equations are developed for analysis of free supported beams loaded with concentrated vertical loads or moments. Beams of any length and flexural rigidity can be analyzed using tables and equations presented in this chapter. Numerical examples confirm the simplicity and accuracy of the method, and show that results of analysis are the same as those obtained by other authors using analytical or numerical exact methods that require much more time and work.

The same equations and tables, as shown by the author of this book, can be used not only for analysis of free supported beams, but also for analysis of complex beams such as beams with various boundary conditions, stepped beams, pin-connected beams, continuous beams, and simple interconnected beams. Chapter 3 also includes a discussion of another simple method for analysis of beams and frames on Winkler foundation by replacing the soil with a series of elastic supports, and using any computer program for analysis of regular statically indeterminate systems. This method of analysis does not require any new software. Any structural system made of line elastic elements and supported on Winkler foundation can be analyzed. Chapter 3 also presents a proposed method of combined analysis of 2D and 3D frames with individual foundations. The soil under the foundation is replaced with a simple beam with both fixed ends or a column with rigidities equivalent to the rigidities of the given soil and foundation.

Chapter 4 describes analysis of beams on elastic half-space developed by Gorbunov-Posadov. All beams are divided into three categories: rigid beams, short beams, and long beams. Tables developed for these three categories of beams allow performing analysis relatively easy. But tables and formulae are limited only to analysis of beams free supported on elastic half-space. Chapter 4 includes a method proposed by the author of this book for analyzing beams with various boundary conditions using the same tables and formulae. The same tables are used for analysis of continuous beams. Several numerical examples illustrate the application of the tables and equations presented in Chapter 4.

Chapter 5 introduces a method of analysis of beams supported on elastic half-space that works under plane strain conditions. The method was developed by Simvulidi and is used for the analysis of long narrow plates in the short (width) direction by cutting off bands from the plate and analzing them as beams on elastic half-space. Simvulidi obtained formulae for beams loaded with concentrated loads, moments, and uniformly distributed loads. Tables and simple formulae allow performing fast and easy analysis. Although the tables and formulae are convenient, they still have limited application since they are used only for plane strain analysis. Several numerical examples solved by the author demonstrate the application of the method to solutions of practical engineering problems. Chapter 5 also includes simple equations and tables for obtaining rotations and settlements of beams supported on elastic half-space. Equations and tables allow analyzing not only beams, but also various statically indeterminate systems such as pin-connected beams, stepped beams, and frames supported on elastic half-space. However, it is important to note that the method does not take into account the influence of the loads applied to the half-space by closely located foundations.

Chapter 6 introduces a numerical method of analysis of beams and 2D as well as 3D frames on elastic half-space. The method was developed by Zhemochkin only for analysis of free supported beams loaded with concentrated vertical loads. Practical limitations of the method can be understood by taking into account that the method was developed a long time ago, when practicing engineers did not use computers, and hand analysis, especially solutions of large systems of linear equations, was simply impossible. However, even today, when computer analysis is a routine procedure, Zhemochkin's method can be successfully used and further developed not only for analysis of simple beams, but also for analysis of more complex problems. In Chapter 6, the method is applied to analysis of beams with various loads including beams with various boundary conditions. The method is applied to analysis of stepped and continuous beams as well as to combined analysis of 2D frames with continuous foundations. It is also used for analysis of interconnected beams supported on elastic half-space.

Chapter 7 describes simplified methods of analysis of mat foundations including infinite mats supported on elastic half-space, elastic layer, and Winkler foundation. Shechter obtained an analytical solution for infinite plates supported on elastic half-space and elastic layer. He also developed tables and equations for practical applications. As shown by Shechter, when the height of the elastic layer is small, results of analysis are practically the same as results obtained from analysis of plates on Winkler foundation. A series of numerical examples presented in Chapter 7 illustrate application of the tables and formulae to analysis of infinite plates. Solutions obtained by Shechter can be used for analysis of finite plates by applying to the edges of the plate loads or deflections that satisfy the actual boundary conditions. Chapter 7 also presents an approximate method of analysis of rectangular finite plates loaded with a series of concentrated vertical loads developed by Gorbunov-Posadov. Numerical examples illustrate application of the method to practical analysis.

Chapter 8 describes a numerical method of analysis of mat foundations supported on Winkler foundation and elastic half-space using the finite element method. Analysis of mat foundations on Winkler foundation is well known, while analysis of finite plates on elastic half-space developed by the author of this book is new. Analysis of mat foundations on elastic half-space is performed by equating settlements of the mat at all finite element nodes to the settlements of the elastic half-space under the same nodes. Two groups of equations are written: one shows the reactions of the plate at each node due to the plate settlements and given loads applied to the plate. The second group shows the settlements of the half-space as functions of the loads applied to the half-space. The total number of unknowns is equal to $2n$ (n reactions plus n settlements). By solving two groups of equations all soil reactions and all settlements are found. Chapter 8 also includes a method of combined analysis of mat foundations and superstructure (frames and walls) as one structural system. A part of Chapter 8 is devoted to analysis of mat foundations supported on piles. It includes mats totally supported on piles and mats partially supported on soil and partially supported on piles. The method is used for two soil models: Winkler foundation and elastic half-space.

Chapter 9 is devoted to analysis of circular and ring foundations. This chapter includes simplified formulae obtained by Beyer and also formulae and tables for analysis of circular plates supported on elastic half-space obtained by Gorbunov-Posadov. All tables and formulae are developed for symmetrically loaded plates. Some numerical examples illustrate the use of the tables and formulae. A numerical method of analysis proposed in Chapter 9 by the author uses the method of finite elements, developed for two soil models: Winkler foundation and elastic half-space. Analysis, in principle, is not different from analysis of rectangular plates. However, the plate is divided into a system of triangular finite elements and the pressure is applied to the soil as a series of circular uniformly distributed loads. Narrow ring foundations are replaced with polygons, and all sides of the polygon are replaced with line elements supported on soil. Any computer program for analysis of 3D statically indeterminate systems can be used for analysis of such types of foundations. Chapter 9 also includes an analytical method of analysis of circular plates loaded with non-symmetrical vertical loads. The method is developed for plates free supported on ring foundations and for plates restrained against any deflections along their perimeter. The method is developed for two soil models: Winkler foundation and elastic half-space.

Chapter 10 is devoted to analysis of some special structures supported on elastic foundation. This chapter includes analysis of composite beams and plates, analysis of walls with continuous foundations, walls supported on frames with individual and continuous foundations, analysis of boxed foundations, including combined analysis of boxed foundations, and 3D frames.

It is important to mention that all methods discussed in this book are recommended for hand as well as for computer analyses. Some of the methods allow obtaining fast and easy solutions of many practical problems. For example, analysis of free supported beams described in Chapters 3 through 5, analysis of beams with various boundary conditions, and others. However, analysis of large continuous beams can be performed using only computer analysis and special software based on methods described in the book. Analysis of many complex problems such as numerical analysis of mat foundations supported on elastic half-space also requires special software. In some cases, complex problems can be solved by using existing well-known computer programs. For example, analysis of 2D and 3D frames with individual foundations supported on Winkler foundation or on elastic half-space can be performed using a standard program for analysis of 3D statically indeterminate systems. The same standard computer programs can be used for analysis of continuous beams, interconnected beams, stepped beams, and various other structures supported on Winkler foundation. It is also important to remember that many problems, such as analysis of composite beams and plates on Winkler foundation, require the use of only one standard program for analysis of regular continuous beams or plates. Such types of programs are widely available. In conclusion, it is also important to note that the content of this book is limited by static analysis only. This book includes mostly foundations and structures that a practicing civil and structural engineer is dealing with.

This book includes formulae and tables developed by some well-known scientists (S. N. Klepikov, O. Y. Shechter, M. I. Gorbunov-Posadov, and I. A. Simvulidi) and detailed descriptions of two classical methods of analysis of finite beams on Winkler foundation—the method of initial parameters and the Hetenyi method. It also includes a description of the Zhemochkin method for analysis of beams on elastic half-space as well as methods of analysis proposed and developed by the author.

The author wishes to thank Professor Samuel Aroni for reviewing the work and making many useful recommendations, Dr. Mark Buhkbinder, P.E., for his helpful criticism and recommended corrections, and Robert Mayer, S.E., for his help and many practical suggestions. The author also wishes to thank all the people at J. Ross Publishing, especially Stephen Buda for his help and cooperation.

Finally, it should be noted that while every care has been taken to avoid errors, it is hard to imagine that all errors in the book were detected. The author would be very grateful for any corrections and suggestions made by the reader.

About the Author

Dr. Edward Tsudik is a licensed Professional Civil Engineer with extensive practical and research experience in structural engineering and structural mechanics. He holds two MS degrees in Hydraulic and Civil Engineering and a PhD in Structural Mechanics from MADI, Department of Bridges and Underground Structures, an internationally recognized engineering school in Moscow. His professional experience includes many years of engineering work combined with teaching and research in the areas of earthquake and foundation engineering. He managed the design and construction of various buildings and other structures, mostly in heavy earthquake zones. He has worked and collaborated with some well-known scientists such as A. A. Umansky and V. A. Kiselev (cofounders of the method of initial parameters), A. P. Sinitsin (coauthor of the well-known Zhemochkin's method for analysis of beams on elastic half-space), O. Y. Shechter, M. I. Gorbunov-Posadov, and others. In 1981, he immigrated to the United States and worked as a structural engineer for Pullman and Kellogg Inc. and Foster-Weller Energy Corporation in Houston, Texas.

Dr. Tsudik is the author of numerous publications including two monographs. His professional interests lie in the areas of structures on elastic foundation and soil-foundation-structure interaction. Material presented in this book reflects his long professional experience in this field. Dr. Tsudik is a member of the American Society of Civil Engineers and the Structural Engineers Association of California. He resides in Santa Monica, California, and can be contacted via e-mail at etsudik@gmail.com.

1

Soils and Soil Models

1.1 Introduction

Analysis of structures and structural elements, supported on soil, requires the knowledge of the properties of the structure, as well as the properties of the soil. While properties of structural materials of the foundation and superstructure are usually well known, obtaining the soil's properties and, especially, evaluating the soil's behavior under applied loads, is very difficult. Various soils react differently to applied loads and, like any bearing material, produce under the same loads different settlements and different stresses. A wide variety of soil and its properties as well as various complex soil conditions require soil investigation and testing in each case. Analysis of beams, plates, walls, frames, and other structures supported on elastic foundations is usually performed by modeling the soil; in other words, by replacing the soil with material that behaves under applied loads like the given real soil. The most popular soil models used by practicing engineers are: Winkler's soil model or Winkler foundation proposed by Winkler (1867) and later used by many scientists like Hetenyi (1946), Umansky (1933), and others; elastic half-space or elastic continuum proposed by Wieghardt (1922), Shechter (1939), Gorbunov-Posadov (1941, 1949), Harr (1966); and elastic layer, developed by Shechter (1948), Giroud (1968, 1972), and Poulos (1967). Some scientists trying to improve the soil models mentioned above recommended the use of new soil models. For example, Pasternak (1954) proposed a soil model with two coefficients of subgrade reaction, and Reissner (1958) recommended a soil model that simplified analysis of foundations supported on elastic half-space. These and other soil models are not discussed in this work. The reader can find a detailed review and description of various soil models in the book written by Selvadurai (1979). Methods of analysis described in this book are developed only for three soil models: Winkler foundation, elastic half-space, and elastic layer.

All tables can be found at the end of the chapter.

1.2 Winkler Foundation

Winkler foundation is based on the following relationship between the pressure applied to the soil and the soil settlement:

$$y = \frac{p}{k} \qquad (1.1)$$

Equation 1.1 is the basic equation for analysis of structures supported on Winkler foundation. In this equation, y is the settlement of the soil, usually in centimeters or in inches, p is the load applied to one square unit of the soil area, usually in $\frac{\text{kip}}{\text{in}^2}$ or $\frac{\text{kg}}{\text{cm}^2}$, and k is the modulus of subgrade reaction in $\frac{\text{kip}}{\text{in}^3}$ or $\frac{\text{kg}}{\text{cm}^3}$.

The modulus of subgrade reaction represents a load that, being applied to one square unit of the soil surface, produces a settlement equal to one unit, and is the only parameter needed to obtain the settlement of the soil. Winkler foundation is based on the following three assumptions:

1. The load applied to the soil surface produces settlements of the soil only under the applied load and does not produce any settlements and stresses outside of the loaded area.
2. The soil can resist compression as well as tension stresses.
3. The shape and size of the foundation do not affect the settlement of the soil.

These assumptions are not always accurate because it is well known that, in many cases, a load applied to the soil produces settlements not only under the applied load, but also outside the loaded area. It is also well known that soil does not resist any, even small, tension stresses. And finally, the settlement of the soil depends not only on the type of soil but also on the shape and size of the foundation. However, numerous experimental and theoretical investigations as shown by many scientists, for example, Klepikov (1967), proved that analysis based on Winkler foundation produces realistic results that are practically close enough to results obtained from soil testing and observations of settlements of real structures. Regarding the assumptions mentioned above, it is important to note that most of the foundation settlements are produced by loads applied directly to the loaded areas, especially for soft soils. Loads applied to the soil outside of the loaded area may affect these settlements only for soils such as hard clay, limestone, rock, and other similar soils that, to a certain degree, follow the rules of the theory of elasticity. Tension stresses between the soil and foundation occur in some cases, not only when Winkler's soil model is used. They occur regardless of what kind of soil model is used. In addition, it is important to remember that in most of the cases, the weight of the structure is very high and tension stresses practically do not take place. Obtaining the modulus of subgrade reaction is a difficult problem. Numerous methods are used in practical applications. Korenev (1962) recommends the use of Table 1.1 developed for various types of soils shown below. However, Table 1.1 can be used only for preliminary calculations.

The modulus of subgrade reaction can be obtained with a higher level of accuracy from testing the soil with a 12 in. square or round plate and plotting a curve p versus y shown in Figure 1.1.

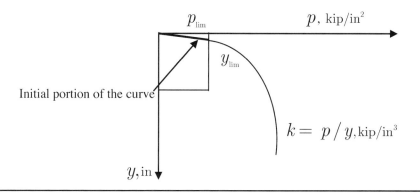

Figure 1.1

In order to extrapolate the test results to a real foundation, Terzaghi (1955) proposed some empirical equations. For example, for clayey soils when the contact pressure is less than one-half of the ultimate bearing capacity, the equation looks as follows:

$$k_S/k_P = B_P/B \tag{1.2}$$

where k_p is the plate load value of the modulus of subgrade reaction using a plate dimension of B_p, and k_S is the value to use under the actual foundation of width B. Terzaghi (1955) proposed for cohesionless soils the use of the following equation:

$$k_S = k_P \left(\frac{B+1}{2B} \right)^2 \tag{1.3}$$

The modulus of subgrade reaction for a rectangular plate of dimensions B and $L = mB$ can be obtained from the soil test performed by a square plate from equation 1.4:

$$k_S = k_P \left(\frac{m+0.5}{1.5} \right) \tag{1.4}$$

Vesic (1961) proposed to find the modulus of subgrade reaction from equation 1.5 using the modulus of elasticity of the soil obtained from a triaxial test.

$$k_S = \frac{0.65}{B} \sqrt[12]{\frac{E_S B^4}{E_b I}} \cdot \frac{E_S}{1 - v^2} \tag{1.5}$$

In equation 1.5, B is the width of the foundation, E_b is the modulus of elasticity of the foundation, E_S is the modulus of elasticity of the soil, I is the moment of inertia of the foundation, and v is the Poisson ratio. Extrapolating the results of the plate load test for the real foundation, Vesic obtained the following equation:

$$k_S = \frac{0.52}{B} \sqrt[12]{\frac{k' B^4}{E_b I}} k' \tag{1.6}$$

where $k' = \dfrac{E_S}{1 - v^2}$

Approximate values of the modulus of elasticity for various soils included in the equations shown above can be obtained from Table 1.2. However, these values cannot be recommended for final analysis, especially for analysis of foundations of complex structures. The modulus of elasticity for final analysis is usually obtained from field soil testing for each job site. The modulus of elasticity of the soil for final analysis is obtained from the same curve shown in Figure 1.1 from the following equation:

$$E = \left(1 - v^2\right)\frac{P}{y_{lim}\, d} \tag{1.7}$$

In equation 1.7, P is the total load applied to the plate taken from the end of the initial portion of the curve, y_{lim} is the settlement of the plate due to load P, and d is the diameter of the plate. If the test is performed with a square plate, the area of the plate (a^2) is replaced with an equivalent area of a circular plate $\left(\dfrac{\pi d^2}{4}\right)$ and diameter d is obtained as $d = a\sqrt{\dfrac{4}{\pi}}$ Equations 1.3–1.6 are developed only for rectangular foundations and cannot be used for foundations of a different shape. Nevertheless, the modulus of subgrade reaction for foundations of different shapes can be found from equation 1.8 shown here:

$$k = \frac{E}{\omega\sqrt{A}\left(1 - v^2\right)} \tag{1.8}$$

where ω is the shape coefficient that is taken from Table 1.3 and A is the area of the foundation. As can be seen, equation 1.8 takes into account the area and shape of the foundation. It also shows that the larger the area of the foundation, the smaller the modulus of subgrade reaction. In other words, equation 1.8 shows that the modulus of subgrade reaction depends not only on the type of soil and applied loads; it depends on the shape and dimensions of the foundation. The reader may find other methods of obtaining the modulus of subgrade reaction recommended by other authors in various publications.

1.3 Modulus of Subgrade Reaction for Individual Foundations

Formulae presented above are developed for analysis of elastic foundations experiencing vertical deflections. The modulus of subgrade reaction is the only soil parameter needed for analyzing elastic beams and plates supported on Winkler foundation. Unlike beams and plates, totally rigid individual foundations, in general, experience rotations and horizontal deflections that, in accordance with Barkan (1962), may be very large. Figure 1.2 shows an individual foundation with three loads: vertical concentrated load N_z applied to the center of the foundation, moment M_x, and a horizontal load N_x applied at the bottom of the foundation. Deflections of the foundation are shown in Figure 1.3.

Deflections of individual footings supported on Winkler foundation can be obtained from the following well-known equations:

$$\Delta_z = \frac{N_z}{K_z A} \quad \varphi_x = \frac{M_x}{K_\varphi I} \quad \Delta_x = \frac{N_x}{K_x A} \tag{1.9}$$

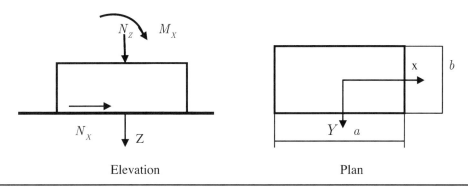

Elevation   Plan

Figure 1.2

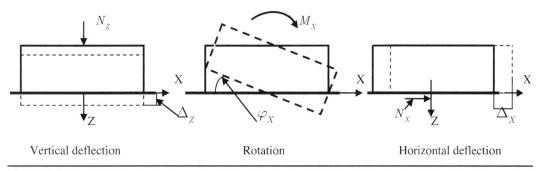

Vertical deflection   Rotation   Horizontal deflection

Figure 1.3

Trying to connect these simple formulae with deflections of an elastic stamp supported on elastic half-space, Klepikov (1967) obtained the following formulae:

$$\Delta_z = \frac{N_z \sqrt{A}\left(1 - v^2\right)}{\omega_z EA} \quad \varphi_x = \frac{M_x \sqrt{A}\left(1 - v^2\right)}{\omega_\varphi EI} \quad \Delta_x = \frac{N_x \sqrt{A}\left(1 - v\right)\left(1 - v\omega_x\right)}{\omega_z EA} \quad (1.10)$$

In formulae 1.9 and 1.10, E is the modulus of elasticity of the soil, v is the Poisson ratio, A is the area of the foundation, and I is the moment of inertia of the foundation about the axis Y.

Dimensionless coefficients, ω_z, ω_φ and ω_x are found depending on a/b from Table 1.4. The vertical load N_z applied to the foundation will produce settlement y and the modulus of subgrade reaction is found as follows:

$$K_z = \frac{N_z}{Ay} \qquad (1.11)$$

Using the first equation from 1.9 and equation 1.11, we obtain the modulus of subgrade reaction of the soil loaded with a vertical load applied to the center of the foundation:

$$K_z = \frac{\omega_z E}{\sqrt{A}\left(1 - v^2\right)} \qquad (1.12)$$

If the foundation is loaded with a moment M, the modulus of subgrade reaction is obtained as follows:

$$K_\varphi = \frac{\omega_\varphi E}{\sqrt{A}\left(1 - v^2\right)} \tag{1.13}$$

If the foundation is loaded with a horizontal load N_X, the friction stress is equal to $N_X/A = p_x = K_x \cdot \Delta$ and the modulus of subgrade reaction is equal to $K_x = N_X/A\Delta$, or using the third equation from 1.10, we have:

$$K_x = \frac{\omega_z E}{\sqrt{A}\left(1 - v\omega_x\right)\left(1 - v\right)} \tag{1.14}$$

As we can see, analysis of a totally rigid individual foundation requires not one modulus of subgrade reaction, but three modules: K_z, K_φ, and K_x.

Comparing equations 1.12–1.14, we can see that the modulus of subgrade reaction depends not only on the soil's properties, but also on the type of the load applied to the foundation. Each modulus also depends on the shape of the foundation. From equations 1.12 and 1.13, we have $K_z/K_\varphi = \omega_z/\omega_\varphi$ that simplifies obtaining the modulus of subgrade reaction K_φ.

$$K_\varphi = \frac{K_z \omega_\varphi}{\omega_z} \tag{1.15}$$

Equations 1.11–1.15 can be used only when the area of the foundation is less than $10m^2$. When the area of the foundation is more than $10m^2$, the modules of subgrade reaction $K_z\,K_\varphi\,K_X$ are found assuming that the area of the foundation is equal to $10m^2$. In this case, as recommended by Barkan (1962), it can be assumed that $K_\varphi = 2K_z$ and $K_X = 0.7\,K_z$.

The actual behavior of individual foundations supported on Winkler foundation and elastic half-space were also studied by other authors such as Weissmann (1961, 1965, 1972).

It is important to mention that foundation settlements obtained from equations shown above do not represent the real foundation settlements. The real total expected foundation settlement is obtained from the following equation:

$$S = S_d + S_c + S_s \tag{1.16}$$

in which S_d is the immediate or distortion settlement, S_c is the consolidation settlement, and S_s is the secondary settlement. The immediate settlement occurs concomitantly with the load application, primarily as a consequence of distortion within the foundation soils.

The distortion settlement, in reality, is not elastic although it is calculated using the elastic theory. Consolidation settlement occurs when the water is expelled from the void spaces in the soil. The secondary settlement is related to the compression of the soil skeleton. It is obvious that the modulus of subgrade reaction obtained from formulae shown above does not take into account the real settlements of the foundation, and settlements obtained from these formulae are different from the real expected settlements. Nevertheless, these formulae allow obtaining deflections of individual foundations as well as the modules of subgrade reaction. The total expected settlement S can be used for obtaining the modulus of subgrade reaction and analysis of various types of foundations:

$$k = p/S \qquad (1.17)$$

In equation 1.17, p is the average soil pressure under the foundation and S is the expected total settlement of the foundation. As shown above, Winkler foundation is a simple soil model and therefore widely used by practicing engineers and developers of engineering structural software. Many methods of analysis discussed in this book are based on Winkler's soil model. However, Winkler foundation is not immune from some problems. The model assumes that loads applied to the soil produce settlements only immediately at the area where the loads are applied. They do not produce settlements outside of the loaded area. If a group of individual foundations is loaded with various loads, each foundation is working independently. The settlement of each foundation is not affected by the settlements of other foundations, even when they are located very close. These assumptions are not acceptable when foundations are supported on soils such as hard clay, limestone, and other similar types of soils.

1.4 Elastic Half-Space or Elastic Continuum

Loads applied to such types of soils usually produce settlements not only under the foundation, but also outside of the loaded area. Settlement of one of the foundations affects the settlements of other neighboring foundations. In other words, the soil, to a certain degree, behaves as an elastic material following the rules of the theory of elasticity and working as an elastic half-space. It is usually understood that elastic half-space is a homogeneous elastic material infinitely large in horizontal and vertical directions. In order to perform analysis of any foundation supported on elastic half-space, we have to know the relationship between the loads, applied to the half-space, and settlements produced by these loads.

If a concentrated vertical load P is applied to the top of the elastic half-space, as shown in Figure 1.4, the settlement of any point k, located at the top of the same

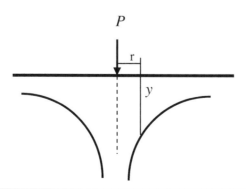

Figure 1.4

half-space at the distance r from that point, in accordance with Boussinesq (1885), can be obtained from the following equation:

$$y = \frac{P(1 - \mu_0^2)}{\pi E_0 r} \tag{1.18}$$

In 1.18, μ_0 is the Poisson ratio of the soil and E_0 is the modulus of elasticity of the soil. As can be seen from 1.18, the settlement under the load P, when $r = 0$, is equal to ∞. However, in reality loads are never applied as concentrated point loads. They are always applied to a certain area of the soil surface. Therefore, it can be assumed that the load P is uniformly distributed over a rectangular area $(b \times c)$ shown in Figure 1.5. Let us find the settlement at point k, due to load P, applied to the area $(b \times c)$. Assuming $P = 1$, we can find the soil reaction per one square unit of the foundation area as equal to:

$$p = \frac{1}{b \times c} \tag{1.19}$$

Now, let us investigate a small area of the bottom of the foundation equal to $d\xi \times d\eta$ with coordinates ξ and η. The total soil reaction applied to this element is equal to:

$$pd\xi \times d\eta = \frac{d\xi \times d\eta}{bc} \tag{1.20}$$

The settlement at point k is equal to:

$$d^2y = \frac{d\xi \times d\eta}{bc} \cdot \frac{(1 - \mu_0^2)}{\pi E_0 r} \tag{1.21}$$

By integrating the expression 1.21 twice, we will find the total settlement at point k, due to the load P applied to the area bc:

$$y = \int_{\xi = x - \frac{c}{2}}^{\xi = x + \frac{c}{2}} 2 \int_{\eta = 0}^{\eta = \frac{b}{2}} \frac{1 - \mu_0^2}{bc\pi E_0 r} d\xi \times d\eta = \frac{2(1 - \mu_0^2)}{bc\pi E_0} \int_{\xi = x - \frac{c}{2}}^{\xi = x + \frac{c}{2}} \int_{\eta = 0}^{\eta = \frac{b}{2}} \frac{d\xi \times d\eta}{\sqrt{\xi^2 + \eta^2}} \tag{1.22}$$

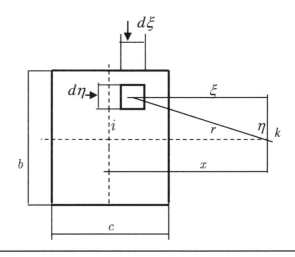

Figure 1.5

After the integration, the following equation is obtained:

$$y = \frac{(1 - \mu_0^2)}{\pi E_0 c} F \qquad (1.23)$$

Where F can be found from 1.24:

$$F = \frac{c}{b} \cdot \left[2\ln\frac{b}{c} - \ln\left(\frac{4x^2}{c^2} - 1\right) - 2\frac{x}{c}\ln\frac{2\frac{x}{c} + 1}{2\frac{x}{c} - 1} + \frac{b}{c}\ln\frac{\left(2\frac{x}{b} + \frac{c}{b}\right) + \sqrt{\left(2\frac{x}{b} + \frac{c}{b}\right)^2 + 1}}{\left(2\frac{x}{b} - \frac{c}{b}\right) + \sqrt{\left(2\frac{x}{b} - \frac{c}{b}\right)^2 + 1}} \right] +$$

$$\frac{c}{b}\left[2\frac{x}{c}\ln\frac{1 + \sqrt{\left(2\frac{x}{b} + \frac{c}{b}\right)^2 + 1}}{1 + \sqrt{\left(2\frac{x}{b} - \frac{c}{b}\right)^2 + 1}} + \ln\left\{1 + \sqrt{\left(2\frac{x}{b} + \frac{c}{b}\right)^2 + 1}\right\} \cdot \left\{1 + \sqrt{\left(2\frac{x}{b} - \frac{c}{b}\right)^2 + 1}\right\} \right]$$

$$(1.24)$$

The settlement of the center of the foundation, when $x = 0$, can be found from equation 1.23, but parameter F in this case should be obtained from equation 1.25 shown below:

$$F = 2\frac{c}{b}\left\{\ln\left(\frac{b}{c}\right) + \frac{b}{c}\ln\left[\frac{c}{b} + \sqrt{\left(\frac{c}{b}\right)^2 + 1}\right] + \ln\left[1 + \sqrt{\left(\frac{c}{b}\right)^2 + 1}\right]\right\} \qquad (1.25)$$

So, if we want to find the settlement at point i, due to the vertical load acting at point k, equations 1.23 and 1.24 are used. When obtaining the settlement at point k due to the load acting at the same point k, equations 1.23 and 1.25 are used. Using 1.23, we can obtain the settlement of the half-space at the center of the applied rectangular load as well as at any distance from that center of the load. In 1.24, x is the distance between the point the load is applied and the point the settlement is obtained, b is the width of the foundation, c is the length of the foundation, and μ_0 is Poisson's ratio of the half-space.

Settlements of an elastic half-space loaded with a circular stamp can be obtained from the same equation 1.23. The equation for a circular stamp looks as shown below. If a uniformly distributed load q is applied to the half-space at point 0, as shown in Figure 1.6, the settlement due to this load at point M is obtained from the following equation 1.26:

$$\Delta_{MO} = \frac{4(1 - \mu^2)q}{\pi E}\left[\int_0^{\frac{\pi}{2}} \sqrt{1 - \left(\frac{a^2}{r^2}\right)\sin^2\theta}\, d\theta - \left(1 - \frac{a^2}{r^2}\right)\int_0^{\frac{\pi}{2}} \frac{d\theta}{\sqrt{1 - \left(\frac{a^2}{r^2}\right)\sin^2\theta}} \right] \qquad (1.26)$$

Solutions of both integrals in formula 1.26 can be found in various mathematical books and handbooks. They look as follows:

$$E = \int_0^{\frac{\pi}{2}} \sqrt{1 - \left(\frac{a^2}{r^2}\right)\sin^2\theta}\, d\theta = \frac{\pi}{2}\left(1 - \frac{1}{2^2}k^2 - \frac{1^2 \cdot 3}{2^2 \cdot 4^2}k^4 - \frac{1 \cdot 3^2 \cdot 5}{2^2 \cdot 4^2 \cdot 6^2}k^6\right) \qquad (1.27)$$

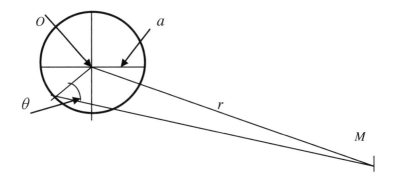

Figure 1.6

$$K = \left(1 - \frac{a^2}{r^2}\right) \int_0^{\frac{\pi}{2}} \frac{d\theta}{\sqrt{1 - \left(\frac{a^2}{r^2}\right)\sin^2\theta}} = \frac{\pi}{2}\left(1 + \frac{1^2}{2^2}k^2 + \frac{1^2 \cdot 3^2}{2^2 \cdot 4^2}k^4 + \frac{1^2 \cdot 3^2 \cdot 5^2}{2^2 \cdot 4^2 \cdot 6^2}k^6\right) \quad (1.28)$$

where $k = \mathrm{Sin}\theta$.

Assuming $r = 0$, we obtain the settlement at the center of the foundation:

$$\Delta_{r=0} = \frac{2(1 - \mu^2)}{E}qa \quad (1.29)$$

Assuming $r = a$, we obtain the settlement at the edge of the foundation:

$$\Delta_{r=a} = \frac{4(1 - \mu^2)}{\pi E}qa \quad (1.30)$$

Equations 1.26–1.30 allow obtaining settlements of circular elastic foundations supported on elastic half-space. As will be shown later, these formulae can be used for analysis of various beams and plates supported on elastic half-space.

Individual foundations usually experience not only settlements but also rotations. Rotations of an individual foundation supported on Winkler's soil model were shown above. Obtaining solutions for an elastic half-space is more difficult. If a rectangular, totally rigid individual foundation, shown in Figure 1.7, is loaded with two moments, m_X and m_Y, applied in directions X and Y, respectively, rotations of the foundation are found from the following two equations obtained by Gorbunov-Posadov (1984):

$$\left.\begin{array}{l} \varphi_X = \dfrac{(1 - \mu_0^2)}{E_0} \cdot \dfrac{K_1 m_X}{a^3} \\[3mm] \varphi_Y = \dfrac{(1 - \mu_0^2)}{E_0} \cdot \dfrac{K_2 m_Y}{b^3} \end{array}\right\} \quad (1.31)$$

In 1.31, K_1 and K_2 are found from Table 1.5. In this Table, $\alpha = a/b$. When the value of α does not coincide with the values given, simple interpolation is used. If $\alpha > 10$, analysis of the foundation is performed as a totally rigid beam supported on elastic

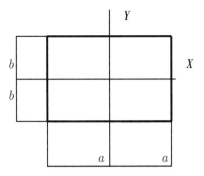

Figure 1.7

half-space. If a circular rigid foundation is loaded with a moment M, the rotation of the foundation is found in equation 1.32, which was obtained by Egorov (1948).

$$tg\alpha = \frac{3(1 - \mu^2)M}{4Er^3} \tag{1.32}$$

where M is the applied moment and r is the radius of the foundation.

It is obvious from the description of Winkler foundation that the soil pressure applied to a totally rigid beam always looks like a straight line. When the same foundation is supported on elastic half-space, the distribution of the soil pressure is more complex. If the average pressure applied to the soil by a circular foundation is equal to p_m, the soil pressure at any point of that foundation can be obtained from the following equation:

$$p_{X,Y} = \frac{p_m}{2\sqrt{1 - \dfrac{\rho^2}{r^2}}} \tag{1.33}$$

In this equation, ρ is the distance from the center of the foundation to any point located under the foundation, and r is the radius of the foundation. It can be seen from 1.33 and Figure 1.8 that when $\rho = 0$, the soil pressure is equal to $p = 0.5 \, p_m$; when $\rho = \frac{r}{2}$, the soil pressure is equal to $p = 0.58 \, p_m$; and when $\rho = r$, $p = \infty$. However, the soil pressure $p = \infty$ never occurs because of the residual plastic strain along the perimeter of the foundation. If the same circular foundation is loaded with an eccentric vertical load, as shown in Figure 1.9, the soil pressure at any point i can be obtained from equation 1.34.

$$p_{i(x,y)} = \frac{3\dfrac{ey}{r^2}}{2\pi r\sqrt{r^2 - x^2 - y^2}} \cdot P \tag{1.34}$$

where e is the eccentricity of the vertical load P applied to the foundation, x and y are the coordinates of the applied load, and r = radius of the circular foundation.

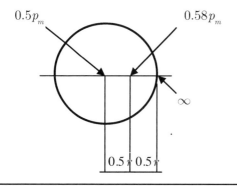

Figure 1.8

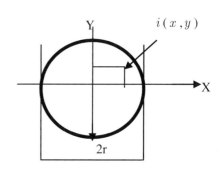

Figure 1.9

Obtaining the soil pressure for a totally rigid rectangular foundation loaded with a vertical load is much more difficult. However, a simplified practical method proposed by Tsytovich (1963) allows finding the soil pressure with a reasonable level of accuracy. If a totally rigid foundation of any shape is supported on an elastic half-space and transfers to the soil a uniformly distributed pressure p, the top of the half-space under the foundation, as shown in Figure 1.10, will experience the same settlement w that can be found from the following equation:

$$w = \frac{1-\mu^2}{E\pi} \int\int_A \frac{p(\zeta,\eta)d\zeta d\eta}{\sqrt{(x-\varepsilon)^2 + (y-\eta)^2}} = \text{Const.} \qquad (1.35)$$

where A is the area of the foundation.

All other parameters in this formula are shown in Figure 1.10. Analysis is performed by dividing the area of the given foundation into a series of small elements and replacing the integral in equation 1.35 with a simple sum as shown below:

$$w = \frac{1}{\pi C} \sum_{\lambda=1}^{\lambda=n} \frac{p_\lambda A_\lambda}{\rho_\lambda(x,y)} \qquad (1.36)$$

where $C = E/(1-\lambda^2)$, p_λ is the unknown average soil pressure on each element, $\rho_\lambda(x,y)$ is the distance from the center of each element to the point the soil pressure has to be found, and A_λ is the area of each element of the foundation.

Equation 1.36 is written for each element of the foundation taking into account that settlements of all elements are the same. If the area of the foundation is divided into n elements, the total number of equations is equal to n. Taking into account the equilibrium of all vertical loads applied to the foundation, the following additional equation is written:

$$\sum_1^n p_\lambda A_\lambda = P \qquad (1.37)$$

In 1.37, P is the resultant of the given vertical loads applied to the center of the foundation. So, the total number of equations is equal to $(n + 1)$. When writing equation

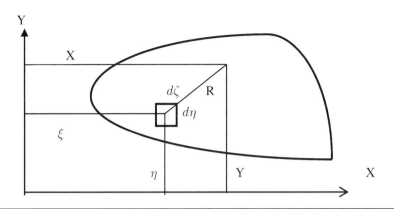

Figure 1.10

1.37, one of the elements will always have $\rho_\lambda = 0$. The settlement of this element can be obtained from the following equation:

$$w_m = \frac{\omega\sqrt{A_\lambda}}{\pi C}p_\lambda \tag{1.38}$$

In 1.38, ω is the coefficient of the shape of the element. This coefficient is equal to 2.97 for a square, 2.95 for a rectangular shape when $a/b = 1{:}1.5$, and 2.89 when $a/b = 1{:}2$.

Example 1.1

A totally rigid foundation, shown in Figures 1.11 and 1.12, is loaded at the center of symmetry (Point 0) with a vertical load $P = 43{,}200$kg and divided into nine elements 40×60 cm. Using equations 1.37 and 1.38, the following equation can be written for point 0:

$$\pi C w_0 = 2.95\sqrt{A_0}\,p_0 + \frac{2A_\lambda}{r_1}p_1 + \frac{4A_\lambda}{r_2}p_2 + \frac{2A_\lambda}{r_3}p_3 \tag{1.39}$$

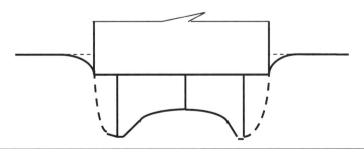

Figure 1.11   The soil pressure diagram

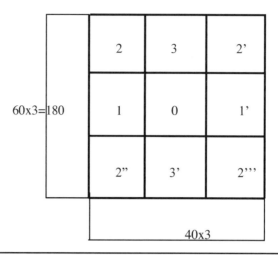

Figure 1.12 Plan of the foundation

From Figure 1.12, we find the distances between point 0 and points 1, 2, and 3 that are equal to $r_1 = 40$ cm, $r_3 = 60$ cm, and $A_0 = 2,400$ cm². Now, introducing the numerical values into equation 1.39 we have:

$$\pi C w_0 = 2.95 \cdot 49 p_0 + \frac{2 \cdot 2,400}{40} p_1 + \frac{4 \cdot 2,400}{72} p_2 + \frac{2 \cdot 2,400}{60} p_3 \qquad (1.40)$$

Analogously similar equations are written for all other points (centers of the elements 1, 2, 3). Finally, we have the following system of four equations:

$$\left.\begin{aligned}
\pi C w_0 &= 145 p_0 + 120 p_1 + 133 p_2 + 80 p_3 \\
\pi C w_0 &= 175 p_1 + 60 p_0 + 128 p_2 + 67 p_3 \\
\pi C w_0 &= 212 p_2 + 64 p_1 + 79 p_3 + 33 p_2 \\
\pi C w_0 &= 165 p_3 + 40 p_0 + 67 p_1 + 158 p_2
\end{aligned}\right\} \qquad (1.41)$$

By simplifying equation 1.40, we find $2,400(p_0 + 2p_1 + 4p_2 + 2p_3) = P = 43,200$ or

$$p_0 + 2p_1 + 4p_2 + 2p_3 = 18 \qquad (1.42)$$

By solving the system (1.41 and 1.42) as one system of five equations, we find $p_0 = 0.86$kg/cm², $p_1 = 1.91$kg/cm², $p_2 = 2.57$kg/cm², $p_3 = 1.49$kg/cm², $\pi C w_0 = 816.5$kg/cm². The settlement of the foundation is:

$$w_{const.} = w_0 = \frac{816.5}{\pi C}\text{cm} \qquad (1.43)$$

Or $p_0 = 0.43 p_m$, $p_1 = 0.95 p_m$, $p_2 = 1.28 p_m$, $p_3 = 0.74 p_m$, where p_m is the average soil pressure equal to $p_m = P/A = 43,200/120 \cdot 180 = 2$kg/cm². The soil pressure diagram and the plan of the foundation are shown in Figures 1.11 and 1.12, respectively.

1.5 Elastic Layer

The elastic layer as a soil model was proposed by Marguerre (1933) and later developed for practical applications by Shechter (1939). The elastic layer takes into account only the upper layer of soil located at the bottom of the foundation, when the soil below this layer does not experience any settlements or when these settlements are negligibly small, and do not have to be taken into account.

Obtaining the height of the elastic layer is a difficult problem, although in some special cases it can be easily obtained. If, for example, a layer of soil is supported on rock, or if a layer of defrosted soil is supported on permafrost, it is obvious that the height of this layer of soil is equal to the height of the elastic layer. But when the foundation is supported on homogeneous soil of unlimited depth or on several layers of various types of soils, obtaining the height of the elastic layer becomes difficult. However, there are some practical recommendations and empirical formulae that allow resolving this problem. Most of the proposed methods for relatively small foundations recommend obtaining the height of the elastic layer as the distance z between the bottom of the foundation and the soil level below, where the following conditions are satisfied:

$$\sigma_z = 0.1 p_z \tag{1.44}$$

$$\sigma_z = 0.2 p_z \tag{1.45}$$

Equations 1.44 and 1.45 are recommended for soft and stiff soils, respectively. In these formulae, σ_z is the pressure produced by the foundation and p_z is the natural vertical soil pressure. Some practical recommendations for obtaining the height of the elastic layer (H) for large foundations are made by various authors. For example, it is recommended to obtain the height of the elastic layer assuming $H = a$ for clayey soils and $H = 2a/3$ for sandy soils, where a is the radius or the width of the foundation. Gorbunov-Posadov and Malikova (1973) recommend finding the height of the elastic layer for homogeneous soils from equation 1.46:

$$H = \zeta' \sqrt[4]{2a} \tag{1.46}$$

and for layered soils from equation 1.47:

$$H = \beta' H_{cl} + (1 - \beta') H_{snd} \tag{1.47}$$

In 1.46, $\zeta' = 6m^{\frac{3}{4}}$ for clayey soils and $\zeta' = 4m^{\frac{3}{4}}$ for sandy soils, $2a$ is the width of the foundation, H_{cl} and H_{snd} are the heights of the clayey and sandy layers, respectively, and β' is the actual percent of clayey soil. For example, if a mat foundation is supported on two layers of clayey and sandy soils and $2a = 20m$, the total height of the elastic layer H is found as follows:

1. $H_{cl} = \zeta' \sqrt[4]{2a} = 6m^{\frac{3}{4}} \cdot \sqrt[4]{20m} = 6m^{\frac{3}{4}} \cdot 2.115m^{\frac{1}{4}} = 12.69m.$

2. $H_{snd} = \zeta' \sqrt[4]{2a} = 4m^{\frac{3}{4}} \cdot \sqrt[4]{20m} = 4m^{\frac{3}{4}} \cdot 2.115m^{\frac{1}{4}} = 8.46m.$

3. $H = \beta' H_{cl} + (1 - \beta') H_{snd} = \dfrac{12.69}{12.69 + 8.46} \cdot 12.69 + (1 - 0.6) \cdot 8.46$

    $= 7.614 + 3.384 = 11m.$

Elastic settlements of circular, rectangular, and strip foundations supported on the elastic layer can be obtained from equation 1.48 recommended by Egorov (1949):

$$S_H = w_0 - w_z \tag{1.48}$$

where w_0 and w_z are settlements of the soil at points 0 (at the bottom of the foundation) and at distance z below the foundation, respectively.

$$w_0 = \frac{2ap(1 - \mu^2)}{E^{(e)}} K_0 \tag{1.49}$$

$$w_z = \frac{2ap(1 - \mu^2)}{E^{(e)}} K_z \tag{1.50}$$

In equations 1.49 and 1.50, p is the pressure applied to the soil, a is half of the width of the rectangular foundation or the radius of the circular foundation, K_z is a dimensionless coefficient that depends on $m = \dfrac{z}{a}$, $n = \dfrac{2b}{2a}$, where b is half of the length of the foundation, and μ is the Poisson ratio.

$K_0 = K_z$, when $z = 0$. $E^{(e)}$ is the modulus of elasticity of the soil obtained from triaxial testing. Using equations 1.48–1.50, we find:

$$S_H = \frac{2ap(1 - \mu^2)}{E^{(e)}} K \tag{1.51}$$

$K = (K_0 - K_z)$ is given in Table 1.6, developed by Egorov (1949), assuming $\mu = 0.3$.
For layered soils the settlement is obtained as follows: $C_i = E_i^{(e)} / (1 - \mu^2)$.

$$S_l = 2ap \sum_{i=1}^{k} \frac{K_i - K_{i-1}}{C_i} \tag{1.52}$$

In 1.52, K_i is a coefficient obtained from Table 1.6 for the layer i, and k is the total number of layers. It is important to note that equation 1.51 is obtained only for settlements of elastic foundations. When the foundation is totally rigid, coefficient K is replaced with K_R and the settlement is found as follows:

$$S_H = \frac{2ap(1 - \mu^2)}{E^{(e)}} K_R \tag{1.53}$$

where K_R for circular and strip foundations is obtained from Table 1.7, assuming that the friction stresses between the elastic layer of soil and the supporting rigid base are equal to zero ($\tau = 0$).

Example 1.2

A rectangular foundation $2a \cdot 2b = 4m \cdot 6m$ is supported on the elastic layer of soil, $H = 10m$, and loaded with a vertical load $P = 300{,}000kg$, $E = 200kg/cm^2$, and $\mu = 0.3$. Find the settlement of the foundation.

Solution:

Find $n = 2b/2a = 6/4 = 1.5$ $m = \dfrac{z}{a} = \dfrac{10}{2} = 5$. From Table 1.7 find the closest $K = 0.759$.

Find the average soil pressure $p = 300,000/400 \cdot 600 = 1.25\text{kg/cm}^2$. Find the settlement of the foundation.

$$S_H = \frac{2ap(1 - \mu^2)}{E^{(e)}}K = \frac{400 \cdot 1.25(1 - 0.3^2)}{200}0.81 = 1.84\text{cm}$$

If the foundation is totally rigid, K is replaced with $K_R = 0.59$ (see Table 1.7) and the settlement is:

$$S_H = \frac{400 \cdot 1.25(1 - 0.3^2)}{200}0.59 = 1.34\text{cm}$$

Similarly, Tables 1.8 and 1.9 for totally rigid rectangular foundations are proposed by Giroud (1972). In accordance with Giroud, the settlement of a rectangular rigid foundation is obtained indifferently by either of the following two equations:

$$\omega = \frac{2ap}{E}P_{Hm} \tag{1.54}$$

$$\omega = \frac{pH}{E}P'_{Hm} \tag{1.55}$$

in which p is the average contact pressure exerted by the foundation on the soil surface, $2a$ is the width of the foundation, H is the thickness of the layer of soil, E is the modulus of elasticity of the soil, P_{Hm} is a dimensionless coefficient depending on μ, $\dfrac{L}{2a}$ and $\dfrac{2a}{H}$, μ is the Poisson ratio of the soil, and L is the length of the rectangular foundation. It is obvious that the same value of ω is obtained either by equations 1.54 or 1.55 and $P'_{Hm} = 2aP_{Hm}/H$. However, in accordance with Giroud (1972), both equations are useful. The first one is more convenient when $H/2a$ is large and the second one when $2a/H$ is large. J. P. Giroud developed Tables 1.8 and 1.9 for coefficients P_{Hm} as well as for coefficients P'_{Hm} for the following Poisson's ratios equal to 0, 0.3, and 0.5.

Example 1.3

Let us find the settlement of a rectangular foundation using equation 1.52 for the following data: $2a = 800\text{cm}$, $L = 1200\text{cm}$, $H = 2000\text{cm}$, $E = 700\text{kg/cm}^2$, $p = 3.3\text{kg/cm}^2$, and $\mu = 0.5$.

Find $L/2a = 12/8 = 1.5$, $H/2a = 20/8 = 2.5$. Using equation 1.52, we have

$$\omega = \frac{2ap}{E}P_{Hm} = \frac{800 \cdot 3.3}{700} \cdot 0.63 = 2.376\text{cm}$$

The same settlement can be obtained using Tables for $P'_{Hm} = 2aP_{Hm}/H = 2 \cdot 400 \cdot 0.63/ 2,000 = 0.252$

$$\omega = \frac{pH}{E}P'_{Hm} = \frac{3.3 \cdot 2,000}{700} \cdot 0.252 = 2.376 cm$$

Individual foundations supported on the elastic layer experience not only settlements but rotations as well. Since an accurate method for obtaining rotations for rectangular foundations supported on the elastic layer has not been developed, it is recommended to use formula 1.31 for an elastic half-space shown earlier. Rotation of a totally rigid circular foundation can be found from the following simple equation:

$$tg\varphi = \frac{1 - \mu^2}{E}k_c \frac{M}{a^3} \tag{1.56}$$

In 1.56, a is the radius of the foundation and k_c is a coefficient obtained from Table 1.10. It is clear from Table 1.10 that rotation of a circular foundation, obtained from equation 1.56, when $H/a < 2$, is smaller than rotation obtained from equation 1.32. When $H/a > 2$, formulae 1.32 and 1.56 produce the same results. As we can see, formula 1.56 can be used for obtaining rotations of a circular foundation supported on elastic half-space as well as on elastic layer. All equations presented above for finding settlements of an elastic layer allow obtaining settlements only of the loaded areas. Finding settlements of the elastic layer at any point outside of the loaded area is more difficult. One of the methods for resolving this problem was developed by Newmark (1942, 1947). He developed influence charts that allow obtaining settlements of the surface of the elastic layer loaded with a uniformly distributed load. Gorbunov-Posadov (1984) recommends the use of the so-called *Method of Corner Points*. However, both methods are not very effective, especially when the number of loads is very large. Using these methods for analysis of mat foundations is practically impossible. One of the best methods for computer analysis of large foundations supported on the elastic layer is replacing the layer with 3D finite elements. Although the use of the finite element method is complex, it allows performing analysis of the mat foundation and obtaining accurate results. The method also allows using variable soil properties and variable heights of the elastic layer. A simplified method of obtaining settlements outside of the loaded area is proposed herein. The method is based on the assumption that the relationship between the settlement of a totally rigid foundation supported on elastic half-space (y_{ii}) and the settlement of the same foundation supported on the elastic layer (Δ_{ii}) is the same as the relationship between the settlement of the half-space (y_{ki}) at any point k located outside of the foundation and the settlement of the elastic layer (Δ_{ki}) at the same point k, as shown in Figure 1.13. In other words:

$$\frac{y_{ii}}{\Delta_{ii}} = \frac{y_{ki}}{\Delta_{ki}} \tag{1.57}$$

From this relationship we find:

$$\Delta_{ki} = \frac{y_{ki}}{y_{ii}}\Delta_{ii} \tag{1.58}$$

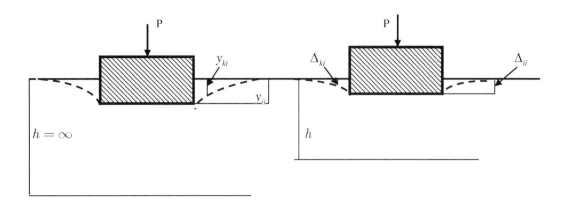

Settlement of the Foundation on Elastic Half-Space   Settlement of the Foundation on Elastic Layer

Figure 1.13

In this equation:

y_{ii} is the settlement of the foundation supported on elastic half-space at point i,
y_{ki} is the settlement of the half-space at point k located outside of the foundation,
Δ_{ii} is the settlement of the elastic layer at point i loaded with the same foundation at the same point, and
Δ_{ki} is the unknown settlement of the elastic layer at point k.

Table 1.1 Modulus of Subgrade Reaction k

General description of the soil	Type of soil	k, in kg/cm³
Lower density soil	Quicksand, filled-up sand, wet soft clay	0.1-0.5
Average density soil	Sandy ballast, loose gravel, wet clay	0.5-5
Compact solid soil	High density sand, gravel, dry clay	5-10
Very solid compacted soil	Compacted sandy clay, stiff clay	10-20
Stiff soil	Soft rocky soil, limestone, sandstone	20-100
Rocky soil	Rock	100-1500
Reinforced soil	Pile foundation	5-15

Table 1.2 Modulus of Elasticity of the Soil

Type of the soil	Modulus of Elasticity of the soil E_s in kg/cm²	
	High density soil	**Average density soil**
Gravel sands regardless on moisture	480	360
Medium sand regardless on moisture	420	310
Fine sand, average saturated	360	250
Sand saturated	310	190
Flour type sand		
a/ With lower content of moisture	210	175
b/ With high content of moisture	175	140
c/Saturated	140	90
Clayey soil	Stiff	Soft
Loam	160-125	90-50
Clay	590-160	160-40
Sandy loam	390-160	160-40

Table 1.3 Coefficients ω

Shape of the foundation	ω
circle	0.89
Square	0.88
Rectangular foundation When a/b is equal	
1.5	0.87
2	0.86
3	0.83
4	0.8
5	0.77
6	0.74
7	0.73
8	0.71
9	0.69
10	0.67

Table 1.4 Coefficients ω_z ω_φ ω_X

Coefficients ω_z, ω_φ, ω_X			
$\dfrac{a}{b}$	ω_z	ω_φ	ω_X
0.2	1.22	1.62	0.53
0.33	1.13	1.65	0.53
0.5	1.09	1.72	0.54
0.66	1.07	1.80	0.53
1	1.06	1.98	0.50
1.5	1.07	2.24	0.45
2	1.09	2.50	0.42
3	1.13	2.97	0.37
5	1.22	3.59	0.29

Table 1.5 Coefficients K_1 and K_2

α	1	1.5	2	3	5	7	10
K_1	0.60	0.81	0.99	1.28	1.70	2.00	2.37
K_2	0.60	0.43	0.339	0.236	0.45	0.105	0.074

Table 1.6 Coefficients *K*

$m = \dfrac{Z}{a}$	Circular foundation $a = R$	Rectangular foundations with the following *n*						Strip foundation when *n* > 10
		1	1.4	1.8	2.4	3.2	5	
0.0	0.000	0.000	0.000	0.000	0.000	0.000	0.000	0.000
0.4	0.090	0.100	0.100	0.100	0.100	0.100	0.100	0.104
0.8	0.179	0.200	0.200	0.200	0.200	0.200	0.200	0.208
1.2	0.266	0.299	0.300	0.300	0.300	0.300	0.300	0.311
1.6	0.348	0.380	0.394	0.397	0.397	0.397	0.397	0.412
2.0	0.411	0.446	0.472	0.482	0.486	0.486	0.486	0.511
2.4	0.461	0.499	0.538	0.556	0.565	0.567	0.567	0.605
2.8	0.501	0.542	0.592	0.618	0.635	0.640	0.640	0.687
3.2	0.532	0.577	0.637	0.671	0.696	0.707	0.709	0.763
3.6	0.558	0.606	0.676	0.717	0.750	0.768	0.772	0.831
4.0	0.579	0.630	0.708	0.756	0.796	0.820	0.830	0.892
4.4	0.596	0.650	0.735	0.789	0.837	0.867	0.883	0.849
4.8	0.611	0.668	0.759	0.819	0.873	0.908	0.932	1.001
5.2	0.624	0.683	0.780	0.884	0.904	0.948	0.977	1.050
5.6	0.635	0.697	0.798	0.867	0.933	0.981	1.018	1.095
6.0	0.645	0.708	0.814	0.887	0.958	1.011	1.056	1.138
6.4	0.653	0.719	0.828	0.904	0.980	1.031	1.090	1.178
6.8	0.661	0.728	0.841	0.920	1.000	1.065	1.122	1.215
7.2	0.668	0.736	0.852	0.935	1.019	1.088	1.152	1.251
7.6	0.674	0.744	0.863	0.948	1.036	1.109	1.180	1.285
8.0	0.679	0.751	0.872	0.960	1.051	1.128	1.205	1.316
8.4	0.684	0.757	0.881	0.970	1.065	1.146	1.229	1.347
8.8	0.689	0.762	0.888	0.980	1.078	1.162	1.251	1.376
9.2	0.693	0.768	0.896	0.989	1.089	1.178	1.272	1.404
9.6	0.697	0.772	0.902	0.998	1.100	1.192	1.291	1.431
10.0	0.700	0.777	0.908	1.005	1.110	1.205	1.309	1.456
11.0	0.705	0.786	0.922	1.022	1.132	1.233	1.349	1.506
12.0	0.710	0.794	0.933	1.037	1.151	1.257	1.384	1.550

Table 1.7 Coefficients K_R for circular and strip foundations

$m = \dfrac{H}{a}$	Strip	Circle		$m = \dfrac{H}{a}$	Strip	Circle	
	$\tau = 0$	$\tau = 0$	$u = 0$		$\tau = 0$	$\tau = 0$	$u = 0$
0	0	0	0	2	0.73	0.52	0.48
0.20	0.10	0.09	0.05	3	0.95	0.60	0.57
0.25	0.12	0.11	0.09	5	1.26	0.67	0.65
0.50	0.23	0.21	0.18	7	1.46	0.70	0.69
1.00	0.43	0.36	0.32	10	1.69	0.73	0.72
1.50	0.59	0.46	0.42	∞	∞	0.79	0.79

Table 1.8 Numerical values of P_{Hm}

$\dfrac{H}{2a}$	L/2a								
	1.0	1.5	2.0	2.5	3. 0	4.0	5.0	10	∞
	P_{Hm}, $\mu = 0$								
0.0	0	0	0	0	0	0	0	0	0
0.1	0.098	0.099	0.099	0.099	0.099	0.099	0.099	0.099	0.098
0.2	0.187	0.190	0.191	0.192	0.193	0.193	0.194	0193	0.191
0.3	0.267	0.274	0.277	0.279	0.281	0.282	0.283	0.284	0.284
0.4	0.338	0.350	0.356	0.359	0.361	0.365	0.366	0.366	0.366
0.5	0.400	0.418	0.427	0.433	0.437	0.441	0.444	0.449	0.446
0.6	0.454	0.479	0.492	0.499	0.504	0.511	0.515	0.522	0.520
0.7	0.500	0.533	0.550	0.560	0.567	0.575	0.580	0.590	0.588
0.8	0.540	0.581	0.602	0.615	0.623	0.634	0.640	0.653	0.651
0.9	0.575	0.624	0.649	0.665	0.675	0.688	0.696	0.711	0.713
1.0	0.604	0.661	0.691	0.710	0.722	0.737	0.746	0.765	0.772
1.1	0.630	0.695	0.729	0.750	0.765	0.782	0.793	0.814	0.821
1.2	0.653	0.724	0.764	0.787	0.803	0.823	0.836	0.860	0.869
1.3	0.672	0.751	0.794	0.821	0.839	0.861	0.875	0.903	0.912
1.4	0.690	0.774	0.822	0.852	0.872	0.897	0.912	0.943	0.956
1.5	0.705	0.796	0.848	0.880	0.902	0.930	0.946	0.980	0.998
2.0	0.762	0.876	0.944	0.990	1.022	1.062	1.086	1.136	1.170
2.5	0.800	0.928	1.010	1.065	1.105	1.158	1.188	1.255	1.308
3.0	0.825	0.963	1.056	1.121	1.168	1.230	1.268	1.350	1.416
4.0	0.858	1.012	1.117	1.194	1.252	1.331	1.382	1.491	1.591
5.0	0.875	1.040	1.155	1.240	1.305	1.395	1.460	1.590	1.743
10	0.900	1.090	1.220	1.320	1.410	1.540	1.630	1.850	2.177
20	0.920	1.120	1.260	1.370	1.480	1.630	1.740	2.030	2.612
∞	0.946	1.148	1.300	1.424	1.527	1.694	1.826	2.246	∞
	P_{Hm}, $\mu = 0.3$								
0.0	0	0	0	0	0	0	0	0	0
0.1	0.075	0.075	0.075	0.075	0.075	0.075	0.075	0.075	0.073
0.2	0.145	0.146	0.147	0.147	0.147	0.147	0.147	0.147	0.145
0.3	0.210	0.214	0.215	0.216	0.217	0.218	0.218	0.218	0.217
0.4	0.271	0.278	0.280	0.282	0.283	0.285	0.286	0.287	0.284
0.5	0.327	0.337	0.342	0.345	0.347	0.350	0.351	0.354	0.349
0.6	0.376	0.391	0.399	0.403	0.406	0.409	0.412	0.416	0.413
0.7	0.419	0.440	0.450	0.456	0.460	0.465	0.468	0.474	0.471
0.8	0.457	0.485	0.498	0.506	0.511	0.518	0.521	0.529	0.525
0.9	0.490	0.525	0.541	0.551	0.558	0.566	0.571	0.580	0.576
1.0	0.519	0.525	0.541	0.551	0.558	0.566	0.571	0.580	0.576
1.1	0.544	0.593	0.617	0.630	0.640	0.651	0.658	0.672	0.670

Table 1.8 Con't

$\dfrac{H}{2a}$	L/2a								
	1.0	**1.5**	**2.0**	**2.5**	**3. 0**	**4.0**	**5.0**	**10**	**∞**
1.2	0.567	0.621	0.650	0.665	0.676	0.690	0.697	0.713	0.711
1.3	0.586	0.647	0.679	0.697	0.709	0.725	0.734	0.751	0.752
1.4	0.604	0.671	0.706	0.727	0.740	0.757	0.768	0.788	0.791
1.5	0.620	0.692	0.731	0.754	0.769	0.788	0.799	0.822	0.825
2.0	0.678	0.772	0.828	0.862	0.886	0.914	0.930	0.964	0.982
2.5	0.715	0.825	0.893	0.938	0.968	1.008	1.030	1.073	1.104
3.0	0.741	0.862	0.939	0.992	1.031	1.079	1.108	1.164	1.206
4.0	0.770	0.909	1.001	1.066	1.115	1.180	1.221	1.301	1.370
5.0	0.785	0.935	1.040	1.110	1.165	1.245	1.295	1.400	1.498
10	0.830	0.980	1.110	1.200	1.280	1.390	1.470	1.650	1.894
20	0.840	1.010	1.145	1.260	1.340	1.485	1.585	1.870	2.293
∞	0.860	1.045	1.183	1.296	1.390	1.542	1.662	2.044	∞
P_{Hm}, $\mu = 0.50$									
0.0	0	0	0	0	0	0	0	0	0
0.1	0.014	0.012	0.011	0.010	0.009	0.009	0.008	0.008	0.007
0.2	0.046	0.040	0.037	0.035	0.034	0.032	0.031	0.028	0.026
0.3	0.088	0.079	0.075	0.072	0.070	0.067	0.065	0.061	0.055
0.4	0133	0.123	0.117	0.113	0.110	0.106	0.104	0.098	0.090
0.5	0.177	0.168	0.161	0.157	0.153	1.49	0.146	0.140	0.129
0.6	0.219	0.211	0.204	0.200	0.196	0.192	0.189	0.182	0.187
0.7	0.258	0.253	0.247	0.243	0.239	0.234	0.231	0.224	0.209
0.8	0.293	0.293	0.288	0.283	0.280	0.275	0.272	0.265	0.248
0.9	0.325	0.329	0.326	0.322	0.319	0.315	0.312	0.305	0.284
1.0	0.354	0.363	0.362	0.359	0.356	0.352	0.350	0.344	0.320
1.1	0.379	0.393	0.395	0.394	0.391	0.387	0.385	0.378	0.353
1.2	0.402	0.422	0.426	0.426	0.424	0.421	0.418	0.412	0.386
1.3	0.422	0.448	0.455	0.456	0.455	0.452	0.450	0.445	0.449
1.4	0.441	0.471	0.481	0.485	0.485	0.482	0.509	0.504	0.476
1.5	0.457	0.493	0.506	0.511	0.512	0.511	0.632	0.630	0.597
2.0	0.518	0.576	0.604	0.618	0.624	0.630	0.728	0.733	0.695
2.5	0.558	0.630	0.673	0.695	0.708	0.723	0.803	0.814	0.778
3.0	0.586	0.671	0.722	0.753	0.773	0.793	0.803	0.814	0.778
4.0	0.622	0.724	0.789	0.834	0.864	0.899	0.918	0.945	0.914
5.0	0.645	0.755	0.830	0.885	0.925	0.975	1.000	1.045	1.021
10	0.675	0.805	0.890	0.960	1.015	1.085	1.140	1.265	1.346
20	0.690	0.830	0.935	1.015	1.075	1.175	1.250	1.455	1.674
∞	0.710	0.861	0.975	1.068	1.145	1.271	1.370	1.685	∞

Table 1.9 Numerical values of P'_{Hm}

$\dfrac{H}{2a}$	L/2a								
	1.0	1.5	2.0	2.5	3. 0	4.0	5.0	10	∞
P'_{Hm}, μ = 0									
0.0	0	0	0	0	0	0	0	0	0
0.1	0.09	0.109	0.122	0.132	0.141	0.154	0.163	0.185	0.218
0.2	0.175	0.208	0.231	0.248	0.261	0.279	0.292	0.318	0.349
0.3	0.251	0.295	0.324	0.345	0.360	0.281	0.393	0.421	0.450
0.4	0.320	0.371	0.404	0.426	0.442	0.463	0.475	0.502	0.522
0.5	0.381	0438	0.472	0.495	0.511	0.531	0.543	0.568	0.585
0.6	0.437	0.496	0.531	0.553	0.568	0.587	0.599	0.622	0.635
0.7	0.486	0.546	0.581	0.602	0.617	0.634	0.645	0.668	0.675
0.8	0.530	0.590	0.623	0.643	0.657	0.674	0.685	0.706	0.712
0.9	0.569	0.628	0.660	0.679	0.692	0.708	0.718	0.7380	0.744
1.0	0.604	0.661	0.691	0.710	0.722	0.737	0.746	0.765	0.772
1.5	0.728	0.774	0.797	0.810	0.820	0.831	0.838	0.852	0.855
2	0.800	0.836	0.854	0.865	0.873	0.882	0.888	0.897	0.892
2.5	0.844	0.874	0.889	0.898	0.903	0.912	0.916	0.923	0.916
3	0.874	0.899	0.912	0.919	0.925	0.931	0.935	0.938	0.929
4	0.912	0.931	0.940	0.946	0.950	0.955	0.957	0.957	0.946
5	0.934	0.988	0.991	0.992	0.992	0.991	0.990	0.989	0.976
20	1.000	1.000	1.000	1.000	1.000	1.000	1.000	1.000	0.986
50	1.000	1.000	1.000	1.000	1.000	1.000	1.000	1.000	0.995
∞	1.000	1.000	1.000	1.000	1.000	1.000	1.000	1.000	1.000
P'_{Hm}, μ = 0.3									
0.0	0	0	0	0	0	0	0	0	0
0.1	0.083	0.098	0.111	0.120	0.128	0.139	0.147	0.165	0189
0.2	0.157	0.187	0.208	0.222	0.233	0.249	0.259	0.280	0.299
0.3	0.226	0.264	0.289	0.306	0.319	0.336	0.345	0.365	0.384
0.4	0.286	0.330	0.357	0.375	0.387	0.403	0.412	0.429	0.442
0.5	0.339	0.386	0.414	0.431	0.443	0.457	0.465	0.482	0.491
0.6	0.385	0.434	0.461	0.477	0.488	0.501	0.508	0.524	0.530
0.7	0.426	0.474	0.499	0.514	0.524	0.536	0.544	0.558	0.563
0.8	0.461	0.508	0.532	0.545	0.554	0.566	0.573	0.586	0.589
0.9	0.492	0.536	0.558	0.571	0.579	0.590	0.596	0.609	0.608
1.0	0.519	0.561	0.581	0.592	0.601	0.610	0.616	0.628	0.624
1.5	0.607	0.636	0.651	0.659	0.664	0.671	0.675	0.683	0.677
2	0.653	0.674	0.684	0.690	0.694	0.699	0.702	0.707	0.698
2.5	0.677	0.694	0.701	0.705	0.708	0.713	0.715	0.717	0.709
3	0.695	0.707	0.713	0.713	0.719	0.722	0.724	0.723	0.715
4	0.713	0.721	0.725	0.729	0.730	0.732	0.732	0.730	0.722

Table 1.9 Con't

$\dfrac{H}{2a}$	L/2a								
	1.0	**1.5**	**2.0**	**2.5**	**3. 0**	**4.0**	**5.0**	**10**	∞
5	0.723	0.729	0.734	0.736	0.737	0.737	0.736	0.734	0.726
10	0.745	0.748	0.749	0.748	0.748	0.746	0.745	0.744	0.734
20	0.752	0.751	0.749	0.748	0.748	0.748	0.748	0.748	0.739
50	0.745	0.744	0.744	0.744	0.744	0.744	0.744	0.744	0.741
∞	0.743	0.743	0.743	0.743	0.743	0.743	0.743	0.743	0.743
P'_{Hm}, $\mu = 0.5$									
0.0	0	0	0	0	0	0	0	0	0
0.1	0.068	0.080	0.089	0.096	0.101	0.108	0.114	0.127	0.135
0.2	0.129	0.151	0.166	0.177	0.185	0.195	0.200	0.209	0.204
0.3	0.180	0.208	0.225	0.235	0.242	0.250	0.254	0.259	0.250
0.4	0.223	0.252	0.269	0.278	0.283	0.289	0.291	0.293	0.278
0.5	0.259	0.288	0.302	0.309	0.312	0.315	0.316	0.315	0.298
0.6	0.288	0.315	0.326	0.330	0.332	0.333	0.332	0.330	0.310
0.7	0.312	0.334	0.342	0.345	0.345	0.344	0.342	0.339	0.318
0.8	0.330	0.348	0.353	0.352	0.352	0.349	0.348	0.343	0.321
0.9	0.344	0.357	0.359	0.358	0.356	0.352	0.350	0.344	0.322
1.0	0.354	0.363	0.362	0.359	0.356	0.352	0.349	0.343	0.320
1.5	0.368	0.359	0.349	0.343	0.337	0.330	0.326	0.315	0.293
2.0	0.353	0.335	0.322	0.313	0.306	0.298	0.292	0.279	0.258
2.5	0.332	0.308	0.298	0.283	0.276	0.266	0.261	0.246	0.225
3.0	0.307	0.281	0.265	0.255	0.248	0.239	0.233	0.218	0.198
4.0	0.265	0.236	0.221	0.211	0.204	0.195	0.189	0.174	0.157
5.0	0.228	0.201	0.186	0.177	0.170	0.161	0.155	0.142	0.130
10	0.135	0.117	0.106	0.098	0.092	0.086	0.083	0.076	0.070
20	0.076	0.061	0.054	0.051	0.048	0.044	0.043	0.042	0.039
50	0.027	0.020	0.018	0.017	0.017	0.017	0.017	0.016	0.015
∞	0	0	0	0	0	0	0	0	0

Table 1.10 Numerical values of k_c

H/a	0.25	0.5	1	2	>2
k_c	0.26	0.43	0.63	0.74	0.75

References

1. Barkan, D. D. 1962. *Dynamic of bases and foundations*. New York: McGraw-Hill.
2. Boussinesq, J. 1885. *Application des potentiels a l'etude de l'equilibre et du mouvement des solides elastique*. Paris: Gauthier-Villars.
3. Egorov, K. E. 1948. Deflections of a circular rigid foundation loaded with a non-symmetrical load. *Journal of Soils and Foundations* (Stroivoenmorizdat, Moscow), ed. 11.
4. Egorov, K. E. 1949. *Methods of obtaining settlements of foundations*. NIIOSP, no. 13. Moscow: Mashstroiizdat.
5. Giroud, J. P. 1968. Settlement of a linearly loaded rectangular area. *Journal of the Soil Mechanics and Foundations Division Proceedings of ASCE* 94:813–831.
6. Giroud, J. P. 1972. Settlement of rectangular foundation on soil layer. *Journal of the Soil Mechanics and Foundations Division: Proceedings of ASCE* 98:149–154.
7. Gorbunov-Posadov, M. I. 1941. *Slabs on elastic foundation*. Moscow: Gosstroiizdat.
8. Gorbunov-Posadov, M. I. 1949. *Beams and slabs on elastic foundation*. Moscow: Mashstroiizdat.
9. Gorbunov-Posadov, M. I. 1984. *Analysis of structures on elastic foundation*. Moscow: Stroiizdat.
10. Gorbunov-Posadov, M. I., and T. A. Malikova. 1973. *Analysis of structures on elastic foundation*. Moscow: Stroiizdat.
11. Harr, M. E. 1966. *Foundations of theoretical soil mechanics*. New York: McGraw-Hill.
12. Hetenyi, M. 1946. *Beams on elastic foundation*. Ann Arbor: University of Michigan Press.
13. Klepikov, S. N. 1967. *Analysis of structures on elastic foundation*. Kiev: Budivelnik.
14. Korenev, B. G. 1962. *Analysis of plates on elastic foundation*. Moscow: Gosstroiizdat.
15. Marguerre, K. 1933. Spannungsverteilung und Wellenaubreitung in der Kontinuierlich Gestutzten Platte. *ING. Arch.* 4:332–353.
16. Newmark, N. M. 1942. Influence charts for computation stresses in elastic foundations. *University of Illinois Bulletin* 40 (12).
17. Newmark, N. M. 1947. Influence charts for computation of vertical displacements in elastic foundations. *University of Illinois Bulletin* 44 (45).
18. Pasternak, P. L. 1954. *On a new method of analysis of an elastic foundation by means of two foundation constants*. Moscow: Gosstroiizdat.
19. Poulos, H. G. 1967. Stresses and displacements in an elastic layer underlain by rigid rough base. *Géotechnique* 17:378–410.
20. Reissner, E. 1958. Deflection of plates on viscoelastic foundation. *Journal of Applied Mechanics ASME* 80:144–145.
21. Selvadurai, A.P.S. 1979. *Elastic analysis of soil-foundation interaction*. Amsterdam/Oxford/New York: Elsevier Scientific.
22. Shechter, O. Y. 1939. *Analysis of an infinite plate resting on an elastic foundation off finite and infinite thickness and loaded by a concentrated force*. Moscow: Fundamentstroi.
23. Shechter, O. Y. 1948. Regarding analysis of plates supported on elastic layer. Collection of Papers, no. 11. *Soil Mechanics and Foundations* (Stroivoenmorizdat, Moscow).
24. Terzaghi, K. 1955. Evaluation of coefficient of subgrade reaction. *Géotechnique* 5:297–326.
25. Tsytovich, N. A. 1963. *Mechanics of Soils*. Moscow: Gosstroiizdat.
26. Umansky, A. A. 1933. *Analysis of beams on elastic foundation*. Leningrad: Central Research Institute of Auto Transportation.
27. Vesic, A. B. 1961. Bending of beams resting on isotropic elastic solid. *Journal of the Engineering Mechanics Division: Proceedings of ASCE* 87 (EM2): 35–53.

28. Weissmann, G. F. 1965. Measuring the modulus of subgrade soil reaction. *Materials Research and Standards* 5(2): 71–75.
29. Weissmann, G. F. 1972. Tilting foundations. *Journal of the Soil Mechanics and Foundations Division Proceedings of ASCE* 98:59–78.
30. Weissmann, G. F., and S. R. White. 1961. Small angular deflexions of rigid foundations. *Géotechnique* 11:186–201.
31. Wieghardt, K. 1922. Uber den Balken auf nachgiebiger Unterlage. *Zeitschrift für Angewandte Mathematik und Mechanik* 2(3): 165–184.
32. Winkler, E. 1867. *Die Lehre von der Elastizitat und Festigkeit*. Prague: Domonicus.

2

Analytical Methods of Analysis of Finite Beams on Winkler Foundation

2.1 Introduction

Analysis of finite beams on Winkler foundation can be performed by two classical analytical methods: method of superposition proposed by Hetenyi (1946) and method of initial parameters proposed by Pouzirevsky (1923), Dutov (1929), Krilov (1930), and later developed in detail by Umansky (1933) and others. Selvadurai (1979) indicates also the strain energy method proposed by Hetenyi (1946) and Christopherson (1956), although the latter method is not well known and not used by practicing engineers and researchers. Some scientists came with new ideas for finite beam analysis. Bowels (1974) recommended the use of the matrix method, White (1963) employed the method of relaxation, and Penzien (1960) used the iterative method. Some other methods were proposed by Malter (1958) and Levinton (1947). In this chapter, only the first two methods are included since both methods are well developed and known by scientists, engineers, and graduate students.

However, even these two methods are not simple. Practical hand analysis is laborious, and therefore rarely used by practicing engineers, especially for analysis of beams with a large number of various loads. Nevertheless, software based on these methods allows analyzing any complex beam supported on Winkler foundation. Chapter 2 also includes some useful improvements of the described methods. As shown below, the method of superposition can be simplified by replacing the use of analytical solutions for infinite beams with solutions for semi-infinite beams. This replacement, shown below, allows obtaining a system of equations that contains two times less unknowns compared to traditional analysis. Analysis of any finite beam can be performed by solving a system with only two linear equations. The method of initial parameters originally developed for analysis of finite beams with various boundary conditions is used by the author for analysis of statically indeterminate systems with elements supported on Winkler foundation. A significant part of this chapter is devoted to analysis

---

All tables can be found at the end of the chapter.

of beams with totally rigid elements at their ends. Solutions obtained by the author (Tsudik 2003, 2006) allow analyzing frames with continuous foundations taking into account the totally rigid elements located at the areas of intersection of the foundations and frame columns. The method is also applied to analysis of interconnected beams on Winkler foundation. It takes into account that areas of intersection of interconnected beams are totally rigid, which significantly reduces the lengths of elastic parts of the beams, final moments, and shear forces. The use of the stiffness method compared to the method of forces reduces the number of unknowns by half.

The method of initial parameters is applied to analysis of piles loaded with horizontal, vertical loads, and moments. This method was developed by Snitco (1970) and is described in this chapter. Numerical examples illustrate practical application of the described methods.

2.2 Analysis of Infinite and Semi-Infinite Beams on Winkler Foundation

If a load applied to the beam is located at the distance more than $(3 \div 4)\lambda$ from one of the ends, the beam is considered infinitely long or *infinite* in the direction of that end, as shown in Figure 2.1, and is called *semi-infinite* overall.

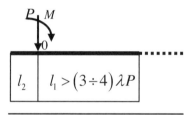

Figure 2.1   Semi-infinite beam

If the point of the load application is located at a distance more than $(3 \div 4)\lambda$ from both ends, as shown in Figure 2.2, the beam is infinitely long in both directions. This type of beam is called *infinite*. Parameter, $\lambda = \sqrt[4]{\dfrac{4EI}{k_0 b}}$ k_0 is the modulus of subgrade reaction, b is the width of the beam, and EI is the flexural rigidity of the beam. Equations shown below are used for analysis of infinite beams on Winkler foundation. If an infinite beam has a vertical load P at the origin 0, as shown in Figure 2.2, the vertical deflection of the beam v_0, rotation of the beam φ_0, moment M_0, and shear force Q_0 at point 0 is obtained from the following simple equations:

$$v_0 = P\frac{\lambda^3}{8EI} \quad \varphi_0 = 0 \quad M_0 = \frac{P\lambda}{4} \quad Q_0 = -\frac{P}{2} \tag{2.1}$$

If an infinite beam is loaded with a moment M at the origin 0, as shown in Figure 2.2, these equations will look as follows:

$$v_0 = 0 \quad \varphi_0 = \frac{M\lambda}{4EI} \quad M_0 = \frac{M}{2} \quad Q_0 = -\frac{M}{2\lambda} \tag{2.2}$$

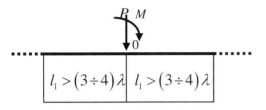

Figure 2.2   Infinite beam

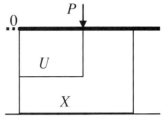

Figure 2.3

The vertical deflection, rotation, moment, and shear force at any point located at distance X to the right from the applied load P, as shown in Figure 2.3, can be found from equations 2.3:

$$
\left.\begin{aligned}
v_x &= P\frac{\lambda^3}{8EI}W_{x-u} \\
\varphi_x &= -\frac{P\lambda^2 V_{x-u}}{4EI} \\
M_x &= \frac{P\lambda U_{x-u}}{4} \\
Q_x &= -\frac{PT_{x-u}}{2}
\end{aligned}\right\}
\tag{2.3}
$$

Functions W_x, V_x, U_x, and T_x are so-called *functions of influence* and can be found from the following equations:

$$
\left.\begin{aligned}
T_x &= T(\xi) = e^{-\xi}\cos\xi \\
U_x &= U(\xi) = e^{-\xi}(\cos\xi - \sin\xi) \\
W_x &= W(\xi) = e^{-\xi}(\sin\xi + \cos\xi) \\
V_x &= V(\xi) = e^{-\xi}\sin\xi
\end{aligned}\right\}
\tag{2.4}
$$

where, $\xi = \dfrac{X}{\lambda}$ and X is the distance between the origin 0 and the point of investigation. When $X = \xi = 0$, functions $T_0 = U_0 = W_0 = 1$ and $V_0 = 0$. Functions 2.4 are interconnected and can be obtained one from another by differentiation as shown below:

$$
\frac{dV_x}{dx} = \frac{1}{\lambda}U_x \quad \frac{dU_x}{dx} = -\frac{2}{\lambda}T_x \quad \frac{dT_x}{dx} = -\frac{1}{\lambda}W_x \quad \frac{dW_x}{dx} = \frac{2}{\lambda}V_x
\tag{2.5}
$$

If an infinite beam, as shown in Figure 2.4, is loaded with a moment M, the vertical deflection, rotation, moment, and shear force at any point located to the right from the applied moment can be found from the following equations:

$$
v_x = \frac{M\lambda^2 V_{x-u}}{4EI} \quad \varphi_x = \frac{M\lambda U_{x-u}}{4EI} \quad M_x = \frac{MT_{x-u}}{2} \quad Q_x = -\frac{MW_{x-u}}{2\lambda}
\tag{2.6}
$$

If the same beam is loaded with a concentrated vertical relative deflection Γ, as shown in Figure 2.5, the same equations will look as follows:

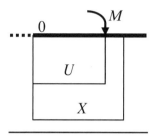

Figure 2.4

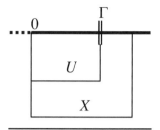

Figure 2.5

$$v_x = \frac{\Gamma T_{x-u}}{2} \quad \varphi_x = -\frac{\Gamma W_{x-u}}{2\lambda} \quad M_x = -\frac{\Gamma EIV_{x-u}}{\lambda^2} \quad Q_x = -\frac{\Gamma EIU_{x-u}}{\lambda^3} \qquad (2.7)$$

If the beam is loaded with a concentrated relative rotation θ, as shown in Figure 2.6, these equations will look as follows:

$$v_x = \frac{\theta\lambda U_{x-u}}{4} \quad \varphi_x = -\frac{\theta T_{x-u}}{2} \quad M_x = -\frac{\theta EIW_{x-u}}{2\lambda} \quad Q_x = \frac{\theta EIV_{x-u}}{\lambda^2} \qquad (2.8)$$

The same equations for a beam loaded with a uniformly distributed load p, as shown in Figure 2.7, look as below:

$$\left. \begin{aligned}
v_x &= p\frac{\lambda^4}{8EI}\left(T_{x-c} - T_{x-d}\right) \\
\varphi_x &= -p\frac{\lambda^3}{8EI}\left(W_{x-c} - Wx-d\right) \\
M_x &= p\frac{\lambda^2}{4}\left(V_{x-c} - V_{x-d}\right) \\
Q_x &= -p\frac{\lambda}{4}\left(U_{x-c} - Ux-d\right)
\end{aligned} \right\} \qquad (2.9)$$

For a triangle-distributed load p', as shown in Figure 2.8, the same equations look as follows:

$$\left. \begin{aligned}
v_x &= p'\frac{\lambda^4}{8EI}\left[\frac{\lambda}{2}(U_{x-c} - U_{x-d}) - (d-c)T_{x-d}\right] \\
\varphi_x &= -p'\frac{\lambda^3}{8EI}\left[\lambda(T_{x-c} - T_{x-d}) - (d-c)W_{x-d}\right] \\
M_x &= p'\frac{\lambda^2}{4}\left[(W_{x-c} - W_{x-d}) - (d-c)V_{x-d}\right] \\
Q_x &= p'\frac{\lambda}{4}\left[\lambda(V_{x-c} - V_{x-d}) + (d-c)U_{x-d}\right]
\end{aligned} \right\} \qquad (2.10)$$

When all loads and deflections mentioned above are applied to the beam, the vertical deflections, rotations, moments, and shear forces can be found from equation 2.11:

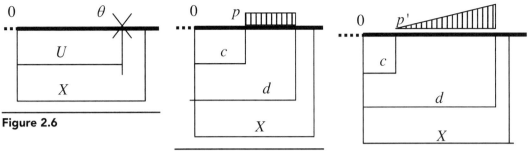

Figure 2.6

Figure 2.7

Figure 2.8

$$v_x = \Gamma \frac{1}{2} T_{x-u} + \theta \frac{\lambda}{4} U_{x-u} + M \frac{\lambda^2}{4EI} V_{x-u} + P \frac{\lambda^3}{8EI} W_{x-u} + p \frac{\lambda^4}{8EI}(T_{x-c} - T_{x-d}) +$$

$$p' \frac{\lambda^4}{8EI}\left[\frac{\lambda}{2}(U_{x-c} - U_{x-d}) - (d-c)T_{x-d}\right]$$

$$\varphi_x = -\theta \frac{1}{2} T_{x-u} + M \frac{\lambda}{4EI} U_{x-u} - P \frac{\lambda^2}{4EI} V_{x-u}$$

$$-\Gamma \frac{1}{2\lambda} W_{x-u} - p \frac{\lambda^3}{8EI}(W_{x-c} - W_{x-d}) - p \frac{\lambda^3}{8EI}[\lambda(T_{x-c} - T_{x-d}) - (d-c)W_{x-d}]$$

$$M_x = M \frac{1}{2} T_{x-u} + P \frac{\lambda}{4} U_{x-u} - \Gamma \frac{EI}{\lambda^2} V_{x-u} - \theta \frac{EI}{2\lambda} W_{x-u} - p \frac{\lambda^2}{4}(V_{x-c} - V_{x-d}) +$$

$$p' \frac{\lambda^2}{4}\left[\frac{\lambda}{2}(W_{x-c} - W_{x-d}) - (d-c)V_{x-d}\right]$$

$$Q_x = -P \frac{1}{2} T_{x-u} - \Gamma \frac{EI}{\lambda^3} U_{x-u} + \theta \frac{EI}{\lambda^2} V_{x-u} - M \frac{1}{2\lambda} W_{x-u}$$

$$-p \frac{\lambda}{4}(U_{x-c} - U_{x-d}) + p' \frac{\lambda}{4}[\lambda(V_{x-c} - V_{x-d}) + (d-c)U_{x-d}]$$

$$\left.\begin{matrix}\\ \\ \\ \\ \\ \\ \\ \\ \\ \\ \\ \\ \end{matrix}\right\} \quad (2.11)$$

It is important to mention that equation 2.11 is written for points of the beam located to the right from the applied loads or deflections. The same equation is used for points located at left from the applied loads or deflections. However, when obtaining φ_x and Q_x, the signs of P and θ have to be changed, and when obtaining v_x and M_x, the signs of M and Γ should also be changed.

As can be seen, equation 2.11 is complex and is not recommended for hand analysis, especially for analysis of beams with many and various types of loads. However, analysis of simple beams is relatively easy.

Equations for analysis of semi-infinite beams with various types of loads are shown in Table 2.1. These formulae are obtained for two types of loads: concentrated vertical loads and concentrated moments. Functions, A_x, B_x, C_x, and D_x are obtained from equations 2.12:

$$A_x = ch\xi \cos\xi \quad B_x = \frac{1}{2}(ch\xi \sin\xi + sh\cos\xi)$$

$$C_x = \frac{1}{2} sh\xi \sin\xi \quad D_x = \frac{1}{4}(ch\xi \sin\xi - sh\xi \cos\xi) \Bigg\} \quad (2.12)$$

The same equation is used for obtaining functions with subscripts u. Table 2.24 for functions T_x, U_x, V_x, and W_x, as well as Table 2.25 for functions A_x, B_x, C_x, and D_x, are given at the end of this chapter. They are helpful for hand analysis of infinite, semi-infinite, and simple finite beams. The use of these tables for computer analysis does not make sense. It is much easier to use equations 2.4 and 2.12 than tables when developing computer software.

2.3 Method of Superposition

When a load applied to the beam is located at a distance $l < (3 \div 4)\lambda$, from both ends of the beam, as shown in Figure 2.9, that type of beam is called *finite*. Exact analysis of finite beams supported on Winkler foundation is a difficult problem. Both analytical methods, the method of superposition and the method of initial parameters, although well known for a long time, are not used by practicing engineers because of their complexity. Both methods are labor and time-consuming and, in most of the cases, their practical application is simply impossible unless software for computer analysis based on these methods is available.

The method of superposition is based on analytical solutions for infinite beams discussed above. The method is explained below by investigating a simple finite beam *AB*, as shown in Figure 2.10. The beam is loaded with a vertical load P, a moment M, and a uniformly distributed load q. Analysis is performed as follows:

> The given finite beam *AB* is extended infinitely in both directions, so the given beam becomes a part of a much larger infinite beam. Immediately outside of the beam ends, A and B, are applied two unknown vertical loads P_A, P_B and two unknown moments M_A and M_B. The purpose of these loads and moments is to cancel out the moments and shear forces produced by the given loads applied to the infinite beam, so the total moment as well as the total shear force at both ends of the beam becomes equal to zero, which reflects the actual boundary conditions of the beam *AB*.

$$|P$$

| $l_1 < (3 \div 4)\lambda$ | $l_2 < (3 \div 4)\lambda$ |

Figure 2.9

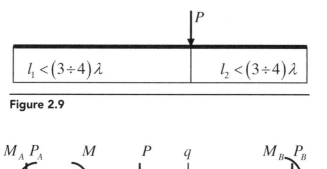

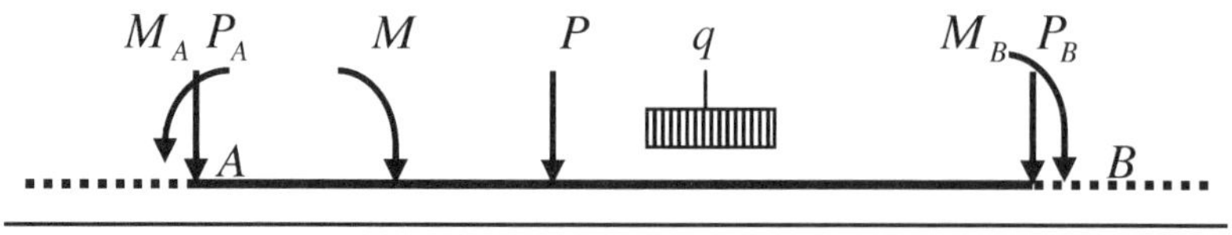

Figure 2.10

Now, using the last two equations from the system 2.11, a system of four linear equations 2.13 is written:

$$
\left.
\begin{aligned}
-\frac{P_A \lambda}{4} + \frac{M_A}{2} - \frac{P_B \lambda U_{AB}}{4} - \frac{M_B T_{AB}}{2} - \frac{P\lambda}{4} U_{An} + \frac{M}{2} T_{Am} - p\frac{\lambda^2}{4}\left(T_{Ad} - T_{Ac}\right) &= 0 \\
\frac{P_A}{2} + \frac{M_A}{2\lambda} + \frac{P_B T_{AB}}{2} - \frac{M_B W_{AB}}{2\lambda} - \frac{MW_{Am}}{2\lambda} - \frac{PT_{An}}{2} + p\frac{\lambda}{4}\left(U_{Ad} - U_{Ac}\right) &= 0 \\
\frac{P_A \lambda U_{AB}}{4} + \frac{M_A T_{AB}}{2} + \frac{P_B \lambda}{4} - \frac{M_B}{2} + \frac{P\lambda U_{Bn}}{4} + \frac{M}{2} T_{Bm} + p\frac{\lambda^2}{4}\left(T_{cB} - T_{dB}\right) &= 0 \\
-\frac{P_A T_{AB}}{2} - \frac{M_A W_{AB}}{2\lambda} - \frac{P_B}{2} - \frac{M_B}{2\lambda} - \frac{MW_{Bm}}{2\lambda} - \frac{PT_{Bn}}{2} - \frac{p\lambda}{4}\left(U_{cB} - U_{dB}\right) &= 0
\end{aligned}
\right\}
\tag{2.13}
$$

In this system, the first and third equations show that each of the total moments at points *A* and *B* are equal to zero. The second and the fourth equations show that each of the shear forces at points *A* and *B* are equal to zero. By solving the system of equations 2.13, all four unknown forces and moments are obtained. By applying the given loads, the found loads, and moments to the infinite beam, the final analysis is performed. Taking into account that all functions included in these equations are complex, building and solution of this system becomes a difficult problem for hand analysis.

Analysis of symmetrical beams can be simplified by reducing the number of unknowns. Let us investigate a symmetrically loaded finite beam *AB*, as shown in Figure 2.11a. The beam *AB*, extended in both directions, is shown in Figure 2.11b. Loads P_A, M_A, P_B, and M_B are applied to both ends of the beam, as shown in Figure 2.11c. Since $P_A = -P_B$ and $M_A = -M_B$, the number of unknown loads is equal only to two.

When the beam is loaded with inverse symmetrical loads, as shown in Figure 2.12, the loads at the ends *A* and *B* have to be inverse symmetrical as well, or $M_A = M_B$ and $P_A = P_B$. The number of unknown forces and moments is also equal to two. So, analysis

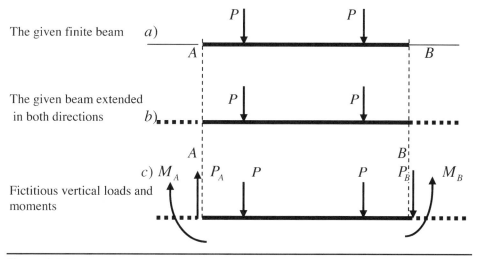

Figure 2.11

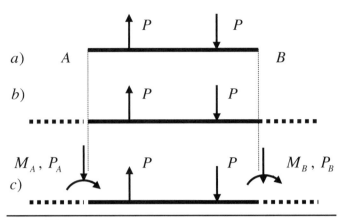

Figure 2.12

of symmetrical beams as well as analysis of inverse symmetrical beams requires solving a system with only two equations that simplifies all calculations.

Both solutions, for symmetrical loads and inverse symmetrical loads, can be used to analyze a beam with any nonsymmetrical loads. In Figure 2.13a, a finite beam loaded with a nonsymmetrical load P is shown. The given load can be replaced with two load combinations: symmetrical and inverse symmetrical, as shown in Figure 2.13b and 2.13c, respectively. Each combination requires the solution of a system with only two linear equations. In other words, the method described above allows replacing a system with four equations with two systems of two equations each that also simplifies analysis, to a certain degree. A more significant simplification proposed by the author of this work can be used if the given finite beam is extended only in one direction and becomes part of a semi-infinite beam. The method is explained below by investigating a finite beam AB, free supported on soil. The beam is loaded with a concentrated vertical load P, as shown in Figure 2.14a. Analysis of the beam is performed as follows:

1. The given beam is extended only in one direction, as shown in Figure 2.14b, and becomes part of a new semi-infinite beam with the origin at the left end of the beam where the boundary conditions are known.
2. The vertical load P_B and moment M_B are applied only to the right end of the given beam.
3. Using equations for analysis of semi-infinite beams, given in Table 2.1, the following equations 2.14 are written:

$$\left.\begin{array}{l} 2\lambda\left(T_lC_l - U_lD_l\right)P_B + \left(-2W_lC_l + 4T_lD_l\right)M_B + \lambda\left(-V_lA_u + W_lB_u\right)P = 0 \\ \left(2T_lB_l - 2U_lC_l\right)P_B + 2\lambda^{-1}\left(-W_lB_l + 2T_lC_l\right)M_B + \left(U_lA_u + 2V_lB_u\right)P = 0 \end{array}\right\} \quad (2.14)$$

Both equations reflect the real boundary conditions at the right end of the given beam. The first equation shows that the total moment at the right end of the beam is equal to zero. The second equation shows that the total shear force at the same end of the beam is equal to zero.

4. By solving this system of equations, moment M_B and the vertical load P_B are found.

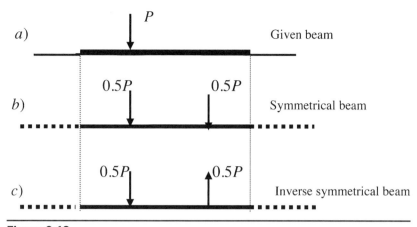

Figure 2.13

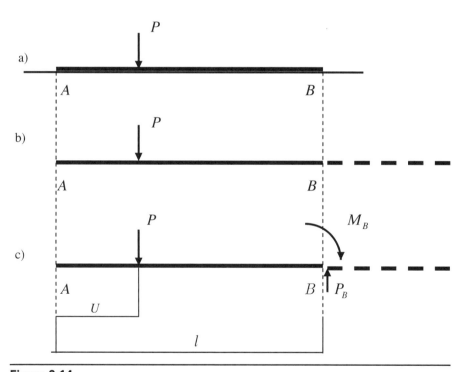

Figure 2.14

5. Final analysis of the beam is performed by applying to the semi-infinite beam the load P, the load P_B, and moment M_B.

The same method can be used for analysis of beams with various boundary conditions. The two equations we write for the right end of the beam reflect the real boundary conditions at that end. If, for example, the beam AB is pinned at the right end and free supported at the left end, the two equations have to show that the total settlement at the end B and total moment at the end B are equal to zero. Using Table 2.1, two equations 2.15 are written:

$$
\left.
\begin{aligned}
& 0.5\lambda^3\left(EI\right)^{-1}\left(T_lA_l - U_lB_l\right)P_B + 0.5\lambda^2\left(EI\right)^{-1}\left(-W_lA_l + 2T_lB_l\right)M_B + \\
& + 0.5\lambda^3\left(EI\right)^{-1}\left(T_lA_u - U_lB_u\right)P = 0 \\
& 2\lambda\left(T_lC_l - U_lD_l\right)P_B + \left(-2W_lC_l + 4T_lD_l\right)M_B + \lambda\left(-V_lA_u + W_lB_u\right)P = 0
\end{aligned}
\right\} \quad (2.15)
$$

For other boundary conditions, the two equations can be written analogously. If, for example, the left end is fixed and the right end is free supported, these two equations will look as shown below:

$$
\left.
\begin{aligned}
& \lambda\left(-V_lA_l + W_lB_l\right)P_B + \left(-U_lA_l - 2V_lB_l\right)M_B + 2\lambda\left(T_lC_u - U_lD_u\right)P = 0 \\
& \left(4V_lD_l + W_lA_l\right)P_B + 2\lambda^{-1}\left(2U_lD_l - V_lA_l\right)M_B + 2\lambda^{-1}\left(-W_lB_u + 2T_lC_u\right)P = 0
\end{aligned}
\right\} \quad (2.16)
$$

Equations 2.16 reflect the actual boundary conditions at the right end of the beam.

When both ends are fixed, the boundary conditions can be expressed by the following two equations:

$$
\left.
\begin{aligned}
& \frac{\lambda^3\left(V_lC_l - W_lD_l\right)}{EI}P_B + \frac{\lambda^2\left(U_lC_l + 2V_lD_l\right)}{EI}M_B + \frac{\lambda^3\left(V_lC_u - W_lD_u\right)}{EI}P = 0 \\
& \frac{\lambda^2\left(V_lB_l - W_lC_l\right)}{EI}P_B + \frac{\lambda\left(U_lB_l + 2V_lC_l\right)}{EI}M_B + \frac{\lambda^2\left(U_lC_u + 2V_lD_u\right)}{EI}P = 0
\end{aligned}
\right\} \quad (2.17)
$$

In this system, the first equation shows that the settlement at point B is equal to zero; the second equation shows that the rotation of the beam at the same right end is also equal to zero. As shown when analyzing the finite beam, the use of analytical closed solutions for semi-infinite beams is much simpler than using solutions for infinite beams.

2.4 Method of Initial Parameters and its Application to Analysis of Regular Beams

The method of initial parameters is a general method of structural mechanics and can be used for analysis of regular simple and continuous beams, including beams with various boundary conditions, stepped beams, and finite beams on Winkler foundation.

Figure 2.15 shows a portion of a beam loaded with a series of various loads and deflections. This portion of the beam is located between the origin 0 and cross section 0' where the settlement of the beam ν_x, the slope of the beam φ_x, the moment M_x, and the shear force Q_x have to be found. For a one-span simple beam, the origin is usually located at the left end. The moment M_0, the shear force Q_0, the settlement ν_0, and the slope φ_0 at point 0 are the initial parameters.

These initial parameters along with the given loads and deflections applied to the beam between points 0 and 0' are used to find ν_x, φ_x, M_x, and Q_x at any point of the beam.

Note that all other given loads and deflections applied to the beam at left, from point 0 and at right from point 0', are not taken into account since they do not affect the unknown deflections ν_x, φ_x, moment M_x, and shear force Q_x at the right end of the

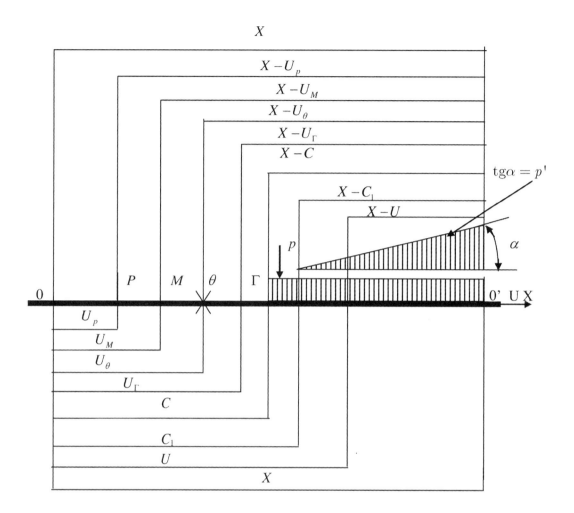

Figure 2.15

beam. Two of four initial parameters are always known. The other two initial parameters can be found using the two known parameters at the right end of the beam.

If the left end of one-span beam is free supported, $M_0 = 0$ and $Q_0 = 0$. If the left end of the beam is pin-supported, $M_0 = 0$ and $v_0 = 0$. If the left end of the beam is guided, $Q_0 = 0$ and $\varphi_0 = 0$. If the left end of the beam is fixed, $v_0 = 0$ and $\varphi_0 = 0$. The same parameters, at the opposite right end of the beam, depending on boundary conditions, can be found analogously. The origin of the beam does not have to be located always at the left end. For symmetrically loaded symmetrical beams, it is better to locate the origin at the center of the beam. When analysis of continuous beams is performed, conditions at intermediate supports also have to be taken into account. A numerical example shown below illustrates analysis of a regular beam by the method of initial parameters.

As can be seen, the beam shown in Figure 2.15 has the following loads: concentrated vertical loads P, concentrated moments M, uniformly distributed vertical loads p and triangle-distributed loads p', concentrated relative rotation θ, and concentrated relative vertical deflection Γ. Uniformly distributed moment m, triangle-distributed moment m', uniformly distributed rotation ϑ, triangle-distributed rotation ϑ', uniformly distributed settlement γ, and triangle-distributed settlement γ' are also applied to the beam. But these types of loads and deflections are not shown in Figure 2.15. They can be shown in Figure 2.15, exactly as uniformly distributed p and triangle-distributed p' loads. Equations 2.18–2.21 include all types of loads and deflections applied to the beam. The number of applied loads and deflections is unlimited. It is assumed that all equations are written for a cross section of the beam, located inside the distributed loads or deflections.

$$Q_x = Q_0 - \sum_0^x P - p(x - c) - p'\frac{(x - c_1)^2}{2} - \dots \int_0^x p(u)\,du \tag{2.18}$$

$$M_x = M_0 + Q_0 x + \sum_0^x M - \sum_0^x P(x - u_p) - p\frac{(x - c)^2}{2} - p'\frac{(x - c_1)^3}{6} + m(x - c) + \dots$$

$$+ m'\frac{(x - c_1)^2}{2} + \dots + \int_0^x m(u)\,du - \int_0^x p(u)(x - u)\,du \tag{2.19}$$

$$\varphi_x = \varphi_0 - M_0\frac{x}{EI} - Q_0\frac{x^2}{2EI} - \sum_0^x \theta - \frac{1}{EI}\sum_0^x M(x - u_M) +$$

$$+ \frac{1}{2EI}\sum_0^x P(x - u_p)^2 - \vartheta(x - c) - m\frac{(x - c)^2}{2EI} + p\frac{(x - c)^3}{6EI} -$$

$$- \vartheta'\frac{(x - c_1)^2}{2} - m'\frac{(x - c_1)^3}{6EI} + p'\frac{(x - c_1)^4}{24EI} - \dots$$

$$- \int_0^x \vartheta(u)\,du - \frac{1}{EI}\int_0^x m(u)(x - u)\,du + \frac{1}{2EI}\int_0^x p(u)(x - u)^2\,du \tag{2.20}$$

$$v_x = v_0 + \varphi_0 x - M_0\frac{x^2}{2EI} - Q_0\left(\frac{x^3}{6EI} - \frac{x}{GF}\right) - \frac{1}{2EI}\sum_0^x M(x - u_M)^2 +$$

$$+ \sum_0^x P\left[\frac{(x - u_p)^3}{6EI} - \frac{(x - u_p)}{GF}\right] + \sum_0^x \Gamma - \sum_0^x \theta(x - u_0)$$

$$+ p\left[\frac{(x - c)^4}{24EI} - \frac{(x - c)^2}{2GF}\right] + \gamma(x - c) - \vartheta\frac{(x - c)^2}{2} - \frac{m}{6EI}(x - c)^3 \tag{2.21}$$

$$+ p'\left[\frac{(x - c_1)^5}{120EI} - \frac{(x - c_1)^3}{6GF}\right] + \gamma'\frac{(x - c_1)^2}{2} - \vartheta'\frac{(x - c_1)^3}{6} - \frac{m'}{24EI}(x - c_1)^4$$

$$+ \int_0^x \gamma(u)\,du - \int_0^x \vartheta(u)(x - u)\,du - \frac{1}{2EI}\int_0^x m(u)(x - u)^2\,du + \int_0^x p(u)\left[\frac{(x - u)^3}{6EI} - \frac{(x - u)}{GF}\right]du$$

If point X is located to the right of the distributed load p, as shown in Figures 2.16 and 2.17, two methods of analysis are available. First method: Q_d, M_d, φ_d, and v_d are found and assuming point d as the new origin, Q_x, M_x, φ_x, and v_x are obtained. The second method is to perform analysis in two steps: first the distributed load p is extended from point d up to point X and the first analysis is performed. Then the beam is loaded with the load $-p$, applied from point d to point X and the second analysis is performed. By summing results of two analyses, final results are obtained.

It is important to mention that in equations 2.18–2.21, all last integrals represent various complex shapes of distributed loads. However, in practical applications, complicated shapes of distributed loads are rarely used. Usually, they are replaced with uniformly and triangle-distributed loads. It is also useful to note that deflections of the beam due to shear forces are usually small and are not taken into account. These deflections can be taken into account only for analysis of relatively short and deep beams.

Now, we can explain the application of the method of initial parameters to analysis of one-span simple beam as well as to the analysis of continuous beams. Figure 2.18 shows a simple beam with one pin-supported and one fixed end with a vertical load P. The known initial parameters are equal to $v_0 = 0$ and $M_0 = 0$. Unknown initial

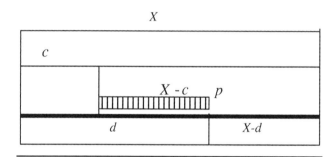

Figure 2.16

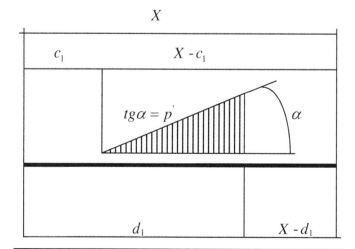

Figure 2.17

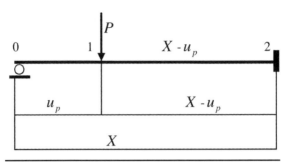

Figure 2.18

parameters are φ_0 and Q_0. Taking into account that $\varphi_2 = 0$ and $v_2 = 0$ and using equations 2.20 and 2.21, the following two equations are written:

$$\left.\begin{array}{l} \varphi_0 - Q_0 \dfrac{l^2}{2EI} + \dfrac{P}{2EI}(l - u_P)^2 = 0 \\[3mm] \varphi_0 l - Q_0 \dfrac{l^3}{6EI} + P\dfrac{(l - u_P)^3}{6EI} = 0 \end{array}\right\} \qquad (2.22)$$

By solving the system 2.22 initial parameters, φ_0 and Q_0 are found and by introducing initial parameters into all four equations 2.18–2.21, v_x, φ_x, M_x, and Q_x are found at any cross section of the beam including the right end of the beam. Now, we can illustrate an application of the method to analyze continuous beams, by using the beam shown in Figure 2.19.

When analyzing continuous beams, we have to take into account not only the boundary conditions at the left and right ends of the beam, but also the conditions at the intermediate supports, at points 1 and 2 in our case.

Taking into account that vertical deflections at points 1, 2, and 3 are equal to zero, and rotation of the beam at point 3 is also equal to zero, using equations 2.20 and 2.21, the following four equations are written:

$$\left.\begin{array}{l} -M_0 \dfrac{l_1^2}{2EI} - Q_0 \dfrac{l_1^3}{6EI} + p' \dfrac{l_1^5}{120EI} = 0 \\[3mm] -M_0 \dfrac{(l_1 + l_2)^2}{2EI} - Q_0 \dfrac{(l_1 + l_2)^3}{6EI} - X_1 \dfrac{l_3^3}{6EI} + p' \dfrac{(l_1 + l_2)^5}{120EI} = 0 \\[3mm] -M_0 \dfrac{l^2}{2EI} - Q_0 \dfrac{l^3}{6EI} - X_1 \dfrac{(l_2 + l_3)^3}{6EI} - X_2 \dfrac{l_3^3}{6EI} + p' \dfrac{l^5}{120EI} = 0 \\[3mm] -M_0 \dfrac{l}{EI} - Q_0 \dfrac{l^2}{2EI} - X_1 \dfrac{(l_2 + l_3)^2}{2EI} - X_2 \dfrac{l_3^2}{2EI} + p' \dfrac{l^4}{24EI} = 0 \end{array}\right\} \qquad (2.23)$$

The first three equations show that the vertical deflections at points 1, 2, and 3 are equal to zero, and the fourth one shows that rotation at point 3 is equal to zero. By solving the system 2.23, initial parameters M_0 and Q_0 and reactions X_1 and X_2 at the intermediate supports 1 and 2 are found. Final analysis of the beam is performed using equations 2.18–2.21. By introducing initial parameters and reactions found above into

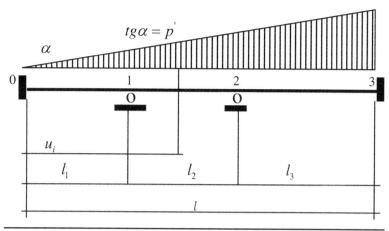

Figure 2.19

equations 2.18–2.21, deflections, moments, and shear forces are found at any point of the beam.

The method of initial parameters can also be applied to analysis of beams supported on springs. For example, if the beam shown above in Figure 2.19 is supported at points 1 and 2 on springs, and the springs' rigidities are equal to C_1 and C_2, respectively, the system of equations will look as follows:

$$
\left.
\begin{aligned}
-M_0 \frac{l_1^2}{2EI} - Q_0 \frac{l_1^3}{6EI} + p' \frac{l_1^5}{120EI} &= C_1 X_1 \\
-M_0 \frac{(l_1+l_2)^2}{2EI} - Q_0 \frac{(l_1+l_2)^3}{6EI} - X_1 \frac{l_2^3}{6EI} + p' \frac{(l_1+l_2)^5}{120EI} &= C_2 X_2 \\
-M_0 \frac{l^2}{2EI} - Q_0 \frac{l^3}{6EI} - X_1 \frac{(l_2+l_3)^3}{6EI} - X_2 \frac{l_3^3}{6EI} + p' \frac{l^5}{120EI} &= 0 \\
-M_0 \frac{l}{EI} - Q_0 \frac{l^2}{2EI} - X_1 \frac{(l_2+l_3)^2}{2EI} - X_2 \frac{l_3^2}{2EI} + p' \frac{l^4}{24EI} &= 0
\end{aligned}
\right\}
\tag{2.24}
$$

It is obvious that the method of initial parameters can also be used for analysis of various types of regular continuous beams, including beams with variable flexural rigidity, various types of supports, and boundary conditions.

The method is also successfully used for analysis of finite beams on Winkler foundation.

2.5 Method of Initial Parameters

Analysis of beams supported on Winkler foundation is based on the following relationship:

$$p = -kv$$

where k is the reaction of the soil, or any other material, per one linear unit of the beam length and

$$k = k_0 b$$

where b is the width of the beam, k_0 is the modulus of subgrade reaction, and v is the vertical deflection of the beam. The fourth-order linear differential equation for a beam supported on Winkler foundation looks as follows:

$$EI \frac{d^4 v}{dx^4} + kv = 0 \qquad (2.25)$$

In order to express the vertical deflections of the beam as a function of x, a new parameter is introduced:

$$\xi = \frac{x}{\lambda} \qquad (2.26)$$

where $\lambda = \sqrt[4]{\dfrac{4EI}{k}}$

As can be seen, parameter λ is related to the rigidity of the beam as well as to the soil characteristic k, but does not depend on the length of the beam. In order to solve the differential equation 2.25, the integral we are looking for is expressed as four, independent linear combinations of trigonometric and hyperbolic functions shown below:

$$\left. \begin{array}{l} A_x = A(\xi) = \cos \xi \times ch\xi \\[2mm] B_x = B(\xi) = \dfrac{1}{2}(\sin \xi \times ch\xi + \cos \xi \times sh\xi) \\[2mm] C_x = C(\xi) = \dfrac{1}{2}\sin \xi \times ch\xi \\[2mm] D_x = D(\xi) = \dfrac{1}{4}(\sin \xi ch\xi - \cos \xi sh\xi) \end{array} \right\} \qquad (2.27)$$

Each of these four functions represents a partial integral of the differential equation 2.25 shown above. Derivatives of these functions can be expressed by the same functions:

$$\frac{d}{dx} = \frac{d}{d\xi} \cdot \frac{d\xi}{dx} = \frac{1}{\lambda} \cdot \frac{d}{d\xi}.$$

By differentiating functions A_x, B_x, C_x, D_x we find:

$$\frac{dA_x}{dx} = -\frac{4}{\lambda}D_x, \quad \frac{dB_x}{dx} = \frac{1}{\lambda}Ax, \quad \frac{dC_x}{dx} = \frac{1}{\lambda}B_x, \quad \frac{D_x}{dx} = \frac{1}{\lambda}C_x$$

Another important property of these functions is when $\xi = 0$, they look as follows:

$$A(0) = 1, B(0) = C(0), = D(0) = 0 \qquad (2.28)$$

Since the general integral of a linear differential equation represents the sum of partial integrals, it can be written:

$$v = H_1 A_x + H_2 B_x + H_3 C_x + H_4 D_x \qquad (2.29)$$

Now, assuming that at point 0, the vertical deflection, the rotation of the beam, the moment and the shear force are equal to v_0, φ_0, M_0, and Q_0, respectively, all four constants H_1, H_2, H_3, and H_4 can be found. Finally, we have:

$$
\left.
\begin{aligned}
v_x &= v_0 A_x + \lambda \varphi_0 B_x - \frac{\lambda^2}{EI} M_0 C_x - \frac{\lambda^3}{EI} Q_0 D_x \\
\varphi_x &= \varphi_0 A_x - \frac{\lambda}{EI} M_0 B_x - \frac{\lambda^2}{EI} Q_0 C_x - \frac{4}{\lambda} v_0 D_x \\
M_x &= M_0 A_x + \lambda Q_0 B_x + k\lambda^2 v_0 C_x + k\lambda^3 \varphi_0 D_x \\
Q_x &= Q_0 A_x + k\lambda v_0 B_x + k\lambda^2 \varphi_0 C_x - \frac{4}{\lambda} M_0 D_x
\end{aligned}
\right\}
\tag{2.30}
$$

Each of these four equations can be obtained one from another by successive differentiation. The equation for M_x is multiplied by EI. By differentiating the fourth equation, we will again obtain the first one multiplied by $\frac{4EI}{\lambda^4} = k$. As can be seen from this system of equations, all coefficients at initial parameters v_0, φ_0, M_0, and Q_0 are numbers of influence for each of the four parameters. For example, λB_x in the first equation is the number of influence for parameter φ_0. In order to find any of parameters v, φ, M, Q, for a given x, we first have to find parameter ξ and using equations 2.27 or Table 2.25, find functions $A(\xi)$, $B(\xi)$, $C(\xi)$, and $D(\xi)$. Two out of four initial parameters v_0, φ_0, M_0, and Q_0, are always known and usually are equal to 0. As an example, if the left end of the beam is fixed, $\varphi_0 = 0$ and $v_0 = 0$. If the left end of the beam is pin-supported, $v_0 = 0$, $M_0 = 0$, and so on. Therefore, the system 2.30 is always simplified: we always have only two unknown initial parameters. These two parameters are found if we write two equations for the opposite right end of the beam. Now, let us investigate the general case when the beam is loaded not only at the ends, but also with various types of loads and deflections between the ends of the beam. In this case, the beam will experience the influence of the initial parameters v_0, φ_0, M_0, and Q_0 as well as the influence of the exterior loads and deflections. As we can see, the beam is loaded with three groups of loads and deflections:

1. Initial parameters v_0, φ_0, M_0, and Q_0
2. Vertical concentrated load P, concentrated moment M, concentrated relative vertical deflection Γ, and concentrated relative rotation θ
3. Distributed loads and deflections including distributed vertical load p, distributed moment m, distributed angles θ, and distributed vertical deflections γ

Taking into account all possible loads and deflections, including initial parameters, the general equations for a beam supported on Winkler foundation can be presented as follows:

$$
\begin{aligned}
v_x ={}& v_0 A_x + \lambda \varphi_0 B_x - \frac{\lambda^2}{EI} M_0 C_x - \frac{\lambda^3}{EI} Q_0 D_x + \\
&+ \sum_0^x \Gamma A_{x-u} - \lambda \sum_0^x \theta \cdot B_{x-u} - \frac{\lambda^2}{EI} \sum_0^x L C_{x-u} + \frac{\lambda^3}{EI} \sum_0^x P D_{x-u} + \\
&+ \int_c^d \gamma A_{x-u}\, du - \lambda \int_c^d \partial B_{x-u}\, du - \frac{\lambda^2}{EI} \int_c^d m C_{x-u}\, du + \frac{\lambda^3}{EI} \int_c^d p D_{x-u}\, du
\end{aligned}
\tag{2.31}
$$

$$\varphi_x = \varphi_0 A_x - \frac{\lambda}{EI} M_0 B_x - \frac{\lambda^2}{EI} Q_0 C_x - \frac{4}{\lambda} v_0 D_x -$$

$$-\sum_0^x \theta A_{x-u} - \frac{\lambda}{EI} \sum_0^x L B_{x-u} + \frac{\lambda^2}{EI} \sum_0^x P C_{x-u} - \frac{4}{\lambda} \sum_0^x \Gamma D_{x-u} -$$

$$-\int_c^d \vartheta A_{x-u} du - \frac{\lambda}{EI} \int_c^d m B_{x-u} du + \frac{\lambda^2}{EI} \int_c^d p C_{x-u} du - \frac{4}{\lambda} \int_c^d \gamma D_{x-u} du \qquad (2.32)$$

$$M_x = M_0 A_x + \lambda Q_0 B_x + k\lambda^2 v_0 C_x + k\lambda^3 \varphi_0 D_x +$$

$$+\sum_0^x L A_{x-u} - \lambda \sum_0^x P B_{x-u} + k\lambda^2 \sum_0^x \Gamma C_{x-u} - k\lambda^3 \sum_0^x \theta D_{x-u} +$$

$$+\int_c^d m A_{x-u} du - \lambda \int_c^d p B_{x-u} du + k\lambda^2 \int_c^d \gamma C_{x-u} du - k\lambda^3 \int_c^d \vartheta D_{x-u} du \qquad (2.33)$$

$$Q_x = Q_0 A_x + k\lambda v_0 B_x + k\lambda^2 \varphi_0 C_x - \frac{4}{\lambda} M_0 D_x -$$

$$-\sum_0^x P A_{x-u} + k\lambda \sum_0^x \Gamma B_{x-u} - k\lambda^2 \sum_0^x \theta C_{x-u} - \frac{4}{\lambda} \sum L D_{x-u} -$$

$$-\int_c^d p A_{x-u} du + k\lambda \int_c^d \gamma B_{x-u} du - k\lambda^2 \int_c^d \vartheta C_{x-u} du - \frac{4}{\lambda} \int_c^d m D_{x-u} du \qquad (2.34)$$

In all four of these equations, the first line shows the influence of initial parameters, the second line shows the influence of all four concentrated loads and deflections, and the third line shows the influence of all four distributed loads and deflections. In order to simplify the calculations, first we have to resolve all integrals located in the third lines of these equations. We denote all functions of distributed loads and deflections as $f(u)$ and their derivatives as $f'(u)$. In general, solutions for all four integrals are presented here:

$$\left.\begin{array}{l}
\displaystyle\int_c^d f(u) A_{x-u} du = -\lambda \int_c^d f(u) \frac{d}{du} B_{x-u} du = -\lambda\{[f(u) B_{x-u}]_c^d - \int_c^d f'(u) B_{x-u} du\} \\[3mm]
\displaystyle\int_c^d f(u) B_{x-u} du = -\lambda\{[f(u) C_{x-u}]_c^d - \int_c^d f'(u) C_{x-u} du \\[3mm]
\displaystyle\int_c^d f(u) C_{x-u} du = -\lambda\{[f(u) D_{x-u}]_c^d - \int_c^d f'(u) D_{x-u} du\} \\[3mm]
\displaystyle\int_c^d f(u) D_{x-u} du = \frac{\lambda}{4}\{[f(u) A_{x-u}]_c^d - \int_c^d f'(u) A_{x-u} du\}
\end{array}\right\} \qquad (2.35)$$

Equations 2.35 can be significantly simplified for uniformly distributed loads and deflections, as shown in Figure 2.20. When $f(u) = g = \text{Const.}$ $f'(u) = 0$, equations 2.36 will look as follows:

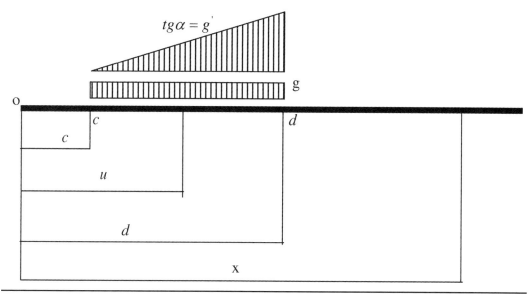

Figure 2.20

$$\int_c^d gA_{x-u}\,du = -g\lambda\,[B_{x-u}]_c^d = g\lambda\,(B_{x-c} - B_{x-d})$$

$$\int_c^d gB_{x-u}\,du = -g\lambda\,[C_{x-u}]_c^d = g\lambda\,(C_{x-c} - C_{x-d})$$

$$\int_c^d gC_{x-c}\,du = -g\lambda\,[D_{x-u}]_c^d = g\lambda\,(D_{x-c} - D_{x-d}) \tag{2.36}$$

$$\int_c^d gD_{x-u}\,du = \frac{g}{4}\lambda\,[A_{x-u}]_c^d = -\frac{g}{4}\lambda\,(A_{x-c} - A_{x-d})$$

When a distributed triangle load, with a slope g', is applied to the beam between points d and c, as shown in Figure 2.20, $(u) = g(u - c)$; $f'(u) = g_1 = \text{Const.}$, equations 2.35 will also look much simpler:

$$\int_c^d g_1\,(u - c)A_{x-u}\,du = g_1\lambda\,[\lambda\,(C_{x-c} - C_{x-d}) - (d - c)B_{x-d}]$$

$$\int_c^d g_1\,(u - c)B_{x-u}\,du = g_1\lambda\,[\lambda\,(D_{x-c} - D_{x-d}) - (d - c)C_{x-d}]$$

$$\int_c^d g_1\,(u - c)C_{x-u}\,du = -g_1\lambda\left[\frac{\lambda}{4}(A_{x-c} - A_{x-d}) + (d - c)D_{x-d}\right] \tag{2.37}$$

$$\int_c^d g_1\,(u - c)D_{x-u}\,du = -g_1\frac{\lambda}{4}\left[\lambda\,(B_{x-c} - B_{x-d}) - (d - c)A_{x-d}\right]$$

If a complex distributed load is applied to the beam, this load can always be replaced with a sum of uniformly distributed and triangle loads that allows using equations 2.36

and 2.37. Two numerical examples solved below illustrate application of the method of initial parameters to analysis of finite beams on Winkler foundation.

Example 2.1

A beam a-b, shown in Figure 2.21, is free supported on soil and loaded with a point load P at the left end a. Find the settlement and rotation at the left end of the beam.

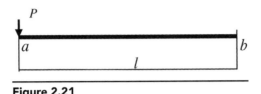

Figure 2.21

Conditions at the right end of the beam can be written as follows:

$$M_0 = 0 \quad Q_0 = -P \quad M_l = 0 \quad Q_l = 0$$

$$\left.\begin{array}{l} -P\lambda B_b + kv_0\lambda^2 C_b + k\varphi_0\lambda^3 D_b = 0 \\ -PA_b + kv_0\lambda B_b + k\varphi_0\lambda^2 C_b = 0 \end{array}\right\} \tag{2.38}$$

Solving this system of equations we will find

$$v_0 = P\frac{B_b C_b - A_b D_b}{k\lambda\,(C_b^2 - B_b D_b)} \quad \varphi_0 = P\frac{A_b C_b - B_b^2}{k\lambda^2\,(C_b^2 - B_b D_b)}$$

Using formulae 2.31–2.34, we will find deflections and forces at any point of the beam.

$$\left.\begin{array}{l} v = v_0 A_x + \lambda\varphi_0 B_x + \dfrac{\lambda^3}{EI}PD_x \\[2mm] \varphi = \varphi_0 A_x + \dfrac{P}{EI}\lambda^2 C_x - \dfrac{4}{\lambda}v_0 D_x \\[2mm] M = -P\lambda B_x + k\lambda^2 v_0 C_x + k\varphi_0\lambda^3 D_x \\[2mm] Q = -PA_x + k\lambda v_0 B_x + k\lambda^2\varphi_0 C_x \end{array}\right\} \tag{2.39}$$

The given data: $l = 10\text{ft}$, $b = 8\text{in}$, $k_0 = 180\dfrac{\text{lb}}{\text{in}^3}$, $k = bk_0 = 8\cdot 180 = 1{,}440\dfrac{\text{lb}}{\text{in}^2}$, $E = 1{,}400{,}000\dfrac{\text{lb}}{\text{in}^2}$, $P = 1$

$$\lambda = \sqrt[4]{\frac{4EI}{k}} = \sqrt[4]{4\cdot\frac{1{,}400{,}000\cdot 240}{1{,}440}} = 31{,}08\text{ in}, \quad \frac{l}{\lambda} = \frac{10\cdot 12}{31.08} = 3.86$$

$A_l = A(3.86) = 17.8751, \quad B_l(3.86) = -16.74335, \quad C_l(3.86) = -7.80685,$
$$D_l(3.86) = 0.5579,$$

$$v_0 = \frac{16.74335\cdot 7.80683 + 17.8751\cdot.3379}{1{,}440\cdot 31.08^2(7.80683^2 + 16.74335\cdot 0.5579)} = 10^{-5}\cdot 4.587\frac{\text{in}}{\text{lb}}$$

$$\varphi_0 = \frac{17.8751 \cdot 7.80683 - 16.74335^2}{1,440 \cdot 31.08^2 \cdot (7.80683^2 + 16.743379 \cdot 0.5579)} = -10^7 \cdot 15.19 \text{lb}^{-1}$$

Introducing $M_0 = 0$, $Q_0 = 0$, $v_0 = 10^5 \cdot 4.5876 \frac{\text{in}}{\text{lb}}$, and $\varphi_0 = -10^{-7} \times 15.19 \text{lb}^{-1}$ into the equations shown above, we find v_x, φ_x, M_x, and Q_x at any section of the beam.

Example 2.2

Let us solve a numerical example shown in Figure 2.22. The length of the beam $l = 2\text{m} \times 3.28 = 76.56\text{ft}$ and the width of the beam $b = 1.25\text{m} = 4.10\text{ft}$. The modulus of elasticity of the material is $E = 10^6 \frac{\text{t}}{\text{m}^2} = 10^6 \times \frac{2240\text{lb}}{10.764\text{ft}^2} = 10^6 \times 208.1 \frac{\text{lb}}{\text{ft}^2}$. $I = 0.256\text{m}^4 = 29.63\text{ft}^4$.

The modulus of subgrade reaction is $k_0 = 3,200 \frac{\text{t}}{\text{m}^3} = 3,200 \frac{2,240\text{lb}}{35.315\text{ft}^3}$ $= 202,973 \frac{\text{lb}}{\text{ft}^3}$

$$P_1 = 10\text{t} = 22,400\text{lb} \quad P_2 = 15\text{t} = 33,600\text{lb} \quad p = 2\frac{\text{t}}{\text{m}} = 2\frac{2,240}{3.28} = 1,366 \frac{\text{lb}}{\text{ft}}$$

$$u_1 = 5\text{m} = 5 \times 3.28 = 16.4\text{ft} \quad u_2 = 12\text{m} = 12 \times 3.28 = 39.36\text{ft}$$
$$c = 6\text{m} = 6 \times 3.28 = 19.68\text{ft} \quad d = 16\text{m} = 16 \times 3.28 = 52.48\text{ft}$$

Let us find $EI = 0.256 \times 10^6 \text{ tm}^2$, $k = k_0, b = 3,2001.25 = 4,000 \frac{\text{t}}{\text{m}^2}$

$$\lambda = \sqrt[4]{\frac{4EI}{k}} = \sqrt[4]{\frac{4 \times 0.256 \times 10^6}{4,000}} = 4\text{m}$$

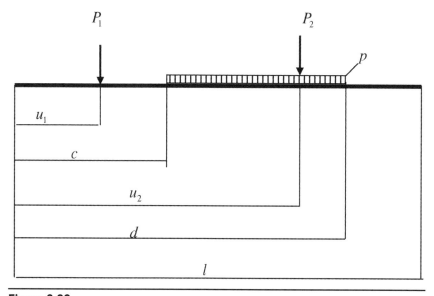

Figure 2.22

Now, using equations from 2.33 and 2.34, we shall find the total moment and shear at the right end of the beam. The following system of equations can be written:

$$\left.\begin{array}{l} k\lambda^2 C_l v_0 + k\lambda^3 D_l \varphi_0 - \lambda P_1 B_{l-u_1} - \lambda P_2 B_{l-u_2} - \lambda p \int_c^d B_{l-u} du = 0 \\ k\lambda^2 C_l \varphi_0 + k\lambda B_l v_0 - P_1 A_{l-u_1} - P_2 A_{l-u_2} - \lambda p (B_{l-c} - B_{l-d}) = 0 \end{array}\right\} \quad (2.40)$$

By solving this system of equations we find:

$$v_0 = \frac{1}{k\lambda^2} \frac{\lambda D_l \overline{Q}_l - C_l \overline{M}_l}{C_l^2 - B_l D_l} \qquad \varphi_0 = \frac{1}{k\lambda^3} \frac{B_l \overline{M}_l - \lambda C_l \overline{Q}_l}{C_l^2 - B_l D_l}$$

where

$$\left.\begin{array}{l} \overline{M}_l = -\lambda P_1 B_{l-u_1} - \lambda P_2 B_{l-u_2} - \lambda p \int_c^d B_{l-u} du = -\lambda P_1 B_{l-u_2} - \lambda P_2 B_{l-u_2} - \lambda^2 p (C_{l-c} - C_{l-d}) \\ \overline{Q}_l = -P_1 A_{l-u_1} - P_2 A_{l-u_2} - \lambda p (B_{l-c} - B_{l-d}) \end{array}\right\} \quad (2.41)$$

$\overline{M}_l$ and $\overline{Q}_l$ are the moment and the shear force at the right end of the beam due to the given loads P_1, P_2, and p. They can be found by introducing the given loads and parameters from Table 2.2 into equations 2.42.

$$\overline{M}_l = -4 \cdot 10 (-14.79715) - 4 \cdot 15 (0.95575) - 4^2 \cdot 2 (-2.9014 - 0.49445) = 643.2 \text{tm}$$

$$\overline{Q}_l = -10 (-17.4552) - 15 (-1.5656) - 4 \cdot 2 (-10.65245 - 0.96675) = 291 \text{t}$$

Now, initial parameters v_0 and φ_0 can be found:

$$v_0 = \frac{1}{k\lambda^2} \frac{\lambda D_l \overline{Q}_l - C_l \overline{M}_l}{C_l^2 - B_l D_l} = \frac{1}{4{,}000 \cdot 4^2} \cdot$$

$$\frac{4 \cdot (-23.0525) \cdot 291.0 - (-35.57745) \cdot 643.2}{688.15} = -0.0008973 \text{m}$$

$$\varphi_0 = \frac{1}{k\lambda^3} \frac{B_l \overline{M}_l - \lambda C_l \overline{Q}_l}{C_l^2 - B_l D_l} = \frac{1}{4{,}000 \cdot 4^2} \cdot$$

$$\frac{(-25.05645) \cdot 643.2 - 4(-35.57745) 291.0}{688.15} = 0.0001436$$

$$C_l^2 - B_l D_l = (-35.57745)^2 - (-25.05645) \cdot (-23.0525) = 688.15$$

Now, using equations 2.31–2.34, the settlements, rotations, moments, and shear forces can be found. The moment diagram is shown in Figure 2.23. Let us show how to find the moment in the middle of the beam. Using equation 2.33, when $x = 0.5l$ we have:

$$M_{0.5l} = k\lambda^2 v_0 C_{0.5l} + k\lambda^3 \varphi_0 D_{0.5l} - \lambda P_1 B_{0.5l-u_1} - p\lambda \int_c^{0.5l} B_{0.5l-u} du \quad (2.42)$$

Since $\int_c^{0.5l} B_{0.5l-u} du = \lambda (C_{0.5l-c} - C_0) = \lambda C_{0.5l-c}$ we can find:

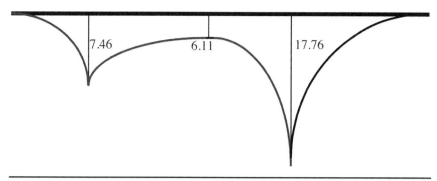

7.46   6.11   17.76

Figure 2.23   Moment diagram

$$M_{0.5l} = 4,000 \cdot 4^2 \left(-0.00008973\right) 1.81045 + 4,000 \cdot 4^3 \cdot 0.0001436$$

$$\cdot 2.12925 - 4 \cdot 10 \cdot 1.1486 - 2 \cdot 4^2 \cdot 0.49445 = 6.11 \text{tm}$$

The settlement of the beam at the same point can be obtained as:

$$v_{0.5l} = v_{0.5l} A_{0.5l} + \lambda \varphi_0 B_{0.5l} + \frac{\lambda^2}{EI} P_1 D_{0.5l - u_1} + \frac{\lambda^3}{EI} p \int_c^{0.5l} D_{0.5l - u} du$$

Taking into account that $\int_c^{0.5l} D_{0.5l - u} du = -\frac{\lambda}{4}\left(A_{0.5l - c} - 1\right)$, we can find the settlement at the middle of the beam as:

$$v_{0.5l} = v_0 A_{0.5l} + \lambda \varphi_0 B_{0.5l} + \frac{\lambda^2}{EI} P_1 D_{0.5l - u_1} - \frac{\lambda^3}{EI} p \frac{\lambda}{4}\left(A_{0.5l - c} - 1\right) = -0.00008997\left(-4.9128\right)$$

$$+ 4 \cdot 0.0001436\left(-0.5885\right) + \frac{4^3}{0.256 \cdot 10^6} \cdot 10 \cdot 0.32175 -$$

$$\frac{4^4}{0.256 \cdot 10^6} \cdot \frac{2}{4}\left(0.8337 - 1\right) = 0.0987 \text{cm}$$

All numerical values of the functions $A(\xi)$, $B(\xi)$, $C(\xi)$, and $D(\xi)$ are taken from Table 2.25, and shown in Tables 2.2 and 2.3.

Moments, shear forces, and deflections of the beam can be obtained analogously at any point. The soil pressure is obtained from this equation $p_{react.i} = k_0 v_i$ where v_i is the settlement of the beam at point i. The soil pressure at the middle of the beam is equal $p_{react.(0.5l)} = 3.2 \cdot 0.0987 = 0.3158 \text{kg/cm}^2 = 4.4823 \text{lb/in}^2$.

As can be seen from this numerical example, hand calculations are difficult and time-consuming. If analysis of the same beam requires several load combinations, analysis is simply impossible. Only computer analysis can produce fast and accurate results for beams with various loads and various boundary conditions. A numerical example shown below illustrates computer analysis of a beam on Winkler foundation.

Example 2.3

A computer program for the analysis of finite beams on Winkler foundation was developed by the author of this work and E. Dmitriyev. The program uses the method of initial parameters. Application of this program to the analysis of a free supported beam loaded with various loads is shown in Figure 2.24.

Data: $l = 20$m $c = 5$m $d = 15$m $u = 5$m $b = 1.25$m $I = 0.256$m^4
$E = 1,000,000$t/m^2 $k_0 = 3,200$t/m^3

$k = k_0 b = 3,200 \cdot 1.25 = 4,000$t/m^2 $P = 10$tn $M = 1$tm $q = 2,000$t/m

$\lambda = (4EI/k)^{1/4} = (4 \times 0.256 \times 10^6/4,000)^{1/4} = 4$m

Perform analysis of the given beam under all three loads separately, find settlements, rotations, moments, and shear forces in the beam with the spacing equal to 0.5m, and build the final moment diagram due to all loads applied to the beam. Results of computer analysis are given in Tables 2.4–2.6. Moment diagrams are shown in Figures 2.25–2.28.

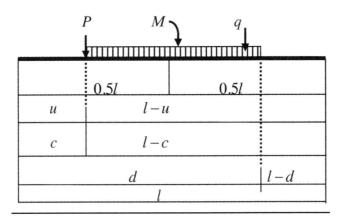

Figure 2.24

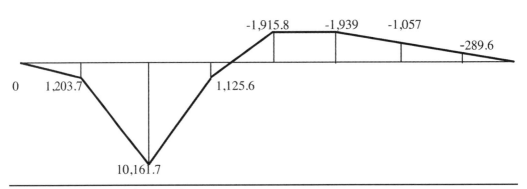

Figure 2.25   Moment diagram due to *P*

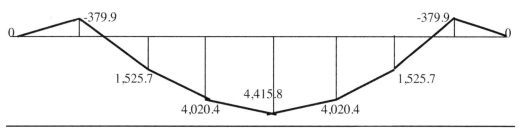

Figure 2.26 Moment diagram due to q

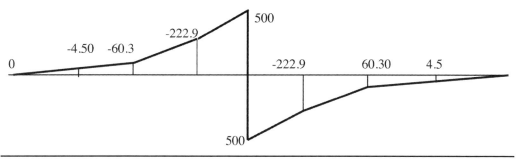

Figure 2.27 Moment diagram due to M

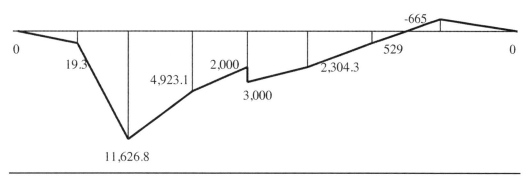

Figure 2.28 Final moment diagram

2.6 Analysis of Continuous Beams and Frames on Winkler Foundation: The Stiffness Method

In many instances, some or all elements of statically indeterminate systems are supported on elastic foundation. For example, a continuous beam shown in Figure 2.29 and a frame with a continuous footing shown in Figure 2.30 are structures supported on elastic foundation.

Methods of structural mechanics, method of forces, stiffness method, and combined method can be used for analysis of beams as well as other structures supported on Winkler foundation. This chapter includes a discussion of the application of the stiffness method to analysis of various beams and frames on Winkler foundation.

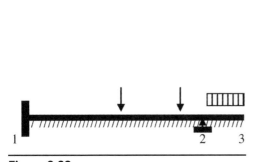

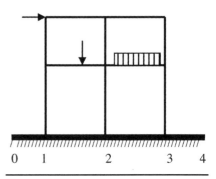

Figure 2.29   **Figure 2.30**

As shown above, the method of initial parameters allows performing analysis of finite beams with various boundary conditions supported on Winkler foundation. When a simple beam is loaded with given loads and boundary conditions are known, analysis is performed as shown above. Deflections, moments, and shear forces at any point of the beam due to initial parameters v_0, φ_0, M_0, and Q_0 can be found from the equations 2.43:

$$
\left.
\begin{aligned}
v_x &= v_0 A_x + \lambda \varphi_0 B_x - \frac{\lambda^2}{EI} M_0 C_x - \frac{\lambda^3}{EI} Q_0 D_x \\
\varphi_x &= \varphi_0 A_x - \frac{\lambda}{EI} M_0 B_x - \frac{\lambda^2}{EI} Q_0 C_x - \frac{4}{\lambda} v_0 D_x \\
M_x &= M_0 A_x + \lambda Q_0 B_x + k\lambda^2 v_0 C_x + k\lambda^3 \varphi_0 D_x \\
Q_x &= Q_0 A_x + k\lambda v_0 B_x + k\lambda^2 \varphi_0 C_x - \frac{4}{\lambda} M_0 D_x
\end{aligned}
\right\}
\tag{2.43}
$$

Equations 2.43 include only deflections and loads applied to the beam ends. They are obtained from equations 2.31–2.34 by excluding all deflections and loads applied to the beam between its ends.

Here we will show how to obtain reactions at the ends of a beam due to deflections applied to the beam ends. Let us find reactions for a beam with one free and one fixed end, as shown in Figure 2.31. Node m is restricted against any deflections. Both ends, e and f, are free supported. If rotation $\varphi_m = 1$ is applied to the node m, the first two equations from the system 2.43 can be rewritten as follows:

$$
\left.
\begin{aligned}
0 &= v_e A_l + \lambda \varphi_e B_l \\
1 &= \varphi_e A_l - \frac{4}{\lambda} v_e D_l
\end{aligned}
\right\}
\tag{2.44}
$$

From equations 2.44, we obtain $v_e = \dfrac{B_l \lambda}{A_l^2 + 4D_l B_l}$ and $\varphi_e = \dfrac{A_l}{A_l^2 + 4D_l B_l}$. By introducing v_e and φ_e into the third and fourth equation of the system 2.43, we find:

$$
M_{me}^\varphi = -\frac{4EI}{\lambda} \cdot \frac{B_l C_l - A_l D_l}{A_l^2 + 4D_l B_l} \qquad Q_{me}^\varphi = -\frac{4EIl}{\lambda} \cdot \frac{B_l^2 - A_l C_l}{A_l^2 + 4D_l B_l}
\tag{2.45}
$$

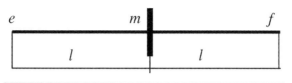

Figure 2.31

Reactions in the beam mf at node m, due to the same rotation $\varphi_m = 1$, are found analogously:

$$M^\varphi_{mf} = -\frac{4EI}{\lambda} \cdot \frac{B_l C_l - A_l D_l}{A_l^2 + 4D_l B_l} \quad Q^\varphi_{mf} = \frac{4EIl}{\lambda} \cdot \frac{B_l^2 - A_l C_l}{A_l^2 + 4D_l B_l} \tag{2.46}$$

By assuming $\alpha_1 = \dfrac{l}{\lambda} \cdot \dfrac{B_l C_l - A_l D_l}{A_l^2 + 4D_l B_l}$ $\alpha_2 = \dfrac{2}{3}\dfrac{l^2}{\lambda^2} \cdot \dfrac{B_l^2 - A_l C_l}{A_l^2 + 4D_l B_l}$ and $\dfrac{EI}{l} = i$ we find:

$$M^\varphi_{me} = -4i\alpha_1, \; Q^\varphi_{me} = -\frac{6i}{l}\alpha_2, \; M^\varphi_{mf} = -4i\alpha_1, \; Q^\varphi_{mf} = \frac{6i}{l}\alpha_2 \tag{2.47}$$

If deflection $v_m = 1$ is applied to the node m, the first two equations from system 2.43 will look as follows:

$$\left.\begin{array}{l} 1 = A_l v_e + \lambda B_l \varphi_e \\ 0 = -\dfrac{4}{\lambda} D_l v_e + A_l \varphi_e \end{array}\right\} \tag{2.48}$$

From 2.48, we have $v_e = \dfrac{A_l}{A_l^2 + 4D_l B_l}$ $\varphi_e = \dfrac{4D_l}{\lambda\left(A_l^2 + 4D_l B_l\right)}$

Now, by introducing v_e and φ_e into the last two equations of the system 2.43, reactions at node m are found:

$$M^v_{me} = \frac{k\lambda^2\left(A_l C_l + 4D_l^2\right)}{A_l^2 + 4D_l B_l} \quad Q^v_{me} = \frac{k\lambda\left(A_l B_l + 4DC_l\right)}{A_l^2 + 4D_l B_l} \tag{2.49}$$

It is easy to show that $A_l C_l + 4D_l^2 = B_l^2 - A_l C_l$. Simplified equations for all reactions at point m due to one-unit vertical deflection are equal:

$$M^v_{me} = \frac{6EI}{l^2}\alpha_2 \; Q^v_{me} = \frac{12EI}{l^3}\beta \; M^v_{mf} = -\frac{6EI}{l^2}\alpha_2, \; Q^v_{mf} = -\frac{12EI}{l^3}\beta \tag{2.50}$$

where $\beta = \dfrac{l^3\left(A_l B_l + 4D_l^2\right)}{3\lambda^3\left(A_l^2 + 4D_l B_l\right)}$

The total moment and shear force at the fixed end of the beam, for example, me, can be found from the following equations:

$$\left.\begin{array}{l} M_{me} = M^0_{me} - 4i_{me}\alpha_1\varphi_m + \dfrac{6i_{me}}{l_{me}}\alpha_2 v_m \\[2mm] Q_{me} = Q^0_{me} - \dfrac{6i_{me}}{l_{me}}\alpha_2\varphi_m + \dfrac{12i_{me}}{l^2_{me}}\beta v_m \end{array}\right\} \tag{2.51}$$

In 2.51, the first members are the moment and shear force due to the given loads, the second members are the moment and shear force due to the rotation φ_m of the node m, and the third members are the moment and the shear force due to the settlement v_m of the node m.

Similar equations can be obtained for a beam with both fixed ends shown in Figure 2.32. Both beams, me and mf, have various loads and deflections applied to their ends. Using the method of initial parameters and the same equations 2.43, we found all reactions at node m due to all loads and deflections.

Reactions at the node m, due to the given applied loads, are found using the method of initial parameters. Reactions at the node m due to one-unit deflections are shown below:

1. One unit rotation applied to the node m will produce:

$$M_{mf} = -\left(EI\right)_{mf}\xi_{mf}, \; Q_{mf} = \left(EI\right)_{mf}\mu_{mf}, \; M_{me} = -\left(EI\right)_{me}\xi_{me}, \; Q_{me} = \left(EI\right)_{me}\mu_{me} \quad (2.52)$$

2. One unit rotation applied to the node f will produce:

$$M_{mf} = -\left(EI\right)_{mf}\eta_{mf}, \; Q_{mf} = -\left(EI\right)_{mf}\gamma_{mf} \quad (2.53)$$

3. One unit rotation applied to the node e will produce:

$$M_{me} = -\left(EI\right)_{me}\eta_{me}, \; Q_{me} = -\left(EI\right)_{me}\gamma_{me} \quad (2.54)$$

4. One unit settlement applied to the node e will produce:

$$M_{me} = -\left(EI\right)_{me}\gamma_{me}, \; Q_{me} = -\left(EI\right)_{me}\sigma_{me} \quad (2.55)$$

5. One unit settlement applied to the node m will produce:

$$M_{mf} = \left(EI\right)_{mf}\mu_{mf} \; Q_{mf} = -\left(EI\right)_{mf}\psi_{mf}, \quad (2.56)$$

$$M_{me} = -\left(EI\right)_{me}\mu_{me} \; Q_{me} = \left(EI\right)_{me}\psi_{me}. \quad (2.57)$$

6. One unit settlement applied to the node f:

$$M_{mf} = \left(EI\right)_{mf}\gamma_{mf} \; Q_{mf} = \left(EI\right)_{mf}\sigma_{mf} \quad (2.58)$$

In equations 2.52–2.58:

$$\left.\begin{array}{l} \xi = \dfrac{BC - AD}{\lambda\left(C^2 - BD\right)}, \mu = \dfrac{AC - B^2}{\lambda^2\left(C^2 - BD\right)}, \eta = \dfrac{D}{\lambda\left(C^2 - BD\right)}, \gamma = \dfrac{C}{\lambda^2\left(C^2 - BD\right)} \\[3mm] \mu = \dfrac{AC + 4D^2}{\lambda^2\left(C^2 - BD\right)}, \psi = \dfrac{4CD + AB}{\lambda^3\left(C^2 - BD\right)}, \sigma = \dfrac{B}{\lambda^3\left(C^2 - BD\right)} \end{array}\right\} \quad (2.59)$$

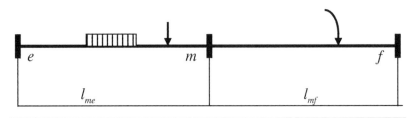

Figure 2.32

Equations 2.51–2.59 are obtained using the same method of initial parameters. The total moment and the total shear force at point m of the beam me are obtained from the equations 2.60:

$$\left.\begin{aligned}
M_{me} &= M^0_{me} - (EI)\xi_{me}\varphi_m - (EI)\eta_{me}\varphi_e + (EI)\xi_{me}v_m - (EI)\gamma_{me}v_e \\
Q_{me} &= Q^0_{me} + (EI)\mu_{me}\varphi_m - (EI)\gamma_{me}\varphi_e + (EI)\psi_{me}v_m - (EI)\sigma_{me}v_e
\end{aligned}\right\} \tag{2.60}$$

The first member of the first equation is the moment due to the given loads, the second member is the moment due to the rotation φ_m, the third member is the moment due to rotation φ_e, the fourth member is the moment due to the settlement v_m, and the fifth member is the moment due to the settlement v_e.

The second equation represents the sum of the shear forces, due to the given loads, and deflections at both ends of the beam me. If, for example, the beam with one fixed and one free supported end, shown in Figure 2.33, has a vertical load P, the moment and shear force at the left end m is obtained from equations 2.33 and 2.34. These two equations will look as follows:

$$\left.\begin{aligned}
0 &= M_0 A_l + \lambda Q_0 B_l - \lambda P B_{l-u} \\
0 &= Q_0 A_l - \frac{4}{\lambda} M_0 D_l - P A_{l-u}
\end{aligned}\right\} \tag{2.61}$$

By solving this system of equations, initial parameters at the left end of the beam are found. These initial parameters are the shear force Q_0 and moment M_0.

$$\left.\begin{aligned}
Q_0 &= Q_{mf} = P\frac{4D_l B_{l-u} + A_l A_{l-u}}{4B_l D_l + A_l^2} \\
M_0 &= M_{mf} = -P\lambda\frac{B_l A_{l-u} - A_l B_{l-u}}{4B_l D_l + A_l^2}
\end{aligned}\right\} \tag{2.62}$$

For other types of loads initial parameters are found analogously. The method of initial parameters allows obtaining equations for beams with various boundary conditions, loaded with various loads and deflections. In order to illustrate application of the equations obtained above to practical analysis, a numerical example is analyzed on the next page:

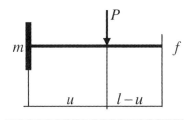

Figure 2.33

Example 2.4

Given the beam shown in Figure 2.34, the loads applied to the beam are:

$$N_1 = 93.89t \ N_2 = 195.00t \ N_3 = 137.11t$$
$$M_1 = 18.78tm \ M_2 = 45.80tm \ M_3 = 46.672tm.$$

The modulus of elasticity $E = 2,650,000t/m^2$, the modulus of subgrade reaction $k = 3,000t/m^3$, $l_1 = 6.4$, $l_2 = 2.95m$, and the moment of inertia of the beam $I = 0.10916m^4$.

Solution:

1. Find parameter $\lambda = \sqrt[4]{\dfrac{4EI}{kb}} = \sqrt[4]{\dfrac{4 \cdot 2,650,000 \cdot 1,091.6}{3,000 \cdot 2 \cdot 10^4}} = 3.727m$

$$\frac{l_1}{\lambda} = \frac{6.4}{3.727} = 1.72, \quad \frac{l_2}{\lambda} = \frac{2.95}{3.727} = 0.80$$

2. Find functions $A_1 = -0.4612$, $B_1 = 1.2195$, $C_1 = 1.34855$, and $D_1 = 0.8263$ for beams 1–2 and 2–3.
3. Find functions $A_2 = 0.9318$, $B_2 = 0.7897$, $C_2 = 0.31855$, and $D_2 = 0.08515$ for beams 0–1 and 3–4.
4. Using equations 2.59, find the following parameters for beams 1–2 and 2–3:

$$\mu = \frac{(-0.4612) \cdot 1.34855 - 1.2195^2}{3.727^2 \cdot (1.34855^2 - 1.2195 \cdot 0.8263)} = -0.19$$

$$\xi = \frac{1.2195 \cdot 1.34855 + 0.4612 \cdot 0.8263}{3.727 \cdot 0.81} = 0.672$$

$$\eta = \frac{0.8263}{3.727 \cdot 0.81} = 0.275, \sigma = \frac{1.2195}{3.727^3 \cdot 0.81} = 0.03, \gamma = \frac{1.34855}{3.727^2 \cdot 0.81} = 0.121$$

$$\psi = \frac{4 \cdot 1.34855 - 0.4612 \cdot 1.2195}{3.727^3 \cdot 0.81} = 0.095$$

Now we can find functions for beams 0–1 and 3–4 using equations 2.46–2.47:

$$\alpha_1 = \frac{0.8(0.7891 \cdot 0.31855 - 0.9318 \cdot 0.08515)}{0.9318^2 + 4 \cdot 0.7891 \cdot 0.08515} = 0.12$$

$$\alpha_2 = \frac{2 \cdot 0.8^2(0.9318 \cdot 0.31855 + 4 \cdot 0.08515^2)}{3 \cdot 1.138} = 0.122$$

Using the stiffness method, we can build a system of linear equations shown in Table 2.7, and by introducing numerical values into Table 2.8, deflections of the nodes 1, 2, and 3 are found:

$$\phi_1 = 0.0991407\frac{t}{m^2}, \phi_2 = 0.0483726\frac{t}{m^2}, \phi_3 = 0.0090306\frac{t}{m^2} \ v_1 = 0.6989112\frac{t}{m},$$

$$v_2 = 1.2632842\frac{t}{m} \ v_3 = 1.0942513\frac{t}{m}$$

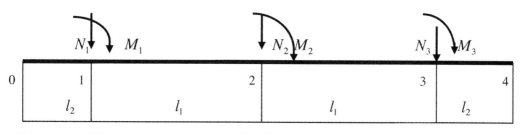

Figure 2.34

Notice that all coefficients of the left part of the system are multiplied by $\dfrac{10^4}{E}$ for convenience. Therefore, all results also have to be multiplied by $\dfrac{10^4}{E}$. Real deflections of the nodes 1, 2, and 3 are equal to:

$$\varphi_1 = 3.741 \cdot 10^{-4} \text{rad}, \quad \varphi_2 = 1.825 \cdot 10^{-4} \text{rad}, \quad \varphi_3 = 0.34 \cdot 10^{-4} \text{rad}$$
$$v_1 = 2.6374 \text{mm}, \quad v_2 = 4.7654 \text{mm}, \quad v_3 = 4.1292 \text{mm}$$

Using deflections found above and equations 2.51 and 2.60, all moments at the ends of the beams are found:

$$M_{1-0} = -4\frac{1{,}091.6}{2.95} \cdot 0.12 \cdot 0.0991407 + \frac{6 \cdot 1{,}091.6}{2.95^2} \cdot 0.122 \cdot 0.6989112 = 46.53 \text{tm}$$

$$M_{1-2} = -1.091.6 \cdot 0.672 \cdot 0.0991407 - 1{,}091.6 \cdot 0.275 \cdot 0.0483726$$
$$-1{,}091.6 \cdot 0.19 \cdot 0.6989112 + 1{,}091.6 \cdot 0.121 \cdot 1.2632842 = -65.31 \text{tm}$$

$$M_{2-1} = -1{,}091.6 \cdot 0.672 \cdot 0.0483726 - 1{,}091.6 \cdot 0.275 \cdot 0.0991407$$
$$+1{,}091.6 \cdot 0.19 \cdot 1.2632842 - 1{,}091.6 \cdot 0.121 \cdot 0.6989112 = 104.45 \text{tm}$$

$$M_{2-3} = -1{,}091.6 \cdot 0.672 \cdot 0.0483726 + 1{,}091.6 \cdot 0.275 \cdot 0.0090306$$
$$-1{,}091.6 \cdot 0.19 \cdot 1.2632842 + 1{,}091.6 \cdot 0.121 \cdot 1.0942513 = -150.25 \text{tm}$$

$$M_{3-2} = -1{,}091.6 \cdot 0.672 \cdot (-0.0090306) - 1{,}091.6 \cdot 0.275 \cdot 0.0483726$$
$$+1{,}091.6 \cdot 0.19 \cdot 1.0942513 - 1{,}091.6 \cdot 0.121 \cdot 1.2632842 = 52.20 \text{tm}$$

$$M_{3-4} = 4\frac{1{,}091.6}{2.95} \cdot 0.12 \cdot 0.0090306 - \frac{6 \cdot 1{,}091.6}{2.95^2} 0.122 \cdot 1.0942513 = -98.87 \text{tm}$$

All shear forces at the same points can be found analogously. Using equations 2.43, deflections, moments, and shear forces are found at any point of the beams 0–1, 1–2, 2–3, and 3–4. The moment diagram is shown in Figure 2.35. The moments are shown only at the ends of the foundation elements at points 0, 1, 2, 3, and 4.

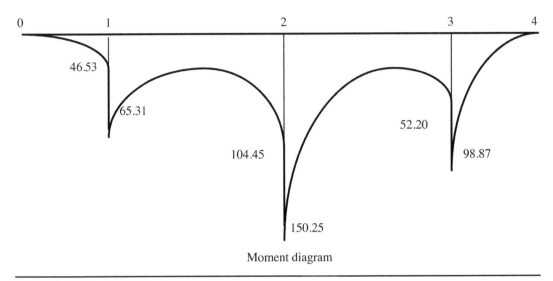

Moment diagram

Figure 2.35

2.7 Exact Analysis of Frames on Winkler Foundation

The stiffness method previously described can also be applied to analysis of frames on elastic foundation. Analysis of a frame, shown in Figure 2.36, in principle, is not different from analysis of any other statically indeterminate system. This frame only differs from other regular frames because some of the frame members (0–1, 1–2, 2–3, 3–4) are supported on elastic foundation.

The stiffness matrix for this frame is built by using well-known equations from structural mechanics and equations for beams supported on Winkler foundation. The system of equations looks as follows:

$$\vec{A}X = \vec{B} \tag{2.63}$$

In this system, $\vec{A}$ is the stiffness matrix of the total system frame-foundation, $\vec{B}$ is the matrix of reactions due to the given loads applied to the frame, and X is the matrix of unknown deflections. The total number of unknowns for the frame shown above is equal to 24. The number of unknowns includes: 6 deflections of the nodes 1, 2, and 3

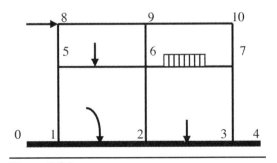

Figure 2.36

and 18 deflections of the nodes 5, 6, 7, 8, 9, and 10. Horizontal deflections of the nodes 1, 2, and 3 are not taken into account.

Once the system 2.63 is solved, it is easy to find the moments and shear forces in all elements of the frame. The shear forces and moments in all elements of the foundation are found using formulae obtained above.

Direct analysis of frames on elastic foundation can only be performed by using computer software that includes solutions for regular line elements as well as solutions for beams supported on Winkler foundation.

Analysis can be simplified by using the second method and is performed as follows:

1. Assuming that the first level columns do not experience any deflections at points 1, 2, and 3, the given frame and foundation are investigated separately.
2. The frame is analyzed successively under the given loads and all support deflections equal one unit.
3. The foundation is analyzed successively under the given loads and deflections of all supports equal one unit.
4. Now, the following system of equations can be built:

$$\left(A_{\text{Frame}} + A_{\text{Found}}\right)X = \left(B_{\text{Frame}} + B_{\text{Found}}\right) \tag{2.64}$$

   where $\left(A_{\text{Frame}} + A_{\text{Found}}\right)$ is the total stiffness matrix of the system frame-foundation and $\left(B_{\text{Frame}} + B_{\text{Found}}\right)$ is the matrix of reactions of all loads applied to the frame and foundation.
5. By solving the system 2.64 deflections at all three points, 1, 2, and 3 are found. The system of equations 2.64 is relatively small. It includes only deflections of the frame supports. Analysis of the frame, shown in Figure 2.36, requires solution of a system with only six unknowns.
6. By applying obtained deflections and the given loads to the frame, the final analysis of the frame is performed.
7. Using equations obtained previously for analysis of beams on Winkler foundation, shear forces and moments are found in all elements of the foundation: 0–1, 1–2, 2–3, and 3–4.

Both methods described above can be used. The second method is easier. It requires the use of standard software for analysis of regular, statically indeterminate systems and software for building and solving a system of linear equations. As shown, the first method requires a standard procedure: building the global stiffness matrix for the total system frame-foundation, solving a system of linear equations and obtaining moments, and shear forces in all elements of the system. The second method avoids creating the global stiffness matrix. Obtained deflections at points of contact between the frame and foundation are used for final analyses of the frame and foundation.

No one of these two methods can be recommended for hand analysis. The use of computer software based on one of the two methods is probably the only practical solution.

It is important to mention that equations of the stiffness method do not take into account that areas of intersection of the frame columns and continuous foundations are totally rigid, especially in multistoried, heavy loaded reinforced concrete frames.

Totally rigid elements include not only the areas of intersection, but also, in many cases, enlargements such as pedestals, as shown in Figures 2.37 and 2.38. Enlargements of these areas are usually designed to reduce high shear stresses in the foundation. Equations of the stiffness method for analysis of beams with totally rigid elements are obtained below. Let us assume we have a beam shown in Figure 2.39.

The beam consists of two beams with fixed ends, *me* and *mf*. Both beams have totally rigid elements at their ends, ee_1, mm_1, mm_2, and ff_2. Elements e_1m_1 and f_2m_2 are elastic.

We shall find the moments and shear forces in these beams, due to the given loads applied to the beam and due to deflections applied to the beam ends. The moments and shear forces at points e_1, m_1, m_2, and f_2 due to the given loads can be obtained by using the method of initial parameters described earlier. Moments and shear forces at the same points, due to node deflections, can be obtained using equations 2.52–2.60. If rotation $\varphi_m = 1$ is applied to the node m, node m_1 will experience the same rotation and vertical deflection equal to a_2 due to this rotation. So, the moment at point m_1 can be found as:

$$M_{m_1 e_1}^{\varphi_m = 1} = -k_{m_1 e_1}\left(\xi_{m_1 e_1} - a_2 \mu_{m_1 e_1}\right) \tag{2.65}$$

The shear force due to the same rotation at point m_1 is equal:

$$Q_{m_1 e_1}^{\varphi_m = 1} = k_{m_1 e_1}\left(\mu_{m_1 e_1} - a_2 \xi_{m_1 e_1}\right) \tag{2.66}$$

If rotation $\varphi_e = 1$ is applied to the node e, the moment at point m_1 due to this rotation is equal:

$$M_{m_1 e_1}^{\varphi_e = 1} = -k_{m_1 e_1}\left(\eta_{m_1 e_1} - a_1 \gamma_{m_1 e_1}\right) \tag{2.67}$$

The shear force at point m_1 due to the same rotation is equal:

$$Q_{m_1 e_1}^{\varphi_e = 1} = -k_{m_1 e_1}\left(\gamma_{m_1 e_1} + a_1 \sigma_{m_1 e_1}\right) \tag{2.68}$$

A vertical deflection of the node m will cause the same deflection of the node m_1. The moment and the shear force at point m_1 due to this settlement are equal, respectively, to:

$$M_{m_1 e_1}^{v_e = 1} = -k_{m_1 e_1} \mu_{m_1 e_1} \tag{2.69}$$

$$Q_{m_1 e_1}^{v_m = 1} = k_{m_1 e_1} \psi_{m_1 e_1} \tag{2.70}$$

The vertical deflection of the node e $v_e = 1$ will produce a moment and shear force at point m_1 equal, respectively:

$$M_{m_1 e_1}^{v_e = 1} = -k_{m_1 e_1} \gamma_{m_1 e_1} \tag{2.71}$$

$$Q_{m_1 e_1}^{v_e = 1} = -k_{m_1 e_1} \sigma_{m_1 e_1} \tag{2.72}$$

Equations for the beam *mf* are obtained analogously. Equations for both beams are shown in Table 2.7. Equations 2.62–2.72 allow finding moments and shear forces at the edges of totally rigid elements in beams with both fixed ends, due to deflections of their ends. Now, the moments and shear forces at the nodes *m*, *f*, and *e* centers of

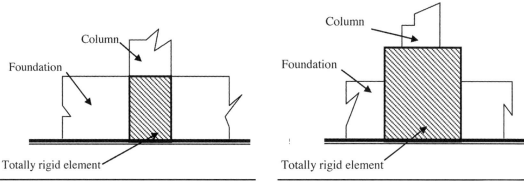

Figure 2.37   **Figure 2.38**

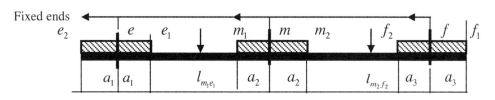

Figure 2.39

totally rigid elements m_1m_2, f_1f_2, and e_1e_2 can be found. The totally rigid element, as shown in Figures 2.40 and 2.41, will experience a settlement v and rotation φ due to applied vertical load P and moment M. The settlement is obtained from equation 2.73:

$$v = \frac{P}{K_z 2ab} \tag{2.73}$$

where $2a$ is the length of the foundation, b is the width of the foundation, and K_z is the modulus of subgrade reaction under the totally rigid element. The moment M will produce rotation of the same element. The soil pressure due to this rotation is equal to:

$$p = K_z z \tag{2.74}$$

where

$$z = \varphi y \tag{2.75}$$

Taking into account the equilibrium of all moments applied to the foundation, the following equation can be written:

$$M - \int_A K_z py dA = 0 \tag{2.76}$$

Since $\int y^2 dA = I$ is the moment of inertia of the foundation, rotation of the foundation can be found as:

$$\varphi = \frac{M}{K_z I} \tag{2.77}$$

As recommended in Chapter 1, K_z is replaced with K_φ and rotation of the foundation is equal to:

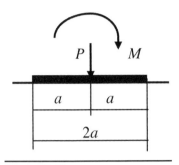

Figure 2.40

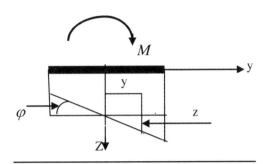

Figure 2.41

$$\varphi = \frac{M}{IK_\varphi} \tag{2.78}$$

where K_φ, in accordance with Barkan (1962), is the modulus of subgrade reaction of a totally rigid foundation experiencing rotation. Now, using equations 2.65–2.72 and 2.73–2.75, the moments and shear forces can be found at the centers of totally rigid elements. Equations for a beam with one fixed and one free supported end, shown in Figure 2.42, can be obtained analogously. Using equations 2.47, 2.50, 2.73, and 2.78, equations for moments and shear forces at points m, m_1, and m_2 can be found. Here are obtained equations for the beam me_1. For the beam mf_2, the equations are found analogously. If rotation $\varphi_m = 1$ is applied to the node m, the beam me_1 will experience at point m_1 the same rotation and a vertical deflection equal to $\varphi_m a = a$. The total moment at this point, due to rotation of the node m $\varphi_m = 1$ is found from equation 2.79:

$$M^{\varphi_m}_{m_1 e_1} = -2i_{m_1 e_1}\left(2\alpha_1 + \frac{3\alpha_2}{l_{m_1 e_1}}a\right) \tag{2.79}$$

The total shear force at the same point is obtained from equation 2.80:

$$Q^{\varphi_m}_{m_1 e_1} = -\frac{6i_{m_1 e_1}}{l_{m_1 e_1}}\left(\alpha_2 + \frac{2\beta}{l_{m_1 e_1}}a\right) \tag{2.80}$$

The settlement of the node m $v_m = 1$ will produce the same settlement of the node m_1, a moment and shear force found from equations 2.81 and 2.82:

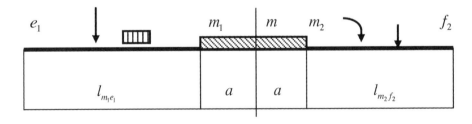

Figure 2.42

$$M^{V_m=1}_{m_1 e_1} = \frac{6i_{m_1 e_1}}{l_{m_1 e_1}} \alpha_2 \tag{2.81}$$

$$Q^{V_m=1}_{m_1 e_1} = \frac{12i_{m_1 e_1}}{l^2_{m_1 e_1}} \beta \tag{2.82}$$

Equations for moments and shear forces at node m, due to the same deflections, can be obtained by taking into account the soil reaction under the totally rigid element and the shear force at node m_1 found above. The moment at point m is equal to:

$$M^{\varphi_m}_{me_1} = M^{\varphi_m}_{m_1 e_1} + Q^{\varphi_m}_{m_1 e_1} a - \frac{K_\varphi I}{2} = -2i_{m_1 e_1}\left(2\alpha_1 + \frac{3\alpha_2}{l_{m_1 e_1}}a\right) - \frac{6i_{m_1 e_1}}{l_{m_1 e_1}}\left(\alpha_2 + \frac{2\beta}{l_{m_1 e_1}}a\right)a - \frac{K_\varphi I}{2}$$

$$= -4i_{m_1 e_1}\left(\alpha_1 + \frac{3a}{l_{m_1 e_1}}\alpha_2 + \frac{3a^2}{l^2_{m_1 e_1}}\beta\right) - \frac{K_\varphi I}{2} \tag{2.83}$$

The shear force at the same point m is equal to:

$$Q^{\varphi_m}_{me_1} = -\frac{6i_{m_1 e_1}}{l_{m_1 e_1}}\left(\alpha_2 + \frac{2\beta}{l_{m_1 e_1}}a\right) - \frac{K_\varphi A_F a}{4} \tag{2.84}$$

The moment and the shear force at point m due to the settlement equal to $v_m = 1$, can be found from the two following equations, respectively:

$$M^{V_m}_{me_1} = \frac{6i_{m_1 e_1}}{l_{m_1 e_1}}\left(\alpha_2 + \frac{2\beta}{l_{m_1 e_1}}a\right) + \frac{K_\varphi A_F a}{4} \tag{2.85}$$

$$Q^{V_m}_{me_1} = \frac{12i_{m_1 e_1}}{l^2_{m_1 e_1}}\beta + \frac{k_z A_F}{4} \tag{2.86}$$

Equations obtained above allow the use of the stiffness method for analysis of beams with totally rigid elements, as well as for analysis of frames, with continuous footings, taking into account totally rigid elements at points of supports. Equations for obtaining the shear forces as well as moments in beams with totally rigid elements are given in Tables 2.9–2.13. These tables allow obtaining moments and shear forces for two types of beams: one with both ends fixed and the second with one fixed and one free supported end. Complete equations for beams with both ends fixed are shown in Table 2.14. Equations for beams with one fixed and one free supported end are given in Table 2.15. In order to illustrate application of these equations, a numerical example shown next is solved.

Example 2.5

Given: A frame with a continuous footing supported on Winkler foundation, as shown in Figure 2.43. All node numbers and rigidities of the frame elements are shown in Figure 2.44. All data are given below:

$$h_1 = 3m \; h_2 = 4.2m \; h_3 = 3.6m \; h_4 = 4.7m \; l = 6.4m \; l_1 = 5.5m \; l_2 = 2.5m \; a = 0.45m$$

$$h_F = 1.2m \; h' = 0.4m \; h_{col} = 1.22m \; q_2 = 6.2\frac{t}{m} \; q_3 = 6.0\frac{t}{m} \; I_F = 0.10916m^4$$

$$S_1 = 3t \; S_2 = 5.9t \; S_3 = 11.7t \; S_4 = 10.6t \; P_1 = 19.52t \; P_2 = 55.84t$$

The modulus of elasticity of the frame and foundation material $E = 2,650,000\frac{t}{m^2}$ and the modulus of subgrade reaction $K_z = 3,000\frac{t}{m^3}$, $K_\phi = 6,000\frac{t}{m^3}$. The cross sections of the columns at the first level are equal to $0.4 \cdot 0.60(h)$ m² at the second, third, and forth levels the cross sections of the columns are equal to $0.4 \cdot 04m^2$. The cross sections of the beams at the first, second, and third levels are equal to $0.3 \cdot 0.63(h)$ m², and the cross section of the beam at the fourth level is equal to $0.4 \cdot 1.1(h)$ m². In order to simplify all calculations, all rigidities of the frame elements, including the rigidity of the soil and foundation, are multiplied by $\frac{10^4}{E}$. All rigidities are found below.

$$i_1 = \frac{4 \cdot 6^3}{3 \cdot 12} = 12 \; i_2 = \frac{4^4}{4.2 \cdot 12} = 5.07 \; i_3 = \frac{4^4}{3.6 \cdot 12} = 5.93$$

$$i_4 = \frac{4^4}{4.7 \cdot 12} = 4.57 \; i_5 = \frac{3 \cdot 6.3^3}{6.4 \cdot 12} = 9.80$$

$$i_6 = \frac{4 \cdot 11^3}{12 \cdot 12.8} = 34.7 \; k_z = 3,000\frac{10^4}{2,650,000} = 11.3$$

$$K_\phi = 6,000\frac{10^4}{2,650,000} = 22.6 \; \lambda = \sqrt[4]{\frac{4 \times 2,650,000 \times 1,091.6}{3,000 \times 2 \times 10^4}} = 3.727$$

$\frac{l_1}{\lambda} = \frac{5.5}{3.727} = 1.48$ (beams 17–18 and 19–20), $\frac{l_2}{\lambda} = \frac{2.5}{3.727} = 0.6708$ (beams 15–16 and 21–22)

Using Table 2.25, we find:

For beams 17–18 and 19–20, $A_1 = 0.2095, B_1 = 1.2448, C_1 = 1.03705, D_1 = 0.5280$

For beams 15–16 and 21–22, $A_2 = 0.9644, B_2 = 0.67515, C_2 = 0.23065, D_2 = 0.0524$

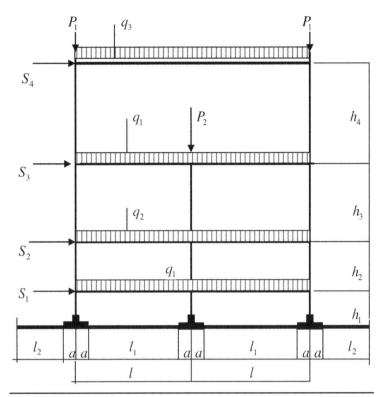

Figure 2.43

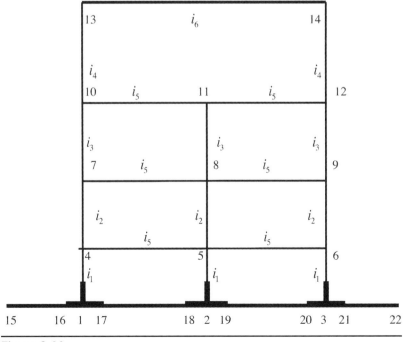

Figure 2.44

For beams 17–18 and 19–20:

$$\xi = \frac{1.2448 \cdot 1.03705 - 0.2099 \cdot 0.528}{(3.727)(1.03705^2 - 1.2448 \cdot 0.528)} = 0.76186,$$

$$\mu = \frac{0.2099 \cdot 1.037705 - 1.2448}{3.727^2 \cdot 0.41822} = -0.23155$$

$$\eta = \frac{0.528}{3.727 \cdot 0.41822} = 0.34087, \ \gamma = \frac{1.03705}{3.727 \cdot 0.41822} = 0.18077,$$

$$\sigma = \frac{1.2448}{3.727 \cdot 0.41822} = 0.05858 \ \psi = \frac{4 \cdot 1.03705 \cdot 0.528 + 0.2099 \cdot 1.2448}{3.727^3 \cdot 0.41822}$$

For beams 15–16 and 21–22 we find:

$$\alpha_1 = 0.68 \frac{0.67515 \cdot 0.23065 - 0.9644 \cdot 0.0524}{0.9644 + 4 \cdot 0.67515 \cdot 0.0524} = 0.06665,$$

$$\alpha_2 = \frac{2}{3} \cdot 0.68^2 \frac{0.9644 \cdot 0.23065 + 4 \cdot 0.524^2}{1.07158} = 0.06715$$

$$\beta = \frac{0.68^3}{3} \cdot \frac{0.9644 \cdot 0.67515 + 4 \cdot 0.23065 \cdot 0.0524}{1.07158} = 0.0672$$

Analysis of the frame and foundation is performed as one statically indeterminate system, using the stiffness method, applying equations for elements of the foundation obtained above and well-known equations from structural mechanics for line elements of the frame. The system of linear equations is written in Tables 2.16–2.18.

All coefficients in the system are found below:

$$a_{11} = a_{33} = 4i_{m_1 e_1}\left(\alpha_1 + \frac{3a}{l_{m_1 e_1}}\alpha_2 + \frac{3a^2}{l_{m_1 e_1}^2}\beta\right) + K_\varphi I + \left(k\xi - 2ak\mu + ka^2\psi\right) +$$

$$+4i_1\left(1 + \frac{3h_p}{h_1} + \frac{3h_p^2}{h_1^2}\right) \ a_{22} = 2\left(k\xi - 2ak\mu + ka^2\psi\right) + K_\varphi I + 4i_1\left(1 + \frac{3h_p}{h_1} + \frac{3h_p^2}{h_1^2}\right)$$

$$a_{44} = a_{66} = 4\left(i_1 + i_2 + i_5\right) \ a_{55} = 4\left(i_1 + i_2 + 2i_5\right) \ a_{77} = a_{99} = 4\left(i_2 + i_3 + i_5\right)$$

$$a_{88} = 4\left(i_2 + i_3 + 2i_5\right) \ a_{10,10} = a_{12,12} = 4\left(i_3 + i_4 + i_5\right) \ a_{11,11} = 4\left(i_3 + 2i_5\right)$$

$$a_{13,13} = a_{14,14} = 4\left(i_4 + i_6\right) \ a_{15,15} = \frac{36i_1}{h_1^2} \ a_{16,16} = \frac{36i_2}{h_2^2} \ a_{17,17} = \frac{36i_3}{h_3^2}$$

$$a_{18,18} = \frac{24i_4}{h_4^2} \ a_{19,19} = \frac{12i_{15-16}}{l_2^2}\beta + A_F k_z + \psi k + \frac{36i_5}{l^2} + \frac{12i_6}{4l^2} \qquad (2.87)$$

$$a_{20,20} = 2\kappa\psi + \frac{72i_5}{l^2} + k_z A_F \ a_{13,14} = 2i_6 \ a_{45} = a_{56} = a_{78} = a_{89} = a_{10,11} = 2i_5$$

$$a_{47} = a_{58} = a_{69} = 2i_2 \ a_{7,10} = a_{8,11} = a_{9,12} = 2i_3 a_{10,13} = a_{12,14} = 2i_4 \ a_{4,15} = a_{5,15} = a_{6,15} = -\frac{6i_1}{h_1}$$

$$a_{4,16} = a_{5,16} = a_{6,16} = a_{7,16} = a_{8,16} = a_{9,16} = -\frac{6i_2}{h_2}$$

$$a_{7,17} = a_{8,17} = a_{9,17} = a_{10,17} = a_{11,17} = a_{12,17} = -\frac{6i_3}{h_3}$$

$$a_{10,18} = a_{12,18} = a_{13,18} = a_{14,18} = -\frac{6i_4}{h_4} \quad a_{1,19} = k(a\psi - \mu) - \frac{6i_{15-16}}{l_2}\left(\alpha_2 + \frac{2a}{l_2}\beta\right)$$

$$a_{3,21} = -k(a\psi - \mu) + \frac{6i_{15-16}}{l_2}\left(\alpha_2 + \frac{2a}{l_2}\beta\right) \quad a_{2,19} = a_{3,20} = -a_{2,21} = k(\gamma + a\sigma)$$

$$a_{13,19} = a_{14,19} = -a_{13,21} = -a_{14,21} = \frac{3i_6}{l}$$

$$a_{4,19} = a_{5,19} = a_{7,19} = a_{8,19} = a_{10,19} = a_{11,19} = a_{6,20} = a_{9,20} = a_{12,20} = \frac{6i_5}{l}$$

$$a_{4,20} = a_{5,21} = a_{6,21} = a_{7,20} = a_{8,21} = a_{9,21} = a_{10,20} = a_{11,21} = a_{12,21} = -\frac{6i_5}{l}$$

The matrix of reactions at the same restraints due to applied loads is shown below. If the stiffness matrix is equal to A, and the matrix of reactions, due to given loads, is equal to B, the following system of linear equations can be written as follows:

$$\vec{A}X = \vec{B} \qquad (2.88)$$

Matrix A is shown in Tables 2.16 and 2.17. The stiffness matrix B is shown in Table 2.18.

Now, by solving this system of equations all deflections can be found, including rotations and settlements of the supports at points 1, 2, and 3. Results of analysis are presented below.

Rotations of the nodes:

$$\varphi_1 = 0.007710572\text{t/m}^2, \ \varphi_2 = 0.0401539\text{t/m}^2, \ \varphi_3 = 0.002197971\text{t/m}^2,$$
$$\varphi_4 = 0.5570987\text{t/m}^2, \ \varphi_5 = 0.03147339\text{t/m}^2, \ \varphi_6 = 0.3196702\text{t/m}^2,$$
$$\varphi_7 = 0.8033868\text{t/m}^2, \ \varphi_8 = 0.3194003\text{t/m}^2, \ \varphi_9 = 0.3140125\text{t/m}^2,$$
$$\varphi_{10} = 0.6881657\text{t/m}^2, \ \varphi_{11} = 0.0772555\text{t/m}^2, \ \varphi_{12} = 0.3504304\text{t/m}^2,$$
$$\varphi_{14} = -0.819296\text{t/m}^2$$

Horizontal deflections of the frame:

$$\delta_1 = 1.029058\text{t/m}, \ \delta_2 = 4.567592\text{t/m}, \ \delta_3 = 2.88701\text{t/m}, \ \delta_4 = 3.597103\text{t/m}$$

Vertical deflections of frame supports:

$$v_1 = 0.7182246\text{t/m}, \ v_2 = 1.170075\text{t/m}, \ v_3 = 1.099772\text{t/m}$$

All deflections found above are not the real deflections. Since all coefficients of the stiffness matrix were multiplied by $\dfrac{10^4}{E}$ in order to find the real deflections, the found deflections are also multiplied by $\dfrac{10^4}{E}$. For example, the real settlement at point 3 is equal to:

$$v_3 = 1.099772 \text{t/m} \cdot \frac{10^4}{E} = 1.099772 \text{t/m} \cdot \frac{10^4}{2,650,000} = 0.00415 \text{m} = 4.15 \text{mm}$$

Now, all moments and shear forces in the frame can be found. We shall find all moments and shear forces in the foundation at the points of support 1, 2, and 3, as well as at the edges of totally rigid elements:

$$M_{1-15} = \left[4i_{15-16}\left(\alpha_1 + \frac{3a}{l_2}\alpha_2 + \frac{3a^2}{l_2^2}\beta\right) + \frac{K_\varphi I_F}{2}\varphi_1 - \left[\frac{6i_{15-16}}{l_2}\left(\alpha_2 + \frac{2a}{l_2}\beta\right) + \frac{k_z Aa}{4}\right]v_1 \right] =$$

$$\left[4 \cdot \frac{1,091.6}{2.5}\left(0.065665 + \frac{3 \cdot 0.45}{2.5}0.06715 + \frac{3 \cdot 0.45^2 0.0672}{2.5^2}\right) + \frac{22.6 \cdot 2 \cdot 0.9^3}{2 \cdot 12}\right]$$

$$\cdot 0.07710572 - \left[6 \cdot \frac{1,091.6}{2.5^2}\left(0.06715 + \frac{0.9}{2.5}0.0672\right) + \frac{11.3 \cdot 2 \cdot 0.9}{4}0.45\right]$$

$$\cdot 0.7182246 = 55.55 \text{tm}$$

$$M_{1-2} = \left[k(2a\mu - \xi - \psi a^2) + \frac{K_\varphi I_F}{2}\varphi_1 + k(\eta + 2a\gamma + a^2\sigma)\varphi_2 + \left[k(\mu - \psi a) + \frac{k_z Aa}{4}\right]\right]$$

$$v_1 - k(\gamma + a\sigma)v_2 = \left(-4 \cdot 271.1576 + \frac{22.6 \cdot 2 \cdot 0.9^3}{2 \cdot 12}\right) \cdot 0.07710572$$

$$+ 562.6384 \cdot 0.04015324 + \left(2 \cdot 154.7135595 + \frac{11.3 \cdot 2 \cdot 0.9}{4}0.45\right) \cdot 0.7182246$$

$$+ 2 \cdot 113.0521 \cdot 1.170075 = -65.65 \text{tm}$$

$$M_{1-4} = 4i_1\left(1 + \frac{3h}{h_1} + \frac{3h^2}{h_1^2}\right)\varphi_1 + 2i_1\left(1 + \frac{3h}{h_1}\right)\varphi_2 - \frac{6i_1}{h_1}\left(1 + \frac{2h}{h_1}\right)\delta_1 =$$

$$-4 \cdot 24\left(1 + \frac{3 \cdot 1.22}{3} + \frac{3 \cdot 1.22^2}{3^2}\right) \cdot 0.07710572 - 2 \cdot 24 \cdot \left(1 + \frac{3 \cdot 1.22}{3}\right) \cdot 0.0401539$$

$$+ \frac{6 \cdot 24}{3}\left(1 + \frac{2 \cdot 1.22}{3}\right)1.029058 = 10.1 \text{tm}$$

$$M_{2-1} = \left[k(2a\mu - \xi - \psi a^2) - \frac{K_\varphi I_F}{2}\varphi_2 - k(\eta + 2a\gamma + a^2\sigma) - k(\gamma + a\sigma)\right]$$

$$v_1 - \left[k(\mu - a\psi) - \frac{k_z Aa}{4}\right]v_2 = -(4 \cdot 271.15764 + 2 \cdot 0.67736) \cdot 0.04015324$$

$$- 562.6384 \cdot 0.07710572 - 226.1042 \cdot 0.7182246$$

$$+ \left(2 \cdot 154.7135595 + \frac{11.3 \cdot 2 \cdot 0.9 \cdot 0.45}{4}\right) \cdot 1.170075 = 115.35 \text{tm}$$

$$M_{2-3} = (4 \cdot 271,15764 + 2 \cdot 0.67736) \cdot 0.040153247$$

$$- 562.6384 \cdot 0.002197971 - \left(2 \cdot 154.7135595 + \frac{11.3 \cdot 2 \cdot 0.90 \cdot 45}{4}\right)$$

$$1.170075 + 2 \cdot 113.0521 \cdot 1.099772 = -160.85 \text{tm}$$

$$M_{2-5} = -2\left(130.3744 \cdot 0.04015324 + 0.3147334 \cdot 53.28 - 43.52 \cdot 1.029058\right)$$
$$= 45.56\text{tm}$$

$$M_{3-6} = -2\left(130.3744 \cdot 0.002197971 + 0.3196702 \cdot 53.28 - 43.52 \cdot 1.029058\right)$$
$$= 54.93\text{tm}$$

$$M_{3-2} = -\left(4 \cdot 271.15764 + 2 \cdot 0.67736\right) \cdot 0.00219797 - 562.6384 \cdot 0.040153324$$
$$- 226.1042 \cdot 1.170075 + \left(2 \cdot 154.7135595 + \frac{11.3 \cdot 2 \cdot 0.9}{4} 0.45\right) 1.099772$$
$$= 53.28\text{tm}$$

$$M_{3-22} = -\left(4 \cdot 47.78588 + 0.67736 \cdot 2\right) \cdot 0.00219797$$
$$-\left(95.72057 + 2.28825\right) \cdot 1.099772 = -108.21\text{tm}$$

We can also obtain the moments at the edges of the totally rigid elements in the foundation. These moments are used for design of the foundations. As we will see, they are much smaller than moments at the centers of the nodes, at points of support. Moments at the edges of the totally rigid elements are found below:

$$M_{16-15} = -2i_{16-15}\left(2\alpha_1 + \frac{3\alpha_2}{l_2}\right)\varphi_1 + \frac{6i_{16-15}\alpha_2}{l_2}v_1$$

$$= -2\frac{1,091.6}{2.5}\left(2 \cdot 0.06665 + \frac{3}{2.5}0.06715 \cdot 0.45\right) \cdot 0.007710572$$

$$+ \frac{6 \cdot 1,091.6}{2.5^2} \cdot 0.06715 \cdot 0.7182246 = 39.12\text{tm}$$

$$M_{17-18} = -k\left(\xi - a\mu\right)\varphi_1 - k\left(\eta + a\gamma\right)\varphi_2 + k\mu v_1 + k\gamma v_2 =$$
$$- 1091.6\left(0.76186 + 0.45 \cdot 0.23155\right) \cdot 0.07710572$$
$$- 1,091.6\left(0.34087 + 0.45 \cdot 0.18077\right) \cdot 0.04015324$$
$$- 1,091.6 \cdot 0.23155 \cdot 0.7182246 + 1,091.6 \cdot 0.18077 \cdot 1.170075 = -42.05\text{tm}$$

$$M_{18-17} = -k\left(\xi - a\mu\right)\varphi_2 - k\left(\eta + a\gamma\right)\varphi_1 - k\mu v_2 - k\gamma v_1$$
$$= -1,091.6 \cdot \left(0.76186 + 0.45 \cdot 0.23155\right) \cdot 0.04015324$$
$$= 1,091.6 \cdot \left(0.34087 + 0.45 \cdot 0.18077\right) \cdot 0.07710572$$
$$+ 1,091.6 \cdot 0.23155 \cdot 1.170075 - 1,091.6 \cdot 0.18077 \cdot 0.7182246 = 80.52\text{tm}$$

$$M_{19-20} = -k\left(\xi - a\mu\right)\varphi_2 - k\left(\eta + a\gamma\right)\varphi_3 + k\mu v_2 + k\gamma v_3 =$$
$$- 1,091.6\left(0.76186 + 0.45 \cdot 0.23155\right) \cdot 0.04015324$$
$$- 1,091.6\left(0.34087 + 0.45 \cdot 0.18077\right) \cdot 0.002197971$$
$$- 1,091.6 \cdot 0.23155 \cdot 1.170075 + 1,091.6 \cdot 0.180077 \cdot 1.099772 = -117.71\text{tm}$$

$$M_{20-19} = -k\left(\xi - a\mu\right)\varphi_3 - k\left(\eta + a\gamma\right)\varphi_2 - k\mu v_3 - k\gamma v_2 =$$
$$- 1,091.6 \cdot \left(0.76186 + 0.45 \cdot 0.23155\right) \cdot 0.002197971$$
$$- 1,091.6 \cdot \left(0.34087 + 0.45 \cdot 0.18077\right) \cdot 0.04015324$$
$$+ 1,091.6 \cdot 0.23155 \cdot 1.099772 - 1,091.6 \cdot 0.18071 \cdot 1.170075 = 26.51\text{tm}$$

$$M_{21-22} = -2i_{21-22}\left(2\alpha_1 + \frac{3\alpha_2 a}{l_2}\right)\varphi_3 - \frac{6i_{21-22}\alpha_2}{l_2}v_3 =$$

$$-2 \cdot \frac{1,091.6}{2.5}\left(2 \cdot 0.06665 + \frac{3 \cdot 0.45 \cdot 0.06715}{2.5}\right) \cdot 0.002197971$$

$$-\frac{6 \cdot 1,091.6}{2.5^2} \cdot 0.06715 \cdot 1.099772 = -71.52$$

In order to evaluate the affect of totally rigid elements on final results, analysis of the same frame was performed a second time without taking into account the totally rigid elements. The moment diagrams of the foundations obtained from both analyses are shown in Figures 2.45 and 2.46. Selected results of analysis are shown and compared

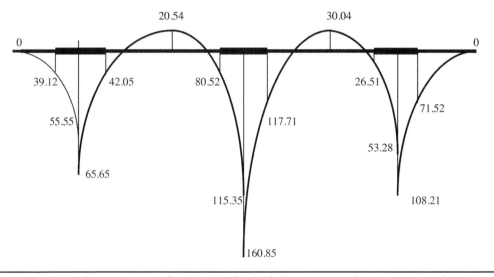

Figure 2.45 Foundation moment diagram (totally rigid elements are taken into account)

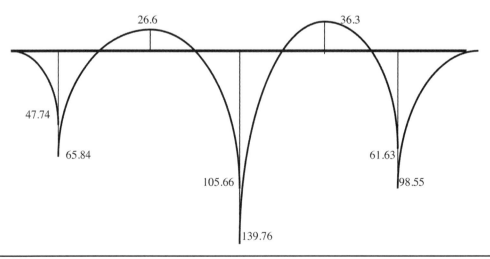

Figure 2.46 Foundation moment diagram (totally rigid elements are not taken into account)

in Tables 2.19 and 2.20. It is easy to note that moments at the edges of totally rigid elements are much smaller than at the centers of these elements.

2.8 Exact Analysis of Interconnected Beams on Winkler Foundation

Foundations of industrial and commercial buildings are often designed as a system of interconnected beams. This type of foundation is usually used for buildings supported on soils with lower or variable bearing capacity. It is even used when the bearing capacity of the soil is high enough, but the loads are large and individual foundations cannot be used. When individual footings are used, it is recommended in many instances to connect them with grade beams, as shown in Figure 2.48. By connecting individual foundations with grade beams, we create a system of interconnected beams supported on elastic foundation. Areas of intersection of interconnected beams can be taken into account as totally rigid elements, since beams running in one direction resist the bending of the beams running in another perpendicular direction. The rigidities of the foundation, in these areas, are also increased by the columns connected with the foundations and usually located at the centers of totally rigid elements. In this case, totally rigid elements are large, much larger than elements of regular interconnected beams, as shown in Figure 2.47. Accurate analysis of a system of interconnected beams

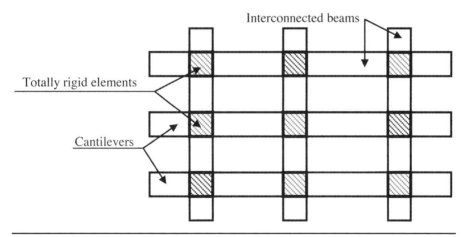

Figure 2.47

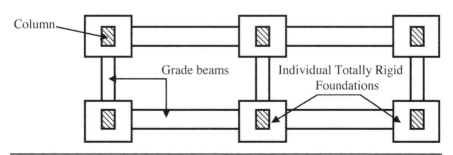

Figure 2.48

supported on Winkler foundation can be performed by using the method of forces or by using the stiffness method. When using the method of forces all beams running in one direction are cut off from all beams running a in perpendicular direction, replacing the given system with a series of simple beams, loaded with the given loads and unknown forces of interaction between the beams, as shown in Figure 2.49. Analysis does not take into account forces and moments of interaction acting in the horizontal plane. Three unknowns, x_1, x_2, and x_3, are applied at each cut. So, the method of forces requires building a system of linear equations with $6n$ unknowns where n is the total number of nodes.

Equations in all cuts are written by equating deflections of the interconnected beams in each cut. For example, three equations written for the cut shown in Figure 2.49 have the following meanings:

1. The angle of torsion φ_A^t of the beam A is equal to the angle of bending φ_B^b of the beam B

$$\varphi_A^t = \varphi_B^b \tag{2.89}$$

2. The angle of bending φ_A^b of the beam A is equal to the angle of torsion φ_B^t of the beam B

$$\varphi_A^b = \varphi_B^t \tag{2.90}$$

3. The settlement Δ_A of the beam A is equal to the settlement of the beam Δ_B

$$\Delta_A = \Delta_B \tag{2.91}$$

By solving this system of equations with $6n$ unknowns, all forces and moments of interaction are found, and each beam, loaded with the given loads and forces of interaction,

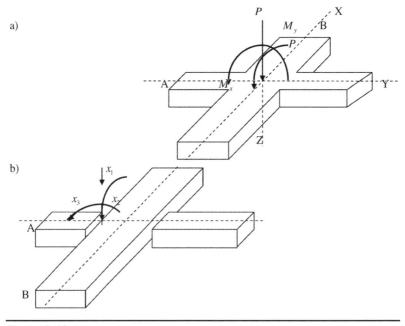

Figure 2.49

is analyzed separately. The use of the method of forces was proposed by Gorbunov-Posadov (1953). The reader can find a more detailed explanation of the method in his book. However, it is important and useful to mention the following:

All areas of intersection of the beams are totally rigid not only because beams running in one direction prevent bending of the beams running in the other direction, but also because they are tied with columns or large column pedestals. Therefore, when replacing a system of interconnected beams with a series of simple beams, the totally rigid areas are still not taken into account. All beams are elastic and analyzed as elastic beams. For example, when final analysis of beam B is performed we do not take into account that the area of intersection with beam A is totally rigid. In order to obtain more accurate results, analysis of beam B should be performed as a beam with totally rigid elements located in the areas of intersection. But this analysis becomes complex and unrealistic. The stiffness method proposed by the author of this book has advantages compared to the method of forces. Figure 2.50 shows a simple system of interconnected beams with vertical loads and moments. All beams under applied loads will experience bending as well as torsion. Equations for obtaining moments and shear forces in beams were shown earlier. Below are given simple equations for obtaining the torsional moments in beams supported on Winkler foundation obtained by Klepikov (1967).

Equations are given for two types of beams: for beams with both ends fixed and beams with one fixed and one free supported end. If rotation $\varphi_x = 1$ is applied to the left end of a beam $a - b$, shown in Figure 2.51 with both ends fixed, a torsional moment will occur at the same end a equal to:

$$M^t_{a-b} = \frac{GI_t\lambda_t}{l \cdot th\lambda_t} \tag{2.92}$$

The torsional moment at the end b is equal to

$$M^t_{b-a} = \frac{GI_t\lambda_t}{l \cdot Sh\lambda_t} \tag{2.93}$$

In equations 2.92 and 2.93, G is the shear modulus of the beam material $G = 0.4E$, l is the length of the beam, and λ_t is the torsional characteristic of the beam found from equation 2.94:

$$\lambda_t = \sqrt[l]{\frac{K_\varphi b^3}{24GI_t}} \tag{2.94}$$

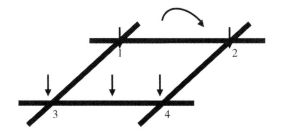

Figure 2.50

Figure 2.51

b is the width of the beam, I_t is the torsional moment of inertia of the beam that is found from equation 2.95:

$$I_t = \alpha b h^3 \tag{2.95}$$

where

K_φ = the modulus of subgrade reaction, h is the height of the beam

α = a coefficient obtained from Table 2.21

If rotation $\varphi_x = 1$ is applied to the fixed left end of a beam $a - b$, with one fixed and one free supported end, as shown in Figure 2.52, a torsional moment will occur at the same end equal to:

$$M^t_{a-b} = \frac{GI_t}{l} \cdot \lambda_t th \lambda_t \tag{2.96}$$

Equation 2.95 is applicable when $h > b$. When $b > h$, formula 2.95 looks as follows:

$$I_{tors} = \alpha h b^3 \tag{2.97}$$

In this case, when using Table 2.21 to find α, h should be replaced by b and b should be replaced by h. It is important to note that formulae 2.95–2.97 are applicable when the cross section of the beam is rectangular. When the shape of the cross section is different, but can be divided into several rectangular elements, the torsional moment of inertia is obtained from the following equation:

$$I_t = I_{1t} + I_{2t} + ... + I_{nt} \tag{2.98}$$

where the right part of the equation is the sum of the torsional moments of inertia of all parts of the cross section of the beam. Equations 2.92–2.98 and equations obtained earlier (2.50–2.60) allow using the stiffness method for analyzing a system of interconnected beams on Winkler foundation. However, these equations do not take into account the totally rigid elements. Exact analysis that takes into account the totally rigid elements can still be performed by using equations 2.92–2.98 and equations given in Tables 2.9–2.13. Using the stiffness method a system of linear equations can be built. The system contains $3n$ unknown deflections: n settlements and $2n$ rotations, while the method of forces usually leads to a system of equations with $6n$ unknowns. Analysis takes into account only settlements and rotations about two horizontal axes x and y as shown in Figure 2.53. The system of equations looks as follows:

$$\begin{vmatrix} a_{11} a_{12} ..... a_{1n} \\ a_{12} a_{22} ..... a_{2n} \\ .... \\ .... \\ a_{1n} a_{2n} .... a_{nn} \end{vmatrix} X \begin{vmatrix} x_1 \\ x_2 \\ ... \\ ... \\ x_n \end{vmatrix} = \begin{vmatrix} p_1 \\ p_2 \\ ... \\ ... \\ p_n \end{vmatrix} \tag{2.99}$$

In 2.99, a_{ij} is the reaction in restraint i due to deflection equal to one unit applied to restraint j, $x_1, x_2, \ldots, x_n$ are unknown deflections, and $p_1, p_2, \ldots, p_n$ are the given exterior vertical loads and moments.

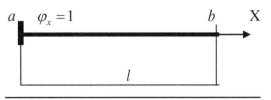

Figure 2.52

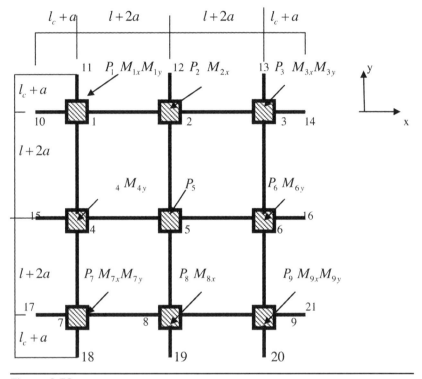

Figure 2.53

In order to illustrate application of the stiffness method and evaluate the affect of totally rigid elements on final results of analysis, a numerical example is solved below.

Example 2.6

Figure 2.53 shows a system of interconnected beams supported on Winkler foundation. All data are given below. $l = 4\text{m}$, $l_c = 2\text{m}$, $a = 1\text{m}$, and the moment of inertia of the cross section of the beam $I = 0.063612\text{m}^4$. The cross section of the foundation is shown in Figure 2.54. The flexural rigidity of the beam $EI = 2,650,0000 \cdot 063612 = 168,667.5\text{tm}^2$.

The characteristic of the foundation $\lambda = \sqrt[4]{\dfrac{4EI}{Kb}} = \sqrt[4]{\dfrac{4 \cdot 168,667.50}{2,000 \cdot 2.0}} = 3.604\dfrac{\text{m}}{\text{t}}$

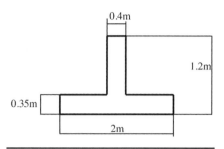

Figure 2.54

The bending characteristic of the beam with $l = 4m$ $\dfrac{l}{\lambda} = \dfrac{4}{3.6} = 1.12$

The bending characteristic of the beam $l_c = 2m$ $\dfrac{l_c}{\lambda} = \dfrac{2}{3.6} = 0.56$

The torsional moment of inertia $I_t = 0.249 \cdot 1 \cdot 0.4^3 + 2 \cdot 0.2535 \cdot 0.8 \cdot 0.3^3 = 0.027 m^4$

The torsional rigidity of the beam is equal $GI_t = 0.40 \cdot 168,667.5 \cdot 0.027 = 28,514 tm^2$

The torsional characteristics of the foundation elements are found below:

1. For elements $4m$ long $\lambda_{t1} = l\sqrt[4]{\dfrac{K_\varphi b^3}{24 GI_{tors}}} = 4\sqrt[4]{\dfrac{4,000 \cdot 2^3}{24 \cdot 28,514}} = 0.86$

2. For elements $2m$ long $\lambda_{t2} = 0.43$

Using Table 2.25, functions A_x, B_x, C_x, D_x are found for beams with characteristics equal to $\dfrac{l}{\lambda} = \dfrac{4}{3.6} = 1.12$ and $\dfrac{l_c}{\lambda} = \dfrac{2}{3.6} = 0.56$

For elements with $\dfrac{l}{\lambda} = 1.12$, $A = 0.7387$, $B = 1.0613$, $C = 0.61625$, $D = 0.23235$

Using equation 2.59 for beams with both ends fixed, find parameters:

$$\xi = \frac{1}{\lambda} \cdot \frac{BC - AD}{C^2 - BD} = \frac{1}{3.6} \cdot \frac{1.0613 \cdot 0.61625 - 0.7387 \cdot 0.23235}{0.61625^2 - 1.0613 \cdot 0.23235} = 1.014$$

$$\mu = \frac{1}{\lambda^2} \cdot \frac{AC - B^2}{C^2 - BD} = \frac{1}{3.6^2} \cdot \frac{0.7387 \cdot 0.61625 - 1.0613^2}{0.61625^2 - 1.0613 \cdot 0.23235} = -0.395$$

$$\eta = \frac{1}{\lambda} \cdot \frac{D}{C^2 - BD} = \frac{1}{3.6} \cdot \frac{0.23235}{0.61625^2 - 1.0613 \cdot 0.23235} = 0.4885$$

$$\gamma = \frac{1}{\lambda^2} \cdot \frac{C}{C^2 - BD} = \frac{1}{3.6^2} \cdot \frac{0.61625}{0.61625^2 - 1.0613 \cdot 0.23235} = 0.3628$$

$$\psi = \frac{1}{\lambda^3} \cdot \frac{4CD + AB}{C^2 - BD} = \frac{1}{3.6^3} \cdot \frac{4 \cdot 0.61625 \cdot 0.23235 + 0.7387 \cdot 1.063}{0.61625^2 - 1.0613 \cdot 0.23235} = 0.2236$$

$$\sigma = \frac{1}{\lambda^3} \cdot \frac{B}{C^2 - BD} = \frac{1}{3.6^3} \cdot \frac{1.0613}{0.61625^2 - 1.0613 \cdot 0.23235} = 0.1749$$

For elements with $\dfrac{l_c}{\lambda} = 0.56$, $A_c = 0.9836$, $B_c = 0.5582$, $C_c = 0.15665$, $D_c = 0.02925$

Using equations 2.46–2.47 for beams with one fixed and one free supported end, find parameters:

$$\alpha_1 = \frac{l_c}{\lambda} \frac{B_c C_c - A_c D_c}{A_c^2 - 4D_c B_c} = \frac{2}{3.6} \cdot \frac{0.5582 \cdot 0,15665 - 0.9836 \cdot 0.02925}{0.9836^2 + 4 \cdot 0.5582} = 0.0318$$

$$\alpha_2 = \frac{2l_c^2}{3\lambda^2} \cdot \frac{B_c^2 - A_c C_c}{A_c^2 + 4B_c D_c} = \frac{2 \cdot 0.56^2}{3} \cdot \frac{0.9836 \cdot 0.15665 + 4 \cdot 0.02925^2}{1.03} = 0.03188$$

$$\beta = \frac{l_c^2 \left(A_c B_c + 4D_c^2 \right)}{3\lambda^3 \left(A_c^2 + 4B_c D_c \right)} = \frac{0.56 \left(0.9836 \cdot 05582 + 4 \cdot 0.15665 \cdot 0.02925 \right)}{3 \cdot 1.03} = 0.03216$$

Loads applied to the foundation:

$$P_1 = P_3 = P_7 = P_9 = 250t \; P_2 = P_8 = 300t \; P_4 = P_{56} = 200t \; M_{1y} = M_{7y} = -M_{3y} = -M_{9y}$$
$$= -15tm$$

$$M_{1x} = M_{3x} = -M_{7x} = -M_{9x} = -30tm \; M_{4y} = -M_{6y} = -25tm \; M_{2x} = -M_{8x} = -40tm$$

The modulus of subgrade reaction $K_z = 2,000\dfrac{t}{m^3}$. The modulus of subgrade reaction under the totally rigid element due to its rotation $K_\varphi = 4,000\dfrac{t}{m^3}$.

As we can see, the system is symmetrical in both directions that allow for the significant simplification of all calculations by minimizing the number of unknowns. The total number of unknowns is equal only to eight.
The unknowns are:

1. Four vertical settlements $v_1 \, v_2 \, v_4 \, v_5$ of the nodes 1, 2, 4, and 5
2. Two rotations $\varphi_{1x} \, \varphi_{1y}$ at node 1
3. One rotation φ_{2x} at node 2
4. One rotation φ_{4y} at node 4

Now, assuming that $\varphi_{1y} = x_1 \; \varphi_{1x} = x_2 \; \varphi_{2x} = x_3 \; \varphi_{4y} = x_4 \; v_1 = x_5 \; v_2 = x_6 \; v_4 = x_7 \; v_5 = x_8$ and taking into account the symmetry of the system, the following system of equations 2.100 is written:

$$\begin{vmatrix} a_{11} & a_{12} & 0 & 0 & a_{15} & a_{16} & 0 & 0 \\ & a_{22} & 0 & 0 & 0 & 0 & a_{27} & a_{28} \\ & & a_{33} & a_{34} & a_{35} & 0 & a_{37} & 0 \\ & & & a_{44} & 0 & a_{46} & 0 & a_{48} \\ & & & & a_{55} & a_{56} & a_{57} & 0 \\ & & & & & a_{66} & 0 & a_{68} \\ & \text{Symmetrical} & & & & & a_{77} & a_{78} \\ & & & & & & & a_{88} \end{vmatrix} \times \begin{vmatrix} x_1 \\ x_2 \\ x_3 \\ x_4 \\ x_5 \\ x_6 \\ x_7 \\ x_8 \end{vmatrix} = \begin{vmatrix} M_{1y} \\ 0.5M_{4y} \\ M_{1x} \\ 0.5M_{2x} \\ P_1 \\ 0.5P_2 \\ 0.50P_3 \\ 0.25P_4 \end{vmatrix} \qquad (2.100)$$

All coefficients in this system of equations 2.100 are found below:

$$a_{11} = EI\left(\xi - 2a\mu + a^2\psi\right) + K_\varphi I_f + \frac{4EI_t}{l_c}\left(\alpha_1 + \frac{3a}{l_c}\alpha_2 + \frac{3a^2}{l_c^2}\beta\right) + \frac{GI_t\lambda_{t1}}{l\cdot th\lambda_{t1}} + \frac{GI_t\lambda_{tc}}{l_c}tgh\lambda_{tc}$$

$$a_{12} = -\frac{GI_t\lambda_{t1}}{l\cdot \sin h\lambda_{t1}} \quad a_{15} = -EI(\mu - a\psi) - \frac{6EI}{l_c^2}\left(\alpha_2 + \frac{2a}{l_c}\beta\right) \quad a_{16} = -EI(\gamma + a\sigma)$$

$$a_{22} = \frac{1}{2}\left[EI\left(\xi - 2a\mu + a^2\psi\right) + K_\varphi I_f + \frac{4EI_t}{l_c}\left(\alpha_1 + \frac{3a}{l_c}\alpha_2 + \frac{3a^2}{l_c^2}\beta\right) + 2\frac{GI_t\lambda}{l\cdot th\lambda}\right] \quad a_{44} = a_{22}$$

$$a_{27} = \frac{1}{2}EI\left[(\mu - a\psi) - \frac{6}{l_c^2}\left(\alpha_2 + \frac{2a}{l_c}\beta\right)\right] \quad a_{28} = -\frac{EI}{2}(\gamma + a\sigma)$$

$$a_{33} = EI\left(\xi - 2a\mu + a^2\psi\right) + K_\varphi I_f + \frac{4EI}{l_c}\left(\alpha_1 + \frac{3a}{l_c}\alpha_2 + \frac{3a^2}{l_c^2}\beta\right) + \frac{GI_{t1}\lambda_{t1}}{l\cdot th\lambda_{t1}} + \frac{GI_{tc}\lambda_{tc}}{l_c}th\lambda_{tc}$$

$$a_{34} = a_{12} \quad a_{35} = -a_{15} \quad a_{37} = -a_{16} \quad a_{44} = a_{22} \quad a_{46} = a_{27} \quad a_{48} = -a_{28}$$

$$a_{55} = \frac{24EI}{l_c^3}\beta + K_z A_f + 2EI\psi \quad a_{56} = a_{57} = -EI\sigma \quad a_{66} = \frac{1}{2}\left(\frac{12EI}{l_c^3}\beta + K_z A_f + 3EI\psi\right)$$

$$a_{68} = -\frac{EI\sigma}{2} \quad a_{77} = a_{66} \quad a_{78} = a_{68} \quad a_{88} = EI\psi + \frac{K_z A_f}{4}$$

All coefficients not shown above are equal to zero. By introducing numerical data into the system 2.100, we will obtain the following system of equations shown in Table 2.22.

By solving this system of equations we find:

$$x_1 = 40.9023\frac{1}{EI} \quad x_2 = 89.28075\frac{1}{EI} \quad x_3 = 30.96219\frac{1}{EI} \quad x_4 = -19.1867\frac{1}{EI}$$

$$x_5 = 1,059.895\frac{1}{EI} \quad x_6 = 1,229.319\frac{1}{EI} \quad x_7 = 950.2162\frac{1}{EI} \quad x_8 = 1,358.019\frac{1}{EI}$$

Below are found all bending and torsional moments at the centers of the nodes:

$$M_{1-2} = -EI_f\left(\xi - 2a\mu + a^2\psi\right)x_1 - \frac{K_\varphi I}{2}x_1 + EI_f(\mu - a\psi)x_5 - \frac{aK_z A_F}{4}x_5 - EI_f(\gamma + a\sigma)x_6$$

$$= -\left(1.014 + 2\cdot 1\cdot 0.395 + 1^2 0.2236 + \frac{4,000\cdot 2^4}{2\cdot 12\cdot 168,667.5}\right)\cdot 40.9023 -$$

$$+\left(-0.395 - 0.2236 - \frac{2,000\cdot 4}{4\cdot 168,667.5}\right)\cdot 1,059.895 - (0.3628 + 0.1749)\cdot 1,229.319$$

$$= -90.794 tm$$

$$M_{1-10} = -\frac{4EI_f}{l_c}\left(\alpha_1 + \frac{3a}{l_c}\alpha_2 + \frac{3a^2}{l_c^2}\beta\right)x_1 - \frac{K_\varphi I_F}{2}x_1 + \frac{6EI_f}{l_c^2}\left(\alpha_1 + \frac{2a\alpha_2}{l_c}\right)x_5$$

$$+\frac{K_z A_F}{4}x_5 = -\frac{4}{2}\left(0.0318 + \frac{3\cdot 1}{2}\cdot 0.03188 + \frac{3\cdot 1^2}{2^2}\cdot 0.03216\right)40.9023$$

$$-\frac{4,000 \cdot 2^4}{12 \cdot 2 \cdot 168,667.5} \cdot 40.9023 + \frac{6}{4}\left(0.03188 + \frac{2 \cdot 0.03216}{2}\right) \cdot 1,059.895$$

$$+\frac{2,000 \cdot 4}{4 \cdot 168,667.5} \cdot 1,059.895 = 105.202 \text{tm}$$

$$M_{1-4}^{tors} = -\frac{GI_{tors}\lambda_{tors1}}{l \cdot th\lambda_{tors1}}x_1 + \frac{GI_{tors}\lambda_{tors1}}{l \cdot sh\lambda_{tors1}}x_4 = -\frac{0.40 \cdot 2,650,000 \cdot 0.027 \cdot 0.86}{4 \cdot 0.6963 \cdot 168,667.5} \cdot 40.9023$$

$$+\frac{0.40 \cdot 2,650,000 \cdot 0.027 \cdot 0.86}{4 \cdot 0.97 \cdot 168,667.5} \cdot 89.28075 = 1.21 \text{tm}$$

$$M_{1-11}^{tors} = -\frac{GI_{tors}\lambda_{tors2}tgh\lambda_{tors2}}{l_c} \cdot x_1 = -\frac{0.40 \cdot 2,650,000 \cdot 0.027 \cdot 0.4053}{4 \cdot 168,667.5} \cdot 40.9023$$

$$=-0.60 \text{tm}$$

$$M_{4-5} = \left(-2.0276 - \frac{0.0316}{2}\right) \cdot 89.28075 + \left(-0.6186 - \frac{0.047431}{4}\right) \cdot 950.2162$$

$$+0.5377 \cdot 1,358.019 = -51.3 \text{tm}$$

$$M_{4-15} = \left(-0.20748 - \frac{0.0316}{2}\right) \cdot 89.28075 + \left(0.09606 + \frac{0.047431}{4}\right) 950.2162$$

$$= 82.60 \text{tm}$$

$$M_{4-1}^{tors} = M_{4-7}^{tors} = -0.05218 \cdot 89.28075 + 0.037459 \cdot 40.9023 = -3.15 \text{tm}$$

$$M_{2-5} = \left(-2.0276 - \frac{0.0316}{2}\right) \cdot (-19.1867) - \left(-0.6186 - \frac{0.047431}{4}\right)$$

$$\cdot 1,229.319 - 0.5377 \cdot 1,358.019 = 84.033 \text{tm}$$

$$M_{2-12} = -\left(0.20748 + \frac{0.0316}{2}\right) \cdot (-19.1667) - \left(0.09606 + \frac{0.047431}{4}\right) 1,229.319$$

$$= -128.355 \text{tm}$$

$$M_{2-1}^{tors} = M_{2-3}^{tors} = -0.05218 \cdot (-19.1867) + 0.037459 \cdot 30.96219 = 2.16 \text{tm}$$

$$M_{1-4} = \left(-2.0276 - \frac{0.0316}{2}\right) \cdot 30.96219 - \left(-0.6186 - \frac{0.047431}{4}\right) \cdot 1,059.895$$

$$-0.5377 \cdot 950.2162 = 94.02 \text{tm}$$

$$M_{1-11} = -\left(0.20748 + \frac{0.0316}{2}\right) \cdot 30.96219 - \left(0.09606 + \frac{0.047431}{4}\right) \cdot 1,059.895$$

$$= -121.194 \text{tm}$$

$$M_{1-2}^{tors} = -0.05218 \cdot 30.96219 + 0.037459 \cdot (-19.1667) = -2.354 \text{tm}$$

$$M_{1-10}^{tors} = -0.0147 \cdot 30.96219 = -0.475 \text{tm}$$

$$M_{2-1} = -(0.4885 + 2 \cdot 1 \cdot 0.3628 + 0.1749) \cdot 40.90230 - 0.5377 \cdot 1,059.895$$

$$-\left(-0.6186 - \frac{0.047431}{4}\right) \cdot 1,229.319 = 148.307 \text{tm}$$

$$M_{5-4} = -(0.4885 + 2 \cdot 1 \cdot 0.3628 + 0.1749) \cdot 89.28075$$

$$-(0.3628 + 0.1749) \cdot 950.2162 - \left(-0.6186 - \frac{0.047431}{4}\right) \cdot 1,358.019$$

$$= 221.233 \text{tm}$$

$$M_{4-1} = \left(-0.6186 - \frac{0.04731}{4}\right) \cdot 950.2162 + 0.5377 \cdot 1,059.895$$

$$-(0.4885 + 2 \cdot 1 \cdot 0.3628 + 0.1749) \cdot 30.96219 = -72.174 \text{tm}$$

$$\left(-0.6186 - \frac{0.047431}{4}\right) \cdot 1,358.019 + 0.5377 \cdot 1,229.319$$

$$-(0.4885 + 2 \cdot 1 \cdot 0.3628 + 0.1749) \cdot (-19.1667) = -168.552 \text{tm}$$

$$M_{5-2} = \left(-0.6186 - \frac{0.047431}{4}\right) \cdot 1,358.019 + 0.5377 \cdot 1,229.319$$

$$-(0.4885 + 2 \cdot 1 \cdot 0.3628 + 0.1749) \cdot (-19.1667) = -168.552 \text{tm}$$

Now, all bending moments at the edges of the totally rigid elements can be obtained.

Node 1:

$$M_{1-10}^{edge} = -2 \cdot 0.5 \cdot \left(2 \cdot 0.0318 + \frac{3 \cdot 0.03188}{2}\right) \cdot 40.9023$$

$$+ \frac{6 \cdot 0.50}{2} \cdot 0.03188 \cdot 1,059.895 = 46.126 \text{tm}$$

$$M_{1-2}^{edge} = -1(1.014 + 0.395) \cdot 40.9023 - 0.395 \cdot 1,095.895$$

$$+ 1 \cdot 0.3628 \cdot 1,229.319 = -30.29 \text{tm}$$

$$M_{1-11}^{edge} = 2 \cdot 0.5 \cdot \left(2 \cdot 0.0318 + \frac{3 \cdot 0.03188}{2}\right) \cdot 30.96219$$

$$- \frac{6 \cdot 0.5}{2} \cdot 0.03188 \cdot 1,059.895 = -47.23 \text{tm}$$

$$M_{1-4}^{edge} = -1 \cdot (1.014 + 0.395) \cdot 30.96219 + 1 \cdot 0.395 \cdot 1,059.895$$

$$- 1 \cdot 0.3628 \cdot 950.2162 = 30.29 \text{tm}$$

Node 2:

$$M_{2-1}^{edge} = -(0.4885 + 2 \cdot 1 \cdot 0.3628 + 0.1749) \cdot 40.9023 - 0.5377 \cdot 1,059.895$$

$$- \left(-0.6186 - \frac{0.047431}{4}\right) \cdot 1,229.319 = 148.307 \text{tm}$$

$$M_{2-5}^{edge} = (1.014 + 0.395) \cdot 19.1667 + 0.395 \cdot 1,229.319 - 0.3628 \cdot 1,358.019$$

$$= 19.89 \text{tm}$$

$$M_{2-12}^{edge} = 2 \cdot 0.5 \cdot \left(2 \cdot 0.0318 + \frac{3}{2} \cdot 0.03188\right) \cdot 19.1667 - \frac{6 \cdot 0.5}{2} \cdot 0.0318 \cdot 1,229.319$$

$$= -56.77 \text{tm}$$

Node 4:

$$M_{4-1}^{edge} = -1(0.3628 + 0.4885) \cdot 30.96219 - 0.395 \cdot 950.2162$$
$$+ 0.3628 \cdot 1,059.895 = -17.17 \text{tm}$$

$$M_{4-5}^{edge} = -(1.014 + 0.395) \cdot 89.2875 - 0.395 \cdot 950.2162 + 0.3628 \cdot 1,358.019$$
$$= -8.45 \text{tm}$$

$$M_{4-15}^{edge} = -2 \cdot 0.5\left(2 \cdot 0.0318 + \frac{3 \cdot 0.03188}{2}\right) \cdot 89.28075$$

$$+ \frac{6 \cdot 0.5}{2} \cdot 0.03188 \cdot 950.2162 = 35.50 \text{tm}$$

Node 5:

$$M_{5-4}^{edge} = (0.4885 + 0.3628)89.28075 + 0.395 \cdot 1,358.019 - 0.3628 \cdot 950.2162$$
$$= 115.679 \text{tm}$$
$$M_{5-2}^{edge} = (0.4885 + 0.3628)19.667 - 0.395 \cdot 1,358.019 + 0.3628 \cdot 1,229.319$$
$$= -73.67 \text{tm}$$

The bending moment diagram for one beam located between nodes 11 and 18 is shown in Figure 2.55. All moments at the nodes 1, 4, and 7, and at the edges of absolutely rigid elements, are shown. As seen, moments at the edges of totally rigid elements are two to four times smaller than moments at the centers of the totally rigid elements. For example, the moment M_{5-2} is about 2.3 times smaller than the moment at the edge or:

$$\frac{M_{5-2}}{M_{5-2}^{edge}} = \frac{168.552}{73.67} = 2.29 \tag{2.101}$$

Results of analysis of the same foundation performed without taking into account the totally rigid elements, shown in Figure 2.56, are significantly different from the results of the exact analysis.

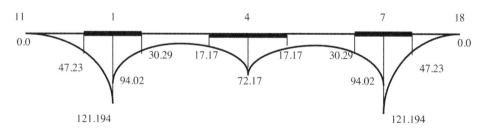

Figure 2.55 The bending moment diagram for the beam 11-18 (the totally rigid elements are taken into account)

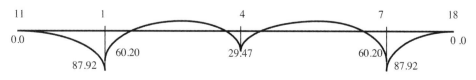

Figure 2.56 The bending moment diagram for the beam 11-18 (the totally rigid elements are not taken into account)

2.9 Analysis of Rigid Piles with Horizontal Loads

A single pile can be considered totally rigid when the following relationship takes place: $\frac{h}{\delta} \leq (10 \div 12)$.

When $\frac{h}{\delta} \geq (10 \div 12)$ the pile is flexible. In this equation h is the length of the embedded in-soil portion of the pile and δ is the diameter of the pile.

A pile with a horizontal load and moment is shown in Figure 2.57. If the soil reaction is equal to p, the following simple relationship between this reaction and horizontal deflection of the pile looks as follows:

$$p = -k_x y \tag{2.102}$$

where k_x can be obtained from equation 2.103:

$$k_x = C_x b \tag{2.103}$$

where C_x is the modulus of subgrade reaction in horizontal direction and b is the width or diameter of the pile. It can be assumed for deep embedded piles, up to about 25 ft, that C_x increases with the depth from zero, at the top of the soil, up to C_h at the bottom of the pile. Approximate values of C_h for various soils are given in Table 2.23.

The modulus of subgrade reaction C_h at any point along the pile is obtained from equation 2.104:

$$C_x = \beta x \tag{2.104}$$

$$\beta = \frac{1}{h} C_h \tag{2.105}$$

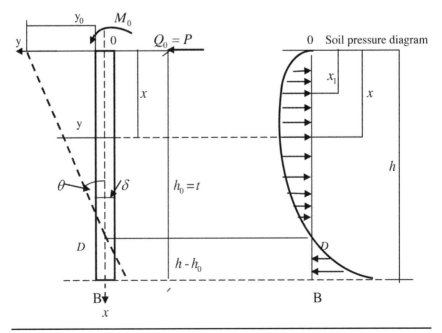

Figure 2.57

Using 2.102, 2.103, and 2.104, equation 2.102 is rewritten:

$$p = -b\beta xy \tag{2.106}$$

It can be also seen from Figure 2.57 that:

$$y = y_0 - \theta x \tag{2.107}$$

where θ is the rotation of the top of the pile. Now, equation 2.106 can be rewritten as follows:

$$p = -b\beta x(y_0 - \theta x) \tag{2.108}$$

Taking into account that $C_h = \beta h$ and $k = C_h b = \beta hb$, equation 2.108 is rewritten:

$$p = -\frac{ky_0}{h}x + \frac{k\theta}{h}x^2 \tag{2.109}$$

Using the method of initial parameters, the shear force at any point is obtained as follows:

$$Q_x = Q_0 + \int_0^x pdx = Q_0 + \int_0^x \left(-\frac{ky_0}{h}x + \frac{k\theta}{h}x^2\right)dx = Q_0 - \frac{ky_0 x^2}{2h} + \frac{k\theta x^3}{3h} \tag{2.110}$$

The moment can be found:

$$M_x = M_0 + Q_x + \int_0^x (pdx_1)(x - x_1) \tag{2.111}$$

where x_1 is the variable coordinate that changes from 0 to x. By introducing 2.109 into 2.111 and integration, we find:

$$M_x = M_0 + Q_0 x + \int_0^x \left(-\frac{k}{h}x_1 + \frac{k\theta}{h}x^2\right)(x - x_1)dx' = M_0 + Q_0 x - \frac{ky_0 x^3}{6h} + \frac{k\theta x^4}{12h} \tag{2.112}$$

Taking into account that at the bottom of the pile $M_{x=h} = Q_{x=h} = 0$, equations 2.109 and 2.112 can be rewritten as follows:

$$\left.\begin{array}{l} \dfrac{ky_0 h}{2} - \dfrac{k\theta h^2}{3} = Q_0 \\[3mm] \dfrac{ky_0 h^2}{6} - \dfrac{k\theta h^3}{12} = M_0 + Q_0 h \end{array}\right\} \tag{2.113}$$

By solving these equations, two initial parameters, ky_0 and $k\theta h$, are found:

$$ky_0 = \frac{24}{h^2}\left(M_0 + \frac{3}{4}Q_0 h\right) \tag{2.114}$$

$$k\theta h = \frac{12}{h}\left(\frac{3M_0}{h} + 2Q_0\right) \tag{2.115}$$

From equations 2.114 and 2.115 we find:

$$h_0 = 2h\frac{H + 0.75}{3H + 2h} \tag{2.116}$$

Equation 2.116 shows the location of the point of rotation of the pile. Using equations 2.109 and 2.111, the shear force and moment diagrams can be built. It is important to mention that the larger the modulus k, the smaller the moment. Therefore, when performing analysis, the smallest possible k should be used in order to obtain the maximum moment.

Example 2.7

Given: A short rigid pile shown in Figure 2.57 embedded in sand $h = 1.9$m, with a lateral load $P = 0.2$t applied at $H = 2$m above the soil diameter of the pile is $b = 30$cm $C_h = 3$kg/cm³.

$$k = C_h b = 3,000 \cdot 0.3 = 900 \frac{t}{m^2}$$

Find the lateral deflection of the top of the pile and location of the center of rotation h_0:

$$M_0 = P \cdot H = 0.2 \cdot 2 = 0.4 \text{tm} \quad Q_0 = 0.2\text{t}$$

$$ky_0 = \frac{24}{1.9^2}\left(0.4 + \frac{2^3}{4}0.2 \cdot 1.9\right) = 4.31\frac{t}{m}$$

$$y_0 = \frac{4.31}{900} = 0.0048\text{m} = \frac{1}{380}h$$

$$k\theta h = \frac{12}{1.9}\frac{12}{6.23}\left(\frac{3 \cdot 0.4}{1.9} + 2 \cdot 0.2\right) = 6.53\frac{t}{m}$$

$$k\theta = \frac{6.53}{1.9} = 3.44\frac{t}{m^2}, \quad h_0 = \frac{ky_0}{k\theta} = \frac{4.31}{3.44} = 1.25\text{m}$$

Now, the location of the maximum moment in the pile can be found from equation 2.110 assuming $Q_x = 0$.

$$0.2 - \frac{4.31x^2}{3.8} + \frac{3.44x^3}{5.7} = 0 \quad x = 0.50\text{m}$$

Now, using equation 2.112 the maximum moment is obtained: $M_{max} = 0.46$tm.

2.10 Analysis of Rigid Piles with Horizontal and Vertical Loads

The analysis described above does not take into account the vertical loads. Analysis of piles with horizontal and vertical loads is presented below. As shown in Figure 2.58, the pile is subjected to a horizontal load Q_0, as well as vertical load P, applied at the top at point A. P_1 is the weight of the pile. The pile moved due to applied loads from position $A0B$ into position $A_10_1B_1$. Rotation of the pile is equal to 0; the horizontal deflection at the soil level is equal to y_0. But these deflections depend not only on lateral loads, they depend on vertical loads P and P_1 as well. By taking into account that

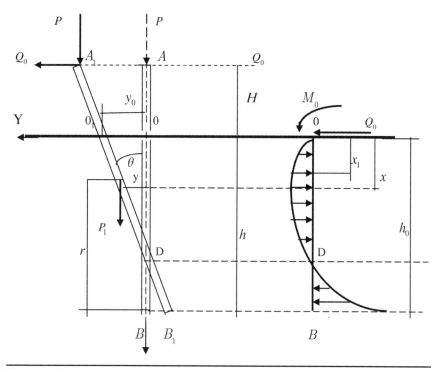

Figure 2.58

the total moment and shear force at point B_1 are equal to zero, the following equation is written:

$$M_0 + Q_0 x + \int_0^x \left(-\frac{ky_0}{h} x_1 + \frac{k\theta}{h} x_1^2 \right)(x - x_1) dx_1 + P(H + h)\theta + P_1 r\theta = 0 \qquad (2.117)$$

After integration it will look as follows:

$$ky_0 \frac{h^2}{6} - k\frac{\theta h^3}{12} = M_0 + Q_0 h + P(H + h)\theta + P_1 r\theta = 0 \qquad (2.118)$$

The second equation is not affected by the vertical loads and looks the same as the first equation:

$$\frac{ky_0 h}{2} - \frac{k\theta h^2}{3} = Q_0 \qquad (2.119)$$

By solving equations 2.118 and 2.119, we find:

$$ky_0 = \frac{M_0 + \frac{3}{4} Q_0 h - \frac{3Q_0}{kh^2}(Ph_P + P_1 r)}{\frac{h^2}{24} - \frac{3}{2kh}(Ph_P + P_1 r)} \qquad (2.120)$$

$$k\theta = \frac{3}{h}\left(\frac{ky_0}{2} - \frac{Q_0}{h} \right) \qquad (2.121)$$

Equations 2.120 and 2.121 allow obtaining deflections ky_0 and $k\theta$ (initial parameters). It is interesting to note, that the pile loses stability when the denominator in formula 2.120 is equal to zero; in other words, when:

$$h^3 - \frac{36}{k}\left(Ph_p + P_1 r\right) = 0 \tag{2.122}$$

From 2.122 the critical vertical load P_{cr} can be obtained:

$$P_{cr} = \frac{1}{H+h}\left(\frac{kh^3}{36} - P_1 r\right) \tag{2.123}$$

That serves as good criteria for pile stability.

2.11 Flexible Piles. Variable Lateral Modulus of Subgrade Reaction

As mentioned above, a pile can be considered a long one, when $\frac{h}{b} \geq 10$. In this case, the pile has to be analyzed as a long flexible beam supported on Winkler foundation as shown in Figure 2.59. The method of initial parameters for long pile analysis, developed by Snitco (1970) is described below. The method takes into account that the modulus of lateral subgrade reaction varies from 0 at the top of the soil, up to k at the depth. It increases in accordance with the following relationship:

$$k_x = k\frac{x}{h} \tag{2.124}$$

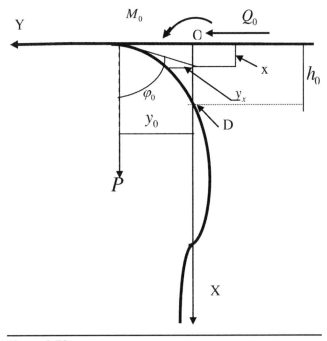

Figure 2.59

In 2.124 k is the modulus of subgrade reaction in horizontal direction at the depth h. The differential equation of the pile looks as follows:

$$y^{IV} = -\alpha xy \tag{2.125}$$

$$\alpha = \frac{k}{EIh} \tag{2.126}$$

Solution of equation 2.125 is obtained in exponential series using the method of initial parameters. Taking into account that, when $x = 0$ $y = y_0$ $y' = \varphi_0$ $y'' = \dfrac{M_0}{EI}$ $y''' = \dfrac{Q_0}{EI}$, the ordinates at any point along the pile are equal to:

$$
\begin{aligned}
y = y_0 &\left(1 - \frac{1\alpha x^5}{5!} + \frac{1 \cdot 6\alpha^2 x^{10}}{10!} - \frac{1 \cdot 6 \cdot 11\alpha^3 x^{15}}{15!} + \cdots\right) + \\
&\varphi_0\left(\frac{x}{1!} - \frac{2\alpha x^6}{6!} + \frac{2 \cdot 7\alpha^2 x^{11}}{11!} - \frac{2 \cdot 7 \cdot 12\alpha^3 x^{16}}{16!} + \cdots\right) \\
&+ \frac{M_0}{EI}\left(\frac{x^2}{2!} - \frac{3\alpha x^7}{7!} + \frac{3 \cdot 8\alpha^2 x^{12}}{12!} - \frac{3 \cdot 8 \cdot 13\alpha^3 x^{17}}{17!} + \cdots\right) \\
&+ \frac{Q_0}{EI}\left(\frac{x^3}{3!} - \frac{4\alpha x^8}{8!} + \frac{4 \cdot 9\alpha^2 x^{13}}{13!} - \frac{4 \cdot 9 \cdot 14\alpha^3 x^{18}}{18!}\right)
\end{aligned} \tag{2.127}
$$

After differentiating expression 2.127 twice and multiplying by EI, the moment and shear force at any point x can be found, as shown below:

$$
\begin{aligned}
M_x = \overline{y}_0 &\left(-\frac{\alpha x^3}{3!} + \frac{1 \cdot 6\alpha^2 x^8}{8!} - \frac{1 \cdot 6 \cdot 11\alpha^3 x^{13}}{13!} + \cdots\right) \\
&+ \overline{\varphi}_0\left(-\frac{2\alpha x^4}{4!} + \frac{2 \cdot 7\alpha^2 x^9}{9!} - \frac{2 \cdot 7 \cdot 12\alpha^3 x^{14}}{14!} + \cdots\right) \\
&+ M_0\left(1 - \frac{3\alpha x^5}{5!} + \frac{3 \cdot 8\alpha^2 x^{10}}{10!} - \frac{3 \cdot 8 \cdot 13\alpha^3 x^{15}}{15!} + \cdots\right) \\
&+ Q_0\left(\frac{x}{1!} - \frac{4\alpha x^6}{6!} + \frac{4 \cdot 9\alpha^2}{11!} - \frac{4 \cdot 9 \cdot 14\alpha^3 x^{16}}{16!} + \cdots\right)
\end{aligned} \tag{2.128}
$$

where $\overline{y}_0 = y_0 EI$ and $\overline{\varphi}_0 = \varphi_0 EI$

$$
\begin{aligned}
Q_x = \overline{y}_0 &\left(-\frac{1\alpha x^2}{2!} + \frac{1 \cdot 6\alpha^2 x^7}{7!} - \frac{1 \cdot 6 \cdot 11\alpha^3 x^{12}}{12!} + \cdots\right) \\
&+ \overline{\varphi}_0\left(-\frac{2\alpha x^3}{3!} + \frac{2 \cdot 7\alpha^2 x^8}{8!} - \frac{2 \cdot 7 \cdot 12\alpha^3 x^{13}}{13!} + \cdots\right) \\
&+ M_0\left(-\frac{3\alpha x^4}{4!} + \frac{3 \cdot 8\alpha^2 x^9}{9!} - \frac{3 \cdot 8 \cdot 13\alpha^3 x^{14}}{14!} + \cdots\right) \\
&+ Q_0\left(1 - \frac{4\alpha x^5}{5!} + \frac{4 \cdot 9\alpha^2 x^{10}}{10!} - \frac{4 \cdot 9 \cdot 14\alpha^3 x^{15}}{15!} + \cdots\right)
\end{aligned} \tag{2.129}
$$

Taking into account that all terms of the series, after the third one, are negligibly small, deflections, moments, and shear forces at any cross section of the pile can be easily obtained. A numerical example that illustrates analysis of long piles loaded with horizontal and vertical loads is shown below:

Example 2.8

Given: $h = 3.5\text{m}$ $C_h = 8.33\dfrac{\text{kg}}{\text{cm}^3}$ $b = 24\text{cm}$ $M_0 = 2.04\text{tm}$

$k = C_h b = 8.33 \cdot 24 = 2{,}000\dfrac{\text{t}}{\text{m}^2}$ $EI = 162.8\dfrac{\text{t}}{\text{m}^2}$ $\alpha = 3.51\dfrac{1}{\text{m}^5}$

Now equations 2.130 and 2.131 can be written for the bottom of the pile, where $x = h$, taking into account that at this point $M_h = 0$ and $Q_h = 0$:

$$\left.\begin{array}{l} 10.937\overline{y}_0 - 9.848\overline{\varphi}_0 - 71.671 = 0 \\ 53.526\overline{y}_0 + 33.484 - 44.191 = 0 \end{array}\right\} \tag{2.130}$$

By solving this system, we find: $\overline{y}_0 = 3.174\text{tm}^3$ $\overline{\varphi}_0 = -3.753\text{tm}^2$.

The real deflections are $y_0 = \dfrac{3.174}{162.8} = 19.5\text{cm}$ $\varphi_0 = -\dfrac{3.753}{162.8} = -0.0231\text{rad}.$

Now, when all initial parameters are known, using formulae 2.127–2.129, lateral deflections, shear forces, and moments can be found at any point of the pile. If a vertical load P is applied to the top of the pile, this load will produce an additional moment equal to:

$$\Delta M = P(y_0 + y_x) \tag{2.131}$$

Table 2.1 Equations for analysis of semi-infinite beams on Winkler foundations

Left end	Deflect, Moment, Shear	$P=1$ (at 0)	$P=1$	$M=1$	$M=1$
Free supported	v_x	$0.5\lambda^3(EI)^{-1}(T_xA_u - U_xB_u)$	$0.5\lambda^3(EI)^{-1}(T_xA_u - U_xB_x)$	$0.5\lambda^2(EI)^{-1}(-U_xA_u - 4T_xD_u)$	$0.5\lambda^2(EI)^{-1}(-W_uA_x + 2T_uB_x)$
	φ_x	$0.5\lambda^2(EI)^{-1}(-W_xA_u - 2T_xB_u)$	$0.5\lambda^2(EI)^{-1}(-4T_xD_x - U_xA_x)$	$\lambda(EI)^{-1}(T_xA_u + 2W_xD_u)$	$\lambda(EI)^{-1}(2W_uD_x + T_uA_x)$
	M_x	$\lambda(-V_xA_u + W_xB_u)$	$2\lambda(T_uC_x - U_uD_x)$	$W_xA_u + 4V_xD_u$	$-2W_uC_x + 4T_uD_x$
	Q_x	$U_xA_u + 2V_xB_u$	$2T_uB_x - 2U_uC_x$	$2\lambda^{-1}(-V_xA_u + 2U_xD_u)$	$2\lambda^{-1}(-W_uB_x + 2T_uC_x)$
Pinned	v_x	$0.125\lambda^3(EI)^{-1}(W_{x+u} + W_{x-u})$	$0.125\lambda^3(EI)^{-1}(-W_{x+u} + W_{x-u})$	$0.25\lambda^2(EI)^{-1}(V_{x+u} + V_{x-u})$	$0.25\lambda^2(EI)^{-1}(V_{x+u} - V_{x-u})$
	φ_x	$0.25\lambda^2(EI)^{-1}(V_{x+u} - V_{x-u})$	$0.25\lambda(EI)^{-1}(V_{x+u} + V_{x-u})$	$\lambda(EI)^{-1}(U_{x+u} + U_{x-u})$	$\lambda(EI)^{-1}(U_{x+u} + U_{x-u})$
	M_x	$0.25\lambda(U_{x+u} - U_{x-u})$	$0.25\lambda(U_{x+u} - U_{x-u})$	$0.5(T_{x+u} + T_{x-u})$	$0.5(T_{x+u} - T_{x-u})$
	Q_x	$0.5(T_{x+u} - T_{x-u})$	$0.5(T_{x+u} - T_{x-u})$	$0.5\lambda(-W_{x+u} - W_{x-u})$	$0.5\lambda(-W_{x+u} - W_{x-u})$
Fixed	v_x	$\lambda^3(EI)^{-1}(V_xC_u - W_xD_u)$	$\lambda^3(EI)^{-1}(V_uC_x - W_uD_x)$	$\lambda^2(EI)(U_xC_u + 2V_xD_u)^{-1}$	$\lambda^2(EI)(U_uC_x + 2V_uD_x)^{-1}$
	φ_x	$\lambda^2(EI)(U_xC_u + 2V_xD_u)^{-1}$	$\lambda^2(EI)^{-1}(V_uB_x - W_uC_x)$	$\lambda(EI)^{-1}(U_xB_x + 2V_xC_x)$	$\lambda(EI)^{-1}(U_uB_x + 2V_uC_x)$
	M_x	$2\lambda(T_uC_u - U_xD_u)$	$\lambda(-V_uA_x + W_uB_x)$	$-U_xA_x - 2V_xB_x$	$-U_uA_x - 2V_uB_x$
	Q_x	$2W_xC_u - 4T_xD_u$	$4V_uD_x + W_uA_x$	$2\lambda^{-1}(2U_xD_x - V_xA_x)$	$2\lambda^{-1}(2U_uD_x - V_uA_x)$
Guided	v_x	$0.125\lambda^3(EI)^{-1}(W_{x+u} + W_{x-u})$	$0.125\lambda^3(EI)^{-1}(W_{x+u} + W_{x-u})$	$0.25\lambda^2(EI)^{-1}(-V_{x+u} + V_{x-u})$	$0.25\lambda^2(EI)^{-1}(-V_{x+u} + V_{x-u})$
	φ_x	$0.25\lambda^2(EI)(V_{x+u} - V_{x-u})$	$0.25\lambda^2(EI)^{-1}(V_{x+u} - V_{x-u})$	$\lambda(EI)^{-1}(-U_{x+u} + U_{x-u})$	$\lambda(EI)^{-1}(-U_{x+u} + U_{x-u})$
	M_x	$0.25\lambda(U_{x+u} + U_{x-u})$	$0.25\lambda(U_{x+u} + U_{x-u})$	$0.5(-T_{x+u} + T_{x-u})$	$0.5(-T_{x+u} + T_{x-u})$
	Q_x	$-0.5(T_{x+u} + T_{x-u})$	$0.5(T_{x+u} + T_{x-u})$	$0.5\lambda^{-1}(W_{x+u} - W_{x-u})$	$0.5\lambda^{-1}(W_{x+u} - W_{x-u})$

Table 2.2

Portion of the beam	Length	ξ	$A(\xi)$	$B(\xi)$	$C(\xi)$	$D(\xi)$
l	20.00	5.00	—	-25.05645	-35.57745	-23.05250
$l - u_1$	15.00	3.75	-17.4552	-14.79715	—	—
$l - u_2$	8.00	2.00	-1.5656	0.95575	—	—
$l - c$	14.00	3.50	—	-10.65245	2.90140	—
$l - d$	4.00	1.00	—	0.96675	0.49014	—

Table 2.3

Portion of the beam	Length	ξ	$A(\xi)$	$B(\xi)$	$C(\xi)$	$D(\xi)$
$0.5l$	10	2.50	−4.9128	−0.5885	1.81045	2.12925
$0.5\,l - u_1$	5	1.25	−	1.1486	−	0.32175
$0.5l - c$	4	1.00	0.8337	−	0.49445	−

Table 2.4 Analysis of the beam loaded with $P = 10tn$

Coordinates m.	Settlement mm	Moment kg-m	Shear force kg
0.00	0.11268	0.0000	0.0000
0.50	0.14108	61.1000	253.8000
1.00	0.16942	263.2000	564.3000
1.50	0.19748	634.8000	931.2000
2.00	0.22491	1,203.7000	1,353.8000
2.50	0.25114	1,997.5000	1,830.1000
3.00	0.27540	3,0423.000	2,357.0000
3.50	0.29667	4,3622.000	2,929.7000
4.00	0.31366	5,978.4000	3,540.9000
4.50	0.32478	7,907.8000	4,180.4000
5.00	**0.32815**	**10,161.4000**	**4,834.8000**
5.00	**0.32815**	**10,161.4000**	**−5,165.2000**
5.50	0.32239	7,742.3000	−4,513.2000
6.00	0.30904	5,644.9000	−3,880.6000
6.50	0.29015	3,856.2000	−3,280.7000
7.00	0.26747	2,357.3000	−2,722.6000
7.50	0.24247	1,125.6000	−2,212.3000
8.00	0.21635	135.4000	−1,753.4000
8.50	0.19008	−636.6000	−1,347.0000
9.00	0.16442	−1,219.3000	−992.7000
9.50	0.13993	−1,637.6000	−688.6000
10.00	0.11703	−1,915.8000	−431.9000
10.50	0.09599	−2,076.9000	−219.2000
11.00	0.07697	−2,141.7000	−46.6000
11.50	0.06004	−2,129.4000	90.1000
12.00	0.04518	−2,057.0000	194.9000
12.50	0.03233	−1,939.1000	272.1000
13.00	0.02137	−1,788.8000	325.5000
13.50	0.01215	−1,616.9000	358.8000
14.00	0.00451	−1,432.8000	375.2000
14.50	−0.00173	−1,244.1000	377.7000
15.00	−0.00675	−1,057.0000	369.1000
15.50	−0.01075	−876.5000	351.4000
16.00	−0.01388	−706.7000	326.7000
16.50	−0.01632	−550.8000	296.4000
17.00	−0.01823	−411.1000	261.7000
17.50	−0.01973	−289.6000	223.7000
18.00	−0.02095	−187.8000	183.0000
18.50	−0.02198	−107.0000	140.0000
19.00	−0.02291	−48.1000	95.1000
19.50	−0.02378	−12.2000	48.4000
20.00	−0.02465	0.0000	0.0000

Table 2.5 Analysis of the beam loaded with $q = 2,000 \text{t/m}$

Coordinates m.	Settlement mm	Moment kg-m	Shear force kg
0.00	-0.09045	0.0000	0.000
0.50	−0.05448	−39.2000	−144.900
1.00	−0.01846	−133.000	−217.900
1.50	0.01769	−245.100	−218.700
2.00	0.05408	−339.600	−147.000
2.50	0.09080	−379.900	−2.100
3.00	0.12789	−329.400	216.500
3.50	0.16528	−151.000	509.600
4.00	0.20281	192.700	877.700
4.50	0.24013	739.200	1,320.700
5.00	0.27671	1,525.700	1,837.700
5.50	0.31180	2,338.900	1,426.500
6.00	0.34462	2,963.600	1,083.400
6.50	0.37456	3,432.700	803.100
7.00	0.40116	3,776.200	579.400
7.50	0.42408	4,020.400	405.300
8.00	0.44307	4,188.400	273.100
8.50	0.45798	4,299.200	174.900
9.00	0.46870	4,367.600	102.200
9.50	0.47516	4,404.300	46.800
10.00	0.47731	4,415.800	0.000
10.50	0.47516	4,404.300	−46.800
11.00	0.46870	4,367.600	−102.200
11.50	0.45798	4,299.200	−174.900
12.00	0.44307	4,188.400	−273.100
12.50	0.42408	4,020.400	−405.300
13.00	0.40116	3,776.200	−579.400
13.50	0.37456	3,432.700	−803.100
14.00	0.34462	2,963.600	−1,083.400
14.50	0.31180	2,338.900	−1,426.500
15.00	0.27671	1,525.700	−1,837.700
15.50	0.24013	739.200	−1,320.700
16.00	0.20281	192.700	−877.700
16.50	0.16528	−151.000	−509.600
17.00	0.12789	−329.400	−216.500
17.50	0.09080	−379.900	2.100
18.00	0.05408	−339.600	147.000
18.50	0.01769	−245.100	218.700
19.00	−0.01846	−133.000	217.900
19.50	−0.05448	−39.2000	144.900
20.00	−0.09045	0.0000	0.000

Table 2.6 Analysis of the beam loaded with a moment $M = 1tm$

Coordinates m.	Settlement mm	Moment kg-m	Shear force kg
0.00	0.00049	0.0000	0.000
0.50	−0.00002	0.2000	0.500
1.00	−0.00053	0.3000	−0.100
1.50	−0.00104	−0.1000	−1.700
2.00	−0.00155	−1.5000	−4.300
2.50	−0.00206	−4.5000	−7.900
3.00	−0.00257	−9.6000	−12.500
3.50	−0.00306	−17.2000	−18.100
4.00	−0.00354	−27.9000	−24.700
4.50	−0.00399	−42.1000	−32.230
5.00	−0.00440	−60.3000	−40.700
5.50	−0.00475	−82.9000	−49.900
6.00	−0.00502	−110.3000	−59.600
6.50	−0.00518	−142.6000	−69.900
7.00	−0.00520	−180.2000	−80.300
7.50	−0.00504	−222.9000	−90.500
8.00	−0.00467	−270.6000	−100.300
8.50	−0.00403	−323.0000	−109.000
9.00	−0.00308	−379.4000	−116.200
9.50	−0.00175	−438.8000	−121.100
10.00	**0.00000**	**−500.000**	**0.000**
10.00	**0.00000**	**500.000**	**0.000**
10.50	0.00175	438.8000	121.100
11.00	0.00308	379.4000	116.200
11.50	0.00403	323.0000	109.000
12.00	0.00467	270.6000	100.300
12.50	0.00504	222.9000	90.500
13.00	0.00520	180.2000	80.300
13.50	0.00518	142.6000	69.900
14.00	0.00502	110.3000	59.600
14.50	0.00475	82.9000	49.900
15.00	0.00440	60.3000	40.700
15.50	0.00399	42.1000	32.230
16.00	0.00354	27.9000	24.700
16.50	0.00306	17.2000	18.100
17.00	0.00257	9.6000	12.500
17.50	0.00206	4.5000	7.900
18.00	0.00155	1.5000	4.300
18.50	0.00104	0.1000	1.700
19.00	0.00053	−0.3000	0.100
19.50	0.00002	−0.2000	−0.500
20.00	−0.00049	0.0000	0.000

Table 2.7

φ_1	φ_2	φ_3	ν_1	ν_2	ν_3			
$EI\xi = 4i_2\alpha_1$	$EI\eta$	0	$-EI\mu - \dfrac{6i_2}{l_2}\alpha_2$	$EI\gamma$	0		φ_1	M_1
	$2EI\xi$	$EI\eta$	$EI\gamma$	0	$-EI\gamma$		φ_2	M_2
		$EI\xi = 4i_2\alpha_1$	0	$EI\gamma$	$EI\mu - \dfrac{6i_2}{l_2}\alpha_2$	$\times$	φ_3	$= M_3$
			$\dfrac{12i_2}{l_2^2}\beta + EI\psi$	$-EI\sigma$	0		ν_1	N_1
	Symmetrical			$EI\psi$	$-EI\sigma$		ν_2	N_2
					$\dfrac{12i_2}{l_2^2}\beta + EI\psi$		ν_3	N_3

Table 2.8

φ_1	φ_2	φ_3	ν_1	ν_2	ν_3			
911.772	300.190	0	115.586	−132.084	0		φ_1	18.780
	1,467.110	300.190	132.084	0	−115.586		φ_2	45.800
		911.172	0	132.084	−115.586	$\times$	φ_3	46.672
			167.992	−32.748	0		ν_1	$=$ 93.980
	Symmetrical			207.404	−32.748		ν_2	189.000
					167.992		ν_3	137.110

Table 2.9 Moments and shear forces at the edges of totally rigid elements for beams with both ends fixed

Load, deflection	$M_{m_2 f_2}$	$Q_{m_2 f_2}$	M_{m,e_1}	Q_{m,e_1}
M^0, Q^0	$M^0_{m_2 f_2}$	$Q^0_{m_2 f_2}$	M^0_{m,e_1}	Q^0_{m,e_1}
Rotation $\varphi_m = 1$	$-k_{m_2 f_2}\left(\xi_{m_2 e_2} - \alpha_2\mu_{m_2 f_2}\right)$	$k_{m_2 f_2}\left(\mu_{m_2 f_2} - \alpha_2\psi_{m_2 f_2}\right)$	$-k_{m,e_1}\left(\xi_{m,e_1} - \alpha_2\mu_{m,e_1}\right)$	$k_{m,e_1}\left(\mu_{m,e_1} - \alpha_2\psi_{m,e_1}\right)$
Rotation $\varphi_f = 1$	$-k_{m_2 f_2}\left(\eta_{m_2 f_2} + \alpha_3\gamma_{m_2 f_2}\right)$	$-k_{m_2 f_2}\left(\gamma_{m_2 f_2} + \alpha_3\sigma_{m_2 f_2}\right)$	0	0
Rotation $\varphi_e = 1$	0	0	$-k_{m,e_1}\left(\eta_{m,e_1} - \alpha_1\gamma_{m,e_1}\right)$	$-k_{m,e_1}\left(\gamma_{m,e_1} + \alpha_1\sigma_{m,e_1}\right)$
Settlement $\nu_m = 1$	$k_{m_2 f_2}\mu_{m_2 f_2}$	$-k_{m_2 f_2}\psi_{m_2 f_2}$	$-k_{m,e_1}\mu_{m,e_1}$	$k_{m,e_1}\psi_{m,e_1}$
Settlement $\nu_f = 1$	$k_{m_2 f_2}\gamma_{m_2 f_2}$	$k_{m_2 f_2}\sigma_{m_2 f_2}$	0	0
Settlement $\nu_e = 1$	0	0	$-k_{m,e_1}\gamma_{m,e_1}$	$-k_{m,e_1}\sigma_{m,e_1}$

1. M^0 and Q^0 are moment and shear force, respectively, at the edges of rigid elements, due to the given loads applied to the beams mf or me
2. All other moments and shear forces in this table are moments and shear forces due to deflections of the nodes m, f, and e
3. Moments and shear forces at the nodes m, f, and e (at the centers of totally rigid elements), due to the same loads and deflections, are shown in Tables 2.10 and 2.11

Table 2.10 Moments and shear forces for the beam *mf* with both ends fixed

Load, deflection	M_{mf}	Q_{mf}
M^0, Q^0	$M^0_{m_2 f_2} + Q^0_{m_2 f_2} \cdot \alpha_2$	$Q^0_{m_2 f_2}$
Rotation $\varphi_m = 1$	$k_{m_2 f_2}\left(2\alpha_2 \mu_{m_2 f_2} - \xi_{m_2 f_2} - \psi_{m_2 f_2}\alpha_2^2\right) - 0.5K_\varphi l_F$	$k_{m_2 f_2}\left(\mu_{m_2 f_2} - \alpha_2 \psi_{m_2 f_2}\right) - 0.25K_\varphi A_F \alpha_2$
Rotation $\varphi_f = 1$	$k_{m_2 f_2}\left(\eta_{m_2 f_2} + \alpha\gamma_{m_2 f_2} + \alpha\gamma_{m_2 f_2} + \alpha_2\alpha_3\sigma_{m_2 f_2}\right)$	$-k_{m_2 f_2}\left(\gamma_{m_2 f_2} + \alpha_2\sigma_{m_2 f_2}\right)$
Rotation $\varphi_e = 1$	0	0
Settlement $\nu_m = 1$	$k_{m_2 f_2}\left(\mu_{m_2 f_2} - \psi_{m_2 f_2}\alpha_2\right) - 0.25K_z A_F \alpha_2$	$-k_{m_2 f_2}\psi_{m_2 f_2} - 0.5K_z A_F$
Settlement $\nu_f = 1$	$k_{m_2 f_2}\left(\gamma_{m_2 f_2} + \alpha_2\sigma_{m_2 f_2}\right)$	$k_{m_2 f_2}\sigma_{m_2 f_2}$
Settlement $\nu_e = 1$	0	0

Table 2.11 Moments and shear forces for the beam *me* with both ends fixed

Load, deflection	M_{me}	Q_{me}
M^0, Q^0	$M^0_{m_1 e_1} + Q^0_{m_1 e_1} \cdot \alpha_2$	$Q^0_{m_1 e_1}$
Rotation $\varphi_m = 1$	$k_{m_1 e_1}\left(2\alpha_2 \mu_{m_1 e_1} - \xi_{m_1 e_1} - \psi_{m_1 e_1}\alpha_2^2\right) - 0.5K_\varphi l_F$	$k_{m_1 e_1}\left(\mu_{m_1 e_1} - \alpha_2 \psi_{m_1 e_1}\right) - 0.25K_\varphi A_F \alpha_2$
Rotation $\varphi_f = 1$	0	0
Rotation $\varphi_e = 1$	$-k_{m_1 e_1}\left(\eta_{m_1 e_1} + \alpha_1\gamma_{m_1 e_1} + \alpha_2\gamma_{m_1 e_1} + \alpha_1\alpha_2\sigma_{m_1 e_1}\right)$	$-k_{m_1 e_1}\left(\gamma_{m_1 e_1} + \alpha_1\sigma_{m_1 e_1}\right)$
Settlement $\nu_m = 1$	$-k_{m_1 e_1}\left(\mu_{m_1 e_1} - \psi_{m_1 e_1}\alpha_2\right) - 0.25K_z A_F \alpha_2$	$k_{m_1 e_1}\psi_{m_1 e_1} + 0.5K_z A_F$
Settlement $\nu_f = 1$	0	0
Settlement $\nu_e = 1$	$-k_{m_1 e_1}\left(\gamma_{m_1 e_1} + \alpha_2\sigma_{m_1 e_1}\right)$	$-k_{m_1 e_1}\sigma_{m_1 e_1}$

Table 2.12 Moments and shear forces in the beam me_1 with one fixed and one free supported end

Type of load or deflection	Moments and shear forces at point m_1		Moments and shear forces at point m	
	$M_{m_1 e_1}$	$Q_{m_1 e_1}$	M_{me_1}	Q_{me_1}
$M^0,\ Q^0$	$M^0_{m_1 e_1}$	$Q^0_{m_1 e_1}$	$M^0_{me_1} + Q^0_{me_1}\alpha$	$Q^0_{me_1}$
Rotation $\varphi_{m=1}$	$-2i_{m_1 e_1}\left(2\alpha_1 + \dfrac{3\alpha_2}{l_{m_1 e_1}}\alpha\right)$	$-\dfrac{6i_{m_1 e_1}}{l_{m_1 e_1}}\left(\alpha_2 + \dfrac{2\beta}{l_{m_1 e_1}}\alpha\right)$	$-4i_{me_1}\left(\alpha_1 + \dfrac{3a}{l_{me_1}}\alpha_2 + \dfrac{3a^2}{l^2_{me_1}}\beta\right) - \dfrac{K_\varphi l}{2}$	$-\dfrac{6i_{me_1}}{l_{me_1}}\left(\alpha_2 + \dfrac{2\beta}{l_{me_1}}\alpha\right) - \dfrac{K_\varphi A_F \alpha}{4}$
Settlement $\nu_{m=1}$	$\dfrac{6i_{m_1 e_1}}{l_{m_1 e_1}}\alpha_2$	$\dfrac{12i_{m_1 e_1}}{l^2_{m_1 e_1}}\beta$	$\dfrac{6i_{me_1}}{l_{me_1}}\left(\alpha_2 + \dfrac{2\beta}{l_{me_1}}a\right) + \dfrac{K_\varphi A_F a}{4}$	$\dfrac{12i_{me_1}}{l^2_{me_1}}\beta + \dfrac{k_z A_F}{4}$

Table 2.13 Moments and shear forces in the beam mf_2 with one fixed and one free supported end

Type of load or deflection	Moments and shear forces at point m_2		Moments and shear forces at point m	
	$M_{m_2 f_2}$	$Q_{m_2 f_2}$	M_{mf_2}	Q_{mf_2}
$M^0,\ Q^0$	$M^0_{m_2 f_2}$	$Q^0_{m_2 f_2}$	$M^0_{mf_2} + Q^0_{mf_2}a$	$Q^0_{mf_2}$
Rotation $\varphi_{m=1}$	$-2i_{m_2 f_2}\left(2\alpha_1 + \dfrac{3\alpha_2}{l_{m_2 f_2}}\alpha\right)$	$-\dfrac{6i_{m_2 f_2}}{l_{m_2 f_2}}\left(\alpha_2 + \dfrac{2\beta}{l_{m_2 f_2}}\alpha\right)$	$-4i_{mf_2}\left(\alpha_1 + \dfrac{3a}{l_{mf_2}}\alpha_2 + \dfrac{3a^2}{l^2_{mf_2}}\beta\right) - \dfrac{K_\varphi l}{2}$	$+\dfrac{6i_{mf_2}}{l_{mf_2}}\left(\alpha_2 + \dfrac{2\beta}{l_{mf_2}}\alpha\right) - \dfrac{K_\varphi A_F \alpha}{4}$
Settlement $\nu_{m=1}$	$\dfrac{6i_{m_2 f_2}}{l_{m_2 f_2}}\alpha_2$	$\dfrac{12i_{m_2 f_2}}{l^2_{m_2 f_2}}\beta$	$\dfrac{6i_{mf_2}}{l_{mf_2}}\left(\alpha_2 + \dfrac{2\beta}{l_{mf_2}}a\right) + \dfrac{K_\varphi A_F a}{4}$	$\dfrac{12i_{mf_2}}{l^2_{mf_2}}\beta + \dfrac{k_z A_F}{4}$

Table 2.14 Equations for obtaining moments and shear forces in beams with totally rigid elements at both ends

$$M_{m_2 f_2} = M^0_{m_2 f_2} + k_{m_2 f_2}\left(\xi_{m_2 e_2} - a_2, \mu_{m_2 f_2}\right)\varphi_m + k_{m_2 f_2}\left(\eta_{m_2 f_2} + a_3 \gamma_{m_2 f_2}\right)\varphi_f + k_{m_2 f_2}\mu_{m_2 f_2} v_m - k_{m_2 f_2}\gamma_{m_2 f_2} v_f$$

$$Q_{m_2 f_2} = Q^0_{m_2 f_2} + k_{m_2 f_2}\left(\mu_{m_2 f_2} - \alpha_2 \psi_{m_2 f_2}\right)\varphi_m - k_{m_2 f_2}\left(\gamma_{m_2 f_2} + \alpha_3 \sigma_{m_2 f_2}\right)\varphi_f - k_{m_2 f_2}\psi_{m_2 f_2} v_m - k_{m_2 f_2}\sigma_{m_2 f_2} v_f$$

$$M_{m_1 e_1} = M^0_{m_1 e_1} + k_{m_1 e_1}\left(\xi_{m_1 e_1} - a_2 \mu_{m_1 e_1}\right)\varphi_m + k_{m_1 e_1}\left(\eta_{m_1 e_1} - a_1 \gamma_{m_1 e_1}\right)\varphi_e - k_{m_1 e_1}\mu_{m_1 e_1} v_m + k_{m_1 e_1}\gamma_{m_1 e_1} v_e$$

$$Q_{m_1 e_1} = Q^0_{m_1 e_1} + k_{m_1 e_1}\left(\mu_{m_1 e_1} - \alpha_2 \psi_{m_1 e_1}\right)\varphi_m + k_{m_1 e_1}\left(\gamma_{m_1 e_1} + \alpha_1 \sigma_{m_1 e_1}\right)\varphi_e - k_{m_1 e_1}\psi_{m_1 e_1} v_m + k_{m_1 e_1}\sigma_{m_1 e_1} v_e$$

$$M_{mf} = M^0_{m_2 f_2} + Q^0_{m_2 f_2}\alpha_2 + \left[k_{m_2 f_2}\left(2\alpha_2\mu_{m_2 e_2} - \xi_{m_2 f_2} - \psi_{m_2 f_2}\alpha_2^2\right) - 0.5 K_\varphi l_F\right]\varphi_m + k_{m_2 f_2}$$
$$\left(\eta_{m_2 f_2} + \alpha\gamma_{m_2 f_2} + \alpha\gamma_{m_2 f_2} + \alpha_2\alpha_3\sigma_{m_2 f_2}\right)\varphi_f + \left[k_{m_2 f_2}\left(\mu_{m_2 f_2} - \psi_{m_2 f_2}\alpha_2\right) - 0.25 K_z A_F \alpha_2\right]v_m - k_{m_2 f_2}\left(\gamma_{m_2 f_2} + \alpha_2\sigma_{m_2 f_2}\right)v_e$$

$$Q_{mf} = Q^0_{m_2 f_2} + \left[k_{m_2 f_2}\left(\mu_{m_2 f_2} - \alpha_2\psi_{m_2 f_2}\right) - 0.25\,K_\varphi A_F \alpha_2\right]\varphi_m - k_{m_2 f_2}\left(\gamma_{m_2 f_2} + \alpha_2\sigma_{m_2 f_2}\right)\varphi_f + \left(k_{m_2 f_2}\psi_{m_2 f_2} + 0.5 K_z A_F\right)v_m - k_{m_2 f_2}\sigma_{m_2 f_2} v_f$$

$$M_{me} = M^0_{m_1 e_1} + Q^0_{m_1 e_1}\alpha_2 + \left[k_{m_1 e_1}\left(2\alpha_2\mu_{m_1 e_1} - \xi_{m_1 e_1} - \psi_{m_1 e_1}\alpha_2^2\right) - 0.5 K_\varphi l_F\right]\varphi_m + k_{m_1 e_1}\left(\eta_{m_1 e_1} + \alpha_1\gamma_{m_1 e_1} + \alpha_2\gamma_{m_1 e_1}\right)$$
$$\varphi_f - \left[k_{m_1 e_1}\left(\mu_{m_1 e_1} - \psi_{m_1 e_1}\alpha_2\right) - 0.25 K_z A_F \alpha_2\right]v_m + k_{m_1 e_1}\left(\gamma_{m_1 e_1} + \alpha_2\sigma_{m_1 e_1}\right)v_e$$

$$Q_{me} = Q^0_{m_1 e_1} + \left[k_{m_1 e_1}\left(\mu_{m_1 e_1} - \alpha_2\psi_{m_1 e_1}\right) - 0.25 K_\varphi A_F \alpha_2\right]\varphi_m - k_{m_1 e_1}\left(\gamma_{m_1 e_1} + \alpha_1\sigma_{m_1 e_1}\right)\varphi_e + \left(k_{m_1 e_1}\psi_{m_1 e_1} + 0.5 K_z A_F\right)v_m - k_{m_1 e_1}\sigma_{m_1 e_1} v_e$$

Table 2.15 Equations for obtaining moments and shear forces in beams with one fixed and one free supported end

$$M_{m_1 e_1} = M^0_{m_1 e_1} + 2i_{m_1 e_1}\left(2\alpha_1 + \frac{3\alpha_2}{I_{m_1 e_1}}a\right)\varphi_m - \frac{6i_{m_1 e_1}}{I_{m_1 e_1}}\alpha_2$$

$$Q_{m_1 e_1} = Q^0_{m_1 e_1} - \frac{6i_{m_1 e_1}}{I_{m_1 e_1}}\left(\alpha_2 + \frac{2\beta}{I_{m_1 e_1}}a\right)\varphi_m + \frac{12i_{m_1 e_1}}{I^2_{m_1 e_1}}\beta v_m$$

$$M_{me_1} = M^0_{m_1 e_1}\alpha + \left[4i_{m_1 e_1}\left(\alpha_1 + \frac{3\alpha}{I_{m_1 e_1}}\alpha_2 + \frac{3\alpha^2}{I^2_{m_1 e_1}}\beta\right) + \frac{K_\varphi l}{2}\right]$$

$$M_{me_1} = M^0_{m_1 e_1}\alpha + \left[4i_{m_1 e_1}\left(\alpha_1 + \frac{3\alpha}{I_{m_1 e_1}}\alpha_2 + \frac{3\alpha^2}{I^2_{m_1 e_1}}\beta\right) + \frac{K_\varphi l}{2}\right]\varphi_m - \left[\frac{6i_{m_1 e_1}}{I_{m_1 e_1}}\left(\alpha_2 + \frac{2\beta}{I_{m_1 e_1}}a\right) + \frac{K_\varphi A_F a}{4}\right]v_m$$

$$Q_{me_1} = Q^0_{m_1 e_1} - \left[\frac{6i_{m_1 e_1}}{I_{m_1 e_1}}\left(\alpha_2 + \frac{2\beta}{I_{m_1 e_1}}a\right) + \frac{K_\varphi A_F a}{4}\right]\varphi_m + \left(\frac{12i_{m_1 e_1}}{I^2_{m_1 e_1}}\beta + \frac{k_z A_F}{4}\right)v_m$$

$$M_{m_2 f_2} = M^0_{m_2 f_2} + 2i_{m_2 f_2}\left(2\alpha_1 + \frac{3\alpha_2}{I_{m_2 f_2}}a\right)\varphi_m - \frac{6i_{m_2 f_2}}{I_{m_2 f_2}}\alpha_2 v_m$$

$$Q_{m_2 f_2} = Q^0_{m_2 f_2} + \frac{6i_{m_2 f_2}}{I_{m_2 f_2}}\left(\alpha_2 + \frac{2\beta}{I_{m_2 f_2}}a\right)\varphi_m + \frac{12i_{m_2 f_2}}{I^2_{m_2 f_2}}\beta v_m$$

$$M_{mf} = M^0_{m_2 f_2} + Q^0_{m_2 f_2}a + \left[4i_{m_2 f_2}\left(\alpha_1 + \frac{3\alpha}{I_{m_2 f_2}}\alpha_2 + \frac{3\alpha^2}{I^2_{m_2 f_2}}\beta\right) + \frac{K_\varphi l}{2}\right]\varphi_m - \left[\frac{6i_{m_2 f_2}}{I_{m_2 f_2}}\left(\alpha_2 + \frac{2\beta}{I_{m_2 f_2}}\alpha\right) + \frac{K_\varphi A_F a}{4}\right]v_m$$

$$Q_{m_2 f_2} = Q^0_{m_2 f_2} + \left[\frac{6i_{m_2 f_2}}{I_{m_2 f_2}}\left(\alpha_2 + \frac{2\beta}{I_{m_2 f_2}}a\right) + \frac{K_\varphi A_F a}{4}\right]\varphi_m + \left(\frac{12i_{m_2 f_2}}{I^2_{m_2 f_2}}\beta + \frac{K_z A_F}{4}\right)v_m$$

Table 2.16 The Stiffness Matrix A

	φ_1	φ_2	φ_3	φ_4	φ_5	φ_6	φ_7	φ_8	φ_9	φ_{10}	φ_{11}	φ_{12}	φ_{13}	φ_{14}	δ_1	δ_2	δ_3	δ_4	ν_1	ν_2	ν_3
φ_1	α_{11}	$\alpha_{1,2}$	0	$\alpha_{1,4}$	0	0	0	0	0	0	0	0	0	0	$\alpha_{1,15}$	0	0	0	$\alpha_{1,19}$	$\alpha_{1,20}$	0
φ_2		α_{22}	α_{23}	0	α_{25}	0	0	0	0	0	0	0	0	0	$\alpha_{2,15}$	0	0	0	$\alpha_{2,19}$	0	$\alpha_{2,21}$
φ_3			α_{33}	0	0	α_{36}	0	0	0	0	0	0	0	0	$\alpha_{3,15}$	0	0	0	0	$\alpha_{3,20}$	$\alpha_{3,21}$
φ_4				α_{44}	α_{45}	0	α_{47}	0	0	0	0	0	0	0	$\alpha_{4,15}$	$\alpha_{4,16}$	0	0	$\alpha_{4,19}$	$\alpha_{4,20}$	0
φ_5					α_{55}	α_{56}	0	$\alpha_{5,8}$	0	0	0	0	0	0	$\alpha_{5,15}$	$\alpha_{5,16}$	0	0	$\alpha_{5,19}$	0	$\alpha_{5,21}$
φ_6						α_{66}	0	0	$\alpha_{6,9}$	0	0	0	0	0	$\alpha_{6,15}$	$\alpha_{6,16}$	0	0	0	$\alpha_{6,20}$	$\alpha_{6,21}$
φ_7							α_{77}	α_{78}	0	$\alpha_{7,10}$	0	0	0	0	0	$\alpha_{7,16}$	$\alpha_{7,17}$	0	$\alpha_{7,19}$	$\alpha_{7,20}$	0
φ_8								α_{88}	$\alpha_{8,9}$	0	$\alpha_{8,11}$	0	0	0	0	$\alpha_{8,16}$	$\alpha_{8,17}$	0	$\alpha_{8,19}$	0	$\alpha_{8,21}$
φ_9									α_{99}	0	0	$\alpha_{9,12}$	0	0	0	$\alpha_{9,16}$	$\alpha_{9,17}$	0	0	$\alpha_{9,20}$	$\alpha_{9,21}$
φ_{10}										$\alpha_{10,10}$	$\alpha_{10,11}$	0	$\alpha_{10,13}$	0	0	0	$\alpha_{10,17}$	$\alpha_{10,18}$	$\alpha_{10,19}$	$\alpha_{10,20}$	0
φ_{11}											$\alpha_{11,11}$	$\alpha_{11,12}$	0	0	0	0	$\alpha_{11,17}$	0	$\alpha_{11,19}$	0	$\alpha_{11,21}$
φ_{12}												$\alpha_{12,12}$	0	$\alpha_{12,14}$	0	0	$\alpha_{12,17}$	$\alpha_{12,18}$	0	$\alpha_{12,20}$	$\alpha_{12,21}$
φ_{13}													$\alpha_{13,13}$	$\alpha_{13,14}$	0	0	0	$\alpha_{13,18}$	$\alpha_{13,19}$	0	$\alpha_{13,21}$
φ_{14}														$\alpha_{14,14}$	0	0	0	$\alpha_{14,18}$	$\alpha_{14,19}$	0	$\alpha_{14,21}$
δ_1															$\alpha_{15,15}$	0	0	0	0	0	0
δ_2																$\alpha_{16,16}$	0	0	0	0	0
δ_3																	$\alpha_{17,17}$	0	0	0	0
δ_4																		$\alpha_{18,18}$	0	0	0
ν_1																			$\alpha_{19,19}$	$\alpha_{19,20}$	$\alpha_{19,21}$
ν_2																				$\alpha_{20,20}$	$\alpha_{21,21}$
ν_3																					$\alpha_{21,21}$

Symmetrical

Table 2.17 The Stiffness Matrix A

Table 2.6	φ1	φ2	φ3	φ4	φ5	φ6	φ7	φ8	φ9	φ10	φ11	φ12	φ13	φ14	δ1	δ2	δ3	δ4	ν1	ν2	ν3
1	1,538.6	564	0	106.60	0	0	0	0	0	0	0	0	0	0	-87.2	0	0	0	185.6	-225.0	0
2		2,431	564.00	0	106.6	0	0	0	0	0	0	0	0	0	-87.2	0	0	0	225.0	0	-225
3			1,538.6	0	0	106.6	0	0	0	0	0	0	0	0	-87.2	0	0	0	0	225	-185.6
4				155.48	19.60	0	10.14	0	0	0	0	0	0	0	-48.0	-7.24	0	0	-9.2	-9.20	0
5					194.68	19.6	0	10.14	0	0	0	0	0	0	-48.0	-7.24	0	0	9.2	0	-9.2
6						155.48	0	0	10.14	0	0	0	0	0	-48.0	-7.24	0	0	0	9.2	-9.2
7							83.20	19.60	0	11.86	0	0	0	0	0	-7.24	-9.88	0	9.2	-9.2	0
8								122.40	19.60	0	11.86	0	0	0	0	-7.24	-9.88	0	9.20	0	-9.2
9									83.20	0	0	11.86	0	0	0	-7.24	-9.88	0	0	9.2	-9.2
10										81.08	19.60	0	9.08	0	0	0	-9.88	-5.8	9.20	-9.20	0
11											102.12	19.60	0	0	0	0	-9.88	0	9.20	0	-9.2
12												81.08	0	9.08	0	0	-9.88	-5.8	0	9.2	-9.2
13													156.9	69.40	0	0	0	-5.8	16.28	0	-16.28
14														156.9	0	0	0	-5.8	16.28	0	16.28
15						Symmetrical									96.00	0	0	0	0	0	0
16																10.34	0	0	0	0	0
17																	16.46	0	0	0	0
18																		4.94	0	0	0
19																			213.76	-72.62	-2.54
20																				289.64	72.62
21																					213.76

Table 2.18 The matrix B

$$
B = \begin{array}{|c|}
\hline
0 \\ \hline
0 \\ \hline
0 \\ \hline
q_1 l^2 \cdot 12_{-1} \\ \hline
0 \\ \hline
-q_1 l^2 \cdot 12_{-1} \\ \hline
q_2 l^2 \cdot 12_{-1} \\ \hline
0 \\ \hline
-q_2 l^2 \cdot 12_{-1} \\ \hline
q_1 l^2 \cdot 12_{-1} \\ \hline
0 \\ \hline
-q_1 l^2 \cdot 12_{-1} \\ \hline
q_3 (2l)^2 \cdot 12_{-1} \\ \hline
-q_3 (2l)^2 \cdot 12_{-1} \\ \hline
S_1 + S_2 + S_3 + S_4 \\ \hline
S_2 + S_3 + S_4 \\ \hline
S_3 + S_4 \\ \hline
S_4 \\ \hline
P_1 + q_1 l + 0.5\, q_2 l + q_3 l \\ \hline
P_2 + 2 q_1 l + q_2 l \\ \hline
P_1 + q_1 l + 0.5\, q_2 l + q_3 l \\ \hline
\end{array}
\quad = \quad
\begin{array}{|c|}
\hline
0 \\ \hline
0 \\ \hline
0 \\ \hline
22.528 \\ \hline
0 \\ \hline
-22.528 \\ \hline
21.163 \\ \hline
0 \\ \hline
-21.163 \\ \hline
22.528 \\ \hline
0 \\ \hline
-22.528 \\ \hline
81.92 \\ \hline
-81.92 \\ \hline
31.20 \\ \hline
28.20 \\ \hline
22.30 \\ \hline
10.60 \\ \hline
120.00 \\ \hline
90.00 \\ \hline
120.00 \\ \hline
\end{array}
$$

Table 2.19 Moments in the foundation

Moments	M_{1-15}	M_{1-2}	M_{2-1}	M_{2-3}	M_{3-2}	M_{3-22}
At the centers of totally rigid elements	55.55	−65.65	115.35	−160.85	53.28	−108.21
Totally rigid elements are not taken into account	47.74	−65.84	105.66	−139.76	61.63	−98.55
Difference %	16.36	0.07	9.17	15.09	15.67	9.80
Moments at the edges of totally rigid elements	M_{16-15}	M_{17-18}	M_{18-17}	M_{19-20}	M_{20-19}	M_{21-22}
	39.12	−42.05	80.52	−117.71	26.51	−71.52
At the centers of totally rigid elements	55.55	−65.65	115.35	−160.85	53.28	−108.21
Difference %	41.9	56.1	43.26	36.65	200.9	51.38

Table 2.20 Moments in the columns

Moments	M_{1-4}	M_{4-1}	M_{2-5}	M_{5-2}	M_{3-6}	M_{6-3}
Totally rigid elements Are taken into account	10.10	12.3	45.5	14.9	54.93	18.47
Totally rigid elements Are not taken into account	18.10	3.47	34.10	23.02	36.92	23.02
Difference %	79.20	354.0	33.43	54.50	48.78	24.63

Table 2.21

$\dfrac{b}{h}$	α	$\dfrac{b}{h}$	α	$\dfrac{b}{h}$	α	$\dfrac{b}{h}$	α
1	0.141	1.8	0.217	3.5	0.274	6.0	0.278
1.2	0.166	2.0	0.229	4.0	0.281	7.0	0.304
1.4	0.187	2.3	0.242	4.5	0.287	8.0	0.307
1.5	0.196	2.5	0.249	5.0	0.291	10.0	0.312
1.6	0.204	3.0	0.263	5.5	0.295	∞	0.333

Table 2.22

x_1	x_2	x_3	x_4	x_5	x_6	x_7	x_8			
2.333560	-0.037459	0.00000	0.00000	0.522540	-0.537700	0.000000	0.000000	x_1		-15.0
	1.185520	0.00000	0.00000	0.000000	0.000000	0.261270	-0.268850	x_2		-12.5
		2.333560	-0.037459	-0.522540	0.000000	0.537700	0.000000	x_3		30.0
			1.185520	0.000000	-0.261270	0.000000	0.268850	x_4	$\times$	20.0
				0.591110	-0.174900	-0.174900	0.000000	x_5	$=$	250.0
					0.383236	0.000000	-0.087450	x_{61}		150.0
						0.383236	-0.087450	x_7		100.0
							0.235546	x_8		100.0

Table 2.23 Approximate values for the modulus of subgrade reactions for pile analysis

Type of soil	C_h kg/cm²
Mud, loose, fine grained sand	20–25
Average size sand	20–30
Soft loam	25–35
Sand supported on hard clay	30–40
High density close sand, clay and coarse gravel	100–150

Table 2.24 Functions for analysis of infinite and semi-infinite beams on Winkler foundation

$\xi = \dfrac{x}{\lambda}$	$T(\xi)$	$U(\xi)$	$V(\xi)$	$W(\xi)$	$\xi = \dfrac{x}{\lambda}$	$T(\xi)$	$U(\xi)$	$V(\xi)$	$W(\xi)$
0.000	1	1	0.0000	1	0.38	0.6351	0.3815	0.2536	0.8887
0.001	0.9990	0.9980	0.0010	1.0000	0.39	0.6262	0.3688	0.2574	0.8836
0.002	0.9980	0.9960	0.0020	1.0000	0.40	0.6174	0.3564	0.2610	0.8784
0.003	0.9970	0.9940	0.0030	1.0000	0.41	0.6087	0.3441	0.2646	0.8732
0.004	0.9960	0.9920	0.0040	1.0000	0.42	0.6000	0.3320	0.2680	0.8679
0.005	0.9950	0.9900	0.0050	1.0000	0.43	0.5913	0.3201	0.2712	0.8625
0.006	0.9940	0.9880	0.0060	1.0000	0.44	0.5827	0.3084	0.2743	0.8570
0.007	0.9930	0.9861	0.0070	0.9999	0.45	0.5742	0.2968	0.2774	0.8515
0.008	0.9920	0.9841	0.0080	0.9999	0.46	0.5657	0.2353	0.2803	0.8459
0.009	0.9910	0.9821	0.0087	0.9999	0.47	0.5573	0.2742	0.2832	0.8403
0.010	0.9900	0.9801	0.0099	0.9999	0.48	0.5489	0.2632	0.2857	0.8346
0.011	0.9890	0.9781	0.0109	0.9999	0.49	0.5406	0.2522	0.2883	0.8289
0.012	0.9880	0.9761	0.0119	0.9999	0.50	0.5323	0.2414	0.2908	0.8231
0.013	0.9870	0.9742	0.0129	0.9998	0..51	0.5241	0.2307	0.2932	0.8173
0.014	0.9860	0.9722	0.0138	0.9998	0.52	0.5159	0.2204	0.2954	0.8113
0.015	0.9850	0.9702	0.0148	0.9998	0.53	0.5079	0.2103	0.2976	0.8054
0.016	0.9840	0.9683	0.0158	0.9997	0.54	0.4998	0.2002	0.2995	0.7994
0.017	0.9830	0.9663	0.0167	0.9997	0.55	0.4918	0.1902	0.3016	0.7934
0.018	0.9820	0.9643	0.0177	0.9997	0.56	0.4839	0.1805	0.3035	0.7873
0.019	0.9810	0.9624	0.0187	0.9996	0.57	0.4761	0.1709	0.3052	0.7813
0.020	0.9800	0.9604	0.0196	0.9996	0.58	0.4683	0.1615	0.3068	0.7752
0.03	0.9700	0.9409	0.0291	0.9991	0.59	0.4606	0.1522	0.3084	0.7690
0.04	0.9600	0.9216	0.0384	0.9984	0.60	0.4529	0.1430	0.3099	0.7628
0.05	0.9501	0.9025	0.0476	0.9976	0.61	0.4453	0.1340	0.3113	0.7566
0.06	0.9401	0.8836	0.0565	0.9966	0.62	0.4378	0.1252	0.3126	0.7503
0.07	0.9302	0.8649	0.0653	0.9954	0.63	0.4301	0.1166	0.3138	0.7442
0.08	0.9202	0.8464	0.0738	0.9940	0.64	0.4230	0.1080	0.3150	0.7379
0.09	0.9103	0.8281	0.0822	0.9924	0.65	0.4156	0.0996	0.3160	0.7315
0.10	0.9003	0.8100	0.0903	0.9906	0.66	0.4083	0.0914	0.3169	0.7252
0.11	0.8904	0.7921	0.0983	0.9887	0.67	0.4011	0.0833	0.3178	0.7189
0.12	0.8806	0.7744	0.1062	0.9867	0.68	0.3940	0.0754	0.3186	0.7126
0.13	0.8707	0.7568	0.1138	0.9844	0.69	0.3869	0.0676	0.3193	0.7062
0.14	0.8606	0.7395	0.1213	0.9821	0.70	0.3798	0.0599	0.3199	0.6997
0.15	0.8510	0.7224	0.1286	0.9796	0.71	0.3729	0.0524	0.3205	0.6933
0.16	0.8413	0.7055	0.1358	0.9770	0.72	0.3659	0.0449	0.3210	0.6869
0.17	0.8315	0.6888	0.1427	0.9742	0.73	0.3591	0.0377	0.3214	0.6805
0.18	0.8218	0.6722	0.1495	0.9713	0.74	0.3524	0.0307	0.3217	0.6741
0.19	0.8121	0.6550	0.1562	0.9683	0.75	03456	0.0237	0.3220	0.6676
0.20	0.8024	0.6398	0.1627	0.9651	0.76	0.3389	0.0168	0.3221	0.6611
0.21	0.7928	0.6238	0.1690	0.9618	0.77	0.3324	0.0101	0.3223	0.6547
0.22	0.7832	0.6080	0.1752	0.9583	0.78	0.3259	0.0035	0.3224	0.6483
0.23	0.7736	0.5924	0.1812	0.9547	0.25 π	0.3224	0.0000	0.3224	0.6448
0.24	0.7641	0.5771	0.1870	0.9511	0.79	0.3195	−0.0030	0.3224	0.6418
0.25	0.7546	0.5619	0.1927	0.9472	0.80	0.3131	−0.0093	0.3223	0.6353
0.26	0.7451	0.5469	0.1982	0.9433	0.81	0.3067	−0.0155	0.3222	0.6289
0.27	0.7357	0.5321	0.2036	0.9393	0.82	0.3004	−0.0217	0.3221	0.6225
0.28	0.7264	0.5176	0.2089	0.9353	0.83	0.2943	−0.0276	0.3219	0.6160
0.29	0.7171	0.5030	0.2140	0.9310	0.84	0.2881	−0.0334	0.3215	0.6096
0.30	0.7078	0.4888	0.2189	0.9267	0.85	0.2821	−0.0391	0.3212	0.6032
0.31	0.6985	0.4748	0.2237	0.9222	0.86	0.2761	−0.0446	0.3207	0.5968
0.32	0.6893	0.4609	0.2284	0.9177	0.87	0.2702	−0.0500	0.3202	0.5904
0.33	0.6801	0.4472	0.2330	0.9130	0.88	0.2643	−0.0554	0.3197	0.5840
0.34	0.6710	0.4337	0.2374	0.9084	0.89	0.2585	−0.0606	0.3191	0.5776
0.35	0.6620	0.4204	0.2416	0.9036	0.90	0.2527	−0.0658	0.3185	0.5712
0.36	0.6530	0.4072	0.2457	0.8986	0.91	0.2470	−0.0708	0.3178	0.5648
0.37	0.6440	0.3943	0.2497	0.8938	0.92	0.2414	−0.0757	0.3169	0.5521

Table 2.24 Con't

$\xi = \frac{x}{\lambda}$	$T(\xi)$	$U(\xi)$	$V(\xi)$	$W(\xi)$	$\xi = \frac{x}{\lambda}$	$T(\xi)$	$U(\xi)$	$V(\xi)$	$W(\xi)$
0.93	0.2359	−0.0805	0.3169	0.5521	1.48	0.0207	−0.2060	0.2267	0.2474
0.94	0.2304	−0.0851	0.3155	0.5459	1.49	0.0183	−0.2064	0.2247	0.2429
0.95	0.2250	−0.0896	0.3146	0.5396	1.50	0.0158	−0.2068	0.2226	0.2384
0.96	0.2196	−0.0941	0.3107	0.5333	1.51	0.0134	−0.2071	0.2205	0.2339
0.97	0.2143	−0.0984	0.3127	0.5270	1.52	0.0111	−0.2073	0.2184	0.2295
0.98	0.2090	−0.1027	0.3117	0.5207	1.53	0.0089	−0.2075	0.2164	0.2252
0.99	0.2038	−0.1069	0.3107	0.5145	1.54	0.0056	−0.2077	0.2143	0.2209
1.00	0.1987	−0.1109	0.3096	0.5083	1.55	0.0044	−0.2078	0.2122	0.2166
1.01	0.1937	−0.1147	0.3085	0.5021	1.56	0.0022	−0.2079	0.2101	0.2123
1.02	0.1888	−0.1185	0.3075	0.4960	1.57	0.0002	−0.2079	0.2081	0.2082
1.03	0.1839	−0.1223	0.3061	0.4899	0.5 π	0.0000	−0.2079	0.2079	0.2079
1.04	0.1790	−0.1259	0.3049	0.4839	1.58	−0.0019	−0.2079	0.2060	0.2041
1.05	0.1742	−0.1294	0.3036	0.4778	1.59	−0.0039	−0.2078	0.2039	0.2000
1.06	0.1694	−0.1328	0.3023	0.4716	1.60	−0.0059	−0.2077	0.2018	0.1960
1.07	0.1647	−0.1362	0.3009	0.4656	1.61	−0.0078	−0.2075	0.1997	0.1919
1.08	0.1601	−0.1394	0.2995	0.4596	1.62	−0.0097	−0.2073	0.1976	0.1879
1.09	0.1555	−0.1426	0.2981	0.4536	1.63	−0.0116	−0.2071	0.1956	0.1840
1.10	0.1509	−0.1458	0.2967	0.4470	1.64	−0.0134	−0.2069	0.1935	0.1801
1.11	0.1464	−0.1488	0.2952	0.4416	1.65	−0.0152	−0.2067	0.1915	0.1793
1.12	0.1420	−0.1516	0.2936	0.4356	1.66	−0.0170	−0.2064	0.1894	0.1725
1.13	0.1378	−0.1543	0.2921	0.4298	1.67	−0.0187	−0.2060	0.1873	0.1686
1.14	0.1350	−0.1570	0.2906	0.4240	1.68	−0.0204	−0.2056	0.1852	0.1648
1.15	0.1293	−0.1597	0.2890	0.4183	1.69	−0.0220	−0.2051	0.1832	0.1612
1.16	0.1252	−0.1622	0.2874	0.4126	1.70	−0.0236	−0.2046	0.1812	0.1576
1.17	0.1241	−0.1647	0.2858	0.4069	1.71	−0.0251	−0.2042	0.1791	0.1540
1.18	0.1171	−0.1671	0.2842	0.4012	1.72	−0.0266	−0.2037	0.1771	0.1505
1.19	0.1131	−0.1694	0.2825	0.3955	1.73	−0.0281	−0.2032	0.1751	0.1470
1.20	0.1091	−0.1716	0.2807	0.3898	1.74	−0.0296	−0.2026	0.1730	0.1435
1.21	0.1053	−0.1737	0.2790	0.3842	1.75	−0.00310	−0.2020	0.1720	0.1400
1.22	0.1014	−0.1758	0.2773	0.3785	1.76	−0.0324	−0.2013	0.1690	0.1365
1.23	0.0977	−0.1778	0.2755	0.3731	1.77	−0.0338	−0.2006	0.1670	01332
1.24	0.0940	−0.1797	0.2737	0.3677	1.78	−0.0351	−0.2000	0.1650	0.1299
1.25	0.0904	−0.1815	0.2719	0.3623	1.79	−0.0364	−0.1993	0.1630	0.1266
1.26	0.0868	−0.1833	0.2701	0.3569	1.80	−0.0376	−0.1985	0.1610	0.1234
1.27	0.0833	−0.1849	0.2683	0.3515	1.81	−0.0388	−0.1978	0.1590	0.1202
1.28	0.0798	−0.1865	0.2664	0.3462	1.82	−0.0400	−0.1970	0.1570	0.1170
1.29	0.0763	−0.1881	0.2645	0.3408	1.83	−0.0412	−0.1962	0.1550	0.1188
1.30	0.0729	−0.1897	0.2626	0.3355	1.84	−0.0423	−0.1953	0.1531	0.1108
1.31	0.0696	−0.1911	0.2607	0.3303	1.85	−0.0434	−0.1945	0.1512	0.1078
1.32	0.0631	−0.1938	0.2569	0.3251	1.86	−0.0444	−0.1936	0.1492	0.1048
1.33	0.0631	−0.1938	0.2569	0.3199	1.87	−0.0454	−0.1927	0.1473	0.1018
1.34	0.0600	−0.1950	0.2550	0.3148	1.88	−0.0464	−0.1917	0.1453	0.0989
1.35	0.0568	−0.1962	0.2530	0.3098	1.89	−0.0474	−0.1908	0.1434	0.0960
1.36	0.0537	−0.1973	0.2510	0.3047	1.90	−0.0484	−0.1899	0.1415	0.0932
1.37	0.0507	−0.1983	0.2490	0.2997	1.91	−0.0493	−0.1889	0.1396	0.0904
1.38	0.0478	−0.1993	0.2470	0.2948	1.92	−0.0501	−0.1879	0.1377	0.0876
1.39	0.0448	−0.2003	0.2450	0.2898	1.93	−0.0510	−0.1869	0.1359	0.0849
1.40	0.0419	−0.2011	0.2430	0.2849	1.94	−0.0519	−0.1859	0.1340	0.0822
1.41	0.0391	−0.2019	0.2410	0.2801	1.95	−0.0527	−0.1849	0.1322	0.0795
1.42	0.0363	−0.2027	0.2390	0.2753	1.96	−0.0535	−0.1838	0.1304	0.0769
1.43	0.0336	−0.2033	0.2370	0.2705	1.97	−0.0543	−0.1827	0.1285	0.0748
1.44	0.0309	−0.2039	0.2349	0.2658	1.98	−0.0550	−0.1816	0.1267	0.0717
1.45	0.0283	−0.2045	0.2329	0.2611	1.99	−0.0556	−0.1804	0.1249	0.0692
1.46	0.0257	−0.2051	0.2308	0.2565	2.00	−0.0563	−0.1793	0.1230	0.0667
1.47	0.0232	−0.2056	0.2288	0.2519	2.01	−0.0569	−0.1782	0.1213	0.0643

Table 2.24 Con't

$\xi = \frac{x}{\lambda}$	$T(\xi)$	$U(\xi)$	$V(\xi)$	$W(\xi)$	$\xi = \frac{x}{\lambda}$	$T(\xi)$	$U(\xi)$	$V(\xi)$	$W(\xi)$
2.02	−0.0576	−0.1771	0.1195	0.0619	2.55	−0.0648	−0.1083	0.0435	−0.0213
2.03	−0.0582	−0.1759	0.1128	0.0595	2.56	−0.0646	−0.1071	0.0425	−0.0221
2.04	−0.0588	−0.1748	0.1160	0.0571	2.57	−0.0644	−0.1058	0.0414	−0.0228
2.05	−0.0594	−0.1737	0.1143	0.0549	2.58	−0.0642	−0.1045	0.0403	−0.0237
2.06	−0.0599	−0.1725	0.1126	0.0526	2.59	−0.0640	−0.1033	0.0394	−0.0246
2.07	−0.0604	−0.1712	0.1108	0.0504	2.60	−0.0637	−0.1020	0.0383	−0.0254
2.08	−0.0609	−0.1700	0.1091	0.0482	2.61	−0.0634	−0.1007	0.0373	−0.0261
2.09	−0.0614	−0.1688	0.1074	0.0460	2.62	−0.0632	−0.0994	0.0363	−0/0269
2.10	−0.0619	−0.1676	0.1057	0.0438	2.63	−0.0629	−0.0982	0.0353	−0.0276
2.11	−0.0620	−0.1663	0.1040	0.0417	2.64	−0.0626	−0.0969	0.0343	−0.0283
2.12	−0.0627	−0.1650	0.1024	0.0397	2.65	−0.0623	−0.0956	0.0334	−0.0289
2.13	−0.0632	−0.1637	0.1007	0.0377	2.66	−0.0620	−0.0944	0.0324	−0.0296
2.14	−0.0634	−0.1625	0.0981	0.0357	2.67	−0.0617	−0.0932	0.0315	−0.0302
2.15	−0.0638	−0.1613	0.0975	0.0337	2.68	−0.0614	−0.0920	0.0306	−0.0308
2.16	−0.0641	−0.1600	0.0959	0.0317	2.69	−0.0611	−0.0908	0.0297	−0.0314
2.17	−0.0645	−0.1587	0.0943	0.0288	2.70	−0.0608	−0.0895	0.0287	−0.0320
2.18	−0.0647	−0.1574	0.0927	0.0280	2.71	−0.0605	−0.0883	0.0279	−0.0326
2.19	−0.0650	−0.1560	0.0911	0.0262	2.72	−0.0601	−0.0871	0.0270	−0.0331
2.20	−0.0652	−0.1547	0.0895	0.0244	2.73	−0.0598	−0.0859	0.0261	−0.0337
2.21	−0.0655	−0.1534	0.0880	0.0226	2.74	−0.0594	−0.0847	0.0253	−0.0342
2.22	−0.0657	−0.1522	0.0865	0.0208	2.75	−0.0591	−0.0835	0.0244	−0.0347
2.23	−0.0660	−0.1509	0.0850	0.0191	2.76	−0.0598	−0.0823	0.0236	−0.0352
2.24	−0.0661	−0.1496	0.0835	0.0174	2.77	−0.0583	−0.0811	0.0228	−0.0356
2.25	−0.0663	−0.1482	0.0820	0.0157	2.78	−0.0581	−0.0799	0.0220	−0.0361
2.26	−0.0664	−0.1469	0.0805	0.0141	2.79	−0.0577	−0.0787	0.0212	−0.0365
2.27	−0.0665	−0.1455	0.0790	0.0125	2.80	−0.0573	−0.0777	0.0204	−0.0369
2.28	−0.0666	−0.1442	0.0776	0.0110	2.81	−0.0570	−0.0765	0.0196	−0.0373
2.29	−0.0667	−0.1429	0.0762	0.0095	2.82	−0.0566	−0.0754	0.0188	−0.0377
2.30	−0.0668	−0.1416	0.0748	0.080	2.83	−0.0562	−0.0742	0.0181	−0.0381
2.31	−0.0669	−0.1403	0.0734	0.0065	2.84	−0.0558	−0.0731	0.0173	−0.0385
2.32	−0.0670	−0.1389	0.0720	0.0050	2.85	−0.0554	−0.0721	0.0167	−0.0388
2.33	−0.0670	−0.1376	0.0706	0.0036	2.86	−0.0550	−0.0710	0.0160	−0.0391
2.34	−0.0671	−0.1362	0.0692	0.0022	2.87	−0.0546	−0.0699	0.0153	−0.0394
2.35	−0.0671	−0.1349	0.0679	0.0008	2.88	−0.0542	−0.0687	0.0145	−0.0397
0.75 π	−0.0671	−0.1342	0.0671	0.0000	2.89	−0.0538	−0.0676	0.0138	−0.0400
2.36	−0.0671	−0.1336	0.0666	−0.0005	2.90	−0.0534	−0.0666	0.0132	−0.0403
2.37	−0.0671	−0.1323	0.0653	−0.0018	2.91	−0.0530	−0.0656	0.0125	−0.0406
2.38	−0.0670	−0.1309	0.0639	−0.0031	2.92	−0.0526	−0.0645	0.0114	−0.0409
2.39	−0.0670	−0.1296	0.0626	−0.0044	2.93	−0.0522	−0.0634	0.0112	−0.0411
2.40	−0.0669	−0.1282	0.0613	−0.0056	2.94	−0.0518	−0.0624	0.0106	−0.0413
2.41	−0.0669	−0.1268	0.0600	−0.0068	2.95	−0.0514	−0.0614	0.0200	−0.0415
2.42	−0.0668	−0.1255	0.0588	−0.0080	2.96	−0.0510	−0.0603	0.0094	−0.0417
2.43	−0.0667	−0.1241	0.0575	−0.0092	2.97	−0.0506	−0.0593	0.0088	−0.0419
2.44	−0.0666	−0.1228	0.0563	−0.0103	2.98	−0.0502	−0.0583	0.0082	−0.0420
2.45	−0.0665	−0.1215	0.0550	−0.0114	2.99	−0.0497	−0.0573	0.0076	−0.0421
2.46	−0.0664	−0.1202	0.0538	−0.0125	3.00	−0.0493	−0.0563	0.0021	−0.0422
2.47	0.0662	−0.1189	0.0526	−0.0135	3.01	−0.0483	−0.0553	0.0065	−0.0423
2.48	−0.0661	−0.1175	0.0514	−0.0146	3.02	−0.0484	−0.0548	0.0059	−0.0424
2.49	−0.0659	−0.1161	0.0503	−0.0156	3.03	−0.0480	−0.0534	0.0054	−0.0425
2.50	−0.0658	−0.1149	0.0492	−0.0166	3.04	−0.0476	−0.0524	0.0049	−0.0426
2.51	−0.0656	−0.1136	0.0480	−0.0176	3.05	−0.0472	−0.0515	0.0043	−0.0427
2.52	−0.0654	−0.1123	0.0464	−0.0185	3.06	−0.0468	−0.0505	0.0039	−0.0428
2.53	−0.0652	−0.1109	0.0457	−0.0195	3.07	−0.0464	−0.0496	0.0034	−0.0429
2.54	−0.0650	−0.1096	0.0446	−0.0204	3.08	−0.0459	−0.0487	0.0029	−0.0430

Table 2.24 Con't

$\xi = \frac{x}{\lambda}$	$T(\xi)$	$U(\xi)$	$V(\xi)$	$W(\xi)$	$\xi = \frac{x}{\lambda}$	$T(\xi)$	$U(\xi)$	$V(\xi)$	$W(\xi)$
3.09	−0.0405	−0.0478	0.0023	−0.0431	3.63	−0.0234	−0.0109	−0.0124	−0.0359
3.10	−0.0450	−0.0469	0.0019	−0.0431	3.64	−0.0231	−0.0105	−0.0125	−0.0356
3.11	−0.0446	−0.0460	0.0015	−0.0431	3.65	−0.0227	−0.0101	−0.0126	−0.0354
3.12	−0.0441	−0.0451	0.0010	−0.0432	3.66	−0.0223	−0.0096	−0.0127	−0.0351
3.13	−0.0437	−0.0442	0.0006	−0.0432	3.67	−0.0220	−0.0092	−0.0128	−0.0348
3.14	−0.0432	−0.0433	0.0001	−0.0432	3.68	0.0217	−0.0088	−0.0129	−0.0346
π	−0.0432	−0.0432	0.0000	−0.0432	3.69	−0.0214	−0.0083	−0.0130	−0.0343
3.15	−0.0428	−0.0424	−0.0004	−0.0432	3.70	−0.0210	−0.0079	−0.0131	−0.0341
3.16	−0.0423	−0.0416	−0.0008	−0.0432	3.71	−0.0207	−0.0075	−0.0132	−0.0338
3.17	−0.0420	−0.0407	−0.0012	−0.0432	3.72	−0.0203	−0.0071	−0.0132	−0.0336
3.18	−0.0415	−0.0399	−0.0016	−0.0431	3.73	−0.0200	−0.0067	−0.0132	−0.0333
3.19	−0.0411	−0.0391	−0.0002	−0.0431	3.74	−0.0197	−0.0063	−0.0133	−0.0330
3.20	−0.0407	−0.0383	−0.0024	−0.0431	3.75	−0.0193	−0.0059	−0.0134	−0.0327
3.21	−0.0403	−0.0375	−0.0029	−0.0430	3.76	−0.0190	−0.0055	−0.0135	−0.0324
3.22	−0.0399	−0.0367	−0.0032	−0.0430	3.77	−0.0187	−0.0051	−0.0136	−0.0322
3.23	−0.0394	−0.0359	−0.0035	−0.0429	3.78	−0.0184	−0.0048	−0.0136	−0.0319
3.24	−0.0390	−0.0351	−0.0039	−0.0428	3.79	−0.0180	−0.0044	−0.0137	−0.0316
3.25	−0.0385	−0.0343	−0.0402	−0.0427	3.80	−0.0177	−0.0040	−0.0137	−0.0314
3.26	−0.0381	−0.0736	−0.0046	−0.0426	3.81	−0.0174	−0.0036	−0.0138	−0.0311
3.27	−0.0377	−0.0328	−0.0049	−0.0425	3.82	−0.0171	−0.0033	−0.0138	−0.0308
3.28	−0.0373	−0.0321	−0.0052	−0.0424	3.83	−0.0168	−0.0030	−0.0138	−0.0305
3.29	−0.0369	−0.0313	−0.0055	−0.0423	3.84	−0.0165	−0.0027	−0.0138	−0.0303
3.30	−0.0365	−0.0306	−0.0058	−0.0422	3.85	−0.0162	−0.0023	−0.0139	−0.0300
3.31	−0.0360	−0.0299	−0.0061	−0.0421	3.86	−0.0159	−0.0020	−0.0139	−0.0297
3.32	−0.0356	−0.0292	−0.0064	−0.0420	3.87	−0.0156	−0.0017	−0.0139	−0.0294
3.33	−0.0352	−0.0285	−0.0067	−0.0419	3.88	−0.0153	−0.0014	−0.0139	−0.0292
3.34	−0.0398	−0.0278	−0.0070	−0.0418	3.89	−0.0150	−0.0011	−0.0139	−0.0289
3.35	−0.0344	−0.0271	−0.0073	−0.0417	3.90	−0.0147	−0.0008	−0.0140	−0.0286
2.36	−0.0340	−0.0264	−0.0075	−0.0415	3.91	−0.0144	−0.0005	−0.0140	−0.0283
3.37	−0.0335	−0.0257	−0.0078	−0.0413	3.92	−0.0141	−0.0002	−0.0140	−0.0280
3.38	−0.0331	−0.0251	−0.0080	−0.0411	1.25 π	−0.0139	0.0000	−0.0140	−0.0278
3.39	−0.0327	−0.0244	−0.0083	−0.0409	3.93	−0.0139	0.0001	−0.0140	−0.0278
3.40	−0.0323	−0.0238	−0.0085	−0.0408	3.94	−0.0136	0.0003	−0.0139	−0.0275
3.41	−0.0319	−0.0231	−0.0088	−0.0406	3.95	−0.0133	0.0005	−0.0139	−0.0272
3.42	−0.0315	−0.0225	−0.0090	−0.0404	3.96	−0.0130	0.0008	−0.0139	−0.0269
3.43	−0.0311	−0.0218	−0.0093	−0.0403	3.97	−0.0128	0.0011	−0.0139	−0.0267
3.44	−0.0307	−0.0212	−0.0094	−0.0401	3.98	−0.0125	0.0014	−0.0139	−0.0264
3.45	−0.0303	−0.0206	−0.0097	−0.0399	3.99	−0.0122	0.0017	−0.0139	−0.0261
3.46	−0.0299	−0.0200	−0.0099	−0.0397	4.00	−0.0120	0.0019	−0.0139	−0.0258
3.47	−0.0295	−0.0194	−0.0101	−0.0395	4.10	−0.0096	0.0040	−0.0136	−0.0231
3.48	−0.0291	−0.0189	−0.0102	−0.0392	4.20	−0.0074	0.0057	−0.0131	−0.0204
3.49	−0.0287	−0.0183	−0.0104	−0.0390	4.30	−0.0055	0.0070	−0.0125	−0.0179
3.50	−0.0283	−0.0177	−0.0106	−0.0388	4.40	−0.0038	0.0079	−0.0117	−0.0155
3.51	−0.0279	−0.0171	−0.0108	−0.0386	4.50	−0.0023	0.0085	−0.0108	−0.0132
3.52	−0.0275	−0.0165	−0.0109	−0.0384	4.60	−0.0012	0.0089	−0.0100	−0.0111
3.53	−0.0271	−0.0160	−0.0111	−0.0382	4.70	−0.0001	0.0090	−0.0091	−0.0092
3.54	−0.0268	−0.0155	−0.0113	−0.0380	1.5 π	0.0000	0.0090	0.0090	−0.0090
3.55	−0.0264	−0.0149	−0.0114	−0.0378	4.80	0.0007	0.0089	−0.0082	−0.0075
3.56	−0.0260	−0.0144	−0.0116	−0.0376	4.90	0.0014	0.0087	−0.0073	−0.0059
3.57	−0.0257	−0.0139	−0.0117	−0.0373	5.00	0.0019	0.0084	−0.0065	−0.0046
3.58	−0.0253	−0.0134	−0.0118	−0.0371	5.10	0.0023	0.0080	−0.0057	−0.0033
3.59	−0.0249	−0.0129	−0.0120	−0.0368	5.20	0.0026	0.0075	−0.0049	−0.0023
3.60	−0.0245	−0.0124	−0.0121	−0.0366	5.30	0.0028	0.0069	−0.0042	−0.0014
3.61	−0.0242	−0.0119	−0.0122	−0.0363	5.40	0.0029	0.0064	−0.0035	−0.0006
3.62	−0.0238	−0.0114	−0.0123	−0.0361	1.75 π	0.0029	0.0058	−0.0029	0.0000

Table 2.24 Con't

$\xi = \frac{x}{\lambda}$	$T(\xi)$	$U(\xi)$	$V(\xi)$	$W(\xi)$	$\xi = \frac{x}{\lambda}$	$T(\xi)$	$U(\xi)$	$V(\xi)$	$W(\xi)$
5.50	0.0029	0.0058	−0.0029	0.0000	6.80	0.0011	0.0004	0.0006	0.0015
5.60	0.0029	0.0052	−0.0023	0.0005	6.90	0.0000	0.0002	0.0006	0.0014
5.70	0.0028	0.0046	−0.0018	0.0009	7.00	0.0007	0.0001	0.0006	0.0013
5.80	0.0027	0.0041	−0.0014	0.0013	2.25 π	0.0006	0.0000	0.0006	0.0012
5.90	0.0026	0.0036	−0.0010	0.0015	7.10	0.0006	0.0000	0.0006	0.0012
6.00	0.0024	0.0031	−0.0007	0.0017	7.20	0.0005	−0.0001	0.0006	0.0011
6.10	0.0022	0.0026	−0.0004	0.0018	7.30	0.0004	−0.0002	0.0006	0.0009
6.20	0.0020	0.0022	−0.0002	0.0019	7.40	0.0003	−0.0003	0.0006	0.0008
2 π	0.0019	0.0019	0.0000	0.0019	7.50	0.0002	−0.0003	0.0006	0.0007
6.30	0.0019	0.0018	0.0001	0.0019	7.60	0.0002	−0.0003	0.0005	0.0007
6.40	0.0017	0.0015	0.0002	0.0018	7.70	0.0002	−0.0004	0.0005	0.0005
6.50	0.0018	0.0012	0.0003	0.0018	7.80	0.0001	−0.0004	0.0005	0.0005
6.60	0.0015	0.0009	0.0004	0.0017	7.90	0.001	−0.0004	0.0004	0.0004
6.70	0.0013	0.0006	0.0005	0.0016	8.00	0.0001	−0.0004	0.0004	0.0003

Table 2.25 Functions for analysis of finite and sem-infinite beams on Winkler foundation

$\xi = \frac{x}{\lambda}$	$A(\xi)$	$B(\xi)$	$C(\xi)$	$D(\xi)$	$\xi = \frac{x}{\lambda}$	$A(\xi)$	$B(\xi)$	$C(\xi)$	$D(\xi)$
1	0	0	0	0	0.35	0.99750	0.34980	0.06125	0.00715
0.001	1.0000	0.00100	0.00000	0.000000	0.36	0.99720	0.35980	0.06480	0.00675
0.002	1.0000	0.00200	0.00000	0.000000	037	0.99600	0.36980	0.06845	0.00845
0.003	1.0000	0.00300	0.000005	0.000000	0.38	0.99650	0.37980	0.07220	0.00915
0.004	1.0000	0.00400	0.000010	0.000000	0.39	0.99610	0.38970	0.07605	0.00990
0.005	1.0000	0.00500	0.000015	0.000000	0.40	0.99570	0.39965	0.08000	0.01070
0.006	1.0000	0.00600	0.000020	0.000000	0.41	0.99530	0.40960	0.0840	0.01150
0.007	1.0000	0.00700	0.000025	0.000000	0.42	0.99480	0.41960	0.08815	0.01235
0.008	1.0000	0.00800	0.00030	0.000000	0.43	0.99430	0.42950	0.09240	0.01325
0.009	1.0000	0.00900	0.000040	0.000000	0.44	0.99380	0.43945	0.09675	0.01420
0.010	1.0000	0.01000	0.000050	0.000000	0.45	0.99320	0.44940	0.10120	0.01520
0.011	1.0000	0.00110	0.000060	0.000000	0.46	0.99250	0.45935	0.10575	0.01620
0.012	1.0000	0.01200	0.000070	0.000000	0.47	0.99190	0.46920	0.11040	0.01730
0.013	1.0000	0.01300	0.000085	0.000000	0.48	0.99110	0.47910	0.11515	0.01840
0.014	1.0000	0.01400	0.00010	0.000000	0.49	0.99040	0.48905	0.11995	0.01960
0.015	1.0000	0.01500	0.000115	0.000000	0.50	0.98950	0.49895	0.12490	0.02080
0.016	1.0000	0.01600	0.000130	0.000000	0.51	0.98870	0.50885	0.12995	0.02210
0.017	1.0000	0.01700	0.000145	0.000000	0.52	0.98780	0.51875	0.13510	0.02340
0.018	1.0000	0.01800	0.000160	0.000000	0.53	0.98690	0.52860	0.14035	0.02480
0.019	1.0000	0.01900	0.00018	0.000000	0.54	0.98580	0.53845	0.14565	0.02620
0.020	1.0000	0.02000	0.00020	0.000000	0.55	0.98470	0.54835	0.15110	0.02770
0.030	1.0000	0.03000	0.00045	0.000005	0.56	0.98360	0.55820	0.15665	0.02925
0.040	1.0000	0.04000	0.00080	0.000010	0.57	0.98240	0.56800	0.16225	0.03025
0.050	1.0000	0.05000	0.00125	0.000000	0.58	0.98110	0.57780	0.16800	0.03085
0.060	1.0000	0.06000	0.00180	0.000005	0.59	0.97980	0.58760	0.17380	0.03250
0.070	1.0000	0.07000	0.00245	0.000005	0.60	0.97840	0.59745	0.17975	0.03600
0.080	1.0000	0.08000	0.00320	0.000010	0.61	0.97690	0.60720	0.18575	0.03780
0.090	1.0000	0.08995	0.00405	0.000010	0.62	0.97540	0.61695	0.19190	0.03970
0.100	1.0000	0.10000	0.00500	0.000150	0.63	0.97380	0.62670	0.19810	0.04165
0.110	1.0000	0.11000	0.00605	0.000200	0.64	0.97210	0.63640	0.20440	0.04365
0.120	1.0000	0.12000	0.00720	0.000300	0.65	0.97030	0.64615	0.21085	0.04570
0.130	0.9999	0.13000	0.00845	0.000350	0.66	0.96840	0.65585	0.21735	0.04790
0.140	0.9999	0.14000	0.00980	0.000450	0.67	0.96640	0.66550	0.22395	0.05010
0.150	0.9999	0.15000	0.01125	0.000550	0.68	0.96440	0.67515	0.23065	0.05240
0.160	0.9999	0.16000	0.01280	0.000700	0.69	0.96230	0.68480	0.23745	0.05470
0.170	0.9999	0.17000	0.01445	0.000800	0.70	0.96000	0.69440	0.24435	0.05710
0.180	0.9998	0.17995	0.01620	0.001000	0.71	0.95770	0.70395	0.25135	0.05960
0.190	0.9998	0.19000	0.01805	0.001150	0.72	0.95520	0.71355	0.25840	0.06210
0.200	0.9997	0.20000	0.02000	0.001350	0.73	0.95270	0.72310	0.26560	0.06475
0.210	0.9997	0.21000	0.02205	0.001550	0.74	0.95010	0.73260	0.27290	0.06745
0.220	0.9996	0.21995	0.02420	0.001800	0.75	0.94730	0.74210	0.28025	0.07020
0.230	0.9995	0.22995	0.02645	0.002000	0.76	0..94440	0.74055	0.28775	0.07300
0.240	0.9995	0.24000	0.02880	0.002300	0.77	0.94150	0.76100	0.29530	0.07595
0.250	0.9993	0.25000	0.03125	0.002600	0.78	0.93840	0.77040	0.30295	0.07895
0.260	0.9992	0.25995	0.03380	0.002900	0.79	0.93510	0.77975	0.31070	0.08200
0.270	0.9991	0.26995	0.036450	0.003200	0.80	0.93180	0.78910	0.31855	0.08515
0.280	0.9990	0.27995	0.02920	0.003000	0.81	0.92830	0.79840	0.32650	0.08840
0.290	0.9988	0.28995	0.04205	0.004100	0.82	0.92870	0.80770	0.33450	0.09170
0.300	0.9987	0.29990	0.04500	0.004500	0.83	0.92100	0.81685	0.34265	0.09510
0.310	0.9985	0.30990	0.04805	0.004950	0.84	0.91710	0.82610	0.35085	0.09855
0.320	0.9983	0.3199	0.05120	0.005450	0.85	0.91310	0.83520	0.35915	0.10210
0.330	0.9980	0.32985	0.05445	0.006000	0.86	0.90900	0.84435	0.36755	0.0570
0.340	0.9978	0.33985	0.05780	0.006600	0.87	0.90470	0.85340	0.37605	0.10945

Table 2.25 Con't

$\xi = \dfrac{x}{\lambda}$	$A(\xi)$	$B(\xi)$	$C(\xi)$	$D(\xi)$	$\xi = \dfrac{x}{\lambda}$	$A(\xi)$	$B(\xi)$	$C(\xi)$	$D(\xi)$
0.88	0.9002	0.86240	0.38460	0.11325	1.41	0.3474	1.22520	0.95055	0.45845
0.89	0.8956	0.87140	0.39330	0.11715	1.42	0.3289	1.22860	0.96480	0.47800
0.90	0.8931	0.88035	0.40205	0.12110	1.43	0.3100	1.23175	0.97510	0.47770
0.91	0.8859	0.88925	0.41090	0.12520	1.44	0.2907	1.23335	0.97645	0.48820
0.92	0.8808	0.89805	0.41985	0.12930	1.45	0.2710	1.23755	0.99980	0.49740
0.93	0.8753	0.90685	0.42885	0.13310	1.46	0.2509	1.24020	1.01220	0.50750
0.94	0.8701	0.91555	0.43795	0.13790	1.47	0.2304	1.24260	1.02460	0.51765
0.95	0.8645	0.92420	0.44715	0.14235	1.48	0.2095	1.24480	1.03705	0.52800
0.96	0.8587	0.93285	0.45645	0.14690	1.49	0.1882	1.24680	1.04950	0.53840
0.97	0.8528	0.94145	0.46585	0.15150	1.50	0.1664	1.24855	1.06195	0.54900
0.98	0.8466	0.94990	0.47530	0.15620	1.51	0.1442	1.25010	1.07445	0.55965
0.99	0.8339	0.95355	0.48485	0.15860	1.52	0.1216	1.25145	1.0870	0.5705
1.00	0.8337	0.96675	0.49445	0.16590	1.53	0.0986	1.25260	1.04950	0.58140
1.01	0.8270	0.97500	0.50415	0.17085	1.54	0.0746	1.25340	1.11205	0.59245
1.02	0.8201	0.98325	0.51395	0.17600	1.55	0.0512	1.25405	1.12405	0.60360
1.03	0.8129	0.99140	0.52380	0.18115	1.56	0.0268	1.25445	1.13710	0.61490
1.04	0.8056	0.99950	0.53375	0.18645	1.57	0.0020	1.25460	1.14965	0.62640
1.05	0.7980	1.00755	0.54380	0.19180	$0.5\,\pi$	0.0000	1.25450	1.15065	0.62730
1.06	0.7902	1.01545	0.55395	0.19730	1.58	−0.0233	1.25450	1.16220	0.63795
1.07	0.7822	1.02330	0.56410	0.20290	1.59	−0.0490	1.25415	1.17475	0.64960
1.08	0.7740	1.03110	0.57440	0.20860	1.60	−0.0753	1.25350	1.18725	0.66145
1.09	0.7655	1.03880	0.58475	0.21340	1.61	−0.1019	1.25260	1.19980	0.67335
1.10	0.7568	1.04645	0.59515	0.22030	1.62	−0.1291	1.25145	1.21235	0.68540
1.11	0.7479	1.05395	0.60565	0.22630	1.63	−0.1568	1.25005	1.22485	0.69760
1.12	0.7387	1.06130	0.61625	0.23235	1.64	−0.1849	1.24835	1.23735	0.70990
1.13	0.7293	1.06870	0.62690	0.23860	1.65	−0.2136	1.24635	1.24980	0.72240
1.14	0.7196	1.07595	0.63760	0.24490	1.66	−0.2427	1.24405	1.26225	0.73490
1.15	0.7097	1.08310	0.64840	0.25135	1.67	−0.2724	1.2415	1.27470	0.74760
1.16	0.6995	1.09015	0.65930	0.25790	1.68	−0.3026	1.23850	1.28710	0.76040
1.17	0.6891	1.09710	0.67020	0.26455	1.69	−0.3332	1.23540	1.29945	0.77060
1.18	0.6784	1.10395	0.68125	0.27130	1.70	−0.3644	1.23190	1.31180	0.78640
1.19	0.6674	1.11065	0.69230	0.27820	1.71	−0.3961	1.22815	1.32410	0.79960
1.20	0.6561	1.11730	0.70345	0.28515	1.72	−0.4284	1.22400	1.33635	0.81290
1.21	0.6446	1.12380	0.71465	0.29225	1.73	−0.4612	1.2195	1.34855	0.82630
1.22	0.6330	1.13060	0.72590	0.29965	1.74	−0.4945	1.2148	1.36075	0.83985
1.23	0.6206	1.13645	0.73725	0.30680	1.75	−0.5284	1.20965	1.37285	0.85350
1.24	0.6082	1.14260	0.74865	0.31420	1.76	−0.5628	1.20420	1.38495	0.86730
1.25	0.5955	1.14800	0.76010	0.32175	1.77	−0.5977	1.19840	1.39695	0.88120
1.26	0.5824	1.15450	0.77160	0.32940	1.78	−0.6833	1.19230	1.40890	0.89525
1.27	0.5691	1.16020	0.78320	0.33720	1.79	−0.6694	1.18570	1.42080	0.90940
1.28	0.5555	1.16590	0.79480	0.34505	1.80	−0.7060	1.17885	1.43260	0.92370
1.29	0.54415	1.17135	0.80650	0.35310	1.81	−0.7433	1.17160	1.44435	0.93805
1.30	0.5272	1.1767	0.81825	0.36120	1.82	−0.7811	1.16400	1.45605	0.95255
1.31	0.5126	1.18190	0.83005	0.36945	1.83	−0.8195	1.15600	1.46765	0.96715
1.32	0.4977	1.18700	0.84190	0.37780	1.84	−0.8584	1.14760	1.47915	0.98190
1.33	0.4824	1.19185	0.85380	0.38630	1.85	−0.8980	1.13885	1.49060	0.99680
1.34	0.4668	1.19660	0.86570	0.39485	1.86	−0.9382	1.12965	1.50195	1.01170
1.35	0.4508	1.20120	0.87770	0.40360	1.87	−0.9790	1.12005	1.51320	1.02680
1.36	0.4345	1.20565	0.88975	0.41240	1.88	−1.0203	1.11005	1.52435	1.0420
1.37	0.4178	1.20985	0.90180	0.42140	1.89	−1.0623	1.09965	1.53540	1.05730
1.38	0.4008	1.21395	0.91395	0.43050	1.90	−1.1049	1.088880	1.54635	1.07270
1.39	0.3833	1.21790	0.92610	0.43965	1.91	−1.1481	1.07755	1.55715	1.08820
1.40	0.3656	1.22165	0.93830	0.44900	1.92	−1.1920	1.06585	1.56790	1.1038

Table 2.25 Con't

$\xi = \frac{x}{\lambda}$	$A(\xi)$	$B(\xi)$	$C(\xi)$	$D(\xi)$	$\xi = \frac{x}{\lambda}$	$A(\xi)$	$B(\xi)$	$C(\xi)$	$D(\xi)$
1.93	−1.23640	1.05375	1.57850	1.11960	2.46	−4.57800	−0.39975	1.83015	2.05640
1.94	−1.28150	1.04110	1.58895	1.13540	2.47	−4.66060	−0.44495	1.82590	2.07470
1.95	−1.32730	1.02810	1.59930	1.15135	2.48	−4.74390	−0.49200	1.82125	2.09293
1.96	−1.37360	1.01455	1.60950	1.16740	2.49	−4.82800	−0.53985	1.81605	2.11110
1.97	−1.42070	1.00065	1.61960	1.18350	2.50	−4.91280	−0.58850	1.81045	2.12925
1.98	−1.46830	0.98615	1.62955	1.19980	2.51	−4.9984	−0.63810	1.80430	2.14735
1.99	−1.51660	0.97125	1.63930	1.21610	2.52	−5.0846	−0.68850	1.79765	2.16535
2.00	−1.56560	0.95575	1.64895	1.23250	2.53	−5.1716	−0.73980	1.79055	2.18330
2.01	−1.61530	0.93990	1.65840	1.24210	2.54	−5.2593	−0.79195	1.78290	2.20120
2.02	−1.66560	0.92350	1.66775	1.26575	2.55	−5.3477	−0.84500	1.77470	2.21895
2.03	−1.71650	0.90660	1.67690	1.28245	2.56	−5.4368	−0.89890	1.76600	2.23665
2.04	−1.76820	0.88915	1.68590	1.29930	2.57	−5.5266	−0.95375	1.75670	2.25430
2.05	−1.82050	0.87125	1.69465	1.31620	2.58	−5.6172	−1.00945	1.74690	2.27180
2.06	−1.87340	0.85280	1.70330	1.33150	2.59	−5.7084	−1.06610	1.73650	2.28920
2.07	−1.92710	0.83375	1.71170	1.35020	2.60	−5.8003	−1.12360	1.72555	2.30650
2.08	−1.98150	0.81420	1.71905	1.36740	2.61	−5.8929	−1.18205	1.71405	2.33770
2.09	−2.03650	0.79390	1.72800	1.38450	2.62	−5.9862	−1.24150	1,70190	2.34080
2.10	−2.09230	0.77350	1.73585	1.40195	2.63	−6.0802	−1.30180	1.68920	2.35775
2.11	−2.14870	0.75230	1.74350	1.41935	2.64	−6.1748	−1.36305	1.67590	2.3746
2.12	−2.20580	0.73055	1.75090	1.43680	2.65	−6.2701	−1.42530	1.66195	2.39125
2.13	−2.226360	0.70815	1.75810	1.45440	2.66	−6.3601	−1.48845	1.64735	2.40780
2.14	−2.32210	0.68525	1.76505	1.47200	2.67	−6.4628	−1.55265	1.63220	2.42420
2.15	−2.38140	0.66175	1.77180	1.48970	2.68	−6.5600	−1.61770	1.61630	2.44045
2.16	−2.44130	0.63760	1.77830	1.50740	2.69	−6.6580	−1.68380	1.60270	2.45656
2.17	−2.50200	0.61290	1.78455	1.52525	2.70	−6.7565	−1.7509	1.58265	2.47245
2.18	−2.56330	0.58760	1.79055	1.54310	2.71	−6.8558	−1.81895	1.56480	2.48820
2.19	−2.62540	0.56160	1.79630	1.56105	2.72	−6.9556	−1.88805	1.54625	2.50370
2.20	−2.68820	0.53510	1.80180	1.57905	2.73	−7.0560	−1.95805	1.52705	2.51910
2.21	−2.75180	0.50785	1.80790	1.59710	2.74	−7.1571	−2.02915	1.50710	2.53430
2.22	−2.81600	0.48005	1.81195	1.61520	2.75	−7.2588	−2.10100	1.48645	2.54925
2.23	−2.88100	0.45155	1.18166	1.63330	2.76	−7.3611	−2.17430	1.46505	2.56400
2.24	−2.94660	0.42240	1.82095	1.65150	2.77	−7.4639	−2.24840	1.44295	2.59285
2.25	−3.01310	0.3926	1.82505	1.66975	2.78	−7.5673	−2.32360	1.42010	2.60695
2.26	−3.08020	0.3621	1.82880	1.68800	2.79	−7.6714	−2.39980	1.39645	2.60695
2.27	−3.14810	0.3310	1.83225	1.70630	2.80	−7.7759	−2.47700	1.37210	2.62080
2.28	−3.21670	0.2992	1.83545	1.72465	2.81	−7.8810	−2.55530	1.34695	2.63440
2.29	−3.28610	0.26665	1.83825	1.74300	2.82	−7.9866	−2.63465	1.3210	2.64770
2.30	−3.35620	0.23345	1.84075	1.76140	2.83	−8.0920	−2.71505	1.29425	2.66080
2.31	−3.42700	0.19955	1.84295	1.77980	2.84	−8.1995	−2.79650	1.2665	2.67360
2.32	−3.49860	0.16485	1.84475	1.79830	2.85	−8.3067	−2.8790	1.2383	2.68615
2.33	−3.57080	0.12955	1.84620	1.81670	2.86	−8.4144	−2.96265	1.20910	2.69840
2.34	−3.64390	0.09350	1.84780	1.83520	2.87	−8.5225	−3.04730	1.17905	2.71030
2.35	−3.71770	0.05655	1.84810	1.85365	2.88	−8.6312	−3.13310	1.14815	2.72195
2.36	−3.77220	0.01910	1.84845	1.87215	2.89	−8.7404	−3.21995	1.11640	2.73330
2.37	−3.86750	−0.01915	1.84850	1.89060	2.90	−8.8471	−3.30790	1.08375	2.74430
2.38	−3.94350	−0.05825	1.84805	1.90910	2.91	−8.9598	−3.39690	1.05025	2.75495
2.39	−4.02020	−0.09805	1.84730	1.92760	2.92	−9.0703	−3.48715	1.01580	2.76530
2.40	−4.09760	−0.13850	1.84610	1.94605	2.93	−9.1811	−3.57835	0.98045	2.77530
2.41	−4.17590	−0.17995	1.84455	1.96450	2.94	−9.2923	−3.67070	0.94425	2.78490
2.42	−4.25480	−0.22210	1.84250	1.98245	2.95	−9.4039	−3.76420	0.90705	2.79415
2.43	−4.33450	−0.26510	1.84010	2.00135	2.96	−9.5158	−3.85880	0.86895	2.80305
2.44	−4.41500	−0.30885	1.83725	2.01980	2.97	−9.6281	−3.95450	0.82985	2.81150
2.45	−4.49610	−0.35340	1.83390	2.03810	2.98	−9.7407	−4.05135	0.78985	2.81960

Table 2.25 Con't

$\xi = \dfrac{x}{\lambda}$	$A(\xi)$	$B(\xi)$	$C(\xi)$	$D(\xi)$	$\xi = \dfrac{x}{\lambda}$	$A(\xi)$	$B(\xi)$	$C(\xi)$	$D(\xi)$
2.99	−9.8536	−4.14930	0.74885	2.82730	3.51	−15.6159	−10.80810	−3.00875	2.38995
3.00	−9.9669	−4.24845	0.70685	2.83460	3.52	−15.7108	−10.9647	−3.11760	2.35930
3.01	−10.0804	−4.34865	0.66385	2.84140	3.53	−15.8046	−11.12230	−3.22800	2.32760
3.02	−10.1943	−4.45005	0.61985	2.84790	3.54	−15.8971	−11.28085	−3.34005	2.29480
3.03	−10.3083	−4.55255	0.52485	2.85380	3.55	−15.9881	−11.44030	−3.45370	2.26075
3.04	−10.4225	−4.65620	0.52885	2.85935	3.56	−16.0780	−11.60070	−3.56890	2.22565
3.05	−10.5317	−4.76105	0.48170	2.86442	3.57	−16.1663	−11.76185	−3.68565	2.18940
3.06	−10.6516	−4.86695	0.43360	2.86900	3.58	−16.2531	−11.92400	−3.80410	2.15195
3.07	−10.7665	−4.97405	0.38440	2.87310	3.59	−16.3384	−12.08695	−3.92415	2.11330
3.08	−10.8815	−5.08230	0.33410	2.87665	3.60	−16.4218	−12.25075	−4.04585	2.0735
3.09	−10.9966	−5.19165	0.28275	2.87980	3.61	−16.5043	−12.41540	−4.16920	2.0324
3.10	−11.1119	−5.30225	0.23030	2.88230	3.62	−16.5847	−12.58080	−4.29420	1.99005
3.11	−11.2272	−5.41300	0.17670	2.88430	3.63	−16.6634	−12.7470	−4.42080	1.94650
3.12	−11.3427	−5.52680	0.12200	2.88585	3.64	−16.7405	−12.91415	−4.54910	1.90170
3.13	−11.4580	−5.64075	0.06615	2.88680	3.65	−16.8155	−13.08135	−4.67910	1.85550
3.14	−11.5736	−5.75595	0.00915	2.88720	3.66	−16.8889	−13.25040	−4.81075	1.80805
π	−11.5919	−5.77435	0.00000	2.88720	3.67	−16.9602	−13.4196	−4.94410	1.75930
3.15	−11.6890	−5.87320	−0.04895	2.88695	3.68	−16.0296	−13.5896	−5.07915	1.70920
3.16	−11.8045	−5.98975	−0.10830	2.88620	3.69	−17.0970	−13.6745	−5.21590	1.70060
3.17	−11.9200	−6.10835	−0.16875	2.88480	3.70	−17.1622	−13.93150	−5.35435	1.60485
3.18	−12.0353	−6.22810	−0.23045	2.88280	3.71	−17.2253	−14.10345	−5.49450	1.55030
3.19	−12.1506	−6.34905	−0.29335	2.88020	3.72	−17.2861	−14.27590	−5.63640	1.49495
3.20	−12.2656	−6.47105	−0.35740	2.87690	3.73	−17.3449	−14.44920	−5.78005	1.43790
3.21	−12.8807	−6.59430	−0.42270	2.87305	3.74	−174022	−14.62285	−5.92540	1.37930
3.22	−12.4956	−6.71875	−0.48935	2.86850	3.75	−17.4552	−14.79715	−6.07250	1.31940
3.23	−12.6101	−6.84420	−0.55710	2.86325	3.76	−17.5067	−14.97195	−6.22135	1.25790
3.24	−12.7373	−6.97095	−0.62670	2.85730	3.77	−17.5557	−15.14725	−6.37195	1.19490
3.25	−12.8388	−7.09880	−0.69660	2.85070	3.78	−17.6024	−15.32315	−6.52430	1.13050
3.26	−129527	−7.22770	−0.76815	2.84340	3.79	−17.6463	−15.49940	−6.67840	1.06445
3.27	−13.0662	−7.35780	−0.84110	2.83540	3.80	−17.6875	−15.67605	−6.83430	0.99690
3.28	−13.1795	−7.48905	−0.91535	2.82660	3.81	−17.7259	−15.85310	−6.99195	0.92775
3.29	−13.2924	−7.62135	−0.99085	2.81710	3.82	−17.7616	−16.03010	−7.15130	0.85705
3.30	−13.4048	−7.75490	−1.06780	2.80675	3.83	−17.7945	−16.20830	−7.31255	0.78470
3.31	−13.5168	−7.88945	−1.14595	2.79570	3.84	−17.8246	−16.38640	−7.47550	0.71080
3.32	−13.6285	−8.02520	−1.22555	2.78385	3.85	−17.8513	−16.56485	−7.64030	0.68520
3.33	−13.7395	−8.16195	−1.30645	2.77120	3.86	−17.8751	−16.74335	−7.80685	0.55790
3.34	−13.8501	−8.30000	−1.38880	2.75770	3.87	−17.8960	−16.82230	−7.97510	0.47910
3.35	−13.9601	−8.43900	−1.47250	2.74340	3.88	−17.9135	−16.10130	−8.14525	0.39845
3.36	−14.0695	−8.57920	−1.55765	2.72820	3,89	−17.9277	−17.28045	−8.31710	0.31610
3.37	−14.1784	−8.72045	−1.64410	2.12200	3.90	−17.9387	−17.45985	−8.49090	0.23210
3.38	−14.2866	−8.86280	−1.73205	2.69535	3.91	−17.9464	−17.63930	−8.66635	0.14635
3.39	−14.3941	−9.00620	−1.82135	2.67760	3.92	−17.9504	−17.81875	−8.84370	0.05870
3.40	−14.5008	−9.15065	−1.91210	2.65890	3.93	−17.9511	−17.99830	−9.02270	−0.03050
3.41	−14.6066	−9.29620	−2.00440	2.63930	3.94	−17.9480	−18.17785	−9.20365	−0.12170
3.42	−14.7118	−9.44270	−2.0980	2.61885	3.95	−17.9412	−18.35720	−9.38630	−0.21720
3.43	−14.8162	−9.59045	−2.19325	2.59740	3.96	−17.9307	−18.53655	−9.57075	−0.30950
3.44	−14.9197	−9.73915	−2.28990	2.57500	3.97	−17.9165	−18.71585	−9.75205	0.40610
3.45	−15.0222	−9.88880	−2.38800	2.55160	3.98	−17.8983	−18.89490	−9.94505	−0.50455
3.46	−15.1238	−10.03955	−2.48760	2.52740	3.99	−17.8761	−19.07380	−10.13495	−0.60495
3.47	−15.2244	−10.19130	−2.58885	2.50180	4.00	−17.8498	−19.25235	−10.32650	−0.70730
3.48	−15.3238	−10.34405	−2.69150	2.47540	4.01	−17.8172	−19.43070	−10.51995	−0.8115
3.49	−15.4224	−10.49775	−2.79570	2.44800	4.02	−17.7850	−19.60875	−10.71510	−0.91760
3.50	−15.5198	−10.65245	−2.90140	2.41950	4.03	−17.7461	−19.78650	−10.91215	−1.02580

Table 2.25 Con't

$\xi = \dfrac{x}{\lambda}$	$A(\xi)$	$B(\xi)$	$C(\xi)$	$D(\xi)$	$\xi = \dfrac{x}{\lambda}$	$A(\xi)$	$B(\xi)$	$C(\xi)$	$D(\xi)$
4.04	−17.7029	−19.96375	−11.11095	−1.13590	4.56	7.2556	−27.24850	−23.61635	−9.99725
4.05	−17.6551	−20.14055	−11.31145	−1.24810	4.57	−6.8510	−27.31915	−23.88920	−10.23480
4.06	−17.6030	−20.31690	−11.51375	−1.36215	4.58	−6.4366	−27.38545	−24.16275	−10.47500
4.07	−17.5461	−20.49255	−11.7178	−1.47830	4.59	−6.0127	−27.44770	−24.43690	−10.71810
4.08	−17.4846	−20.66770	−11.92355	−1.59655	4.60	−5.5791	−27.50565	−24.71165	−10.96800
4.09	−17.4185	−20.84225	−12.13110	−1.71680	4.61	−5.1358	−27.55925	−24.98695	−11.21230
4.10	−17.3472	−21.0160	−12.34040	−1.83920	4.62	−4.8237	−27.60860	−25.26295	−11.46355
4.11	−17.2712	−21.18905	−12.55135	−1.96360	4.63	−4.2189	−27.65310	−25.53920	−11.71750
4.12	−17.1900	−21.36140	−12.76415	−2.0902	4.64	−3.7450	−27.69280	−25.81585	−11.97425
4.13	−17.1040	−21.53290	−12.97785	−2.21890	4.65	−3.2607	−27.72770	−26.09285	−12.23380
4.14	−16.026	−21.70345	−13.1948	−2.34980	4.66	−2.7663	−27.75810	−26.37500	−12.49615
4.15	−16.9160	−21.87310	−13.41265	−2.48280	4.67	−2.2611	−27.78310	−26.64810	−12.76120
4.16	−16.8139	−22.0417	−13.63220	−2.61800	4.68	−1.7449	−27.80315	−26.92620	−13.02930
4.17	−16.7064	−22.20940	−13.85355	−2.75550	4.69	−1.2187	−27.81805	−27.20420	−13.29980
4.18	−16.5934	−22.37590	−14.07650	−2.89515	4.70	−0.6812	−27.82740	−27.48230	−13.57315
4.19	−16.4748	−22.54125	−14.30105	−3.28700	4.71	−0.1327	−27.83165	−27.76080	−13.84950
4.20	−16.3505	−22.70545	−14.52735	−3.18120	$1.5\,\pi$	0.0000	−27.83170	−27.82720	−13.01585
4.21	−16.2203	−22.86815	−14.75505	−3.32750	4.72	0.4268	−27.83005	−28.03900	−14.1284
4.22	−16.0842	−23.02985	−14.98470	−3.47630	4.73	0.9976	−27.82280	−28.31715	−14.41015
4.23	−15.9423	−23.18995	−15.21575	−3.62720	4.74	1.5799	−27.81005	−28.59550	−14.6948
4.24	−15.7939	−23.34850	−15.44835	−3.78055	4.75	2.1731	−27.79130	−28.87340	−14.98205
4.25	−15.6398	−23.50585	−15.68270	−3.93620	4.76	2.7782	−27.76675	−29.15135	−15.27230
4.26	−15.4793	−23.66155	−15.91865	−4.09420	4.77	3.3951	−27.73565	−29.42875	−15.56520
4.27	−15.3122	−23.66155	−16.15090	−4.25455	4.78	4.0236	−27.69875	−29.70610	−15.86090
4.28	−15.1387	−23.81530	−16.39400	−4.41740	4.79	4.6638	−27.65530	−29.98275	−16.15925
4.29	−14.9587	−23.96765	−16.63525	−4.5325	4.80	5.3164	−27.60515	−30.25890	−16.46040
4.30	−14.7722	−24.11805	−16.87730	−4.75010	4.81	5.9811	−27.54875	−30.53480	−16.76450
4.31	−14.5788	−24.26685	−17.120065	−4.92000	4.82	6.6574	−27.48590	−30.81015	−17.07120
4.32	−14.3786	−24.41360	−17.36550	−5.09245	4.83	7.3466	−27.41555	−31.08445	−17.38060
4.33	−14.1714	−24.55840	−17.61185	−5.26735	4.84	8.0477	−27.33890	−31.35835	−17.69280
4.34	−13.9570	−24.84170	−17.85945	−5.44170	4.85	8.7623	−27.25465	−31.63135	−18.00790
4.35	−13.7357	−24.98015	−18.10855	−5.62450	4.86	9.4890	−27.16340	−31.90350	−18.32565
4.36	−13.5070	−25.11635	−18.35905	−5.80690	4.87	10.2282	−27.06495	−32.17465	−18.64600
4.37	−13.2712	−25.2500	−18.61100	−5.99155	4.88	10.9806	−26.95885	−32.44475	−18.96910
4.38	−13.0276	−25.38185	−18.86415	−6.17920	4.89	11.7458	−26.84515	−32.71370	−19.29480
4.39	−12.7766	−25.51075	−19.11850	−6.36900	4.90	12.5239	−26.72385	−3298140	−19.62320
4.40	−12.5180	−25.63725	−19.37425	−6.56150	4.91	13.3158	−26.59460	−33.24815	−19.95445
4.41	−12.2517	−25.76120	−19.6313	−6.75655	4.92	14.1202	−26.45775	−33.51350	−20.28820
4.42	−11.9776	−25.88235	−19.88745	−6.95410	4.93	14.9388	−26.31225	−33.77735	−20.62480
4.43	−11.6625	−26.00065	−20.14885	−7.15430	4.94	15.7704	−26.15875	−34.03965	−20.96380
4.44	−11.4051	−26.11610	−20.40950	−7.35710	4.95	16.6157	−25.99670	−34.30025	−21.30535
4.45	−11.1069	−26.20735	−20.67115	−7.55170	4.96	17.4750	−25.82620	−34.55945	−21.64975
4.46	−10.8003	−26.33840	−20.93410	−7.77050	4.97	18.3478	−25.64720	−34.81680	−21.99660
4.47	−10.4851	−26.44475	−21.19805	−7.98120	4.98	19.2348	−25.45940	−35.07200	−22.34620
4.48	−10.1615	−26.54795	−21.46295	−8.19450	4.99	20.1356	−25.26230	−35.32590	−22.69805
4.49	−9.8295	−26.64790	−21.72885	−8.41040	5.00	21.0504	−25.05645	−35.57745	−23.05250
4.50	−9.4890	−26.74465	−21.99590	−8.62900	5.01	21.9800	−24.84125	−35.82715	−23.40970
4.51	−9.1392	−26.83770	−22.26385	−8.85035	5.02	22.8474	−24.61700	−36.07445	−23.76910
4.52	−8.7805	−26.92720	−22.53265	−9.07440	5.03	23.8815	−24.38265	−36.31925	−24.13110
4.53	−8.4133	−27.01320	−22.80225	−9.30095	5.04	24.8537	−24.13915	−36.56190	−24.49540
4.54	−8.0368	−27.09565	−23.07295	−9.53040	5.05	25.8407	−23.88595	−36.80225	−24.86230
4.55	−7.6509	−27.17395	−23.34415	−9.76240	5.06	26.8427	−23.62250	−37.03975	−25.23150

Table 2.25 Con't

$\xi = \frac{x}{\lambda}$	$A(\xi)$	$B(\xi)$	$C(\xi)$	$D(\xi)$	$\xi = \frac{x}{\lambda}$	$A(\xi)$	$B(\xi)$	$C(\xi)$	$D(\xi)$
5.07	27.8598	−23.34890	−37.27475	−25.60325	5.59	102.9739	8.71480	−42.7695	−47.1281
5.08	28.8914	−23.06505	−37.50675	−25.97710	5.60	104.8687	9.75435	−42.67745	−47.5558
5.09	29.9377	−22.77110	−37.73595	−26.35325	5.61	106.7790	10.81250	−42.57440	−47.9818
5.10	30.9997	−22.46605	−37.96185	−26.73170	5.62	108.7074	11.89025	−42.46090	−48.4071
5.11	32.0766	−22.15085	−38.1852	−27.11260	5.63	110.6512	12.98645	−42.33660	−48.8309
5.12	33.1687	−2182455	−38.40505.	−27.49550	5.64	112.6133	14.10285	−42.20130	−49.2538
5.13	34.2762	−21.48740	−38.62160	−27.88060	5.65	114.5922	15.23895	−42.05470	−49.6752
5.14	35.3991	−21.13910	−38.8348	−28.26790	5.66	116.5866	16.39495	−41.89590	−50.0944
5.15	36.5377	−20.77950	−39.04450	−28.65740	5.67	118.5994	17.57055	−41.72675	−50.5130
5.16	37.6913	−20.40835	−39.25015	−29.04860	5.68	120.6277	18.66600	−41.54485	−50.9292
5.17	38.8617	−20.0254	−39.45245	−29.44230	5.69	122.6730	19.98350	−41.35065	−51.3434
5.18	40.0474	−19.63095	−39.65090	−29.83790	5.70	124.1352	21.21990	−41.14535	−51.7563
5.19	41.2485	−19.22475	−39.84525	−30.23540	5.71	126.8144	22.47845	−40.92650	−52.1667
5.20	42.4661	−18.80570	−40.0350	−30.63460	5.72	128.9091	23.75705	−40.69520	−52.5746
5.21	43.6994	−18.37535	−40.22135	−31.03610	5.73	131.0207	25.05680	−40.45135	−52.9806
5.22	44.9485	−17.93215	−40.40280	−31.43910	5.74	133.1478	26.28100	−40.13650	−53.3359
5.23	46.2148	−17.47580	−40.57960	−31.84400	5.75	135.2903	27.71920	−39.92380	−53.7842
5.24	47.4958	−17.00730	−40.75205	−3225040	5.76	137.4497	29.08315	−39.63960	−54.1819
5.25	48.7949	−16.52580	−40.91965	−33.65900	5.77	139.6260	30.46925	−39.3416	−54.5770
5.26	50.1091	−16.03165	−41.08260	−33.06900	5.78	141.8144	31.87550	−39.03040	−54.9689
5.27	51.4399	−15.52395	−41.24035	−33.48060	5.79	144.0228	33.30525	−38.70410	−55.3574
5.28	52.7876	−15.00295	−41.39320	−33.89890	5.80	146.2448	34.75640	−38.36395	−55.7429
5.29	54.1511	−14.46840	−4154050	−34.30840	5.81	148.4819	36.23005	−38.0089	−56.1246
5.30	55.5317	−13.92010	−41.68255	−34.72455	5.82	150.7340	37.72555	−37.63945	−56.5029
5.31	56.9296	−13.35735	−41.81870	−35.14210	5.83	153.0028	39.24490	−37.25450	−56.8776
5.32	58.3438	−12.78080	−41.94930	−35.56090	5.84	155.2847	40.78585	−36.85455	−57.2481
5.33	59.7745	−12.19030	−42.07415	−35.98100	5.85	157.5988	42.35035	−36.43845	−57.6143
5.34	61.2218	−11.58560	−42.19315	−36.40230	5.86	159.8947	43.93780	−3600770	−57.9772
5.35	62.68690	−10.96595	−42. 30605	−36.8250	5.87	162.2208	45.54840	−35.56010	−58.3349
5.36	64.1678	−10.33210	−42.41270	−37.2485	5.88	164.5613	47.18250	−35.09635	−58.6882
5.37	65.6657	−9.68225	−42.51240	−37.6731	5.89	166.9145	48.83940	−34.61605	−59.0363
5.38	67.1818	−9.01835	−42.60600	−38.0986	5.90	169.2837	50.52030	−34.11980	−59.3805
5.39	68.7140	−8.33900	−42.69280	−38.5251	5.92	174.0609	53.95415	−33.07460	−60.0521
5.40	70.2637	−7.64400	−42.77270	−38.9324	5.93	176.0704	55.70670	−32.52680	−60.3806
5.41	71.8308	−6.93360	−42.84585	−39.3808	5.94	178.8917	57.48330	−31.86085	−60.7030
5.42	73.4144	−6.20760	−42.91170	−39.8100	5.95	181.3266	59.28520	−31.37640	−61.0195
5.43	750158	−5.46515	−42.96100	−40.2390	5.96	183.7730	61.73030	−30.77505	−61.0201
5.44	76.6338	−4.70715	−43.02100	−40.6691	5.97	186.2326	63.30865	−30.154550	−61.4608
5.45	78.2687	−3.93280	−43.06415	−41.0993	5.98	188.7034	64.83465	−29.51550	−61.9332
5.46	79.9216	−3.14180	−43.09965	−41.5303	5.99	191.1870	66.73440	−28.85750	−62.2251
5.47	81.5916	−2.33395	−43.12675	−41.9613	6.00	193.6813	68.65775	−28.21160	−62.5106
5.48	83.2786	−1.50950	−43.14585	−42.3926	6.01	196.1881	70.60790	−27.48455	−62.7889
5.49	84.9829	−0.66825	−43.15680	−42.8241	6.02	198.7051	72.58215	−26.76890	−63.0603
5.50	86.7044	0.19005	−43.15925	−43.2557	6.03	201.2322	74.58170	−26.03295	−63.3241
5.51	88.4432	1.06560	−43.15310	−43.6879	6.04	203.7710	76.60665	−25.27740	−63.5810
5.52	90.1996	1.95885	−43.13810	−44.1189	6.05	206.3194	78.65470	−24.50085	−63.8299
5.53	91.9722	2.86925	−43.11405	−44.5500	6.06	208.8770	80.73305	−23.70405	−64.0708
5.54	93.7637	3.79840	−43.08065	−44.9812	6.07	211.4435	82.83495	−22.88545	−64.3032
5.55	95.5716	4.74530	−43.03775	−45.4117	6.08	214.0209	84.96215	−22.04690	−64.5282
5.56	97.3960	5.70950	−42.98575	−45.8418	6.09	216.6066	87.11495	−21.1870	−64.7447
5.57	99.2383	6.69270	−42.92375	−45.2714	6.10	219.2004	89.29465	−20.30425	−64.9518
5.58	101.0984	7.69500	−42.85155	−46.7003	6.11	221.8019	92.49915	−19.40050	−65.1503

Table 2.25 Con't

$\xi = \frac{x}{\lambda}$	$A(\xi)$	$B(\xi)$	$C(\xi)$	$D(\xi)$	$\xi = \frac{x}{\lambda}$	$A(\xi)$	$B(\xi)$	$C(\xi)$	$D(\xi)$
6.12	224.4109	93.72995	−18.47425	−65.3395	7.40	358.7306	546.9343	367.56875	94.1019
6.13	227.0292	95.98710	−17.52625	−65.5200	7.50	313.3700	580.6710	423.9858	133.6503
6.14	229.6542	98.27085	−16.55505	−65.6906	7.60	251.0334	609.0402	483.52330	179.0035
6.15	232.2833	100.5538	−15.56015	−65.8372	7.70	169.3472	630.2295	545.55570	230.4412
6.16	234.9208	102.9168	−14.54245	−65.0010	7.80	65.8475	642.1835	609.25955	288.1681
6.17	237.5639	105.2793	−13.50155	−66.1413	2.5 π	0.0000	643.9927	643.99255	321.9964
6.18	240.2122	107.6680	−12.43700	−66.2711	7.90	−62.0375	642.5872	673.60570	352.3123
6.19	242.8654	110.0831	−11.34845	−66.3901	8.0	−216.8647	628.8779	737.3101	422.8713
6.20	245.5231	112.5249	−10.23560	−66.4981	8.1	−401.1674	598.2344	798.8179	499.7008
6.21	248.1847	114.9934	−9.09795	−66.5947	8.2	−617.4142	547.5808	856.2877	582.4975
6.22	250.8499	117.4888	−7.93515	−66.6796	8.3	−867.9091	473.5998	907.5542	670.7544
6.23	253.5208	120.0113	−6.74805	−66.7538	8.4	−1154.658	372.7866	950.1158	763.7226
6.24	256.1917	122.5599	−5.53500	−66.8150	8.5	−1479.370	241.41360	981.0984	860.3917
6.25	258.8649	125.1346	−4.2969	−66.8642	8.6	−1843.288	75.60880	997.2527	959.4484
6.26	261.5398	127.7369	−3.03205	−66.9005	8.7	−2247.040	−128.5824	994.9377	1059.2289
6.27	264.2159	130.3657	−1.74135	−66.9242	8.8	−2690.485	−375.1167	970.1255	1157.6839
6.28	266.8926	133.0195	−0.4257	−66.9354	8.9	−3172.602	−667.9794	818.8664	1252.3561
2 π	267.7468	133.8725	0.00000	−66.9362	9.0	−3691.482	1010.8700	834.8607	1340.3007
6.30	272.2887	138.4120	2.28855	−66.9175	9.1	4243.5551	−1407.3690	714.40845	1418.09300
6.40	298.8909	166.9722	17.53620	−65.9486	9.2	−4824.0587	−1860.5365	551.49275	1481.76105
6.50	324.7861	198.1637	35.77125	−63.2105	9.3	5426.5154	−2372.9486	340.30910	1526.7834
6.60	349.2554	231.8801	57.25280	−58.6871	9.4	−6042.3167	−2946.2708	74.887500	1548.0229
6.70	371.4244	267.9374	82.22550	−51.7430	3 π	−6195.8239	−3097.9119	0.0000000	1548..9560
6.80	390.2947	306.0558	110.90870	−42.1181	9.5	−6660.9594	−3581.4756	−250.99590	1539.7419
6.90	404.7145	347.34985	143.49270	−30.1819	9.6	7269.3664	−4278.1693	−643.4861	1495.5985
7.00	413.3762	386.80715	180.11910	−13.2842	9.7	−7851.7063	−5034.4714	−1108.6183	1408.6174
7.10	414.8263	428.2849	220.87175	6.7295	9.8	−8389.5687	−5847.0360	−1652.2517	1271.2663
7.20	407.4216	469.4772	265.76635	31.0281	9.9	−8860.9431	−6710.2070	−2279.7354	1075.3680
7.30	389.3783	509.4156	314.72645	60.0189	10.0	−9240.8733	−7616.1462	−2995.7095	812.36360

References

1. Barkan, D. D. 1962. *Dynamics of bases and foundations*. New York: McGraw Hill.
2. Bowels, J. E. 1974. *Analytical and computer methods in foundation engineering*. New York: McGraw Hill.
3. Christopherson, D. G. 1956. *The design and construction of engineering foundations*. London.
4. Dutov, G. D. 1929. *Analysis of beams on elastic foundation*. L. Kubuch.
5. Gorbunov-Posadov, M. I. 1953. *Analysis of structures on elastic foundation*. Moscow: Gosstroiizdat.
6. Hetenyi, M. 1946. *Beams on elastic foundation*. Ann Arbor: University of Michigan Press.
7. Klepikov, S. N. 1967. *Analysis of structures on elastic foundation*. Kiev: Budivelnik.
8. Krilov, A. N. 1930. *Regarding analysis of beams on elastic foundation*. Academy of Sciences USSR.
9. Levinton, Z. 1947. Elastic foundations analysis by the method of redundant reactions. *Journal of the Structural Division: Proceedings of ASCE*, pp. 1529–1541.
10. Malter, H. 1958. Numerical solutions for beams on elastic foundations. *Journal of the Structural Division: Proceedings of ASCE* 84 (ST2), Paper no. 1562.
11. Penzien, J. 1960. Discontinuity stresses in beams on elastic foundation. *Journal of the Structural Division: Proceedings of ASCE*, Paper no. 2545.
12. Pouzirevsky, N. P. 1923. *Soils and Foundations*. Moscow.
13. Selvadurai, A.P.S. 1979. *Elastic analysis of soil-foundation interaction*. Amsterdam/Oxford/New York: Elsevier Scientific.
14. Snitco, N. K. 1970. *Static and dynamic soil pressure and analysis of retaining walls*, L. 1970, pp. 176–184.
15. Tsudik, E. A. 2003. Specified analysis of beams and frames on elastic foundation. In *Proceedings of first international conference on foundations*, UK, Scotland.
16. Tsudik, E. A. 2006. *Analysis of beams and frames on elastic foundation*. Canada/UK: Trafford.
17. Umansky, A. A. 1933. *Analysis of beams on elastic foundation*. Leningrad: Central Research Institute of Auto Transportation.
18. White, R. N. 1963. Optimum solution techniques for finite difference equations. 3rd Conference on Electronic Computation. *Journal of the Structural Division: Proceedings ASCE* 89 (ST4): 115–136.

3

Simplified Analysis of Beams and Frames on Winkler Foundation

3.1 Introduction

Chapter 3 is devoted completely to practical analysis of beams and frames supported on Winkler foundation. It contains simple tables and equations for hand analysis of free supported beams with concentrated loads and moments. As shown by the author of this book, the same tables and equations can be used for analysis of beams with various boundary conditions as well as for analysis of stepped beams, pin-connected beams, interconnected, and various continuous beams. Computer analysis of large complex beams and frames with continuous foundations is recommended to be performed by modeling the soil with individual elastic supports that allows the use of widely available software developed for analysis of regular statically indeterminate systems. Chapter 3 also includes a simple method of computer analysis of 2D and 3D frames with individual foundations. Analysis is performed by modeling the system soil-foundation with simple line elements that allows analyzing frames with individual foundations as regular statically indeterminate systems using available computer software. Numerous examples illustrate the application of the described methods to practical analyses.

3.2 Analysis of Free Supported Beams

Tables 3.1–3.14 presented at the end of this chapter were developed by Klepikov (1967) for hand analysis of free supported beams on Winkler foundation. All beams are divided into three categories: rigid beams when $\lambda < 1$, short beams when $1 \leq \lambda \leq 6$, and long beams when $\lambda \geq 7$. Parameter λ is obtained from the following equation:

$$\lambda = lm \tag{3.1}$$

where $m = \sqrt[4]{\dfrac{kb}{4EI}}$ and l is the length of the beam, b is the width of the beam and EI is the flexural rigidity of the beam. The soil pressure diagram under a rigid beam can be assumed as a straight line, and analysis of such beams can be performed as an analysis

---

All tables can be found at the end of the chapter.

of a totally rigid foundation. The soil pressure diagram under short and long beams is more complex, and analysis of these beams is performed as an analysis of elastic beams on Winkler foundation. Long beams are analyzed as infinite or semi-infinite beams. All tables are developed for various numerical values of λ (1, 2, 3, 4, 5, 6, and 7) and for two types of loads: vertical concentrated loads N and concentrated moments M. Distributed loads are replaced with a series of concentrated vertical loads.

Tables are developed for loads shown in Figures 3.1 and 3.2. The beam is divided into 10 equal sections. Vertical concentrated loads N and moments M can be applied at any of the 11 points of the beam (0, 1, 2, 3, . . ., 10). If the given load, N or M, is applied between these points, simple interpolation is used. If several loads are applied to the beam, analysis of the beam is performed separately for each load (N and M) and final results are obtained by simple superposition. If λ is not equal to the exact values given in Tables 3.1–3.14, the closest value of λ from these tables is used. If, for example, $\lambda = 4.2$, the closest $\lambda = 4$ is used. If $\lambda = 4.85$, the closest $\lambda = 5$ is used, and so on. The given vertical loads are positive when they are applied downward; the moments are positive when they are applied counterclockwise. Loads applied to the beam and shown in Figures 3.1 and 3.2 are located at the left half of the beam. When the loads are applied to the right half of the beam, the same tables are used, but the numbering of the points starts at the right end of the beam. It is important to note that when a positive moment is applied to the right half of the beam, all data taken from the tables change their sign. Analysis of a short beam supported on Winkler foundation is performed in the following order:

1. Find coefficient λ using equation 3.1.
2. If a vertical concentrated load N is applied to the beam, use Tables 3.1–3.7 for obtaining $\overline{p}$, $\overline{M}$, and $\overline{Q}$.

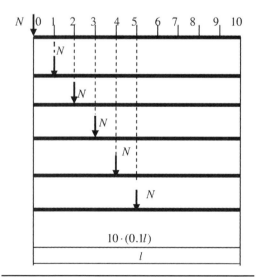

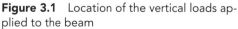

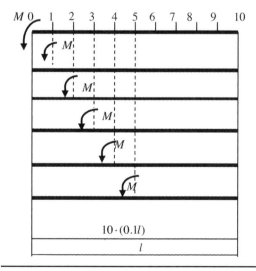

Figure 3.1   Location of the vertical loads applied to the beam

Figure 3.2   Location of concentrated moments applied to the beam

3. If a concentrated moment M is applied to the beam, use Tables 3.8–3.14 for obtaining $\overline{p}$, $\overline{M}$, and $\overline{Q}$.
4. Find the soil pressure p, moments M, and shear forces Q at all points of the beam using equations 3.2 and 3.3 shown below:

$$p = \overline{p}\frac{N_i}{l},\ M = \overline{M}N_i l,\ Q = \overline{Q}N_i \tag{3.2}$$

$$p = \overline{p}\frac{M_i}{l^2},\ M = \overline{M}M_i,\ Q = \overline{Q}\frac{M_i}{l} \tag{3.3}$$

All $\overline{M}$ and $\overline{Q}$ that are shown **bold** in the tables belong to the moments and shear forces located at left from the cross section where the moment M_i or the vertical load N_i are applied. The moment $\overline{M}_R$ and the shear $\overline{Q}_R$ at the right of the same cross section are found from the following equations:

$$\overline{M}_R = \overline{M}_{Left} - 1,\ \overline{Q}_R = \overline{Q}_{Left} - 1 \tag{3.4}$$

The settlements of the beam at any point i are obtained from equation 3.5:

$$y_i = \frac{p_i}{kb} \tag{3.5}$$

It is important to note that if parameter $\lambda < 1$, tables for $\lambda = 1$ are used. Analysis of infinite long beams and semi-infinite long beams, when $\lambda > 7$, is performed using tables for $\lambda = 7$. However, in this case, equations 3.2 and 3.3 are replaced with equations 3.6 and 3.7 shown below. When a vertical load N_i is applied to a long beam, equations 3.6 are used:

$$p = \overline{p}\frac{N_i m}{7}\ \ M = \overline{M}\frac{7N_i}{m}\ \ Q = \overline{Q}N_i \tag{3.6}$$

When a moment M_i is applied to a long beam, equations 3.7 are used:

$$p = \overline{p}\frac{Mm^2}{49}\ \ M = \overline{M}M_i\ \ Q = \overline{Q}\frac{M_i}{7} \tag{3.7}$$

It is important to remember, when performing analysis of infinite and semi-infinite long beams, that the length of the beam cannot be simply divided into 10 sections because the length is infinitely large. In this case, coordinates of the cross sections of the beam are equal to $\xi = xm$. For analysis of infinite and semi-infinite long beams, the spacing between the points cannot be equal to $0.1l$; it is equal to $0.1m$. Analysis of infinite and semi-infinite beams is performed by using only two Tables, 3.7 and 3.14, when $\lambda = 7$. It is also important to mention, that when a vertical load is applied to an infinite long beam, only column 5 from Table 3.7 ($\lambda = 7$) is used. When a vertical load is applied to a semi-infinite long beam, columns 0, 1, 2, 3, and 4 from Table 3.7 are used. When a moment is applied to an infinite long beam, column 5 from Table 3.14 is used. When a moment is applied to a semi-infinite beam, columns 0, 1, 2, 3, and 4 from Table 3.14 are used. It is also important to mention that all coefficients given in Tables 3.1–3.14 are dimensionless. Therefore, different types of units can be used.

If the data are given in meters and metric tons, the soil pressure will be obtained in $\frac{ton}{m}$, the moments will be obtained in tm, and shear forces will be obtained in tons. If the data are given in pounds and inches, the soil pressure is obtained in $\frac{lb}{in}$, the moments are obtained in lb · in, and the shear forces in pounds. Some numerical examples presented below illustrate the application of Tables 3.1–3.14 to analysis of beams free supported on Winkler foundation.

Example 3.1

Given: A concrete beam supported on soil and loaded with a vertical load, as shown in Figure 3.3. The length of the beam $l = 240$ in, the width of the beam $b = 12$in, the moment of inertia $I = 1.56 · 10^4$ in^4, the modulus of subgrade reaction $k = 100\frac{lb}{in^3}$ $kb = 1,200$ psi, and modulus of elasticity of the beam $E = 4 \times 10^6$psi $l = 240$in.

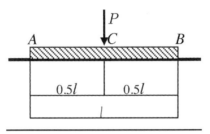

Figure 3.3

Find the maximum deflection and the moment at the center of the beam at point C.

Solution:

1. Find parameter $m = \sqrt[4]{\dfrac{kb}{4EI}} = \sqrt[4]{\dfrac{100 \times 12}{4 \times 4 \times 10^6 \times 1.56 \times 10^4}} = 0.008327\dfrac{1}{in}$.

2. Find parameter $\lambda = lm = 0.008327 \times 240 = 2.0$.

3. Using Table 3.2 for $\lambda = 2$, find $\overline{p} = 1.1807$ and $\overline{M} = 0.1152$.

4. Using equations 3.2, find the soil pressure and the settlement at the center C of the beam

$$p = \overline{p}\frac{P}{l} = 1.1807\frac{P}{240} = 0.0049195P \; y_c = \frac{p}{kb} = \frac{0.0049195P}{1,200} = 4.10 · 10^{-6}P.$$

5. Using equations 3.2, find the moment $M_c = \overline{M}Pl = 0.1152 \times 240P = 27,648P.$ Results of analysis of this beam performed by R. F. Scott (1981) are the same.

Example 3.2

Given: A concrete beam supported on Winkler foundation shown in Figure 3.4. The length of the beam $l = 48$ft, the width of the beam $b = 3.30$ft, the height of the beam

$h = 4.0\text{ft}$, the modulus of subgrade reaction $k = 125,000\dfrac{\text{lb}}{\text{ft}^3}$, and the modulus of elasticity of the beam material $E = 4.3 \times 10^8 \dfrac{\text{lb}}{\text{ft}^2}$. The beam is loaded with two moments $M = 721,600\text{lbft}$, as shown in Figure 3.4. Find the soil pressure, moment, and shear force at points 0, k, and 5.

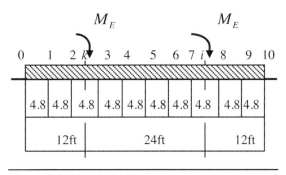

Figure 3.4

Solution:

1. Find parameter $m = \sqrt[4]{\dfrac{kb}{4EI}} = \sqrt[4]{\dfrac{125,000 \times 3.3}{4 \times 4.3 \times 10^8 \times 20.37}} = 0.0586\dfrac{1}{\text{ft}}$.

2. Find parameter $\lambda = lm = 0.0586 \times 52.50 = 3.077$.

As can be seen, the given moments are not applied at the points dividing the beam into 10 sections. The moments are applied between points 2 and 3 and points 7 and 8. Therefore, when using the tables for $\lambda = 3$, we shall use interpolation between the values shown in point 2 and values shown in point 3, taking into account that point k is located at the center of section 2–3 and the direction of the moment is counterclockwise:

$$p_0 = -\left[0.5(13.7134 + 10.3536) + 0.5(0.7851 + 1.4484)\right]\dfrac{721,600}{52.50} = -3,444.05\dfrac{\text{lb}}{\text{ft}}$$

$$p_k = -\left[0.5\left(\dfrac{3.5716 - 0.6718}{2} + \dfrac{4.59 + 1.0626}{2}\right)\right.$$
$$\left. + 0.5\left(\dfrac{2.5371 + 3.3063}{2} + \dfrac{2.913 + 3.4946}{2}\right)\right] \times \dfrac{721,600}{52.50^2} = 1,361.58\dfrac{\text{lb}}{\text{ft}}.$$

$$p_5 = \left[0.5(3.8307 + 3.3711) + 0.5(-3.8307 - 3.3711)\right]\dfrac{721,600}{52.5^2} = 0.$$

$$M_k^{left} = -\left[\dfrac{0.3369 + 0.2078}{2} + 0.50 \times \left(\dfrac{0.0389 + 0.02078}{2} + \dfrac{0.0275 + 0.0748}{2}\right)\right]$$

$\times 721,600 = 240,260.321\text{lb} \times \text{ft}$, $M_k^{right} = 721,600 - 240,260.321 = 481,339.68\text{lb} \times \text{ft}$.

$M_0 = M_5 = 0 \quad Q_0 = 0.$

$$Q_k = -\left[0.5\left(\frac{1.7394 + 1.8844}{2} + \frac{1.5032 + 1.7859}{2}\right)\right.$$

$$\left. +0.5\left(\frac{0.3333 + 0.6255}{2} + \frac{0.4378 + 0.7582}{2}\right)\right] \times \frac{721,600}{52.5} = -31,200 \, \text{lb}$$

$$Q_5 = -0.5(2 \cdot 1.3648 + 2 \cdot 1.4764)\frac{721,600}{52.5} = -39,183 \text{lb}$$

Our results are the same as obtained by Kornevits and Ender (1932).

Example 3.3

Let us analyze a beam shown in Figure 3.5.

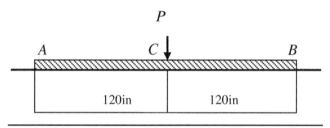

Figure 3.5

Given: The length of the beam l = 240in, the width of the beam b = 10in, the moment of inertia of the beam I = 426.7in⁴, the modulus of elasticity of the beam material $E = 1.5 \cdot 10^6 \frac{\text{lb}}{\text{in}^2}$, and the modulus of subgrade reaction $k = 200\frac{\text{lb}}{\text{in}^3}$, P = 5,000 lb, $m = \sqrt[4]{\frac{kb}{4EI}} = \sqrt[4]{\frac{200 \cdot 10}{4 \cdot 1.5 \cdot 10^6 \cdot 426.7}} = 0.02973 \, \text{in}^{-1}$ $\lambda = lm = 0.02973 \cdot 240$ = 7.13. Find the soil pressure and settlements of the beam.

Solution:

Since λ = 7.13 and the load P is applied to the center of the beam, this beam has to be analyzed as an infinite long beam. Assuming that λ = 7, using Table 3.7 and equation 3.5, the soil pressure and settlements are found. The soil pressure and settlements are found in Table 3.15. The soil pressure p is obtained in $\frac{\text{lb}}{\text{in}}$; the settlements are obtained in inches using equation 3.5.

Example 3.4

All given data are the same as in Example 3.3 except the point of application of the load P, as shown in Figure 3.6. This beam belongs to the long semi-infinite beams and has to be analyzed using Table 3.7, column 3.

Find the soil pressure, the settlements, moments, and shear forces.

1. Find $\bar{p}$, $\overline{M}$, and $\overline{Q}$ from Table 3.7.

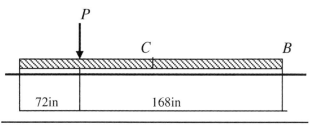

Figure 3.6

2. Find the real soil pressure using the first equation from 3.6:

$$p = \bar{p}\frac{Nm}{7} = \bar{p}\frac{5,000 \cdot 0.02973}{7} = \bar{p} \cdot 21.2357 \text{lb/in.}$$

3. Find the settlements of the beam using equation 3.5.
4. Find the moments using the second equation from 3.6:

$$M = \bar{M}\frac{7N}{m} = \bar{M}\frac{7 \cdot 5,000}{0.02973} = \bar{M} \, 1,177,262.$$

5. Find the shear forces using the third equation from 3.6:

$$Q = N\bar{Q} = 5,000\bar{Q}.$$

All calculations are performed in Table 3.16. The soil pressure is obtained in $\frac{\text{lb}}{\text{in}}$, settlements are obtained in inches, moments are obtained in lb · in, and shear forces are obtained in pounds.

For rigid beams, the soil pressure diagram always looks as a straight line or very close to a straight line, no matter where the vertical load or moment is applied. When parameter $\lambda \le 1$, tables for $\lambda = 1$ are used. This type of beam is working as a totally rigid foundation.

It is easy to see from Tables 3.1 and 3.8 with $\lambda = 1$, that if a beam is loaded at the center with a vertical load P, the soil pressure at the center is equal $p_{center} = \bar{p}\frac{P}{l} = 1.0125\frac{P}{l}$, and the soil pressure at the end of the beam is equal to $p_{end} = 0.9815\frac{P}{l}$. The soil pressure at the center of the foundation is close to the soil pressure at the edge of the beam. The difference is only 3.16%. If the same beam is loaded with a moment applied at the center of the beam, the soil pressure at both ends of the beam is the same $p_{end} = 0.9815\frac{P}{l}$. The soil pressure applied to one-half of the beam is positive while the soil pressure applied to the other half of the beam is negative. However, the soil pressure diagram is practically a straight line or very close to a straight line. And, of course, if a rigid foundation is loaded with a vertical load and a moment, the soil pressure diagram also looks like a straight line.

3.3 Analysis of Beams with Various Boundary Conditions

As mentioned earlier, Tables 3.1–3.14 are developed only for analysis of simple free supported beams. The same tables can be used for analysis of beams with various boundary conditions. Table 3.17 shows beams supported on Winkler foundation with various boundary conditions. By removing all restraints and replacing them with unknown reactions, as shown in Table 3.18, any beam with any boundary conditions can always be replaced with a free supported beam loaded with the given loads and unknown reactions. Figure 3.7 shows a beam on Winkler foundation with a free supported left end at point A and pin-supported right end at point B. By replacing the support at point B with the unknown reaction X, as shown in Figure 3.8, and taking into account that the total settlement of the beam at point B is equal to zero, the following equation is written:

$$\omega_{BB}X + \omega_{BP} + \omega_{BM} = 0 \qquad (3.8)$$

In this equation ω_{BB} is the settlement of the beam at point B due to $X = 1$ and ω_{BP} and ω_{BM} are settlements of the beam at point B due to the given loads P and M, respectively. From 3.8 the reaction at point B is found:

$$X = -\frac{\omega_{BP} + \omega_{BM}}{\omega_{BB}} \qquad (3.9)$$

Now, the final analysis of the beam can be performed. By applying P, M, and X to the beam and using Tables 3.1–3.14, the soil pressure, moments, and shear forces are obtained. Each load is applied to the beam separately and results are found using simple superposition. It is important to mention that the beam settlements at point B cannot be found directly from Tables 3.1–3.14. However, from these tables the soil pressure at point B can be found. If the soil pressure at point B, due to the load P, is obtained and is equal to $\overline{p}_{BP}$, the settlement at this point can be found as $\omega_{BP} = \dfrac{\overline{p}_{BP}P}{kb}$. The settlement at the same point B due to the moment M is equal to $\omega_{BM} = \dfrac{\overline{p}_{BM}M}{kb}$, and settlement at the same point due to reaction X is equal to $\omega_{BB} = \dfrac{\overline{p}_{BB}X}{kb}$.

If the same beam AB is supported on individual supports at both ends, as shown in Figure 3.9, analysis is performed as follows. By replacing both end supports, A and B, with unknown reactions, we obtain a beam supported on soil and loaded with given loads P and M and unknown reactions X_1 and X_2, as shown in Figure 3.10.

Taking into account that settlements at points A and B are equal to zero, the following two equations are written:

$$\left. \begin{array}{l} \omega_{AA}X_1 + \omega_{AB}X_2 + \omega_{AP} + \omega_{AM} = 0 \\ \omega_{BA}X_1 + \omega_{BB}X_2 + \omega_{BP} + \omega_{BM} = 0 \end{array} \right\} \qquad (3.10)$$

In this system, ω_{AA} and ω_{BB} are vertical deflections of the beam at points A and B due to support reactions $X_1 = 1$ and $X_2 = 1$, respectively; ω_{AP} and ω_{BP} are settlements of the beam at points A and B due to the given load P, respectively; and ω_{AM} and ω_{BM} are settlements of the beam at points A and B, respectively, due to the given moment M. ω_{AB} and ω_{BA} are settlements at points A and B, respectively, due to reactions $X_1 = 1$ and $X_2 = 1$. From the system of equations 3.10, unknown reactions X_1 and X_2 are found and final analysis of the beam is performed. As we can see, analysis of beams with one

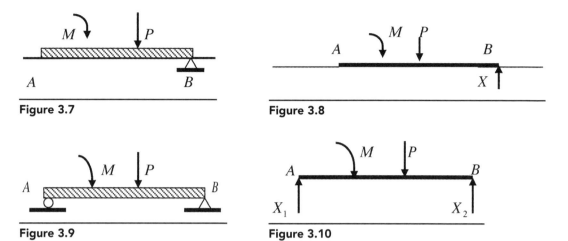

Figure 3.7

Figure 3.8

Figure 3.9

Figure 3.10

and two end supports is performed in two steps. First, support reactions are found and then final analysis of the beam is performed. Two numerical examples presented next illustrate application of the described method to practical analyses.

Example 3.5

Let us solve an example shown in Figure 3.11. The left end of the beam is free supported on soil at point A and the right end is supported on a non-yielding support at point B. The length of the beam $l = 240$in, the moment of inertia $l = 1.56 \cdot 10^4$in^4, modulus of subgrade reaction $k = 100\dfrac{\text{lb}}{\text{in}^3}$, modulus of elasticity of the beam material $E = 4 \cdot 10^6$psi, $P = 20{,}000$ lb, and the width of the beam $b = 12$in. Find the reaction at point B and build the moment diagram.

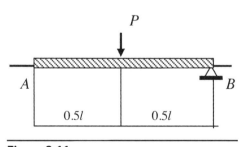

Figure 3.11

Solution:

1. Find parameter $m = \sqrt[4]{\dfrac{kb}{4EI}} = \sqrt[4]{\dfrac{100 \cdot 12}{4 \cdot 4 \cdot 10^6 \cdot 1.56 \cdot 10^4}} = 0.008327\text{in}^{-1}$.

2. Find parameter $\lambda = lm = 0.008327 \cdot 240 = 2$.

3. Replace support at point B with unknown reaction X. The given beam will look as shown in Figure 3.12.

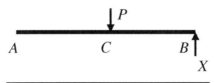

Figure 3.12

4. Analyze the beam loaded with $X = 1$ lb applied to the right end, and find the settlement of the beam at point B using Table 3.2 and taking into account the direction of the reaction X $\bar{p} = -4.5441$

$$p = \bar{p}\frac{N}{l} = -4.5441\frac{1}{240} = -189.337 \cdot 10^{-4}\text{lb/in},$$

$$\omega_{BB} = \frac{-p}{kb} = -\frac{189.337 \cdot 10^{-4}\dfrac{\text{lb}}{\text{in}}}{1,200 \cdot \dfrac{\text{lb}}{\text{in}^2}} = -0.1578 \cdot 10^{-4}\text{in}.$$

5. The soil pressure of the beam at point B due to the given load P can be found from the same Table 3.2 $\bar{p} = 0.7368$, $\bar{p} = 0.7368\dfrac{20,000}{240} = 61.40\text{lb/in}.$

6. The settlement of the beam is $\omega_{BP} = \dfrac{p}{kb} = \dfrac{61.4\dfrac{\text{lb}}{\text{in}}}{1,200\dfrac{\text{lb}}{\text{in}^2}} = 0.05117\text{in}.$

Find the reaction at point B from the following equation:

$$-\omega_{BB}X + \omega_{BP} = 0 \qquad (3.11)$$

By introducing numerical values into 3.11 we have:

$$-0.1578 \cdot 10^{-4} X + 511.7 \cdot 10^{-4} = 0$$

$$X = 3,242.7\text{lb}$$

Now, analysis of the beam AB with the given load $P = 20,000$lb and reaction X can be performed. Let us build two moment diagrams: one due to the given load P and the second due to the reaction X. All moments are given in k-ft.

The second equation from equations 3.2 is used to obtain moments given in Tables 3.19 and 3.20. For example, the moment at point 4, shown in Table 3.19, due to applied load $P = 20,000$lb is equal to:

$$M = \overline{M}Nl = 0.0711 \times 20.0 \times 240/12 = 28.44\text{k-ft}$$

The moment at the same point, due to reaction X applied at point B, is equal to:

$$M = \overline{M}Nl = 0.0819 \times 3.24271 \times 20 = 5.312\text{k-ft}$$

The moment diagrams due to the load P and reaction X are shown in Figures 3.13 and 3.14, respectively. The final moment diagram, shown in Figure 3.15, is obtained by simple superposition. All moments are obtained in kip-ft.

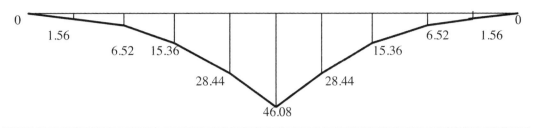

0

1.56

6.52   15.36

28.44

46.08

6.52   1.56

15.36

28.44

Figure 3.13   Moment diagram for the beam *AB* due to load *P*

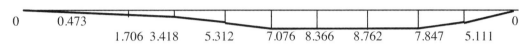

0   0.473

1.706 3.418   5.312   7.076 8.366   8.762   7.847   5.111

0

Figure 3.14   Moment diagram for the beam *AB* due to reaction *X*

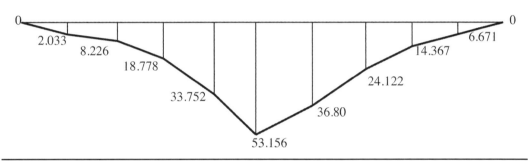

0

2.033

8.226

18.778

33.752

53.156

36.80

24.122

14.367

6.671

0

Figure 3.15   The final moment diagram for the beam *AB*

Example 3.6

The beam shown in Figure 3.16 is supported on soil. Both ends of the beam are supported on non-yielding supports at points A and B. $P = 300$kips, $L = 30$ft, the modulus of subgrade reaction $k = 400\dfrac{\text{kips}}{\text{ft}^3}$, $b = 3$ft, $h = 2.5$ft, $I = \dfrac{bh^3}{12} = \dfrac{3 \cdot 2.5^3}{12} = 3.906 \text{ ft}^4$,

$E = 600{,}000\dfrac{\text{kip}}{\text{ft}^2}$, $m = \sqrt[4]{\dfrac{1{,}200}{4 \cdot 600{,}000 \cdot 3.906}} = 0.106\dfrac{1}{\text{ft}}$, $\lambda = lm = 0.106 \cdot 30 = 3.18$.

Find the soil pressure and support reactions at points A and B.

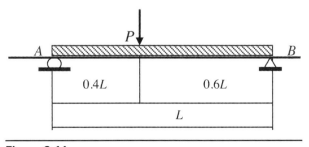

P

A   *B*

0.4L   0.6L

L

Figure 3.16

Solution:

By replacing both supports with unknown reactions, we have a beam supported on soil with the given vertical load P and reactions X_1 and X_2 shown in Figure 3.17. Now, the following system of equations can be written:

$$\left.\begin{array}{c} \omega_{AA}X_1 + \omega_{AB}X_2 + \omega_{AP} = 0 \\ \omega_{BA}X_1 + \omega_{BB}X_2 + \omega_{BP} = 0 \end{array}\right\} \tag{3.12}$$

The equations show that settlements of both supports are equal to zero. All coefficients in this system of equations are settlements of the beam at points A and B due to one-unit support reactions and given loads applied to the beam. ω_{ij} is the settlement of the beam at point i due to one unit load applied at point j, and ω_{iP} is the settlement at point i due the given load applied to the beam. Since Tables 3.1–3.14 do not contain settlements, we shall find first the soil pressures and then the settlements.

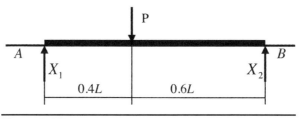

Figure 3.17

From Table 3.3, obtain the soil pressure and then settlements at points A and B.

$$p_{AA} = \overline{p}_{AA}\frac{-X_1}{L} = -6.0163\frac{1}{30} = -0.200543\frac{\text{kips}}{\text{ft}}$$

$$\omega_{AA} = \frac{p_A}{kb} = \frac{-0.200543}{400 \cdot 3} = -167.1 \cdot 10^{-3}\ \text{ft}$$

$$p_{BA} = \overline{p}_{BA}\frac{1}{L} = 0.6768\frac{1}{30} = 22.56 \cdot 10^{-3}\frac{\text{kips}}{\text{ft}}$$

$$\omega_{BA} = \frac{p_{BA}}{kb} = \frac{0.02256}{400 \cdot 3} = 0.0188 \cdot 10^{-3}\ \text{ft}$$

Find the soil pressure and the vertical deflections at points A and B due to the given load P.

$$p_{AP} = \overline{p}_{AP}\frac{P}{L} = 0.6831 \cdot \frac{300}{30} = 6.831\frac{\text{kips}}{\text{ft}},$$

$$p_{BP} = \overline{p}_{BP}\frac{P}{L} = -0.2567 \cdot \frac{300}{30} = -2.567\frac{\text{kips}}{\text{ft}}$$

$$\omega_{AP} = \frac{p_{AP}}{kb} = \frac{6.831}{1,200} = 0.0056925\ \text{ft},\quad \omega_{BP} = \frac{p_{BP}}{kb} = \frac{-2.567}{1,200} = -0.002139\ \text{ft}$$

Find reactions X_1 and X_2.

By introducing all numerical values into the system 3.12 and taking into account that $\omega_{AB} = \omega_{BA}$, $\omega_{AA} = \omega_{BB}$, the system of equations 3.13 is written:

$$\left.\begin{array}{c} -167.1X_1 + 18.8X_2 + 5,692.50 = 0 \\ +18.8X_1 - 167.1X_2 - 2,139.00 = 0 \end{array}\right\} \qquad (3.13)$$

From this system of equations we have $X_1 = 33.045$kips and $X_2 = -9.083$kips.

Now, analysis of the beam, loaded with the given load $P = 300$ kips and support reactions X_1 and X_2 can be performed.

Analysis of the beam is performed in Table 3.22. In this table $\bar{p}$ is taken from Table 3.3, p is obtained from equation 3.2. Now, the equilibrium of all loads applied to the beam can be checked. The total active load applied to the beam is equal to 300kips. The sum of the applied load and all soil reactions (concentrated reactions X_1 and X_2 at points A, B and uniformly distributed reactions at all other points from 1 to 9) is $300 - 33.045 + 9.083 - 3 \times (5.185 + 9.971 + 13.796 + 15.779 + 15.151 + 12.81 + 9.717 + 6.426 + 3.177) = 0$ which confirms the equilibrium of all given loads and reactions applied to the beam. The soil pressure diagram is shown in Figure 3.18.

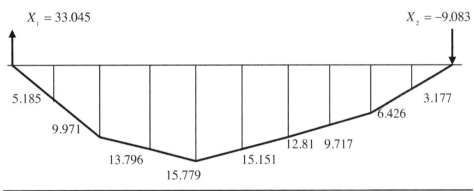

Figure 3.18   Soil pressure diagram for the beam AB

Example 3.7

A beam supported on Winkler foundation, as shown in Figure 3.19, is loaded with a moment $M = 723.285$kip-ft applied to the center of the beam. The modulus of elasticity of the beam $E = 204.87 \cdot 10^6 \dfrac{\text{lb}}{\text{ft}^2}$, the moment of inertia $I = 29.66\text{ft}^4$, the modulus of subgrade reaction $k = 199,774.71 \dfrac{\text{lb}}{\text{ft}^3}$, the length of the beam is $l = 65.616$ft, and the width $b = 4.10$ft. Find the moments and show the moment diagram.

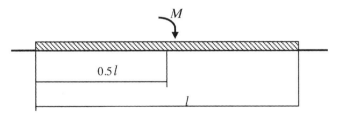

Figure 3.19

Solution:

$$EI = 204.82 \cdot 10^6 \cdot 29.66 = 6{,}074.96 \cdot 10^6 \text{lbft}^2, \; kb = 199{,}774.71 \cdot 4.1 = 819{,}076.3\frac{\text{lb}}{\text{ft}^2}$$

$$\text{Parameter } m = \sqrt[4]{\frac{kb}{4EI}} = \sqrt[4]{\frac{819{,}076.3}{4 \cdot 6{,}074.96 \cdot 10^6}} = 0.07622\frac{1}{\text{ft}} \; \lambda = ml = 0.07622 \cdot 65.616 = 5.0$$

Now, using Table 3.12 for $\lambda = 5$, moments at all points of the beam are found. Moments are obtained in Table 3.23. The moment diagram is shown in Figure 3.20.

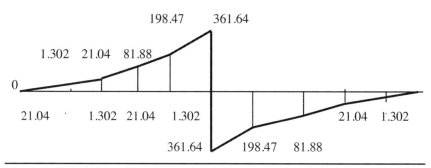

Figure 3.20   Moment diagram

3.4 Analysis of Interconnected Beams

Tables 3.1–3.14 can also be used for the analysis of interconnected beams supported on Winkler foundation. A simple system of two interconnected beams is shown in Figure 3.21.

Simplified analysis of a system of interconnected beams is based on the assumption that beams running in one direction are supported on beams running in another direction. In our case, we assume that the beam AB is supported on beam CD at point O. Now, the given system of two interconnected beams can be replaced with two independently working beams AB and CD, as shown in Figure 3.22. Taking into account that the vertical deflections of both beams at the point of intersection are the same, the following equation is written:

$$\left(\omega_{0,AB} + \omega_{0,CD}\right)X + \omega_{0P_1} + \omega_{0P_2} = 0 \tag{3.14}$$

From this equation the loads of interaction X are found:

$$X = -\frac{\omega_{0P_1} + \omega_{0P_2}}{\left(\omega_{0,AB} + \omega_{0,CD}\right)} \tag{3.15}$$

In equations 3.14 and 3.15, $\omega_{0,AB}$ and $\omega_{0,CD}$ are settlements at point 0 due to $X = 1$ applied to the beams AB and CD, respectively; ω_{0P_1} and ω_{0P_2} are settlements at point 0 of the beams AB and CD due to the given loads P_1 and P_2, respectively. Now, each beam is analyzed independently under the given loads P_1, P_2, and X.

Simplified analysis of interconnected beams requires solving of a system of linear equations with a number of unknowns equal to the number of points of intersection. When the number of such points is equal to 4–5, hand calculations can be performed.

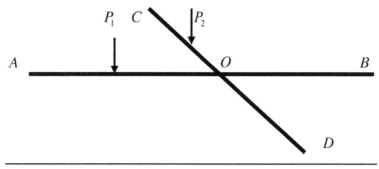

Figure 3.21

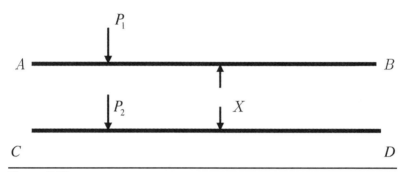

Figure 3.22

When the number of unknowns is larger, computer analysis is the best solution. An example shown below illustrates the application of the method to the solution of practical problems.

Example 3.8

Let us analyze a simple system of interconnected beams, shown in Figure 3.23, assuming that $EI = 0.256 \cdot 10^6 tm^2$, $l_{AB} = l_{CD} = 20m$, $k = 3,200t/m^3$, $b = 1.25m$, $P = 10t$, $OE = OF = 4m$, and $0E = 0F = 4m$. We also assume that the point of intersection 0 is located at the middle of both beams and the beam AB is supported on beam CD.

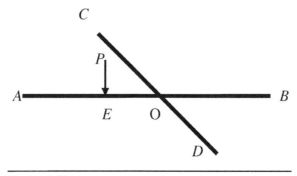

Figure 3.23

Solution:

Find parameter $m = \sqrt[4]{\dfrac{k}{4EI}} = \sqrt[4]{\dfrac{4{,}000}{4 \cdot 0.256 \cdot 10^6}} = 0.25m^{-1}$ and $\lambda = ml = 0.25 \cdot 20 = 5$

Assuming that the forces of interaction between both beams at point O are equal to X, the given system of two interconnected beams is replaced with two simple beams shown in Figures 3.24 and 3.25. The soil pressure and the vertical deflection of the beam AB at point O are equal to:

$$p_{OP} = \bar{p}_{OP}\frac{P}{l} = 1.3491 \cdot \frac{10}{20} = 0.67455\frac{t}{m} \qquad \omega_{OP} = \frac{p_{OP}}{k} = \frac{0.67455}{4{,}000} = 168.6 \cdot 10^{-6}m$$

$$p_{OX} = -\bar{p}_{OX}\frac{X}{l} = -2.6417 \cdot \frac{X}{20} = -0.132085X\frac{t}{m} \qquad \omega_{OX} = -\frac{0.132085X}{4{,}000} = -33 \cdot 10^{-6}X$$

The vertical deflection of the beam CD at point O, due to X, is equal to $\omega_{OX} = 33 \cdot 10^{-6}$ X. Taking into account that the settlement of the beam AB at point O is equal to the settlement of the beam CD, at the same point the following equation can be written:

$$\left(\omega_{OX} + \omega_{OX}\right)X + \omega_{OP} = 0 \tag{3.16}$$

From 3.16 we have $X = -\dfrac{\omega_{OP}}{2\omega_{OX}} = -\dfrac{168.6 \cdot 10^{-6}}{2 \cdot 33 \cdot 10^{-6}} = -2.5545t$.

Now, each beam is analyzed separately. Analysis of beam AB is performed in Table 3.24 and analysis of beam CD is performed in Table 3.25. We shall obtain only the moments in both beams and build a moment diagram for beam AB using the second equation from the system of equations 3.2. The moment diagram is shown in Figure 3.26.

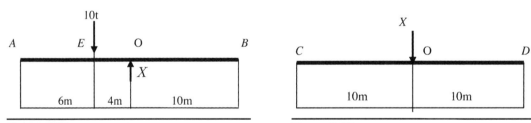

Figure 3.24

Figure 3.25

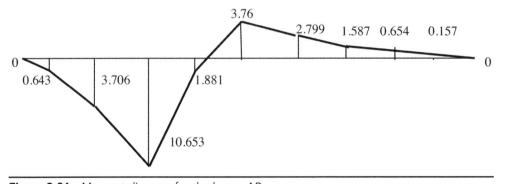

Figure 3.26   Moment diagram for the beam AB

3.5 Analysis of Pin-Connected Beams

Tables 3.1–3.14 can be used for analysis of a system of pin-connected beams. Let us investigate a system of two beams shown in Figure 3.27. The system has three vertical loads and one moment. The left end is free supported and the right end is pin-supported. Analysis of this beam is performed as follows:

By assuming that the forces of interaction at point 2 are equal to X_1 and the reaction at point 3 is equal to X_2, the given system can be divided into two beams as shown in Figure 3.28. Taking into account that the settlements of both beams, a and b, at point 2 are the same and the settlement of the beam at point 3 is equal to zero, the following two equations are written:

$$\left(\omega_{21}^{1-2} + \omega_{21}^{2-3} \right) X_1 + \omega_{22}^{2-3} X_2 + \omega_{2P_1}^{1-2} + \omega_{2P_2}^{1-2} + \omega_{2P_3}^{2-3} + \omega_{2M}^{2-3} = 0$$

$$\omega_{3}^{2-3} X_1 + \omega_{32}^{2-3} X_2 + \omega_{3M}^{2-3} + \omega_{3P_1}^{2-3} = 0 \qquad (3.17)$$

In this system, the first equation shows that the settlement of the beam a at point 2 is equal to the settlement of the beam b at the same point. The second equation shows that the total settlement of the beam at point 3 is equal to zero. In this system, ω_{ij}^{m} is the settlement of the beam m at point i due to reaction $X_j = 1$ at point j and ω_{iG}^{m} is the settlement of the beam m at point i due to the given load G applied to the beam.

All settlements, due to the given loads, as well as settlements due to reactions $X_1 = 1$ and $X_2 = 1$, are found from Tables 3.1–3.14. Since settlements cannot be obtained directly from Tables 3.1–3.14, the soil pressure is obtained instead, and settlements are found from this equation: $\delta_i = \dfrac{p_i}{kb}$ where p_i is the soil pressure, k is the modulus of subgrade reaction, and b is the width of the beam. By solving the system 3.17, X_1 and X_2 can be found, and final analysis of the beam performed. Example 3.9 presented below illustrates practical application of the described method.

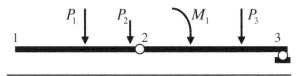

Figure 3.27

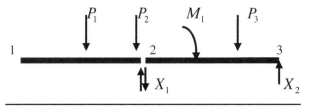

Figure 3.28

Example 3.9

Given: Two pin-connected beams have two vertical loads $P_1 = 100,000\text{lb}$ and $P_2 = 75,000\text{lb}$, as shown in Figure 3.29. The modulus of elasticity of both beams is $E = 200 \cdot 10^6 \dfrac{\text{lb}}{\text{ft}^2}$ $l_1 = 30\text{ft}$ $l_2 = 18\text{f}$ $b_{1-2} = 4\text{ft}$ $b_{2-3} = 3\text{ft}$. The modulus of subgrade reaction is $k = 200,000\dfrac{\text{lb}}{\text{ft}^3}$ and the moments of inertia of the beams 1–2 and 2–3 are equal to $I_{1-2} = 20\text{ft}^4$ and $I_{2-1} = 15\text{ft}^4$, successively. Find the soil pressure for the beams 1–2 and 2–3 at point 2.

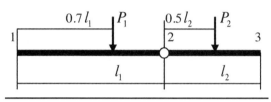

Figure 3.29

Solution:

By replacing the hinge with the forces of interaction X, we have two beams supported on the elastic foundation shown in Figure 3.30. By taking into account that the soil pressure at the right end of the beam 1–2 is equal to the soil pressure at the left end of the beam 2–3, the following equation is written:

$$p_{(1-2)2P_1} + p_{(1-2)2X} = p_{(2-3)2P_2} + p_{(2-3)2X} \tag{3.18}$$

In equation 3.18, the first term shows the soil pressure at the right end of the beam 1–2, due to the given load P_1; the second term shows the soil pressure at the same point due to unknown reaction X. The right part of this equation shows the total soil pressure at the same point 2, due to the given load P_2 and unknown reaction X applied to the beam 2–3.

Parameters of the beam 1–2 are equal to $m_{1-2} = \sqrt[4]{\dfrac{(k_0)_{1-2}b_{1-2}}{4EI_{1-2}}} = \sqrt[4]{\dfrac{400,000 \cdot 4}{4 \cdot 200 \cdot 10^6 \cdot 20}}$ $= 0.10$ $\lambda_{1-2} = m_{1-2} \cdot l_1 = 0.1 \cdot 30 = 3$.

Parameters of the beam 2–3 are equal to $m_{2-1} = \sqrt[4]{\dfrac{(k_0)_{2-3}b_{2-3}}{4EI_{2-3}}} = \sqrt[4]{\dfrac{600,000 \cdot 3}{4 \cdot 200 \cdot 10^6 \cdot 15}}$ $= 0.11$ $\lambda_{2-3} = m_{2-3} \cdot l_2 = 0.11 \cdot 18 = 1.98$.

Equation 3.18 can be rewritten as follows:

$$p_{(1-2)2P_1} - \overline{p}_{(1-2)2}X = p_{(2-3)2P_2} + \overline{p}_{(2-3)2}X \tag{3.19}$$

From 3.19 we find:

$$X = \frac{p_{(1-2)2P_1} - p_{(2-3)2P_2}}{\overline{p}_{(2-3)2} + \overline{p}_{(1-2)2}} \tag{3.20}$$

In 3.19 and 3.20, $\overline{p}_{(1-2)2}$ and $\overline{p}_{(2-3)2}$ are the soil pressures at point 2 at beams 1–2 and 2–3 due to $X = 1$, respectively. From Table 3.3 for $\lambda = 3$ we find $\overline{p} = 1.5564$, and using the first equation from equations 3.2 we find $p_{(1-2)2P_1} = \overline{p}\dfrac{P_1}{l_1} = 1.5564\dfrac{100,000}{30} = 5,188\text{lb/ft}$.

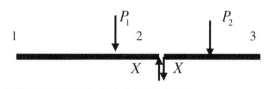

Figure 3.30

The soil pressure at the same point 2, due to reaction $X = 1$, is equal to:

$$\overline{p}_{(1-2)2} = 6.0163\frac{-1}{30} = -0.20\text{lb/ft}$$

Using Table 3.2 for $\lambda = 2$ we find $\overline{p} = 0.7368$, and using the first equation from 3.2 we obtain the soil pressure at the left end of the beam 2–3:

$$p_{(1-2)2P_1} = \overline{p}\frac{P_2}{l_{2-3}} = 0.7368\frac{75,000}{18} = 3,070\text{lb/ft}$$

$$\overline{p}_{(1-2)2P_2} = 4.5441\frac{1}{18} = 0.252\text{lb/ft}$$

Now using equation 3.22 we can find:

$$X = \frac{5,188 - 3,070}{0.252 + 0.20} = 4,685\text{lb}$$

By introducing X into equation 3.18, the equilibrium of the soil pressures at point 2 is checked:

$$5,188 - 0.20 \cdot 4,685 = 3,070 + 0.252 \cdot 4,685 \quad \text{or} \quad 4.251\frac{\text{lb}}{\text{ft}} = 4.251\frac{\text{lb}}{\text{ft}}$$

In other words, the soil pressure at the right end of the beam 1–2 is equal to the soil pressure at the left end of the beam 2–3. Now, each beam 1–2 and 2–3 can be analyzed separately. Beam 1–2 is shown in Figure 3.31 and beam 2–3 is shown in Figure 3.32. Analysis of the beam 1–2 is performed in Tables 3.26–3.28. Analysis of the beam is performed twice: under the load P_1 and under the reaction X. The total soil pressure for the beam 1–2 is estimated and shown in Table 3.28. Analysis of the beam 2–3 is not shown but can be performed analogously. It is important to mention that although

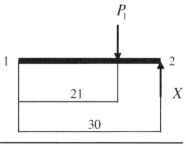

Figure 3.31

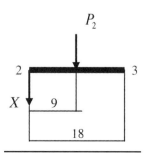

Figure 3.32

the soil pressures for both beams at point 2 are the same, the soil pressures per one square unit are different. The soil pressure at the right end of the beam 1–2 is equal to $\frac{4,250}{1\cdot4} = 1,062.5\frac{lb}{ft^2}$, while the soil pressure at the left end of the beam 2–3 is different and equal to $\frac{4,250}{1\cdot3} = 1,416.7\frac{lb}{ft^2}$.

3.6 Analysis of Continuous Beams

The method used for analysis of beams, with one or two end supports, can also be applied to analysis of continuous beams supported on a series of individual supports. Analysis of continuous beams is not different, in principle, from analysis of simple, one-span beams. The number of unknowns is equal to the number of supports or to the total number of restraints. The beam shown in Figure 3.33 is analyzed by replacing all supports with unknown reactions, as shown in Figure 3.34. Taking into account that the settlements at points 1, 2, and 3 are equal to zero, and using the method of forces, the following system of equations is written:

$$\left.\begin{array}{l} \omega_{11}X_1 + \omega_{12}X_2 + \omega_{13}X_3 + \omega_{1P} = 0 \\ \omega_{21}X_1 + \omega_{22}X_2 + \omega_{23}X_3 + \omega_{2P} = 0 \\ \omega_{31}X_1 + \omega_{32}X_2 + \omega_{33}X_3 + \omega_{3P} = 0 \end{array}\right\} \tag{3.21}$$

In this system of equations, ω_{ij} is the settlement of the beam at point i due to reaction $X_j = 1$ at any point j, and ω_{iP} is the settlement at the same point due to all given loads.

By solving the system 3.21, reactions X_1, X_2, and X_3 can be found. Now, analysis of the beam can be performed. Analysis of the beam loaded with each load is performed separately. The final results of analysis are obtained by simple superposition. Various types of continuous beams supported on Winkler foundation can be analyzed by using the method of forces and Tables 3.1–3.14. The same method is used for analysis

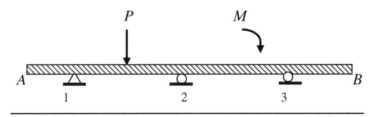

Figure 3.33

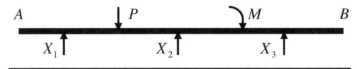

Figure 3.34

of beams on Winkler foundation with non-yielding and elastic supports. For example, if the same beam is supported at point 2 on a spring with rigidity equal to C_2, as shown in Figures 3.35 and 3.36, analysis of this beam can be performed using the same method. However, unlike the analysis explained above, in this case the settlement of the spring at point 2 is not equal to zero. By replacing all three supports with unknown reactions X_1, X_2, and X_3 the following system of equations is written:

$$\left. \begin{aligned} \omega_{11}X_1 + \omega_{12}X_2 + \omega_{13}X_3 + \omega_{1PM} &= 0 \\ \omega_{21}X_1 + \left(\omega_{22} + \frac{1}{C_2}\right)X_2 + \omega_{23}X_3 + \omega_{2PM} &= 0 \\ \omega_{31}X_1 + \omega_{32}X_2 + \omega_{33}X_3 + \omega_{3PM} &= 0 \end{aligned} \right\} \qquad (3.22)$$

By solving this system of equations, all reactions are found and final analysis of the beam with the given loads and obtained reactions is performed.

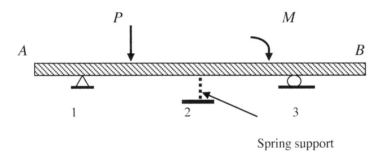

Figure 3.35

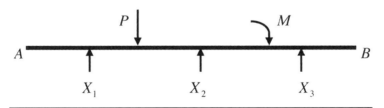

Figure 3.36

3.7 Analysis of Stepped Beams

Analysis of stepped beams can be performed using the method of initial parameters described earlier in the Chapter 2, but Tables 3.1–3.14 also allow performing the same analysis. For example, a stepped beam shown in Figure 3.37 consists of two simple beams, *a* and *b*. Taking into account that the settlements and rotations of both beams at the point of contact are the same, the following two equations are written:

$$\left. \begin{aligned} \left(\omega_{a1} + \omega_{b1}\right)X_1 + \left(\omega_{a2} + \omega_{b2}\right)X_2 + \omega_{aP} + \omega_{bP} &= 0 \\ \left(\varphi_{a1} + \varphi_{b1}\right)X_1 + \left(\varphi_{a2} + \varphi_{b2}\right)X_2 + \varphi_{aP} + \varphi_{bP} &= 0 \end{aligned} \right\} \qquad (3.23)$$

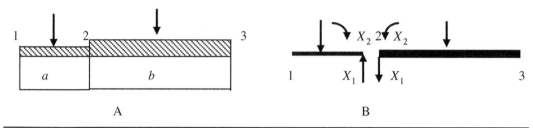

Figure 3.37

In this system of equations:

$(\omega_{a1} + \omega_{b1})$ is the total vertical deflection of both beams due to $X_1 = 1$

$(\omega_{a2} + \omega_{b2})$ is the total vertical deflection of both beams due to $X_2 = 1$

$(\varphi_{a1} + \varphi_{b1})$ is the total rotation of both beams due to $X_1 = 1$

$(\varphi_{a2} + \varphi_{b2})$ is the total rotation of both beams due to $X_2 = 1$

The last two terms in both equations show the settlements and rotations due to the given loads applied to both beams. As we can see, the system of equations 3.23 requires not only the values of settlements due to one-unit loads, but also rotations due to these loads. Tables 3.1–3.14 do not contain such types of information. Simple equations and Tables 3.29 and 3.30, developed by Korenev (1962), allow obtaining not only settlements but rotations as well. Table 3.29 contains equations for obtaining settlements and rotations of finite beams free supported on Winkler foundation, and with concentrated vertical loads $P = 1$ and moments $M = 1$ applied to the left and right ends of the beam. Equations in Table 3.29 contain coefficients ω_P^0, ω_M^0, φ_P^0 and φ_M^0. Numerical values of these coefficients are given in Table 3.30 for various $\lambda = lm$ $(0.8 < \lambda < 5.0)$, where l is the length of the beam and coefficient $m = \sqrt[4]{\dfrac{kb}{4EI}}$.

When $\lambda > 5.0$, a vertical load or moment applied to one of the ends does not affect the opposite end. Therefore, beams with $\lambda > 5.0$ are not included in Table 3.30. All signs of the numerical values in Table 3.30 reflect the actual directions of forces and moments. Vertical loads are positive when directed downward; moments are positive when they act clockwise. Linear deflections in the upward direction are positive and clockwise rotations are positive.

Point 0 can be located at the left or at the right ends of the beam. In Table 3.30, values located in columns (point 0) are used to obtain deflections produced by loads applied at the same point 0. Values located in columns (point l) are used to obtain deflections produced by loads applied to the opposite end of the beam. A numerical example shown below illustrates the application of both Tables 3.29 and 3.30 to practical analysis.

Example 3.10

A simple beam is loaded with two vertical loads and two moments, as shown in Figure 3.38. $P_1 = 50tm$ $P_2 = 70tm$ $M_1 = 10tm$ $M_2 = 15tm$ $l = 16m$. $E = 2 \cdot 10^6 t/m^2$ $b = 3m$

$h = 1.2m$ $EI = 2,000,000 \cdot \dfrac{3 \cdot 1.2^3}{12} = 864,000tm^2.$

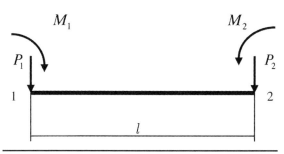

Figure 3.38

Find the settlements and rotations of the beam at points 1 and 2 due to each load applied to the beam.

Solution:

Find parameter $m = \sqrt[4]{\dfrac{kb}{4EI}} = \sqrt[4]{\dfrac{30 \cdot 3.0}{4 \cdot 864,000}} = 0.127\dfrac{1}{m}$ and $ml = 0.127 \cdot 16 = 2.03$.

Using Tables 3.29 and 3.30 we find:

Settlement at point 1 due to P_1 $\omega_{1P_2} = -\omega_P^0 \dfrac{m}{kb} P_2 = -(-0.800)\dfrac{0.127}{30 \cdot 3} \cdot 70 = 0.79\text{cm}$

Settlement at point 1 due to P_2 $\omega_{1P_2} = -\omega_P^0 \dfrac{m}{kb} P_2 = -(-0.800)\dfrac{0.127}{30 \cdot 3} \cdot 70 = 0.79\text{cm}$

Settlement at point 1 due to M_1 $\omega_{1M_1} = \omega_M^0 \dfrac{m^2}{kb} M_1 = 2.268 \cdot \dfrac{0.127^2}{30 \cdot 3} \cdot 10 = 0.0406\text{cm}$

Settlement at point 1 due to M_2 $\omega_{1M_2} = -\omega_M^0 \dfrac{m^2}{kb} M_2 = -(-1.070) \cdot \dfrac{0.127^2}{30 \cdot 3} \cdot (-15)$
$$= -0.02876\text{cm}$$

Rotation at point 1 due to P_1 $\varphi_{1P_1} = -\varphi_P^0 \dfrac{m^2}{kb} P_1 = -2.268 \cdot \dfrac{0.127^2}{30 \cdot 3} \cdot 50 = -2.03 \cdot 10^{-3}\text{rad}$

Rotation at point 1 due to P_2 $\varphi_{1P_2} = \varphi_P^0 \dfrac{m^2}{kb} P_2 = 1.070 \cdot \dfrac{0.127^2}{30 \cdot 3} \cdot 70 = 1.3422 \cdot 10^{-3}\text{rad}$

Rotation at point 1 due to M_1 $\varphi_{1M_1} = \varphi_M^0 \dfrac{m^3}{kb} M_1 = 4.305 \cdot \dfrac{0.127^3}{30 \cdot 3} \cdot 10 = 0.098 \cdot 10^{-3}\text{rad}$

Rotation at point 1 due to M_2 $\varphi_{1M_2} = \varphi_M^0 \dfrac{m^3}{kb} M_2 = -0.620 \cdot \dfrac{0.127^3}{30 \cdot 3} \cdot 15$
$$= -0.0211 \cdot 10^{-3}\text{rad}$$

Settlement at point 2 due to P_1 $\omega_{2P_1} = -\omega_P^0 \dfrac{m}{kb} P_1 = -(-0.800) \cdot \dfrac{0.127}{30 \cdot 3} \cdot 50$
$$= 0.56444\text{cm}$$

Settlement at point 2 due to P_2 $\omega_{2P_2} = -\omega_P^0 \dfrac{m}{kb} P_2 = -2.275\dfrac{0.127}{30 \cdot 3} \cdot 70 = -2.247\text{cm}$

Settlement at point 2 due to M_1 $\omega_{2M_1} = \omega_M^0 \dfrac{m}{kb} M_1 = -1.070 \cdot \dfrac{0.127^2}{30 \cdot 3} \cdot 10$

$$= -0.01917\,\text{cm}$$

Settlement at point 2 due to M_2 $\omega_{2M_2} = -\omega_M^0 \dfrac{m^2}{kb} M_2 = -2.268 \cdot \dfrac{0.127^2}{30 \cdot 3} \cdot (-15)$

$$= 0.061\,\text{cm}$$

Rotation at point 2 due to P_1 $\varphi_{2P_1} = -\varphi_P^0 \dfrac{m^2}{kb} P_1 = -1.070 \cdot \dfrac{0.127^2}{30 \cdot 3} \cdot 50$

$$= -0.959 \cdot 10^{-3}\,\text{rad}$$

Rotation at point 2 due to P_2 $\varphi_{2P_2} = \varphi_P^0 \dfrac{m^2}{kb} P_2 = 2.268 \cdot \dfrac{0.127^2}{30 \cdot 3} \cdot 70 = 2.845 \cdot 10^{-3}\,\text{rad}$

Rotation at point 2 due to M_1 $\varphi_{2M_1} = \varphi_M^0 \dfrac{m^3}{kb} M_1 = 0.620 \cdot \dfrac{0.127^3}{30 \cdot 3} \cdot 10$

$$= 0.0141 \cdot 10^{-3}\,\text{rad}$$

Rotation at point 2 due to M_2 $\varphi_{2M_2} = \varphi_M^0 \dfrac{m^3}{kb} M_2 = 4.305 \cdot \dfrac{0.127^3}{30 \cdot 3} \cdot (-15)$

$$= -0.1469 \cdot 10^{-3}\,\text{rad}$$

Deflections at the ends of the beam, due to loads applied between both ends 3.24–3.29, are shown below. They are obtained using the method of initial parameters discussed in Chapter 2 for three types of loads: vertical concentrated load, concentrated moment, and uniformly distributed load. A vertical concentrated load P shown in Figure 3.39 applied to the beam, will produce a settlement ω_{left}^P and rotation φ_{left}^P at the left end equal to:

$$\omega_{\text{left}}^P = \frac{Pm}{k} \cdot \frac{D_l A_{l-u} - C_l B_{l-u}}{C_l^2 - B_l D_l} \qquad \varphi_{\text{left}}^P = \frac{Pm^2}{k} \cdot \frac{B_l B_{l-u} - C_l A_{l-u}}{C_l^2 - B_l D_l} \tag{3.24}$$

The settlement and rotation at the right end of the beam are equal to:

$$\omega_{\text{right}}^P = \omega_{\text{left}}^P A_l + \varphi_{\text{left}}^P \frac{B_l}{m} + \frac{PD_{l-u}}{m^3 EI} \qquad \varphi_{\text{right}}^P = \varphi_{\text{left}}^P A_l - 4m\omega_{\text{left}}^P + \frac{PC_{l-u}}{m^2 EI} \tag{3.25}$$

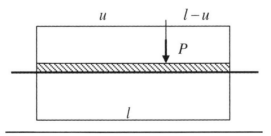

Figure 3.39

If moment M is applied to the beam, as shown in Figure 3.40, the settlement of the beam and rotation at the left end are equal to:

$$\omega^M_{left} = -\frac{Mm^2}{k} \cdot \frac{4D_l D_{l-u} + C_l A_{l-u}}{C_l^2 - B_l D_l} \qquad \varphi^M_{left} = \frac{Mm^3}{k} \cdot \frac{B_l A_{l-u} + 4C_l D_{l-u}}{C_l^2 - B_l D_l} \qquad (3.26)$$

The settlement and rotation at the right end are equal to:

$$\omega^M_{right} = \omega^M_{left} A_l + \varphi^M_{left} \frac{B_l}{m} - \frac{MC_{l-u}}{m^2 EI} \qquad \varphi^M_{right} = \varphi^M_{left} A_l - 4m\omega^M_{left} D_l - \frac{MB_{l-u}}{mEI} \qquad (3.27)$$

If a uniformly distributed load is applied to the beam, as shown in Figure 3.41, the settlement and the rotation at the left end are equal:

$$\left. \begin{aligned}
\omega^P_{left} &= -\frac{p}{k} \cdot \frac{\left(C_{l-c} - C_{l-d}\right)C_l - \left(B_{l-c} - B_{l-d}\right)D_l}{C_l^2 - B_l D_l} \\
\varphi^P_{left} &= -\frac{pm}{k} \cdot \frac{\left(B_{l-c} - B_{l-d}\right)C_l - \left(C_{l-c} - C_{l-d}\right)B_l}{C_l^2 - B_l D_l}
\end{aligned} \right\} \qquad (3.28)$$

The settlement and rotation at the right end of the beam are equal:

$$\left. \begin{aligned}
\omega^P_{right} &= \omega^P_{left} A_l + \varphi^P_{left} B_l + \frac{p}{4m^4 EI}\left(A_{l-d} - A_{l-c}\right) \\
\varphi^P_{right} &= \varphi^P_{left} A_l - 4m\omega^P_{left} D_l + \frac{p}{m^3 EI}\left(D_{l-c} - D_{l-d}\right)
\end{aligned} \right\} \qquad (3.29)$$

Equations for obtaining functions A, B, C, and D and Table 2.25 are given in Chapter 2.

Using Tables 3.1–3.14 and the equations shown above, any beam supported on Winkler foundation with various end conditions and various loads at any point can be analyzed.

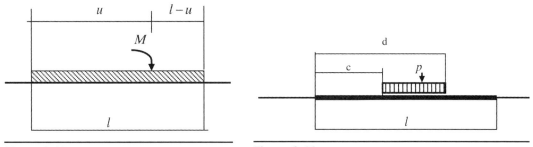

Figure 3.40   **Figure 3.41**

Example 3.11

A simple beam, free supported on Winkler foundation is shown in Figure 3.42. The length of the beam is $l = 118.1$in, the width of the beam $b = 7.88$in, $E = 1,419,355.0$lb/in², $I = 240.25$in, and the modulus of subgrade reaction $k = 180.33$lb/in³. The beam is loaded with one vertical concentrated load $P = 1$ applied to the left end of the beam. Find the settlement and the slope at the left end.

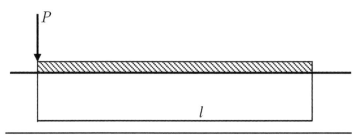

Figure 3.42

1. Find parameter $m = \sqrt[4]{\dfrac{k_0 b}{4EI}} = \sqrt[4]{\dfrac{180.33 \cdot 7.88}{4 \cdot 1,419,355 \cdot 240.25}} = 0.031948 \dfrac{1}{in}.$

2. Find parameter $\lambda = lm = 0.031948 \cdot 118.11 = 3.773.$
3. From Tables 3.29 and 3.30, use the following equations for settlements and rotations for $\lambda = 3.773 \approx 3.8$:

$$\omega_P = -\omega_P^0 \frac{m}{k_0 b} = -2.002 \cdot \frac{0.031948}{180.33 \cdot 7.88} = -10^{-4} \cdot 0.45 \frac{in}{lb}$$

$$\varphi_P = -\varphi_P^0 \frac{m^2}{k_0 b} = -2.003 \frac{0.031948^2}{180.33 \cdot 7.88} = -10^{-7} 14 \frac{1}{lb}$$

Example 3.12

We will analyze the beam from Example 3.11 shown in Figure 3.43, applying a vertical load $P = 1,000lb$ at the center of the beam $l = 118.11in$, the width of the beam $b = 7.88in$, $E = 1,419,355.0lb/in^2$, $I = 240.25in^4$, the modulus of subgrade reaction $k = 180.33lb/in^3$ $m = 0.031948$, and $\lambda = lm \approx 3.8$. Find the deflections at the left end of the beam.

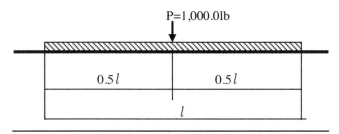

P=1,000.0lb

0.5*l*   0.5*l*

l

Figure 3.43

Solution:

Deflections at the left end of the beam are derived from equations 3.24:

$$\omega_{left}^P = \frac{Pm}{k} \cdot \frac{D_l A_{l-u} - C_l B_{l-u}}{C_l^2 - B_l D_l}, \quad \varphi_{left}^P = \frac{Pm^2}{k} \cdot \frac{B_l B_{l-u} - C_l A_{l-u}}{C_l^2 - B_l D_l}$$

Functions A, B, C, and D are found from Table 2.25 in Chapter 2.

In these formulae $\xi = \lambda = lm$.

Functions A, B, C, and D for the total beam are equal:

$$A_l = -17.6875, \ B_l = -15.67605, \ C_l = -6.8343, \ D_l = 0.9969$$

The same functions for half of the length of the beam when $\xi = \lambda = 0.5lm = 1.9$.

$$A_{0.5l} = -1.1049, \ B_{0.5l} = 1.0888, \ C_{0.5l} = 1.54635, \ D_{0.5l} = 1.0727$$

Now, deflections at the left end of the beam can be found:

$$\omega_{left}^P = \frac{Pm}{k} \cdot \frac{D_l A_{l-u} - C_l B_{l-u}}{C_l^2 - B_l D_l}$$

$$= \frac{1000 \cdot 0.031948}{180.33 \cdot 7.88} \cdot \frac{0.9969 \cdot (-1.1049) - (-6.8343) \cdot 1.0888}{(-6.8343)^2 - (-15.67605) \cdot 0.9969}$$

$$\varphi_{left}^P = \frac{Pm^2}{k} \cdot \frac{B_l B_{l-u} - C_l A_{l-u}}{C_l^2 - B_l D_l}$$

$$= \frac{1000 \cdot 0.031948^2}{180.33 \cdot 7.88} \cdot \frac{(-15.67605) 1.0888 - (-6.8343) \cdot (-1.1049)}{(-6.8343)^2 - (-15.67605) \cdot 0.9969}$$

$$\omega_{left}^P = 2.2865 \cdot 10^{-3} \text{in}, \quad \varphi_{left}^P = -0.2836 \cdot 10^{-3} \text{rad}$$

Deflections of the beam at the right end can be found using equations 3.25. Taking into account the symmetry of the beam, deflections at the right end of the beam are equal:

$$\omega_{left}^P = 2.2865 \cdot 10^{-3} \text{in}, \quad \varphi_{left}^P = -0.2836 \cdot 10^{-3} \text{rad}$$

Now, using the equations obtained above, Tables 3.1–3.14, and the formulae from these Tables, any beam with any boundary conditions can be analyzed. In order to illustrate analysis of beams with fixed and guided ends, some numerical examples are solved below.

Example 3.13

Let us analyze a beam shown in Figure 3.44. All data are the same as in the previous example except the boundary conditions. The given beam is fixed at the left end and free supported at the right end. Find reactions at the left end of the beam.

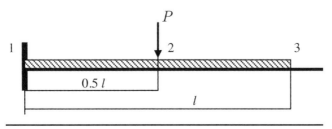

Figure 3.44

Solution:

Analysis is performed in the following order: the two restraints at the left end are removed and replaced with two unknown reactions. The restraint against the vertical settlement is replaced with a vertical reaction X_1, and the restraint against rotation is replaced with a moment X_2. So the given beam with the left fixed and right free supported ends, loaded with a vertical load P, is replaced with a beam with both free ends, loaded with the same load P, and two unknown reactions X_1 and X_2, as shown in Figure 3.45. Unknown reactions X_1 and X_2 can be found from the following system of equations:

$$\left.\begin{array}{l} \omega_{11}X_1 + \omega_{12}X_2 + \omega_{1P} = 0 \\ \varphi_{21}X_1 + \varphi_{22}X_2 + \varphi_{2P} = 0 \end{array}\right\} \tag{3.30}$$

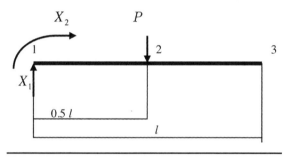

Figure 3.45

Deflections of the left end of the beam, due to reaction $X_1 = -1$ are equal:

$$\omega_{11} = \omega_1^0 \frac{m}{k_0 b} = 2.002 \cdot \frac{0.031948}{180.33 \cdot 7.88} = 0.45 \cdot 10^{-4} \frac{in}{lb}$$

$$\varphi_{21} = \varphi_2^0 \frac{m^2}{k_0 b} = 2.003 \cdot \frac{0.031948^2}{180.33 \cdot 7.88} = 14 \cdot 10^{-7} \frac{1}{lb}$$

$$\omega_{12} = \omega_2^0 \frac{m^2}{kb} = 2.003 \cdot \frac{0.031948^2}{180.33 \cdot 7.88} = 14 \cdot 10^{-7} \frac{1}{lb}$$

$$\varphi_{22} = \varphi_2^0 \frac{m^3}{kb} = 4.011 \cdot \frac{0.031948^3}{180.33 \cdot 7.88} = 0.92 \cdot 10^{-7} \frac{1}{lb \times in}$$

Deflections of the left end of the beam, due to the given load $P = 1,000lb$ are equal:

$$\omega_P = 22,865 \cdot 10^{-7} \quad \varphi_P = 22,836 \cdot 10^{-7}$$

By introducing all deflections into the system 3.30, we have:

$$\left.\begin{array}{l} 450X_1 + 14X_2 + 22,865 = 0 \\ 14X_1 + 0.92X_2 + 2,836 = 0 \end{array}\right\} \tag{3.31}$$

By solving this system of equations, we find $X_1 = 85.633943lb$ and $lb \times in$. The further analysis of the beam is simple. Using Tables 3.1–3.14, we perform analysis of the beam due to the load P and reactions X_1 and X_2 found above.

The method of forces used above for analysis of a simple stepped beam can be applied to analysis of complex continuous beams with boundary conditions as shown in Figures 3.46 and 3.47. The same method can also be applied to analysis of beams with totally rigid elements.

This type of foundation is usually used in multistory buildings where areas of intersection of the foundations and columns work as totally rigid elements. These areas are usually large and can significantly affect the final results of analysis. A beam with totally rigid elements is shown in Figure 3.48.

Analysis of such type of foundations, in principle, is not different from analysis of the stepped foundations discussed previously. The only difference is the method of obtaining deflections of the totally rigid elements BC and DE. Taking into account that these two elements are totally rigid, it is recommended to obtain the settlements and rotations using the following, simple equations:

$$\varphi = \frac{\Sigma M}{k_\varphi I_x} \tag{3.32}$$

$$y = \frac{\Sigma N}{kba} \tag{3.33}$$

The totally rigid elements are shown in Figures 3.49 and 3.50. The total settlement of that element at its edges can be found from the following equation:

$$Y = y \pm \varphi \frac{a}{2} \tag{3.34}$$

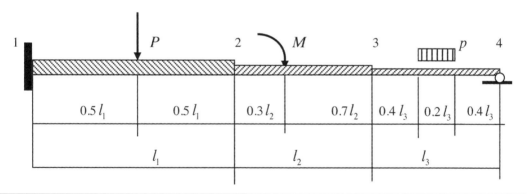

Figure 3.46

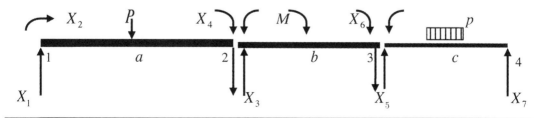

Figure 3.47

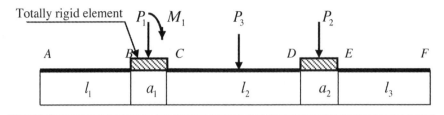

Figure 3.48

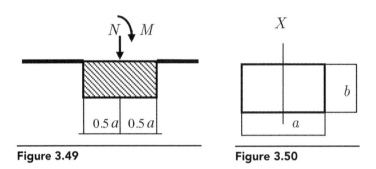

Figure 3.49   **Figure 3.50**

In this equation:

$\sum M$ is the total moment applied to the totally rigid element of the foundation

$\sum N$ is the total vertical load applied to the same element of the foundation

k_φ is the modulus of subgrade reaction of the soil of the same element under rotation

I_x is the moment of inertia of the same element of the foundation

Analysis of a beam with totally rigid elements, shown in Figure 3.48, is explained next. The beam is divided into five simple beams: three elastic beams, *AB*, *CD*, and *EF*, and two totally rigid beams, *BC* and *DE*, as shown in Figure 3.51. Two totally rigid beams have the given loads P_1, P_2, and moment M_1. Element *CD* is loaded with a vertical load P_3. Each beam is loaded with the given loads, unknown forces, and moments of interaction $X_1, X_2, \ldots, X_8$.

The total number of unknowns is equal to eight. By solving this system of equations, we find all forces and moments of interaction $X_1, X_2, \ldots, X_8$ between the elements

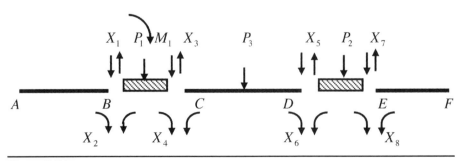

Figure 3.51

of the given beam. By applying these forces and moments to beams *AB*, *CD*, and *EF*, the soil pressure, the shear forces, and moments can be found at any point of these beams. Analysis of the same beam using the stiffness method introduced in Chapter 2 requires solution of a system with only four unknowns or two times less than the method of forces. However, equations of the stiffness method are much more complex compared to the method of forces. Both methods can be used for hand analysis, but only for small beams that require the solution of a small system of linear equations.

Example 3.14

The given beam is shown in Figure 3.52. All data are shown below:

$$l_{AB} = 200\text{cm} \quad l_{BC} = 300\text{cm} \quad b = 20\text{cm} \quad k = 5\text{kg/cm}^3 \quad I = 10^4\text{cm}^4$$

$$EI_{AB} = 2 \cdot 10^9\text{kgcm}^2 \quad EI_{BC} = 10^9\text{kgcm}^2 \quad P = 30,000\text{kg}$$

Solution:

1. Divide the given stepped beam into two beams, as shown in Figure 3.53. Forces and moments of interaction are shown in Figure 3.54.
2. Using the method of forces, the following system of equations is written:

$$\left.\begin{array}{l} \omega_{11}X_1 + \omega_{12}X_2 + \omega_{1P} = 0 \\ \varphi_{21}X_1 + \varphi_{22}X_2 + \varphi_{2P} = 0 \end{array}\right\} \qquad (3.35)$$

where ω_{11} and ω_{12} are the total vertical deflection at point *B* due to forces X_1 and X_2 equal to one unit, respectively, φ_{21} and φ_{22} are total rotations at the node *B* due to forces X_1 and X_2 equal to one unit, respectively, ω_{1P} and φ_{2P} are settlement and rotation at point *B* due to the given load *P*.

3. Find parameters $m_{AB} = \sqrt[4]{\dfrac{100}{4 \cdot 2 \cdot 10^9}} = 0.0106$; $m_{BC} = \sqrt[4]{\dfrac{100}{4 \cdot 10^9}} = 0.0125$; $l_{AB}m$

= $200 \cdot 0.0106 = 2.12$; $l_{BC}m = 300 \cdot 0.0125 = 3.75$.

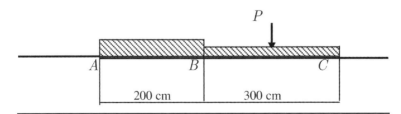

Figure 3.52

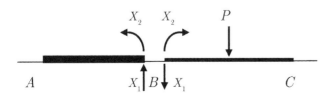

Figure 3.53

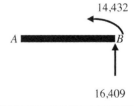

Figure 3.54

4. Now, using equations from Table 3.29 and numerical values from Table 3.30, find deflections due to one-unit vertical loads and moments applied to the ends of both beams, AB and BC, at point B:

$$\omega_{11} = \left(\omega_{11}^{AB} + \omega_{11}^{BC}\right) = \left(\omega_{1}^{0AB}\frac{m_{AB}}{kb} - \omega_{1}^{0BC}\frac{m_{BC}}{kb}\right) = 2.169\frac{0.0106}{100} - 2.002\frac{0.0125}{100} = -203\cdot10^{-7}\,\mathrm{cm}$$

$$\omega_{12} = \left(\omega_{12}^{AB} + \omega_{12}^{BC}\right) = \left(-\omega_{12}^{0AB}\frac{m_{AB}^{2}}{kb} - \omega_{12}^{0BC}\frac{m_{BC}^{2}}{kb}\right) = -2.136\frac{0.0106^{2}}{100} + 2.003\frac{0.0125^{2}}{100} = -7\cdot10^{-7}\,\mathrm{cm}$$

$$\varphi_{22} = \left(\varphi_{22}^{AB} + \varphi_{22}^{BC}\right) = \left(\varphi_{22}^{0AB}\frac{m_{AB}^{3}}{kb} + \varphi_{22}^{0BC}\frac{m_{BC}^{3}}{kb}\right) = 4.140\frac{0.0106^{3}}{100} + 4.011\frac{0.0125^{3}}{100} = 1.27\cdot10^{-7}\,\mathrm{rad}$$

$$\varphi_{21} = \left(\varphi_{21}^{AB} + \varphi_{21}^{BC}\right) = \left(-\varphi_{21}^{0AB}\frac{m_{AB}^{2}}{kb} - \varphi_{21}^{0BC}\frac{m_{BC}^{2}}{kb}\right) = -2.136\frac{0.0106^{2}}{100} + 2.003\frac{0.0125^{2}}{100} = -7\cdot10^{-7}\,\mathrm{rad}$$

Deflections of the beam BC due to load P are found using equations 3.24:

$$\omega_{B,P} = \frac{Pm}{kb}\cdot\frac{D_{l_2}A_{0.5l_2} - C_{l_2}B_{0.5l_2}}{C_{l_2}^{2} - B_{l_2}D_{l_2}}$$

$$= \frac{30{,}000\cdot0.0106}{100}\cdot\frac{0.9969\cdot(-1.1049) - (-6.8343)\cdot1.0888}{(-6.8343)^{2} - (-15.67605)\cdot0.9969} = 0.323\,\mathrm{cm}$$

$$\varphi_{B,P} = \frac{Pm^{2}}{kb}\cdot\frac{B_{l_2}B_{0.5l_2} - C_{l_2}A_{0.5l_2}}{C_{l_2}^{2} - B_{l_2}D_{l_2}}$$

$$= \frac{30{,}000\cdot(0.0106)^{2}}{100}\cdot\frac{(-15.67605)\,1.0888 - (-6.8343)\cdot(-1.1049)}{62.33511} = 0.1332\,\mathrm{rad}$$

By introducing all numerical values into the system 3.35 we have:

$$\left.\begin{array}{r} -203\cdot10^{-7}X_{1} + 7\cdot10^{-7}X_{2} + 0.323 = 0 \\ 7\cdot10^{-7}X_{1} + 1.27\cdot10^{-7}X_{2} - 0.01332 = 0 \end{array}\right\} \tag{3.36}$$

From 3.36 we find $X_{1} = 16{,}409\,\mathrm{kg}$ and $X_{2} = 14{,}432\,\mathrm{kgm}$.

Now, each part of the stepped beam can be analyzed separately: beam AB loaded with two reactions X_{1}, X_{2}, as shown in Figure 3.54, and beam BC loaded with the same reactions acting in opposite directions and given load P, as shown in Figure 3.55. The moments at the centers of beam AB and BC:

$$M_{CenterAB} = \overline{M}_{N}X_{1}l_{AB} + \overline{M}_{m}X_{2} = 0.1091\cdot16{,}409\cdot2 + 0.4261\cdot14{,}432 = 9{,}730\,\mathrm{kgm}$$

$$M_{CenterBC} = \overline{M}_{N}X_{1}l_{BC} + \overline{M}_{m}X_{2} + \overline{M}_{N}P = -0.0311\cdot16{,}409\cdot3 + 0.0690\cdot14{,}432$$
$$+ 0.0662\cdot30{,}000\cdot3 = 5{,}422.8\,\mathrm{kgm}$$

The moment diagram is shown in Figure 3.56.

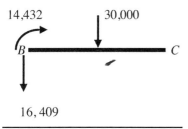

Figure 3.55

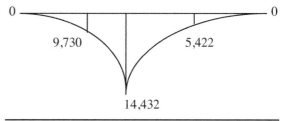

Figure 3.56   Moment diagram for the beam *BC*

3.8 Analysis of Complex Beams on Winkler Foundation

Hand analysis of beams free supported on Winkler foundation, as shown above, is easy, accurate, and can be recommended for practical applications. However, analysis of continuous, stepped, and even simple beams with various boundary conditions is more complex; it requires building and solving a system of linear equations and, in most cases, it is very time consuming. A computer program based on the described method or analytical methods described in Chapter 2 can resolve this problem. Development of such type of software will also require a lot of work and time. The simplest and most practical method for computer analysis of beams on Winkler foundation is modeling the soil with a series of independently working elastic supports; in other words, replacing the given beam on Winkler foundation with a regular statically indeterminate system. For example, the given beam shown in Figure 3.57 will look as shown in Figure 3.58.

The given beam continuously supported on soil is replaced with a beam supported on a series of elastic supports. The spacing of elastic supports along the beam is usually the same, although it can vary depending on the nature of the problem, type of

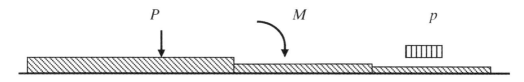

Figure 3.57

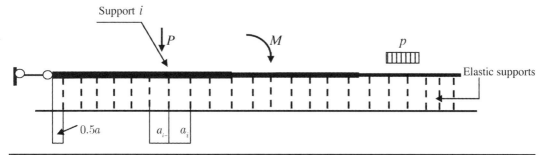

Figure 3.58

beam, and expected accuracy of analysis. The rigidity of each support at any point i is obtained from the following equation: $C_i = k_i a_i b_i$, where k_i is the modulus of subgrade reaction at point i, a_i is the spacing of supports (when the spacing at the left and right from point i is the same), and b_i is the width of the beam at point i. When the spacing of supports at the left and right from point i is not the same, the rigidity of this support is found as $C_i = k_i b_i \cdot (a_{i-1} + a_i)0.5$, where a_{i-1} is the spacing at the left from point i and a_i is the spacing at the right from point i. The soil can be replaced not only with elastic springs as shown above but also with elastic columns. The rigidity of the column, assuming that the length of the column $h = 1$ at any point i is equal to $E_i A_i = k_i a_i b_i$. Horizontal deflections of beams supported on the elastic foundation usually are not taken into account. The restraint against horizontal deflections is shown at the left end of the beam in Figure 3.58. Any computer program developed for analysis of regular statically indeterminate systems can be used for analysis of beams supported on Winkler foundation. The same method is applied to analysis of frames with continuous foundations (Figures 3.59 and 3.60) and even to analysis of interconnected beams and 3D frames with continuous foundations. Any statically indeterminate system with all or some of its elements supported on Winkler foundation can be analyzed by modeling the soil with independently working elastic supports.

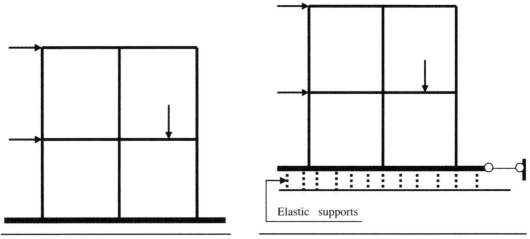

Elastic supports

Figure 3.59   **Figure 3.60**

3.9 Combined Analysis of Frames and Individual Foundations

Modeling the soil with elastic supports can also be applied to analysis of 2D and 3D frames with individual totally rigid foundations. However, analysis of frames with individual foundations is different from analysis of frames with continuous foundations. Continuous foundations are working as beams on the elastic foundation. They experience mostly bending and shear stresses due to settlements, applied loads, and soil pressure. Individual rigid foundations are much more sensitive to applied loads and soil reactions. They experience much larger differential settlements, rotations, and horizontal linear deflections. Large deflections produce additional stresses in all elements of the superstructure. Combined analysis of 2D and 3D frames with individual foundations can significantly improve the accuracy of the final results and lead to more economical and realistic design. A method of analysis is introduced below that allows replacing a frame with individual foundations with a simple statically indeterminate system, and analyzing that system using usually available software developed for analysis of regular frames.

Let us assume that a 2D frame is supported on individual foundations as shown in Figure 3.61.

By replacing the soil with a series of elastic supports and replacing the foundations with totally rigid T-elements, the same frame will look as shown in Figure 3.62. Horizontal supports shown in Figure 3.62 indicate that foundations are restrained against horizontal deflections and the rigidities of all horizontal supports are equal to ∞. If the foundation is not restrained completely against horizontal deflections, the actual rigidity of the horizontal support can be taken into account by using elastic supports. The rigidity of this support is equal to $K_X A$, where K_X is the modulus of subgrade reaction of the soil in the horizontal direction and A is the area of the foundation. K_X is obtained from the third equation 1.9 given in Chapter 1. The frame shown in Figure 3.62 can be analyzed as a simple statically indeterminate system using any available computer software for analysis of regular frames.

Two other methods described below are developed and recommended for analysis of 2D and 3D frames with individual foundations by modeling the system soil-foundation with line elements. In Figure 3.63, individual foundations of a 2D frame are replaced with simple beams restrained at both ends against any deflections. It is assumed that the center of rotation of the foundation is located at its bottom. The height of the foundation h is added to the height of the column and is totally rigid. The beam replacing the actual foundation is shown in Figure 3.64. Rigidities of the beam and soil are found by equating deflections of the individual foundation to the deflections of the equivalent beam. Deflections of an individual foundation of a 2D frame, as shown in Chapter 1, can be obtained from the following equations:

$$S_z = \frac{N_z}{AK_z} \quad \varphi_x = \frac{M_x}{I_x K_{\varphi x}} \quad \Delta_x = \frac{N_x}{AK_x} \tag{3.37}$$

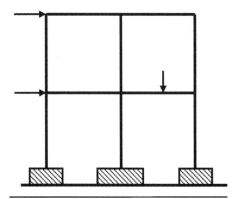

Figure 3.61

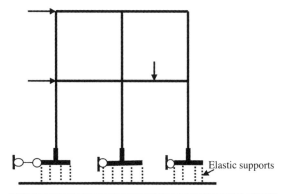

Elastic supports

Figure 3.62

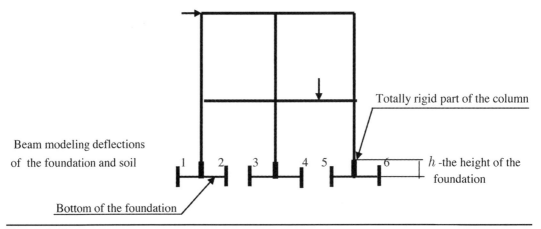

Totally rigid part of the column

Beam modeling deflections
of the foundation and soil

h -the height of the
foundation

Bottom of the foundation

Figure 3.63

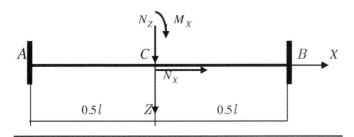

Figure 3.64

In these equations N_Z, M_X, and N_X are the vertical load, moment, and horizontal load applied to the foundation, respectively; A is the area of the foundation; I_X is the moment of inertia of the foundation; K_Z is the modulus of subgrade reaction of the soil under uniform vertical pressure; $K_{\varphi X}$ is the modulus of subgrade reaction of the soil under rotation in the X direction; and K_X is the modulus of subgrade reaction of the soil under horizontal deflection of the foundation in the X direction. Deflections of a beam loaded with the same loads applied to its center, as shown in Figure 3.64, are found from the following equations:

$$\Delta_z = \frac{N_z l^3}{192\,(EI)_X} \quad \varphi_X = \frac{M_X l}{16\,(EI)_X} \quad \Delta_X = \frac{N_X l}{4EA_B} \tag{3.38}$$

In these equations l is the length of the beam, $(EI)_X$ is the flexural rigidity of the beam in the vertical direction, and A_B is the area of the cross section of the beam. By equating deflections of the foundation and deflections of the beam, we find:

$$\frac{N_Z}{AK_Z} = \frac{N_z l^3}{192\,(EI)_X} \quad \frac{M_X}{I_X K_{\varphi X}} = \frac{M_x l}{16\,(EI)_X} \quad \frac{N_X}{AK_X} = \frac{N_x l}{4EA_B} \tag{3.39}$$

From the first two equations the length and flexural rigidity of the beam are found:

$$l = a\sqrt{\frac{K_{\varphi X}}{K_Z}} \tag{3.40}$$

$$\left(EI \right)_{X} = \frac{I_{X} K_{\varphi X} l}{16} \tag{3.41}$$

Now, from the third equation 3.39 the tension-compression rigidity of the beam is obtained:

$$EA_{B} = \frac{AK_{X} l}{4} \tag{3.42}$$

Coefficients K_Z, $K_{\varphi X}$, and K_X are obtained from equations 1.11–1.13 in Chapter 1. Now, using the length of the beam and rigidities obtained from equations 3.40–3.42, the frame shown in Figure 3.63 is analyzed as a regular statically indeterminate system.

The same method can be applied to analysis of 3D frames. A simple 3D frame is shown in Figure 3.65. Unlike foundations of 2D frames, foundations of 3D frames experience not three but six deflections: three linear deflections and three rotations. In addition to deflections discussed above, individual foundations experience linear deflection along axis Y (Δ_Y), rotation along axis Y (φ_Y), and rotation about the vertical axis Z (φ_Z). The last one is usually small and not taken into account. The first two deflections are found as follows:

$$\Delta_{Y} = \frac{N_{Y}}{AK_{Y}} \quad \varphi_{Y} = \frac{M_{Y}}{I_{Y} K_{\varphi Y}} \tag{3.43}$$

Horizontal deflection in the Y direction Δ_Y and rotation of the beam φ_Y in the Y direction, as shown in Figure 3.66, are found from equations:

$$\Delta_{Y} = \frac{N_{Y} l^{3}}{192 \left(EI \right)_{Y}} \quad \varphi_{Y} = \frac{M_{Y} l}{4 GI_{T}} \tag{3.44}$$

In the second equation 3.44, G is the shear modulus and I_T is the moment of inertia of the beam under torsion in the Y direction. Now, by equating deflections of the foundation to the deflections of the beam, we have two additional equations:

$$\frac{N_{Y}}{AK_{Y}} = \frac{N_{Y} l^{3}}{192 EI_{Y}} \quad \frac{M_{Y}}{I_{Y} K_{\varphi Y}} = \frac{M_{Y} l}{4 GI_{T}} \tag{3.45}$$

From equations 3.45 the flexural and torsional rigidities of the beam in direction Y are found:

$$EI_{Y} = \frac{AK_{Y} l^{3}}{192} \quad GI_{T} = \frac{I_{Y} K_{\varphi Y} l}{4} \tag{3.46}$$

Now, when all rigidities and the length of the beam are obtained, analysis of the 3D frame shown in Figure 3.65 can be performed.

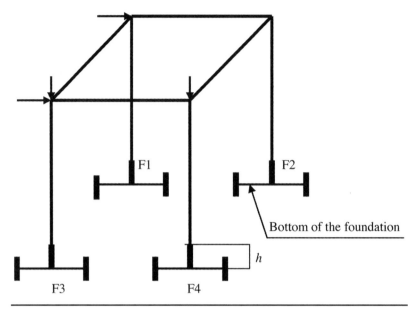

Figure 3.65

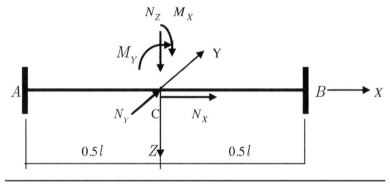

Figure 3.66

3.10 The Second Method of Modeling and the Iterative Method

Combined analysis of 2D and 3D frames with individual foundations can also be performed by modeling the system soil-foundation with simple elastic columns, as shown in Figure 3.67. The column shown in Figure 3.68 is modeling the actual system soil-foundation. The column can replace the actual foundation only when deflections of the column at its top are the same as deflections of the actual foundation under the same applied loads.

Rigidities of these columns can be found by equating the deflections of the columns to the deflections of the actual foundations. Let us assume that a simple vertical cantilever column, shown in Figure 3.68, has three loads: a vertical concentrated load N_Z,

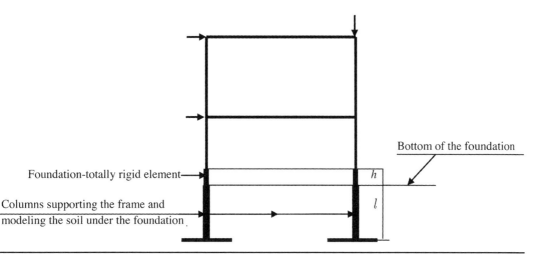

Figure 3.67

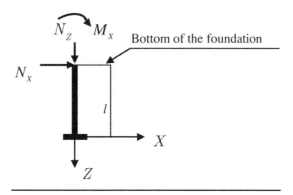

Figure 3.68

horizontal concentrated load N_X, and moment M_X. Deflections of the column tops are shown below:

$$S_c = \frac{N_z l}{E_c A_c} \quad \varphi_c = \frac{M_X l}{E_c I_{cX}} \quad \Delta_c = \frac{N_X l^3}{3 E_c I_{cX}} \tag{3.47}$$

In formulae 3.47, E_c is the modulus of elasticity of the column material, A_c is the cross area of the column, and I_c is the moment of inertia of the same cross section of the column $E_i A_i = k_i b_i \cdot (a_{i-1} + a_i) \cdot 0.5$ in the X direction. By equating deflections of the column to the deflections of the foundation, we find:

$$\frac{N_z}{AK_z} = \frac{N_z l}{E_c A_c} \quad \frac{M_X}{I_X K_{\varphi}} = \frac{M_X l}{E_c I_c} \quad \frac{N_X}{AK_X} = \frac{N_X l^3}{3 E_c I_c} \tag{3.48}$$

or

$$\frac{1}{AK_z} = \frac{l}{E_c A_c} \quad \frac{1}{I_X K_{\varphi}} = \frac{l}{E_c I_c} \quad \frac{1}{AK_X} = \frac{l^3}{3 E_c I_c} \tag{3.49}$$

From 3.49 the length and rigidities of the column is found:

$$l = \sqrt{\frac{3I_x K_\varphi}{AK_x}} \quad E_c A_c = AK_z l \quad E_c I_c = I_x K_\varphi l \tag{3.50}$$

Now, using equations 3.50, individual foundations of the frame shown in Figure 3.61 are replaced with elastic columns as shown in Figure 3.67.

Since all deflections are applied to the bottom of the foundation, the height of the frame column also includes the height of the foundation h. This portion of the column is assumed totally rigid.

Modeling the individual foundations for 3D frames is performed analogously. The only difference is the number of rigidities. The column modeling an individual foundation experiences six deflections: three linear deflections and three rotations. As shown in Figure 3.70, the column has six loads: three concentrated loads and three moments. Rotation of the foundation about the vertical axis Z is not taken into account. Moment M_Z about the axis Z is not shown in Figure 3.70.

Rotation of the column in Y direction can be obtained by equating rotation of the column to rotation of the foundation. This equation looks as follows:

$$\frac{M_Y}{I_Y K_\varphi} = \frac{M_Y l}{E_c I_{cY}} \tag{3.51}$$

From 3.51 we find the flexural rigidity of the column in the Y direction:

$$E_c I_{cY} = ll_Y K_\varphi \tag{3.52}$$

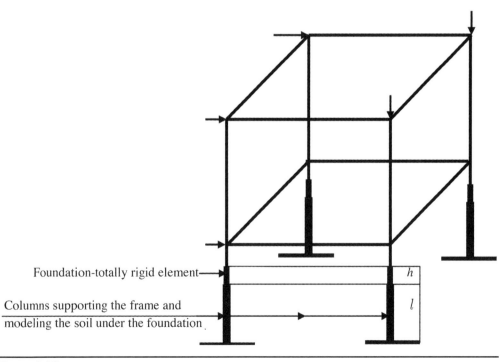

Foundation-totally rigid element——

Columns supporting the frame and modeling the soil under the foundation

h

l

Figure 3.69

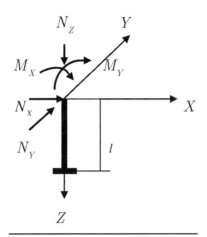

Figure 3.70

Now, when the length l and rigidities of the columns, $E_c A_c$, $E_c I_{cX}$ and $E_c I_{cY}$ are found, analysis of the frame can be performed using any computer software developed for analysis of statically indeterminate systems. Both methods of modeling the soil and individual foundations with elastic beams and columns are simple: they allow performing computer analysis of the frame taking into account deflections of the foundations. However, the described methods do not produce final results for the total system frame-foundation. Obtained results are used only for design of the frame, but not for final design of the foundations. Final analysis of the foundation is performed after analysis of the frame is performed by applying to the foundations loads found from analysis of the frame.

Nevertheless, analysis that produces final results for the frame and foundations can be performed by using the iterative method. Analysis is performed in several steps-iterations.

> **Step 1:** Assuming that all columns of the frame do not experience any deflections at all points of support, analysis of the frame loaded with the given loads is performed.

> **Step 2:** Reactions in the columns, at all points of support acting in the opposite direction, are applied to the foundations, and analysis of the foundations is performed.

> **Step 3:** Deflections of the foundations obtained in Step 2 are applied to the frame along with the given loads, and analysis of the frame is performed.

> **Step 4:** New reactions at the columns acting in the opposite direction along with the given loads applied to the foundations will produce new deflections of the foundation.

These deflections are used to repeat Step 3. The process stops when loads or deflections found in Step (i) are close enough to loads or deflections obtained in Step (i-1).

This method is simple. It requires the use of two computer programs: one for frame analysis and a second for foundations analysis. Both programs are widely used and available. Analysis can be simplified by combining both programs into one, which requires some programming work.

Both methods of analysis presented above can be recommended for practical applications. However, it is important to understand the difference between these two methods. When modeling the foundation with a beam or a column, the frame is analyzed taking into account the deflections of the foundation. Analysis of the foundations is performed separately using the loads applied to the foundations obtained from analysis of the frame. When the iterative method is used, both analyses of the frame and foundations are performed simultaneously and final results of the frame and foundations' analyses are found.

Table 3.1 Tables for analysis of beams on Winkler foundation with a vertical concentrated load N

	$\lambda = 1$					
	Point of application of the vertical load N					
Point #	0	1	2	3	4	5
			$\overline{p}$			
0	4.0374	3.4175	2.8005	2.1885	1.5822	0.9815
1	3.4172	2.9294	2.4414	1.9548	1.4708	0.9898
2	2.8000	2.4412	2.0817	1.7205	1.3589	0.9978
3	2.1881	1.9545	1.7204	1.4846	1.2446	1.0050
4	1.5818	1.4705	1.3588	1.2458	1.1301	1.0104
5	0.9811	0.9895	0.9977	1.0049	1.0104	1.0125
6	0.3854	0.5115	0.6375	0.7630	0.8876	1.0104
7	−0.2065	0.0359	0.2783	0.5207	0.7631	1.0050
8	−0.7959	−0.4381	−0.0801	0.2784	0.6376	0.9978
9	−1.3841	−0.9112	−0.4380	0.0360	0.5117	0.9898
10	−1.9718	−1.3840	−0.7958	−0.2063	0.3857	0.9815
			$\overline{M}$			
0	0.0000	0.0000	0.0000	0.0000	0.0000	0.0000
1	−0.0808	0.0163	0.0134	0.0106	0.0077	0.0049
2	−0.1275	−0.0382	0.0512	0.0407	0.0302	0.0197
3	−0.1462	−0.0682	0.0099	0.0880	0.0662	0.0445
4	−0.1429	−0.0786	−0.0143	0.0501	0.1147	0.0794
5	−0.1239	−0.0744	−0.0249	0.0247	0.0744	0.1243
6	−0.0950	−0.0603	−0.0255	0.0093	0.0443	0.0794
7	−0.0623	−0.0410	−0.0197	0.0016	0.0230	0.0445
8	−0.0316	−0.0214	−0.0111	−0.0009	0.0094	0.0197
9	−0.0089	−0.0061	−0.0034	−0.0006	0.0021	0.0049
10	0.0000	0.0000	0.0000	0.0000	0.0000	0.0000
			$\overline{Q}$			
0	**0.0000**	0.0000	0.0000	0.0000	0.0000	0.0000
1	−0.6273	**0.3173**	0.2621	0.2072	0.1526	0.0986
2	−0.3164	−0.4141	**0.4883**	0.3909	0.2941	0.1979
3	−0.0670	−0.1943	−0.3216	**0.5512**	0.4244	0.2981
4	0.1215	−0.0231	−0.1676	−0.3123	**0.5432**	0.3989
5	0.2496	0.0999	−0.0499	−0.1998	−0.3498	**0.5000**
6	0.3180	0.1750	0.0319	−0.1114	−0.2549	−0.3989
7	0.3269	0.2023	0.0777	−0.0472	−0.1724	−0.2981
8	0.2768	0.1822	0.0876	−0.0072	−0.1023	−0.1979
9	0.1678	0.1148	0.0617	0.0085	−0.0449	−0.0986
10	0.0000	0.0000	0.0000	0.0000	0.0000	0.0000

Analysis is performed in the following order

1. Find $m = \sqrt[4]{\dfrac{kb}{4EI}}$ 2. Find $\lambda = lm$ 3. Find $p = \overline{p}\dfrac{N_i}{l}$, $M = \overline{M}N_i l$ and $Q = \overline{Q}N_i$

k-Modulus of Subgrade Reaction, b-width of the beam, E-Modulus of Elasticity of the beam material, I-Moment of Inertia of the beam

Table 3.2

	λ = 2					
	Point of application of the vertical load *N*					
Point #	0	1	2	3	4	5
	$\overline{p}$					
0	4.5441	3.6508	2.8040	2.0317	1.3437	0.7368
1	3.6466	3.0550	2.4599	1.8841	1.3470	0.8537
2	2.7976	2.4573	2.1058	1.7298	1.3453	0.9678
3	2.0245	1.8806	1.7283	1.5494	1.3256	1.0709
4	1.3369	1.3428	1.3431	1.3248	1.2664	1.1489
5	0.7310	0.8500	0.9657	1.0699	1.1486	1.1807
6	0.1945	0.3992	0.6027	0.8013	0.9875	1.1489
7	−0.2897	−0.0178	0.2547	0.5283	0.8019	1.0709
8	−0.7400	−0.4120	−0.0816	0.2556	0.6043	0.9678
9	−1.1730	−0.7942	−0.4112	−0.0160	0.4021	0.8537
10	−1.6008	−1.1726	−0.7386	−0.2869	0.1988	0.7368
	$\overline{M}$					
0	0.0000	0.0000	0.0000	0.0000	0.0000	0.0000
1	−0.0788	0.0173	0.0134	0.0099	0.0067	0.0039
2	−0.1210	−0.0349	0.0515	0.0387	0.0269	0.0163
3	−0.1351	−0.0625	0.0105	0.0847	0.0605	0.0384
4	−0.1289	−0.0712	−0.0132	0.0461	0.1073	0.0711
5	−0.1091	−0.0664	−0.0234	0.0207	0.0667	0.1152
6	−0.0819	−0.0531	−0.0240	0.0060	0.0375	0.0711
7	−0.0527	−0.0357	−0.0185	−0.0007	0.0181	0.0384
8	−0.0263	−0.0184	−0.0104	−0.0021	0.0067	0.0163
9	−0.0073	−0.0052	−0.0031	−0.0010	0.0013	0.0039
10	0.0000	0.0000	0.0000	0.0000	0.0000	0.0000
	$\overline{Q}$					
0	0.0000	**0.0000**	0.0000	0.0000	0.0000	0.0000
1	−0.5905	**0.3353**	0.2632	0.1958	0.1345	0.1706
2	−0.2683	−0.3891	**0.4915**	0.3765	0.2691	0.1706
3	−0.0272	−0.1722	−0.3168	**0.5405**	0.4027	0.2725
4	0.1409	−0.0110	−0.1633	−0.3158	**0.5323**	0.3835
5	0.2443	0.0986	−0.0478	−0.1961	−0.3470	**0.5000**
6	0.2906	0.1611	0.0306	−0.1025	−0.2402	−0.3835
7	0.2858	0.1801	0.0735	−0.0360	−0.1507	−0.2725
8	0.2343	0.1587	0.0821	0.0032	−0.0804	−0.1706
9	0.1387	0.0983	0.0575	0.0151	−0.0300	−0.0795
10	0.0000	0.0000	0.0000	0.0000	0.0000	0.0000

Analysis is performed in the following order

1. Find $m = \sqrt[4]{\dfrac{kb}{4EI}}$ 2. Find $\lambda = lm$ 3. Find $p = \overline{p}\dfrac{N_i}{l}$, $M = \overline{M}N_i l$ and $Q = \overline{Q}N_i$

k-Modulus of Subgrade Reaction, *b*-width of the beam, *E*-Modulus of Elasticity of the beam material, *I*-Moment of Inertia of the beam

Table 3.3

			$\lambda = 3$			
		Point of application of the vertical load N				
Point #	0	1	2	3	4	5
				$\overline{p}$		
0	6.0163	4.2818	2.7602	1.5564	0.6831	0.1003
1	4.2622	3.3981	2.5052	1.6840	1.0070	0.4911
2	2.7325	2.4945	2.2001	1.7812	1.3153	0.8769
3	1.5281	1.6691	1.7755	1.7674	1.5620	1.2369
4	0.6586	0.9920	1.3077	1.5593	1.6586	1.5224
5	0.0814	0.4786	0.8697	1.2337	1.5216	1.6450
6	−0.2696	0.1104	0.4920	0.8721	1.2314	1.5224
7	−0.4652	−0.1477	0.1765	0.5162	0.8742	1.2369
8	−0.5687	−0.3350	−0.0915	0.1792	0.4973	0.8769
9	−0.6289	−0.4869	−0.3328	−0.1423	0.1194	0.4911
10	−0.6768	−0.6280	−0.5652	−0.4577	−0.2567	0.1003
				$\overline{M}$		
0	0.0000	0.0000	0.0000	0.0000	0.0000	0.0000
1	−0.0728	0.0199	0.0134	0.0080	0.0040	0.0012
2	−0.1027	−0.0262	0.0517	0.0328	0.0180	0.0072
3	−0.1047	−0.0472	0.0119	0.0752	0.0450	0.0220
4	−0.0908	−0.0513	−0.0103	0.0349	0.0874	0.0490
5	−0.0699	−0.0452	−0.0193	0.0101	0.0460	0.0910
6	−0.0478	−0.0341	−0.0196	−0.0025	0.0196	0.0490
7	−0.0281	−0.0217	−0.0148	−0.0063	0.0054	0.0220
8	−0.0129	−0.0107	−0.0082	−0.0049	−0.0001	0.0072
9	0.0033	−0.0290	−0.0024	−0.0018	−0.0007	0.0012
10	0.0000	0.0000	0.0000	0.0000	0.0000	0.0000
				$\overline{Q}$		
0	**0.0000**	0.0000	0.0000	0.0000	0.0000	0.0000
1	−0.4861	**0.3840**	0.2633	0.1620	0.0845	0.0296
2	−0.1363	−0.3214	**0.4985**	0.3353	0.2006	0.0980
3	0.0767	−0.1132	−0.3027	**0.5127**	0.3445	0.2037
4	0.1860	0.0199	−0.1485	−0.3210	0.5055	0.3416
5	0.2230	0.0934	−0.0396	−0.1813	−0.3355	**0.5000**
6	0.2136	0.1228	0.0284	−0.0760	−0.1978	−0.3416
7	0.1769	0.1210	0.0619	−0.0066	−0.0925	−0.2037
8	0.1252	0.0968	0.0661	0.0282	−0.0240	−0.0980
9	0.6530	0.0557	0.0449	0.0300	0.0069	−0.0296
10	0.0000	0.0000	0.0000	0.0000	0.0000	0.0000

Analysis is performed in the following order

1. Find $m = \sqrt[4]{\dfrac{kb}{4EI}}$ 2. Find $\lambda = lm$ 3. Find $p = \overline{p}\dfrac{N_i}{l}$, $M = \overline{M}N_i l$ and $Q = \overline{Q}N_i$

k-Modulus of Subgrade Reaction, b-width of the beam, E-Modulus of Elasticity of the beam material, I-Moment of Inertia of the beam

Table 3.4

Point #	0	1	2	3	4	5
λ = 4						
Point of application of the vertical load *N*						
$\overline{p}$						
0	7.9516	4.9423	2.5174	0.8761	−0.0598	−0.4763
1	4.8865	3.7757	2.5396	1.4354	0.6275	0.1258
2	2.4484	2.5144	2.4110	1.9281	1.3007	0.7408
3	0.8160	1.4043	1.9160	2.1500	1.8816	1.3646
4	−0.1031	0.6005	1.2870	1.8768	2.1611	1.9170
5	−0.5028	0.1065	0.7293	1.3594	1.9157	2.1799
6	−0.5801	−0.1434	0.3181	0.9331	1.3966	1.9170
7	−0.4885	−0.2319	0.0493	0.3944	0.9363	1.3646
8	−0.3267	−0.2316	−0.1174	0.0525	0.3255	0.7408
9	−0.1450	−0.1932	−0.2298	−0.2264	−0.1318	0.1258
10	0.0391	−0.1449	−0.3253	−0.4830	−0.5664	−0.4763
$\overline{M}$						
0	0.0000	0.0000	0.0000	0.0000	0.0000	0.0000
1	−0.0654	0.0228	0.0126	0.0053	0.0008	−0.0014
2	−0.0808	−0.0169	0.0504	0.0249	0.0079	−0.0015
3	−0.0704	−0.0311	0.0117	0.0663	0.0279	0.0058
4	−0.0507	−0.0308	−0.0081	0.0223	0.0662	0.0267
5	−0.0311	−0.0240	−0.0149	−0.0003	0.0252	0.0662
6	−0.0160	−0.0156	−0.0142	−0.0093	0.0029	0.0267
7	−0.0065	−0.0085	−0.0100	−0.0098	−0.0055	0.0058
8	−0.0017	−0.0035	−0.0052	−0.0062	−0.0055	−0.0015
9	−0.0001	−0.0008	−0.0015	−0.0020	−0.0021	−0.0014
10	0.0000	0.0000	0.0000	0.0000	0.0000	0.0000
$\overline{Q}$						
0	0.0000	0.0000	0.0000	0.0000	0.0000	0.0000
1	−0.3581	**0.4359**	0.2528	0.1156	0.0284	−0.0175
2	0.0087	−0.2496	**0.5004**	0.2838	0.1248	0.0258
3	0.1719	−0.0537	−0.2832	**0.4877**	0.2839	0.1311
4	0.2075	0.0466	−0.1230	−0.3110	**0.4860**	0.2952
5	0.1772	0.0819	−0.0222	−0.1492	−0.3101	**0.5000**
6	0.1231	0.0801	0.0302	−0.0396	−0.1445	−0.2952
7	0.0696	0.0613	0.0485	−0.0218	−0.0329	−0.1311
8	0.0289	0.0381	0.0451	0.0442	0.0252	−0.0258
9	0.0053	0.0169	0.0278	0.0355	0.0349	−0.0175
10	0.0000	0.0000	0.0000	0.0000	0.0000	0.0000

Analysis is performed in the following order

1. Find $m = \sqrt[4]{\dfrac{kb}{4EI}}$ 2. Find $\lambda = lm$ 3. Find $p = \overline{p}\dfrac{N_i}{l}$, $M = \overline{M}N_i l$ and $Q = \overline{Q}N_i$

k-Modulus of Subgrade Reaction, *b*-width of the beam, *E*-Modulus of Elasticity of the beam material, *I*-Moment of Inertia of the beam

Table 3.5

			$\lambda = 5$			
		Point of application of the vertical load N				
Point #	0	1	2	3	4	5
			$\overline{p}$			
0	9.8937	5.3402	2.0196	0.1672	−0.5770	−0.6843
1	5.2185	4.0639	2.5511	1.2106	0.3412	−0.0893
2	1.8898	2.5059	2.7556	2.1645	1.1286	0.5683
3	0.0740	1.1624	2.1477	2.6466	2.1858	1.3549
4	−0.6283	0.3066	1.2685	2.1798	2.6498	2.1874
5	−0.7045	−0.1082	0.5559	1.3491	2.1859	2.6417
6	−0.5243	−0.2333	0.1210	0.6341	1.3615	2.1874
7	−0.2976	−0.2105	−0.0811	1.7067	0.6372	1.3549
8	−0.1064	−0.1331	−0.1399	−0.0791	0.1273	0.5683
9	0.0438	−0.0450	−0.1337	−0.2092	−0.2257	−0.0893
10	0.1766	0.0424	−0.1100	−0.3011	−0.5214	−0.6843
			$\overline{M}$			
0	0.0000	0.0000	0.0000	0.0000	0.0000	0.0000
1	−0.0583	0.0246	0.0110	0.0026	−0.0014	−0.0024
2	−0.0622	−0.0107	0.0469	0.0171	0.0007	−0.0056
3	−0.0447	−0.0205	0.0091	0.0525	0.0156	−0.0030
4	−0.0246	−0.0179	−0.0077	0.0128	0.0517	0.0133
5	−0.0097	−0.0115	−0.0116	−0.0058	0.0126	0.0509
6	−0.0014	−0.0057	−0.0094	−0.0106	−0.0051	0.0133
7	0.0017	−0.0020	−0.0057	−0.0087	−0.0091	−0.0030
8	0.0017	−0.0003	−0.0025	−0.0047	−0.0064	−0.0056
9	0.0007	0.0001	0.0006	−0.0014	−0.0021	−0.0024
10	0.0000	0.0000	0.0000	0.0000	0.0000	0.0000
			$\overline{Q}$			
0	**0.0000**	0.0000	0.0000	0.0000	0.0000	0.0000
1	−0.2444	**0.4702**	0.2285	0.0689	−0.0118	−0.0387
2	0.1110	−0.2013	**0.4939**	0.2376	0.0696	−0.0147
3	0.2092	−0.0179	−0.2610	**0.4782**	0.2432	0.0814
4	0.1815	0.0556	−0.0901	−0.2805	**0.4850**	0.2585
5	0.1149	0.0655	0.0011	−0.1040	−0.2733	**0.5000**
6	0.0534	0.0484	0.0349	−0.0049	−0.0959	−0.2585
7	0.0123	0.0262	0.0369	0.0353	0.0041	−0.0814
8	−0.0079	0.0090	0.0259	0.0399	0.0423	0.0147
9	−0.0110	0.0001	0.0122	0.0255	0.0374	0.0387
10	0.0000	0.0000	0.0000	0.0000	0.0000	0.0000

Analysis is performed in the following order

1. Find $m = \sqrt[4]{\dfrac{kb}{4EI}}$ 2. Find $\lambda = lm$ 3. Find $p = \overline{p}\dfrac{N_i}{l}$, $M = \overline{M}N_i l$ and $Q = \overline{Q}N_i$

k-Modulus of Subgrade Reaction, b-width of the beam, E-Modulus of Elasticity of the beam material, I-Moment of Inertia of the beam

Table 3.6

	$\lambda = 6$					
	Point of application of the vertical load N					
Point #	0	1	2	3	4	5
			$\overline{p}$			
0	11.8071	5.4781	1.3560	−0.4487	−0.8247	−0.6133
1	5.2533	4.3068	2.5664	1.0109	0.1304	−0.1935
2	1.1517	2.4975	3.2124	2.4172	1.2106	0.3568
3	−0.5654	0.9482	2.3961	3.1828	2.4219	1.2246
4	−0.8680	0.0947	1.1915	2.4156	3.1288	2.3722
5	−0.6169	−0.2067	0.3461	1.2193	2.3708	3.0931
6	−0.2961	−0.2197	−0.0540	0.3666	1.2028	2.3722
7	−0.0804	−0.1384	−0.1586	−0.0435	0.3690	1.2246
8	−0.0219	−0.0560	−0.1281	−0.1584	−0.0503	0.3568
9	0.0588	0.0060	−0.0581	−0.1412	−0.2186	−0.1935
10	0.0751	0.0573	0.0165	−0.0902	−0.3060	−0.6133
			$\overline{M}$			
0	0.0000	0.0000	0.0000	0.0000	0.0000	0.0000
1	−0.0519	0.0254	0.0088	0.0002	−0.0025	−0.0024
2	−0.0472	−0.0071	0.0423	0.0104	−0.0036	−0.0065
3	−0.0269	−0.0143	0.0055	0.0437	0.0078	−0.0064
4	−0.0100	−0.0108	−0.0080	0.0063	0.0424	0.0063
5	−0.0008	−0.0054	−0.0089	−0.0077	0.0060	0.0420
6	0.0023	−0.0016	−0.0057	−0.0039	−0.0075	0.0063
7	0.0023	0.0001	−0.0025	−0.0057	−0.0083	−0.0064
8	0.0013	0.0004	−0.0006	−0.0024	−0.0048	−0.0065
9	0.0003	0.0002	0.0000	−0.0005	−0.0014	−0.0024
10	0.0000	0.0000	0.0000	0.0000	0.0000	0.0000
			$\overline{Q}$			
0	**0.0000**	0.0000	0.0000	0.0000	0.0000	0.0000
1	−0.1470	**0.4892**	0.1961	0.0281	−0.0347	−0.0403
2	−0.1733	−0.1705	**0.4851**	0.1995	0.0323	−0.0322
3	0.2026	0.0017	−0.0235	**0.4795**	0.2140	0.0469
4	0.1309	0.0539	−0.0551	−0.2406	**0.4915**	0.2267
5	0.0567	0.0483	0.0217	−0.0588	−0.2335	**0.5000**
6	0.0110	0.0270	0.0364	0.0205	−0.0549	−0.2267
7	−0.0078	0.0091	0.0257	0.0366	0.0237	−0.0469
8	−0.0107	−0.0007	0.0114	0.0265	0.0397	0.0322
9	−0.0067	−0.0032	0.0021	0.0116	0.0262	0.0403
10	0.0000	0.0000	0.0000	0.0000	0.0000	0.0000

Analysis is performed in the following order

1. Find $m = \sqrt[4]{\dfrac{kb}{4EI}}$ 2. Find $\lambda = lm$ 3. Find $p = \overline{p}\dfrac{N_i}{l}$, $M = \overline{M}N_i l$ and $Q = \overline{Q}N_i$

k-Modulus of Subgrade Reaction, b-width of the beam, E-Modulus of Elasticity of the beam material, I-Moment of Inertia of the beam

Table 3.7

			$\lambda = 7$			
		Point of application of the vertical load N				
Point #	0	1	2	3	4	5
			$\overline{p}$			
0	13.6867	5.4001	0.6381	−0.8856	−0.8370	−0.4126
1	5.0300	4.5599	2.6006	0.8234	−0.0333	−0.2349
2	0.3561	2.5065	3.7409	2.6175	1.0442	0.1298
3	−1.0065	0.7518	2.5937	3.7068	2.5692	1.0150
4	−0.8575	−0.0640	1.0261	2.5633	3.6142	2.5068
5	−0.3976	−0.2401	0.1221	1.0107	2.5056	3.5794
6	−0.0844	−0.1676	−0.1700	0.1161	0.9917	2.5068
7	0.0380	−0.0676	−0.1639	−0.1732	0.1176	1.0150
8	0.0516	−0.0085	−0.0825	−0.1647	−0.1688	0.1298
9	0.0279	0.0158	−0.0103	−0.0717	0.1714	−0.2349
10	−0.0019	0.0275	0.0488	0.0291	−0.1011	−0.4126
			$\overline{M}$			
0	0.0000	0.0000	0.0000	0.0000	0.0000	0.0000
1	−0.0460	0.0256	0.0065	−0.0016	−0.0028	−0.0018
2	−0.0351	−0.0052	0.0376	0.0052	−0.0056	−0.0056
3	−0.0150	−0.0105	0.0023	0.0370	0.0029	−0.0072
4	−0.0025	−0.0067	−0.0078	0.0022	0.0363	0.0023
5	0.0019	−0.0024	−0.0065	−0.0077	0.0022	0.0362
6	0.0021	−0.0001	−0.0030	−0.0064	−0.0075	0.0023
7	0.0011	0.0005	−0.0007	−0.0030	−0.0063	−0.0072
8	0.0003	0.0004	0.0002	−0.0007	−0.0028	−0.0056
9	0.0000	0.0001	0.0001	0.0000	−0.0006	−0.0018
10	0.0000	0.0000	0.0000	0.0000	0.0000	0.0000
			$\overline{Q}$			
0	**0.0000**	0.0000	0.0000	0.0000	0.0000	0.0000
1	−0.0642	**0.4980**	0.1619	−0.0031	−0.0435	−0.0324
2	0.2051	−0.1487	**0.4790**	0.1689	0.0070	−0.0376
3	0.1726	0.0142	−0.2043	**0.4852**	0.1877	0.0196
4	0.0794	0.0486	−0.0233	−0.2013	**0.4969**	0.1957
5	0.0167	0.0334	0.0341	−0.0226	−0.1971	**0.5000**
6	−0.0071	0.0130	0.0317	0.0337	−0.0223	−0.1957
7	−0.0098	0.0013	0.0150	0.0308	0.0332	−0.0196
8	−0.0053	−0.0025	0.0027	0.0139	0.0306	0.0376
9	−0.0013	−0.0022	−0.0019	−0.0021	0.0136	0.0324
10	0.0000	0.0000	0.0000	0.0000	0.0000	0.0000

Analysis is performed in the following order

1. Find $m = \sqrt[4]{\dfrac{kb}{4EI}}$ 2. Find $\lambda = lm$ 3. Find $p = \overline{p}\dfrac{N_i}{l}$, $M = \overline{M}N_i l$ and $Q = \overline{Q}N_i$

k-Modulus of Subgrade Reaction, b-width of the beam, E-Modulus of Elasticity of the beam material, I-Moment of Inertia of the beam

Table 3.8 Tables for analysis of beams on winkler foundation loaded with a concentrated moment M

			$\lambda = 1$			
			Point of application of the moment M			
Point #	0	1	2	3	4	5
0	6.2047	6.1885	6.1468	6.0921	6.0343	5.9809
1	4.8770	4.8802	4.8758	4.8546	4.8252	4.7946
2	3.5880	3.5906	3.6036	3.6158	3.6149	3.6070
3	2.3346	2.3367	2.3469	2.3726	2.4002	2.4152
4	1.1123	1.1139	1.1214	1.1407	1.1769	1.2147
5	−0.0843	−0.0833	−0.0784	−0.0655	−0.0407	0.0000
6	−1.2611	−1.2606	−1.2583	−1.2519	−1.2384	−1.2147
7	−2.4239	−2.4240	−2.4243	2.4242	−2.4220	−2.4152
8	−3.5782	−3.5788	−3.5817	−3.5870	−3.5970	−3.6070
9	−4.7282	−4.7294	−4.7349	−4.7474	−4.7677	−4.7946
10	−5.8770	−5.8788	−5.8869	−5.9057	−5.9371	−5.9809
0	**0.0000**	0.0000	0.0000	0.0000	0.0000	0.0000
1	−0.9712	**0.0288**	0.0286	0.0284	0.0282	0.0279
2	−0.8935	−0.8936	**0.1060**	0.1053	0.1046	0.1038
3	−0.7800	−0.7801	−0.7806	**0.2184**	0.2171	0.2157
4	−0.6430	−0.6431	−0.6436	−0.6447	**0.3536**	0.3518
5	−0.4948	−0.4949	−0.4954	−0.4960	−0.4980	**0.5000**
6	−0.3475	−0.3476	−0.3479	−0.3488	−0.3501	−0.3518
7	−0.2127	−0.2128	−0.2130	−0.2136	−0.2145	−0.2157
8	−0.1022	−0.1022	−0.1024	−0.1027	−0.1031	−0.1038
9	−0.0275	−0.0275	−0.0275	−0.0276	−0.0277	−0.0279
10	0.0000	0.0000	0.0000	0.0000	0.0000	0.0000
0	0.0000	0.0000	0.0000	0.0000	0.0000	0.0000
1	0.5541	0.5534	0.5511	0.5473	0.5430	0.5388
2	0.9773	0.9770	0.9751	0.9709	0.9650	0.9589
3	1.2735	1.2733	1.2726	1.2703	1.2657	1.2600
4	1.4458	1.4459	1.4460	1.4459	1.4446	1.4415
5	1.4972	1.4974	1.4982	1.4997	1.5014	1.5022
6	1.4299	1.4302	1.4314	1.4338	1.4374	1.4415
7	1.2457	1.2460	1.2472	1.2500	1.2544	1.2600
8	0.9456	0.9458	0.9469	0.9494	0.9535	0.9589
9	0.5303	0.5304	0.5311	0.5327	0.5352	0.5388
10	0.0000	0.0000	0.0000	0.0000	0.0000	0.0000

Analysis is performed in the following order

1. Find $m = \sqrt[4]{\dfrac{kb}{4EI}}$ 2. Find $\lambda = lm$ 3. Find $p = \overline{p}\dfrac{M_i}{l^2}$, $M = \overline{M}M_i$ and $Q = \overline{Q}\dfrac{M_i}{l}$

k-Modulus of Subgrade Reaction, b-width of the beam, E-Modulus of Elasticity of the beam material, I-Moment of Inertia of the beam

Table 3.9

Point #	0	1	2	3	4	5
			$\lambda = 2$			
		Point of application of the moment M				
			$\overline{p}$			
0	9.0168	8.7647	8.1254	7.3057	6.4609	5.6994
1	5.8971	5.9523	5.8957	5.5837	5.1555	4.7148
2	3.3888	3.4317	3.6392	3.8371	3.8280	3.7104
3	1.4291	1.4607	1.6158	2.0090	2.4270	2.6396
4	−0.0670	−0.0456	0.0617	0.3406	0.8764	1.4318
5	−1.1945	−1.1822	−1.1180	−0.9439	−0.5948	0.0000
6	−2.0487	−2.0445	−2.0191	−1.9405	−1.7638	−1.4318
7	−2.7175	−2.7207	−2.7309	−2.7403	−2.7241	−2.6396
8	−3.2762	−3.2863	−3.3297	−3.4223	−3.5584	−3.7104
9	−3.7826	−3.7993	−3.8750	−4.0481	−4.3322	−4.7148
10	−4.2738	−4.2971	−4.4046	−4.6574	−5.0881	−5.6994
			$\overline{M}$			
0	**0.0000**	0.0000	0.0000	0.0000	0.0000	0.0000
1	−0.9601	**0.0391**	0.0369	0.0337	0.0301	0.0269
2	−0.8602	−0.8617	**0.1327**	0.1231	0.1118	0.1008
3	−0.7256	−0.7273	−0.7347	**0.2508**	0.2316	0.2118
4	−0.5758	−0.5776	−0.5851	−0.6011	**0.3754**	0.3489
5	−0.4261	−0.4277	−0.4343	−0.4490	−0.4719	**0.5000**
6	−0.2879	−0.2891	−0.2943	−0.3059	−0.3246	−0.3489
7	−0.1699	−0.1707	−0.1741	−0.1819	−0.1946	−0.2118
8	−0.0789	−0.0793	−0.0810	−0.0850	−0.0917	−0.1008
9	−0.0206	−0.0207	−0.0211	−0.0223	−0.0242	−0.0269
10	0.0000	0.0000	0.0000	0.0000	0.0000	0.0000
			$\overline{Q}$			
0	0.0000	0.0000	0.0000	0.0000	0.0000	0.0000
1	0.7457	0.7359	0.7011	0.6445	0.5808	0.5207
2	1.2100	1.2051	1.1778	1.1155	1.0300	0.9420
3	1.4509	1.4497	1.4405	1.4078	1.3427	1.2595
4	1.5190	1.5204	1.5244	1.5254	1.5079	1.4630
5	1.4559	1.4590	1.4716	1.4952	1.5220	1.5346
6	1.2938	1.2977	1.3147	1.3510	1.4041	1.4630
7	1.0554	1.0594	1.0772	1.1169	1.1797	1.2595
8	0.7558	0.7591	0.7742	0.8088	0.8655	0.9420
9	0.4028	0.4048	0.4140	0.4353	0.4710	0.5207
10	0.0000	0.0000	0.0000	0.0000	0.0000	0.0000

Analysis is performed in the following order

1. Find $m = \sqrt[4]{\dfrac{kb}{4EI}}$ 2. Find $\lambda = lm$ 3. Find $p = \overline{p}\dfrac{M_i}{l^2}$, $M = \overline{M}M_i$ and $Q = \overline{Q}\dfrac{M_i}{l}$

k-Modulus of Subgrade Reaction, b-width of the beam, E-Modulus of Elasticity of the beam material, I-Moment of Inertia of the beam

Table 3.10

λ = 3						
Point of application of the moment *M*						
Point #	0	1	2	3	4	5
			$\overline{p}$			
0	17.7377	16.5575	13.7134	10.3536	7.1863	4.5828
1	8.5336	8.8557	8.7520	7.5591	5.9588	4.3913
2	2.3080	2.5230	3.5716	4.5916	4.6055	4.1148
3	−1.4524	−1.3247	−0.6718	1.0626	2.8279	3.5391
4	−3.3555	−3.2932	−2.9455	−1.9410	0.1818	2.3190
5	−3.9775	−3.9601	−3.8307	−3.3711	−2.2419	0.0000
6	3.7958	−3.8074	−3.8242	−3.7488	−3.3602	−2.3190
7	−3.1659	−3.1951	−3.3063	−3.4946	−3.6474	−3.5391
8	−2.3239	−2.3637	−2.5371	−2.9130	−3.4759	−4.1148
9	−1.4048	−1.4520	−1.6721	−2.1974	−3.1034	−4.3913
10	−0.4693	−0.5229	−0.7851	−1.4484	−2.6769	−4.5828
			$\overline{M}$			
0	**0.0000**	0.0000	0.0000	0.0000	0.0000	0.0000
1	−0.9267	**0.0700**	0.0603	0.0471	0.0339	0.0226
2	−0.7630	−0.7693	**0.2078**	0.1695	0.1272	0.0890
3	−0.5722	−0.5791	−0.6075	**0.3369**	0.2658	0.1960
4	−0.3928	−0.3990	−0.4262	−0.4842	**0.4312**	0.3373
5	−0.2448	−0.2498	−0.2721	−0.3221	−0.4012	**0.5000**
6	−0.1352	−0.1387	−0.1547	−0.1919	−0.2538	−0.3373
7	−0.0629	−0.0650	−0.0748	−0.0982	−0.1386	−0.1960
8	−0.0218	−0.0228	−0.0275	−0.0389	−0.0591	−0.0890
9	−0.0039	−0.0042	−0.0054	−0.0085	−0.0141	−0.0226
10	0.0000	0.0000	0.0000	0.0000	0.0000	0.0000
			$\overline{Q}$			
0	0.0000	0.0000	0.0000	0.0000	0.0000	0.0000
1	1.3136	1.2707	1.1233	0.8956	0.6573	0.4487
2	1.8556	1.8396	1.7394	1.5032	1.1855	0.8740
3	1.8984	1.8995	1.8844	1.7859	1.5571	1.2567
4	1.6580	1.6686	1.7036	1.7420	1.7076	1.5496
5	1.2914	1.3060	1.3648	1.4764	1.6046	1.6656
6	0.9027	0.9176	0.9820	1.1204	1.3245	1.5496
7	0.5546	0.5675	0.6255	0.7582	0.9741	1.2567
8	0.2801	0.2895	0.3333	0.4378	0.6180	0.8740
9	0.0937	0.0987	0.1229	0.1823	0.2890	0.4487
10	0.0000	0.0000	0.0000	0.0000	0.0000	0.0000

Analysis is performed in the following order

1. Find $m = \sqrt[4]{\dfrac{kb}{4EI}}$ 2. Find $\lambda = lm$ 3. Find $p = \overline{p}\dfrac{M_i}{l^2}$, $M = \overline{M}M_i$ and $Q = \overline{Q}\dfrac{M_i}{l}$

k-Modulus of Subgrade Reaction, b-width of the beam, E-Modulus of Elasticity of the beam material, I-Moment of Inertia of the beam

Table 3.11

Point #	λ = 4					
	Point of application of the moment *M*					
Point #	0	1	2	3	4	5
	$\overline{p}$					
0	31.2088	27.8610	20.3743	12.6287	6.4262	2.2373
1	10.7220	11.8807	12.1621	9.6790	6.4862	3.6660
2	−0.8716	−0.2380	2.9440	6.0511	6.1611	4.9247
3	−5.9699	−5.7103	−4.2153	0.2014	4.4669	5.4915
4	−7.0464	−7.0127	−6.6047	−4.8603	−0.1758	4.3587
5	−6.0674	−6.1442	−6.3198	−6.1648	−4.6123	0.0000
6	−4.3303	−4.4417	−4.8485	−5.4577	−5.6753	−4.3587
7	−2.5315	−2.6349	−3.0628	−3.9050	−4.9273	−5.4915
8	−0.9259	−1.0016	−1.3464	−2.1360	−3.3968	−4.9247
9	−0.4956	0.4543	0.2319	−0.3862	−1.6279	−3.6660
10	1.8417	1.8360	1.7447	1.3283	0.1766	−2.2373
	$\overline{M}$					
0	**0.0000**	0.0000	0.0000	0.0000	0.0000	0.0000
1	−0.8781	**0.1127**	0.0882	0.0582	0.0322	0.0136
2	−0.6342	−0.6494	**0.2963**	0.2121	0.1287	0.0635
3	−0.3881	−0.4028	−0.4627	**0.4228**	0.2845	0.1615
4	−0.1950	−0.2063	−0.2559	−0.3632	**0.4800**	0.3117
5	−0.0690	−0.0764	−0.1107	−0.1915	−0.3259	**0.5000**
6	−0.0024	−0.0065	−0.0267	−0.0781	−0.1723	−0.3117
7	0.0210	0.0192	0.0093	−0.0179	−0.0724	−0.1615
8	0.0188	0.0182	0.0146	0.0036	−0.0205	−0.0635
9	0.0070	0.0069	0.0062	0.0038	−0.0021	−0.0136
10	0.0000	0.0000	0.0000	0.0000	0.0000	0.0000
	$\overline{Q}$					
0	0.0000	0.0000	0.0000	0.0000	0.0000	0.0000
1	2.0966	1.9871	1.6268	1.1154	0.6456	0.2952
2	2.5891	2.5692	2.3821	1.9019	1.2780	0.7247
3	2.2470	2.2718	2.3186	2.2145	1.8094	1.2455
4	1.5962	1.6357	1.7776	1.9816	2.0239	1.7380
5	0.9405	0.9778	1.1313	1.4303	1.7845	1.9560
6	0.4206	0.4485	0.5729	0.8492	1.2701	1.7380
7	0.0775	0.0947	0.1774	0.3811	0.7400	1.2455
8	−0.0953	−0.0871	−0.0431	0.0790	0.3238	0.7247
9	−0.1169	−0.1145	−0.0988	−0.0471	0.0726	0.2952
10	0.0000	0.0000	0.0000	0.0000	0.0000	0.0000

Analysis is performed in the following order

1. Find $m = \sqrt[4]{\dfrac{kb}{4EI}}$ 2. Find $\lambda = lm$ 3. Find $p = \overline{p}\dfrac{M_i}{l^2}$, $M = \overline{M}M_i$ and $Q = \overline{Q}\dfrac{M_i}{l}$

k-Modulus of Subgrade Reaction, *b*-width of the beam, *E*-Modulus of Elasticity of the beam material, *I*-Moment of Inertia of the beam

Table 3.12

Point #	0	1	2	3	4	5
			$\lambda = 5$			
		Point of application of the moment M				
			$\overline{p}$			
0	47.9693	40.6688	25.5801	12.1977	3.5276	−0.7612
1	10.5336	13.5694	15.2089	11.1958	6.3057	2.5761
2	−6.6002	−5.2843	1.7619	8.4811	8.3972	5.8160
3	−10.9751	−10.7008	−8.4335	−0.0915	6.6584	8.2088
4	−9.2832	−9.4821	−9.6921	−7.9439	0.0273	6.6065
5	−5.8570	−6.1747	−7.2560	−8.5202	−7.5648	0.0000
6	−2.8212	−3.0895	−4.1801	−6.2102	−8.1959	−6.6065
7	−0.8148	0.9835	−1.7487	−3.4558	−5.9649	−8.2088
8	0.2912	0.2187	−0.1631	−1.1869	−3.1046	−5.8160
9	0.8849	0.8938	0.8643	0.5740	−0.3987	−2.5761
10	1.3142	1.3970	1.6967	2.1225	2.1529	0.7612
			$\overline{M}$			
0	**0.0000**	0.0000	0.0000	0.0000	0.0000	0.0000
1	−0.8225	**0.1582**	0.1106	0.0593	0.0223	0.0018
2	−0.5059	−0.5342	**0.3682**	0.2277	0.1064	0.0291
3	−0.2340	−0.2570	−0.3512	**0.4712**	0.2699	0.1132
4	−0.0618	−0.0758	−0.1400	−0.2850	**0.4984**	0.2744
5	0.0205	0.0141	−0.0196	−0.1086	−0.2727	**0.5000**
6	0.0436	0.0418	0.0293	−0.0125	−0.1079	−0.2744
7	0.0368	0.0371	0.0354	0.0221	−0.0203	−0.1132
8	0.0203	0.0209	0.0225	0.0214	0.0088	−0.0291
9	0.0059	0.0061	0.0071	0.0080	0.0065	−0.0018
10	0.0000	0.0000	0.0000	0.0000	0.0000	0.0000
			$\overline{Q}$			
0	0.0000	0.0000	0.0000	0.0000	0.0000	0.0000
1	2.9251	2.7119	2.0395	0.1697	0.4917	0.0907
2	3.1218	3.1262	2.8880	2.1535	1.2268	0.5103
3	2.2431	2.3269	2.5544	2.5700	2.0296	1.2116
4	1.2301	1.3178	1.6481	2.1712	2.4139	2.0023
5	0.4731	0.5349	0.8007	1.3480	2.0370	2.3827
6	0.0392	0.0717	0.2289	0.6115	1.2490	2.0023
7	−0.1426	−0.1319	−0.0675	0.1281	0.5409	1.2116
8	−0.1688	−0.1702	−0.1631	−0.1042	0.0875	0.5103
9	−0.1100	−0.1145	−0.1280	−0.1348	−0.0877	0.0907
10	0.0000	0.0000	0.0000	0.0000	0.0000	0.0000

Analysis is performed in the following order

1. Find $m = \sqrt[4]{\dfrac{kb}{4EI}}$ 2. Find $\lambda = lm$ 3. Find $p = \overline{p}\dfrac{M_i}{l^2}$, $M = \overline{M}M_i$ and $Q = \overline{Q}\dfrac{M_i}{l}$

k-Modulus of Subgrade Reaction, b-width of the beam, E-Modulus of Elasticity of the beam material, I-Moment of Inertia of the beam

Table 3.13

			$\lambda = 6$			
		Point of application of the moment M				
Point #	0	1	2	3	4	5
				$\overline{p}$		
0	67.7867	54.2986	28.5526	9.2938	−0.3078	−3.1256
1	7.3100	13.7745	18.1966	12.3163	5.6191	1.3386
2	−14.1562	−12.0611	0.5797	12.1504	10.8738	6.1678
3	−15.1097	−15.1893	−12.9776	0.0616	11.9574	10.8546
4	−9.3884	−10.1046	−11.9782	−11.6090	0.2963	11.7264
5	−3.8922	−4.5209	−6.9131	−10.6065	−11.4020	0.0000
6	−0.6464	−1.0007	−2.6153	−6.1271	−10.5781	−11.7264
7	0.6238	0.4932	−0.2503	−2.3415	−6.2117	−10.8546
8	0.7811	0.7758	0.6156	−0.1677	−2.2796	−6.1678
9	0.5104	0.5644	0.7391	0.8926	0.5039	−1.3386
10	0.1483	0.2389	0.6546	1.5679	2.7494	3.1256
				$\overline{M}$		
0	**0.0000**	0.0000	0.0000	0.0000	0.0000	0.0000
1	−0.7619	**0.2040**	0.1255	0.0515	0.0083	−0.0032
2	−0.3856	−0.4299	**0.4209**	0.2209	0.1717	−0.0024
3	−0.1167	−0.1465	−0.2712	**0.4918**	0.2369	0.0649
4	0.0122	−0.0012	−0.0688	−0.2359	**0.5005**	0.2343
5	0.0468	0.0437	0.0206	−0.0585	−0.2331	**0.5000**
6	0.0388	0.0401	0.0396	0.0185	−0.0598	−0.2343
7	0.0210	0.0231	0.0292	0.0332	0.0136	0.0649
8	0.0076	0.0089	0.0139	0.0217	0.0242	0.0024
9	0.0013	0.0017	0.0034	0.0067	0.0100	0.0082
10	0.0000	0.0000	0.0000	0.0000	0.0000	0.0000
				$\overline{Q}$		
0	0.0000	0.0000	0.0000	0.0000	0.0000	0.0000
1	3.7548	3.4037	2.3375	1.0805	0.2656	−0.0894
2	3.4125	3.4893	3.2763	2.3038	1.0902	0.2860
3	1.9492	2.1268	2.6564	2.9144	2.2318	1.1371
4	0.7243	0.8621	1.4086	2.3371	2.8445	2.2661
5	0.0603	0.1308	0.4640	1.2263	2.2892	2.8525
6	−0.1666	−0.1452	−0.0124	0.3896	1.1902	2.2661
7	−0.1678	−0.1706	−0.1557	−0.0338	0.3507	1.1371
8	−0.0975	−0.1072	−0.1374	−0.1593	−0.0739	0.2860
9	−0.0329	−0.0402	−0.0697	0.1230	−0.1627	−0.0894
10	0.0000	0.0000	0.0000	0.0000	0.0000	0.0000

Analysis is performed in the following order

1. Find $m = \sqrt[4]{\dfrac{kb}{4EI}}$ 2. Find $\lambda = lm$ 3. Find $p = \overline{p}\dfrac{M_i}{l^2}$, $M = \overline{M}M_i$ and $Q = \overline{Q}\dfrac{M_i}{l}$

k-Modulus of Subgrade Reaction, b-width of the beam, E-Modulus of Elasticity of the beam material, I-Moment of Inertia of the beam

Table 3.14

Point #	0	1	2	3	4	5
			λ = 7			
		Point of application of the moment *M*				
			$\overline{p}$			
0	90.2681	68.0632	28.9367	4.7606	−3.7246	−4.0516
1	0.7264	12.6498	21.5576	13.2153	4.5964	0.1493
2	-22.3793	−19.7542	−0.1465	17.0102	13.0055	5.6014
3	−17.2399	−18.2686	−17.6916	0.3824	16.8947	12.7938
4	−7.4519	−8.9028	−13.4450	−16.1346	0.3413	16.4644
5	−1.2930	−2.1374	−5.7472	−12.3983	−16.1660	0.0000
6	0.9232	0.6488	−0.9506	−5.3562	−12.4750	−16.4644
7	1.0544	1.0612	0.7592	−0.9308	−5.4805	−12.7938
8	0.5753	0.6525	0.8376	0.6823	−0.9784	−5.6014
9	0.1015	0.1598	0.4051	0.8440	1.0506	−0.1493
10	−0.3010	−0.2816	−0.0940	0.6106	2.1474	4.0516
			$\overline{M}$			
0	**0.0000**	0.0000	0.0000	0.0000	0.0000	0.0000
1	−0.6979	**0.2480**	0.1324	0.0379	−0.0048	−0.0133
2	−0.2778	−0.3392	**0.4565**	0.2002	0.0366	−0.0229
3	−0.0344	−0.0675	−0.2140	**0.4985**	0.2005	0.0263
4	0.0443	0.0347	−0.0250	−0.1991	**0.4992**	0.1976
5	0.0425	0.0435	0.0352	−0.0244	−0.1985	**0.5000**
6	0.0211	0.0244	0.0332	0.0319	−0.0243	−0.1976
7	0.0055	0.0077	0.0165	0.0303	0.0307	−0.0263
8	−0.0005	0.0003	0.0047	0.0147	0.0268	0.0229
9	−0.0008	−0.0007	0.0040	0.0034	0.0089	0.0133
10	0.0000	0.0000	0.0000	0.0000	0.0000	0.0000
			$\overline{Q}$			
0	0.0000	0.0000	0.0000	0.0000	0.0000	0.0000
1	4.5497	4.0357	2.5247	0.8988	0.0436	−0.1951
2	3.4671	3.6804	3.5953	2.4101	0.9237	0.0924
3	1.4861	1.7793	2.7034	3.2797	2.4187	1.0122
4	0.2515	0.4207	1.1465	2.4921	3.2805	2.4751
5	−0.1857	−0.1313	0.1869	1.0654	2.4893	3.2983
6	−0.2042	−0.2057	−0.1480	0.1777	1.0572	2.4751
7	−0.1053	−0.1202	−0.1575	−0.1366	0.1594	1.0122
8	−0.0239	−0.0345	−0.0777	−0.1490	−0.1635	0.0924
9	0.0100	0.0061	−0.0156	−0.0727	−0.1599	−0.1951
10	0.0000	0.0000	0.0000	0.0000	0.0000	0.0000

Analysis is performed in the following order

1. Find $m = \sqrt[4]{\dfrac{kb}{4EI}}$ 2. Find $\lambda = lm$ 3. Find $p = \overline{p}\dfrac{M_i}{l^2}$, $M = \overline{M}M_i$ and $Q = \overline{Q}\dfrac{M_i}{l}$

k-Modulus of Subgrade Reaction, b-width of the beam, E-Modulus of Elasticity of the beam material, I-Moment of Inertia of the beam

Table 3.15

#	0	1	2	3	4	5	6	7	8	9	10
$\overline{p}$	−0.4126	−0.2349	0.1298	1.0150	2.5068	3.5794	2.5068	1.0150	0.1298	−0.2349	−0.4126
p	−8.7618	−4.9883	2.7564	21.554	53.234	76.0111	53.234	21.554	2.7564	−4.9883	−8.7618
y	−0.0044	−0.0025	0.0014	0.0108	0.0266	0.0380	0.0266	0.0108	0.0014	−0.0025	−0.0044

Table 3.16

#	0	1	2	3	4	5	6	7	8	9	10
$\overline{p}$	−0.8856	0.8234	2.6175	3.7068	2.5633	1.0107	0.1161	−0.1732	−0.1647	−0.0717	0.0291
p	−18.8063	17.4850	55.5844	78.716	54.433	21.4630	2.4650	3.6773	−3.4975	−1.5226	0.6180
y	−0.0094	0.00874	0.0278	0.0394	0.0272	0.0107	0.0012	0.0018	−0.0017	−0.0008	0.0003
$\overline{M}$	0	−0.0016	0.0052	0.0370	0.0022	−0.0077	−0.0064	−0.0030	−0.0007	0.0000	0
M	0	−1,883.6	6,121.8	43,558.7	2,590.0	−9,065.0	−7,534.5	−3,531.8	−824.10	0.0000	0
$\overline{Q}$	0	−0.0031	0.1689	0.4852 −0.5148	−0.2013	−0.0226	0.0337	0.0308	0.0139	−0.0021	0
Q	0	−15.50	844.50	2,426 −2,574	−1,006.5	−113.0	168.5	154.0	69.50	−10..5	0

Table 3.17

Left end \ Right end	Free Supported	Pin Supported	Guided	Fixed	Elastic Vertical Support	Elastic Rotational Support
Free Supported	1	2	3	4	5	6
Pin Supported	7	8	9	10	11	12
Guided	13	14	15	16	17	18
Fixed	19	20	21	22	23	24
Elastic Vertical	25	26	27	28	29	30
Elastic Rotational	31	32	33	34	35	36

Table 3.18

Left end \ Right end	Free Supported	Pin Supported	Guided	Fixed	Elastic Vertical Support	Elastic Rotational Support
Free Supported	1	2	3	4	5	6
Pin Supported	7	8	9	10	11	12
Guided	13	14	15	16	17	18
Fixed	19	20	21	22	23	24
Elastic Vertical	25	26	27	28	29	30
Elastic Rotational	31	32	33	34	35	36

Table 3.19

Point	0	1	2	3	4	5	6	7	8	9	10
$\overline{M}$	0	0.0039	0.0163	0.0384	0.0711	0.1152	0.0711	0.0384	0.0163	0.0039	0
M	0	1.5600	6.5200	15.3600	**28.440**	46.080	28.440	15.360	6.520	1.560	0

Table 3.20

Point	0	1	2	3	4	5	6	7	8	9	10
$\overline{M}$	0	0.0073	0.0263	0.0527	0.0819	0.1091	0.129	0.1351	0.1210	0.0788	0
M	0	0.473	1.706	3.418	5.312	7.076	8.366	8.762	7.847	5.111	0

Table 3.21

Moments										
0	1	2	3	4	5	6	7	8	9	10
0	2.033	8.226	18.778	33.752	53.156	36.806	24.122	14.367	6.671	0

Table 3.22

Point	0	1	2	3	4	5	6	7	8	9	10
$\overline{p}_{x_1}$	−6.0163	-42622	−2.7325	−1.5281	−0.6586	−0.0814	0.2696	0.4652	0.5687	0.6289	0.6768
p_{x_1}	−6.626	−4.695	−3.010	−1.683	−0.725	−0.090	0.297	0.512	0.626	0.693	0.745
$\overline{p}_{x_2}$	−0.6768	−0.6289	−0.5687	−0.4652	−0.2696	0.0814	0.6586	1.5281	2.7325	4.2622	6.0163
p_{x_2}	−0.205	−0.190	−0.172	−0.141	−0.082	0.025	0.199	0.463	0.827	1.290	1.820
$\overline{p}_p$	0.6831	1.0070	1.3153	1.5620	1.6586	1.5216	1.2314	0.8742	0.4973	0.1194	−0.2567
p_p	6.831	10.070	13.153	15.620	16.586	15.216	12.314	8.742	4.973	1.194	−2.567
$\sum p$	0	5.185	9.971	13.796	15.779	15.151	12.810	9.717	6.426	3.177	0

Table 3.23

Point	0	1	2	3	4	5	6	7	8	9	10
$\overline{M}$	0	−0.0018	−0.0291	−0.1132	−0.2744	0.5/−0.5	0.2744	0.1132	0.0291	0.0018	0
$\overline{M}M$	0	−1.3020	−21.04	−81.876	−198.469	$\dfrac{-361.64}{361.64}$	198.469	81.876	21.04	1.3020	0

Table 3.24

Point	0	1	2	3	4	5	6	7	8	9	10
$\overline{M}_P$	0	0.0026	0.0171	0.0525	0.0128	−0.0058	−0.0106	−0.0087	−0.0047	−0.0014	0
M_P	0	0.5200	3.420	10.500	2.5600	−1.160	−2.1200	−1.7400	−0.940	−0.2800	0
$\overline{M}_X$	0	0.0024	0.0056	0.003	−0.0133	−0.0509	−0.0133	0.0030	0.0056	0.0024	0
M_X	0	0.1230	0.286	0.153	−0.6790	−2.600	−0.6790	0.1530	0.2860	0.1230	0
$M_P + M_X$	0	0.6430	3.706	10.653	1.8810	−3.760	−2.7990	−1.5870	−0.6540	−0.157	0

Table 3.25

Point	0	1	2	3	4	5	6	7	8	9	10
$\overline{M}_X$	0	−0.0024	−0.0056	−0.003	0.0133	0.0509	0.0133	−0.003	−0.0056	−0.0024	0
M	0	−0.1230	−0.2860	−0.153	0.679	2.600	0.679	−0.153	−0.2860	−0.1230	0

Table 3.26

Point	0	1	2	3	4	5	6	7	8	9	10
$\overline{p}$	−0.4577	−0.1423	0.1792	0.5162	0.8721	1.2337	1.5593	1.7674	1.7812	1.6840	1.5564
p	−1,526	−474	597	1,721	2,907	4,112	5,198	5,891	5,937	5,613	5,189

Table 3.27

Point	0	1	2	3	4	5	6	7	8	9	10
$\bar{p}$	0.6768	0.6289	0.5687	0.4652	0.2606	−0.0814	−0.6586	−1.5281	−2.7325	−4.2622	−6.0163
p	105.00	98.21	88.81	72.65	40.70	−12.71	−102.85	−238.62	−426.73	−665.61	−939.00

Table 3.28

Point	0	1	2	3	4	5	6	7	8	9	10
p	−1,421	−375.8	685.81	1,793.7	2,947.7	4.099	5,095.15	5,652.4	5,510.3	4,947.4	4,250

Table 3.29 Settlements and rotations of free supported finite beams

Beam and load	$P=1$	$P=1$	$M=1$	$M=1$
Settlement	$\omega_P = -\omega_P^0 \dfrac{m}{kb}$	$\omega_P = -\omega_P^0 \dfrac{m}{kb}$	$\omega_M = \omega_M^0 \dfrac{m^2}{kb}$	$\omega_M = -\omega_M^0 \dfrac{m^2}{kb}$
Slope	$\varphi_P = -\varphi_P^0 \dfrac{m^2}{kb}$	$\varphi_P = \varphi_P^0 \dfrac{m^2}{kb}$	$\varphi_M = \varphi_M^0 \dfrac{m^3}{kb}$	$\varphi_M = \varphi_M^0 \dfrac{m^3}{kb}$

Table 3.30 Numerical values of coefficients φ_M^0 φ_P^0

Coefficient	Point 0	Point *l*	Point 0	Point	Point 0	Point	Point 0	Point *l*
	$\lambda = 0.8$		$\lambda = 1.0$		$\lambda = 1.2$		$\lambda = 1.4$	
φ_M^0	24.624	23.028	13.480	11.451	8.713	6.340	6.422	3.681
φ_P^0	9.509	9.296	6.208	5.877	4.465	3.992	3.463	2.827
ω_M^0	9.509	−9.296	6.208	−5.877	4.465	−3.992	3.463	−2.827
ω_P^0	5.519	−2.485	4.038	−1.972	3.398	−1.618	2.959	−1.353
Coefficient	$\lambda = 1.6$		$\lambda = 1.8$		$\lambda = 2.0$		$\lambda = 2.2$	
φ_M^0	5.243	2.159	4.630	1.224	4.305	0.620	4.140	0.223
φ_P^0	2.860	2.045	2.492	1.487	2.268	1.070	2.136	0.750
ω_M^0	2.860	−2.045	2.492	−1.487	2.268	−1.070	2.136	−0.750
ω_P^0	2.649	−1.139	2.430	−0.959	2.275	−0.800	2.169	−0.658
Coefficient	$\lambda = 2.4$		$\lambda = 2.6$		$\lambda = 2.8$		$\lambda = 3.0$	
φ_M^0	4.062	−0.038	4.029	−0.202	4.018	−0.297	4.015	−0.340
φ_P^0	2.062	0.502	2.062	0.310	2.007	0.164	2.001	0.056
ω_M^0	2.062	−0.529	2.062	−0.310	2.007	−0.164	2.001	−0.056
ω_P^0	2.099	0.000	2.099	−0.414	2.028	−0.313	2.013	−0.226
Coefficient	$\lambda = 3.2$		$\lambda = 3.4$		$\lambda = 3.6$		$\lambda = 3.8$	
φ_M^0	4.015	−0.347	4.014	−0.327	4.013	−0.292	4.011	−0.252
φ_P^0	2.000	−0.018	2.001	−0.068	2.002	−0.097	2.003	−0.109
ω_M^0	2.000	0.018	2.001	0.068	2.002	0.097	2.003	−0.109
ω_P^0	2.006	−0.154	2.003	−0.095	2.002	0.049	2.002	−0.017
Coefficient	$\lambda = 4.0$		$\lambda = 4.2$		$\lambda = 4.4$		$\lambda = 4.6$	
φ_M^0	4.008	−0.207	4.006	−0.166	4.004	−0.120	4.003	−0.088
φ_P^0	2.003	−0.112	2.003	−0.106	2.002	−0.093	2.002	−0.080
ω_M^0	2.003	0.112	2.003	0.106	2.002	0.093	2.002	0.080
ω_P^0	2.002	0.007	2.001	0.022	2.001	0.032	2.001	0.036
Coefficient	$\lambda = 4.8$		$\lambda = 5.0$					
φ_M^0	4.002	−0.062	4.000	−0.032				
φ_P^0	2.001	0.066	2.001	−0.051				
ω_M^0	2.001	0.066	2.001	0.051				
ω_P^0	2.001	0.035	2.001	0.032				

References

1. Klepikov, S. N. 1967. *Analysis of structures on elastic foundation*. Kiev: Budivelnik, pp. 57–73.
2. Korenev, G. B., and E. I. Chernigovskaya. 1962. *Analysis of plates on elastic foundation*. Moscow: Gosstroiizdat, pp. 190–201.
3. Kornevits, E. F., and G. F. Ender. 1932. *Formulae for analysis of beams on elastic foundation*. Moscow: Gosstroiizdat.
4. Scott, R. F. 1981. *Foundation analysis*. Civil Engineering and Engineering Mechanics Series. Englewood Cliffs, NJ: Prentice Hall, pp. 137–138.

4

Analysis of Beams on Elastic Half-Space

4.1 Introduction

As mentioned earlier in Chapter 1, some soils under applied pressure behave as elastic material following the rules of the theory of elasticity, and soil models such as elastic half-space and elastic layer may produce more realistic practical results compared to results obtained by using Winkler foundation. But analysis of beams supported on elastic half-space is a difficult problem and few works are devoted to this subject. Most known publications are scientific and research papers written by academics and researchers that cannot be used by practicing engineers. Nevertheless, there are some methods that allow performing analysis of simple beams supported on elastic half-space as well as on elastic layer. These methods can be used for hand calculations and computer analysis as well. Chapter 4 presents equations and tables for analysis of free supported beams on elastic half-space developed by Borowicka (1938, 1939) and Gorbunov-Posadov (1940, 1949, 1953, 1984). Theoretical bases of the method are not discussed in this book. The reader can find a detailed explanation of the method in the original publications mentioned above and a good review of the method by Selvadurai (1979). This chapter also includes the analysis of complex beams, such as beams with various boundary conditions, various continuous beams, stepped beams developed by the author of this book (Tsudik 2006), and analysis of frames on elastic half-space including direct methods as well as iterative methods. Tables are developed for beams that meet the following two requirements: $a/b \geq 7$ and the width of the beam is narrow enough so the bending in the transverse direction can be ignored; a is half of the length of the beam and b is half of the width of the beam, as shown in Figures 4.1 and 4.2. Tables are developed for three categories of beams: 1. rigid beams, 2. short beams, and 3. long beams.

In order to find out if the beam belongs to rigid beams, the following parameter is obtained:

$$t = \pi E_0 a^3 b / \left[2(1 - v^2) E_1 I \right] \qquad (4.1)$$

All tables can be found at the end of the chapter.

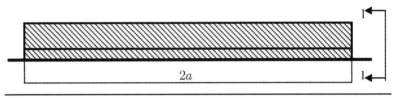

Figure 4.1

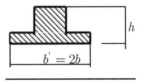

Figure 4.2

where E_0 is the modulus of elasticity of the soil, v is Poisson's ratio of the soil, E_1 is the modulus of elasticity of the beam material, and I is the moment of inertia of the beam. The beam belongs to rigid beams when $t \leq 0.5$. The beam belongs to rigid beams also when $\alpha = a/b < 20$ and $0.5 \leq t \geq 1$.

In order to find out if the beam belongs to long beams, parameter L of the system beam-soil is found as follows: $L = \sqrt[3]{\dfrac{2E_1 I(1 - v^2)}{b' E_0}}$ where $b' = 2b$ and coefficients λ and β are obtained as $\lambda = \dfrac{a}{L}$ and $\beta = \dfrac{b}{L}$. The beam belongs to long beams when $\beta < 0.15$ and $\lambda > 1$ or when $\beta \leq 0.30$ and $\lambda > 2$ or when $\beta \leq 0.50$ and $\lambda > 3.5$. If parameters of the beam do not meet requirements for rigid and long beams, the beam belongs to the short beams and analysis of such beams is performed using tables for analysis of short beams. Tables for analysis of short and long beams are shown at the end of the chapter. Now, we can start with the analysis of rigid beams. Tables and equations presented in Chapter 4 are developed only for analysis of simple free supported beams.

The author of this book (Tsudik 2006) proposed to use the method of forces for analysis of complex beams that includes simple beams with various boundary conditions and continuous beams, including beams with various intermediate and end supports.

4.2 Analysis of Rigid Beams

Tables 4.1–4.5 are developed for analysis of rigid beams on elastic half-space for two types of loads: concentrated vertical loads and moments applied to the center of the beam. If a concentrated vertical load P_0 is applied to the center of the beam, the following equations are used:

$$p = \overline{P}_0 \frac{P_0}{b' \cdot a}, \; Q = \pm \overline{Q}_0 \cdot P_0, \; M = \overline{M}_0 \cdot P_0 \cdot a, \; Y = \overline{Y}_0 \frac{1 - v^2}{E_0} \cdot \frac{P_0}{a}, \; tg\varphi = 0 \quad (4.2)$$

In equations 4.2, p is the soil pressure, Q is the shearing force, M is the moment, and Y is the settlement of the beam. Coefficients $\overline{P}_0$, $\overline{Q}_0$, $\overline{M}_0$, and $\overline{Y}_0$ are dimensionless and taken from Table 4.2. Rotation of the beam for this type of load, $tg\varphi = 0$. If the beam is loaded with a concentrated moment m_0 equations 4.3 are used:

$$
\left.
\begin{aligned}
p &= \pm \overline{P}_1 \frac{m_0}{b'a^2}, \quad Q = \overline{Q}_1 \frac{m_0}{a}, \quad M = \pm \overline{M}_1 m_0 \\
Y &= \overline{tg}\varphi_1 \frac{1-v^2}{E_0} x \frac{m_0}{a^2}, \quad tg\varphi = \pm \overline{tg}\varphi_1 \frac{1-v^2}{E_0} \cdot \frac{m_0}{a_0}
\end{aligned}
\right\}
\tag{4.3}
$$

In these equations, $x = x'/a$, where x' is the distance of the cross section from the center of the beam and is always positive regardless of where the cross section is located, whether at the left or right half of the beam. Two signs ($\pm$) mean that the upper sign belongs to the right half and the lower sign belongs to the left half of the beam. Tables 4.2 and 4.4 do not contain information for beams with $7 \leq \alpha < 10$. In order to obtain the soil pressure, shearing forces, and moments in this case, it is recommended to use data for $\alpha = 10$. However, settlements and rotations of the beam, when $\alpha < 10$, cannot be obtained from these tables and have to be found from equations 4.4 shown below:

$$
\left.
\begin{aligned}
\omega_p &= \frac{1-v^2}{E_0} K_0 \frac{P_0}{\sqrt{A}} \\
tg\varphi &= \frac{1-v^2}{E_0} \cdot \frac{K_1 m_x}{a^3}
\end{aligned}
\right\}
\tag{4.4}
$$

In 4.4, coefficient K_0 is dimensionless and is obtained from Table 4.1; A is the area of the foundation. Coefficient K_1 is also dimensionless and obtained from Table 4.1; K_x is the moment applied to the beam in the direction of the length. Table 4.1 is built using specified values of these two coefficients. Equations 4.4 are obtained for totally rigid individual rectangular foundations and used here for beam analysis when $\alpha < 10$. Analysis of the beam is performed as follows. Each half of the beam is divided into 10 sections, as shown in Figure 4.3. Since the loads are symmetrical or asymmetrical, coefficients are given only for the right half of the beam starting from point 0.

If the beam has a series of vertical concentrated loads, they have to be replaced with one load and one moment applied to the center of the beam. If the beam has

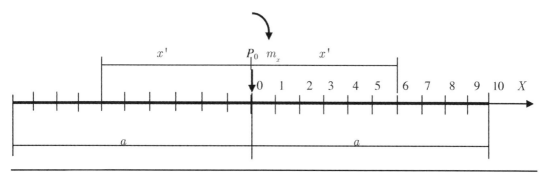

Figure 4.3

a distributed load located at any part of the beam, this load is replaced with a series of concentrated vertical loads and all these loads, in turn, are also replaced with one moment and one vertical load applied to the center of the beam. The total moment applied to the center of the beam m_0 is equal to the sum of all moments produced by vertical loads P_i and moments m_i applied to the beam. The total vertical load P_0 is equal to the sum of all vertical loads P_i including distributed loads $q(x)$. The moment is positive if applied clockwise. When P_0 and m_0 are found, the soil pressure is obtained as follows:

$$p = \overline{p}_0 \frac{P_0}{b'a} \pm \overline{p}_1 \frac{m_0}{b'a^2} \qquad (4.5)$$

Coefficients $\overline{p}_0$ and $\overline{p}_1$ are obtained from Tables 4.2 and 4.4, respectively, taking into account the actual value of parameters $\alpha = a/b$ and $x = x'/a$ at each point of the beam. Vertical concentrated loads P_i, distributed loads $q(x)$, and moments applied to the beam will produce shear forces $Q(x)$. The shearing force at any point of the beam is obtained from equation 4.6:

$$Q(x) = \pm \overline{Q}_0 P_0 + \overline{Q}_1 \frac{m_0}{a} - Q_{ext} \qquad (4.6)$$

where $\overline{Q}_0$ and $\overline{Q}_1$ are found from Tables 4.2 and 4.4, respectively, and Q_{ext} for the right half of the beam is equal to $\sum P_x - P_0$. For the left half of the beam $Q_{ext} = \sum P_x$, where $\sum P_x$ includes all vertical loads located at distance x between the left end of the beam and the point the shear is obtained. Vertical concentrated loads P_i, distributed loads $q(x)$, and moments m_i will also produce moments that are obtained from equation 4.7:

$$M = \overline{M}_0 P_0 a \pm \overline{M}_1 m_0 + M_{ext} \qquad (4.7)$$

In 4.7 coefficients $\overline{M}_0$ and $\overline{M}_1$ are obtained from Tables 4.2 and 4.4, respectively. M_{ext} for the right half of the beam are found from equation 4.8:

$$M_{ext} = \sum M_x + \sum m_x + P_0 ax - m_0 \qquad (4.8)$$

where $\sum M_x$ is the sum of all moments produced by all vertical concentrated (P_i) and distributed loads $q(x)$ located at left from point x; $\sum m_x$ is the sum of all concentrated moments applied to the beam and located at left from point x.

For the left half of the beam:

$$M_{ext} = \sum M_x + \sum m_x \qquad (4.9)$$

The settlement of the beam is obtained from equation 4.10:

$$Y = \left(\overline{Y}_0 P_0 \pm tg\varphi_1 x \frac{m_0}{a} \right) \frac{(1 - \nu^2)}{aE_0} \qquad (4.10)$$

The soil pressure, shear forces, moments, settlements, and rotations of the beam can be found using equations presented above and Tables 4.1–4.5. A simple numerical example that illustrates analysis of rigid beams supported on elastic half-space is shown next.

Example 4.1

Given: A reinforced concrete beam loaded with a vertical load $P = 200t$ is shown in Figure 4.4. The length of the beam is $2a = 10m$, the cross section is equal to $b' \cdot h = 2m \cdot 1.5m(h) = 3m^2$ ($b' = 2b$), and the moment of inertia of the beam section is equal to $I = \dfrac{b'h^3}{12} = \dfrac{2 \cdot 1.5^3}{12} = 0.5625m^3$. The beam is loaded with a vertical concentrated load equal to 200t. The modulus of elasticity of the soil is $E = 300kg/cm^2 = 3000t/m^2$; Poisson's ratio $\lambda = 0.4$ and the modulus of elasticity of the concrete is $E_1 = 3,000,000t/m^2$. Find the soil pressure, shearing forces, and moments. Build the soil pressure, shear, and moment diagrams.

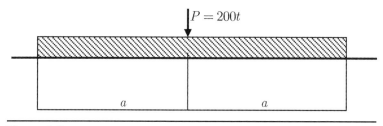

Figure 4.4

Solution:

1. Find parameter $t = \pi E a^3 b/[2(1 - \lambda^2)EI] = 3.14 \cdot 3,000 \cdot 5^3 \cdot 1/(1.68 \cdot 0.5625 \cdot 3,000,000) = 0.415$. Since $t < 0.5$, the beam is rigid.
2. Find $\alpha = \dfrac{a}{b} = \dfrac{5}{1} = 5$ and coefficients $\overline{P}_0$ from Table 4.2. As can be seen, the closest parameter $\alpha = 10$. So all coefficients are obtained for $\alpha = 10$. The soil pressure diagram is shown in Figure 4.5.
3. Find the soil pressure. The soil pressure is obtained, for convenience, only in all other points of the beam.

$p_0 = \overline{P}_0 P_0/b \cdot a = 0.439 \cdot 200/1 \cdot 5 = 17.56t/m$; $p_{0.1} = 0.440 \cdot 40 = 17.6t/m$;
$p_{0.2} = 0.442 \cdot 40 = 17.68t/m$; $p_{0.3} = 0.446 \cdot 40 = 17.84t/m$; $p_{0.4} = 0.445 \cdot 40 = 18.2t/m$;
$p_{0.5} = 0.462 \cdot 40 = 18.48t/m$; $p_{0.6} = 0.475 \cdot 40 = 19t/m$; $p_{0.7} = 0.498 \cdot 40 = 19.92t/m$;
$p_{0.8} = 0.541 \cdot 40 = 21.64t/m$; $p_{0.9} = 0.632 \cdot 40 = 25.28t/m$; $p_{1.0} = 0.842 \cdot 40 = 33.68t/m$

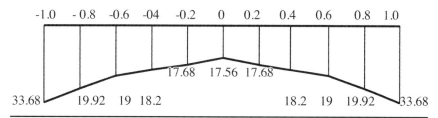

Figure 4.5   The soil pressure diagram

The total soil pressure applied to the beam is equal to:

$\sum\limits_{P_{0.1}}^{P_{1.0}} p_i + 0.5 p_0 = 198.2t$, that is close to the load applied to the beam $P_0 = 200t$. The difference is only 0.9%. The shear diagram is built analogously, as shown in Figure 4.6. Shear forces are found below using the second equation from 4.2:

$Q_0 = -\overline{Q}_0 \cdot P_0 = -0.5 \cdot 200 = -100t$; $Q_{0.1} = -0.456 \cdot 200 = -91.2t$;
$Q_{0.2} = -0.412 \cdot 200 = -52.4t$; $Q_{0.3} = -0.367 \cdot 200 = -73.4t$; $Q_{0.4} = -0.322 \cdot 200 = -64.4t$;
$Q_{0.5} = -0.277 \cdot 200 = -55.4t$; $Q_{0.6} = -0.230 \cdot 200 = -46.0t$; $Q_{0.7} = -0.182 \cdot 200 = -36.4t$;
$Q_{0.8} = -0.130 \cdot 200 = -26.4t$; $Q_{0.9} = -0.072 \cdot 200 = -14.4t$; $Q_{1.0} = 0$.

Obtained moments and the moment are shown below. The moment diagram is shown in Figure 4.7.

$M_0 = \overline{M}_0 \cdot P_0 \cdot a = 0.2703 \cdot 200 \cdot 5 = 270.3tm$; $M_{0.1} = 0.2225 \cdot 200 \cdot 5 = 222.5tm$;
$M_{0.2} = 0.1791 \cdot 200 \cdot 5 = 179.1tm$; $M_{0.3} = 0.1401 \cdot 200 \cdot 5 = 140.1tm$;
$M_{0.4} = 0.1056 \cdot 200 \cdot 5 = 105.6tm$; $M_{0.5} = 0.0756 \cdot 200 \cdot 5 = 75.6tm$;
$M_{0.6} = 0.0502 \cdot 200 \cdot 5 = 50.2tm$; $M_{0.7} = 0.0295 \cdot 200 \cdot 5 = 29.5tm$;
$M_{0.8} = 0.0139 \cdot 200 \cdot 5 = 13.9tm$; $M_{0.9} = 0.0037 \cdot 200 \cdot 5 = 3.7tm$; $M_{1.0} = 0$

Taking into account that $\alpha = 5 < 10$, the settlement of the beam is obtained from the first equation 4.4:

$$\omega_p = \frac{1 - v^2}{E_0} K_0 \frac{P_0}{\sqrt{A}} = \frac{1 - 0.4^2}{3,000} \cdot 0.77 \cdot \frac{200}{\sqrt{10 \cdot 2}} = 0.964cm$$

The soil pressure, shear, and moment diagrams are shown in Figures 4.5, 4.6, and 4.7, respectively.

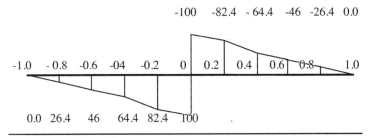

Figure 4.6   Shear diagram

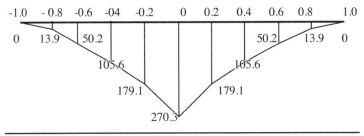

Figure 4.7   Moment diagram

Example 4.2

The given beam is loaded with asymmetrical loads as shown in Figure 4.8, assuming $P_1 = 100t$ $M_1 = 10t$ $x_1' = 3m$ $x_2' = 2m$. Find the shear forces and rotation of the beam.

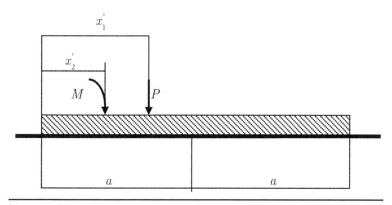

Figure 4.8

Solution:

1. The total moment applied to the center of the beam is equal to:

$$m_0 = M - P(a - x_1') = 25 - 100(5 - 3) = -175 \text{tm}$$

2. The total vertical concentrated load applied to the center of the beam is equal to:

$$P_0 = P = 100t$$

3. The shear forces are obtained below. For the right half of the beam we have:

$$Q_0 = \overline{Q}_0 P_0 + \overline{Q}_1 \frac{m_0}{a} - Q_{ext} = -0.5 \cdot 100 + (-0.709) \cdot \frac{(-175)}{5} = -25.185t$$

$$Q_{0.1} = \overline{Q}_0 P_0 + \overline{Q}_1 \frac{m_0}{a} - Q_{ext} = -0.456 \cdot 100 + (-0.704) \cdot \frac{(-175)}{5} = -20.96t$$

$$Q_{0.2} = \overline{Q}_0 P_0 + \overline{Q}_1 \frac{m_0}{a} - Q_{ext} = -0.412 \cdot 100 + (-0.686) \cdot \frac{(-175)}{5} = -17.19t$$

$$Q_{0.3} = \overline{Q}_0 P_0 + \overline{Q}_1 \frac{m_0}{a} - Q_{ext} = -0.367 \cdot 100 + (-0.658) \cdot \frac{(-175)}{5} = -13.67t$$

$$Q_{0.4} = \overline{Q}_0 P_0 + \overline{Q}_1 \frac{m_0}{a} - Q_{ext} = -0.322 \cdot 100 + (-0.617) \cdot \frac{(-175)}{5} = -10.605t$$

$$Q_{0.5} = \overline{Q}_0 P_0 + \overline{Q}_1 \frac{m_0}{a} - Q_{ext} = -0.277 \cdot 100 + (-0.563) \cdot \frac{(-175)}{5} = -7.995t$$

$$Q_{0.6} = \overline{Q}_0 P_0 + \overline{Q}_1 \frac{m_0}{a} - Q_{ext} = -0.230 \cdot 100 + (-0.496) \cdot \frac{(-175)}{5} = -5.64t$$

$$Q_{0.7} = \overline{Q}_0 P_0 + \overline{Q}_1 \frac{m_0}{a} - Q_{ext} = -0.182 \cdot 100 + (-0.413) \cdot \frac{(-175)}{5} = -3.545t$$

$$Q_{0.8} = \overline{Q}_0 P_0 + \overline{Q}_1 \frac{m_0}{a} - Q_{ext} = -0.130 \cdot 100 + (-0.311) \cdot \frac{(-175)}{5} = -2.115t$$

$$Q_{0.9} = \overline{Q}_0 P_0 + \overline{Q}_1 \frac{m_0}{a} - Q_{ext} = -0.072 \cdot 100 + (-0.180) \cdot \frac{(-175)}{5} = -0.900t$$

$$Q_1 = 0$$

For the left half of the beam:

$$Q_0 = \overline{Q}_0 P_0 + \overline{Q}_1 \frac{m_0}{a} - Q_{ext} = 0.5 \cdot 100 - (0.709) \cdot \frac{(-175)}{5} - 100 = -25.185t$$

$$Q_{-0.1} = \overline{Q}_0 P_0 + \overline{Q}_1 \frac{m_0}{a} - Q_{ext} = 0.456 \cdot 100 - (0.704) \cdot \frac{(-175)}{5} - 100 = -29.76t$$

$$Q_{-0.2} = \overline{Q}_0 P_0 + \overline{Q}_1 \frac{m_0}{a} - Q_{ext} = 0.412 \cdot 100 + (-0.686) \frac{(-175)}{5} - 100 = -35.79t$$

$$Q_{-0.3} = \overline{Q}_0 P_0 + \overline{Q}_1 \frac{m_0}{a} - Q_{ext} = 0.367 \cdot 100 + (-0.658) \frac{(-175)}{5} - 100 = -40.27t$$

$$Q_{-0.4} = \overline{Q}_0 P_0 + \overline{Q}_1 \frac{m_0}{a} - Q_{ext} = 0.322 \cdot 100 + (-0.617) \frac{(-175)}{5} - 100 = -46.205t$$

$$Q_{-0.5} = \overline{Q}_0 P_0 + \overline{Q}_1 \frac{m_0}{a} - Q_{ext} = 0.277 \cdot 100 + (-0.563) \frac{(-175)}{5} = 47.401t$$

$$Q_{-0.6} = \overline{Q}_0 P_0 + \overline{Q}_1 \frac{m_0}{a} - Q_{ext} = 0.230 \cdot 100 + (-0.496) \frac{(-175)}{5} = 40.36t$$

$$Q_{-0.7} = \overline{Q}_0 P_0 + \overline{Q}_1 \frac{m_0}{a} - Q_{ext} = 0.182 \cdot 100 + (-0.413) \frac{(-175)}{5} = 32.655t$$

$$Q_{-0.8} = \overline{Q}_0 P_0 + \overline{Q}_1 \frac{m_0}{a} - Q_{ext} = 0.130 \cdot 100 + (-0.311) \frac{(-175)}{5} = 23.885t$$

$$Q_{-0.9} = \overline{Q}_0 P_0 + \overline{Q}_1 \frac{m_0}{a} - Q_{ext} = 0.072 \cdot 100 + (-0.180) \frac{(-175)}{5} = 13.5t$$

$$Q_{-1.0} = 0.$$

Taking into account that $\alpha = 5 < 10$, rotation of the foundation is obtained using the second equation from 4.4.

The shear diagram is shown in Figure 4.9.

$$tg\varphi = \frac{1 - v^2}{E_0} \cdot \frac{K_1 m_x}{a^3} = \frac{1 - 0.4^2}{3,000} \cdot 1.45 \cdot \frac{(-175)}{2^3} = -0.00888$$

$m_0 = m_x = -175tm$. $K_1 = 1.45$ is obtained from Table 4.1 for $\alpha = 5$.

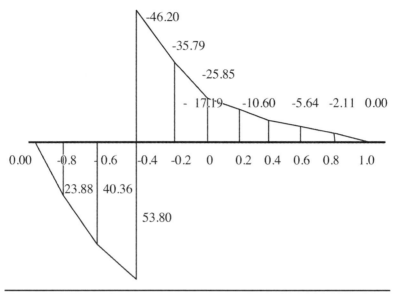

Figure 4.9   Shear diagram

4.3 Short Beam Analysis

Analysis of short beams is performed using Tables 4.6–4.13 that are developed for analysis of beams with concentrated vertical loads P and moments m. Table 4.14 is developed for analysis of short beams with a uniformly distributed load q along the length of the beam. The beam is divided into 20 sections of equal length. The length of each section is equal to $0.1\,a$ where a is half of the beam's length, as shown previously in Figure 4.3. Although the tables are developed for beams with $a/b = 10$, they produce good practical results when $7 \leq a/b \geq 15$.

Tables 4.6–4.9 and equations 4.11 allow obtaining moments, soil pressures, shear forces, and settlements of the beam due to applied vertical load P. These equations look as follows:

$$M = \overline{M}Pa, \quad p = \overline{p}\frac{P}{b'a} \quad Q = \overline{Q}P \quad \omega = \overline{\omega}\frac{(1-v^2)P}{E_0 a} \tag{4.11}$$

In order to use Tables 4.6–4.9 and equations 4.11, we have to obtain parameter t from equation 4.1 and two other parameters $\delta = \dfrac{d}{a}$ and $x = \dfrac{x'}{a}$, where d is the distance between the point the load is applied and the center of the beam, and x' is the distance from the center of the beam and the point in which M, p, Q, and ω are obtained. It is important to mention that coefficients $\overline{M}$ and $\overline{Q}$ have to be multiplied by 10^{-3}; coefficients $\overline{p}$ and $\overline{\omega}$ have to be multiplied by 10^{-2}.

Tables 4.10–4.13 and equations 4.12 allow calculating moments, shear forces, soil pressures, and settlements of the beam due to applied concentrated moments m:

$$M = \overline{M}m, \quad Q = \overline{Q}\frac{m}{a}, \quad p = \overline{p}\frac{m}{b'a^2}, \quad \omega = \overline{\omega}\frac{(1-v^2)m}{E_0 a^2} \tag{4.12}$$

The use of Tables 4.10–4.13 and equations 4.12 requires the following data: parameter t that is obtained from equation 4.1, parameters $\delta = \dfrac{d}{a}$ and $x = \dfrac{x'}{a}$, where d is the distance between the center of the beam and the point the moment m is applied, and x' is the distance between the center of the beam and the point in which M, Q, p, and ω have to be found. If the distance d is not divisible by $0.1a$, the concentrated load (vertical load or moment) should be moved to the closest point the beam is divided. All tables are developed for the following numerical values of t: 1, 2, 5, and 10. If $t < 0.75$ analysis is performed using tables for rigid beams (Tables 4.1–4.5), when $t > 10$ analysis is performed using Tables 4.15–4.19 for analysis of long beams.

All numerical values of $\overline{M}$ and $\overline{Q}$ in the tables, for convenience, are multiplied by 1000 and all numerical values of $\overline{p}$ and $\overline{Y}$ are multiplied by 100. If, for example, in tables for $\overline{M}$ or $\overline{Q}$ the values are shown as equal to –045, the actual value of this coefficient is equal to –0.045, while in tables for $\overline{p}$ and $\overline{Y}$ it is equal to –0.45.

In Tables 4.6–4.9, the shearing force $\overline{Q}$ is shown in bold at all points of the beam, where the vertical load P is applied. This shear is equal to the shear $\overline{Q}_L$ at left from the point the load is applied. The shear at right from this point is found as follows: $\overline{Q}_R = \overline{Q}_L - 1$. When the beam is loaded at any point with a moment m as shown in tables 4.10–4.13, the moment shown **bold** is the moment at left from the point where the moment is applied and is equal to $\overline{M}_L$. The moment at right from that point is equal to $\overline{M}_R = \overline{M}_L + 1$. Two numerical examples illustrate the use of Tables 4.6–4.9 and equations 4.11 and Tables 4.10–4.13 and equations 4.12.

Example 4.3

The given beam loaded with a vertical concentrated load is shown in Figure 4.10. The data are: the modulus of elasticity of the soil $E_0 = 3,000t/m^2$, the total length of the beam $2a = 12m$, $b' = 2b = 1.6m$, Poisson's ratio $\nu = 0.40$, the modulus of elasticity of the beam material $E_1 = 2,600,000t/m^2$, the height of the beam $h = 1m$, the load $P = 100t$ applied to the beam, and $x' = 2.4m$ at right from the center. Build the moment and shear diagrams.

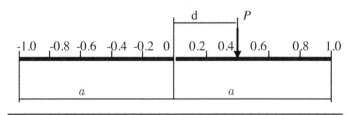

Figure 4.10

To find out the category the beam belongs to, we find parameter t:

$$t = \pi E_0 a^3 b / \left[2(1 - v^2) E_1 I \right]$$

$$= 3.14 \cdot 2,300 \cdot 6^3 \cdot 0.8 / 1.68 \cdot 2,600,000 \cdot \frac{1.6 \cdot 1^3}{12} = 2.14 > 0.5$$

Since $t = 2.14 > 0.5$, the beam is not rigid. Since $L = \sqrt[3]{\dfrac{2 \cdot 3,000,000 \dfrac{1.6 \cdot 1^3}{12}}{1.6 \cdot 2,300}} = 6.013$,

and $\beta = b/L = 0.8/6.013 = 0.133 < 0.15$, but $\lambda = a/L = 6/6.013 = 0.997 < 1$, the beam belongs to short beams.

Solution:

Using Table 4.7 for $t = 2$ and $\delta = d/a = 2.4/6 = 0.4$, we can find the moments, shear forces, soil pressure, and settlements in all 21 points of the beam. Below, the moments are obtained using the first equation from 4.11:

$M = \overline{M}Pa = \overline{M} \cdot 100 \cdot 6 \cdot 10^{-3} = 0.6\overline{M}$.
$M_{-1} = 0$, $M_{-0.9} = 0.6$tm, $M_{-0.8} = -1.2$tm, $M_{-0.7} = -1.8$tm, $M_{-0.6} = -2.4$tm, $M_{-0.5} = -1.8$tm,
$M_{-0.4} = 0$, $M_{-0.3} = 3.0$tm, $M_{-0.2} = 8.4$tm, $M_{-0.1} = 16.2$tm, $M_{-0.0} = 27.0$tm,
$M_{0.1} = 40.80$tm, $M_{0.2} = 58.20$tm, $M_{0.3} = 79.20$tm, $M_{0.4} = 105.0$tm, $M_{0.5} = 74.40$tm,
$M_{0.6} = 49.20$tm, $M_{0.7} = 28.80$tm, $M_{0.8} = 13.80$tm, $M_{0.9} = 3.60$tm, $M_{1.0} = 0$.

The moment diagram is shown in Figure 4.11.

The shear diagram can be built analogously. Shear forces are obtained using the third equation from 4.11:

$Q = \overline{Q}P = 100\overline{Q} \cdot 10^{-3} = 0.1\overline{Q}$
$Q_{-1} = 0$, $Q_{-0.9} = -11 \cdot 0.1 = -1.1$t, $Q_{-0.8} = -15 \cdot 0.1 = -1.5$t, $Q_{-0.7} = -11 \cdot 0.1 = -1.1$t,
$Q_{-0.6} = 0 \cdot 0.1 = 0$, $Q_{-0.5} = 17 \cdot 0.1 = 1.7$t, $Q_{-0.4} = 42 \cdot 0.1 = 4.2$t, $Q_{-0.3} = 72 \cdot 0.1 = 7.2$t,
$Q_{-0.2} = 109 \cdot 0.1 = 10.9$t, $Q_{-0.1} = 152 \cdot 0.1 = 15.2$t, $Q_0 = 202 \cdot 0.1 = 20.2$t,
$Q_{0.1} = 258 \cdot 0.1 = 25.8$t, $Q_{0.2} = 320 \cdot 0.1 = 32.0$t, $Q_{0.3} = 338 \cdot 0.1 = 33.8$t,
$Q_{0.4} = 461 \cdot 0.1 = 46.1$t, $Q_{0.5} = -462 \cdot 0.1 = -46.2$t, $Q_{0.6} = -382 \cdot 0.1 = -38.2$t,
$Q_{0.7} = -299 \cdot 0.1 = -29.9$t, $Q_{0.8} = -213 \cdot 0.1 = -21.3$t, $Q_{0.9} = -117 \cdot 0.1 = -11.7$t,
$Q_1 = 0$.

The shear diagram is shown in Figure 4.12. Numerical values of the moments and shear forces are found for all 21 points of the beam, but shown, for convenience, in both diagrams only at all other points.

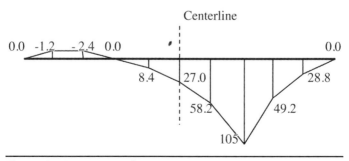

Centerline

0.0 -1.2 -2.4 0.0   0.0

8.4   27.0   28.8

58.2   49.2

105

Figure 4.11 Moment diagram

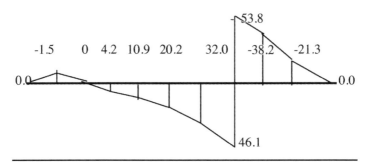

Figure 4.12   Shear diagram

Example 4.4

The given beam is shown in Figure 4.13. Data: $t = 2$, $\delta = \dfrac{d}{a} = \dfrac{2.7}{5.4} = 0.5$, $m = 240\text{tm}$. Find the soil pressure and settlements of the beam.

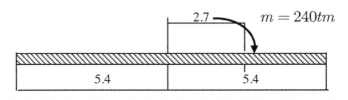

Figure 4.13

By introducing the given numerical data into equations 4.12 we find:

$$M = \overline{M}m = 240\overline{M}, Q = \overline{Q}\frac{m}{a} = \frac{240}{5.4}\overline{Q} = 44.44\overline{Q}, p = \overline{p}\frac{m}{b'a^2} = \overline{p}\frac{240}{1.4 \cdot 5.4^2} = 5.88\overline{p},$$

$$\omega = \overline{\omega}\frac{(1-v^2)m}{E_0 a^2} = \frac{0.84 \cdot 240}{2,100 \cdot 5.4^2}\overline{\omega} = 0.0032921\overline{\omega}$$

Now, the soil pressure and settlements of the beam can be found.

$$p_{-1.0} = -5.88 \cdot 2.1 = -12.3480\text{t/m}, \; p_{-0.9} = -5.88 \cdot 1.43 = -8.4080\text{t/m},$$
$$p_{-0.8} = -5.88 \cdot 1.08 = -6.3504\text{t/m}, \; p_{-0.7} = -5.88 \cdot 0.87 = -5.1156\text{t/m},$$
$$p_{-0.6} = -5.88 \cdot 0.73 = -4.2924\text{t/m}, \; p_{-0.5} = -5.88 \cdot 0.60 = -3.5280\text{t/m},$$
$$p_{-0.4} = -5.88 \cdot 0.49 = -2.8800\text{t/m}, \; p_{-0.3} = -5.88 \cdot 0.38 = -2.2344\text{t/m},$$
$$p_{-0.2} = -5.88 \cdot 0.27 = -1.5876\text{t/m}, \; p_{-0.1} = -5.88 \cdot 0.17 = -1.0000\text{t/m},$$
$$p_{-0.0} = -5.88 \cdot 0.07 = -0.4116\text{t/m}, \; p_{0.1} = 5.88 \cdot 0.04 = 0.2352\text{t/m},$$
$$p_{0.2} = 5.88 \cdot 0.15 = 0.8820\text{t/m}, \; p_{0.3} = 5.88 \cdot 0.28 = 1.6464\text{t/m},$$
$$p_{0.4} = 5.88 \cdot 0.43 = 2.5284\text{t/m}, \; p_{0.5} = 5.88 \cdot 0.58 = 3.4104\text{t/m},$$
$$p_{0.6} = 5.88 \cdot 0.75 = 4.4100\text{t/m}, \; p_{0.7} = 5.88 \cdot 0.95 = 5.5860\text{t/m},$$
$$p_{0.8} = 5.88 \cdot 1.21 = 7.1148\text{t/m}, \; p_{0.9} = 5.88 \cdot 1.60 = 9.4080\text{t/m},$$
$$p_{1.0} = 5.88 \cdot 2.32 = 13.6416\text{t/m}.$$

Settlements of the beam are found analogously:

$$\omega = \overline{\omega}\frac{(1-v^2)m}{E_0a^2} = \frac{0.84 \cdot 240}{2,100 \cdot 5.4^2}\overline{\omega} = 0.0032921\overline{\omega}m = 0.329\text{cm}$$

$\omega_{-1.0} = -0.32921 \cdot 0.87 = -0.286\text{cm}$, $\omega_{-0.9} = -0.32921 \cdot 0.90 = -0.296\text{cm}$,
$\omega_{-0.8} = -0.32921 \cdot 0.94 = -0.395\text{cm}$, $\omega_{-0.7} = -0.32921 \cdot 0.96 = -0.316\text{cm}$,
$\omega_{-0.6} = -0.32921 \cdot 0.98 = -0.322\text{cm}$, $\omega_{-0.5} = -0.32921 \cdot 0.99 = -0.326\text{cm}$,
$\omega_{-0.4} = -0.32921 \cdot 1.00 = -0.329\text{cm}$, $\omega_{-0.3} = -0.32921 \cdot 0.98 = -0.322\text{cm}$,
$\omega_{-0.2} = -0.32921 \cdot 0.94 = -0.395\text{cm}$, $\omega_{-0.1} = -0.32921 \cdot 0.87 = -0.286\text{cm}$,
$\omega_{0.00} = -0.32921 \cdot 0.76 = -0.250\text{cm}$, $\omega_{0.1} = -0.32921 \cdot 0.60 = -0.198\text{cm}$,
$\omega_{0.2} = -0.32921 \cdot 0.39 = -0.128\text{cm}$, $\omega_{0.3} = -0.32921 \cdot 0.12 = -0.0395\text{cm}$,
$\omega_{0.4} = 0.32921 \cdot 0.23 = 0.076\text{cm}$, $\omega_{0.5} = 0.32921 \cdot 0.67 = 0.221\text{cm}$,
$\omega_{0.6} = 0.32921 \cdot 1.14 = 0.375\text{cm}$, $\omega_{0.7} = 0.32921 \cdot 1.58 = 0.520\text{cm}$,
$\omega_{0.8} = 0.32921 \cdot 2.02 = 0.665\text{cm}$, $\omega_{0.9} = 0.32921 \cdot 2.44 = 0.665\text{cm}$,
$\omega_{1.0} = 0.32921 \cdot 2.87 = 0.945\text{cm}$.

Soil pressure and settlements' diagrams are shown in Figures 4.14 and 4.15, respectively. Numerical values of the soil pressure and settlements, for convenience, are not shown at all points of the beam. Moment and shear diagrams can be built analogously.

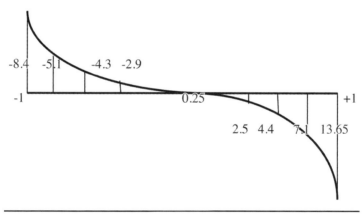

Figure 4.14 Soil pressure diagram

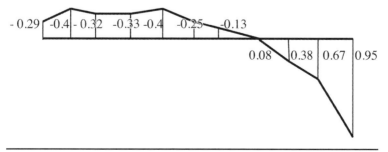

Figure 4.15 Diagram of the settlements

4.4 Analysis of Short Beams with Uniformly Distributed Loads

Tables and equations for such type of analysis are developed only for beams loaded completely from the left to the right ends, as shown in Figure 4.16.

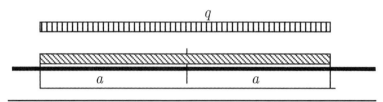

Figure 4.16

Equations for obtaining M, Q, p, and Y are shown below:

$$M = \overline{M}a^2 q, \; Q = \overline{Q}aq, \; p = \overline{p}q, \; Y = \overline{Y}\frac{1-v^2}{E_0}q \tag{4.13}$$

Coefficients $\overline{M}$, $\overline{Q}$, $\overline{p}$, and $\overline{Y}$ are taken from Table 4.14. Since the load applied to the beam is symmetrical, all coefficients are given only for the right half of the beam. Coefficient $\overline{Q}$ for the left half of the beam should be taken with the opposite sign.

Parameter x has the same meaning as in the tables shown earlier and is equal to $x = \dfrac{x'}{a}$, where x' is the distance from the center of the beam and the point coefficients $\overline{M}$, $\overline{Q}$, $\overline{p}$, and $\overline{Y}$ are found. A numerical example is shown below.

It is important to mention that rotations of short beams at all points, except the left and right ends, are found from the following formula: $tg\varphi = (Y_{i+1} - Y_{i-1})/0.2a$. Rotations of the beam at the ends are found as follows:

$$tg\varphi = (Y_i - Y_{i-1})/0.1a$$

where Y_i is the settlement of the beam at point i.

Example 4.5

The given beam is 10m long, parameter $t = 10$m. The beam is loaded with a uniformly distributed load $q = 100$t/m, coefficient $t = 10$. Build the moment and soil pressure diagrams.

Table 4.14 and equations 4.13 are used to build the moment and soil pressure diagrams shown in Figures 4.17 and 4.18, respectively. Because of the beam symmetry both diagrams are built only for the right half of the beam.

Let us check the equilibrium of vertical loads applied to the beam. The total uniformly distributed load applied to the beam is equal to $100 \cdot 5 = 500$t. The total soil pressure is equal to:

$$98 \cdot 0.25 + 2 \cdot 97 \cdot 0.5 + 3 \cdot 96 \cdot 0.5 + 93 \cdot 0.5 + 95 \cdot 0.5 + 100 \cdot 0.5 + 110 \cdot 0.5$$
$$+ 134 \cdot 0.25 = 498t$$

As shown, the difference between the total applied load and total soil pressure is negligible: $(498 \approx 500)$.

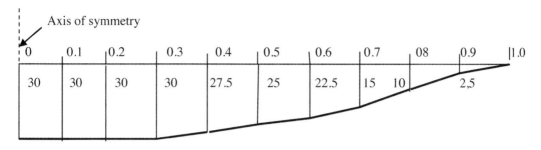

Figure 4.17 Moment diagram

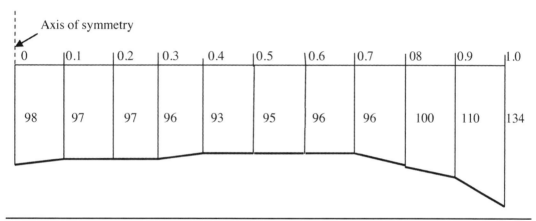

Figure 4.18 Soil pressure diagram

4.5 Analysis of Long Beams

As mentioned earlier, the beam belongs to the category of long beams when:

$\beta < 0.15$ and $\lambda > 1$ or $\beta \leq 0.3$ and $\lambda > 2$ or $\beta \leq 0.5$ and $\lambda > 3.5$, where $\beta = \dfrac{b}{L}$ and $\lambda = \dfrac{a}{L}$ $L = \sqrt[3]{\dfrac{2E_1I(1-\lambda^2)}{b'E_0}}$. Tables are developed only for analysis of beams with concentrated vertical loads. If the beam is loaded with distributed loads, these loads have to be replaced with a series of concentrated vertical loads. When a moment is applied to the beam, the beam should be analyzed as a short beam. Analysis is performed using Tables 4.17–4.21 and the following equations:

$$p = \overline{p}\frac{P}{L}, \quad M = \overline{M}PL, \quad Q = \overline{Q}P, \quad Y = \overline{Y}\frac{1-\lambda^2}{E_0} \cdot \frac{P}{L} \tag{4.14}$$

Coefficients $\overline{p}$, $\overline{M}$, $\overline{Q}$, and $\overline{Y}$ are obtained from Tables 4.17–4.21 that are developed for the following numerical values of β: 0.025, 0.075, 0.15, 0.3, 0.5. Analysis is performed in several steps:

Step 1: Choose the right table for analysis taking into account the following:

When $0.01 \leq \beta \leq 0.04$, use Table 4.17 for $\beta = 0.025$.
When $0.04 < \beta \leq 0.10$, use Table 4.18 for $\beta = 0.075$.

When $0.10 < \beta \le 0.20$, use Table 4.19 for $\beta = 0.15$.
When $0.20 < \beta \le 0.40$, use Table 4.20 for $\beta = 0.30$.
When $0.40 < \beta \le 0.70$, use Table 4.21 for $\beta = 0.50$.

If the beam has several vertical loads P_i, as shown in Figure 4.19, analysis is performed for each load and final results are obtained by superposition. Analysis of long beams is based on analysis of infinite and semi-infinite beams.

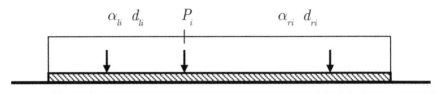

Figure 4.19

Step 2: After the table for analysis is chosen, the so-called conditional distances between each load and the left and right ends of the beam have to be found. Conditional distances, α_{li} and α_{ri}, are obtained as follows:

$$\alpha_{li} = \frac{d_{li}}{L} \quad \alpha_{ri} = \frac{d_{ri}}{L}$$

where d_{li} and d_{ri} are the actual distances between the load and the left and right ends, respectively. Both types of distances are shown in Figure 4.19. When performing analysis of long beams two cases are possible:

1. When $\beta \le 0.2$ and one of two values $(\alpha_{li}, \alpha_{ri})$ is less than 1, or when $\beta > 0.2$ and one of the same values is less than 2. In this case analysis of the beam is performed using the smallest of the two α. The beam is analyzed as a semi-infinite beam.
2. When $\beta \le 0.2$ and one of two values $(\alpha_{li}, \alpha_{ri})$ is larger than 1, or when $\beta > 0.2$ and one of the same two values is larger than 2. In this case analysis is performed using $\alpha = \infty$. The beam is analyzed as an infinite beam.

Example 4.6

Given: A beam shown in Figure 4.20, $L = 2\text{m}$ $\beta = 0.075$ $\alpha_l = 0.4$ $\alpha_r = 2.2$ $E_0 = 180\text{kg/cm}^2$ $P = 60\text{t}$ $v_0 = 0.3$ $l = 6.2\text{m}$.

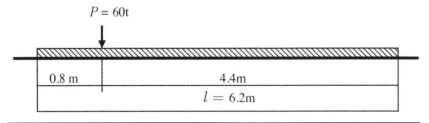

Figure 4.20

Find the soil pressure and settlements of the beam.

Solution:

$$p = \overline{p}\frac{P}{L} = \overline{p}\frac{60000}{200} = 300\overline{p} \quad Y = \overline{Y}\frac{1 - \lambda_0^2}{E_0}\frac{P}{L} = \overline{Y}\frac{1 - 0.3^2}{180} \cdot \frac{60,000}{200} = 1.5$$

Analysis is performed in Table 4.15.

The soil pressure and settlements' diagrams are shown in Figures 4.21 and 4.22, respectively. The soil pressure is shown in kg/m, the settlements in cm.

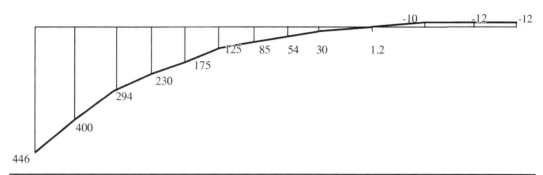

Figure 4.21   Soil pressure diagram

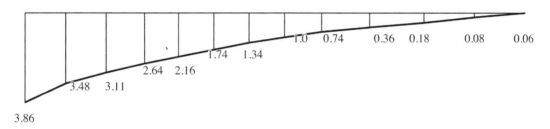

Figure 4.22   Settlements diagram

Example 4.7

The given beam is shown in Figure 4.23.

$$L = 3\text{m}, \ \beta = 0.3 \ \alpha_l = \frac{3.6}{3} = 1.2, \ \alpha_r = \frac{4.2}{3} = 1.4, \ Q = \overline{Q}P = \overline{Q}\cdot 80$$

Find the shear forces and build the shear diagram.

Analysis is performed in Table 4.16. Using Table 4.20 for $\beta = 0.3$ and $\alpha_l = 1.2$, numerical values of $\overline{Q}$ are found and final shear forces are obtained. The shear diagram is shown in Figure 4.24. The soil pressure, moments, and settlements of the beam can be obtained analogously.

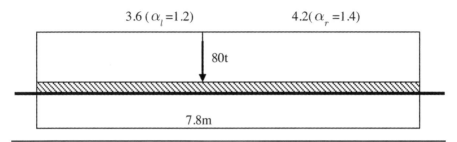

Figure 4.23

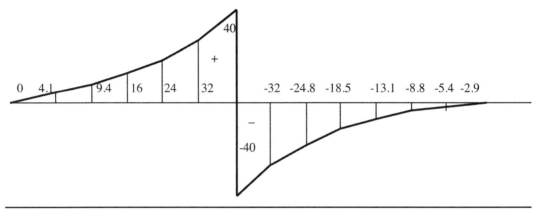

Figure 4.24   Shear diagram

4.6 Analysis of Complex Beams

All tables presented and used in this chapter are developed for analysis of free supported beams. When the ends of the beam are restrained against various deflections, analysis becomes more complex. For example, analysis of the beam shown in Figure 4.25 is performed easily using one of the tables presented in this chapter. Analysis of the beam shown in Figure 4.26, with one restraint against vertical deflection at the left end and two other restraints against vertical deflection and rotation at the right end, requires solving a system of three equations that reflect the actual boundary conditions. Analysis, in principle, is not different from the analysis used in Chapter 3 for finite beams on Winkler foundation. The same method is applied to analysis of beams supported on elastic half-space using tables and equations introduced above. Analysis of the beam shown in Figure 4.26 is performed as follows: After removal of all three restraints and replacing them with unknown reactions X_1, X_2, and X_3, the beam will look as shown in Figure 4.27.

Figure 4.25

Figure 4.26

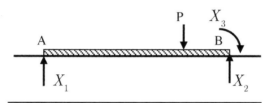

Figure 4.27

Now, the beam is analyzed applying the given load P and all three unknown reactions successively assuming that all unknown reactions are equal to one unit. All analyses are performed using tables and equations presented earlier. Using the results of these analyses, the following system of equations is written:

$$\left.\begin{array}{l} \delta_{11}X_1 + \delta_{12}X_2 + \delta_{13}X_3 + \delta_{1P} = 0 \\ \delta_{21}X_1 + \delta_{22}X_2 + \delta_{23}X_3 + \delta_{2P} = 0 \\ \delta_{31}X_1 + \delta_{32}X_2 + \delta_{33}X_3 + \delta_{3P} = 0 \end{array}\right\} \qquad (4.15)$$

In this system, δ_{ik} is deflection of the beam at point i due to one unit load applied at point k, δ_{iP} is the deflection of the beam at point i due to the given load P. By solving system 4.15, reactions X_1, X_2, and X_3 are found and final analysis of the beam is performed by applying to the beam all given loads and reactions. The same method can be used for analysis of continuous beams, including beams with spring supports. For example, analysis of the beam shown in Figure 4.28 is performed as follows: After removing all restraints-supports and replacing them with unknown reactions, the given beam will look as shown in Figure 4.29. The total number of unknowns in this case is equal to the number of removed restraints. The system of equations will look as follows:

$$\left.\begin{array}{l} \delta_{11}X_1 + \delta_{12}X_2 + \delta_{13}X_3 + \delta_{14}X_4 + \delta_{15}X_5 + \delta_{1P} = 0 \\ \delta_{21}X_1 + \delta_{22}X_2 + \delta_{23}X_3 + \delta_{24}X_4 + \delta_{25}X_5 + \delta_{2P} = 0 \\ \delta_{31}X_1 + \delta_{32}X_2 + \left(\delta_{33} + \dfrac{1}{C_D}\right)X_3 + \delta_{34}X_4 + \delta_{35}X_5 + \delta_{3P} = 0 \\ \delta_{41}X_1 + \delta_{42}X_2 + \delta_{43}X_3 + \delta_{44}X_4 + \delta_{45}X_5 + \delta_{4P} = 0 \\ \delta_{51}X_1 + \delta_{52}X_2 + \delta_{53}X_3 + \delta_{54}X_4 + \delta_{55}X_5 + \delta_{5P} = 0 \end{array}\right\} \qquad (4.16)$$

In this system, C_D is the rigidity of an elastic support D. All other coefficients in this system have the same meaning as in system 4.15. By solving the system of equations 4.16 all unknown reactions are found. Final analysis of the beam is performed by applying to the beam the given loads and obtained reactions.

As shown, analysis of these types of beams is much more time consuming compared to analysis of beams with free supported ends. Moreover, the method can be used only when the given beam is replaced with one free supported beam with the given loads and unknown reactions, as shown in Figure 4.29. When analysis requires replacement of the given complex beam with several simple beams, the method cannot be used because each simple beam supported on elastic half-space produces settlements not

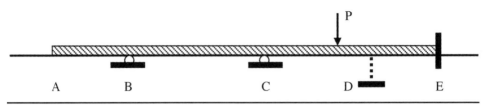

Figure 4.28

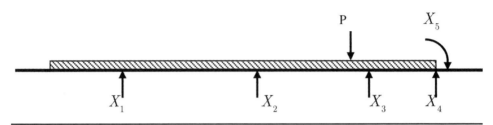

Figure 4.29

only under its loaded area, but also under other neighboring beams that are not taken into account. This method of analysis is justified for beams on Winkler foundation where all beams work independently without affecting each other, but when the beam is supported on elastic half-space, this method cannot be applied directly. However, we propose a simple method that can resolve this problem. Let us assume we have to analyze a pin-connected beam shown in Figure 4.30. Analysis is performed in the following steps:

Step 1: The given beam is divided into two simple beams, as shown in Figure 4.31, and unknown forces of interaction X_1 are applied to both beams at points 2. Taking into account that settlements of the beams 1–2 and 2–3 at point 2 are the same, equation 4.17 is written:

$$\left(\omega_{1(1-2)} + \omega_{1(2-3)}\right)X_1 + \omega_{1P_1} + \omega_{1P_2} = 0 \qquad (4.17)$$

where $\omega_{1(1-2)}$ and $\omega_{1(2-3)}$ are settlements in direction X_1 of the beams 1–2 and 2–3 due to one unit loads applied at point 2; ω_{1P_1} and ω_{1P_2} are settlements of the beams 1–2 and 2–3 due to the given loads P_1 and P_2. Reactions X_1 obtained from 4.17 are used for separate analyses of both beams.

Step 2: Both beams are analyzed separately as two simple beams.

Step 3: In order to specify analysis of one of the beams, let's say beam 1–2, by taking into account **the soil pressure produced** by the beam 2–3, the soil pressure under beam 2–3 is replaced with individual loads-stamps applied to the soil in 21 points. For example, the load applied at point i can be found from equation 4.18:

$$P_i = \frac{p_i l}{20} \qquad (4.18)$$

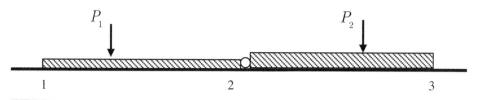

Figure 4.30

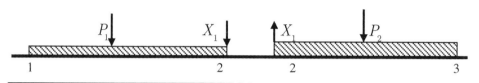

Figure 4.31

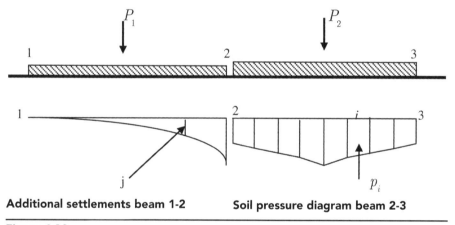

Additional settlements beam 1-2   Soil pressure diagram beam 2-3

Figure 4.32

where p_i is the soil pressure per one unit of the beam length 2–3 and l is the length of the beam. By applying loads P_i to the half-space under beam 2–3 and using equations 1.23 and 1.25 from Chapter 1, additional settlements under beam 1–2 are found. These settlements will produce additional moments and shear forces in beam 1–2. Assuming that beam 1–2 is supported at all 21 points on non-yielding supports, as shown in Figure 4.33, and applying the settlements to all supports, analysis of beam 1–2 is performed.

Now, by summing obtained results with the results of the original analysis of beam 1–2, final moments and shear forces in beam 1–2 are found. The same method is used for analysis of beam 2–3.

It is obvious that this method can be applied to analysis of stepped beams and groups of close located foundations.

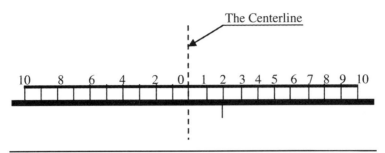

Figure 4.33   Beam 1–2 supported on 21 supports

4.7 Analysis of Frames on Elastic Half-Space

Analysis of continuous foundations supported on the elastic half-space discussed above can take into account the rigidity of the superstructure. A method of analysis of foundations and 2D frames using the frame shown in Figure 4.34 is explained below. Analysis is performed in the following order:

The given system frame-foundation is divided into two parts: frame and foundation, as shown in Figure 4.35. All frame supports are restrained against any deflections. By applying successively to the frame the given loads and one unit deflections at all points of support, the following system of equations is written:

$$\left.\begin{aligned}
X_1 &= r_{11}\omega_1 + r_{12}\omega_2 + \ldots + r_{1n}\omega_n + r_{1P} \\
X_2 &= r_{21}\omega_1 + r_{22}\omega_2 + \ldots + r_{2n}\omega_n + r_{2P} \\
&\ldots\ldots\ldots\ldots\ldots\ldots\ldots\ldots\ldots\ldots\ldots\ldots\ldots\ldots \\
X_n &= r_{n1}\omega_1 + r_{n2}\omega_2 + \ldots + r_{nn}\omega_n + r_{nP}
\end{aligned}\right\} \tag{4.19}$$

In this system of equations:

> X_i is the unknown reaction at point i
> r_{ij} is the reaction at point i due to one unit load applied at point j
> ω_i is the settlement or rotation at point i
> r_{iP} is the reaction at point i due to the given loads applied to the frame.

Now, by applying the given loads, and unknown reactions acting in the opposite direction to the foundation, all deflections of the foundation can be expressed as follows:

$$\left.\begin{aligned}
\omega_1 &= \Delta_{11}X_1 + \Delta_{12}X_2 + \ldots + \Delta_{1n}X_n + \omega_{1P} \\
\omega_2 &= \Delta_{21}X_1 + \Delta_{22}X_2 + \ldots + \Delta_{2n}X_n + \omega_{2P} \\
&\ldots\ldots\ldots\ldots\ldots\ldots\ldots\ldots\ldots\ldots\ldots\ldots\ldots\ldots \\
\omega_n &= \Delta_{n1}X_1 + \Delta_{n2}X_2 + \ldots + \Delta_{nn}X_n + \omega_{nP}
\end{aligned}\right\} \tag{4.20}$$

In this system of equations, Δ_{ij} is deflection of the foundation at point i due to reaction equal to one unit and applied to point j, and ω_{iP} is the deflection of the foundation at point i due to the given loads applied to the foundation.

The number of unknowns can be reduced by introducing reactions X_i from 4.19 into the system of equations 4.20 that leads to a system of equations with only n unknown deflections. In this case, the stiffness method is used. If deflections ω_i from the

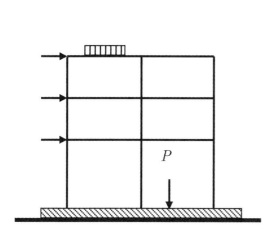

Figure 4.34

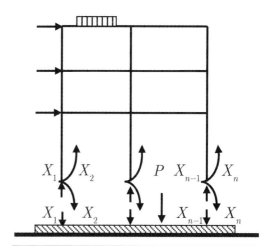

Figure 4.35

system of equations 4.20 are introduced into the system of equations 4.19, we again obtain a system of equations with only n unknown reactions. In this case, the method of forces is used.

By solving both systems of equations 4.19 and 4.20 as one system, we find $2n$ unknowns: n deflections and n reactions. In this case we use the so-called combined method of analysis. Final analysis of the frame is performed by applying to the frame the given loads and deflections of all supports. Final analysis of the foundation is performed by applying to the foundation the given loads and obtained reactions acting in the opposite direction. It should be mentioned that the iterative method is another way to perform analysis. This method was discussed in Chapter 3 in the analysis of frames on Winkler foundation. The only difference between the analyses described in Chapter 3 and analysis in this chapter is the method of obtaining deflections of the foundation.

Table 4.1 Coefficients K_0 and K_1

α	1	2	3	4	5	6	7	8	9	10
K_0	0.88	0.86	0.82	0.79	0.77	0.74	0.73	0.71	0.69	0.67
K_1	0.52	0.85	1.10	1.30	1.45	1.60	1.70	1.80	1.90	2.00

Table 4.2 Dimensionless coefficients for analysis of rigid beams loaded with a vertical concentrated load P_0 applied at the center of the beam

α	x										
	0.0	0.1	0.2	0.3	0.4	0.5	0.6	0.7	0.8	0.9	1.0
	$\overline{P}_0$										
10	0.439	0.440	0.442	0.446	0.455	0.462	0.475	0.498	0.541	0.632	0.842
15	0.447	0.448	0.450	0.454	0.459	0.468	0.480	0.500	0.537	0.611	0.773
20	0.452	0.453	0.455	0.458	0.464	0.471	0.483	0.502	0.535	0.600	0.737
30	0.458	0.459	0.460	0.463	0.468	0.475	0.486	0.503	0.532	0.586	0.699
50	0.464	0.465	0.466	0.468	0.473	0.480	0.489	0.503	0.538	0.574	0.665
100	0.469	0.470	0.471	0.473	0.477	0.483	0.481	0.504	0.525	0.561	0.634
	$\overline{Q}_0$										
10	−0.500	−0.456	−0.412	−0.367	−0.322	−0.277	−0.230	−0.182	−0.130	−0.072	0
15	−0.500	−0.455	−0.411	−0.365	−0.320	−0.273	−0.225	−0.177	−0.125	−0.068	0
20	−0.500	−0.454	−0.410	−0.363	−0.317	−0.271	−0.223	−0.174	−0.122	−0.066	0
30	−0.500	−0.454	−0.408	−0.362	−0.315	−0.268	−0.220	−0.171	−0.119	−0.063	0
50	−0.500	−0.453	−0.407	−0.360	−0.313	−0.265	−0.217	−0.168	−0.116	−0.061	0
100	−0.500	−0.453	−0.406	−0.359	−0.311	−0.263	−0.214	−0.165	−0.113	−0.058	0
	$\overline{M}_0$										
10	0.2703	0.2225	0.1791	0.1401	0.1056	0.0756	0.0502	0.0295	0.0139	0.0037	0
15	0.2672	0.2194	0.1761	0.1373	0.1931	0.0735	0.0486	0.0284	0.0132	0.0035	0
20	0.2654	0.2176	0.1744	0.1357	0.1017	0.0722	0.0475	0.0277	0.0128	0.0034	0
30	0.2634	0.2156	0.1725	0.1340	0.1002	0.0710	0.0465	0.0270	0.0125	0.0033	0
50	0.2615	0.2137	0.1707	0.1323	0.0986	0.0697	0.0455	0.0263	0.0120	0.0031	0
100	0.2596	0.2120	0.1690	0.1307	0.0972	0.0685	0.0446	0.0256	0.0117	0.0030	0

Table 4.3 Dimensionless coefficients $\overline{Y}_0$ for rigid beams loaded with a vertical concentrated load P_0 applied at the center of the beam

α	10	15	20	30	50	100
$\overline{Y}_0$	1.081	1,210	1.302	1.431	1.595	1.814

Table 4.4 Dimensionless coefficients for analysis of rigid beams loaded with a concentrated moment m_0 applied at the center of the beam

α	0.0	0.1	0.2	0.3	0.4	0.5	0.6	0.7	0.8	0.9	1.0
						$\overline{P}_0$						
10	0	0.114	0.229	0.347	0.470	0.600	0.746	0.918	1.148	1.506	2.155
15	0	0.119	0.239	0.362	0.490	0.625	0.772	0.943	1.162	1.483	2.024
20	0	0.122	0.245	0.371	0.501	0.638	0.786	0.957	1.169	1.449	1.957
30	0	0.125	0.252	0.381	0.513	0.652	0.802	0.971	1.176	1.415	1.885
50	0	0.129	0.258	0.390	0.525	0.667	0.817	0.984	1.181	1.414	1.820
100	0	0.133	0.265	0.400	0.538	0.680	0.831	0.996	1.86	1.415	1.761
						$\overline{Q}_0$						
10	−0.709	−0.704	−0.686	−0.658	−0.617	−0.563	−0.496	−0.413	−0.311	−0.180	0
15	−0.716	−0.710	−0.692	−0.662	−0.619	−0.564	−0.494	−0.408	−0.304	−0.173	0
20	−0.719	−0.713	−0.695	−0.664	−0.621	−0.564	−0.493	−0.406	−0.300	−0.169	0
30	−0.723	−0.717	−0.698	−0.667	−0.622	−0.564	−0.491	−0.403	−0.296	−0.165	0
50	−0.727	−0.721	−0.701	-.0669	−0.623	−0.564	−0.490	−0.400	−0.292	−0.162	0
100	−0.731	−0.724	−0.705	−0.671	−0.624	−0.564	−0.488	−0.397	−0.288	−0.158	0
						$\overline{M}_0$						
10	0.5	0.4293	0.3597	0.2924	0.2285	0.1694	0.1163	0.0707	0.0343	0.0095	0
15	0.5	0.4285	0.3584	0.2907	0.2265	0.1672	0.1142	0.0690	0.0332	0.0091	0
20	0.5	0.4283	0.3577	0.2897	0.2253	0.1660	0.1130	0.0679	0.0324	0.0088	0
30	0.5	0.4297	0.3570	0.2887	0.2241	0.1647	0.1119	0.0670	0.0319	0.0084	0
50	0.5	0.4275	0.3563	0.2876	0.2229	0.1635	0.1107	0.0660	−0.0313	0.0083	0
100	0.5	0.4271	0.3556	0.2867	0.2217	0.1622	01095	0.0651	−0.0306	0.0081	0

Table 4.5 Coefficients $\overline{tg}\varphi_1$ for rigid beams loaded with a concentrated moment m_0

α	10	15	20	30	50	100
$\overline{tg}\varphi_1$	2.088	2.466	3.737	3.122	3.606	4.311

Table 4.6 Dimensionless coefficients for analysis of short beams loaded with a concentrated vertical load P

$t = 1$

Coef.	δ	x																				
		1	0.9	0.8	0.7	0.6	0.5	0.4	0.3	0.2	0.1	0	-0.1	-0.2	-0.3	-0.4	-0.5	-0.6	-0.7	-0.8	-0.9	-1
$\bar{M}$	0	0	003	010	022	038	059	085	116	152	193	241	193	152	116	085	059	038	022	010	003	0
$\bar{Q}$		0	-050	-096	-140	-186	-234	-283	-335	-338	-444	500*	444	388	335	283	234	186	140	096	050	0
$\bar{P}$		055	047	045	045	046	048	050	053	055	056	056	056	055	053	050	048	046	045	045	047	055
$\bar{\omega}$		087	092	097	102	107	111	116	119	122	124	125	124	122	119	116	111	107	102	097	092	087
$\bar{M}$	0.1	0	003	007	015	027	043	063	088	117	152	192	152	117	088	063	043	027	015	007	002	0
$\bar{Q}$		0	-069	-127	-181	-235	-290	-346	-403	-460	482*	426	372	319	269	223	180	139	102	068	034	0
$\bar{P}$		077	062	056	0.54	0.54	055	056	057	057	057	056	054	051	048	045	042	038	036	034	033	0.36
$\bar{\omega}$		106	109	113	116	119	122	124	125	127	126	124	121	116	112	106	101	095	088	082	076	069
$\bar{M}$	0.2	0	004	017	036	061	093	131	176	226	182	144	112	084	061	043	029	017	009	004	001	0
$\bar{Q}$		0	-085	-158	-224	-288	-350	-412	-473	467*	408	352	300	250	205	165	128	095	066	040	019	0
$\bar{P}$		096	077	069	065	063	062	061	061	059	057	055	051	047	043	039	035	031	027	023	020	019
$\bar{\omega}$		129	131	131	132	134	134	134	132	130	127	122	117	111	104	097	089	081	074	066	059	050
$\bar{M}$	0.3	0	005	020	044	074	112	157	208	166	129	099	073	053	036	024	014	008	003	001	000	0
$\bar{Q}$		0	-106	-192	-270	-330	-410	-481	454*	393	335	280	230	185	145	109	079	053	032	015	005	0
$\bar{P}$		121	094	081	075	071	069	066	053	060	056	052	048	043	038	033	028	024	019	014	008	002
$\bar{\omega}$		152	151	150	148	146	144	142	137	132	126	120	112	104	095	087	078	069	060	052	043	033
$\bar{M}$	0.4	0	007	024	052	088	132	184	142	107	079	055	037	023	012	005	001	-001	-001	-001	000	0
$\bar{Q}$		0	-127	-229	-319	402	-479	448*	380	318	260	208	162	121	085	056	031	012	000	-007	-007	0
$\bar{P}$		145	112	095	086	080	075	070	065	060	055	049	044	038	033	027	022	016	010	003	-003	-010
$\bar{\omega}$		178	174	169	165	160	155	149	142	134	124	116	106	096	086	076	066	056	046	036	025	015

		1	2	3	4	5	6	7	8	9	10	11	12	13	14	15	16	17	18	19	20	21
0.5	$\overline{M}$	0	008	029	061	102	153*	111	077	050	028	012	000	-007	-011	-013	-012	-010	-007	-004	-001	0
	$\overline{Q}$	0	-148	-267	-370	-462	453*	376	307	244	188	138	095	058	028	003	-015	-027	-032	-031	-021	0
	$\overline{p}$	170	131	109	097	088	080	073	066	059	053	046	040	034	027	021	015	009	002	-005	-015	-027
	$\overline{\omega}$	204	197	189	181	174	165	155	145	134	122	112	101	089	078	066	055	044	033	022	011	0
0.6	$\overline{M}$	0	009	033	070	118	075	041	014	-006	-020	-029	-034	-035	-034	-030	-024	-018	-012	-006	-002	0
	$\overline{Q}$	0	-172	-309	-425	474*	384	303	232	169	115	068	029	-003	-029	-047	-059	-065	-063	-052	-032	0
	$\overline{p}$	198	151	125	108	096	085	075	067	068	051	043	036	029	022	015	009	002	-006	-015	-026	-039
	$\overline{\omega}$	233	221	210	199	186	174	161	148	134	120	107	094	081	068	056	044	031	020	007	-004	-016
0.7	$\overline{M}$	0	010	038	080	034	-002	-029	-049	-061	-068	-070	-069	-063	-056	-047	-037	-026	-017	-008	-002	0
	$\overline{Q}$	0	-198	-352	519*	409	313	230	158	096	044	000	-036	-064	-085	-097	-103	-102	-092	-074	-044	0
	$\overline{p}$	229	172	140	119	103	089	077	067	057	048	040	032	024	017	009	002	-005	-014	-024	-036	-052
	$\overline{\omega}$	262	246	230	214	199	182	165	149	133	118	102	088	074	060	046	033	020	007	-006	-018	-030
0.8	$\overline{M}$	0	012	043	-010	-050	-079	-099	-111	-116	-116	-111	-103	-091	-078	-064	-049	-035	-021	-010	-003	0
	$\overline{Q}$	0	-224	603*	462	344	243	157	084	023	-028	-069	-101	-124	-139	-147	-147	-139	-122	-095	-055	0
	$\overline{p}$	259	194	155	129	109	093	079	067	056	046	036	027	019	011	004	-004	-012	-022	-033	-047	-065
	$\overline{\omega}$	293	272	252	231	211	190	170	151	132	115	097	081	066	050	036	022	008	-006	-020	-033	-047
0.9	$\overline{M}$	0	013	-052	-100	-133	-156	-169	-173	-171	-164	-152	-136	-119	-100	-08	-061	-043	-026	-013	-003	0
	$\overline{Q}$	0	748*	558	404	278	172	084	010	-050	-099	-137	-165	-187	-194	-196	-190	-175	-151	-116	-067	0
	$\overline{p}$	293	212	170	138	115	097	081	067	054	043	033	023	014	006	-002	-010	-019	-029	-041	-057	-079
	$\overline{\omega}$	325	298	272	247	222	198	174	152	131	111	092	074	058	041	025	011	-004	-018	-033	-047	-062
1.0	$\overline{M}$	0	085	147	-189	-217	-232	-233	-235	-226	-213	-192	-170	-147	-122	-097	-073	-051	-031	-015	-004	0
	$\overline{Q}$	1000*	721	512	347	212	102	011	-064	-123	-170	-205	-229	-244	-249	-245	-233	-212	-180	-137	-079	0
	$\overline{p}$	326	238	184	148	121	100	082	066	053	041	029	019	010	001	-008	-017	-026	-037	-050	-067	-092
	$\overline{\omega}$	356	325	293	263	233	206	179	154	130	108	088	068	050	032	016	-001	-016	-031	-046	-061	-076

Table 4.7 Dimensionless coefficients for analysis of short beams loaded with a concentrated vertical load P

$t = 2$

Coef.	δ											x										
		−1	−0.9	−0.8	−0.7	−0.6	−0.5	−0.4	−0.3	−0.2	−0.1	0	0.1	0.2	0.3	0.4	0.5	0.6	0.7	0.8	0.9	1
$\overline{M}$	0	0	002	007	016	030	047	070	098	132	173	219	173	132	098	070	047	030	016	007	002	0
$\overline{Q}$		0	036	072	111	155	202	254	310	371	434	500*	−434	−371	−310	−254	−202	−155	−111	072	036	0
$\overline{p}$		039	035	037	041	045	049	054	058	062	065	066	065	062	058	054	049	045	041	037	035	039
$\overline{\omega}$		071	080	089	097	106	114	121	127	133	136	137	136	133	127	121	114	106	097	089	080	071
$\overline{M}$	0.1	0	001	004	010	020	033	050	072	099	132	171	217	169	128	093	064	041	023	011	003	0
$\overline{Q}$		0	021	046	076	110	150	195	246	301	361	424	489*	−445	−381	−318	−259	−203	−152	−103	−055	0
$\overline{p}$		020	023	027	031	037	042	048	053	058	062	064	065	065	064	061	057	053	049	048	050	−61
$\overline{\omega}$		057	066	075	085	094	103	111	119	126	132	136	138	137	134	129	124	117	110	102	095	087
$\overline{M}$	0.2	0	000	002	005	010	019	031	047	068	094	126	164	208	159	117	082	053	030	014	004	0
$\overline{Q}$		0	007	022	043	070	103	140	183	232	287	348	412	478*	−454	−387	−322	−258	−196	−134	−071	0
$\overline{p}$		005	011	018	024	030	035	040	046	052	058	063	066	067	067	066	065	063	062	062	065	078
$\overline{\omega}$		040	050	060	071	081	091	101	110	119	127	133	138	141	142	140	137	133	128	123	118	113
$\overline{M}$	0.3	0	000	−001	000	002	007	014	025	040	059	084	114	151	194	145	102	067	039	019	005	0
$\overline{Q}$		0	005	001	014	034	059	089	126	169	219	274	335	400	468*	−461	−389	−317	−244	−171	−093	0
$\overline{p}$		−010	001	010	017	022	028	034	040	046	053	058	063	067	069	071	072	072	072	074	083	106
$\overline{\omega}$		026	036	047	058	068	079	089	100	110	119	128	136	142	146	148	148	147	145	142	140	137
$\overline{M}$	0.4	0	001	−002	−003	−004	−003	000	005	014	027	045	068	097	132	175	124	082	048	023	006	0
$\overline{Q}$		0	−011	−015	−011	000	017	042	072	109	152	202	258	320	388	461*	−462	−382	−299	−213	−117	0
$\overline{p}$		−014	−007	000	007	014	021	027	034	040	046	053	059	065	071	075	079	081	084	090	103	134
$\overline{\omega}$		010	021	032	043	054	066	077	088	099	110	121	131	140	148	155	159	162	163	164	165	165

206

The table below lists influence values $\bar{M}$, $\bar{Q}$, $\bar{p}$, $\bar{\omega}$ for parameter values 0.5 through 1.0 (values ×1000; `*` marks the critical/maximum ordinate).

Param	Var	1	2	3	4	5	6	7	8	9	10	11	12	13	14	15	16	17	18	19	20	21
0.5	$\bar{M}$	0	007	027	058	098	148	106	071	044	023	007	−004	−011	−014	−015	−014	−011	−008	−004	−001	0
	$\bar{Q}$	0	−141	−256	−357	−451	462*	381	308	243	184	133	088	051	020	−003	−021	−032	−037	−035	−023	0
	$\bar{p}$	161	125	107	097	090	084	077	069	062	055	048	041	034	027	021	014	008	001	−006	−017	−031
	$\bar{\omega}$	196	191	186	181	175	168	159	149	138	126	114	102	090	078	066	055	043	032	021	−010	−001
0.6	$\bar{M}$	0	009	033	069	117	074	040	014	−006	−020	−028	−033	−034	−032	−028	−023	−017	−011	−006	−002	0
	$\bar{Q}$	0	−170	−305	−422	475*	382	300	228	165	111	064	026	−005	−029	−046	−057	−061	−059	−049	−003	0
	$\bar{p}$	195	149	124	109	098	087	077	067	059	050	042	035	027	020	014	008	001	−006	−014	−025	−035
	$\bar{\omega}$	231	221	211	200	180	176	163	149	134	120	106	093	079	067	054	042	031	019	008	−003	−015
0.7	$\bar{M}$	0	010	039	081	036	001	−025	−043	−055	−061	−063	−061	−056	−049	−041	−032	−023	−014	−007	−002	0
	$\bar{Q}$	0	−202	−358	510*	398	301	218	148	089	039	−002	−034	−059	−077	−087	−091	−089	−081	−064	−038	0
	$\bar{p}$	234	175	142	121	104	089	076	064	054	045	037	029	021	014	007	001	−005	−012	−021	−032	−043
	$\bar{\omega}$	265	249	233	217	199	182	164	147	130	114	098	084	070	057	044	032	021	009	−002	−013	−024
0.8	$\bar{M}$	0	012	045	−009	−044	−071	−089	−099	−103	−102	−097	−088	−078	−066	−054	−041	−029	−018	−009	−002	0
	$\bar{Q}$	0	−234	585*	440	321	221	138	069	013	−032	−068	−094	−112	−123	−128	−126	−117	−102	−078	−045	0
	$\bar{p}$	272	203	160	131	109	091	075	062	050	040	030	022	014	007	001	−005	−012	−019	−028	−039	−052
	$\bar{\omega}$	304	281	258	235	211	188	166	145	125	107	089	074	059	045	032	020	008	−004	−015	−026	−037
0.9	$\bar{M}$	0	014	−048	−093	−123	−142	−152	−154	−150	−142	−130	−116	−100	−083	−066	−050	−034	−021	−010	−003	0
	$\bar{Q}$	0	730*	522	370	243	140	057	−010	−063	−103	−133	−159	−165	−170	−168	−158	−145	−123	−093	−053	0
	$\bar{p}$	317	231	177	141	114	093	075	059	046	035	025	016	008	001	−005	−011	−018	−025	−034	−046	−061
	$\bar{\omega}$	344	313	281	251	221	193	167	142	120	100	081	064	049	034	021	009	003	−014	−025	−036	−047
1.0	$\bar{M}$	0	−084	−142	−180	−203	−217	−215	−210	−198	−182	−164	−164	−122	−100	−079	−058	−040	−024	−012	−003	0
	$\bar{Q}$	1000*	695	471	300	165	059	−025	−089	−138	−174	−198	−212	−218	−216	−208	−193	−172	−144	−108	−061	0
	$\bar{p}$	361	258	194	151	119	094	074	056	043	029	019	010	002	−005	−012	−018	−024	−032	−041	−053	−070
	$\bar{\omega}$	385	344	305	267	231	198	168	140	115	093	073	055	038	024	010	−002	−014	−025	−036	−046	−057

207

Table 4.8 Dimensionless coefficients for analysis of short beams loaded with a concentrated vertical load P

$t = 5$

Coef.	δ	-1	-0.9	-0.8	-0.7	-0.6	-0.5	-0.4	-0.3	-0.2	-0.1	0	0.1	0.2	0.3	0.4	0.5	0.6	0.7	0.8	0.9	1
$\overline{M}$	0	0	000	002	007	014	026	043	066	096	134	180	134	096	066	043	026	014	007	002	000	0
$\overline{Q}$		0	011	029	058	096	143	199	263	336	416	500*	-416	-336	-263	-199	-143	-096	-058	-029	-011	0
$\overline{p}$		010	014	023	033	043	051	060	069	077	082	084	082	077	069	060	051	043	033	023	014	010
$\overline{\omega}$		042	058	073	088	102	117	130	142	152	159	162	159	152	142	130	117	102	088	073	058	042
$\overline{M}$	0.1	0	-001	000	002	006	014	025	042	065	096	133	179	133	096	066	043	026	014	006	002	0
$\overline{Q}$		0	-005	009	028	057	095	142	199	265	336	417	500*	-418	-336	-264	-198	-141	-096	-060	-031	0
$\overline{p}$		007	011	015	024	033	043	041	050	061	077	081	083	081	077	071	062	051	040	031	029	035
$\overline{\omega}$		036	050	064	078	092	106	120	132	143	152	159	160	157	149	139	126	112	097	081	066	050
$\overline{M}$	0.2	0	-012	002	002	001	003	010	022	039	061	091	128	174	128	090	060	037	020	009	002	0
$\overline{Q}$		0	-021	-011	003	026	056	094	139	195	260	335	415	498*	-419	-340	-266	-199	-139	-087	-042	0
$\overline{p}$		-018	-005	008	019	027	033	041	050	061	070	078	082	083	080	077	071	064	056	048	043	044
$\overline{\omega}$		024	037	051	065	079	093	106	120	133	144	154	160	163	160	154	144	133	120	107	093	080
$\overline{M}$	0.3	0	-011	-003	-006	-006	-005	-002	005	017	033	056	085	122	167	121	088	052	029	013	003	0
$\overline{Q}$		0	-028	-024	-016	001	023	052	080	137	193	259	332	411	493*	-424	-342	-264	-193	-129	-069	0
$\overline{p}$		-031	-017	005	013	019	026	033	042	052	061	070	076	080	083	083	080	075	067	061	062	081
$\overline{\omega}$		016	028	041	054	067	080	093	106	120	132	144	153	160	164	162	156	148	138	127	116	105
$\overline{M}$	0.4	0	-001	-003	-006	-009	-010	-010	-007	000	010	026	048	076	112	156	109	070	041	019	005	0
$\overline{Q}$		0	-017	-027	-029	-022	-007	015	046	084	130	185	249	320	309	484*	-428	-341	-257	-178	-097	0
$\overline{p}$		-020	-014	-006	002	011	019	027	034	042	050	059	068	076	082	087	088	086	081	079	086	113
$\overline{\omega}$		001	013	025	038	050	063	076	090	104	117	131	143	154	163	168	168	165	160	153	146	139

0.5 $\bar{M}$	0.5 $\bar{Q}$	0.5 $\bar{p}$	0.5 $\bar{\omega}$	0.6 $\bar{M}$	0.6 $\bar{Q}$	0.6 $\bar{p}$	0.6 $\bar{\omega}$	0.7 $\bar{M}$	0.7 $\bar{Q}$	0.7 $\bar{p}$	0.7 $\bar{\omega}$	0.8 $\bar{M}$	0.8 $\bar{Q}$	0.8 $\bar{p}$	0.8 $\bar{\omega}$	0.9 $\bar{M}$	0.9 $\bar{Q}$	0.9 $\bar{p}$	0.9 $\bar{\omega}$	1.0 $\bar{M}$	1.0 $\bar{Q}$	1.0 $\bar{p}$	1.0 $\bar{\omega}$
0	0	143	176	0	0	187	223	0	0	244	268	0	0	296	326	0	0	365	384	0	1000*	433	444
006	-124	111	178	008	-162	144	218	011	207	179	253	013	-254	219	299	-016	693*	258	341	-081	640	298	384
024	-230	101	180	031	-296	124	211	040	368	147	238	049	552*	171	270	-041	470	191	298	-132	388	212	327
052	-330	099	181	067	-414	113	204	083	495*	127	221	002	399	136	241	-080	303	145	257	-161	207	154	273
090	-428	097	180	114	477*	103	195	040	378	109	202	-032	277	109	212	-103	176	111	218	175	074	112	225
138	478*	092	176	071	379	092	183	007	278	091	182	-055	179	087	184	-116	079	084	183	-177	-021	080	183
094	389	085	169	038	293	081	168	-016	196	074	162	-068	102	068	157	-120	007	062	151	-171	-088	055	144
059	308	076	158	012	218	070	152	-032	128	060	142	-076	041	053	133	-118	-046	044	123	-160	-132	035	112
032	236	067	145	-006	153	059	136	-042	074	049	123	-077	-006	040	111	-111	-083	030	098	-145	-160	020	055
012	173	058	132	-019	099	049	119	-047	030	039	106	-075	-040	029	092	-102	-107	019	077	-128	-174	008	063
-002	119	050	118	-026	054	040	104	-048	-005	031	090	-069	-064	020	074	-090	-121	010	060	-111	-178	000	045
-012	073	041	104	-030	018	032	089	-047	-032	023	075	-062	-080	012	059	-078	-127	003	045	-093	-175	-006	030
-017	036	033	090	-030	-009	023	075	-42	-051	015	062	-054	-089	006	046	-065	-127	-003	033	-076	-166	-011	019
-019	006	025	077	-028	-029	016	062	-037	-063	008	051	-045	-092	001	035	-052	-123	-007	022	-060	-153	-014	010
-019	-015	018	065	-025	-042	010	051	-030	-068	003	040	-035	-091	-003	025	-040	-114	-010	014	-045	-137	-017	002
-017	-030	011	053	-020	-050	005	039	-023	-069	-001	031	-026	-087	-006	017	-030	-103	-012	007	-032	-120	-018	-004
-013	-039	006	042	-015	-054	000	029	-016	-006	-004	022	-018	-078	-010	009	-020	-089	-015	000	-022	-100	-020	-009
-009	-042	000	031	-009	-051	-005	019	-010	-059	-009	013	-011	-066	-015	001	-012	-073	-018	-006	-013	-079	-022	-013
-005	-039	-007	021	-005	-042	-013	009	-005	-046	-016	-006	-005	-049	-020	-006	-005	-053	-022	-011	-006	-056	-025	-016
-001	-027	-019	010	-001	-025	-022	-001	-001	-026	-024	-002	-001	-026	-024	-012	-001	-028	-027	-016	-001	-029	-028	-020
0	0	-036	000	0	0	-026	010	0	0	-025	-010	0	0	-026	-019	0	0	-030	-021	0	0	-030	-024

209

Table 4.9 Dimensionless coefficients for analysis of short beams loaded with a concentrated vertical load P

$t = 10$

Coef.	δ	x = -1	-0.9	-0.8	-0.7	-0.6	-0.5	-0.4	-0.3	-0.2	-0.1	0	0.1	0.2	0.3	0.4	0.5	0.6	0.7	0.8	0.9	1
$\overline{M}$	0	0	000	-001	-001	002	009	022	040	066	101	146	101	066	040	022	009	002	-001	-008	000	00
$\overline{Q}$		0	-007	-003	016	048	093	150	221	304	399	500*	-399	-304	-221	-150	-093	-048	-016	003	007	00
$\overline{p}$		-008	-003	011	026	039	051	064	077	090	090	102	099	090	077	064	051	039	026	011	-003	-008
$\overline{\omega}$		020	039	059	079	098	118	137	154	169	179	184	179	169	154	137	118	098	079	059	039	020
$\overline{M}$	0.1	0	-001	-003	-004	-004	-001	006	019	038	065	101	147	103	068	042	024	013	006	003	001	0
$\overline{Q}$		0	-017	-018	-007	015	050	096	156	230	315	408	507*	-395	-301	-216	-143	-087	-048	-027	-014	0
$\overline{p}$		-026	-009	005	017	028	040	053	067	080	090	097	099	097	090	080	065	047	029	015	011	021
$\overline{\omega}$		025	042	058	074	091	108	125	142	157	170	179	181	175	162	145	125	104	081	059	036	013
$\overline{M}$	0.2	0	-002	-005	-008	-009	-008	-004	003	016	035	062	098	145	101	066	040	022	011	004	001	0
$\overline{Q}$		0	-025	-033	-024	-005	021	055	090	156	230	316	412	512*	-391	-299	-218	-147	-089	-046	-018	0
$\overline{p}$		-032	002	002	015	023	029	038	050	065	080	092	099	100	095	087	076	064	051	036	021	017
$\overline{\omega}$		017	032	047	062	077	093	109	126	143	158	172	181	184	179	167	151	132	112	091	070	049
$\overline{M}$	0.3	0	-002	-005	-009	-012	-013	-012	-008	-001	013	032	060	097	143	099	064	038	021	009	002	0
$\overline{Q}$		0	-003	-041	-034	-020	-002	022	056	103	164	237	322	413	509*	-394	-300	-215	-144	-090	-048	0
$\overline{p}$		-047	-019	001	011	016	021	029	040	054	067	079	088	094	097	096	090	079	063	046	041	063
$\overline{\omega}$		016	028	040	052	065	079	094	109	125	141	157	170	179	182	177	165	150	131	112	092	072
$\overline{M}$	0.4	0	-001	-003	-007	-010	-014	-015	-014	-011	-002	011	030	057	093	138	093	058	033	015	004	0
$\overline{Q}$		0	-016	-031	-038	-036	-025	-006	022	058	104	162	231	311	403	502*	-397	-300	-215	-143	-078	0
$\overline{p}$		-016	-016	-011	003	006	015	023	032	041	052	063	075	086	096	101	100	092	078	066	067	095
$\overline{\omega}$		-003	009	021	033	046	059	073	088	104	121	138	154	168	178	183	179	170	157	142	126	109

Influence coefficients (values ×10⁻³). The first value in each row is the support (x = 0) coefficient; the remaining 21 values are the ordinates along the span. Bold starred values mark the maxima.

λ		C0	1	2	3	4	5	6	7	8	9	10	11	12	13	14	15	16	17	18	19	20
0.5	$\overline{M}$	0	0	005	020	045	080	126	082	048	022	003	-010	-017	-021	-022	-020	-017	-013	-009	-005	-001	0
	$\overline{Q}$	0	0	-106	-200	-299	-404	**491***	391	300	222	155	100	054	019	-006	-024	-034	-040	-042	-040	-029	0
	$\overline{p}$	123	-041	096	095	102	106	104	096	085	073	061	050	040	030	021	014	008	004	000	-006	-018	-041
	$\overline{\omega}$	153	006	163	173	181	187	187	180	168	153	136	119	103	088	074	062	051	041	031	023	014	006
0.6	$\overline{M}$	0	0	008	030	064	110	067	034	010	-007	-018	-024	-026	-026	-024	-020	-016	-012	-007	-003	-001	0
	$\overline{Q}$	0	0	-151	-280	-402	**482***	376	283	203	137	083	040	007	-016	-030	-038	-043	-044	-042	-034	-018	0
	$\overline{p}$	176	-013	136	124	119	112	100	086	073	060	048	038	028	018	011	006	003	000	-005	-012	-019	-013
	$\overline{\omega}$	212	006	212	212	210	204	192	176	157	138	119	101	085	070	057	046	035	026	018	010	008	006
0.7	$\overline{M}$	0	0	011	046	084	042	012	-10	-024	-032	-036	-036	-034	-030	-026	-020	-015	-010	-006	-003	000	0
	$\overline{Q}$	0	0	-208	-372	**486***	361	250	173	109	059	020	-009	-031	-045	-051	-052	-049	-045	-040	-030	-015	0
	$\overline{p}$	250	-007	179	151	134	115	094	073	056	044	034	026	018	010	003	-002	-004	-004	-007	-012	-017	-007
	$\overline{\omega}$	265	003	253	241	226	206	184	161	139	118	099	082	068	056	046	037	030	024	018	013	008	003
0.8	$\overline{M}$	0	0	014	052	008	022	041	052	056	056	052	047	041	034	027	021	015	010	006	002	000	0
	$\overline{Q}$	0	0	-270	**524***	416	364	240	144	071	062	049	030	016	006	-002	-011	-011	-010	-012	-014	-013	0
	$\overline{p}$	313	-004	234	181	147	117	083	062	045	031	020	011	004	-001	-005	-007	-008	-008	-010	-013	-014	-004
	$\overline{\omega}$	344	-008	313	282	248	214	180	150	123	099	079	062	048	037	028	020	014	009	005	001	-001	-008
0.9	$\overline{M}$	0	0	018	-035	-067	-085	-092	-092	-086	-078	-068	-058	-047	-038	-029	-021	-015	-009	-005	-002	000	0
	$\overline{Q}$	0	0	**657***	234	181	117	105	073	049	030	016	001	-010	-014	-013	-012	-011	-011	-014	-013	-014	0
	$\overline{p}$	412	-006	285	282	261	214	171	135	104	079	059	043	031	022	015	010	007	005	001	000	-008	-006
	$\overline{\omega}$	421	-003	336	313	313	227	203	181	147	120	095	073	055	040	029	020	013	008	004	000	-003	-003
1.0	$\overline{M}$	0	0	0	-078	-122	-148	-137	-135	-159	-167	-163	-152	-136	-118	-099	-081	-064	-048	-033	-020	-008	0
	$\overline{Q}$	**1000***	0	585	308	153	101	068	035	015	001	008	014	-017	-018	-019	-018	-016	-015	-014	-011	-011	0
	$\overline{p}$	507	-004	585	336	227	153	120	101	086	068	035	015	001	008	014	-015	-014	-011	-011	-002	-002	-004
	$\overline{\omega}$	504	-001	422	345	275	214	162	120	086	059	039	024	014	006	002	-001	-002	-002	-002	-002	-001	-001

Table 4.10 Dimensionless coefficients for analysis of short beams loaded with a concentrated moment m

$t = 1$

Coef.	δ	-1	-0.9	-0.8	-0.7	-0.6	-0.5	-0.4	-0.3	-0.2	-0.1	0	0.1	0.2	0.3	0.4	0.5	0.6	0.7	0.8	0.9	1
$\overline{M}$	0	0	-009	-032	-066	-110	-162	-220	-285	-353	-426	**500***	426	353	285	220	162	110	066	032	009	0
$\overline{Q}$		0	-162	-292	-396	-581	-552	-614	-667	-709	-736	-745	-736	-709	-667	-614	-552	-481	-396	-292	-162	0
$\overline{p}$		-180	-145	-115	-092	-077	-066	-058	-048	-035	-018	0	018	035	048	058	066	077	093	115	145	180
$\overline{\omega}$		-190	-175	-159	-143	-127	-109	-091	-071	-050	-026	0	026	050	071	091	109	127	143	159	175	190
$\overline{M}$	0.1	0	-008	-029	-061	-102	-151	-207	-268	-335	-406	-479	**-553**	373	302	235	173	119	072	035	010	0
$\overline{Q}$		0	-146	-270	-371	-454	-524	-586	-641	-685	-723	-742	-743	-725	-693	-644	-583	-510	-422	-316	-180	0
$\overline{p}$		-155	-136	-112	-091	-075	-066	-059	-052	-041	-027	-010	008	026	041	055	067	080	096	119	154	210
$\overline{\omega}$		-199	-184	-168	-153	-136	-120	-102	-083	-062	-039	-013	015	045	071	095	119	141	162	183	204	224
$\overline{M}$	0.2	0	-007	-026	-056	-096	-142	-196	-255	-319	-388	-460	-534	**-608**	319	249	184	126	077	037	010	0
$\overline{Q}$		0	-135	-250	-349	-431	-502	-564	-619	-667	-706	-732	-744	-739	-716	-675	-615	-540	-448	-338	-196	0
$\overline{p}$		-145	-125	-107	-090	-076	-066	-058	-051	-043	-033	-020	-004	014	032	051	068	083	100	123	163	236
$\overline{\omega}$		-189	-176	-162	-148	-134	-120	-104	-087	-069	-048	-025	001	030	060	088	115	140	164	188	212	236
$\overline{M}$	0.3	0	-006	-025	-053	-090	-135	-186	-243	-306	-373	-443	-516	-590	**-665**	263	196	135	082	040	011	0
$\overline{Q}$		0	-129	-236	-328	-409	-481	-544	-600	-648	-688	-719	-739	-745	-773	-701	-548	-573	-477	-360	-211	259
$\overline{p}$		-143	-116	-099	-086	-076	-067	-060	-052	-044	-036	-025	-013	002	021	043	065	085	105	130	173	259
$\overline{\omega}$		-170	-159	-148	-137	-125	-113	-101	-087	-071	-053	-033	-010	016	046	076	106	134	164	188	214	241
$\overline{M}$	0.4	0	-006	-024	-052	-087	-130	-180	-236	-297	-362	-431	-503	-577	-652	**725**	205	141	086	041	011	1
$\overline{Q}$		0	-127	-229	-315	-394	-466	-530	-586	-633	-673	-706	-731	-745	-743	728	-673	-600	-502	-378	-219	0
$\overline{p}$		-152	-111	-092	-082	-076	-069	-060	-051	-044	-037	-025	-020	-006	012	011	060	086	111	139	182	265
$\overline{\omega}$		-152	-143	-134	-125	-116	-107	-096	-085	-071	-057	-039	-019	004	031	062	095	127	157	186	216	246

Ratio																						
0.5	$\overline{M}$	0	-009	-032	-063	-099	-142	-198	-244	-302	-364	-429	-497	-568	-540	-711	-781	154	096	048	014	0
	$\overline{Q}$	0	-175	-270	-338	-399	-457	-510	-557	-599	-636	-668	-696	-715	-721	-710	-677	-618	-534	-420	-258	0
	$\overline{p}$	-248	-122	-076	-063	-059	-056	-050	-044	-039	-035	-030	-024	-014	001	022	046	071	098	133	191	340
	$\overline{\omega}$	-142	-134	-127	-120	-112	-104	-095	-085	-073	-059	-043	-025	-003	022	051	085	120	155	190	223	257
0.6	$\overline{M}$	0	-007	-026	-054	-088	-130	-178	-231	-289	-352	-418	-488	-560	-533	-707	-778	-844	096	047	013	0
	$\overline{Q}$	0	-138	-234	-313	-383	-417	-506	-559	-606	-647	-682	-710	-729	-735	-724	-693	-636	-548	-423	-249	0
	$\overline{p}$	-173	-112	-086	-073	-067	-062	-056	-050	-014	-038	-032	-024	-013	001	020	043	071	105	147	206	300
	$\overline{\omega}$	-146	-139	-131	-123	-116	-107	-098	-087	-076	-062	-046	-028	-006	019	048	081	120	161	201	241	281
0.7	$\overline{M}$	0	-005	-019	-041	-072	-110	-155	-201	-264	-326	-392	-462	-535	-610	-685	-758	-828	-892	054	015	0
	$\overline{Q}$	0	-094	-184	-268	-346	-418	-483	-543	-596	-644	-684	-717	-739	-749	-745	-722	-675	-598	-474	-286	0
	$\overline{p}$	-093	-093	-087	-081	-075	-068	-062	-057	-051	-044	-036	-027	-017	-003	013	034	061	099	151	220	353
	$\overline{\omega}$	-151	-143	-135	-127	-119	-110	-101	-090	-078	-064	-048	-030	-008	017	046	080	118	162	208	254	300
0.8	$\overline{M}$	0	-006	-023	-048	-081	-121	-167	-219	-276	-339	-405	-475	-547	-621	-694	-766	-834	-896	-949	014	0
	$\overline{Q}$	0	-116	-212	-294	-366	-430	-491	-547	-599	-645	-683	-712	-731	-737	-739	-705	-657	-578	-455	-271	0
	$\overline{p}$	-128	-105	-088	-076	-068	-062	-058	-054	-049	-042	-034	-024	-013	000	-015	-035	062	099	150	222	326
	$\overline{\omega}$	-157	-148	-140	-132	-123	-113	-104	-092	-080	-066	-049	-030	-008	018	047	081	120	164	214	267	320
0.9	$\overline{M}$	0	-006	-022	-048	-081	-120	-166	-218	-276	-338	-404	-474	-546	-620	-694	-766	-834	-896	-948	**-985**	0
	$\overline{Q}$	0	-115	-210	-293	-365	-431	-491	-547	-598	-644	-682	-712	-731	-738	-731	-706	-657	-578	-456	-274	0
	$\overline{p}$	-126	-104	-088	-077	-069	-063	-058	-054	-048	-042	-034	-025	-014	000	016	036	062	098	148	221	335
	$\overline{\omega}$	-153	-146	-137	-129	-121	-112	-102	-092	-080	-066	-049	-031	-009	016	045	079	117	161	210	265	324
1.0	$\overline{M}$	0	-006	-022	-048	-081	-120	-167	-219	-276	-338	-404	-474	-547	-620	-694	-766	-834	-896	-948	-993	-10^3
	$\overline{Q}$	0	-116	-211	-292	-365	-432	-492	-548	-599	-644	-682	-711	-731	-739	-732	-706	-656	-576	-455	-274	0
	$\overline{p}$	-129	-104	-087	-077	-069	-063	-058	-053	-048	-041	-034	-025	-014	-001	016	036	063	098	147	220	336
	$\overline{\omega}$	-150	-142	-134	-127	-118	-110	-101	-090	-079	-065	-049	-031	-010	005	044	077	115	158	207	263	324

Table 4.11 Dimensionless coefficients for analysis of short beams loaded with a moment m

$t = 2$

Coef.	δ											x										
		-1	-0.9	-0.8	-0.7	-0.6	-0.5	-0.4	-0.3	-0.2	-0.1	0	0.1	0.2	0.3	0.4	0.5	0.6	0.7	0.8	0.9	1
$\overline{M}$	0	0	-008	-030	-063	-106	-156	-214	-278	-348	-423	**-500**	423	348	278	214	156	106	063	030	008	0
$\overline{Q}$		0	-151	-279	-382	-466	-540	-609	-673	-731	-764	-777	-764	-731	-673	-609	-540	-466	-382	-279	-151	0
$\overline{p}$		-157	-141	-115	-092	-078	-071	-067	-060	-046	-025	0	025	046	000	067	071	078	092	115	141	157
$\overline{\omega}$		-171	-160	-149	-137	-126	-113	-097	-079	-057	-031	0	031	057	079	097	113	126	137	149	160	171
$\overline{M}$	0.1	0	-009	-032	-068	-112	-164	-223	-286	-354	-424	-495	**-567**	362	294	229	169	116	070	034	009	0
$\overline{Q}$		0	-168	-298	-402	-486	-555	-611	-656	-689	-710	-719	-714	-697	-667	-624	-569	-499	-414	-309	-176	0
$\overline{p}$		-192	-147	-115	-093	-076	-063	-050	-039	-027	-015	-002	011	024	036	049	062	077	094	117	151	205
$\overline{\omega}$		-192	-181	-169	-159	-146	-132	-116	-099	-078	-053	-023	013	049	080	108	132	154	175	195	214	233
$\overline{M}$	0.2	0	-009	-032	-067	-111	-162	-220	-283	-350	-420	-491	-563	**-633**	298	232	172	118	071	034	009	0
$\overline{Q}$		0	-166	-294	-397	-481	-550	-606	-651	-684	-706	-718	-715	-700	-673	-631	-576	-506	-420	-314	-179	0
$\overline{p}$		-190	-144	-114	-092	-076	-062	-050	-039	-028	-016	-005	008	021	034	048	062	078	095	118	153	210
$\overline{\omega}$		-175	-167	-159	-151	-142	-132	-120	-106	-090	-069	-045	-014	023	061	094	125	153	179	204	228	253
$\overline{M}$	0.3	0	-008	-031	-066	-109	-161	-218	-280	-347	-416	-487	-559	-629	**-698**	236	174	100	072	035	010	0
$\overline{Q}$		0	-164	-291	-392	-476	-546	-602	-647	-680	-703	-714	-714	-702	-676	-637	-583	-513	-426	-319	-183	0
$\overline{p}$		-190	-142	-113	-092	-076	-063	-050	-039	-028	-017	-006	006	019	032	046	062	078	096	112	155	216
$\overline{\omega}$		-140	-136	-133	-129	-125	-120	-113	-104	-093	-078	-059	-034	-004	033	073	108	141	172	201	230	260
$\overline{M}$	0.4	0	-008	-031	-065	-109	-159	-217	-279	-345	-414	-484	-557	-627	-696	**-762**	176	121	073	035	010	0
$\overline{Q}$		0	-164	-289	-389	-473	-542	-598	-643	-677	-699	-711	-712	-702	-679	-641	-589	-519	-432	-323	-184	0
$\overline{p}$		-191	-141	-111	-091	-076	-063	-051	-039	-028	-017	-007	009	017	030	045	061	078	097	122	158	217
$\overline{\omega}$		-104	-105	-106	-106	-106	-106	-104	-099	-093	-083	-069	-050	-025	006	045	087	125	162	197	232	267

Group		1	2	3	4	5	6	7	8	9	10	11	12	13	14	15	16	17	18	19	20	21
0.5	$\overline{M}$	0	010	037	075	124	**-821**	-759	-693	-625	-554	-484	-414	-345	-280	-219	-162	-111	-068	-033	-009	0
	$\overline{Q}$	0	-193	-331	-438	-523	-590	-639	-674	-696	-705	-704	-692	-670	-637	-594	-540	-474	-394	-297	-173	0
	$\overline{p}$	232	160	121	095	075	058	043	028	015	004	-007	-017	-027	-038	-049	-060	-073	-087	-108	-143	-210
	$\overline{\omega}$	287	244	202	158	114	067	023	-012	-039	-060	-076	-087	-094	-098	-100	-099	-098	-096	-094	-090	-087
0.6	$\overline{M}$	0	010	036	075	**-876**	-820	-758	-692	-623	-552	-481	-411	-343	-217	-216	-159	-109	-067	-032	-009	0
	$\overline{Q}$	0	-191	-333	-442	-527	-593	-643	-677	-699	-708	-706	-694	-671	-637	-593	-538	-470	-388	-290	-165	0
	$\overline{p}$	225	162	124	096	075	057	042	027	015	004	-007	-018	-026	-039	-050	-062	-074	-089	-110	-141	-195
	$\overline{\omega}$	332	278	225	170	113	061	018	-017	-043	-064	-080	-091	-099	-103	-105	-106	-105	-104	-102	-100	-097
0.7	$\overline{M}$	0	010	038	**-922**	-872	-815	-753	-686	-617	-546	-475	-405	-337	-272	-211	-155	-105	-063	-030	-008	0
	$\overline{Q}$	0	-199	-344	-453	-536	-600	-648	-681	-701	-710	-707	-693	-668	-633	-588	-530	-461	-378	-278	-156	0
	$\overline{p}$	236	168	125	095	073	055	040	027	014	003	-008	-019	-030	-040	-051	-063	-076	-091	-110	-137	178
	$\overline{\omega}$	369	303	237	171	109	057	015	-019	-046	-067	-083	-095	-102	-108	-111	-112	-112	-111	-111	-109	-108
0.8	$\overline{M}$	0	010	**-963**	-923	-874	-817	-755	-689	-620	-549	-478	-408	-340	-275	-213	-157	-107	-065	-031	-008	0
	$\overline{Q}$	0	-196	-340	-448	-532	-596	-644	-678	-700	-709	-707	-693	-669	-635	-590	-534	-465	-384	-285	-161	0
	$\overline{p}$	230	166	124	095	073	056	041	027	015	003	-009	-019	-029	-040	-050	-062	-074	-090	-110	-140	-186
	$\overline{\omega}$	416	328	246	176	113	061	018	-017	-045	-067	-084	-097	-106	-112	-116	-118	-120	-120	-120	-120	-120
0.9	$\overline{M}$	0	**-990**	-962	-923	-874	-817	-755	-689	-620	-549	-478	-408	-340	-275	-213	-157	-107	-054	-031	-008	0
	$\overline{Q}$	0	-196	-340	-448	-532	-596	-645	-679	-700	-709	-706	-693	-669	-635	-590	-534	-466	-384	-284	-160	0
	$\overline{p}$	232	166	124	095	073	056	041	027	015	003	-008	-019	-029	-040	-050	-062	-074	-090	-110	-139	-185
	$\overline{\omega}$	416	326	241	169	108	056	014	-020	-048	-069	-085	-096	-104	-110	-113	-115	-116	-116	-116	-115	-114
1.0	$\overline{M}$	**-10³**	**-990**	-962	-923	-874	-817	-755	-689	-620	-549	-478	-408	-340	-275	-213	-157	-107	-064	-031	-008	0
	$\overline{Q}$	**0**	-196	-339	-448	-532	-596	-645	-679	-700	-709	-706	-693	-669	-635	-590	-534	-466	-384	-284	-168	0
	$\overline{p}$	232	166	124	095	074	056	041	027	015	003	-008	-019	-029	-040	-050	-062	-075	-090	-110	-139	-186
	$\overline{\omega}$	416	320	236	164	103	053	011	-022	-049	-069	-085	-095	-103	-108	-111	-111	-111	-111	-109	-108	-106

Table 4.12 Dimensionless coefficients for analysis of short beams loaded with a concentrated moment m

$t = 5$

Coef.	δ	$x=-1$	-0.9	-0.8	-0.7	-0.6	-0.5	-0.4	-0.3	-0.2	-0.1	0	0.1	0.2	0.3	0.4	0.5	0.6	0.7	0.8	0.9	1
$\bar{M}$	0	0	−008	−026	−055	−094	−141	−196	−260	−333	−414	**−500**	−414	−333	−260	−196	−141	−094	−055	−026	007	0
$\bar{Q}$		0	−118	−243	−344	−427	−506	−594	−689	−777	−840	−864	−840	−777	−689	−594	−506	−427	−344	−243	−118	0
$\bar{p}$		−092	−130	−114	−089	−078	−083	−093	−094	−077	−045	000	045	077	094	093	083	078	089	114	130	092
$\bar{\omega}$		−113	−117	−120	−122	−122	−120	−113	−099	−078	0045	000	045	078	099	113	120	122	122	120	117	113
$\bar{M}$	0.1	0	−004	−016	−039	−070	−109	−155	−210	−276	−351	−433	**−519**	395	314	241	176	120	073	036	011	0
$\bar{Q}$		0	−064	−175	−274	−351	−424	−507	−604	−705	−792	−848	−863	−837	−777	−696	−606	−514	−420	−314	−181	0
$\bar{p}$		−002	−103	−109	−086	−077	−076	−091	−101	−097	−074	−037	006	045	073	087	092	092	097	116	154	210
$\bar{\omega}$		−187	−184	−181	−178	−173	−166	−155	−140	−118	−087	−045	011	068	112	146	172	193	210	224	237	250
$\bar{M}$	0.2	0	−001	−009	−025	−050	−083	−123	−172	−229	−297	−373	−456	**−543**	370	286	210	144	089	044	013	0
$\bar{Q}$		0	−033	−118	−208	−289	−364	−443	−532	−627	−721	−802	−856	−876	−857	−800	−714	−610	−500	−383	−236	0
$\bar{p}$		015	−069	−092	−087	−077	−076	−083	−093	−097	−089	−070	−039	000	039	073	097	108	112	126	179	305
$\bar{\omega}$		−161	−161	−163	−164	−164	−162	−158	−150	−136	−115	−084	−042	−015	073	120	157	188	214	237	259	280
$\bar{M}$	0.3	0	−001	−007	−020	−040	−067	−101	−144	−196	−257	−326	−405	−490	**−580**	330	244	167	102	050	014	0
$\bar{Q}$		0	−039	−095	−161	−232	−308	−389	−473	−561	−653	−741	−822	−882	−908	−880	−822	−715	−584	−442	−273	0
$\bar{p}$		−026	−049	−062	−069	−074	−078	−082	−086	−090	−091	−086	−072	−045	−005	043	089	122	137	149	202	372
$\bar{\omega}$		−084	−094	−104	−114	−123	−130	−136	−138	−136	−127	−111	−083	−043	012	071	118	158	193	224	254	283
$\bar{M}$	0.4	0	−001	−007	−017	−033	−055	−086	−125	−171	−226	−280	−363	−445	−535	**−630**	276	189	114	055	014	0
$\bar{Q}$		0	−048	−078	−123	−189	−267	−348	−427	−505	−589	−681	−778	−868	−933	−952	−914	−817	−676	−504	−298	0
$\bar{p}$		−077	−030	−034	−056	−074	−080	−083	−085	−086	−088	−096	−096	−080	−045	008	068	122	158	185	236	388
$\bar{\omega}$		−005	−024	−062	−060	−078	−095	−110	−122	−130	−133	−128	−114	−089	−049	007	068	120	166	208	248	288

		0	017	064	133	218	-688	-589	-493	-405	-325	-257	-198	-148	-106	-071	-044	-025	-013	-005	-001	0
0.5	$\bar{M}$	0	017	064	133	218	**-688**	-589	-493	-405	-325	-257	-198	-148	-106	-071	-044	-025	-013	-005	-001	0
	$\bar{Q}$	0	-347	-590	-775	-906	-976	-981	-931	-842	-737	-633	-540	-459	-385	-310	-232	-156	-094	-057	-038	0
	$\bar{p}$	435	282	211	159	102	037	-025	-073	-100	-107	-099	-086	-079	-079	-077	-073	-071	-050	-025	-019	-068
	$\bar{\omega}$	325	270	213	155	092	022	-041	-086	-116	-132	-138	-136	-128	-114	-098	-079	-058	-037	-016	-006	028
0.6	$\bar{M}$	0	021	076	154	**-753**	-654	-551	-461	-376	-300	-234	-178	-132	-93	-063	-040	-023	-013	-006	-001	0
	$\bar{Q}$	0	-389	-683	-870	-971	-999	-971	-902	-810	-708	-606	-510	-423	-343	-267	-196	-133	-084	-054	-034	0
	$\bar{p}$	469	337	233	142	062	-003	-051	-083	-099	-103	-099	-091	-083	-078	-074	-068	-057	-039	-022	-021	-056
	$\bar{\omega}$	428	347	266	181	092	011	-049	-092	-120	-136	-143	-142	-135	-124	-110	-094	-077	-059	-041	-023	-004
0.7	$\bar{M}$	0	026	089	**-824**	-726	-625	-527	-435	-352	-278	-215	-161	-117	-081	-054	-034	-019	-009	-004	-001	0
	$\bar{Q}$	0	-475	-774	-939	-965	-998	-954	-878	-786	-687	-586	-488	-396	-312	-238	-173	-119	-075	-041	-016	0
	$\bar{p}$	580	379	226	110	027	-029	-064	-085	-097	-100	-100	-095	-088	-079	-070	-059	-049	-039	-029	-020	-010
	$\bar{\omega}$	510	403	294	183	082	004	-054	-095	-122	-139	-146	-146	-142	-133	-122	-110	-096	-081	-066	-052	-037
0.8	$\bar{M}$	0	028	**-903**	-813	-712	-611	-514	-424	-342	-270	-207	-153	-110	-076	-050	-031	-018	-009	-003	-001	0
	$\bar{Q}$	0	-514	-820	-972	-018	-997	-938	-860	-772	-679	-582	-483	-386	-297	-220	-159	-112	-074	-039	-008	0
	$\bar{p}$	640	400	220	092	007	-044	-071	-084	-091	-095	-098	-099	-094	-084	-069	-053	-041	-035	-034	-025	017
	$\bar{\omega}$	601	462	322	194	093	014	-045	-088	-117	-136	-146	-149	-148	-143	-136	-126	-116	-105	-094	-083	-072
0.9	$\bar{M}$	0	**-971**	-902	-810	-711	-610	-512	-422	-340	-267	-204	-152	-108	-075	-049	-029	-016	-007	-002	000	0
	$\bar{Q}$	0	-528	-826	-970	-996	-998	-943	-864	-774	-677	-577	-478	-384	-298	-224	-163	-112	-068	-002	-002	0
	$\bar{p}$	681	396	210	088	009	-040	-069	-085	-094	-099	-099	-097	-090	-080	-068	-056	-046	-040	-031	-020	021
	$\bar{\omega}$	624	456	302	178	079	002	-055	-095	-123	-139	-147	-148	-145	-138	-129	-118	-106	-094	-081	-068	-056
1.0	$\bar{M}$	-10^3	-971	-902	-812	-712	-611	-513	-422	-340	-258	-205	-153	-110	-076	-049	-029	-016	-007	-002	-000	0
	$\bar{Q}$	**0**	-528	-820	-965	-1114	-103	-947	-868	-774	-674	-573	-436	-386	-303	-229	-165	-111	-066	-031	-007	0
	$\bar{p}$	691	391	207	090	014	-036	-069	-088	-098	-101	-099	-094	-087	-078	-069	-059	-049	-040	-030	-017	005
	$\bar{\omega}$	623	440	289	166	068	-006	-062	-101	-126	-141	-147	-146	-141	-132	-121	-108	-094	-080	-065	-050	036

Table 4.13 Dimensionless coefficients for analysis of short beams loaded with a moment m

$t = 10$

δ	Coef.	-1	-0.9	-0.8	-0.7	-0.6	-0.5	-0.4	-0.3	-0.2	-0.1	0 (x)	0.1	0.2	0.3	0.4	0.5	0.6	0.7	0.8	0.9	1
0	$\overline{M}$	0	-006	-019	-044	-078	-119	-170	-234	-312	-402	**-500**	-402	-312	-234	-170	-119	-078	-044	-019	-006	0
	$\overline{Q}$	0	-070	-193	-293	-370	-456	-570	-709	-849	-953	-991	-953	-849	-709	-570	-456	-370	-293	-193	-070	0
	$\overline{p}$	011	-119	-116	-085	-076	-098	-129	-145	-128	-075	0	075	128	145	129	098	076	085	116	119	011
	$\overline{\omega}$	-029	-052	-074	-095	-113	-126	-132	-126	-106	-066	0	066	106	126	132	126	113	095	074	052	029
0.1	$\overline{M}$	0	000	-005	-021	-015	-077	-116	-168	-234	-315	-407	**-505**	399	309	231	166	112	069	035	012	0
	$\overline{Q}$	0	010	-104	-208	-250	-350	-449	-584	-736	-873	-961	-983	-037	-841	-718	-594	-483	-386	-289	-167	0
	$\overline{p}$	156	-086	-120	-085	-065	-081	-118	-148	-150	-117	057	013	075	114	127	119	103	093	105	143	191
	$\overline{\omega}$	-208	-210	-213	-215	-216	-214	-207	-192	-167	-126	066	020	107	168	210	236	254	262	266	268	270
0.2	$\overline{M}$	0	045	-030	-126	-208	-283	-370	-481	-612	-751	-873	-959	**-514**	-962	-877	-754	-618	-492	-381	-248	0
	$\overline{Q}$	154	-037	-096	-090	-076	-078	-098	-123	-138	-134	-107	-061	-002	058	107	134	134	116	111	170	350
	$\overline{p}$	-151	-195	-195	-195	-200	-203	-204	-199	-187	-164	-125	-066	-020	107	169	212	242	252	277	288	298
	$\overline{\omega}$	0	003	004	-003	-020	-044	-077	-119	-174	-242	-324	-416	-514	388	296	214	-145	090	016	014	0
0.3	$\overline{M}$	0	014	-014	-067	-134	-212	-300	-398	-508	-631	-760	-884	-983	**-537**	-1015	-924	-778	-614	-463	-302	0
	$\overline{Q}$	017	-011	-043	-061	-072	-083	-093	-104	-117	-127	-129	-116	-078	-018	056	123	161	160	145	197	459
	$\overline{p}$	-044	-062	-081	-101	-120	-138	-155	-168	-175	-174	-161	-131	-080	000	081	140	182	213	237	257	276
	$\overline{\omega}$	102	068	034	001	-033	-067	-099	-129	-154	-173	-185	-178	-156	-110	-035	046	107	154	194	229	263
0.4	$\overline{M}$	0	002	000	000	-005	-017	-037	-067	-104	-151	-210	282	-369	-470	**-579**	311	210	125	059	015	0
	$\overline{Q}$	0	-019	-004	-020	-079	-162	-250	-334	-421	-523	-649	-797	-945	-1061	-1109	-1068	-941	-757	-553	-333	0
	$\overline{p}$	076	015	005	-039	-075	-081	-085	-089	-092	-113	-136	-153	-138	-087	-005	087	162	199	207	248	464
	$\overline{\omega}$	102	068	034	001	-033	-067	-099	-129	-154	-173	-185	-178	-156	-110	-035	046	107	154	194	229	263

0.5																				
$\overline{M}$ 0	002	002	004	003	-005	-022	-046	-077	-117	-167	-231	-311	-407	-517	**-632**	255	155	074	019	0
$\overline{Q}$ 0	010	018	010	-013	-123	-204	-270	-353	-443	-564	-716	-884	-1038	-1139	-1157	-1078	-915	-689	-407	0
$\overline{p}$ 073	028	016	-013	-071	-073	-078	-079	-079	-104	-137	-164	-167	-133	-063	030	124	198	251	323	528
$\overline{\omega}$ 154	117	079	042	001	-035	-072	-109	-142	-170	-181	-201	-197	-172	-122	-038	054	129	195	255	316
0.6																				
$\overline{M}$ 0	000	000	-001	-001	-003	-014	-032	-058	-092	-138	-197	-271	-360	-464	-579	**-697**	190	093	025	0
$\overline{Q}$ 0	-011	-011	-013	-019	-074	-142	-217	-300	-399	-519	-661	-816	-970	-1100	-1178	-1176	-1071	-845	-491	0
$\overline{p}$ 056	018	017	-015	-046	-064	-071	-078	-090	-109	-131	-151	-158	-146	-108	-043	050	163	289	420	567
$\overline{\omega}$ 082	052	022	-007	-037	-067	-097	-125	-152	-175	-192	-200	-195	-173	-128	-054	057	181	293	399	502
0.7																				
$\overline{M}$ 0	001	003	005	005	003	-005	-018	-040	-071	-114	-169	-239	-322	-421	-532	-651	**-772**	117	034	0
$\overline{Q}$ 0	019	024	015	-009	-048	-103	-174	-262	-367	-488	-621	-765	-912	-1050	-1162	-1219	-1182	-1002	-625	0
$\overline{p}$ 028	011	-002	-016	-031	-047	-063	-080	-096	-113	-127	-139	-146	-145	-129	-090	-017	100	269	492	765
$\overline{\omega}$ 007	-014	-034	-055	-076	-097	-119	-140	-160	-178	-191	-196	-191	-170	-129	-061	041	184	345	498	648
0.8																				
$\overline{M}$ 0	001	004	005	066	004	-001	-012	-032	-062	-104	-159	-226	-307	-401	-508	-628	-752	**-870**	039	0
$\overline{Q}$ 0	000	022	010	-003	-027	-076	-151	-249	-363	-484	-608	-737	-872	-1011	-1142	-1233	-1235	-1082	-695	0
$\overline{p}$ 082	000	-014	-021	-016	-035	-062	-088	-107	-118	-123	-126	-132	-138	-138	-117	-057	063	257	529	874
$\overline{\omega}$ -067	-080	-092	-105	-118	-131	-144	-158	-170	-181	-188	-188	-177	-152	-108	-038	064	206	396	610	821
0.9																				
$\overline{M}$ 0	002	006	009	010	008	002	-010	-030	-060	-102	-155	-222	-303	-398	-506	-625	-749	-866	**-959**	0
$\overline{Q}$ 0	042	037	019	-005	-038	-087	-157	-247	-354	-474	-603	-739	-879	-1018	-1142	-1226	-1229	-1091	-723	0
$\overline{p}$ 087	-010	-016	-021	-027	-040	-059	-080	-099	-114	-125	-133	-139	-141	-135	-109	-052	059	236	520	956
$\overline{\omega}$ -036	-051	-066	-081	-097	-114	-131	-148	-164	-178	-189	-193	-187	-167	-127	-062	035	172	357	597	865
1.0																				
$\overline{M}$ 0	002	005	009	010	008	001	-012	-032	-062	-103	-156	-222	-304	-400	-509	-628	-751	-867	**-960**	**-10^3**
$\overline{Q}$ 0	033	037	024	-003	-042	-096	-164	-248	-349	-466	-597	-740	-887	-1028	-1148	-1224	-1219	-1081	-723	**0**
$\overline{p}$ -055	-015	-005	-020	-033	-046	-061	-076	-092	-109	-124	-138	-146	-146	-134	-102	-042	060	230	510	977
$\overline{\omega}$ 004	-015	-034	-053	-073	-094	-115	-136	-156	-175	-189	-196	-194	-177	-141	-080	014	148	330	567	864

219

Table 4.14 Dimensionless coefficients for analysis of beams loaded with a uniformly distributed load

t	x										
	0.0	0.1	0.2	0.3	0.4	0.5	0.6	0.7	0.8	0.9	1.0
						$\overline{M}$						
1	032	032	031	028	0.25	021	016	012	007	002	0
2	0.27	0.27	0.27	0.24	0.21	0.19	0.15	0.11	0.06	0.02	0
5	020	020	019	017	0.16	014	011	008	004	002	0
10	012	012	012	012	011	010	009	006	004	001	0
						$\overline{Q}$						
1	0	−009	−019	−028	−035	−043	−049	−052	−049	−036	0
2	0	−008	−016	−022	−030	−036	−042	−045	−046	−032	0
5	0	−004	−010	−015	−020	−024	−030	−033	−033	−023	0
10	0	−0.02	−005	−008	−010	−015	−019	−023	−026	−020	0
						$\overline{p}$						
1	090	090	091	091	092	093	095	100	107	122	153
2	092	092	092	092	093	094	096	098	116	147	198
5	096	095	095	095	095	094	094	093	102	114	141
10	098	097	097	096	093	095	096	096	100	110	134
						$\overline{Y}_x$						
1	221	220	220	219	218	217	217	216	215	213	212
2	222	222	221	220	220	218	217	214	212	211	208
5	226	226	225	224	223	218	217	213	210	205	201
10	230	230	229	227	224	221	216	212	207	202	196

Table 4.15 Soil pressure and settlements

ξ	0	0.2	0.4	06	0.8	1.0	1.2	1.4	1.6	2.0	2.4	2.8	3.0
$\overline{p}$	1.487	1.322	0.981	0.772	0.582	0.418	0.284	0.179	0.100	0.004	−0.034	−0.042	−0.042
p	446	400	294	230	175	125	85	54	30	1.2	−10	−12	−12
$\overline{Y}$	2.57	2.32	2.07	1.76	1.44	1.16	0.89	0.67	0.49	0.24	0.12	0.05	0.04
Y	3.86	3.48	3.11	2.64	2.16	1.74	1.34	1.00	0.74	0.36	0.18	0.08	0.06

Table 4.16 Shear forces

ξ	0.0	0.2	0.4	0.6	0.8	1.0	1.2	1.4	1.6	1.8	2.0	2.2	2.4	2.6
$\overline{Q}$	0.00	0.051	0.118	0.200	0.296	0.397	0.500	−0.400	−0.310	−0.231	−0.164	−0.11	−0.067	−0.036
Q	0.0	4.1	9.4	16	24	32	40/−40	−32	−24.8	−18.5	−13.1	−8.8	−5.4	−2.9

Table 4.17 Dimensionless coefficients for analysis of long beams loaded with a concentrated vertical load P

$\beta = 0.025$

ζ	α											
	0	0.1	0.2	0.3	0.4	0.5	0.6	0.7	0.8	0.9	1.0	∞
						$\bar{p}$						
0.0	3.227	2.747	2.259	1.828	1.402	1.101	0.867	0.723	0.570	0.434	0.273	**0.761**
0.1	2.626	**2.296**	1.963	1.644	1.335	1.057	0.825	0.643	0.482	0.340	0.219	0.733
0.2	2.105	1.805	**1.683**	1.465	1.248	1.034	0.829	0.643	0.480	0.342	0.226	0.668
0.3	1.658	1.538	1.421	**1.291**	1.159	1.106	0.846	0.680	0.537	0.395	0.284	0.602
0.4	1.282	1.229	1.182	1.121	**1.059**	0.967	0.857	0.727	0.588	0.480	0.371	0.552
0.5	0.963	0.959	0.962	0.958	0.947	**0.911**	0.851	0.765	0.669	0.569	0.471	0.509
0.6	0.698	0.730	0.768	0.802	0.831	0.839	**0.825**	0.775	0.723	0.649	0.573	0.456
0.7	0.481	0.538	0.603	0.660	0.715	0.755	0.775	**0.772**	0.750	0.711	0.659	0.388
0.8	0.306	0.376	0.454	0.528	0.601	0.665	0.714	0.745	**0.757**	0.751	0.724	0.310
0.9	0.165	0.246	0.328	0.413	0.494	0.575	0.645	0.703	0.742	**0.759**	0.736	0.233
1.0	0.057	0.141	0.230	0.314	0.401	0.484	0.563	0.634	0.693	0.737	**0.762**	0.164
1.2	-0.085	-0.003	0.079	0.160	0.243	0.325	0.409	0.494	0.576	0.650	0.712	0.065
1.4	-0.152	-0.083	-0.012	0.057	0.127	0.199	0.275	0.353	0.432	0.507	0.583	0.013
1.6	-0.170	-0.115	-0.062	-0.006	0.047	0.106	0.167	0.233	0.300	0.367	0.438	-0.009
1.8	-0.158	-0.121	-0.083	-0.040	-0.003	0.043	0.089	0.141	0.192	0.243	0.301	-0.016
2.0	-0.134	-0.109	-0.084	-0.055	-0.030	0.002	0.035	0.073	0.110	0.149	0192	-0.016
2.2	-0.106	-0.091	-0.074	-0.058	-0.042	-0.023	-0.002	0.024	0.051	0.079	0.111	-0.014
2.4	-0.076	-0.068	-0.059	-0.051	-0.043	-0.034	-0.019	-0.008	0.013	0.028	0.050	-0.012
2.6	-0.052	-0.050	-0.045	-0.042	-0.039	-0.035	-0.031	-0.017	-0.013	-0.009	-0.003	-0.009
2.8	-0.032	-0.033	-0.033	-0.033	-0.035	-0.035	-0.034	-0.032	-0.031	-0.031	-0.031	-0.007
3.0	-0.021	-0.020	-0.020	-0.024	-0.028	-0.034	-0.040	-0.044	-0.039	-0.039	-0.030	-0.005

Table 4.17 Con't

β = 0.025

$\overline{M}$

ζ	0	0.1	0.2	0.3	0.4	0.5	0.6	0.7	0.8	0.9	1.0	∞
						α						
0.0	0	0	0	0	0	0	0	0	0	0	0	**0.176**
0.1	0.085	**0.013**	0.011	0.007	0.007	0.006	0.006	0.005	0.004	0.003	0.002	0.130
0.2	-0.142	-0.050	**0.041**	0.033	0.028	0.022	0.018	0.015	0.011	0.007	0.005	0.091
0.3	-0.181	-0.094	-0.011	**0.072**	0.059	0.048	0.039	0.031	0.022	0.017	0.011	0.058
0.4	-0.202	-0.124	-0.050	0.026	**0.104**	0.085	0.068	0.054	0.041	0.030	0.020	0.032
0.5	-0.209	-0.142	-0.076	-0.007	0.059	**0.131**	0.105	0.085	0.065	0.046	0.031	0.011
0.6	-0.207	-0.150	-0.092	-0.035	0.022	0.085	**0.152**	0.122	0.092	0.070	0.048	-0.006
0.7	-0.198	-0.150	-0.102	-0.054	-0.004	0.050	0.107	**0.168**	0.131	0.100	0.070	-0.016
0.8	-0.185	-0.144	-0.105	-0.067	-0.024	0.020	0.068	0.120	**0.176**	0.137	0.100	-0.023
0.9	-0.168	-0.135	-0.105	-0.072	-0.039	-0.002	0.039	0.081	0.130	**0.179**	0.137	-0.027
1.0	-0.150	-0.124	-0.100	-0.074	-0.048	-0.018	0.015	0.050	0.089	0.135	**0.181**	-0.029
1.2	-0.113	-0.098	-0.085	-0.070	-0.056	-0.039	-0.018	0.004	0.030	0.059	0.092	-0.028
1.4	-0.080	-0.072	-0.067	-0.061	-0.054	-0.044	-0.033	-0.020	-0.007	0.011	0.031	-0.024
1.6	-0.050	-0.054	-0.048	-0.048	-0.046	-0.043	-0.039	-0.033	-0.028	-0.018	-0.007	-0.020
1.8	-0.028	-0.030	-0.033	-0.035	-0.035	-0.037	-0.037	-0.035	-0.034	-0.031	-0.026	-0.016
2.0	-0.013	-0.017	-0.020	-0.024	-0.026	-0.031	-0.031	-0.032	-0.035	-0.035	-0.035	-0.012
2.2	-0.004	-0.007	-0.011	-0.015	-0.018	-0.024	-0.024	-0.026	-0.030	-0.033	-0.035	-0.010
2.4	0.002	0.000	-0.006	-0.007	-0.011	-0.017	-0.017	-0.018	-0.024	-0.028	-0.031	-0.006
2.6	0.005	0.002	-0.002	-0.003	-0.005	-0.009	-0.009	-0.013	-0.017	-0.020	-0.024	-0.006
2.8	0.006	0.004	0.000	0.000	-0.002	-0.004	-0.004	-0.006	-0.011	-0.015	-0.018	-0.005
3.0	0.006	0.004	0.002	0.001	0.000	-0.001	-0.002	-0.002	-0.005	-0.009	-0.013	-0.004

$\bar{Q}$

x												
0.0	**0.000**	0.000	0.000	0.000	0.000	0.000	0.000	0.000	0.000	0.000	0.00	**-0.500**
0.1	-0.708	**0.252**	0.220	0.174	0.137	0.106	0.083	0.066	0.050	0.038	0.026	-0.425
0.2	-0.472	-0.539	**0.393**	0.330	0.266	0.211	0.166	0.128	0.091	0.071	0.047	-0.355
0.3	-0.284	-0.367	-0.452	**0.467**	0.387	0.313	0.250	0.194	0.146	0.107	0.072	-0.292
0.4	-0.138	-0.229	-0.322	-0.413	**0.497**	0.412	0.335	0.264	0.202	0.150	0.103	-0.234
0.5	-0.026	-0.120	-0.215	-0.309	-0.402	**0.506**	0.420	0.340	0.266	0.202	0.144	-0.181
0.6	0.056	-0.036	-0.129	-0.221	-0.314	-0.407	**0.504**	0.417	0.386	0.263	0.197	-0.132
0.7	0.113	0.027	-0.061	-0.149	-0.237	-0.328	-0.416	**0.495**	0.413	0.333	0.261	-0.090
0.8	0.153	0.073	-0.009	-0.089	-0.171	-0.256	-0.340	-0.427	**0.488**	0.407	0.330	-0.055
0.9	0.176	0.103	0.030	-0.042	-0.116	-0.194	-0.272	-0.354	-0.438	**0.482**	0.402	-0.028
1.0	0.187	0.122	0.058	-0.006	-0.072	-0.141	-0.212	-0.288	-0.365	-0.442	**0.482**	-0.008
1.2	0.183	0.135	0.088	0.040	-0.008	-0.060	-0.114	-0.174	-0.237	-0.303	-0.373	0.014
1.4	0.159	0.126	0.093	0.061	0.028	-0.009	-0.047	-0.090	-0.137	-0.188	-0.243	0.021
1.6	0.146	0.106	0.086	0.065	0.048	0.021	-0.004	-0.033	-0.064	-0.100	-0.141	0.021
1.8	0.092	0.091	0.070	0.060	0.049	0.036	0.022	-0.004	-0.016	-0.040	-0.067	0.018
2.0	0.062	0.058	0.053	0.050	0.046	0.040	0.034	0.025	0.019	-0.001	-0.018	0.014
2.2	0.039	0.038	0.038	0.039	0.039	0.038	0.037	0.035	0.029	0.022	0.013	0.011
2.4	0.020	0.022	0.024	0.027	0.030	0.032	0.034	0.036	0.035	0.032	0.028	0.009
2.6	0.008	0.011	0.014	0.018	0.022	0.025	0.029	0.032	0.033	0.033	0.032	0.007
2.8	-0.001	0.002	0.006	0.010	0.014	0.018	0.022	0.027	0.029	0.029	0.029	0.005
3.0	-0.005	-0.003	0.000	0.005	0.008	0.012	0.016	0.020	0.022	0.022	0.022	0.004

Table 4.17 Con't

$\beta = 0.025$

ζ	0	0.1	0.2	0.3	0.4	0.5 $\bar{y}$	0.6	0.7	0.8	0.9	1.0	∞
0	**6.92**	6.00	5.12	4.28	3.52	2.84	2.20	1.72	1.24	0.88	0.60	2.14
0.1	6.00	**5.32**	4.64	3.96	3.32	2.76	2.28	1.80	1.40	1.08	0.80	2.11
0.2	5.12	4.64	**4.16**	3.68	3.20	2.88	2.34	1.92	1.56	1.24	1.00	2.02
0.3	4.28	3.96	3.68	**3.44**	3.08	2.72	2.40	2.08	1.76	1.48	1.2	1.90
0.4	3.52	3.32	3.20	3.08	**2.88**	2.64	2.40	2.16	1.88	1.64	1.40	1.76
0.5	2.84	2.76	2.88	2.72	2.64	**2.56**	2.43	2.24	2.04	1.80	1.60	1.60
0.6	2.20	2.28	2.34	2.40	2.43	2.44	2.44	2.24	2.12	1.96	1.78	1.44
0.7	1.72	1.80	1.92	2.08	2.16	2.24	**2.36**	**2.24**	2.16	2.04	1.92	1.28
0.8	1.24	1.40	1.56	1.76	1.88	2.04	2.24	2.16	**2.20**	2.04	2.04	1.13
0.9	0.88	1.08	1.24	1.48	1.64	1.80	2.12	2.04	2.12	**2.16**	2.12	0.99
1.0	0.60	0.80	1.00	1.20	1.40	1.60	1.96	1.92	2.12	2.12	**2.16**	0.85
1.2	0.16	0.36	0.56	0.80	1.00	1.20	1.76	1.56	2.04	1.52	2.04	0.62
1.4	-0.08	0.08	0.28	0.48	0.68	0.88	1.40	1.24	1.76	1.60	1.76	0.43
1.6	-0.16	-0.04	0.12	0.24	0.44	0.60	1.04	0.92	1.40	1.24	1.40	0.28
1.8	-0.20	-0.12	0.00	0.16	0.28	0.40	0.76	0.64	1.08	0.96	1.08	0.16
2.0	-0.20	-0.12	-0.04	0.08	0.16	0.28	0.52	0.44	0.80	0.64	0.80	0.07
2.2	-0.12	-0.12	-0.04	0.04	0.12	0.16	0.36	0.32	0.56	0.48	0.56	0.00
2.4	-0.08	-0.08	-0.04	0.04	0.08	0.12	0.24	0.20	0.40	0.32	0.36	-0.06
2.6	-0.04	-0.04	0.00	0.04	0.04	0.08	0.16	0.16	0.28	0.20	0.24	-0.11
2.8	0.00	0.00	0.04	0.04	0.04	0.08	0.12	0.12	0.16	0.12	0.16	-0.15
3.0	0.04	0.04	0.04	0.04	0.04	0.04	0.08	0.08	0.12	0.08	0.08	-0.18

α

Table 4.18 Dimensionless coefficients for analysis of long beams loaded with a concentrated vertical load P

$\beta = 0.075$

ζ						$\bar{p}$	α					
	0	0.1	0.2	0.3	0.4	0.5	0.6	0.7	0.8	0.9	1.0	∞
0.0	**2.800**	2.481	2.131	1.803	1.487	1.238	1.015	0.851	0.697	0.547	0.405	**0.620**
0.1	2.295	**2.064**	1.814	1.575	1.332	1.125	0.927	0.762	0.615	0.469	0.341	0.605
0.2	1.876	1.716	**1.549**	1.387	1.208	1.041	0.873	0.726	0.589	0.641	0.361	0.572
0.3	1.525	1.421	1.315	**1.203**	1.092	0.970	0.844	0.714	0.594	0.485	0.388	0.536
0.4	0.229	1.170	1.111	1.046	**0.981**	0.902	0.813	0.714	0.616	0.521	0.432	0.494
0.5	0.979	0.945	0.930	0.903	0.875	**0.832**	0.779	0.712	0.638	0.561	0.484	0.460
0.6	0.768	0.769	0.771	0.772	0.772	0.761	**0.739**	0.702	0.664	0.596	0.534	0.417
0.7	0.589	0.610	0.632	0.654	0.674	0.687	0.689	**0.678**	0.655	0.623	0.581	0.369
0.8	0.439	0.474	0.510	0.546	0.582	0.612	0.635	0.646	**0.648**	0.637	0.614	0.318
0.9	0.314	0.359	0.404	0.450	0.496	0.539	0.577	0.608	0.627	**0.633**	0.627	0.266
1.0	0.209	0.261	0.313	0.365	0.418	0.469	0.516	0.555	0.591	0.616	**0.629**	0.216
1.2	0.055	0.112	0.170	0.233	0.284	0.341	0.397	0.452	0.505	0.553	0.592	0.132
1.4	-0.043	0.012	0.068	0.123	0.179	0.234	0.292	0.348	0.401	0.458	0.510	0.071
1.6	-0.099	-0.050	0.000	0.050	0.100	0.151	0.204	0.256	0.309	0.362	0.415	0.131
1.8	-0.126	-0.084	-0.042	0.001	0.043	0.088	0.132	0.179	0.225	0.272	0.320	0.008
2.0	-0.132	-0.098	-0.064	-0.029	0.004	0.040	0.077	0.117	0.155	0.195	0.237	0.002
2.2	-0.125	-0.099	-0.073	-0.046	-0.020	0.007	0.036	0.067	0.099	0.133	0.167	-0.009
2.4	-0.111	-0.092	-0.072	-0.061	-0.034	-0.013	0.006	0.025	0.054	0.081	0.109	-0.011
2.6	-0.094	-0.080	-0.067	-0.054	-0.041	-0.026	-0.011	0.007	0.021	0.038	0.055	-0.011
2.8	-0.076	-0.068	-0.059	-0.051	-0.042	-0.034	-0.025	-0.016	-0.006	0.003	0.014	-0.010
3.0	-0.058	-0.054	-0.049	-0.046	-0.042	-0.040	-0.038	-0.035	-0.029	-0.021	-0.011	-0.008

Table 4.18 Con't

$\beta = 0.075$

ζ	α					$\overline{M}$						
	0	0.1	0.2	0.3	0.4	0.5	0.6	0.7	0.8	0.9	1.0	∞
0.0	**0.000**	0.000	0.000	0.000	0.000	0.000	0.000	0.000	0.000	0.000	0.000	**0.214**
0.1	-0.087	**-0.011**	0.009	0.007	0.006	0.006	0.006	0.005	0.004	0.003	0.002	0.167
0.2	-0.150	-0.057	**0.039**	0.033	0.028	0.022	0.020	0.018	0.015	0.011	0.007	0.127
0.3	-0.195	-0.106	-0.019	**0.070**	0.061	0.050	0.043	0.037	0.030	0.024	0.017	0.091
0.4	-0.224	-0.152	-0.061	0.022	**0.104**	0.087	0.074	0.063	0.052	0.039	0.030	0.060
0.5	-0.242	-0.169	-0.093	0.017	0.057	**0.138**	0.113	0.096	0.078	0.054	0.046	0.036
0.6	-0.248	-0.182	-0.115	-0.046	0.020	0.087	**0.151**	0.137	0.113	0.089	0.069	0.016
0.7	-0.246	-0.189	-0.130	-0.070	-0.009	0.050	0.115	**0.185**	0.154	0.122	0.096	-0.001
0.8	-0.241	-0.191	-0.139	-0.085	-0.033	0.019	0.078	0.139	**0.202**	0.163	0.130	-0.013
0.9	-0.230	-0.185	-0.141	-0.096	-0.050	-0.004	0.044	0.098	0.154	**0.209**	0.171	-0.023
1.0	-0.217	-0.180	-0.141	-0.119	-0.063	-0.024	0.019	0.065	0.113	0.163	**0.215**	-0.029
1.2	-0.183	-0.158	-0.132	-0.138	-0.076	-0.050	-0.019	0.015	0.050	0.085	0.128	-0.036
1.4	-0.148	-0.135	-0.109	-0.096	-0.080	-0.061	-0.041	-0.017	0.007	0.031	0.061	-0.038
1.6	-0.113	-0.104	-0.094	-0.083	-0.074	-0.063	-0.050	-0.037	-0.020	-0.004	0.013	-0.036
1.8	-0.083	-0.076	-0.074	-0.068	-0.065	-0.059	-0.052	-0.044	-0.035	-0.026	-0.015	-0.034
2.0	-0.057	-0.056	-0.056	-0.054	-0.052	-0.052	-0.048	-0.044	-0.041	-0.037	-0.032	-0.031
2.2	-0.037	-0.037	-0.039	-0.041	-0.041	-0.043	-0.043	-0.041	-0.039	-0.039	-0.039	-0.027
2.4	-0.022	-0.024	-0.026	-0.028	-0.032	-0.033	-0.035	-0.035	-0.035	-0.037	-0.039	-0.022
2.6	-0.011	-0.013	-0.017	-0.014	-0.022	-0.022	-0.026	-0.026	-0.028	-0.032	-0.035	-0.018
2.8	-0.004	-0.006	-0.009	-0.011	-0.013	-0.017	-0.017	-0.019	-0.020	-0.024	-0.028	-0.015
3.0	0.000	-0.002	-0.002	-0.006	-0.006	-0.007	-0.009	-0.011	-0.011	-0.015	-0.020	-0.011

$\overline{Q}$

0.0	**0.000**	0.000	0.000	0.000	0.000	0.000	0.000	0.000	0.000	0.000	0.000	**-0.500**
0.1	-0.744	**0.228**	0.196	0.168	0.141	0.117	0.097	0.079	0.065	0.052	0.039	-0.439
0.2	-0.535	-0.584	**0.364**	0.316	0.274	0.288	0.187	0.156	0.124	0.103	0.075	-0.391
0.3	-0.365	-0.426	-0.492	**0.445**	0.383	0.326	0.274	0.226	0.184	0.146	0.112	-0.325
0.4	-0.228	-0.298	-0.372	-0.443	**0.480**	0.419	0.356	0.297	0.244	0.196	0.147	-0.274
0.5	-0.117	-0.192	-0.269	-0.345	-0.421	**0.506**	0.435	0.387	0.307	0.250	0.197	-0.225
0.6	0.031	-0.107	-0.184	-0.261	-0.338	-0.414	**0.511**	0.439	0.371	0.308	0.248	-0.181
0.7	0.037	-0.035	-0.115	-0.190	-0.266	-0.342	-0.418	**0.508**	0.436	0.369	0.305	-0.142
0.8	0.089	0.017	-0.057	-0.130	-0.203	-0.277	-0.351	-0.425	**0.503**	0.432	0.365	-0.106
0.9	0.125	0.058	-0.012	-0.080	-0.150	-0.219	-0.289	-0.362	-0.433	**0.496**	0.426	-0.079
1.0	0.152	0.008	0.024	-0.040	-0.104	-0.169	-0.236	-0.304	-0.372	-0.441	**0.491**	-0.054
1.2	0.177	0.125	0.071	0.019	-0.034	-0.088	-0.145	-0.203	-0.262	-0.317	-0.388	-0.020
1.4	0.177	0.131	0.095	0.054	0.012	-0.032	-0.076	-0.124	-0.172	-0.224	-0.277	0.004
1.6	0.163	0.132	0.101	0.071	0.039	0.007	-0.027	-0.063	-0.101	-0.141	-0.185	0.010
1.8	0.140	0.118	0.096	0.075	0.055	0.031	0.006	-0.020	-0.047	-0.078	-0.111	0.013
2.0	0.113	0.100	0.085	0.074	0.057	0.043	0.027	0.009	-0.009	-0.031	-0.055	0.015
2.2	0.088	0.079	0.072	0.065	0.056	0.048	0.039	0.028	0.016	-0.002	-0.015	0.013
2.4	0.064	0.061	0.057	0.054	0.050	0.047	0.042	0.037	0.031	0.027	0.012	0.011
2.6	0.044	0.044	0.043	0.044	0.043	0.043	0.041	0.040	0.038	0.034	0.028	0.009
2.8	0.027	0.029	0.030	0.033	0.035	0.036	0.038	0.039	0.039	0.041	0.036	0.007
3.0	0.013	0.017	0.020	0.023	0.026	0.030	0.032	0.033	0.036	0.037	0.036	0.004

Table 4.18 Con't

β = 0.075

ζ	0	0.1	0.2	0.3	0.4	0.5	0.6	0.7	0.8	0.9	1.0	∞
						α $\overline{\gamma}$							
0	**4.31**	3.84	3.40	2.97	2.57	2.2	1.85	1.55	1.27	1.04	0.85	1.37
0.1	3.84	**3.47**	3.11	2.79	2.44	2.12	1.83	1.54	1.29	1.09	0.93	1.35
0.2	3.40	3.11	**2.85**	2.59	2.32	2.05	1.80	1.56	1.35	1.16	1.00	1.32
0.3	2.97	2.79	2.59	**2.40**	2.20	1.99	1.78	1.57	1.40	1.23	1.09	1.27
0.4	2.57	2.44	2.32	2.20	**2.07**	1.92	1.76	1.59	1.44	1.29	1.16	1.21
0.5	2.20	2.12	2.05	1.99	1.92	**1.81**	1.71	1.59	1.47	1.35	1.23	1.13
0.6	1.85	1.83	1.80	1.78	1.76	1.71	**1.65**	1.56	1.49	1.39	1.29	1.05
0.7	1.55	1.54	1.56	1.57	1.59	1.59	1.56	**1.52**	1.48	1.43	1.35	0.97
0.8	1.27	1.29	1.35	1.40	1.44	1.47	1.49	1.48	**1.48**	1.45	1.41	0.89
0.9	1.04	1.09	1.16	1.23	1.29	1.35	1.39	1.43	1.45	**1.44**	1.42	0.80
1.0	0.85	0.93	1.00	1.09	1.16	1.23	1.29	1.35	1.41	1.42	**1.44**	0.73
1.2	0.49	0.59	0.68	0.80	0.89	0.97	1.08	1.15	1.25	1.31	1.32	0.60
1.4	0.27	0.36	0.45	0.57	0.67	0.76	0.87	0.96	1.07	1.15	1.24	0.48
1.6	0.09	0.20	0.29	0.40	0.49	0.57	0.69	0.77	0.88	0.96	1.08	0.35
1.8	0.00	0.09	0.17	0.27	0.35	0.44	0.53	0.61	0.72	0.80	0.89	0.26
2.0	-0.05	0.03	0.09	0.17	0.24	0.32	0.40	0.47	0.57	0.64	0.73	0.17
2.2	-0.07	-0.01	0.04	0.11	0.17	0.23	0.29	0.36	0.44	0.51	0.57	0.10
2.4	-0.07	-0.03	0.01	0.07	0.12	0.16	0.21	0.27	0.33	0.39	0.44	0.04
2.6	-0.06	-0.03	0.00	0.05	0.08	0.11	0.15	0.19	0.24	0.38	0.38	0.00
2.8	-0.05	-0.01	0.00	0.04	0.05	0.08	0.10	0.13	0.17	0.20	0.23	-0.03
3.0	-0.03	0.00	0.00	0.03	0.04	0.05	0.07	0.08	0.11	0.13	0.15	-0.07

Table 4.19 Dimensionless coefficients for analysis of long beams loaded with a concentrated vertical load P

β = 0.15

$\bar{\rho}$

ζ	0	0.1	0.2	0.3	0.4	0.5	0.6	0.7	0.8	0.9	1.0	∞
						α						
0.0	**2.732**	2.435	2.128	1.841	1.552	1.316	1.107	0.946	0.794	0.640	0.507	**0.555**
0.1	2.200	**1.981**	1.773	1.562	1.351	1.159	0.975	0.827	0.687	0.557	0.446	0.548
0.2	1.781	1.635	**1.488**	1.339	1.190	1.043	0.898	0.763	0.639	0.523	0.425	0.529
0.3	1.444	1.349	1.258	**1.154**	1.055	0.949	0.838	0.727	0.619	0.520	0.432	0.503
0.4	1.173	1.115	1.057	0.998	**0.938**	0.868	0.790	0.706	0.619	0.535	0.456	0.476
0.5	0.948	0.948	0.890	0.861	0.831	**0.793**	0.745	0.687	0.624	0.557	0.490	0.452
0.6	0.761	0.753	0.746	0.741	0.733	0.720	**0.698**	0.666	0.624	0.575	0.522	0.426
0.7	0.604	0.613	0.623	0.634	0.644	0.649	0.647	**0.636**	0.615	0.586	0.550	0.393
0.8	0.471	0.493	0.515	0.538	0.561	0.581	0.594	0.603	**0.597**	0.590	0.570	0.352
0.9	0.357	0.389	0.419	0.453	0.484	0.516	0.543	0.565	0.578	**0.582**	0.576	0.302
1.0	0.264	0.302	0.339	0.380	0.416	0.454	0.486	0.520	0.545	0.564	**0.573**	0.252
1.2	0.118	0.163	0.208	0.252	0.297	0.341	0.385	0.428	0.469	0.508	0.539	0.155
1.4	0.019	0.065	0.111	0.157	0.203	0.249	0.299	0.340	0.385	0.433	0.471	0.081
1.6	−0.046	−0.003	0.041	0.085	0.129	0.173	0.217	0.261	0.305	0.330	0.394	0.035
1.8	−0.083	−0.045	−0.007	0.034	0.072	0112	0.153	0.194	0.234	0.278	0.316	0.009
2.0	−0.102	−0.069	−0.037	−0.002	0.031	0.065	0.101	0.137	0.173	0.210	0.246	−0.003
2.2	−0.109	−0.081	−0.054	−0.026	0.002	0.030	0.059	0.090	0.121	0.153	0.186	−0.009
2.4	−0.105	−0.083	−0.061	−0.039	−0.016	0.005	0.028	0.051	0.077	0.105	0.132	−0.011
2.6	−0.097	−0.083	−0.064	−0.047	−0.030	−0.013	0.004	0.024	0.042	0.062	0.080	−0.011
2.8	−0.087	−0.081	−0.063	−0.050	−0.039	−0.027	−0.014	−0.001	0.012	0.024	0.036	−0.010
3.0	−0.076	−0.075	−0.061	−0.053	−0.046	−0.040	−0.032	−0.025	−0.017	−0.007	−0.003	−0.008

Table 4.19 Con't

β = 0.15

$\overline{M}$

ζ	α											
	0	0.1	0.2	0.3	0.4	0.5	0.6	0.7	0.8	0.9	1.0	∞
0.0	0.000	0.000	0.000	0.000	0.000	0.000	0.000	0.000	0.000	0.000	0.000	0.230
0.1	-0.087	0.012	0.009	0.009	0.008	0.005	0.005	0.004	0.003	0.003	0.003	0.183
0.2	-0.152	-0.056	0.037	0.033	0.029	0.024	0.020	0.017	0.014	0.012	0.010	0.141
0.3	-0.200	-0.110	-0.021	0.070	0.061	0.052	0.044	0.037	0.031	0.026	0.020	0.105
0.4	-0.232	-0.148	-0.065	0.020	0.105	0.090	0.077	0.066	0.055	0.046	0.036	0.072
0.5	-0.253	-0.176	-0.099	-0.021	0.058	0.137	0.116	0.100	0.085	0.070	0.056	0.046
0.6	-0.265	-0.194	-0.124	-0.053	0.019	0.091	0.165	0.139	0.120	0.100	0.081	0.024
0.7	-0.269	-0.205	-0.142	-0.077	-0.012	-0.053	0.120	0.191	0.163	0.137	0.110	0.006
0.8	-0.267	-0.210	-0.153	-0.096	-0.037	0.021	0.081	0.145	0.210	0.178	0.147	-0.006
0.9	-0.259	-0.209	-0.158	-0.108	-0.057	-0.005	0.049	0.106	0.164	0.225	0.190	-0.018
1.0	-0.249	-0.205	-0.161	-0.116	-0.071	-0.026	0.022	0.124	0.124	0.179	0.238	-0.025
1.2	-0.221	-0.188	-0.156	-0.122	-0.088	-0.054	-0.019	0.020	0.060	0.103	0.147	-0.033
1.4	-0.188	-0.165	-0.142	-0.118	-0.093	-0.069	-0.044	-0.015	0.014	0.047	0.080	-0.035
1.6	-0.154	-0.139	-0.124	-0.107	-0.091	-0.074	-0.056	-0.036	-0.016	0.007	0.031	-0.033
1.8	-0.122	-0.133	-0.103	-0.093	-0.083	-0.072	-0.061	-0.047	-0.034	-0.018	-0.002	-0.030
2.0	-0.093	-0.088	-0.083	-0.077	-0.072	-0.066	-0.059	-0.051	-0.043	-0.033	-0.022	-0.026
2.2	-0.068	-0.066	-0.065	-0.062	-0.059	-0.057	-0.053	-0.049	-0.044	-0.039	-0.032	-0.022
2.4	-0.047	-0.048	-0.048	-0.047	-0.047	-0.046	-0.045	-0.042	-0.041	-0.038	-0.035	-0.018
2.6	-0.031	-0.033	-0.034	-0.034	-0.035	-0.036	-0.035	-0.031	-0.034	-0.033	-0.032	-0.015
2.8	-0.013	-0.021	-0.023	-0.023	-0.024	-0.025	-0.026	-0.026	-0.026	-0.026	-0.026	-0.013
3.0	-0.010	-0.011	-0.014	-0.014	-0.015	-0.016	-0.017	-0.017	-0.017	-0.019	-0.019	-0.010

$\bar{Q}$

$\bar{y}$												
0.0	**0.000**	0.000	0.000	0.000	0.000	0.000	0.000	0.000	0.000	0.000	0.000	-0.500
0.1	-0.755	**0.219**	0.196	0.168	0.144	0.124	0.104	0.089	0.074	0.062	0.046	-0.445
0.2	-0.556	-0.600	**0.358**	0.313	0.271	0.234	0.198	0.168	0.140	0.116	0.090	-0.391
0.3	-0.395	-0.454	-0.505	**0.439**	0.384	0.334	0.285	0.242	0.203	0.167	0.133	-0.339
0.4	-0.265	-0.328	-0.390	-0.454	**0.483**	0.424	0.366	0.313	0.264	0.219	0.177	-0.290
0.5	-0.159	-0.226	-0.293	-0.293	-0.428	**0.507**	0.443	0.382	0.326	0.273	0.224	-0.244
0.6	-0.075	-0.143	-0.211	-0.281	-0.350	-0.417	**0.515**	0.450	0.388	0.329	0.275	-0.200
0.7	-0.007	-0.075	-0.143	-0.213	-0.280	-0.349	-0.418	**0.515**	0.451	0.388	0.329	-0.159
0.8	0.047	-0.020	-0.086	-0.154	-0.221	-0.287	-0.356	-0.423	**0.512**	0.477	0.385	-0.123
0.9	0.088	0.024	-0.040	-0.104	-0.169	-0.232	-0.299	-0.364	-0.429	**0.506**	0.442	-0.090
1.0	0.119	0.059	-0.002	-0.063	-0.123	-0.184	-0.248	-0.311	-0.373	-0.436	**0.501**	-0.060
1.2	0.157	0.105	0.052	0.000	-0.053	-0.105	-0.160	-0.215	-0.272	-0.330	-0.389	-0.021
1.4	0.170	0.126	0.083	0.040	-0.003	-0.047	-0.093	-0.139	-0.187	-0.236	-0.287	0.003
1.6	0.167	0.132	0.099	0.064	0.030	-0.005	-0.042	-0.079	-0.118	-0.158	-0.201	0.017
1.8	0.153	0.128	0.102	0.075	0.050	0.023	-0.005	-0.034	-0.064	-0.096	-0.130	0.020
2.0	0.134	0.116	0.097	0.078	0.060	0.041	0.021	-0.001	-0.023	-0.047	-0.074	0.019
2.2	0.114	0.102	0.088	0.076	0.063	0.061	0.037	0.022	0.016	-0.011	-0.030	0.018
2.4	0.092	0.084	0.076	0.069	0.061	0.054	0.045	0.036	0.026	0.014	0.001	0.015
2.6	0.072	0.068	0.064	0.060	0.056	0.053	0.048	0.043	0.037	0.030	0.022	0.012
2.8	0.053	0.052	0.052	0.050	0.049	0.049	0.047	0.045	0.043	0.039	0.034	0.910
3.0	0.037	0.037	0.039	0.040	0.041	0.042	0.043	0.043	0.042	0.040	0.038	0.009

Table 4.19 Con't

β = 0.15

ζ	0	0.1	0.2	0.3	0.4	0.5	0.6	0.7	0.8	0.9	1.0	∞
						α						
0	**3.04**	2.75	2.48	2.21	1.96	1.72	1.50	1.30	1.11	0.95	0.79	1.07
0.1	2.75	**2.53**	2.31	2.09	1.87	1.67	1.47	1.3	1.13	0.98	0.84	1.06
0.2	2.48	2.31	**2.12**	1.95	1.77	1.60	1.44	1.29	1.14	1.00	0.88	1.04
0.3	2.21	2.09	1.95	**1.81**	1.67	1.51	1.41	1.27	1.15	1.03	0.92	1.01
0.4	1.96	1.87	1.77	1.67	**1.59**	1.49	1.38	1.27	1.17	1.06	0.97	0.98
0.5	1.72	1.67	1.60	1.51	1.49	**1.41**	1.33	1.25	1.17	1.08	1.00	0.94
0.6	1.50	1.47	1.44	1.41	1.38	1.33	**1.29**	1.24	1.17	1.11	1.06	0.85
0.7	1.30	1.30	1.29	1.27	1.27	1.25	1.24	**1.21**	1.17	1.11	1.07	0.80
0.8	1.11	1.13	1.14	1.15	1.17	1.17	1.17	1.17	**1.14**	1.11	1.07	0.80
0.9	0.95	0.98	1.00	1.03	1.06	1.08	1.11	1.11	1.11	**1.11**	1.09	0.76
1.0	0.79	0.84	0.88	0.92	0.97	1.00	1.03	1.06	1.07	1.09	**1.09**	0.65
1.2	0.55	0.61	0.66	0.72	0.78	0.83	0.89	0.93	0.97	1.01	1.05	0.55
1.4	0.36	0.43	0.49	0.55	0.62	0.68	0.75	0.81	0.86	0.91	0.97	0.46
1.6	0.22	0.29	0.35	0.41	0.49	0.55	0.62	0.68	0.74	0.80	0.86	0.36
1.8	0.12	0.18	0.24	0.31	0.37	0.43	0.50	0.56	0.62	0.68	0.75	0.28
2.0	0.05	0.11	0.17	0.22	0.28	0.34	0.40	0.46	0.51	0.57	0.63	0.15
2.2	0.01	0.06	0.11	0.16	0.21	0.26	0.32	0.37	0.42	0.47	0.53	0.15
2.4	-0.01	0.03	0.07	0.11	0.16	0.20	0.25	0.29	0.33	0.37	0.43	0.10
2.6	-0.02	0.01	0.05	0.08	0.11	0.14	0.19	0.23	0.26	0.29	0.34	0.05
2.8	-0.02	0.00	0.03	0.05	0.09	0.11	0.14	0.17	0.19	0.22	0.26	0.02
3.0	-0.01	-0.01	0.02	0.04	0.06	0.08	0.11	0.12	0.13	0.15	0.18	-0.01

Table 4.20 Dimensionless coefficients for analysis of long beams loaded with a concentrated vertical load P

$\beta = 0.3$

$\bar{\rho}$

ζ	α											
	0	0.1	0.2	0.3	0.4	0.5	0.6	0.7	0.8	0.9	1.0	∞
0.0	**2.25**	1.81	1.34	0.95	0.54	0.34	0.24	0.28	0.31	0.28	0.13	0.49
0.2	1.65	**1.55**	1.14	0.89	0.64	0.44	0.29	0.20	0.13	0.07	0.03	0.47
0.4	1.18	1.06	**0.94**	0.80	0.67	0.52	0.37	0.23	0.11	0.03	0.02	0.43
0.6	0.82	0.78	0.75	**0.70**	0.65	0.56	0.45	0.30	0.18	0.09	0.02	0.36
0.8	0.54	0.56	0.58	0.59	**0.60**	0.57	0.50	0.39	0.28	0.19	0.11	0.30
1.0	0.34	0.37	0.44	0.49	0.53	**0.54**	0.52	0.45	0.38	0.30	0.22	0.23
1.2	0.19	0.25	0.32	0.39	0.47	0.49	**0.51**	0.49	0.45	0.39	0.33	0.17
1.4	0.08	0.15	0.22	0.30	0.37	0.43	0.48	**0.50**	0.49	0.46	0.42	0.13
1.6	0.00	0.07	0.14	0.22	0.29	0.36	0.43	0.47	**0.50**	0.50	0.48	0.09
1.8	-0.04	0.02	0.08	0.15	0.22	0.30	0.37	0.43	0.49	**0.50**	0.51	0.06
2.0	-0.08	-0.02	0.04	0.10	0.16	0.23	0.30	0.37	0.43	0.48	**0.51**	0.04
2.2	-0.09	-0.03	0.01	0.06	0.12	0.18	0.24	0.31	0.38	0.44	0.48	0.02
2.4	-0.09	-0.05	-0.02	0.03	0.08	0.13	0.18	0.25	0.31	0.38	0.44	0.01
2.6	-0.09	-0.06	-0.03	0.01	0.05	0.09	0.14	0.19	0.25	0.32	0.38	0.00
2.8	-0.09	-0.06	-0.03	0.00	0.03	0.06	0.10	0.14	0.20	0.25	0.32	0.00
3.0	-0.08	-.005	-0.04	-0.01	0.01	0.04	0.07	0.10	0.15	0.20	0.25	0.00
3.2	-0.07	-0.05	-0.04	-0.02	0.00	0.02	0.04	0.07	0.11	0.15	0.19	-0.01
3.4	-0.06	-0.05	-0.03	-0.02	-0.01	0.01	0.03	0.05	0.08	0.10	0.14	-0.01
3.6	-0.05	-0.04	-0.03	-0.02	-0.01	0.00	0.01	0.03	0.05	0.07	0.10	-0.01
3.8	-0.04	-0.03	-0.03	-0.02	-0.02	-0.01	0.00	0.02	0.03	0.05	0.07	-0.01
4.0	-0.03	-0.03	-0.03	-0.02	-0.01	-0.01	0.00	0.01	0.02	0.03	0.05	-0.01

Table 4.20 Con't

β = 0.3

$\overline{M}$

ζ	0	0.1	0.2	0.3	0.4	0.5	0.6	0.7	0.8	0.9	1.0	∞
0.0	**0.000**	0.000	0.000	0.000	0.000	0.000	0.000	0.000	0.000	0.000	0.000	0.282
0.2	-0.159	**0.034**	0.026	0.019	0.012	0.008	0.005	0.005	0.005	0.004	0.003	0.192
0.4	-0.252	-0.077	**0.096**	0.073	0.049	0.028	0.022	0.019	0.016	0.012	0.007	0.120
0.6	-0.296	-0.145	0.005	**0.159**	0.112	0.079	0.053	0.041	0.031	0.021	0.012	0.066
0.8	-0.310	-0.182	-0.057	0.073	**0.202**	0.146	0.103	0.075	0.053	0.033	0.017	0.026
1.0	-0.300	-0.196	-0.095	0.011	0.115	**0.237**	0.172	0.125	0.087	0.054	0.026	-0.002
1.2	-0.280	-0.194	-0.115	-0.032	0.050	0.149	**0.262**	0.193	0.136	0.086	0.044	-0.022
1.4	-0.246	-0.183	-0.123	-0.067	0.003	0.081	0.172	**0.281**	0.202	0.133	0.077	-0.033
1.6	-0.212	-0.166	-0.120	-0.075	-0.036	0.030	0.101	0.189	**0.289**	0.200	0.124	-0.039
1.8	-0.178	-0.145	-0.114	-0.081	-0.049	-0.006	0.047	0.115	0.195	**0.288**	0.191	-0.041
2.0	-0.146	-0.124	-0.110	-0.82	-0.060	-0.031	0.010	0.058	0.120	0.191	**0.277**	-0.041
2.2	-0.117	-0.104	-0.091	-0.078	-0.065	-0.045	-0.020	0.018	0.062	0.116	0.189	-0.039
2.4	-0.090	-0.084	-0.078	-0.072	-0.065	-0.054	-0.037	-0.012	0.020	0.059	0.110	-0.037
2.6	-0.068	-0.067	-0.065	-0.064	-0.062	-0.058	-0.047	-0.031	-0.010	0.017	0.055	-0.033
2.8	-0.050	-0.051	-0.054	-0.055	-0.057	-0.056	-0.052	-0.043	-0.031	-0.013	0.013	-0.030
3.0	-0.035	-0.039	-0.043	-0.047	-0.051	-0.053	-0.053	-0.049	-0.043	-0.032	-0.015	-0.027
3.2	-0.023	-0.027	-0.033	-0.039	-0.044	-0.048	-0.050	-0.051	-0.049	-0.044	-0.034	-0.023
3.4	-0.013	-0.019	-0.025	-0.031	-0.037	-0.042	-0.047	-0.050	-0.050	-0.049	-0.044	-0.020
3.6	-0.007	-0.013	-0.019	-0.025	-0.031	-0.036	-0.041	-0.046	-0.049	-0.050	-0.049	-0.018
3.8	-0.002	-0.008	-0.013	-0.019	-0.025	-0.030	-0.036	-0.041	-0.045	-0.049	-0.050	-0.015
4.0	0.001	-0.004	-0.009	-0.015	-0.020	-0.025	-0.030	-0.035	-0.040	-0.045	-0.048	-0.013

α

$\bar{Q}$

0.0	**0.000**	0.000	0.000	0.000	0.000	0.000	0.000	0.000	0.000	0.000	0.000	**-0.500**
0.2	-0.612	**0.321**	0.246	0.184	0.119	0.078	0.051	0.046	0.040	0.032	0.020	-0.403
0.4	-0.331	-0.435	**0.454**	0.353	0.224	0.174	0.118	0.099	0.062	0.041	0.021	-0.313
0.6	-0.133	-0.253	-0.377	**0.503**	0.384	0.256	0.200	0.148	0.091	0.052	0.020	-0.234
0.8	-0.001	-0.120	-0.244	-0.368	**0.510**	0.397	0.296	0.210	0.137	0.079	0.032	-0.168
1.0	0.088	-0.026	-0.142	-0.260	-0.376	**0.508**	0.397	0.294	0.203	0.128	0.065	-0.115
1.2	0.139	0.036	-0.066	-0.173	-0.277	-0.388	**0.500**	0.390	0.283	0.197	0.121	-0.074
1.4	0.165	0.076	-0.013	-0.105	-0.195	-0.295	-0.400	**0.489**	0.381	0.284	0.195	-0.043
1.6	0.173	0.097	0.023	-0.053	-0.129	-0.215	-0.310	-0.411	**0.481**	0.380	0.285	-0.025
1.8	0.162	0.105	0.046	-0.016	-0.078	-0.149	-0.231	-0.323	-0.421	**0.481**	0.384	0.006
2.0	0.156	0.106	0.057	0.009	-0.039	-0.096	-0.164	-0.243	-0.330	-0.421	**0.386**	0.006
2.2	0.140	0.101	0.064	0.026	-0.011	-0.056	-0.110	-0.174	-0.248	-0.329	-0.417	0.012
2.4	0.120	0.092	0.064	0.037	0.009	-0.025	-0.067	-0.118	-0.181	-0.246	-0.323	0.016
2.6	0.101	0.082	0.062	0.040	0.022	-0.004	-0.036	-0.076	-0.125	-0.177	-0.241	0.017
2.8	0.084	0.070	0.057	0.042	0.029	0.012	-0.042	-0.042	-0.078	-0.121	-0.176	0.017
3.0	0.066	0.058	0.051	0.040	0.033	0.022	0.005	-0.016	-0.043	-0.076	-0.116	0.016
3.2	0.052	0.047	0.044	0.037	0.034	0.027	0.016	0.001	-0.014	-0.041	-0.071	0.015
3.4	0.039	0.038	0.036	0.033	0.033	0.030	0.023	0.014	-0.001	0.015	-0.037	0.014
3.6	0.029	0.029	0.029	0.029	0.031	0.030	0.027	0.022	0.013	0.002	-0.014	0.013
3.8	0.020	0.022	0.023	0.025	0.027	0.030	0.029	0.028	0.022	0.014	0.003	0.011
4.0	0.015	0.016	0.017	0.021	0.024	0.028	0.029	0.030	0.026	0.022	0.014	0.010

Table 4.20 Con't

$\beta = 0.3$

ζ	α					ȳ						
	0	0.1	0.2	0.3	0.4	0.5	0.6	0.7	0.8	0.9	1.0	∞
0.0	**2.29**	1.92	1.56	1.25	1.56	1.25	1.01	0.78	0.57	0.40	0.12	**0.78**
0.2	1.92	**1.66**	1.40	1.14	0.99	0.80	0.62	0.46	0.35	0.27	0.22	0.77
0.4	1.56	1.40	**1.24**	1.09	0.97	0.82	0.66	0.52	0.40	0.33	0.29	0.72
0.6	1.25	1.14	1.09	**1.03**	0.95	0.84	0.72	0.61	0.50	0.43	0.38	0.66
0.8	1.01	0.99	0.97	0.95	**0.92**	0.86	0.77	0.67	0.58	0.43	0.38	0.66
1.0	0.78	0.80	0.82	0.84	0.86	**0.84**	0.80	0.73	0.65	0.59	0.54	0.53
1.2	0.57	0.62	0.66	0.72	0.77	0.80	**0.79**	0.76	0.71	0.65	0.61	0.45
1.4	0.40	0.46	0.52	0.61	0.67	0.73	0.76	**0.77**	0.75	0.71	0.67	0.39
1.6	0.28	0.35	0.40	0.50	0.58	0.65	0.71	0.75	**0.76**	0.75	0.72	0.32
1.8	0.19	0.27	0.33	0.43	0.51	0.59	0.65	0.71	0.75	0.77	0.77	0.26
2.0	0.12	0.22	0.29	0.38	0.46	0.54	0.61	0.67	0.72	0.77	**0.79**	0.21
2.2	0.07	0.16	0.22	0.31	0.38	0.45	0.52	0.59	0.66	0.72	0.77	0.16
2.4	0.03	0.12	0.17	0.25	0.31	0.38	0.45	0.52	0.59	0.66	0.72	0.12
2.6	0.00	0.08	0.14	0.21	0.26	0.33	0.39	0.46	0.53	0.60	0.66	0.08
2.8	-0.01	0.06	0.11	0.17	0.22	0.27	0.33	0.40	0.46	0.53	0.60	0.05
3.0	-0.01	0.04	0.09	0.14	0.18	0.23	0.28	0.34	0.40	0.46	0.53	0.02
3.2	-0.01	0.02	0.07	0.12	0.15	0.20	0.24	0.29	0.34	0.40	0.46	0.00
3.4	-0.01	0.01	0.05	0.10	0.13	0.17	0.20	0.25	0.29	0.34	0.39	-0.02
3.6	-0.01	0.00	0.04	0.09	0.11	0.14	0.17	0.21	0.25	0.29	0.34	-0.04
3.8	0.00	0.00	0.03	0.07	0.10	0.12	0.15	0.18	0.21	0.25	0.28	-0.05
4.0	0.00	0.00	0.02	0.06	0.09	0.11	0.13	0.16	0.18	0.21	0.24	-0.07

Table 4.21 Dimensionless coefficients for analysis of long beams loaded with a concentrated vertical load P

β = 0.5

ζ	0	0.1	0.2	0.3	0.4	0.5 $\bar{p}$	0.6	0.7	0.8	0.9	1.0	∞
						α						
0.0	**2.27**	1.86	1.43	1.04	0.64	0.39	0.24	0.22	0.21	0.18	0.14	**0.44**
0.2	1.63	**1.39**	1.15	0.92	0.68	0.48	0.31	0.19	0.11	0.05	0.01	0.43
0.4	1.15	0.96	**0.92**	0.80	0.67	0.53	0.39	0.25	0.14	0.05	0.00	0.40
0.6	0.79	0.75	0.72	**0.68**	0.64	0.56	0.46	0.33	0.22	0.12	0.05	0.36
0.8	0.53	0.54	0.55	0.56	**0.57**	0.55	0.50	0.41	0.32	0.23	0.14	0.31
1.0	0.32	0.37	0.42	0.46	0.50	**0.52**	0.51	0.47	0.40	0.33	0.25	0.25
1.2	0.20	0.25	0.30	0.37	0.42	0.47	**0.49**	0.49	0.46	0.41	0.34	0.20
1.4	0.09	0.15	0.22	0.28	0.35	0.41	0.46	**0.48**	0.49	0.46	0.42	0.14
1.6	0.02	0.09	0.15	0.21	0.28	0.34	0.41	0.46	**0.48**	0.49	0.47	0.10
1.8	-0.02	0.04	0.09	0.15	0.22	0.28	0.35	0.41	0.46	**0.48**	0.49	0.07
2.0	-0.05	0.00	0.06	0.11	0.16	0.22	0.29	0.35	0.40	0.46	**0.49**	0.04
2.2	-0.07	-0.03	0.03	0.07	0.12	0.17	0.23	0.29	0.35	0.41	0.46	0.02
2.4	-0.08	-0.04	0.01	0.04	0.09	0.14	0.18	0.23	0.29	0.36	0.41	0.00
2.6	-0.08	-0.05	-0.01	0.02	0.06	0.10	0.14	0.19	0.25	0.30	0.35	0.00
2.8	-0.07	-0.05	-0.02	0.01	0.04	0.06	0.10	0.14	0.19	0.24	0.30	0.00
3.0	-0.07	-0.05	-0.02	0.00	0.02	0.05	0.07	0.11	0.15	0.19	0.24	-0.01
3.2	-0.07	-0.05	-0.03	-0.01	0.01	0.03	0.05	0.08	0.11	0.15	0.19	-0.01
3.4	-0.06	-0.04	-0.03	-0.01	0.00	0.02	0.04	0.06	0.08	0.10	0.14	-0.01
3.6	-0.05	-0.04	-0.03	-0.02	-0.01	0.01	0.02	0.04	0.06	0.08	0.10	-0.01
3.8	-0.04	-0.03	-0.03	-0.02	-0.01	0.00	0.01	0.03	0.04	0.06	0.08	-0.01
4.0	-0.03	-0.03	-0.03	-0.02	-0.01	0.00	0.01	0.02	0.03	0.04	0.06	0.00

Table 4.21 Con't

β = 0.5

$\bar{M}$

ζ	α											
	0	0.1	0.2	0.3	0.4	0.5	0.6	0.7	0.8	0.9	1.0	∞
0.0	**0.000**	0.000	0.000	0.000	0.000	0.000	0.000	0.000	0.000	0.000	0.000	**0.300**
0.2	−0.159	**0.034**	0.026	0.020	0.013	0.009	0.005	0.004	0.003	0.002	0.001	0.207
0.4	−0.253	−0.076	**0.099**	0.077	0.053	0.036	0.023	0.017	0.011	0.008	0.004	0.135
0.6	−0.300	−0.144	0.009	**0.165**	0.121	0.085	0.057	0.039	0.025	0.019	0.006	0.075
0.8	−0.316	−0.183	−0.052	0.081	**0.217**	0.156	0.108	0.074	0.048	0.027	0.012	0.033
1.0	−0.310	−0.199	−0.091	0.019	0.128	**0.250**	0.180	0.126	0.083	0.049	0.023	0.001
1.2	−0.291	−0.201	−0.113	−0.025	0.064	0.162	**0.272**	0.197	0.134	0.084	0.043	−0.020
1.4	−0.264	−0.192	−0.123	−0.053	0.016	0.094	0.183	**0.287**	0.204	0.135	0.078	−0.033
1.6	−0.232	−0.177	−0.124	−0.071	−0.018	0.042	0.113	0.197	**0.293**	0.204	0.129	−0.041
1.8	−0.200	−0.159	−0.119	−0.080	−0.041	0.005	0.058	0.124	0.201	**0.293**	0.198	−0.044
2.0	−0.169	−0.139	−0.111	−0.082	−0.055	−0.022	0.018	0.068	0.128	0.201	**0.288**	−0.045
2.2	−0.140	−0.120	−0.101	−0.081	−0.062	−0.039	−0.011	0.026	0.071	0.127	0.196	−0.044
2.4	−0.114	−0.101	−0.089	−0.077	−0.065	−0.050	−0.030	−0.006	0.028	0.070	0.126	−0.043
2.6	−0.091	−0.083	−0.077	−0.070	−0.064	−0.055	−0.043	−0.025	−0.003	0.028	0.067	−0.029
2.8	−0.070	−0.068	−0.066	−0.063	−0.061	−0.057	−0.051	−0.040	−0.024	−0.004	0.024	−0.017
3.0	−0.053	−0.054	−0.055	−0.055	−0.056	−0.056	−0.053	−0.047	−0.039	−0.025	−0.006	−0.012
3.2	−0.039	−0.042	−0.045	−0.048	−0.051	−0.053	−0.054	−0.051	−0.046	−0.039	−0.027	−0.009
3.4	−0.028	−0.032	−0.036	−0.041	−0.045	−0.048	−0.051	−0.051	−0.050	−0.046	−0.039	−0.009
3.6	−0.019	−0.024	−0.029	−0.034	−0.038	−0.043	−0.047	−0.019	−0.050	−0.046	−0.008	−0.008
3.8	−0.012	−0.017	−0.022	−0.028	−0.033	−0.038	−0.042	−0.046	−0.049	−0.050	−0.049	−0.008
4.0	−0.007	−0.012	−0.017	−0.022	−0.027	−0.032	−0.037	−0.041	−0.045	−0.048	−0.049	−0.008

	$\bar{Q}$											
0.0	**0.000**	0.000	0.000	0.000	0.000	0.000	0.000	0.000	0.000	0.000	0.000	**-0.500**
0.2	-0.614	**0.323**	0.258	0.196	0.133	0.088	0.055	0.039	0.029	0.020	0.014	-0.411
0.4	-0.339	-0.436	**0.484**	0.367	0.269	0.189	0.126	0.083	0.053	0.030	0.013	-0.329
0.6	-0.147	-0.258	-0.372	**0.514**	0.400	0.299	0.211	0.142	0.088	0.082	0.036	-0.186
0.8	-0.017	-0.127	-0.245	-0.362	**0.522**	0.410	0.307	0.218	0.142	0.082	0.036	-0.186
1.0	0.068	-0.040	-0.149	-0.260	-0.371	**0.516**	0.408	0.305	0.214	0.137	0.075	-0.130
1.2	0.120	0.021	-0.075	-0.178	-0.278	-0.384	**0.509**	0.401	0.302	0.211	0.134	-0.085
1.4	0.149	0.061	-0.026	-0.113	-0.202	-0.296	-0.396	**0.499**	0.397	0.299	0.211	-0.050
1.6	0.160	0.084	0.010	-0.064	-0.140	-0.221	-0.310	-0.406	**0.494**	0.395	0.300	-0.027
1.8	0.160	0.096	0.034	-0.028	-0.090	-0.159	-0.235	-0.320	-0.412	**0.493**	0.397	-0.011
2.0	0.152	0.100	0.048	-0.001	-0.053	-0.108	-0.171	-0.243	-0.325	-0.413	**0.495**	0.000
2.2	0.139	0.097	0.056	0.016	-0.025	-0.069	-0.119	-0.179	-0.247	-0.325	-0.410	0.006
2.4	0.124	0.092	0.059	0.028	-0.004	-0.039	-0.080	-0.126	-0.182	-0.248	-0.322	0.009
2.6	0.109	0.083	0.060	0.035	0.011	-0.016	-0.047	-0.086	-0.130	-0.183	-0.244	0.010
2.8	0.092	0.074	0.056	0.038	0.021	0.001	-0.023	-0.051	-0.087	-0.129	-0.178	0.009
3.0	0.077	0.064	0.052	0.039	0.027	0.012	-0.006	-0.028	0.053	-0.086	-0.125	0.008
3.2	0.063	0.055	0.047	0.038	0.031	0.019	0.007	-0.009	-0.028	-0.052	-0.082	0.007
3.4	0.050	0.045	0.041	0.035	0.031	0.024	0.016	0.004	-0.009	-0.027	-0.049	0.005
3.6	0.040	0.037	0.035	0.032	0.030	0.026	0.021	0.014	0.004	-0.008	-0.025	0.004
3.8	0.031	0.030	0.029	0.029	0.028	0.027	0.025	0.021	0.014	0.005	-0.007	0.003
4.0	0.023	0.023	0.023	0.025	0.025	0.027	0.027	0.025	0.021	0.014	0.006	0.002

Table 4.21 Con't

β = 0.5

ζ	0	0.1	0.2	0.3	0.4	0.5 $\bar{y}$	0.6	0.7	0.8	0.9	1.0	∞
0.0	**1.69**	1.43	1.21	1.01	0.82	0.66	0.52	0.40	0.31	0.24	0.20	**0.60**
0.2	1.43	**1.25**	1.08	0.94	0.78	0.66	0.54	0.44	0.36	0.29	0.24	0.59
0.4	1.21	1.08	**0.97**	0.87	0.76	0.66	0.56	0.48	0.40	0.34	0.28	0.56
0.6	1.01	0.94	0.87	**0.80**	0.73	0.66	0.59	0.51	0.45	0.39	0.32	0.52
0.8	0.82	0.78	0.76	0.73	**0.71**	0.66	0.61	0.54	0.50	0.44	0.39	0.47
1.0	0.66	0.66	0.66	0.66	0.66	**0.65**	0.62	0.57	0.54	0.49	0.44	0.43
1.2	0.52	0.54	0.56	0.59	0.61	0.62	**0.62**	0.60	0.57	0.53	0.48	0.38
1.4	0.40	0.44	0.48	0.51	0.54	0.57	0.60	**0.60**	0.59	0.56	0.53	0.33
1.6	0.31	0.36	0.40	0.45	0.50	0.54	0.57	0.59	**0.60**	0.59	0.57	0.28
1.8	0.24	0.29	0.34	0.39	0.44	0.49	0.53	0.56	0.59	**0.61**	0.60	0.24
2.0	0.20	0.24	0.28	0.32	0.39	0.44	0.48	0.53	0.57	0.60	**0.61**	0.20
2.2	0.16	0.19	0.24	0.29	0.34	0.39	0.43	0.48	0.52	0.57	0.60	0.16
2.4	0.13	0.15	0.20	0.24	0.29	0.34	0.39	0.43	0.48	0.52	0.57	0.12
2.6	0.10	0.13	0.16	0.21	0.25	0.30	0.34	0.39	0.44	0.48	0.53	0.09
2.8	0.08	0.11	0.14	0.18	0.22	0.26	0.30	0.34	0.39	0.44	0.49	0.05
3.0	0.07	0.09	0.12	0.15	0.19	0.23	0.26	0.30	0.34	0.39	0.44	0.03
3.2	0.06	0.08	0.10	0.13	0.16	0.20	0.23	0.27	0.30	0.35	0.39	0.00
3.4	0.05	0.07	0.09	0.12	0.14	0.17	0.20	0.23	0.27	0.31	0.35	-0.03
3.6	0.04	0.06	0.08	0.10	0.12	0.15	0.18	0.20	0.23	0.27	0.30	-0.05
3.8	0.03	0.05	0.07	0.09	0.11	0.13	0.15	0.18	0.20	0.23	0.27	-0.07
4.0	0.02	0.04	0.06	0.08	0.10	0.12	0.14	0.16	0.18	0.20	0.23	-0.09

α

References

1. Borowicka, H. 1938. *The distribution of pressure under a uniformly loaded strip resting on elastic isotropic ground*. 2nd Congress International Association for Bridge and Structural Engineering (Berlin), Final Report, VIII, 3.
2. Borowicka, H. 1939. Druckverteilung unterelastischen Platten. *ING. Arch.* (Berlin) 10(2): 113–125.
3. Gorbunov-Posadov, M. I. 1940. Analysis of beams and plates on elastic half-space. *Applied Mechanics and Mathematics* 4(3): 60–80.
4. Gorbunov-Posadov, M. I. 1949. *Beams and slabs on elastic foundation*. Moscow: Mashstroiizdat.
5. Gorbunov-Posadov, M. I. 1953. *Analysis of structures on elastic foundation*. Moscow: Gosstroiizdat.
6. Gorbunov-Posadov, M. I. 1984. *Analysis of structures on elastic foundation*. 3rd ed. Moscow: Stroiizdat, pp. 252–298.
7. Selvadurai, A.P.S. 1979. *Elastic analysis of soil-foundation interaction*. Amsterdam/Oxford/New York: Elsevier Scientific, pp. 11–121.
8. Tsudik E. A. 2006. *Analysis of beams and frames on elastic foundation*. Canada/UK: Trafford, pp. 29–47.

5

Analysis of Beams on Elastic Half-Space, Plane Strain Analysis

5.1 Introduction

Methods of analysis of beams on elastic half-space described in Chapter 4 were developed for analysis of free supported beams. The method is used usually for design of continuous foundations of buildings or other similar structures. This type of foundation is working as a 3D structural system supported on elastic half-space. However, in some cases, beams supported on elastic half-space do not work as isolated foundations. They represent elements of much larger foundations and work under different conditions. For example, analysis of a very long slab in the short direction is usually performed by cutting off a band, as shown Figure 5.1, and analyzing this band as a beam on an elastic foundation. The same method is used for analysis of foundations of retaining walls (shown in Figure 5.2), foundations of concrete dams, and other structures. The same method is usually used by practicing engineers for simplified analysis of mat foundations.

When cutting off a band, including the soil, and analyzing the band as a beam on elastic half-space, we assume that all neighboring bands are working under the same conditions known as *plane strain conditions*. A simple practical method for analysis of such type of beams was developed by Simvulidi (1949, 1958, 1973).

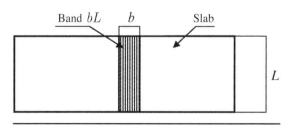

Figure 5.1

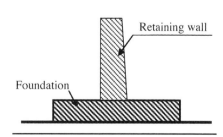

Figure 5.2

All tables can be found at the end of the chapter.

In Chapter 5, the reader will find a detailed description of the method as well as a simplified hand analysis using simple equations and tables. Additional information related to this method can be found in the books mentioned above and other publications such as Flamant (1892) and Gersevanov (1933, 1934). The method was used by the author of this work for analysis of multistoried frames on elastic half-space (Tsudik 1975, 2006). The method allows analyzing free supported beams with three types of loads: uniformly distributed loads, concentrated vertical loads, and concentrated moments.

5.2 The Method of Analysis

The differential equation of the elastic line of the beam supported on elastic foundation looks as follows:

$$EI\frac{d^4y}{dx^4} + p_x = \psi_x \tag{5.1}$$

where EI is the flexural rigidity of the beam, y is the unknown vertical deflection of the beam, p_x is the unknown distributed soil reaction and ψ_x is the given active load.

In order to obtain both unknowns, y and p_x, and to establish the relationship between them a second additional equation is needed. In accordance with Simvulidi the soil reaction looks as follows:

$$p_x = a_0 + \frac{2a_1}{L}\left(x - \frac{L}{2}\right) + \frac{4a_2}{L^2}\left(x - \frac{L}{2}\right)^2 + \frac{8a_3}{L^3}\left(x - \frac{L}{2}\right)^3 \tag{5.2}$$

where L is the length of the beam, and a_0, a_1, a_2, and a_3 are unknown parameters that depend on the rigidity, the length of the beam, the modulus of elasticity of the soil, and the type applied and location of the load, and x is the distance of the load from the left end of the beam.

Let us replace the given load ψ_x with three types of active loads shown in Figure 5.3.

$$\psi_x = \sum \Gamma_{l_{Bi}}^{l_{ki}} f(z) + \sum \Gamma_{l_{2i}}'' M_i + \Gamma_{3i}' P_i \tag{5.3}$$

where P_i is the vertical concentrated load, M_i is the moment, $f(z)$ is the variable distributed load, l_{Bi} is the distance between the left end of the beam and the beginning of the distributed load, l_{ki} is the distance between the left end of the beam and the end of the distributed load, l_{2i} is the distance between the left end of the beam and the moment M_i and l_{3i} is the distance between the left end of the beam of the load P_i.

Γ are special stepped functions introduced by Gersevanov (1934) for analysis of various problems in structural mechanics and used by Simvulidi in his work mentioned above. By introducing ψ_x and p_x into equation 5.3, we have:

$$EI\frac{d^4y}{dx^4} = -a_0 - \frac{2a_1}{L}\left(x - \frac{L}{2}\right) - \frac{4a_2}{L^2}\left(x - \frac{L}{2}\right)^2 - \frac{8a_3}{L^3}\left(x - \frac{L}{2}\right)^3$$
$$+ \sum \Gamma_{Bi}^{ki} f(z) - \sum \Gamma_{l_{2i}}'' M_i + \sum \Gamma_{l_{3i}}' P_i \tag{5.4}$$

By integration of equation 5.4 four times, we find:

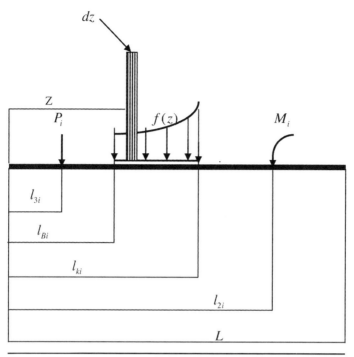

Figure 5.3

$$EI\frac{d^3y}{dx^3} = -a_0 x - \frac{2a_1}{L}\frac{\left(x-\frac{L}{2}\right)^2}{2!} - \frac{8a_2}{L^2}\frac{\left(x-\frac{L}{2}\right)^3}{3!} - \frac{48a_3}{L^3}\frac{\left(x-\frac{L}{2}\right)^4}{4!} +$$

$$+\sum\Gamma_{l_{Bi}}\int_{l_{Bi}}^{x} f(z)\,dz - \sum\Gamma_{l_{ki}}\int_{l_{ki}}^{x} f(z)\,dz + \sum\Gamma'_{l_{2i}}M_i + \sum\Gamma_{l_{3i}}P + D_3 \qquad (5.5)$$

$$I\frac{d^2y}{dx^2} = -\frac{a_0}{2!}x - \frac{2a_1}{L}\frac{\left(x-\frac{L}{2}\right)^3}{3!} - \frac{8a_2}{L^2}\frac{\left(x-\frac{L}{2}\right)^4}{4!} - \frac{48a_3}{L^3}\frac{\left(x-\frac{L}{2}\right)^5}{5!}$$

$$+\sum\Gamma_{l_{Bi}}\int_{l_{Bi}}^{x} f(z)(x-z)\,dz - \sum\Gamma_{l_{ki}}\int_{l_{ki}}^{x} f(z)(x-z)\,dz + \sum\Gamma'_{l_{2i}} M_i$$

$$+\sum\Gamma_{l_{3i}}P(x-l_{3i}) + D_3 x + D_2 \qquad (5.6)$$

$$EI\frac{dy}{dx} = -\frac{a_0 x^3}{3!} - \frac{2a_1}{L}\frac{\left(x-\frac{L}{2}\right)^4}{4!} - \frac{8a_2}{L^2}\frac{\left(x-\frac{L}{2}\right)^5}{5!} - \frac{48a_3}{L^3}\frac{\left(x-\frac{L}{2}\right)^6}{6!} +$$

$$+\sum\Gamma_{l_{Bi}}\int_{l_{Bi}}^{x} f(z)\frac{(x-z)^2}{2!}dz - \sum\Gamma_{l_{ki}}\int_{l_{ki}}^{x} f(z)\frac{(x-z)^2}{2!}dz + \sum\Gamma'_{l_{2i}}M_i\left(x-l_{2i}\right) +$$

$$+\sum\Gamma_{l_{3i}}P\frac{(x-l_{3i})^2}{2!} + D_3\frac{x^2}{2!} + D_2 x + D_1 \qquad (5.7)$$

$$Ely = -\frac{a_0 x^4}{4!} - \frac{2a_1}{L}\cdot\frac{\left(x-\frac{L}{2}\right)^5}{5!} - \frac{8a_2}{L^2}\cdot\frac{\left(x-\frac{L}{2}\right)^6}{6!} - \frac{48a_3}{L^3}\cdot\frac{\left(x-\frac{L}{2}\right)^7}{7!} +$$

$$+\sum\Gamma_{l_{Bi}}\int_{l_{Bi}}^{x} f(z)\frac{(x-z)^3}{3!}dz - \sum\Gamma_{l_{ki}}\int_{l_{ki}}^{x} f(z)\frac{(x-z)^3}{3!}dz + \sum\Gamma_{l_{2i}}M_i\frac{(x-l_{2i})^2}{2!} +$$

$$+\sum\Gamma_{l_{3i}}P_i\frac{(x-l_{3i})^3}{3!} + D_3\frac{x^3}{3!} + D_2\frac{x^2}{2!} + D_1 X + D_0 \qquad (5.8)$$

Equation 5.8 is the general equation of the elastic line of the beam supported on the elastic foundation. It includes four parameters: a_0, a_1, a_2, a_3 and four constants of integration: D_0, D_1, D_2, and D_3. In order to obtain all eight unknowns, eight additional equations have to be written.

The first two equations take into account the equilibrium of the vertical loads and moments applied to the beam. The equilibrium of all vertical loads applied to the beam looks as follows:

$$\int_0^L \left[a_0 + \frac{2a_1}{L}\left(x - \frac{L}{2}\right) + \frac{4a_2}{L^2}\left(x - \frac{L}{2}\right)^2 + \frac{8a_3}{L^3}\left(x - \frac{L}{2}\right)^3 \right] dx =$$

$$= \sum \int_{l_{Bi}}^L f(z)\,dz - \sum \int_{l_{ki}}^L f(z)\,dz + \sum P_i \qquad (5.9)$$

By integrating the left part of this equation and by designating the right part of the equation as AL, we have the following **first equation**:

$$a_0 + \frac{a_2}{3} = A \qquad (5.10)$$

where

$$A = \frac{1}{L}\left[\sum \int_{l_{Bi}}^L f(z)\,dz - \sum \int_{l_{ki}}^L f(z)\,dz + \sum P_i \right] \qquad (5.11)$$

The equilibrium of moments applied to the beam is shown below in equation 5.12:

$$\int_0^L \left[a_0 + \frac{2a_1}{L}\left(x - \frac{L}{2}\right) + \frac{4a_2}{L^2}\left(x - \frac{L}{2}\right)^2 + \frac{8a_3}{L^3}\left(x - \frac{L}{2}\right)^3 \right] x\,dx =$$

$$= \sum \int_{l_{Bi}}^L f(z)\,z\,dz - \sum \int_{l_{ki}}^L f(z)\,z\,dz + \sum P_i l_{3i} - \sum M_i \qquad (5.12)$$

Now, by designating the right part of this equation as L^2C and by integrating its left part of the same equation, the **second equation** is obtained:

$$\frac{a_0}{2} + \frac{a_1}{6} + \frac{a_2}{6} + \frac{a_3}{10} = C \qquad (5.13)$$

where

$$L^2C = \sum \int_{l_{Bi}}^L f(z)\,z\,dz - \sum \int_{l_{ki}}^L f(z)\,z\,dz + \sum P_i l_{3i} - \sum M_i \qquad (5.14)$$

The third equation reflects the boundary conditions at the left end of the beam when $x = 0$ and $y'' = 0$, which allows writing the **third equation**:

$$D_2 = -\frac{L}{6!}(30a_1 - 15a_2 + 9a_3) \qquad (5.15)$$

The fourth equation reflects the boundary conditions at the right end of the beam when $x = L$ and $y'' = 0$, which allows writing the **fourth equation**:

$$D_3 = \frac{a_0 L}{2} + \frac{L}{5!}(10a_1 + 3a_3) + K \tag{5.16}$$

where

$$K = -\frac{1}{L}\left[\sum \int_{l_{Bi}}^{L} f(z)(L-z)\,dz - \sum \int_{l_{ki}}^{L} f(z)(L-z)\,dz + \sum M_i + \sum P_i(L - l_{3i})\right] \tag{5.17}$$

The next four equations show the following:

- The settlements of the beam and soil at the left end of the beam are the same.
- The settlement of the beam and soil at the middle point of the beam are the same.
- The area of the diagram of the settlements of the beam is equal to the area of the diagram of the settlements of the soil under the centerline of the beam.
- The third order derivatives of the beam and soil curves, at the middle point of the beam, are the same.

The curve of the surface of the half-space with a distributed load by can be expressed by the following equation (Simvulidi, 1973) :

$$v = -\frac{(1 - \mu_0^2)L}{12\pi E_0} \times$$

$$\times \left\{ \begin{aligned} &24a_0\left[\frac{x}{L}\ln\frac{x}{L} + \frac{L-x}{L}\ln\frac{L-x}{L}\right] + 24a_1\left[\frac{(L-x)x}{L^2}\ln\frac{L-x}{x} - \frac{x}{L}\right] + \\ &+ 8a_2\left[\frac{L^3 + (2x-L)^3}{2L^3}\ln\frac{x}{L} + \frac{L^3 - (2x-L)^3}{2L^3}\ln\frac{L-x}{L} + 4\frac{(L-x)x}{L^2}\right] + \\ &+ a_3\left[3\frac{L^4 - (2x-L)^4}{L^4}\ln\frac{L-x}{x} - 6\frac{L^3 + (2x-L)^3}{L^3} - 4\frac{x}{L}\right] \end{aligned} \right\} \tag{5.18}$$

where μ_0 is Poisson's ratio of the soil and E_0 is the modulus of elasticity of the soil. Now, the fifth equation can be obtained. Taking into account that:

$$y_{x=0} = v_{x=0} \tag{5.19}$$

and using equations 5.8 and equation 5.18, we have:

$$EIy = \frac{a_1 L^4}{16\cdot 5!} - \frac{a_2 L^4}{8\cdot 6!} + \frac{3a_3 L^4}{8\cdot 7!} + D_0 \tag{5.20}$$

From 5.20 the **fifth equation** is obtained:

$$D_0 = -\frac{L^4}{8!}(21a_1 - 7a_2 + 3a_3) \tag{5.21}$$

The sixth equation shows that:

$$y_{x=\frac{L}{2}} = v_{x=\frac{L}{2}} \tag{5.22}$$

By introducing $x = \dfrac{L}{2}$ into equation 5.18, we find:

$$v_{x=\frac{L}{2}} = \left[\frac{2a_0 L}{\pi E_0} \ln 2 + \frac{a_1 L}{\pi E_0} - \frac{2a_2 L}{3\pi E_0}(1 - \ln 2) + \frac{2a_3 L}{3\pi E_0} \right](1 - \mu_0^2) \tag{5.23}$$

Now, from equation 5.8 using $x = \dfrac{L}{2}$ we find:

$$y_{x=\frac{L}{2}} = \frac{L^4}{EI}\left\{ \frac{a_0}{128} + W + \frac{K}{48L} + \frac{1}{8!}\left(-161a_1 + 112a_2 - 45a_3 + D_1 \frac{1}{2L^3}\right) \right\} \tag{5.24}$$

where

$$W = \frac{1}{L^4}\left[\sum \Gamma_{\frac{L}{2}}^{\frac{L}{2}} \int_0^{l_{Bi}} f(z) \frac{\left(\frac{L}{2} - z\right)^3}{3!} dz - \sum \Gamma_{0}^{\frac{L}{2}} \int_0^{l_{ki}} f(z) \frac{\left(\frac{L}{2} - z\right)^3}{3!} dz + \right.$$

$$\left. + \sum \Gamma_{0}^{\frac{L}{2}} M_i \frac{\left(\frac{L}{2} - l_{2i}\right)^2}{2!} + \sum \Gamma_{0}^{\frac{L}{2}} P \frac{\left(\frac{L}{2} - l_{3i}\right)^3}{3!} \right] \tag{5.25}$$

It is important to mention the following: Taking into account equation 5.20, it can be concluded that when obtaining W it is necessary to take into consideration the influence of the loads located at the left half of the beam. Therefore, the first term in equation 5.25 is always equal to zero when the left end of the distributed load is located outside of the left half of the beam or $l_{Bi} \geq \dfrac{L}{2}$. The second term is also equal to zero when the end of the distributed load is not located at the left half of the beam or $l_{ki} \geq \dfrac{L}{2}$.

The third term in this equation is equal to zero when a concentrated moment is applied outside of the left half of the beam, and the fourth term is equal to zero when the vertical concentrated load is applied outside of the left half of the beam or $l_{3i} \geq \dfrac{L}{2}$. Now, by introducing 5.23 and 5.24 into equation 5.22 we find the **sixth equation**:

$$D_1 = 2L^3\left\{ \frac{a_0}{128a}(256\ln 2 - \alpha) + \frac{1}{8!}\left[161a_1\alpha - 112a_2\alpha + 45a_3\alpha + \right.\right.$$

$$\left.\left. + 40,320a_1 - 26,880(1 - \ln 2)a_2 + 26,880a_3 \right] - W - KL^{-1}0.0208333 \right\} \tag{5.26}$$

where α is the index of flexibility obtained from equation 5.25:

$$\alpha = \frac{\pi E_0 L^3}{(1 - \mu_0^2)EI} \tag{5.27}$$

When analyzing beams cut off from plates, the flexural rigidity of the plate should be specified as:

$$\frac{EI}{1 - \mu^2}$$

where μ is Poisson's ratio of the beam material.

By introducing $\dfrac{EI}{1-\mu^2}$ into 5.27, we have:

$$\alpha = \frac{1-\mu^2}{1-\mu_0^2}\cdot\frac{\pi E_0 L^3}{EI} \tag{5.28}$$

In 5.28 the width of the beam is equal to one unit; when the width of the beam is equal to b, equation 5.28 becomes:

$$\alpha = \frac{1-\mu^2}{1-\mu_0^2}\cdot\frac{\pi E_0 bL^3}{EI} \tag{5.29}$$

The seventh equation shows that:

$$F_S = F_B \tag{5.30}$$

where F_S is the area of the soil settlement diagram and F_B is the area of the beam settlement diagram. By integration of equation 5.8, the area F_B is obtained as shown below:

$$F_B = \frac{1}{EI}\Bigg[-\frac{a_0 L^5}{5!}-\frac{a_2 L^5}{8!}+\sum\int_{l_{Bi}}^{L}f(z)\frac{(L-z)^4}{4!}dz-\sum\int_{l_{ki}}^{L}f(z)\frac{(L-z)^4}{4!}dz+$$
$$+M_i\frac{(L-l_{2i})^3}{3!}+P_i\frac{(L-l_{3i})^4}{4!}+D_3\frac{L^4}{4!}+D_2\frac{L^3}{3!}+D_1\frac{L^2}{2!}+D_0 L\Bigg] \tag{5.31}$$

Now, by integration of equation 5.18 and some simplifications, F_S is found:

$$\frac{1}{1-\mu_0^2}F_S = -\frac{2a_0 L}{\pi E_0}\int_0^{L}\left[\frac{x}{L}\ln\frac{x}{L}+\frac{L-x}{L}\ln\frac{L-x}{L}\right]dx-$$
$$-\frac{2a_1 L}{\pi E_0}\int_0^{L}\left[\frac{(L-x)x}{L^2}\ln\frac{2L-x}{x}-\frac{x}{L}\right]dx-$$
$$-\frac{2a_2 L}{3\pi E_0}\int_0^{L}\left[\frac{4(L-x)x}{L^2}+\frac{L^3+(2x-L)^3}{2L^3}\ln\frac{x}{L}+\frac{L^3-(2x-L)^3}{2L^3}\ln\frac{(L-x)}{L}\right]dx-$$
$$-\frac{a_3 L}{12\pi E_0}\int_0^{L}\left[3\frac{L^4-(2x-L)^4}{L^4}\ln\frac{L-x}{x}-6\frac{L^3+(2x-L)^3}{L^3}-4\frac{x}{L}\right]dx=$$
$$=\frac{a_0 L^2}{\pi E_0}+\frac{a_1 L^2}{\pi E_0}+\frac{2a_3 L^2}{3\pi E_0} \tag{5.32}$$

By introducing equations 5.31 and 5.32 into 5.30, the **seventh equation** is obtained:

$$\frac{a_0}{640\alpha}(640-3\alpha-1,280\ln 2)-\frac{a_2}{8!\alpha}\left[34\alpha-26,880(1-\ln 2)\right]=B \tag{5.33}$$

where B is equal to:

$$B=\frac{1}{L^5}\Bigg[\sum\int_{Bi}^{L}f(z)\frac{(L-z)^4}{4!}dz-\sum\int_{ki}^{L}f(z)\frac{(L-z)^4}{4!}dz+$$
$$+\sum M_i\frac{(L-l_{2i})^3}{3!}+\sum P_i\frac{(L-l_{3i})^4}{4!}-WL^5+\frac{L^4}{48}K\Bigg] \tag{5.34}$$

The eighth equation is obtained by equating the third order derivatives of the settlements of the beam and soil at the middle of the beam:

$$\left(\frac{d^3 y}{dx^3}\right)_{x=\frac{L}{2}} = \left(\frac{d^3 v}{dx^3}\right)_{x=\frac{L}{2}} \tag{5.35}$$

Now, by differentiating equations 5.8 and 5.18 three times, the following two equations are found:

$$\left(\frac{d^3 v}{dx^3}\right)_{x=\frac{L}{2}} = -\left[\frac{32a_1}{\pi E_0 L^2} - \frac{32a_3}{\pi E_0 L^2}\right](1 - \mu_0^2) \tag{5.36}$$

$$\left(\frac{d^3 y}{dx^3}\right)_{x=\frac{L}{2}} = \frac{1}{EI}\left[\begin{array}{c} -\dfrac{a_0 L}{2} + \sum \Gamma_0^{\frac{L}{2}} \int_{Bi} f(z)\,dz - \sum \Gamma_0^{\frac{L}{2}} \int_{ki} f(z)\,dz + \\[2mm] +\Gamma_0^{\frac{L}{2}} \sum P_i + \dfrac{a_0 L}{2} + \dfrac{L}{5!}(10a_1 + 3a_3) + K \end{array}\right] \tag{5.37}$$

By equating the right parts of these two equations, we obtain the **eighth equation** that looks as follows:

$$\frac{1}{EI}\left[NL + \frac{L}{120}(10a_1 + 3a_3)\right] = -\left[\frac{32a_1}{\pi E_0 L^2} - \frac{32a_3}{\pi E_0 L^2}\right](1 - \mu_0^2) \tag{5.38}$$

where N is obtained from equation 5.39:

$$N = \frac{1}{L}\left[\sum \Gamma_0^{\frac{L}{2}} \int_{Bi} f(z)\,dz - \sum \Gamma_0^{\frac{L}{2}} \int_{ki} f(z)\,dz + \sum \Gamma_0^{\frac{L}{2}} P_i + K\right] \tag{5.39}$$

Now, by solving equations 5.10, 5.13, 5.33, and 5.38 we find:

$$\left.\begin{array}{l} a_0 = \dfrac{(8,252 - 34\alpha)A - 13,440B\alpha}{13,440 + 29\alpha} \\[3mm] a_1 = 3 \cdot \dfrac{(2C - A)(1,280 - \alpha) - 8N\alpha}{2,048 + \alpha} \\[3mm] a_2 = 3 \cdot \dfrac{(5,188 + 63\alpha)A + 13,440B\alpha}{13,440 + 29\alpha} \\[3mm] a_3 = 10 \cdot \dfrac{(2C - A)(384 + \alpha) + 4N\alpha}{2,048 + \alpha} \end{array}\right\} \tag{5.40}$$

Equations for obtaining the shear forces and moments are found from equations 5.5 and 5.6:

$$Q = \left[\frac{a_0}{2}(2x - L) - \frac{L}{120}(10a_1 + 3a_3) - K\right]$$

$$+ \frac{2a_1}{L} \cdot \frac{\left(x - \dfrac{L}{2}\right)^2}{2!} + \frac{8a_2}{L^2} \cdot \frac{\left(x - \dfrac{L}{2}\right)^3}{3!} + \frac{48a_3}{L^3} \cdot \frac{\left(x - \dfrac{L}{2}\right)^4}{4!} -$$

$$- \sum \Gamma_{l_{3i}} P_i - \sum \Gamma_{l_{Bi}} \int_{l_{Bi}}^{x} f(z)\,dz + \sum \Gamma_{l_{ki}} \int_{l_{ki}}^{x} f(z)\,dz \tag{5.41}$$

$$M = \frac{L^2}{240} \cdot (10a_1 - 5a_2 + 3a_3) + \left[\frac{a_0}{2}(x - L) - \frac{L}{120}(10a_1 + 3a_3) - K\right]x +$$

$$+ \frac{2a_1}{L} \cdot \frac{\left(x - \frac{L}{2}\right)^3}{3!} + \frac{8a_2}{L^2} \cdot \frac{\left(x - \frac{L}{2}\right)^4}{4!} + \frac{48a_3}{L^3} \cdot \frac{\left(x - \frac{L}{2}\right)^5}{5!} x -$$

$$-\sum \Gamma_{l_{2i}} M_i - \sum \Gamma_{l_{3i}} P_i (x - l_{3i}) - \sum \Gamma_{l_{Bi}} \int\limits_{l_{Bi}}^{x} f(z)(x - z)\,dz + \sum \Gamma_{l_{ki}} \int\limits_{l_{ki}}^{x} f(z)(x - z)\,dz \quad (5.42)$$

All equations shown above allow analyzing beams working under plane strain conditions and can be applied to solutions of practical problems. However, it is obvious that the method cannot be used for hand calculations. In order to simplify the application of his method, Simvulidi developed a series of tables and simple equations for hand analysis. Tables 5.1–5.3 were developed for analysis of beams with uniformly distributed loads.

Tables 5.4–5.6 were developed for analysis of beams with concentrated vertical loads and Tables 5.7–5.9 were developed for analysis of beams with concentrated moments. Results of analysis are the soil pressures, shear forces, and moments. We shall describe and explain the application of these tables by solving a series of numerical examples. In order to perform analysis of a beam, the following data are needed:

1. E_0—Modulus of elasticity of the soil
2. E—Modulus of elasticity of the beam material
3. I—Moment of inertia of the cross section of the beam
4. L—The length of the beam
5. The loads applied to the beam and their coordinates, as shown in Figure 5.3

For each concentrated vertical load P_i and moment M_i, one coordinate is given: the distance from the left end of the beam l_{2i} for a moment and l_{3i} for a vertical load. Two coordinates are given for all uniformly distributed loads: the first coordinate l_{Bi} is the distance between the left end of the beam and the left end of the distributed load. The second coordinate l_{ki} is the distance between the left end of the beam and the right end of the load, as shown in Figure 5.3. It is important to note that the tables are developed in such a way that the right end of the uniformly distributed load is located at the right end of the beam. If the right end of the load is not located at the right end of the beam, analysis can still be performed. We will illustrate that below by performing analysis of a practical example.

5.3 Analysis of Beams with Uniformly Distributed Loads

If a beam has a uniformly distributed load q, the soil pressure p, the shear force Q, and the moment M at any point are obtained from the following equations:

$$\left.\begin{array}{l} p = \overline{p}q \\ Q = \overline{Q}qbL \\ M = \overline{M}qbL^2 \end{array}\right\} \quad (5.43)$$

Where $\overline{p}$, $\overline{Q}$, and $\overline{M}$ are coefficients given in Tables 5.1, 5.2, and 5.3, respectively. All given coefficients are dimensionless. Tables allow using metric units as well as English units. If the load q is given in $\frac{t}{m^2}$, the length L and the width of the beam b in meters, the soil pressure will be obtained in $\frac{t}{m^2}$, the shear force in tons and the moment in tm. If the load q is given in $\frac{lb}{ft^2}$, the length L and the width of the beam b in ft, the soil pressure will be obtained in $\frac{lb}{ft^2}$, the shear in lb, and the moment in lb · ft. The beam is always divided into 10 sections and all coefficients are given at 11 points equally spaced including the left and the right ends of the beam (points 0 and 1.0). Coefficient ξ, shown in the top row of the table is equal to $\xi = \frac{x}{L}$, where x is the distance between the left end of the beam and the point of investigation. Analysis is performed in the following order:

1. Find the index of flexibility α using equation 5.29.
2. Find parameter $\beta = \frac{l_B}{L}$.
3. Go to Tables 5.1–5.3 and find coefficients $\overline{p}$, $\overline{Q}$, and $\overline{M}$ using equations 5.43 find p, Q, and M.

Tables 5.1–5.3 are used for analysis of beams with uniformly distributed loads. Table 5.1 is used for obtaining the soil pressure, Table 5.2 is used for obtaining the shear forces, and Table 5.3 is used for obtaining the moments.

Example 5.1

Given: A beam with a uniformly distributed load located at the right part of the beam, as shown in Figure 5.4. Find the soil pressure, shear forces, and moments:

$$\alpha = 150, q = 2,000\text{lb/ft}^2, L = 20\text{ft}, l_B = 8\text{ft}, b = 3\text{ft}.$$

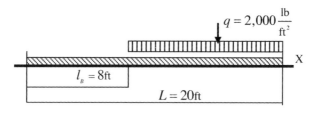

Figure 5.4

Solution:

1. Find $\beta = \frac{l_B}{L} = \frac{8}{20} = 0.4.$
2. Find $\overline{p}$ from Table 5.1 and using the first equation from 5.43, $p = \overline{p}q$, find the soil pressure.
3. Find $\overline{Q}$ from Table 5.2 and using the second equation from 5.43, $Q = \overline{Q}qbL$, find the shear forces.

4. Find $\overline{M}$ from Table 5.3 and using the third equation from 5.43, $M = \overline{M}qbL^2$, find the moments.

The soil pressures, shear forces, and moments are found in 11 points $\xi = 0, 0.1, 0.2, ..., 1$. All calculations are performed in Table 5.10. The shear forces diagram is shown in Figure 5.5. The soil pressure diagram and the moment diagram can be built analogously.

Example 5.2

This example illustrates analysis of a beam with uniformly distributed loads when the load is located at the central area of the beam, as shown in Figure 5.6. All data are the same as in Example 5.1 except the location of the load $l_B = 4ft$, $l_k = 16ft$.

Solution:

Since the right end of the given load q is not located at the right end of the beam, analysis is performed in three steps.

> **Step 1:** The given load is extended up to the right end of the beam as shown in Figure 5.7, and analysis of the beam is performed exactly as it was performed in Example 5.1.

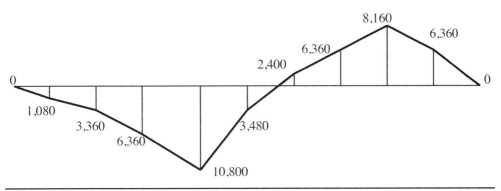

Figure 5.5   Shear forces diagram

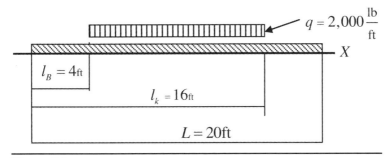

Figure 5.6

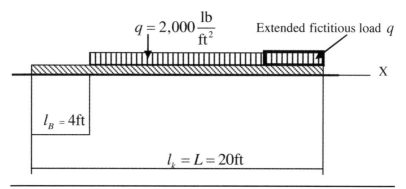

Figure 5.7

Step 2: A fictitious load $(-q)$ is applied to the beam in the opposite direction between the right end of the given load and the right end of the beam as shown in Figure 5.8, and analysis of the beam is performed.

Step 3: By summing the results of both analyses, final results are found.

Analysis of the beam shown in Figure 5.8 is performed below:

Coefficients $\overline{p}$ are obtained from Table 5.1 for $\beta = \dfrac{l_B}{L} = \dfrac{4}{20} = 0.20$, and using the first equation from 5.43, $p = \overline{p}q$, the soil pressure is found.

Coefficients $\overline{Q}$ are obtained from Table 5.2, and using the second equation from 5.43, $Q = \overline{Q}qbL$, the shear forces are found.

Coefficients $\overline{M}$ are obtained from Table 5.3, and using the third equation from 5.43, $M = \overline{M}qbL^2$, all moments are found.

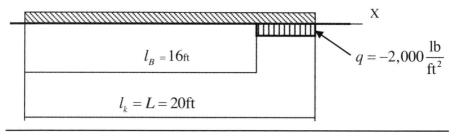

Figure 5.8

Analysis of the beam is performed in Tables 5.11–5.14. Analysis of the beam loaded with the given and fictitious loads is shown in Table 5.11. Analysis of the beam loaded with the fictitious load $(-q)$ is shown in Table 5.12. The final soil pressures and final moments are obtained in Tables 5.13 and 5.14, respectively.

The final soil pressure and moment diagrams are shown in Figures 5.9 and 5.10.

If the left end of the uniformly distributed load is located at the left end of the beam, as shown in Figure 5.11, analysis is performed exactly as analysis in Example 5.1. The only difference is that the numbering of the points (cross sections of the beam) is made from right to left. In Example 5.1, the load is located at the right end of the beam. In the beam shown in Figure 5.11, the same load is located at the left end of the beam. By comparing the shear diagrams shown in Figures 5.5 and 5.12, it can be seen that both analyses produce asymmetrical results.

5.4 Analysis of Beams with Concentrated Vertical Loads

If a beam has a concentrated vertical load P, as shown in Figure 5.13, the following three equations are used to find the soil pressure p, the shear forces Q, and moments:

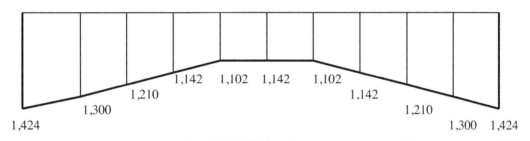

Figure 5.9　Soil pressure diagram

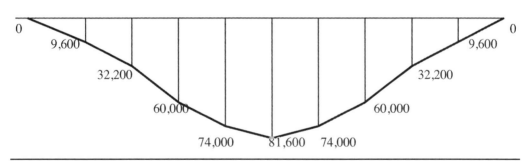

Figure 5.10　Moment diagram

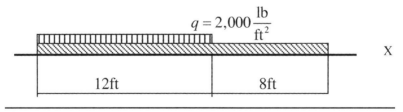

Figure 5.11

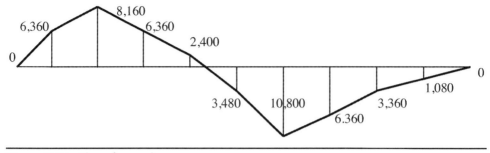

Figure 5.12   Shear forces diagram

$$\left.\begin{array}{l} p = \overline{p}\dfrac{P}{bL} \\[2mm] Q = \overline{Q}P \\[2mm] M = \overline{M}PL \end{array}\right\} \qquad (5.44)$$

In these equations $\overline{p}$, $\overline{Q}$, and $\overline{M}$ are coefficients given in Tables 5.4, 5.5, and 5.6, respectively. All given coefficients are dimensionless. Table 5.4 is used for obtaining soil pressure, Table 5.5 is used for obtaining shear forces, and Table 5.6 is used for obtaining moments.

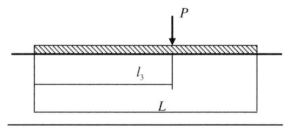

Figure 5.13

Example 5.3

A beam 17 ft long, loaded with a vertical load $P = 220$kips, located at 7 ft from the left end of the beam, is shown in Figure 5.14.

$$E_0 = 1,230\frac{\text{kips}}{\text{ft}^2}, \ E = 440,000\frac{\text{kips}}{\text{ft}^2}, \ I = 2.970\text{ft}^4, \ b = 3.3\text{ft}$$

$$\beta = \frac{7}{17} = 0.4, \ \alpha = \frac{\pi E_0 bl^3}{EI(1 - \mu_0^2)} = \frac{3.141 \cdot 1,220 \cdot 3.3 \cdot 17^3}{440,000 \cdot 3.1 \cdot (1 - 0.3^2)} = 50.$$

Find the soil pressure, shear forces, and moments.

Solution:

1. Find the soil pressure coefficients $\overline{p}$ from Table 5.4, and using the first equation from 5.44, $p = \overline{p}\dfrac{P}{bL}$, obtain the actual soil pressure.

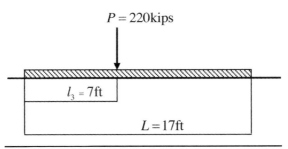

Figure 5.14

2. Find the shear forces coefficients $\overline{Q}$ from Table 5.5, and using the second equation from 5.44, $Q = \overline{Q}P$, obtain the actual shear forces.
3. Find the moments coefficients $\overline{M}$ from Table 5.6, and using the third equation from 5.44, $M = \overline{M}PL$, find the actual moments.

Analysis is performed in Table 5.15. The soil pressure, shear, and moment diagrams are shown in Figures 5.15, 5.16, and 5.17, respectively.

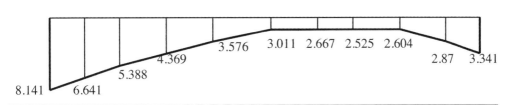

Figure 5.15 Soil pressure diagram

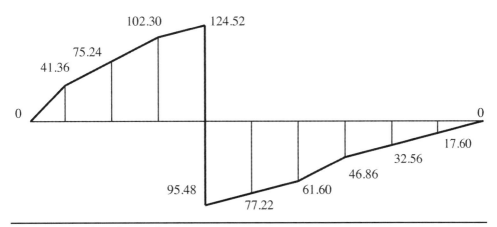

Figure 5.16 Shear forces diagram

0   41.14 138.38   291.32   486.20   336.60   220.66   127.16 59.84   18.70   0

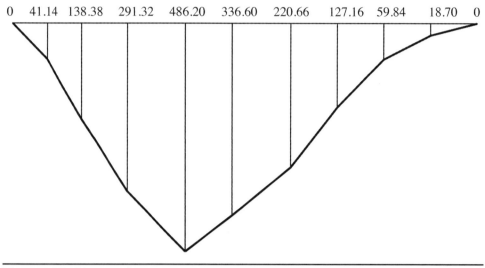

Figure 5.17 Moment diagram

5.5 Analysis of Beams Loaded with Concentrated Moments

If a moment M_A is applied to the beam at any point A, as shown in Figure 5.18, the soil pressure p, the shear forces Q, and moments M can be obtained from the following formulae:

$$p = \overline{p}\frac{M_A}{bL^2}$$
$$Q = \overline{Q}\frac{M_A}{L} \qquad (5.45)$$
$$M = \overline{M}M_A$$

Coefficients $\overline{p}$, $\overline{Q}$, and $\overline{M}$ are given in Tables 5.7, 5.8, and 5.9, respectively, depending on parameters α and β. Table 5.7 is used for obtaining the soil pressure, Table 5.8 is used for obtaining the shear forces, and Table 5.9 is used for obtaining the moments. It is important to note that the moment is positive if applied counterclockwise and negative if applied clockwise.

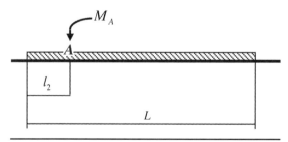

Figure 5.18

Example 5.4

Let us perform analysis of the beam shown in Figure 5.19 loaded with a moment $M_A = 200\text{kip} \times \text{ft}$. The beam is 30 ft long, 5 ft wide, and 4 ft deep. The modulus of elasticity of the soil is $E_0 = 1,200\dfrac{\text{kip}}{\text{ft}^2}$, Poisson's ratio $\mu_0 = 0.40$, and modulus of elasticity of the beam $E = 440,000\dfrac{\text{kip}}{\text{ft}^2}$

$$I = \frac{bh^3}{12} = \frac{5 \cdot 4^3}{12} = 26.67\text{ft}^4, \ \alpha = \frac{\pi E_0 bL^3}{EI(1 - 0.4^2)} = \frac{3.14 \cdot 1,200 \cdot 5 \cdot 30^3}{440,000 \cdot 26.67 \cdot 0.84} = 51.6$$

$$\beta = \frac{6}{30} = 0.2$$

Find the soil pressure, shear force, and moments.

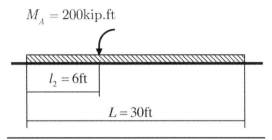

$$M_A = 200\text{kip.ft}$$

$$l_2 = 6\text{ft}$$

$$L = 30\text{ft}$$

Figure 5.19

Solution:

1. Find the soil pressure coefficients $\overline{p}$ from Table 5.7, and using the first equation from 5.45, $p = \overline{p}\dfrac{M_A}{bL^2}$, find the soil pressure.

2. Find the shear forces coefficients $\overline{Q}$ from Table 5.8, and using the second equation from 5.45, $Q = \overline{Q}\dfrac{M_A}{L}$, find the shear forces.

3. Find the moments coefficients $\overline{M}$ from Table 5.9, and using the third equation from 5.45, $M = \overline{M}M_A$, find the moments.

Analysis is performed in Table 5.16. The soil pressure, shear forces, and moment diagrams are given in Figures 5.20, 5.21, and 5.22, respectively.

In conclusion, it is important to mention that in many occasions analysis requires not only obtaining moments, shear forces, and soil pressures, but also deflections of the beam. Simvulidi obtained equations for settlements and rotations of beams supported on elastic half-space. He also developed tables that significantly simplify practical application of these equations. Equations and tables are introduced for obtaining settlements and rotations for the following three types of loads: uniformly distributed loads q, concentrated vertical loads P, and concentrated moments M.

Tables 5.17 and 5.18 were developed and used for analysis of beams with uniformly distributed loads. Tables 5.19 and 5.20 were developed and used for beams with verti-

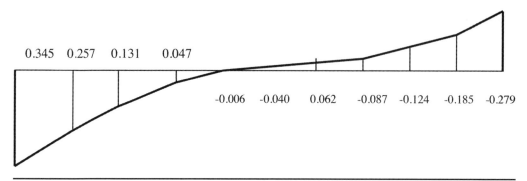

Figure 5.20   Soil pressure diagram

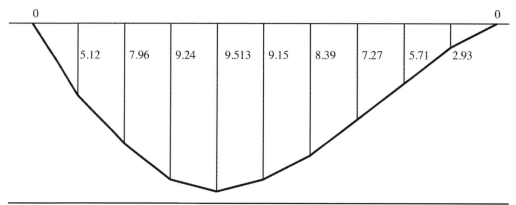

Figure 5.21   Shear diagram

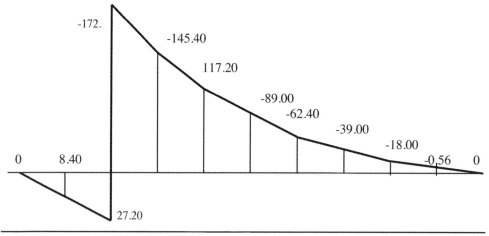

Figure 5.22   Moment diagram

cal concentrated loads. Tables 5.21 and 5.22 are developed and used for beams loaded with concentrated moments.

5.6 Deflections of Beams with Uniformly Distributed Loads

Equations for obtaining rotations and vertical deflections of a beam loaded with a uniformly distributed load q, as shown in Figure 5.23, look as follows:

$$\varphi = \overline{\varphi}\frac{q}{\pi E_0} \tag{5.46}$$

$$y = \overline{y}\frac{qL}{\pi E_0} \tag{5.47}$$

Coefficients $\overline{\varphi}$ and $\overline{y}$ are taken from Tables 5.17 and 5.18, respectively, depending on α, β, and ξ. Two examples presented below illustrate the practical application of Tables 5.17 and 5.18.

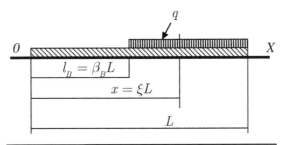

Figure 5.23

Example 5.5

The given beam, as shown in Figure 5.24, has a uniformly distributed load q. Find the angle or rotation at the left end of the beam and the settlement at the center of the beam.

Given: $l_B = 0$, $\alpha = 300$, $b = 1m$, $L = 5m$, $E_0 = 300kg/cm^2$, $q = 10t/m^2$.

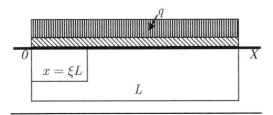

Figure 5.24

Solution:

As we can see, $\beta = \beta_B = \dfrac{l_B}{L} = 0$, $\dfrac{q}{\pi E_0} = \dfrac{1}{3.14 \cdot 300} = 0.001062$, $\dfrac{qL}{\pi E_0} = \dfrac{500}{3.14 \cdot 300}$

$= 0.531cm$.

From Table 5.17 for $\alpha = 300$ and $\beta = 0$, we find $\overline{\varphi} = 1.864$.

The angle of rotation at the left end of the beam is equal to $\varphi = \overline{\varphi} \cdot \dfrac{q}{\pi E_0} = $ $1.864 \cdot 0.001062 = 0.00198$. Now, the settlement of the beam at the center can be found. From Table 5.18 for $\alpha = 300$ and $\beta = 0$, coefficient $\overline{y} = 0.641$ is obtained and the settlement is found as:

$$y = \overline{y}\frac{qL}{\pi E_0} = 0.641 \cdot 0.5310 = 0.3404 \text{cm}$$

Settlements at all other points are found analogously.

Example 5.6

The beam shown in Figure 5.25 is partially loaded with a uniformly distributed load. Find angles of rotation at points $\xi = 0$, $\xi = 0.5$, and $\xi = 1.0$ and settlement of the beam at point $\xi = 0.5$.

Given: $\alpha = 250$, $q = 10\text{t/m}^2$, $b = 1\text{m}$, $l_B = 1.4\text{m}$, $L = 7\text{m}$, and $E_0 = 200\text{kg/cm}^2$.

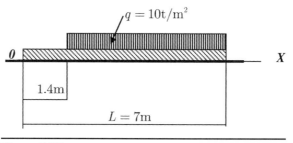

Figure 5.25

Solution:

$$\beta = \beta_H = \frac{l_H}{L} = \frac{1.4}{7} = 0.2, \quad \frac{q}{\pi E_0} = \frac{1}{3.14 \cdot 200} = 0.001592, \quad \frac{qL}{\pi E_0} = \frac{1}{3.14 \cdot 200} = 0.01.$$

Now, from Table 5.17 for $\alpha = 250$ and $\beta = 0.2$ we obtain:

$$\varphi_{\xi=0} = \overline{\varphi}_{\xi=0}\frac{q}{\pi E_0} = 2.967 \cdot 0.001592 = 0.004708$$

$$\varphi_{\xi=0.5} = \overline{\varphi}_{\xi=0.5}\frac{q}{\pi E_0} = 0.957 \cdot 0.001592 = 0.001524$$

$$\varphi_{\xi=1.0} = \overline{\varphi}_{\xi=1.0}\frac{q}{\pi E_0} = -0.935 \cdot 0.001592 = 0.001489$$

Settlement at point $\xi = 0.5$ is equal to $y_{\xi=0.5} = \overline{y}_{\xi=0.5}\dfrac{qL}{\pi E_0} = 1.169 \cdot \dfrac{1\text{kg/cm}^2 \cdot 700\text{cm}}{3.14 \cdot 200\text{kg/cm}^2}$

$= 1.303\text{cm}$. $\overline{y}_{\xi=0.5}$ is taken from Table 5.18 for $\alpha = 250$, $\beta = 0.2$, and $\xi = 0.5$. Settlements at any other point of the beam can be obtained analogously.

5.7 Deflections of Beams with Vertical Concentrated Loads

When a beam is loaded with a concentrated vertical load P, as shown in Figure 5.26, rotations of the beam and settlements are found from the following formulae, respectively:

$$\varphi = \overline{\varphi} \frac{P}{\pi E_0 bL} \tag{5.48}$$

$$y = \overline{y} \frac{P}{\pi E_0 b} \tag{5.49}$$

Coefficients $\overline{\varphi}$ and $\overline{y}$ are given in Tables 5.19 and 5.20, respectively. A numerical example illustrates application of these tables and equations to practical analysis.

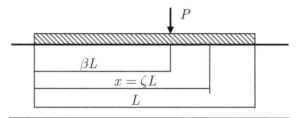

Figure 5.26

Example 5.7

Given: The beam shown in Figure 5.27 is loaded with a load $P = 100,000$kg, $E_0 = 300$kg/cm, $b = 1$m, $L = 5$m, and $\alpha = 350$. Find rotations of the beam at the left end φ_l, rotation of the beam at the right end φ_r, and rotation of the beam at the point the load P is applied φ_P. Also find the relative settlements of the beam at all 11 points, from point $0, 0.1, \ldots,$ up to 1.

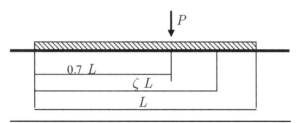

Figure 5.27

Solution:

From Table 5.19, using $\alpha = 350$, $\beta = 0.7$ and $\zeta = 0$, $\zeta = 0.7$ and $\zeta = 1.0$, we find $\overline{\varphi}_l = 5.171$, $\overline{\varphi}_r = -2.903$, and $\overline{\varphi}_p = 0.245$. Using equation 5.48 we find:

$$\varphi_l = \frac{P}{\pi E_0 bL} \overline{\varphi}_l = \frac{100,000}{3.14 \cdot 300 \cdot 500} \cdot 5.171 = 0.512 \cdot 5.171 = 2.648,$$

$$\varphi_r = 0.512 \cdot (-2.903) = -1.486, \ \varphi_p = 0.512 \cdot 0.245 = 0.1254.$$

Now, using equation 5.49 and Table 5.20 for $\alpha = 350$, $\beta = 0.7$, and all values of ζ from 0 to 1.0, we find all coefficients $\overline{y}$ and all relative settlements. For example, when $\overline{y}_{\zeta=0.5} = 2.467$, the actual settlement at this point is equal to $y_{\zeta=0.5} = \overline{y}_{\zeta=0.5} \cdot \dfrac{P}{\pi E_0 b} =$

$2.467 \cdot \dfrac{100,000}{3.14 \cdot 300 \cdot 100} = 2.62\,\text{cm}.$

The actual relative settlements of the beam are found in Table 5.23.

The diagram of settlements is shown in Figure 5.28. Settlements are shown only in four points.

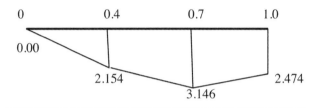

Figure 5.28   The diagram of settlements

5.8 Deflections of Beams Loaded with Concentrated Moments

Deflections of a beam loaded with a concentrated moment M applied at any point, as shown in Figure 5.29, is performed using equations 5.50–5.51:

$$\varphi = \overline{\varphi}\frac{M}{\pi E_0 b L^2} \tag{5.50}$$

$$y = \overline{y}\frac{M}{\pi E_0 b L} \tag{5.51}$$

Coefficients $\overline{\varphi}$ and $\overline{y}$ are given in Tables 5.21 and 5.22, respectively. By introducing these coefficients into equations 5.50 and 5.51, rotations and relative settlements are found. Analysis, in principle, is not different from analysis of beams with vertical distributed or concentrated vertical loads.

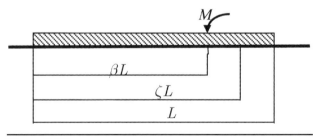

Figure 5.29

Example 5.8

Let us find the rotation and relative settlement at the center of the beam shown in Figure 5.29, assuming $E_0 = 300 \text{kg/cm}^2$, $b = 1\text{m}$, $L = 5\text{m}$, $\alpha = 350$, $M = 50\text{tm} = 5{,}000{,}000 \text{kgcm}$, $\zeta = 0.50$, and $\beta = 0.7$.

Solution:

From Table 5.21 find $\overline{\varphi} = -12.024$; from Table 5.22 find $\overline{y} = 0.194$.

$$\varphi = \overline{\varphi}\frac{M}{\pi E_0 bL^2} = -12.024 \cdot \frac{5000000}{3.14 \cdot 300 \cdot 100 \cdot 500} = 1.276$$

$$y = \overline{y}\frac{M}{\pi E_0 bL} = 0.194 \cdot \frac{5000000}{3.14 \cdot 300 \cdot 100 \cdot 500} = 0.0206\text{cm}$$

Deflections of the beam at any other point are found analogously. If the beam is loaded with several various loads, analysis of the beam is performed under each load separately and final results are obtained by simple superposition. It is important to mention that settlements obtained from this analysis are relative settlements, not real settlements. However, these settlements allow obtaining the differential settlements between various points of the beam that can be used for combined analysis of continuous foundations and the superstructure.

5.9 Combined Analysis of 2D Frames and Continuous Foundations

When a mat foundation supported on elastic half-space is supporting a 3D frame, as shown in Figures 5.30 and 5.31, analysis of the foundation can be performed by taking into account the rigidity of the superstructure. This analysis specifies the actual behavior of the foundation as well as the behavior of the frame. Analysis is performed in the following order: A band $b \cdot L$ including a 2D frame is cut off from the given 3D structural system and that frame is analyzed as a 2D frame with a continuous foundation supported on elastic half-space. Two practical methods of analysis are introduced and explained below: the direct method and the iterative method.

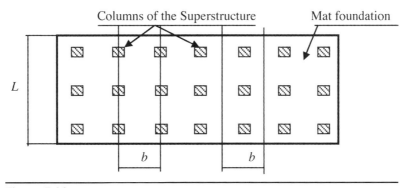

Figure 5.30

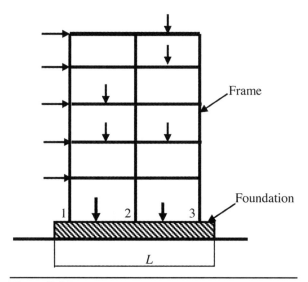

Figure 5.31

Basic classical methods of structural mechanics (the stiffness method, the method of forces, and the combined method) can be used for analysis of continuous foundations and frames. The total structural system frame-foundation consists of two parts: frame and foundation. The frame is loaded with the given loads and unknown deflections of the first floor columns at their points of support. The foundation is loaded with the given loads and unknown frame reactions acting in the opposite direction and applied to the foundation. In order to perform the final analysis of the frame, deflections of the columns at points of support have to be known. To perform final analysis of the foundation, loads applied to the foundation have to be known. Each column of the frame experiences two deflections: one settlement and one rotation. The foundation at each of the same points is loaded with two loads: one vertical load and one moment. The total number of unknowns is equal to $4n$: $2n$ deflections and $2n$ loads. Below, Figures 5.32 and 5.33 show the frame and the foundation. The frame is loaded with the given loads and unknown deflections. The foundation is loaded with the given loads and reactions acting in the opposite direction.

Now, by applying to the frame the given loads and to the frame supports one unit deflections, respectively, the following system of equations is written:

$$\left.\begin{aligned}
r_1 &= r_{11}\delta_1 + r_{12}\delta_2 + \ldots + r_{1n}\delta_n + r_{1P} \\
r_2 &= r_{21}\delta_1 + r_{22}\delta_2 + \ldots + r_{2n}\delta_n + r_{2P} \\
&\qquad\cdots\cdots\cdots\cdots\cdots\cdots\cdots\cdots \\
r_n &= r_{n1}\delta_1 + r_{n2}\delta_2 + \ldots + r_{nn}\delta_n + r_{nP}
\end{aligned}\right\} \tag{5.52}$$

In this equation, r_i is the total unknown reaction at restraint i, r_{ik} is the reaction at restraint due to one unit deflection applied to the restraint k, r_{iP} is the reaction at restraint i due to the given loads applied to the frame and δ_i is the unknown deflection at restraint i.

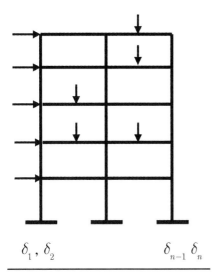

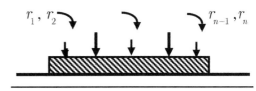

Figure 5.33

δ_1, δ_2 $\qquad\qquad$ $\delta_{n-1}\ \delta_n$

Figure 5.32

Deflections of the foundation are obtained from the following system of equations:

$$\delta_1 = \delta_{11} r_1 + \delta_{12} r_2 + \dots + \delta_{1n} r_n + \delta_{1P}$$
$$\delta_2 = \delta_{21} r_1 + \delta_{22} r_2 + \dots + \delta_{2n} r_n + \delta_{2P}$$
$$\dots\dots\dots\dots\dots\dots\dots\dots\dots\dots \tag{5.53}$$
$$\delta_n = \delta_{n1} r_1 + \delta_{n2} r_2 + \dots + \delta_{nn} r_n + \delta_{nP}$$

In this system, δ_{ik} is the deflection of the foundation at restraint i due to the one unit load applied to the foundation at restraint k, δ_i is the unknown deflection of the foundation at restraint i, r_i is the unknown reaction at restraint i of the foundation and δ_{iP} is the unknown deflection at the restraint i due to the given loads applied to the foundation.

By solving both systems of equations all reactions and all deflections are found. Final analysis of the frame is performed by applying to the frame the given loads and found deflections. Final analysis of the foundation is performed by applying to the foundation the given loads and found reactions. The soil pressure, shear forces, and moments are found using Tables 5.1–5.9. Tables 5.17–5.22 are used for obtaining the deflections of the foundation. As shown, the total number of unknowns in both systems, 5.52 and 5.53, is equal to $2n$: n reactions and n deflections. This method of analysis allows us to find deflections and reactions and perform final analyses of the frame and foundation.

Analysis can be simplified by reducing the number of unknowns. By introducing reactions $r_1, r_2, \dots, r_n$ from the system 5.52 into system 5.53, we obtain a system of equations with only n unknown deflections. Now, analysis of the frame can be performed by applying the given loads and deflections to the frame. By applying the given loads and reactions to the foundation, final analysis is performed.

If deflections $\delta_1, \delta_2, \dots, \delta_n$ from the system of equations 5.53 are introduced into the system of equations 5.52, we obtain again a system of only n equation reactions. Obtained reactions along with the given loads are applied to the foundation and final analysis of the foundation is performed. Deflections found from this analysis and the given loads applied to the frame allow performing the final analysis of the frame.

The same analysis can be performed by using the iterative method. Analysis is performed in several steps-iterations as follows:

First iteration: Analysis of the frame is performed assuming that all columns are restrained against any deflections. Obtained reactions are applied to the foundation and deflections of the foundation are found.

Second iteration: Analysis of the frame is performed again by applying to the frame the given loads and deflections found in the first iteration. Obtained reactions acting in opposite directions are applied to the foundation and deflections of the foundation are found.

Third iteration: Repeats the same procedures described above in the second one and so on. Analysis is stopped when reactions obtained in the iteration $(n + 1)$ are close enough to the reactions found in iteration n, or when deflections found in iteration $(n + 1)$ are close enough to deflections found in iteration n. After 3–4 iterations, reasonable results can be obtained.

Table 5.1 Uniformly distributed load q. Coefficients $\bar{p}$ for obtaining the soil pressure $p = \bar{p}q$

α	β	ξ										
		0.0	0.1	0.2	0.3	0.4	0.5	0.6	0.7	0.8	0.9	1.0
0	0.0	1.772	1.355	1.031	0.0799	0.660	0.614	0.660	0.799	1.031	1.355	1.772
	0.1	1.257	0.998	0.790	0.642	0.560	0.553	0.630	0.798	1.066	1.440	1.931
	0.2	0.818	0.690	0.580	0.500	0.466	0.491	0.590	0.778	1.070	1.478	2.018
	0.3	0.453	0.432	0.401	0.378	0.380	0.430	0.544	0.742	1.043	1.464	2.027
	0.4	0.164	0.223	0.252	0.270	0.302	0.368	0.490	0.688	0.986	1.403	1.964
	0.5	-0.049	0.065	0.133	0.183	0.232	0.07	0.428	0.617	0.897	1.291	1.821
	0.6	-0.192	-0.048	0.045	0.111	0.170	0.246	0.358	0.529	0.779	1.132	1.608
	0.7	-0.255	-0.109	-0.012	0.058	0.116	0.184	0.28	0.422	0.630	0.923	1.319
	0.8	-0.246	-0.123	-0.039	0.021	0.070	0.123	0.194	0.299	0.451	0.665	0.954
	0.9	-0.159	-0.085	-035	0.002	0.031	0.061	0.101	0.158	0.241	0.357	0.515
25	0.0	1.732	1.337	1.029	0.810	0.678	0.634	0.678	0.810	1.029	1.337	1.732
	0.1	1.202	0.973	0.789	0.656	0.584	0.579	0.652	0.810	1.063	1.417	1.882
	0.2	0.759	0.664	0.578	0.515	0.492	0.519	0.614	0.793	1.066	1.454	1.965
	0.3	0.404	0.410	0.399	0.391	0.402	0.455	0.566	0.755	1.041	1.440	1.976
	0.4	0.132	0.206	0.248	0.277	0.318	0.387	0.508	0.701	0.986	1.386	1.920
	0.5	-0.056	0.055	0.128	0.181	0.238	0.317	0.440	0.629	0.902	1.281	1.788
	0.6	-0.188	-0.049	0.043	0.108	0.170	0.247	0.360	0.532	0.781	1.131	1.600
	0.7	-0.244	-0.104	-0.012	0.055	0.112	0.179	0.276	0.419	0.631	0.926	1.328
	0.8	-0.231	-0.116	-0.037	0.016	0.063	0.114	0.185	0.294	0.451	0.674	0.975
	0.9	-0.150	-0.081	0.033	0.000	0.026	0.055	0.094	0.154	0.241	0.363	0.530
50	0.0	1.696	1.320	1.028	0.819	0.694	0.652	0.694	0.819	1.028	1.320	1.696
	0.1	1.151	0.950	0.788	0.669	0.606	0.603	0.672	0.823	1.060	1.396	1.837
	0.2	0.704	0.640	0.577	0.529	0.515	0.545	0.637	0.805	1.063	1.430	1.916
	0.3	0.356	0.386	0.395	0.402	0.423	0.478	0.587	0.768	1.041	1.422	1.932
	0.4	0.104	0.191	0.244	0.284	0.331	0.404	0.525	0.712	0.988	1.369	1.880
	0.5	-0.060	0.048	0.122	0.181	0.242	0.326	0.452	0.639	0.906	1.272	1.756
	0.6	-0.182	-0.048	0.040	0.106	0.168	0.247	0.362	0.534	0.784	1.130	1.594
	0.7	-0.234	-0.101	-0.013	0.051	0.106	0.173	0.270	0.417	0.633	0.935	1.342
	0.8	-0.220	-0.109	-0.036	0.014	0.057	0.107	0.179	0.290	0.450	0.681	0.992
	0.9	-0.141	-0.078	-0.032	-0.004	0.022	0.049	0.088	0.150	0.240	0.368	0.545

Table 5.1 Con't

α	β	ξ										
		0.0	0.1	0.2	0.3	0.4	0.5	0.6	0.7	0.8	0.9	1.0
100	0.0	1.634	1.292	1.025	0.835	0.721	0.683	0.721	0.835	1.025	1.292	1.634
	0.1	1.066	0.915	0.786	0.694	0.642	0.644	0.708	0.840	1.054	1.357	1.758
	0.2	0.614	0.601	0.576	0.555	0.554	0.589	0.674	0.825	1.058	1.387	1.830
	0.3	0.279	0.351	0.393	0.422	0.458	0.518	0.622	0.788	1.037	1.383	1.849
	0.4	0.054	0.163	0.236	0.295	0.354	0.434	0.554	0.733	0.990	1.343	1.810
	0.5	-0.065	0.035	0.111	0.175	0.247	0.341	0.473	0.659	0.915	1.257	1.701
	0.6	-0.176	-0.051	0.035	0.102	0.167	0.249	0.367	0.540	0.789	1.129	1.580
	0.7	-0.215	-0.092	-0.011	0.047	0.099	0.173	0.280	0.413	0.633	0.940	1.355
	0.8	-0.194	-0.097	-0.032	0.009	0.046	0.093	0.166	0.279	0.450	0.689	1.022
	0.9	-0.122	-0.064	-0.029	-0.005	0.012	0.038	0.078	0.141	0.239	0.378	0.570
150	0.0	1.584	1.269	1.023	0.848	0.743	0.708	0.743	0.848	1.023	1.269	1.584
	0.1	0.990	0.880	0.786	0.713	0.675	0.679	0.737	0.857	1.050	1.326	1.694
	0.2	0.535	0.566	0.575	0.577	0.588	0.626	0.706	0.843	1.053	1.354	1.761
	0.3	0.219	0.32	0.389	0.438	0.486	0.551	0.652	0.808	1.035	1.354	1.784
	0.4	0.018	0.143	0.231	0.303	0.372	0.458	0.578	0.749	0.991	1.319	1.750
	0.5	-0.065	0.025	0.100	0.171	0.252	0.354	0.492	0.677	0.924	1.243	1.649
	0.6	-0.166	-0.050	0.032	0.099	0.167	0.250	0.369	0.545	0.792	1.126	1.566
	0.7	-0.200	-0.085	-0.011	0.041	0.091	0.157	0.257	0.411	0.634	0.949	1.372
	0.8	-0.177	-0.085	-0.030	0.006	0.037	0.082	0.155	0.272	0.448	0.703	1.049
	0.9	-0.112	-0.059	-0.026	-0.008	0.007	0.030	0.069	0.136	0.238	0.387	0.592
200	0.0	1.540	1.248	1.022	0.860	0.762	0.730	0.762	0.860	1.022	1.248	1.540
	0.1	0.927	0.853	0.783	0.730	0.761	0.708	0.761	0.870	1.047	1.301	1.641
	0.2	0.469	0.538	0.573	0.595	0.732	0.657	0.732	0.857	1.049	1.326	1.703
	0.3	0.156	0.294	0.388	0.453	0.677	0.580	0.677	0.823	1.032	1.326	1.724
	0.4	-0.013	0.124	0.227	0.309	0.600	0.479	0.600	0.765	0.933	1.298	1.697
	0.5	-0.062	0.017	0.091	0.166	0.508	0.365	0.508	0.694	0.931	1.231	1.602
	0.6	-0.157	-0.050	0.029	0.095	0.375	0.251	0.375	0.551	0.795	1.124	1.553
	0.7	-0.186	-0.079	-0.010	0.038	0.252	0.151	0.252	0.408	0.634	0.953	1.382
	0.8	-0.163	-0.077	-0.028	0.003	0.146	0.073	0.146	0.265	0.448	0.711	1.071
	0.9	-0.101	-0.052	-0.026	-0.011	0.061	0.022	0.061	0.129	0.238	0.396	0.613

		1.502	1.231	1.020	0.869	0.779	0.749	0.779	0.869	1.020	1.231	1.502
250	0.0	1.502	1.231	1.020	0.869	0.779	0.749	0.779	0.869	1.020	1.231	1.502
	0.1	0.870	0.829	0.783	0.745	0.725	0.734	0.783	0.883	1.043	1.277	1.594
	0.2	0.411	0.512	0.573	0.610	0.642	0.684	0.754	0.870	1.045	1.302	1.653
	0.3	0.109	0.272	0.386	0.465	0.533	0.604	0.699	0.735	1.030	1.304	1.675
	0.4	-0.039	0.109	0.221	0.314	0.401	0.497	0.617	0.778	0.995	1.281	1.651
	0.5	-0.059	0.010	0.080	0.161	0.256	0.375	0.524	0.709	0.940	1.220	1.559
	0.6	-0.149	-0.50	0.025	0.091	0.162	0.252	0.378	0.555	0.799	1.122	1.541
	0.7	-0.173	-0.073	-0.010	0.034	0.081	0.145	0.247	0.404	0.634	0.959	1.393
	0.8	-0.151	-0.071	-0.025	0.000	0.025	0.065	0.137	0.260	0.447	0.719	1.091
	0.9	-0.094	-0.047	-0.023	-0.013	-0.003	0.016	0.055	0.125	0.237	0.401	0.630
300	0.0	1.468	1.215	1.019	0.878	0.794	0.766	0.794	0.878	1.019	1.215	1.468
	0.1	0.822	0.808	0.784	0.758	0.745	0.756	0.801	0.892	1.040	1.256	1.554
	0.2	0.359	0.490	0.571	0.625	0.664	0.708	0.774	0.879	1.043	1.280	1.609
	0.3	0.066	0.251	0.382	0.476	0.551	0.625	0.717	0.846	1.030	1.287	1.634
	0.4	-0.062	0.095	0.216	0.318	0.412	0.513	0.634	0.792	0.998	1.265	1.610
	0.5	-0.051	0.005	0.072	0.155	0.257	0.383	0.537	0.723	0.946	1.211	1.519
	0.6	-0.142	-0.050	0.021	0.087	0.160	0.253	0.382	0.561	0.803	1.120	1.530
	0.7	-0.166	-0.072	-0.010	0.032	0.077	0.141	0.243	0.402	0.637	0.964	1.402
	0.8	-0.141	-0.064	-0.025	-0.001	0.020	0.058	0.130	0.253	0.447	0.726	1.109
	0.9	-0.086	-0.041	-0.021	-0.014	-0.007	0.010	0.049	0.120	0.235	0.407	0.646
350	0.0	1.440	1.202	1.018	0.886	0.806	0.780	0.806	0.886	1.018	1.202	1.440
	0.1	0.781	0.791	0.783	0.770	0.763	0.775	0.817	0.900	1.037	1.239	1.519
	0.2	0.313	0.469	0.570	0.636	0.682	0.728	0.792	0.890	1.049	1.263	1.575
	0.3	0.030	0.235	0.381	0.486	0.568	0.644	0.734	0.856	1.027	1.269	1.594
	0.4	-0.080	0.084	0.213	0.322	0.422	0.527	0.650	0.802	0.999	1.250	1.572
	0.5	-0.044	0.000	0.064	0.149	0.260	0.390	0.550	0.737	0.954	1.202	1.484
	0.6	-0.132	-0.048	0.019	0.084	0.157	0.253	0.385	0.564	0.805	1.118	1.520
	0.7	-0.156	-0.067	-0.010	0.030	0.074	0.137	0.240	0.400	0.636	0.967	1.408
	0.8	-0.135	-0.061	-0.024	-0.004	0.015	0.052	0.125	0.250	0.448	0.733	1.127
	0.9	-0.079	-0.037	-0.019	-0.014	-0.011	0.005	0.043	0.116	0.235	0.411	0.659

Table 5.1 Con't

						ξ					
α	β	0.0	0.1	0.2	0.3	0.4	0.5	0.6	0.7	0.8	0.9	1.0
400	0.0	1.414	1.190	1.017	0.892	0.818	0.793	0.818	0.892	1.017	1.190	1.414
	0.1	0.743	0.774	0.778	0.780	0.781	0.792	0.830	0.907	1.034	1.224	1.489
	0.2	0.272	0.451	0.570	0.647	0.700	0.747	0.806	0.897	1.038	1.247	1.540
	0.3	0.000	0.221	0.380	0.494	0.582	0.660	0.748	0.864	1.026	1.253	1.560
	0.4	-0.097	0.072	0.208	0.323	0.429	0.539	0.663	0.813	1.002	1.240	1.541
	0.5	-0.035	-0.003	0.056	0.143	0.257	0.396	0.559	0.715	0.930	1.195	1.500
	0.6	-0.127	-0.050	0.015	0.079	0.155	0.254	0.389	0.569	0.809	1.118	1.511
	0.7	-0.146	-0.062	-0.016	0.028	0.070	0.133	0.236	0.398	0.636	0.970	1.414
	0.8	-0.126	-0.056	-0.022	-0.005	0.011	0.046	0.117	0.245	0.446	0.740	1.142
	0.9	-0.075	-0.034	-0.017	-0.014	-0.012	0.001	0.038	0.112	0.233	0.416	0.671
450	0.0	1.392	1.180	1.016	0.898	0.828	0.804	0.828	0.898	1.016	1.180	1.392
	0.1	0.711	0.762	0.783	0.791	0.794	0.807	0.842	0.913	1.031	1.210	1.461
	0.2	0.238	0.437	0.571	0.658	0.714	0.763	0.818	0.904	1.035	1.231	1.510
	0.3	-0.032	0.206	0.378	0.501	0.594	0.675	0.762	0.873	1.026	1.240	1.532
	0.4	-0.110	0.063	0.204	0.326	0.437	0.550	0.675	0.22	1.004	1.229	1.510
	0.5	-0.027	-0.007	0.049	0.130	0.257	0.402	0.571	0.768	0.967	1.187	1.419
	0.6	-0.120	-0.050	0.012	0.077	0.153	0.255	0.391	0.573	0.812	1.116	1.500
	0.7	-0.142	-0.061	-0.010	0.026	0.066	0.130	0.234	0.398	0.638	0.973	1.422
	0.8	-0.118	-0.051	-0.019	-0.006	0.008	0.041	0.112	0.240	0445	0.743	1.154
	0.9	-0.069	-0.029	-0.016	-0.015	-0.015	-0.003	0.033	0.107	0.232	0.419	0.681
500	0.0	1.372	1.171	1.015	0.903	0.836	0.814	0.836	0.903	1.015	1.171	1.372
	0.1	0.680	0.749	0.784	0.800	0.807	0.821	0.853	0.918	1.028	1.197	1.436
	0.2	0.212	0.427	0.573	0.667	0.729	0.777	0.831	0.909	1.031	1.215	1.480
	0.3	-0.056	0.196	0.377	0.508	0.605	0.688	0.773	0.880	1.025	1.228	1.514
	0.4	-0.116	0.057	0.202	0.327	0.443	0.559	0.685	0.830	1.004	1.219	1.480
	0.5	-0.019	-0.011	0.041	0.131	0.255	0.407	0.581	0.745	0.945	1.183	1.510
	0.6	-0.110	-0.048	0.011	0.073	0.151	0.255	0.393	0.577	0.813	1.114	1.480
	0.7	-0.134	-0.057	-0.010	0.024	0.064	0.127	0.232	0.396	0.638	0.975	1.426
	0.8	-0.108	-0.044	-0.016	-0.006	0.006	0.037	0.108	0.236	0.442	0.744	1.160
	0.9	-0.063	-0.026	-0.014	-0.014	-0.010	-0.006	0.030	0.104	0.230	0.422	0.687

Table 5.2 Uniformly distributed load q. Coefficients $\overline{Q}$ for obtaining shear forces $Q = \overline{Q}qbL$

α	β	ξ										
		0.0	0.1	0.2	0.3	0.4	0.5	0.6	0.7	0.8	0.9	1.0
0	0.0	0	0.055	0.074	0.065	0.037	0.000	-0.037	-0.065	-0.074	-0.055	0
	0.1	0	0.113	0.102	0.073	0.032	-0.013	-0.054	-0.083	-0.092	-0.067	0
	0.2	0	0.076	0.140	0.093	0.041	-0.012	-0.059	-0.091	-0.100	-0.074	0
	0.3	0	0.044	0.085	0.124	0.062	0.002	-0.050	-0.086	-0.099	-0.074	0
	0.4	0	0.020	0.043	0.069	0.098	0.031	-0.026	-0.069	-0.087	-0.068	0
	0.5	0	0.002	0.012	0.028	0.049	0.075	0.011	-0.038	-0.062	-0.054	0
	0.6	0	-0.011	-0.013	-0.004	0.011	0.031	0.061	0.004	-0.031	-0.037	0
	0.7	0	-0.019	-0.025	-0.022	-0.013	0.002	0.025	0.060	0.011	-0.011	0
	0.8	0	-0.018	-0.025	-0.026	-0.022	-0.012	0.004	0.028	0.065	0.020	0
	0.9	0	-0.011	-0.017	-0.018	-0.017	-0.013	-0.005	0.008	0.027	0.057	0
25	0.0	0	0.052	0.070	0.061	0.035	0.000	-0.035	-0.061	-0.070	-0.052	0
	0.1	0	0.108	0.096	0.069	0.030	-0.013	-0.052	-0.079	-0.086	-0.064	0
	0.2	0	0.071	0.134	0.088	0.038	-0.012	-0.056	-0.086	-0.094	-0.069	0
	0.3	0	0.040	0.081	0.120	0.060	0.002	-0.048	-0.082	-0.095	-0.070	0
	0.4	0	0.016	0.040	0.066	0.097	0.031	-0.025	-0.066	-0.082	-0.064	0
	0.5	0	0.000	0.010	0.025	0.046	0.073	0.011	-0.037	-0.060	-0.052	0
	0.6	0	-0.011	-0.012	-0.004	0.011	0.031	0.061	0.004	-0.030	-0.035	0
	0.7	0	-0.018	-0.024	-0.021	-0.012	0.002	0.024	0.059	0.010	-0.012	0
	0.8	0	-0.017	-0.023	-0.025	-0.021	-0.012	0.003	0.027	0.063	0.019	0
	0.9	0	-0.011	-0.016	-0.017	-0.017	-0.013	-0.005	0.007	0.026	0.056	0
50	0.0	0	0.050	0.066	0.059	0.033	0.000	-0.033	-0.059	-0.066	-0.050	0
	0.1	0	0.105	0.092	0.063	0.028	-0.013	-0.050	-0.075	-0.082	-0.061	0
	0.2	0	0.068	0.128	0.084	0.036	-0.012	-0.054	-0.082	-0.090	-0.066	0
	0.3	0	0.037	0.077	0.116	0.057	0.002	-0.045	-0.078	-0.089	-0.067	0
	0.4	0	0.015	0.038	0.063	0.095	0.031	-0.023	-0.063	-0.079	-0.061	0
	0.5	0	0.000	0.008	0.023	0.045	0.073	0.011	-0.035	-0.058	-0.050	0
	0.6	0	-0.011	-0.011	-0.004	0.011	0.031	0.061	0.005	-0.030	-0.035	0
	0.7	0	-0.016	-0.022	-0.020	-0.012	0.002	0.024	0.058	0.010	-0.014	0
	0.8	0	-0.016	-0.023	-0.023	-0.020	-0.012	0.002	0.025	0.061	0.018	0
	0.9	0	-0.011	-0.016	0.017	-0.016	-0.013	-0.006	0.006	0.026	0.055	0

Table 5.2 Con't

α	β	ξ										
		0.0	0.1	0.2	0.3	0.4	0.5	0.6	0.7	0.8	0.9	1.0
100	0.0	0	0.046	0.061	0.053	0.031	0.000	-0.031	-0.053	-0.061	-0.046	0
	0.1	0	0.098	0.083	0.057	0.024	-0.013	-0.046	-0.069	-0.074	-0.054	0
	0.2	0	0.060	0.119	0.075	0.031	-0.012	-0.049	-0.075	-0.081	-0.060	0
	0.3	0	0.031	0.069	0.109	0.054	0.002	-0.042	-0.071	-0.081	-0.061	0
	0.4	0	0.012	0.032	0.058	0.092	0.030	-.021	-0.058	-0.072	-0.056	0
	0.5	0	-0.001	0.007	0.021	0.042	0.071	0.011	-0.033	0.055	-0.047	0
	0.6	0	-0.011	-0.011	-0.005	0.010	0.030	0.060	0.005	-0.029	-0.033	0
	0.7	0	-0.014	-0.020	-0.018	0.011	0.002	0.023	0.056	0.008	-0.014	0
	0.8	0	-0.014	-0.021	-0.021	-0.019	-0.012	0.001	0.023	0.059	0.014	0
	0.9	0	-0.009	-0.014	-0.016	-0.015	-0.013	-0.007	0.004	0.022	0.053	0
150	0.0	0	0.042	0.056	0.049	0.028	0	-0.028	-0.049	-0.056	-0.042	0
	0.1	0	0.093	0.076	0.052	0.020	-0.012	-0.042	-0.062	-0.068	-0.049	0
	0.2	0	0.055	0.112	0.069	0.028	-0.012	-0.046	-0.069	-0.074	-0.055	0
	0.3	0	0.028	0.063	0.104	0.050	0.002	-0.038	-0.066	-0.075	-0.056	0
	0.4	0	0.009	0.028	0.053	0.090	0.029	-0.020	-0.053	-0.068	-0.053	0
	0.5	0	-0.002	0.004	0.018	0.039	0.069	0.011	-0.030	-0.052	-0.044	0
	0.6	0	-0.011	-0.011	-0.005	0.008	0.029	0.060	0.005	-0.029	-0.033	0
	0.7	0	-0.014	-0.018	-0.017	-0.010	0.002	0.022	0.055	0.006	-0.014	0
	0.8	0	-0.013	-0.020	-0.019	-0.018	-0.012	0.000	0.021	0.056	0.013	0
	0.9	0	-0.007	-0.013	-0.013	-0.014	-0.012	-0.008	0.003	0.021	0.051	0
200	0.0	0	0.039	0.052	0.045	0.026	0	-0.026	-0.045	-0.052	-0.039	0
	0.1	0	0.090	0.071	0.047	0.018	-0.012	-0.040	-0.057	-0.063	-0.046	0
	0.2	0	0.050	0.107	0.065	0.026	-0.011	-0.042	-0.063	-0.069	-0.050	0
	0.3	0	0.023	0.057	0.099	0.048	0.002	-0.036	-0.061	-0.069	-0.051	0
	0.4	0	0.006	0.024	0.050	0.088	0.029	-0.018	-0.050	-0.062	-0.048	0
	0.5	0	-0.002	0.003	0.016	0.037	0.068	0.011	-0.030	-0.049	-0.041	0
	0.6	0	-0.010	-0.011	-0.005	0.008	0.029	0.060	0.005	-0.028	-0.033	0
	0.7	0	-0.012	-0.017	-0.016	-0.010	0.002	0.022	0.054	0.005	-0.016	0
	0.8	0	-0.011	-0.017	-0.018	-0.016	-0.011	0.000	0.020	0.055	0.011	0
	0.9	0	-0.007	-0.011	-0.012	-0.013	-0.012	-0.009	0.002	0.019	0.051	0

250	0.0	0	0.036	0.048	0.042	0.024	0	-0.024	-0.042	-0.048	-0.036	0
	0.1	0	0.085	0.066	0.042	0.015	-0.012	-0.037	-0.054	-0.058	-0.043	0
	0.2	0	0.046	0.101	0.060	0.023	-0.011	-0.039	-0.058	-0.063	-0.046	0
	0.3	0	0.019	0.053	0.095	0.045	0.002	-0.033	-0.057	-0.065	-0.047	0
	0.4	0	0.004	0.021	0.048	0.086	0.028	-0.017	-0.048	-0.059	-0.046	0
	0.5	0	-0.003	0.002	0.014	0.035	0.066	0.011	-0.028	-0.046	-0.039	0
	0.6	0	-0.010	-0.011	-0.005	0.007	0.028	0.059	0.005	-0.027	-0.032	0
	0.7	0	-0.012	-0.015	-0.015	-0.009	0.002	0.021	0.053	0.003	-0.016	0
	0.8	0	-0.011	-0.015	-0.017	-0.015	-0.011	-0.001	0.018	0.053	0.011	0
	0.9	0	-0.006	-0.010	-0.012	-0.013	-0.012	-0.009	0.000	0.018	0.048	0
300	0.0	0	0.034	0.045	0.039	0.023	0	-0.023	-0.039	-0.045	-0.034	0
	0.1	0	0.082	0.061	0.038	0.013	-0.012	-0.035	-0.050	-0.053	-0.040	0
	0.2	0	0.043	0.096	0.055	0.021	-0.011	-0.037	-0.055	-0.060	-0.043	0
	0.3	0	0.017	0.049	0.092	0.043	0.002	-0.031	-0.054	-0.061	-0.045	0
	0.4	0	0.002	0.018	0.044	0.084	0.027	-0.015	-0.044	-0.056	-0.044	0
	0.5	0	-0.003	0.001	0.012	0.033	0.065	0.011	-0.026	-0.043	-0.036	0
	0.6	0	-0.010	-0.011	-0.006	0.007	0.027	0.059	0.006	-0.027	-0.032	0
	0.7	0	-0.011	-0.015	-0.014	-0.008	0.002	0.021	0.052	0.003	-0.017	0
	0.8	0	-0.009	-0.014	-0.017	-0.014	-0.011	-0.002	0.018	0.050	0.009	0
	0.9	0	-0.006	-0.009	-0.011	-0.012	-0.012	-0.010	-0.001	0.017	0.048	0
350	0.0	0	0.032	0.042	0.037	0.021	0	-0.021	-0.037	-0.042	-0.032	0
	0.1	0	0.079	0.058	0.036	0.010	-0.011	-0.032	-0.046	-0.050	-0.037	0
	0.2	0	0.040	0.092	0.052	0.019	-0.011	-0.035	-0.052	-0.056	-0.042	0
	0.3	0	0.014	0.045	0.088	0.041	0.002	-0.029	-0.050	-0.057	-0.042	0
	0.4	0	0.000	0.015	0.043	0.083	0.027	-0.014	-0.043	-0.053	-0.040	0
	0.5	0	-0.003	0.001	0.010	0.031	0.063	0.010	-0.026	-0.041	-0.034	0
	0.6	0	-0.009	-0.011	-0.006	0.007	0.027	0.059	0.006	-0.027	-0.031	0
	0.7	0	-0.011	-0.015	-0.013	-0.008	-0.002	0.020	0.051	0.003	-0.017	0
	0.8	0	-0.009	-0.014	-0.015	-0.014	-0.011	-0.002	0.016	0.050	0.008	0
	0.9	0	-0.005	-0.008	-0.009	-0.011	-0.011	-0.010	-0.001	0.016	0.047	0

Table 5.2 Con't

α	β						ξ					
		0.0	0.1	0.2	0.3	0.4	0.5	0.6	0.7	0.8	0.9	1.0
400	0.0	0	0.030	0.040	0.034	0.020	0	-0.020	-0.034	-0.040	-0.030	0
	0.1	0	0.076	0.054	0.034	0.010	-0.011	-0.030	-0.044	-0.046	0.034	0
	0.2	0	0.036	0.088	0.049	0.017	-0.011	-0.033	-0.049	-0.052	-0.039	0
	0.3	0	0.012	0.042	0.086	0.040	0.002	-0.028	-0.048	-0.054	-0.040	0
	0.4	0	-0.001	0.013	0.040	0.082	0.026	-0.014	-0.040	-0.049	-0.037	0
	0.5	0	-0.003	0.000	0.010	0.030	0.062	0.010	-0.026	-0.040	-0.033	0
	0.6	0	-0.008	-0.011	-0.006	0.006	0.026	0.058	0.006	-0.026	-0.030	0
	0.7	0	-0.010	-0.014	-0.013	-0.008	0.002	0.020	0.051	0.002	-0.018	0
	0.8	0	-0.008	-0.013	-0.014	-0.013	-0.011	-0.003	0.015	0.049	0.006	0
	0.9	0	-0.004	-0.007	-0.008	-0.010	-0.011	-0.010	-0.002	0.015	0.046	0
450	0.0	0	0.028	0.038	0.033	0.019	0	-0.019	-0.033	-0.038	-0.028	0
	0.1	0	0.074	0.051	0.029	0.009	-0.011	-0.029	-0.043	-0.045	-0.032	0
	0.2	0	0.035	0.085	0.047	0.017	-0.010	-0.031	-0.045	-0.049	-0.037	0
	0.3	0	0.010	0.039	0.083	0.038	0.002	-0.026	-0.045	-0.051	-0.038	0
	0.4	0	-0.002	0.012	0.038	0.081	0.026	-0.013	-0.038	-0.047	-0.036	0
	0.5	0	-0.003	0.000	0.010	0.028	0.061	0.010	-0.024	-0.038	-0.030	0
	0.6	0	-0.008	-0.010	-0.006	0.006	0.026	0.058	0.006	-0.025	-0.030	0
	0.7	0	-0.010	-0.013	-0.012	-0.008	0.002	0.020	0.050	0.001	-0.018	0
	0.8	0	-0.008	-0.011	-0.013	-0.012	-0.010	-0.003	0.014	0.047	0.006	0
	0.9	0	-0.004	-0.007	-0.009	-0.010	-0.011	-0.010	-0.004	0.013	0.046	0
500	0.0	0	0.026	0.036	0.031	0.018	0	0.018	-0.031	-0.036	-0.026	0
	0.1	0	0.072	0.049	0.028	0.008	-0.011	-0.028	-0.042	-0.045	-0.032	0
	0.2	0	0.031	0.083	0.045	0.015	-0.010	-0.029	-0.043	-0.047	-0.035	0
	0.3	0	0.008	0.037	0.081	0.037	0.002	-0.025	-0.043	-0.049	-0.036	0
	0.4	0	-0.003	0.011	0.037	0.080	0.025	-0.018	-0.037	-0.045	-0.035	0
	0.5	0	-0.003	0.000	0.009	0.027	0.060	0.009	-0.023	-0.036	-0.028	0
	0.6	0	-0.008	-0.009	-0.006	0.005	0.025	0.057	0.006	-0.025	-0.030	0
	0.7	0	-0.009	-0.013	-0.012	-0.008	0.002	0.020	0.050	0.001	-0.019	0
	0.8	0	-0.008	-0.011	-0.012	-0.011	-0.010	-0.003	0.013	0.047	0.006	0
	0.9	0	-0.004	-0.007	-0.008	-0.010	-0.011	-0.010	-0.004	0.011	0.046	0

Table 5.3 Uniformly distributed load q. Coefficients $\overline{M}$ for obtaining moments $M = \overline{M}qbL^2$

α	β	ξ										
		0.0	0.1	0.2	0.3	0.4	0.5	0.6	0.7	0.8	0.9	1.0
0	0.0	0	0.003	0.010	0.017	0.022	0.024	0.022	0.017	0.010	0.003	0
	0.1	0	0.006	0.017	0.026	0.030	0.031	0.028	0.021	0.013	0.004	0
	0.2	0	0.005	0.015	0.026	0.033	0.034	0.031	0.023	0.014	0.004	0
	0.3	0	0.002	0.008	0.019	0.028	0.031	0.029	0.022	0.013	0.005	0
	0.4	0	0.001	0.005	0.010	0.018	0.024	0.024	0.019	0.011	0.005	0
	0.5	0	0.000	0.001	0.002	0.006	0.012	0.016	0.014	0.009	0.003	0
	0.6	0	0.000	-0.001	-0.002	-0.002	-0.001	0.004	0.007	0.005	0.002	0
	0.7	0	0.000	-0.001	-0.005	-0.007	-0.008	-0.007	-0.003	-0.001	0.001	0
	0.8	0	0.000	-0.003	-0.006	-0.009	-0.010	-0.011	-0.009	-0.004	-0.001	0
	0.9	0	0.000	-0.002	-0.004	-0.005	-0.007	-0.008	-0.009	-0.007	-0.001	0
25	0.0	0	0.003	0.009	0.016	0.021	0.023	0.021	0.016	0.009	0.003	0
	0.1	0	0.006	0.017	0.024	0.029	0.030	0.027	0.019	0.011	0.004	0
	0.2	0	0.004	0.014	0.025	0.032	0.033	0.029	0.022	0.013	0.004	0
	0.3	0	0.002	0.008	0.018	0.027	0.030	0.028	0.021	0.013	0.004	0
	0.4	0	0.001	0.004	0.009	0.017	0.023	0.023	0.018	0.011	0.004	0
	0.5	0	0.000	0.001	0.002	0.005	0.011	0.015	0.013	0.008	0.003	0
	0.6	0	0.000	-0.001	-0.002	-0.002	-0.001	0.004	0.007	0.005	0.002	0
	0.7	0	0.000	-0.001	-0.005	-0.007	-0.008	-0.007	-0.003	-0.001	0.001	0
	0.8	0	0.000	-0.003	-0.006	-0.008	-0.009	-0.010	-0.009	-0.004	-0.001	0
	0.9	0	0.000	-0.002	-0.004	-0.006	-0.007	-0.008	-0.009	-0.007	-0.002	0
50	0.0	0	0.003	0.009	0.016	0.020	0.021	0.020	0.016	0.009	0.003	0
	0.1	0	0.006	0.015	0.023	0.027	0.028	0.025	0.018	0.010	0.003	0
	0.2	0	0.004	0.013	0.023	0.030	0.031	0.027	0.020	0.012	0.004	0
	0.3	0	0.002	0.007	0.017	0.025	0.028	0.027	0.019	0.012	0.004	0
	0.4	0	0.001	0.004	0.009	0.016	0.022	0.022	0.018	0.011	0.004	0
	0.5	0	0.000	0.000	0.002	0.005	0.011	0.015	0.013	0.008	0.003	0
	0.6	0	0.000	-0.001	-0.002	-0.002	-0.001	0.004	0.007	0.005	0.002	0
	0.7	0	0.000	-0.001	-0.005	-0.007	-0.007	-0.007	-0.003	-0.001	0.001	0
	0.8	0	0.000	-0.003	-0.006	-0.008	-0.009	-0.010	-0.009	-0.004	-0.001	0
	0.9	0	0.000	-0.002	-0.003	-0.005	-0.007	-0.008	-0.008	-0.007	-0.002	0

Table 5.3 Con't

α	β	0.0	0.1	0.2	0.3	0.4	0.5	0.6	0.7	0.8	0.9	1.0
100	0.0	0	0.002	0.008	0.014	0.018	0.020	0.018	0.014	0.008	0.002	0
	0.1	0	0.006	0.015	0.022	0.026	0.026	0.023	0.016	0.010	0.003	0
	0.2	0	0.003	0.012	0.022	0.027	0.028	0.025	0.019	0.011	0.003	0
	0.3	0	0.002	0.006	0.015	0.024	0.027	0.024	0.018	0.011	0.003	0
	0.4	0	0.001	0.003	0.007	0.015	0.021	0.021	0.016	0.010	0.003	0
	0.5	0	0.000	0.000	0.001	0.005	0.010	0.014	0.013	0.008	0.002	0
	0.6	0	0.000	-0.001	-0.002	-0.002	-0.001	0.004	0.007	0.005	0.002	0
	0.7	0	0.000	-0.001	-0.005	-0.006	-0.007	-0.006	-0.002	0.000	0.001	0
	0.8	0	0.000	-0.003	-0.005	-0.007	-0.009	-0.010	-0.008	-0.004	-0.001	0
	0.9	0	0.000	-0.001	-0.003	-0.005	-0.006	-0.007	-0.007	-0.006	-0.002	0
150	0.0	0	0.002	0.007	0.013	0.017	0.018	0.017	0.013	0.007	0.002	0
	0.1	0	0.005	0.014	0.020	0.024	0.024	0.021	0.015	0.009	0.003	0
	0.2	0	0.003	0.011	0.020	0.025	0.026	0.024	0.017	0.009	0.003	0
	0.3	0	0.001	0.005	0.014	0.022	0.024	0.022	0.016	0.010	0.003	0
	0.4	0	0.001	0.002	0.007	0.014	0.019	0.019	0.015	0.009	0.002	0
	0.5	0	0.000	0.000	0.001	0.004	0.009	0.013	0.012	0.007	0.002	0
	0.6	0	0.000	-0.001	-0.002	-0.002	-0.001	0.003	0.006	0.005	0.002	0
	0.7	0	0.000	-0.001	-0.005	-0.006	-0.006	-0.005	-0.002	-0.001	0.001	0
	0.8	0	0.000	-0.002	-0.005	-0.006	-0.008	-0.009	-0.008	-0.004	-0.001	0
	0.9	0	0.000	-0.001	-0.002	-0.004	-0.006	-0.007	-0.007	-0.006	-0.002	0
200	0.0	0	0.002	0.007	0.011	0.015	0.017	0.015	0.011	0.007	0.002	0
	0.1	0	0.005	0.013	0.019	0.022	0.022	0.019	0.014	0.008	0.003	0
	0.2	0	0.003	0.011	0.019	0.024	0.024	0.022	0.016	0.008	0.003	0
	0.3	0	0.001	0.005	0.013	0.020	0.022	0.021	0.015	0.009	0.003	0
	0.4	0	0.001	0.002	0.005	0.012	0.017	0.018	0.014	0.009	0.003	0
	0.5	0	0.000	0.000	0.000	0.003	0.009	0.012	0.011	0.007	0.002	0
	0.6	0	0.000	-0.001	-0.002	-0.002	-0.001	0.003	0.006	0.005	0.002	0
	0.7	0	0.000	-0.001	-0.004	-0.005	-0.006	-0.005	-0.002	-0.001	0.001	0
	0.8	0	0.000	-0.002	-0.004	-0.006	-0.007	-0.008	-0.007	-0.004	-0.001	0
	0.9	0	0.000	-0.001	-0.002	-0.004	-0.005	-0.006	-0.007	-0.006	-0.002	0

ξ

250	0.0	0	0.002	0.006	0.010	0.014	0.015	0.014	0.010	0.006	0.002	0
	0.1	0	0.004	0.012	0.017	0.020	0.020	0.017	0.013	0.007	0.002	0
	0.2	0	0.002	0.010	0.017	0.022	0.022	0.019	0.015	0.008	0.003	0
	0.3	0	0.001	0.004	0.012	0.018	0.020	0.019	0.015	0.009	0.003	0
	0.4	0	0.001	0.002	0.005	0.012	0.017	0.017	0.013	0.008	-0.002	0
	0.5	0	0.000	0.000	0.000	0.002	0.008	0.012	0.011	0.007	0.002	0
	0.6	0	0.000	-0.001	-0.002	-0.002	-0.001	0.003	0.006	0.005	0.002	0
	0.7	0	0.000	-0.001	-0.004	-0.005	-0.005	-0.004	-0.002	-0.001	0.001	0
	0.8	0	0.000	-0.002	-0.004	-0.005	-0.006	-0.007	-0.007	-0.004	-0.001	0
	0.9	0	0.000	-0.001	-0.002	-0.004	-0.005	-0.006	-0.007	-0.006	-0.003	0
300	0.0	0	0.002	0.006	0.010	0.013	0.014	0.013	0.010	0.006	0.002	0
	0.1	0	0.004	0.011	0.016	0.018	0.018	0.016	0.012	0.006	0.002	0
	0.2	0	0.002	0.010	0.016	0.021	0.021	0.018	0.014	0.007	0.003	0
	0.3	0	0.001	0.003	0.011	0.017	0.019	0.018	0.014	0.008	0.003	0
	0.4	0	0.000	0.001	0.005	0.012	0.016	0.016	0.012	0.007	0.002	0
	0.5	0	0.000	0.000	0.000	0.003	0.007	0.010	0.010	0.006	0.002	0
	0.6	0	0.000	-0.001	-0.003	-0.003	-0.001	0.003	0.006	0.005	0.002	0
	0.7	0	0.000	-0.001	-0.003	-0.004	-0.005	-0.004	-0.002	-0.001	0.001	0
	0.8	0	0.000	-0.002	-0.004	-0.005	-0.006	-0.007	-0.006	-0.004	-0.001	0
	0.9	0	0.002	-0.001	-0.002	-0.003	-0.004	-0.005	-0.006	-0.006	-0.003	0
350	0.0	0	0.002	0.006	0.010	0.013	0.015	0.015	0.011	0.006	0.002	0
	0.1	0	0.004	0.010	0.015	0.018	0.017	0.017	0.013	0.007	0.003	0
	0.2	0	0.002	0.010	0.015	0.019	0.020	0.017	0.013	0.007	0.003	0
	0.3	0	-0.001	0.003	0.010	0.016	0.018	0.017	0.012	0.006	0.002	0
	0.4	0	-0.001	0.001	0.004	0.010	0.015	0.015	0.010	0.006	0.002	0
	0.5	0	-0.001	-0.001	0.000	0.003	0.007	0.010	0.006	0.005	0.002	0
	0.6	0	-0.001	-0.002	-0.003	-0.003	-0.001	0.003	-0.002	-0.001	0.002	0
	0.7	0	-0.002	-0.003	-0.005	-0.005	-0.005	-0.004	-0.006	-0.004	0.001	0
	0.8	0	-0.001	-0.002	-0.005	-0.006	-0.006	-0.007	-0.006	-0.004	-0.001	0
	0.9	0	0.000	-0.001	-0.003	-0.004	-0.004	-0.005	-0.006	-0.006	-0.003	0

Table 5.3 Con't

α	β						ξ					
		0.0	0.1	0.2	0.3	0.4	0.5	0.6	0.7	0.8	0.9	1.0
400	0.0	0	0.002	0.006	0.009	0.012	0.013	0.012	0.009	0.006	0.002	0
	0.1	0	0.004	0.010	0.015	0.016	0.016	0.015	0.010	0.006	0.002	0
	0.2	0	0.001	0.008	0.014	0.018	0.018	0.016	0.012	0.006	0.002	0
	0.3	0	-0.001	0.003	0.009	0.015	0.017	0.016	0.012	0.007	0.002	0
	0.4	0	-0.001	0.000	0.003	0.009	0.014	0.014	0.012	0.006	0.002	0
	0.5	0	-0.001	-0.001	0.000	0.003	0.0078	0.010	0.009	0.006	0.002	0
	0.6	0	-0.001	-0.002	-0.003	-0.003	-0.001	0.003	0.006	0.005	0.002	0
	0.7	0	-0.001	-0.002	-0.003	-0.004	-0.005	-0.004	-0.002	-0.001	0.001	0
	0.8	0	-0.001	-0.002	-0.003	-0.004	-0.005	-0.006	-0.006	-0.004	-0.001	0
	0.9	0	0.000	-0.001	-0.002	-0.003	-0.004	-0.005	-0.006	-0.006	-0.003	0
450	0.0	0	0.002	0.006	0.009	0.012	0.013	0.012	0.009	0.006	0.002	0
	0.1	0	0.004	0.010	0.015	0.016	0.016	0.015	0.010	0.006	0.002	0
	0.2	0	0.001	0.008	0.014	0.018	0.018	0.016	0.012	0.006	0.002	0
	0.3	0	-0.001	0.003	0.009	0.015	0.017	0.016	0.012	0.007	0.002	0
	0.4	0	-0.001	0.000	0.003	0.009	0.014	0.014	0.012	0.006	0.002	0
	0.5	0	-0.001	-0.001	0.000	0.003	0.0078	0.010	0.009	0.006	0.002	0
	0.6	0	-0.001	-0.002	-0.003	-0.003	-0.001	0.003	0.006	0.006	0.002	0
	0.7	0	-0.001	-0.002	-0.003	-0.004	-0.005	-0.004	-0.002	-0.001	0.001	0
	0.8	0	-0.001	-0.002	-0.003	-0.004	-0.005	-0.006	-0.006	-0.004	-0.001	0
	0.9	0	0.000	-0.001	-0.002	-0.003	-0.004	-0.005	-0.006	-0.006	-0.003	0
500	0.0	0	0.002	0.005	0.008	0.010	0.011	0.010	0.008	0.005	0.002	0
	0.1	0	0.004	0.010	0.014	0.015	0.015	0.013	0.009	0.005	0.002	0
	0.2	0	0.001	0.007	0.013	0.016	0.016	0.014	0.010	0.006	0.002	0
	0.3	0	-0.001	0.002	0.008	0.013	0.016	0.014	0.011	0.006	0.002	0
	0.4	0	-0.001	0.000	0.002	0.008	0.013	0.013	0.011	0.005	0.002	0
	0.5	0	-0.001	-0.001	0.000	0.002	0.006	0.009	0.008	0.005	0.002	0
	0.6	0	-0.001	-0.002	-0.003	-0.003	-0.001	0.003	0.006	0.004	0.002	0
	0.7	0	-0.001	-0.002	-0.003	-0.003	-0.004	-0.003	-0.002	-0.001	0.001	0
	0.8	0	-0.001	-0.002	-0.003	-0.003	-0.004	-0.005	-0.005	-0.003	-0.001	0
	0.9	0	0.000	-0.001	-0.002	-0.003	-0.004	-0.005	-0.006	-0.006	-0.003	0

Table 5.4 Vertical load P. Coefficients $\bar{p}$ for obtaining the soil pressure $p = \bar{p}\dfrac{Pb}{L}$

α	β						ξ					
		0.0	0.1	0.2	0.3	0.4	0.5	0.6	0.7	0.8	0.9	1.0
	0	5.522	3.812	2.560	1.669	1.050	0.614	0.270	-0.071	-0.498	-1.102	-1.978
	0.1	4.772	3.323	2.255	1.495	0.972	0.614	0.348	0.103	-0.193	-0.613	-1.228
	0.2	4.017	2.828	1.948	1.321	0.894	0.614	0.426	0.277	0.114	-0.118	-0.473
	0.3	3.272	2.333	1.643	1.147	0.816	0.614	0.504	0.451	0.419	0.371	0.272
	0.4	2.517	1.844	1.336	0.973	0.738	0.614	0.582	0.625	0.726	0.866	1.027
0	0.5	1.772	1.355	1.031	0.799	0.660	0.614	0.660	0.799	1.031	1.355	1.772
	0.6	1.027	0.866	0.726	0.625	0.582	0.614	0.738	0.973	1.336	1.844	2.517
	0.7	0.272	0.371	0.419	0.451	0.504	0.614	0.816	1.147	1.643	2.333	3.272
	0.8	-0.473	-0.118	0.114	0.277	0.426	0.614	0.894	1.321	1.948	2.828	4.017
	0.9	-1.228	-0.613	-0.193	0.103	0.348	0.614	0.972	1.495	2.255	3.323	4.772
	1.0	-1.978	-1.102	-0.498	-0.071	0.270	0.614	1.050	1.669	2.560	3.812	5.522
	0	5.739	3.900	2.553	1.604	0.959	0.523	0.201	-0.100	-0.477	-1.022	-1.831
	0.1	4.868	3.363	2.253	1.468	0.931	0.572	0.315	0.086	0.185	-0.575	-1.156
	0.2	3.999	2.827	1.954	1.332	0.905	0.620	0.427	0.272	0.106	-0.127	-0.479
	0.3	3.125	2.288	1.653	1.191	0.871	0.663	0.535	0.459	0.401	0.332	0.223
	0.4	2.290	1.767	1.355	1.046	0.830	0.695	0.634	0.636	0.693	0.795	0.930
25	0.5	1.584	1.269	1.023	0.848	0.743	0.708	0.743	0.848	1.023	1.269	1.584
	0.6	0.930	0.795	0.693	0.626	0.634	0.695	0.830	1.046	1.355	1.767	2.290
	0.7	0.223	0.332	0.401	0.459	0.535	0.663	0.871	1.191	1.653	2.288	3.125
	0.8	-0.479	-0.127	0.106	0.272	0.427	0.62	0.905	1.332	1.954	2.827	3.999
	0.9	-1.156	-0.575	-0.185	0.086	0.315	0.572	0.931	1.468	2.253	3.363	4.868
	1.0	-1.831	-1.022	-0.477	-0.100	0.201	0.523	0.959	1.604	2.553	3.900	5.739

Table 5.4 Con't

						ξ					
α	β	0.0	0.1	0.2	0.3	0.4	0.5	0.6	0.7	0.8	0.9	1.0
50	0.0	5.948	3.981	2.547	1.544	0.876	0.440	0.138	-0.126	-0.457	-0.951	-1.708
	0.1	4.956	3.399	2.252	1.444	0.895	0.534	0.285	0.072	-0.178	-0.541	-1.092
	0.2	3.970	2.819	1.959	1.340	0.912	0.625	0.428	0.270	0.101	-0.129	-0.470
	0.3	3.002	2.247	1.666	1.233	0.921	0.707	0.563	0.463	0.380	0.293	0.170
	0.4	2.076	1.694	1.374	1.114	0.912	0.768	0.680	0.644	0.664	0.732	0.852
	0.5	1.414	1.190	1.017	0.892	0.818	0.793	0.818	0.892	1.017	1.190	1.414
	0.6	0.852	0.732	0.664	0.644	0.680	0.768	0.912	1.114	1.374	1.694	2.076
	0.7	0.170	0.293	0.380	0.463	0.563	0.707	0.921	1.233	1.666	2.247	3.002
	0.8	-0.470	-0.129	0.101	0.270	0.428	0.625	0.912	1.340	1.959	2.819	3.970
	0.9	-1.092	-0.541	-0.178	0.072	0.285	0.534	0.895	1.444	2.252	3.399	4.956
	1.0	-1.708	-0.951	-0.457	-0.126	0.138	0.440	0.876	1.544	2.547	3.981	5.948
100	0.0	6.303	4.117	2.530	1.437	0.728	0.297	0.034	-0.169	-0.418	-0.823	-1.491
	0.1	5.115	3.463	2.252	1.399	0.831	0.468	0.233	0.047	-0.166	-0.485	-0.987
	0.2	3.924	2.809	1.968	1.358	0.928	0.634	0.428	0.262	0.090	-0.135	-0.460
	0.3	2.767	2.168	1.686	1.306	1.011	0.783	0.607	0.468	0.348	0.232	0.101
	0.4	1.697	1.571	1.411	1.237	1.060	0.895	0.755	0.653	0.605	0.623	0.723
	0.5	1.120	1.055	1.005	0.969	0.947	0.940	0.947	0.969	1.005	1.055	1.120
	0.6	0.723	0.623	0.605	0.653	0.755	0.895	1.060	1.237	1.411	1.571	1.697
	0.7	0.101	0.232	0.348	0.468	0.607	0.783	1.011	1.306	1.686	2.168	2.767
	0.8	-0.460	-0.135	0.090	0.262	0.428	0.634	0.928	1.358	1.968	2.809	3.924
	0.9	-0.987	-0.485	-0.166	0.047	0.233	0.468	0.831	1.399	2.252	3.463	5.115
	1.0	-1.491	-0.823	-0.418	-0.169	0.034	0.297	0.728	1.437	2.530	4.117	6.303
150	0	6.618	4.235	2.516	1.345	0.603	0.177	-0.051	-0.201	-0.384	-0.721	-1.326
	0.1	5.242	3.513	2.247	1.361	0.776	0.413	0.190	0.029	-0.153	-0.433	-0.894
	0.2	3.880	2.798	1.978	1.374	0.943	0.642	0.427	0.254	0.080	-0.140	-0.448
	0.3	2.563	2.104	1.708	1.377	1.087	0.847	0.643	0.463	0.316	0.178	0.049
	0.4	1.351	1.457	1.445	1.346	1.188	1.000	0.812	0.654	0.553	0.541	0.645
	0.5	0.875	0.942	0.995	1.032	1.055	1.062	1.055	1.032	0.995	0.942	0.875
	0.6	0.645	0.541	0.553	0.654	0.812	1.000	1.188	1.346	1.445	1.457	1.351

	0.7	0.049	0.178	0.316	0.463	0.643	0.847	1.087	1.377	1.708	2.104	2.563
	0.8	-0.448	-0.140	0.080	0.254	0.427	0.642	0.943	1.374	1.978	2.798	3.880
	0.9	-0.894	-0.433	-0.153	0.029	0.190	0.413	0.776	1.361	2.247	3.513	5.242
	1.0	-1.326	-0.721	-0.384	-0.201	-0.051	0.177	0.603	1.345	2.516	4.235	6.618
200	0	6.890	4.334	2.500	1.261	0.495	0.075	-0.123	-0.223	-0.352	-0.632	-1.190
	0.1	5.374	3.558	2.243	1.327	0.730	0.366	0.154	0.013	-0.141	-0.392	-0.838
	0.2	3.845	2.793	1.988	1.388	0.956	0.648	0.424	0.246	0.068	-0.145	-0.437
	0.3	2.386	2.050	1.736	1.434	1.157	0.901	0.669	0.464	0.280	0.132	0.010
	0.4	1.061	1.369	1.485	1.446	1.302	1.091	0.858	0.648	0.501	0.463	0.575
	0.5	0.666	0.846	0.987	1.087	1.147	1.167	1.147	1.087	0.987	0.846	0.066
	0.6	0.575	0.463	0.501	0.648	0.858	1.091	1.302	1.446	1.485	1.369	1.061
	0.7	0.010	0.132	0.280	0.464	0.669	0.901	1.157	1.434	1.736	2.050	2.386
	0.8	-0.437	-0.145	0.068	0.246	0.424	0.648	0.956	1.388	1.988	2.793	3.845
	0.9	-0.838	-0.392	-0.141	0.013	0.154	0.366	0.730	1.327	2.243	3.558	5.374
	1.0	-1.190	-0.632	-0.352	-0.223	-0.123	0.075	0.495	1.261	2.500	4.334	6.890
250	0.0	7.128	4.419	2.483	1.189	0.399	-0.012	-0.181	-0.241	-0.321	-0.557	-1.080
	0.1	5.456	3.596	2.239	1.298	0.689	0.326	0.125	0.002	-0.131	-0.356	-0.760
	0.2	3.804	2.783	1.995	1.403	0.968	0.654	0.424	0.237	0.061	-0.147	-0.420
	0.3	2.227	2.004	1.755	1.489	1.217	0.948	0.691	0.457	0.253	0.092	-0.019
	0.4	0.778	1.287	1.518	1.537	1.401	1.168	0.895	0.637	0.456	0.403	0.540
	0.5	0.486	0.764	0.979	1.134	1.226	1.257	1.226	1.134	0.979	0.764	0.486
	0.6	0.540	0.403	0.456	0.637	0.895	1.168	1.401	1.537	1.518	1.287	0.778
	0.7	-0.019	0.092	0.253	0.457	0.691	0.948	1.217	1.489	1.755	2.004	2.227
	0.8	-0.420	-0.147	0.061	0.237	0.424	0.654	0.968	1.403	1.995	2.783	3.804
	0.9	-0.760	-0.356	-0.131	0.002	0.125	0.326	0.689	1.298	2.239	3.596	5.456
	1.0	-1.080	-0.557	-0.321	-0.241	-0.181	-0.012	0.399	1.189	2.483	4.419	7.128

Table 5.4 Con't

α	β	0.0	0.1	0.2	0.3	0.4	0.5	0.6	0.7	0.8	0.9	1.0
	0.0	7.337	4.492	2.467	1.120	0.316	-0.088	-0.230	-0.252	-0.293	-0.490	-0.985
	0.1	5.546	3.630	2.235	1.270	0.653	0.291	0.099	-0.008	-0.121	-0.326	-0.710
	0.2	3.777	2.780	2.004	1.416	0.979	0.659	0.421	0.230	0.050	-0.152	-0.412
	0.3	2.074	1.959	1.774	1.539	1.271	0.989	0.709	0.449	0.228	0.061	-0.030
	0.4	0.546	1.217	1.552	1.619	1.490	1.235	0.924	0.625	0.410	0.351	0.514
300	0.5	0.330	0.692	0.973	1.174	1.295	1.335	1.295	1.174	0.973	0.692	0.330
	0.6	0.514	0.351	0.410	0.625	0.924	1.235	1.490	1.619	1.552	1.217	0.546
	0.7	-0.030	0.061	0.228	0.449	0.709	0.989	1.271	1.539	1.774	1.959	2.074
	0.8	-0.710	-0.326	-0.121	-0.008	0.099	0.291	0.653	1.270	2.235	3.630	5.546
	0.9	-0.710	-0.326	-0.121	-0.008	0.099	0.291	0.653	1.270	2.235	3.630	5.546
	1.0	-0.985	-0.490	-0.293	-0.252	-0.230	-0.088	0.316	1.120	2.467	4.492	7.337
	0	7.534	4.561	2.451	1.061	0.241	-0.155	-0.273	-0.263	-0.267	-0.435	-0.914
	0.1	5.621	3.658	2.230	1.246	0.620	0.260	0.078	-0.116	-0.112	-0.296	-0.661
	0.2	3.742	2.771	2.011	1.427	0.988	0.663	0.418	0.223	0.043	-0.151	-0.394
	0.3	1.946	1.926	1.798	1.587	1.322	1.024	0.720	0.436	0.198	0.030	-0.042
	0.4	0.327	1.162	1.589	1.697	1.573	1.294	0.945	0.609	0.363	0.298	0.497
350	0.5	0.194	0.629	0.968	1.210	1.355	1.403	1.355	1.210	0.968	0.629	0.194
	0.6	0.497	0.298	0.363	0.609	0.945	1.294	1.573	1.697	1.589	1.162	0.327
	0.7	-0.042	0.030	0.198	0.436	0.720	1.024	1.322	1.587	1.798	1.926	1.946
	0.8	-0.394	-0.151	0.043	0.223	0.418	0.663	0.988	1.427	2.011	2.771	3.742
	0.9	-0.661	-0.296	-0.112	-0.016	0.078	0.260	0.62	1.246	2.230	3.658	5.621
	1.0	-0.914	-0.435	-0.267	-0.263	-0.273	-0.155	0.241	1.061	2.451	4.561	7.534
	0	7.699	4.617	2.434	1.005	0.173	-0.214	-0.309	-0.267	-0.240	-0.383	-0.843
	0.1	5.698	3.687	2.226	1.225	0.591	0.233	0.059	-0.023	-0.104	-0.275	-0.630
	0.2	3.707	2.763	2.016	1.438	0.998	0.667	0.416	0.216	0.038	-0.151	-0.375
	0.3	1.816	1.886	1.814	1.628	1.366	1.056	0.732	0.430	0.178	0.010	-0.040
400	0.4	0.121	1.100	1.617	1.768	1.646	1.347	0.964	0.592	0.327	0.262	0.491
	0.5	0.072	0.573	0.963	1.241	1.408	1.464	1.408	1.241	0.963	0.573	0.072
	0.6	0.491	0.262	0.327	0.592	0.964	1.347	1.646	1.768	1.617	1.100	0.121
	0.7	-0.040	0.010	0.178	0.430	0.732	1.056	1.366	1.628	1.814	1.886	1.816

	0.8	3.707	2.763	2.016	1.438	0.998	0.667	0.416	0.216	0.038	-0.151	-0.375
	0.9	5.698	3.687	2.226	1.225	0.591	0.233	0.059	-0.023	-0.104	-0.275	-0.630
	1.0	7.699	4.617	2.434	1.005	0.173	-0.214	-0.309	-0.267	-0.240	-0.383	-0.843
450	0.0	-0.804	-0.344	-0.218	-0.270	-0.340	-0.266	0.112	0.954	2.420	4.674	7.868
	0.1	-0.598	-0.254	-0.096	-0.028	0.041	0.209	0.565	1.206	2.222	3.710	5.762
	0.2	-0.364	-0.155	0.026	0.206	0.413	0.670	1.084	1.450	2.026	2.763	3.684
	0.3	-0.033	-0.008	0.148	0.419	0.742	1.084	1.406	1.669	1.838	1.854	1.697
	0.4	0.495	0.227	0.289	0.573	0.978	1.393	1.714	1.835	1.649	1.049	-0.067
	0.5	-0.036	0.523	0.959	1.269	1.456	1.518	1.456	1.269	0.959	0.523	-0.036
	0.6	-0.067	1.049	1.649	1.835	1.714	1.393	0.978	0.573	0.289	0.227	0.495
	0.7	1.697	1.854	1.838	1.669	1.406	1.084	0.742	0.419	0.148	-0.008	-0.033
	0.8	3.684	2.763	2.026	1.450	1.007	0.670	0.413	0.206	0.026	-0.155	-0.364
	0.9	5.762	3.710	2.222	1.206	0.565	0.209	0.041	-0.028	-0.096	-0.254	-0.598
	1.0	7.868	4.674	2.420	0.954	0.112	-0.266	-0.340	-0.270	-0.218	-0.344	-0.804
500	0.0	-0.762	-0.303	-0.196	-0.272	-0.367	-0.313	0.057	0.906	2.406	4.719	8.014
	0.1	-0.570	-0.236	-0.089	-0.032	0.028	0.187	0.542	1.186	2.219	3.732	5.822
	0.2	-0.350	-0.155	0.020	0.199	0.409	0.673	1.015	1.461	2.032	2.757	3.658
	0.3	-0.027	-0.026	0.134	0.407	0.748	1.109	1.444	1.707	1.848	1.826	1.591
	0.4	0.513	0.200	0.251	0.556	0.990	1.435	1.776	1.896	1.676	1.000	-0.253
	0.5	-0.132	0.479	0.955	1.294	1.498	1.566	1.498	1.294	0.955	0.479	-0.132
	0.6	-0.253	1.000	1.676	1.896	1.776	1.435	0.990	0.556	0.251	0.200	0.513
	0.7	1.591	1.826	1.848	1.707	1.444	1.109	0.748	0.407	0.134	-0.026	-0.027
	0.8	3.658	2.757	2.032	1.461	1.015	0.673	0.409	0.199	0.020	-0.155	-0.350
	0.9	5.822	3.732	2.219	1.186	0.542	0.187	0.028	-0.032	-0.089	-0.236	-0.570
	1.0	8.014	4.719	2.406	0.906	0.057	-0.313	-0.367	-0.272	-0.196	-0.303	-0.762

Table 5.5 Vertical load P. Coefficients $\overline{Q}$ for obtaining shear forces $Q = \overline{Q}P$

α	β	ξ										
		0.0	0.1	0.2	0.3	0.4	0.5	0.6	0.7	0.8	0.9	1.0
0	0.0	−1	−0.538	−0.222	−0.013	0.13	0.203	0.247	0.257	0.230	0.152	0
	0.1	0	$\frac{0.400}{-0.600}$	−0.323	−0.138	−0.016	0.062	0.110	0.132	0.129	0.090	0
	0.2	0	0.339	$\frac{0.577}{-0.423}$	−0.62	−0.152	−0.078	−0.026	0.008	0.029	0.029	0
	0.3	0	0.278	0.476	$\frac{0.614}{-0.386}$	−0.290	−0.219	−0.164	−0.116	−0.072	−0.032	0
	0.4	0	0.216	0.374	0.489	$\frac{0.573}{-0.427}$	−0.360	−0.301	−0.241	−0.174	−0.094	0
	0.5	0	0.155	0.274	0.365	0.437	$\frac{0.500}{-0.500}$	−0.437	−0.365	−0.274	−0.155	0
	0.6	0	0.094	0.174	0.241	0.301	0.360	$\frac{0.427}{-0.573}$	−0.489	−0.374	−0.216	0
	0.7	0	0.032	0.072	0.164	0.219	0.290	0.290	$\frac{0.386}{-0.614}$	−0.476	−0.278	0
	0.8	0	−0.029	−0.029	−0.008	0.026	0.078	0.152	0.262	$\frac{0.423}{-0.577}$	−0.339	0
	0.9	0	−0.090	−0.129	−0.132	−0.110	−0.062	0.016	0.138	0.323	$\frac{0.400}{-0.600}$	0
	1.0	0	−0.152	−0.230	−0.257	−0.247	−0.203	−0.121	0.013	0.222	0.538	1
25	0.0	−1	−0.522	−0.204	0.001	0.128	0.200	0.236	0.241	0.214	0.140	0
	0.1	0	$\frac{0.408}{-0.592}$	−0.314	−0.130	−0.012	0.062	0.106	0.126	0.122	0.086	0
	0.2	0	0.338	$\frac{0.575}{-0.425}$	−0.262	−0.152	−0.077	−0.026	0.010	0.029	0.029	0
	0.3	0	0.270	0.466	$\frac{0.605}{-0.395}$	−0.292	−0.216	−0.156	−0.107	−0.064	−0.028	0
	0.4	0	0.202	0.358	0.476	$\frac{0.569}{-0.431}$	−0.355	−0.289	−0.226	−0.160	−0.086	0
	0.5	0	0.142	0.256	0.349	0.428	$\frac{0.500}{-0.500}$	−0.428	−0.349	−0.256	−0.142	0
	0.6	0	0.086	0.160	0.226	0.289	0.355	$\frac{0.431}{-0.569}$	−0.476	−0.358	−0.202	0
	0.7	0	0.028	0.064	0.107	0.156	0.216	0.292	$\frac{0.395}{-0.605}$	−0.466	−0.270	0
	0.8	0	−0.029	−0.029	−0.010	0.026	0.077	0.152	0.262	$\frac{0.425}{-0.575}$	−0.338	0
	0.9	0	−0.086	−0.122	−0.126	−0.106	−0.062	0.012	0.013	0.314	$\frac{0.592}{-0.408}$	0
	1.0	0	0.140	−0.214	−0.241	−0.236	−0.200	−0.128	−0.001	0.204	0.522	1

Table 5.5 Con't

α	β	ξ 0.0	0.1	0.2	0.3	0.4	0.5	0.6	0.7	0.8	0.9	1.0
	0.0	−1	−0.507	−0.186	0.015	0.134	0.198	0.226	0.227	0.198	0.131	0
	0.1	0	$\frac{0.414}{-0.586}$	−0.306	−0.125	−0.009	0.061	0.101	0.119	0.114	0.080	0
	0.2	0	0.336	$\frac{0.574}{-0.426}$	−0.263	−0.152	−0.076	0.024	0.011	0.030	0.029	0
	0.3	0	0.260	0.454	$\frac{0.599}{-0.401}$	−0.295	−0.214	−0.151	−0.101	−0.058	−0.024	0
	0.4	0	0.188	0.342	0.465	$\frac{0.566}{-0.434}$	−0.351	−0.280	−0.213	−0.148	−0.080	0
50	0.5	0	0.130	0.240	0.334	0.419	$\frac{0.500}{-0.500}$	−0.419	−0.334	−0.240	−0.13	0
	0.6	0	0.080	0.148	0.213	0.280	0.351	$\frac{0.434}{-0.566}$	−0.465	−0.342	−0.188	0
	0.7	0	0.024	0.058	0.101	0.151	0.214	0.295	$\frac{0.401}{-0.599}$	−0.454	−0.260	0
	0.8	0	−0.029	−0.030	−0.011	0.024	0.076	0.152	0.263	$\frac{0.426}{-0.574}$	−0.336	0
	0.9	0	−0.080	−0.114	−0.119	−0.101	−0.061	0.009	0.125	0.306	$\frac{0.586}{-0.414}$	0
	1.0	0	−0.131	−0.198	−0.227	−0.226	−0.198	−0.134	−0.015	0.186	0.507	1
	0.0	−1	−0.485	−0.156	0.038	0.144	0.193	0.208	0.202	0.174	0.113	0
	0.1	0	$\frac{0.425}{-0.575}$	−0.292	−0.113	−0.004	0.060	0.094	0.109	0.102	0.071	0
	0.2	0	0.334	$\frac{0.570}{-0.430}$	−0.265	−0.152	−0.075	0.022	0.013	0.030	0.029	0
	0.3	0	0.245	0.438	$\frac{0.587}{-0.473}$	−0.298	−0.209	−0.140	−0.085	0.046	−0.017	0
	0.4	0	0.163	0.313	0.445	$\frac{0.559}{-0.441}$	−0.343	−0.261	−0.191	−0.129	−0.067	0
100	0.5	0	0.109	0.212	0.310	0.406	$\frac{0.500}{-0.500}$	−0.406	−0.310	−0.212	−0.109	0
	0.6	0	0.067	0.129	0.191	0.261	0.343	$\frac{0.441}{-0.559}$	−0.445	−0.313	−0.163	0
	0.7	0	0.017	0.046	0.085	0.140	0.209	0.298	$\frac{0.413}{-0.587}$	−0.438	−0.245	0
	0.8	0	−0.029	−0.030	−0.013	0.022	0.075	0.152	0.265	$\frac{0.430}{-0.570}$	−0.334	0
	0.9	0	−0.071	−0.102	−0.109	−0.094	−0.060	0.004	0.113	0.292	$\frac{0.575}{-0.425}$	0
	1.0	0	−0.113	−0.174	−0.202	−0.208	−0.193	−0.144	−0.038	0.156	0.485	1

Table 5.5 Con't

α	β	ξ										
		0.0	0.1	0.2	0.3	0.4	0.5	0.6	0.7	0.8	0.9	1.0
150	0.0	−1	−0.464	−0.131	0.058	0.152	0.189	0.194	0.180	0.153	0.100	0
	0.1	0	$\frac{0.434}{-0.566}$	−0.282	−0.105	0.000	0.058	0.088	0.099	0.092	0.064	0
	0.2	0	0.331	$\frac{0.567}{-0.433}$	−0.266	−0.151	−0.073	−0.019	0.014	0.030	0.029	0
	0.3	0	0.233	0.423	$\frac{0.578}{-0.422}$	−0.300	−0.204	−0.130	−0.074	−0.035	−0.011	0
	0.4	0	0.142	0.288	0.428	$\frac{0.555}{-0.445}$	−0.335	−0.245	−0.172	−0.112	−0.058	0
	0.5	0	0.091	0.188	0.289	0.394	$\frac{0.500}{-0.500}$	−0.394	−0.289	−0.188	−0.091	0
	0.6	0	0.058	0.112	0.172	0.245	0.335	$\frac{0.445}{-0.555}$	−0.428	−0.288	−0.142	0
	0.7	0	0.011	0.035	0.074	0.130	0.204	0.300	$\frac{0.422}{-0.578}$	−0.423	−0.233	0
	0.8	0	−0.029	−0.030	−0.014	0.019	0.073	0.151	0.266	$\frac{0.433}{-0.567}$	−0.331	0
	0.9	0	−0.064	−0.092	−0.099	−0.088	−0.058	0.000	0.105	0.282	$\frac{0.566}{-0.434}$	0
	1.0	0	−0.100	−0.153	−0.180	−0.194	−0.189	−0.152	−0.058	0.131	0.464	1
200	0.0	−1	−0.445	−0.109	0.074	0.159	0.185	0.181	0.164	0.135	0.089	0
	0.1	0	$\frac{0.442}{-0.558}$	−0.272	−0.098	0.004	0.057	0.082	0.090	0.084	0.058	0
	0.2	0	0.330	$\frac{0.567}{-0.433}$	−0.266	−0.150	−0.071	−0.018	0.015	0.031	0.028	0
	0.3	0	0.223	0.410	$\frac{0.569}{-0.431}$	−0.301	−0.199	−0.121	−0.065	−0.028	−0.007	0
	0.4	0	0.123	0.268	0.415	$\frac{0.553}{-0.447}$	−0.327	−0.229	−0.155	−0.098	−0.051	0
	0.5	0	0.076	0.168	0.272	0.384	$\frac{0.500}{-0.500}$	−0.384	−0.272	−0.168	−0.076	0
	0.6	0	0.051	0.098	0.155	0.229	0.327	$\frac{0.447}{-0.553}$	−0.415	−0.268	−0.123	0
	0.7	0	0.007	0.028	0.065	0.121	0.199	0.301	$\frac{0.431}{-0.569}$	−0.410	−0.223	0
	0.8	0	−0.028	−0.031	−0.015	0.018	0.071	0.150	0.266	$\frac{0.433}{-0.567}$	−0.330	0
	0.9	0	−0.058	−0.084	−0.090	−0.082	−0.057	−0.004	0.098	0.272	$\frac{0.588}{-0.442}$	0
	1.0	0	−0.089	−0.135	−0.164	−0.181	−0.185	−0.159	−0.074	0.109	0.445	1

Table 5.5 Con't

α	β	ξ										
		0.0	0.1	0.2	0.3	0.4	0.5	0.6	0.7	0.8	0.9	1.0
250	0.0	−1	−0.429	−0.090	0.088	0.164	0.181	0.170	0.148	0.120	0.079	0
	0.1	0	$\frac{0.449}{-0.551}$	−0.264	−0.091	0.007	0.056	0.077	0.083	0.078	0.053	0
	0.2	0	0.328	$\frac{0.565}{-0.435}$	−0.267	−0.149	−0.069	−0.015	0.017	0.031	0.028	0
	0.3	0	0.211	0.399	$\frac{0.561}{-0.439}$	−0.303	−0.195	−0.113	−0.055	−0.021	−0.003	0
	0.4	0	0.107	0.249	0.402	$\frac{0.551}{-0.449}$	−0.320	0.217	−0.140	−0.087	−0.045	0
	0.5	0	0.063	0.151	0.257	0.375	$\frac{0.500}{-0.500}$	−0.375	−0.257	−0.151	−0.063	0
	0.6	0	0.045	0.087	0.140	0.217	0.320	$\frac{0.449}{-0.551}$	−0.402	−0.249	−0.107	0
	0.7	0	0.003	0.021	0.055	0.113	0.195	0.303	$\frac{0.439}{-0.561}$	−0.399	−0.211	0
	0.8	0	−0.028	−0.031	−0.017	0.015	0.069	0.149	0.267	$\frac{0.435}{-0.565}$	−0.328	0
	0.9	0	−0.053	−0.078	−0.083	−0.077	−0.056	−0.007	0.091	0.264	$\frac{0.551}{-0.449}$	0
	1.0	0	−0.079	−0.120	−0.148	−0.170	−0.181	−0.164	−0.088	0.090	0.429	1
300	0.0	−1	−0.415	−0.075	0.101	0.169	0.177	0.159	0.135	0.109	0.071	0
	0.1	0	$\frac{0.455}{-0.545}$	−0.257	−0.085	0.008	0.054	0.072	0.077	0.071	0.049	0
	0.2	0	0.326	$\frac{0.563}{-0.437}$	−0.268	−0.149	−0.068	−0.015	0.018	0.031	0.028	0
	0.3	0	0.201	0.389	$\frac{0.555}{-0.445}$	−0.304	−0.191	−0.106	−0.049	−0.015	−0.001	0
	0.4	0	0.091	0.232	0.394	$\frac{0.548}{-0.452}$	−0.314	−0.206	−0.128	−0.078	−0.041	0
	0.5	0	0.052	0.136	0.244	0.368	$\frac{0.500}{-0.500}$	−0.368	−0.244	−0.136	−0.052	0
	0.6	0	0.041	0.078	0.128	0.206	0.314	$\frac{0.452}{-0.548}$	−0.394	−0.232	−0.091	0
	0.7	0	0.001	0.015	0.049	0.106	0.191	0.304	$\frac{0.445}{-0.555}$	−0.389	−0.201	0
	0.8	0	−0.028	−0.031	−0.018	0.015	0.068	0.149	0.268	$\frac{0.437}{-0.563}$	−0.326	0
	0.9	0	−0.049	−0.071	−0.077	−0.072	−0.054	−0.008	0.085	0.257	$\frac{0.545}{-0.455}$	0
	1.0	0	−0.071	−0.109	−0.135	−0.159	−0.177	−0.169	−0.101	0.075	0.415	1

Table 5.5 Con't

α	β	ξ 0.0	0.1	0.2	0.3	0.4	0.5	0.6	0.7	0.8	0.9	1.0
	0.0	−1	−0.400	−0.064	0.111	0.173	0.174	0.151	0.123	0.094	0.068	0
	0.1	0	$\frac{0.460}{-0.540}$	−0.250	−0.080	0.011	0.053	0.069	0.072	0.066	0.046	0
	0.2	0	0.325	$\frac{0.562}{-0.438}$	−0.268	−0.148	−0.067	−0.014	0.018	0.031	0.027	0
	0.3	0	0.193	0.382	$\frac{0.551}{-0.449}$	−0.303	−0.186	−0.099	−0.041	−0.010	0.001	0
	0.4	0	0.076	0.218	0.384	$\frac{0.549}{-0.451}$	−0.307	−0.195	−0.116	−0.070	−0.038	0
350	0.5	0	0.042	0.123	0.232	0.361	$\frac{0.500}{-0.500}$	−0.361	−0.232	−0.123	−0.042	0
	0.6	0	0.038	0.070	0.116	0.195	0.307	$\frac{0.451}{-0.549}$	−0.384	−0.218	−0.076	0
	0.7	0	−0.001	0.010	0.041	0.099	0.186	0.303	$\frac{0.449}{-0.551}$	−0.382	−0.193	0
	0.8	0	−0.027	−0.031	−0.018	0.014	0.067	0.148	0.268	$\frac{0.438}{-0.562}$	−0.325	0
	0.9	0	−0.046	−0.066	−0.072	−0.069	−0.053	−0.011	0.080	0.250	$\frac{0.540}{-0.460}$	0
	1.0	0	−0.068	−0.094	−0.123	−0.151	−0.174	−0.173	−0.111	0.064	0.400	1
	0.0	−1	−0.391	−0.046	0.121	0.176	0.170	0.142	0.113	0.088	0.059	0
	0.1	0	$\frac{0.464}{-0.536}$	−0.245	−0.076	0.013	0.052	0.065	0.068	0.061	0.042	0
	0.2	0	0.323	$\frac{0.559}{-0.441}$	−0.269	−0.148	−0.066	−0.012	0.019	0.031	0.027	0
	0.3	0	0.187	0.372	$\frac{0.546}{-0.454}$	−0.305	−0.183	−0.093	−0.036	−0.006	0.003	0
	0.4	0	0.065	0.204	0.378	$\frac{0.549}{-0.451}$	−0.301	−0.185	−0.106	−0.062	−0.035	0
400	0.5	0	0.033	0.111	0.222	0.355	$\frac{0.500}{-0.500}$	−0.355	−0.222	−0.111	0.033	0
	0.6	0	0.035	0.062	0.106	0.185	0.301	$\frac{0.451}{-0.549}$	−0.378	−0.204	−0.065	0
	0.7	0	−0.003	0.006	0.036	0.093	0.183	0.305	$\frac{0.454}{-0.546}$	−0.372	−0.187	0
	0.8	0	−0.027	−0.031	−0.019	0.012	0.066	0.148	0.269	$\frac{0.441}{-0.559}$	−0.323	0
	0.9	0	−0.042	−0.061	−0.068	−0.065	−0.052	−0.013	0.076	0.245	$\frac{0.536}{-0.464}$	0
	1.0	0	−0.059	−0.088	−0.113	−0.142	−0.170	−0.176	−0.121	0.046	0.391	1

Table 5.5 Con't

α	β	ξ 0.0	0.1	0.2	0.3	0.4	0.5	0.6	0.7	0.8	0.9	1.0
	0.0	−1	−0.382	−0.035	0.127	0.178	0.166	0.134	0.103	0.079	0.054	0
	0.1	0	$\frac{0.468}{-0.532}$	−0.240	−0.071	0.014	0.051	0.062	0.063	0.056	0.040	0
	0.2	0	0.322	$\frac{0.558}{-0.442}$	−0.270	−0.147	−0.064	−0.011	0.020	0.032	0.026	0
	0.3	0	0.178	0.364	$\frac{0.541}{-0.459}$	−0.305	−0.180	−0.089	−0.031	−0.004	0.004	0
	0.4	0	0.054	0.192	0.370	$\frac{0.549}{-0.451}$	−0.294	−0.175	−0.098	−0.056	−0.034	0
450	0.5	0	0.026	0.101	0.213	0.350	$\frac{0.500}{-0.500}$	−0.350	−0.213	−0.101	−0.026	0
	0.6	0	0.034	0.056	0.098	0.175	0.294	$\frac{0.451}{-0.549}$	−0.370	−0.192	−0.054	0
	0.7	0	−0.004	0.004	0.031	0.089	0.180	0.305	$\frac{0.459}{-0.541}$	−0.364	−0.178	0
	0.8	0	−0.026	−0.032	−0.020	0.0011	0.064	0.147	0.270	$\frac{0.442}{-0.558}$	−0.322	0
	0.9	0	−0.040	−0.056	−0.063	−0.062	−0.051	−0.014	0.071	0.240	$\frac{0.532}{-0.468}$	0
	1.0	0	−0.054	−0.079	−0.103	−0.134	−0.166	−0.178	−0.127	0.035	0.382	1
	0.0	−1	−0.373	−0.024	0.136	0.179	0.163	0.127	0.094	0.072	0.049	0
	0.1	0	$\frac{0.472}{-0.528}$	−0.235	−0.067	0.015	0.050	0.059	0.059	0.053	0.038	0
	0.2	0	0.321	$\frac{0.558}{-0.442}$	−0.270	−0.147	−0.063	−0.009	0.020	0.032	0.025	0
	0.3	0	0.173	0.359	$\frac{0.537}{-0.463}$	−0.305	−0.176	−0.083	−0.025	−0.001	0.005	0
	0.4	0	0.041	0.181	0.363	$\frac{0.547}{-0.453}$	−0.290	−0.169	−0.091	0.053	−0.033	0
500	0.5	0	0.019	0.091	0.205	0.346	$\frac{0.500}{-0.500}$	−0.346	−0.205	−0.091	−0.019	0
	0.6	0	0.033	0.053	0.091	0.169	0.290	$\frac{0.453}{-0.547}$	−0.363	−0.181	−0.041	0
	0.7	0	−0.005	0.001	0.025	00.083	0.176	0.305	$\frac{0.463}{-0.537}$	−0.359	−0.173	0
	0.8	0	−0.025	−0.032	−0.020	0.009	0.063	0.147	0.270	$\frac{0.442}{-0.558}$	−0.321	0
	0.9	0	−0.038	−0.053	−0.059	−0.059	−0.050	−0.015	0.067	0.235	$\frac{0.528}{-0.472}$	0
	1.0	0	−0.049	−0.072	−0.094	−0.127	−0.163	−0.179	−0.163	0.024	0.373	1

Table 5.6 Vertical load P. Coefficients $\overline{M}$ for obtaining moments $M = \overline{M}PL$

α	β	ξ 0.0	0.1	0.2	0.3	0.4	0.5	0.6	0.7	0.8	0.9	1.0
0	0.0	0	-0.076	-0.112	-0.123	-0.117	-0.100	-0.077	-0.52	-0.027	-0.008	0
	0.1	0	0.022	-0.024	-0.046	-0.053	-0.050	-0.042	-0.030	-0.017	-0.005	0
	0.2	0	0.018	0.065	0.031	0.011	-0.001	-0.006	-0.007	-0.005	-0.001	0
	0.3	0	0.014	0.053	0.106	0.074	0.049	0.030	0.017	0.006	0.002	0
	0.4	0	0.012	0.042	0.085	0.138	0.099	0.066	0.039	0.018	0.008	0
	0.5	0	0.008	0.030	0.061	0.102	0.149	0.102	0.061	0.030	0.008	0
	0.6	0	0.008	0.018	0.039	0.066	0.099	0.138	0.085	0.042	0.012	0
	0.7	0	0.002	0.006	0.017	0.030	0.049	0.074	0.106	0.065	0.018	0
	0.8	0	-0.001	-0.005	-0.007	-0.006	-0.001	0.011	0.031	0.065	0.018	0
	0.9	0	-0.005	-0.017	-0.030	-0.042	-0.050	-0.053	-0.046	-0.024	0.002	0
	1.0	0	-0.008	-0.027	-0.052	-0.077	-0.100	-0.117	-0.123	-0.112	-0.076	0
25	0.0	0	-0.076	-0.111	-0.120	-0.113	-0.096	-0.074	-0.050	-0.027	-0.007	0
	0.1	0	0.022	-0.023	-0.045	-0.051	-0.049	-0.040	-0.028	-0.016	-0.005	0
	0.2	0	0.018	0.064	0.031	0.010	-0.001	-0.006	-0.007	-0.005	-0.001	0
	0.3	0	0.014	0.052	0.105	0.072	0.046	0.028	0.015	0.005	0.002	0
	0.4	0	0.011	0.039	0.081	0.134	0.095	0.063	0.037	0.018	0.005	0
	0.5	0	0.008	0.028	0.057	0.097	0.143	0.097	0.057	0.028	0.008	0
	0.6	0	0.005	0.018	0.037	0.063	0.095	0.134	0.081	0.039	0.011	0
	0.7	0	0.002	0.006	0.015	0.028	0.046	0.072	0.105	0.052	0.014	0
	0.8	0	-0.001	-0.005	-0.007	-0.006	-0.001	0.010	0.031	0.064	0.018	0
	0.9	0	-0.005	-0.016	-0.028	-0.040	-0.049	-0.051	-0.045	-0.023	0.022	0
	1.0	0	-0.007	-0.027	-0.050	-0.074	-0.096	-0.113	-0.120	-0.111	-0.076	0
50	0.0	0	-0.074	-0.107	-0.115	-0.107	-0.090	-0.069	-0.046	-0.024	-0.007	0
	0.1	0	0.022	-0.022	-0.043	-0.049	-0.046	-0.038	-0.027	-0.015	-0.005	0
	0.2	0	0.018	0.064	0.030	0.010	-0.001	-0.006	-0.007	-0.005	-0.001	0
	0.3	0	0.013	0.050	0.104	0.068	0.043	0.025	0.012	0.005	0.001	0
	0.4	0	0.011	0.037	0.078	0.130	0.090	0.059	0.034	0.016	0.005	0
	0.5	0	0.007	0.025	0.054	0.092	0.138	0.092	0.054	0.025	0.007	0
	0.6	0	0.005	0.016	0.034	0.059	0.090	0.130	0.078	0.037	0.011	0

Group												
	0.7	0	0.001	0.005	0.012	0.025	0.043	0.068	0.104	0.050	0.013	0
	0.8	0	-0.001	-0.005	-0.007	-0.006	-0.001	0.010	0.030	0.064	0.018	0
	0.9	0	-0.005	-0.015	-0.027	-0.038	-0.016	-0.049	-0.043	-0.022	0.022	0
	1.0	0	-0.007	-0.024	-0.046	-0.069	-0.090	-0.107	-0.116	-0.107	-0.074	0
100	0.0	0	-0.073	-0.103	-0.108	-0.099	-0.081	-0.061	-0.040	-0.021	-0.007	0
	0.1	0	0.023	-0.020	-0.030	-0.045	-0.042	-0.034	-0.023	-0.013	-0.004	0
	0.2	0	0.018	0.064	0.030	0.009	-0.002	-0.007	-0.007	-0.005	-0.001	0
	0.3	0	0.012	0.047	0.098	0.063	0.038	0.021	0.010	0.003	0.001	0
	0.4	0	0.009	0.032	0.070	0.121	0.082	0.052	0.029	0.014	0.004	0
	0.5	0	0.005	0.021	0.047	0.083	0.128	0.083	0.047	0.021	0.005	0
	0.6	0	0.004	0.014	0.029	0.052	0.082	0.121	0.070	0.032	0.009	0
	0.7	0	0.001	0.003	0.010	0.021	0.038	0.063	0.098	0.047	0.012	0
	0.8	0	-0.001	-0.005	-0.007	-0.007	-0.002	0.009	0.030	0.064	0.018	0
	0.9	0	-0.004	-0.013	-0.023	-0.034	-0.042	-0.045	-0.039	-0.020	0.023	0
	1.0	0	-0.007	-0.021	-0.040	-0.061	-0.081	-0.099	-0.108	-0.103	-0.073	0
150	0.0	0	-0.071	-0.098	-0.101	-0.090	-0.073	-0.053	-0.034	-0.018	-0.005	0
	0.1	0	0.023	-0.018	-0.037	-0.042	-0.038	-0.031	-0.021	-0.012	-0.003	0
	0.2	0	0.018	0.063	0.029	0.008	-0.003	-0.007	-0.007	-0.005	-0.001	0
	0.3	0	0.012	0.045	0.095	0.059	0.034	0.018	0.008	0.003	0.001	0
	0.4	0	0.007	0.029	0.065	0.114	0.075	0.047	0.026	0.012	0.003	0
	0.5	0	0.005	0.019	0.042	0.077	0.121	0.077	0.042	0.019	0.005	0
	0.6	0	0.003	0.012	0.026	0.047	0.075	0.114	0.065	0.029	0.007	0
	0.7	0	0.001	0.003	0.008	0.018	0.034	0.059	0.095	0.045	0.012	0
	0.8	0	-0.001	-0.005	-0.007	-0.007	-0.003	0.008	0.029	0.063	0.018	0
	0.9	0	-0.001	-0.012	-0.021	-0.031	-0.038	-0.042	-0.037	-0.018	0.023	0
	1.0	0	-0.005	-0.018	-0.034	-0.053	-0.073	-0.090	-0.101	-0.098	-0.071	0

Table 5.6 Con't

α	β						ξ					
		0.0	0.1	0.2	0.3	0.4	0.5	0.6	0.7	0.8	0.9	1.0
200	0.0	0	-0.070	-0.096	-0.097	-0.084	-0.067	-0.048	-0.031	-0.016	-0.005	0
	0.1	0	0.024	-0.017	-0.034	-0.039	-0.035	-0.028	-0.019	-0.011	-0.003	0
	0.2	0	0.018	0.063	0.029	0.008	-0.003	-0.007	-0.007	-0.005	-0.001	0
	0.3	0	0.012	0.044	0.093	0.057	0.032	0.016	0.007	0.003	0.001	0
	0.4	0	0.007	0.026	0.060	0.109	0.070	0.043	0.024	0.011	0.003	0
	0.5	0	0.004	0.015	0.037	0.070	0.114	0.070	0.037	0.015	0.004	0
	0.6	0	0.003	0.011	0.024	0.043	0.070	0.109	0.060	0.026	0.007	0
	0.7	0	0.001	0.003	0.007	0.016	0.032	0.057	0.093	0.044	0.012	0
	0.8	0	-0.001	-0.005	-0.007	-0.007	-0.003	0.008	0.029	0.063	0.018	0
	0.9	0	-0.003	-0.011	-0.019	-0.028	-0.035	-0.039	-0.034	-0.017	0.024	0
	1.0	0	-0.005	-0.016	-0.031	-0.048	-0.067	-0.084	-0.097	-0.096	-0.070	0
250	0.0	0	-0.069	-0.094	-0.093	-0.080	-0.062	-0.044	-0.028	-0.015	-0.004	0
	0.1	0	0.024	-0.016	-0.032	-0.036	-0.033	-0.026	-0.018	-0.010	-0.003	0
	0.2	0	0.017	0.063	0.028	0.007	-0.003	-0.007	-0.007	-0.005	-0.001	0
	0.3	0	0.011	0.041	0.090	0.053	0.028	0.013	0.005	0.001	0.000	0
	0.4	0	0.005	0.023	0.055	0.103	0.065	0.038	0.021	0.010	0.003	0
	0.5	0	0.003	0.013	0.034	0.065	0.109	0.065	0.034	0.013	0.003	0
	0.6	0	0.003	0.010	0.021	0.038	0.065	0.103	0.055	0.023	0.005	0
	0.7	0	0.000	0.001	0.005	0.013	0.028	0.053	0.090	0.041	0.011	0
	0.8	0	-0.001	-0.005	-0.007	-0.007	-0.003	0.007	0.028	0.063	0.017	0
	0.9	0	-0.003	-0.010	-0.018	-0.026	-0.033	-0.036	-0.032	-0.016	0.024	0
	1.0	0	-0.004	-0.015	-0.028	-0.044	-0.062	-0.080	-0.093	-0.094	-0.069	0
300	0.0	0	-0.069	-0.092	-0.089	-0.075	-0.057	-0.041	-0.026	-0.014	-0.003	0
	0.1	0	0.025	-0.014	-0.030	-0.034	-0.030	-0.024	-0.016	-0.009	-0.002	0
	0.2	0	0.017	0.062	0.028	0.007	-0.003	-0.007	-0.007	-0.005	-0.001	0
	0.3	0	0.010	0.039	0.087	0.050	0.025	0.011	0.003	0.000	0.000	0
	0.4	0	0.005	0.020	0.052	0.099	0.061	0.035	0.019	0.009	0.003	0
	0.5	0	0.003	0.012	0.030	0.061	0.104	0.061	0.030	0.012	0.003	0

Group	x											
	0.6	0	0.003	0.009	0.019	0.035	0.061	0.099	0.052	0.020	0.005	0
	0.7	0	0.000	0.000	0.003	0.011	0.025	0.050	0.087	0.039	0.010	0
	0.8	0	-0.001	-0.005	-0.007	-0.007	-0.003	0.007	0.028	0.062	0.017	0
	0.9	0	-0.002	-0.009	-0.016	-0.024	-0.030	-0.034	-0.030	-0.014	0.025	0
	1.0	0	-0.003	-0.014	-0.026	-0.041	-0.057	-0.075	-0.089	-0.092	-0.069	0
350	0.0	0	-0.068	-0.089	-0.085	-0.070	-0.053	-0.037	-0.023	-0.012	-0.003	0
	0.1	0	0.025	-0.013	-0.029	-0.032	-0.028	-0.022	-0.015	-0.008	-0.002	0
	0.2	0	0.017	0.062	0.027	0.006	-0.004	-0.008	-0.007	-0.005	-0.001	0
	0.3	0	0.017	0.062	0.027	0.006	-0.004	-0.008	-0.007	-0.005	-0.001	0
	0.4	0	0.004	0.018	0.048	0.094	0.057	0.032	0.017	0.008	0.002	0
	0.5	0	0.002	0.010	0.027	0.057	0.100	0.057	0.027	0.010	0.002	0
	0.6	0	0.002	0.008	0.017	0.032	0.057	0.094	0.048	0.018	0.004	0
	0.7	0	0.000	0.000	0.002	0.009	0.023	0.047	0.085	0.038	0.009	0
	0.8	0	-0.001	-0.005	-0.007	-0.008	-0.004	0.006	0.027	0.062	0.017	0
	0.9	0	-0.002	-0.008	-0.015	-0.022	-0.028	-0.032	-0.029	-0.013	0.025	0
	1.0	0	-0.003	-0.012	-0.023	-0.037	-0.053	-0.070	-0.085	-0.089	-0.068	0
400	0.0	0	-0.067	-0.087	-0.082	-0.067	-0.049	-0.034	-0.021	-0.011	-0.003	0
	0.1	0	0.025	-0.013	-0.028	-0.030	-0.027	-0.021	-0.014	-0.008	-0.002	0
	0.2	0	0.017	0.062	0.027	0.006	-0.004	-0.008	-0.007	-0.005	-0.001	0
	0.3	0	0.009	0.037	0.083	0.046	0.021	0.008	0.002	0.000	0.000	0
	0.4	0	0.002	0.016	0.045	0.091	0.054	0.030	0.016	0.008	0.002	0
	0.5	0	0.001	0.008	0.025	0.053	0.096	0.053	0.025	0.008	0.001	0
	0.6	0	0.002	0.008	0.016	0.030	0.054	0.091	0.054	0.016	0.002	0
	0.7	0	0.000	0.000	0.002	0.008	0.021	0.046	0.083	0.037	0.009	0
	0.8	0	-0.001	-0.005	-0.007	-0.008	-0.004	0.006	0.027	0.062	0.017	0
	0.9	0	-0.002	-0.008	-0.014	-0.021	-0.027	-0.030	-0.028	-0.013	0.025	0
	1.0	0	0.003	-0.011	-0.021	-0.034	-0.049	-0.067	-0.082	-0.087	-0.067	0

Table 5.6 Con't

α	β	0.0	0.1	0.2	0.3	0.4	ξ 0.5	0.6	0.7	0.8	0.9	1.0
450	0.0	0	-0.066	-0.085	-0.079	-0.063	-0.046	-0.031	-0.019	-0.010	-0.003	0
	0.1	0	0.025	-0.012	-0.027	-0.029	0.025	-0.020	-0.013	-0.007	-0.002	0
	0.2	0	0.017	0.061	0.026	0.006	-0.004	-0.008	-0.007	-0.005	-0.001	0
	0.3	0	0.009	0.036	0.081	0.043	0.020	0.006	0.001	0.000	0.000	0
	0.4	0	0.002	0.014	0.042	0.088	0.051	0.029	0.016	0.007	0.002	0
	0.5	0	0.001	0.006	0.022	0.050	0.092	0.050	0.022	0.006	0.001	0
	0.6	0	0.002	0.007	0.016	0.029	0.051	0.088	0.042	0.014	0.002	0
	0.7	0	0.000	0.000	0.001	0.006	0.020	0.043	0.081	0.036	0.009	0
	0.8	0	-0.001	-0.005	-0.007	-0.008	-0.004	0.006	0.026	0.061	0.017	0
	0.9	0	-0.002	-0.007	-0.113	-0.020	-0.025	-0.029	-0.027	0.012	0.025	0
	1.0	0	0.003	-0.010	-0.019	-0.031	-0.046	-0.063	-0.079	-0.085	-0.066	0
500	0.0	0	-0.066	-0.084	-0.077	-0.061	-0.044	-0.029	-0.018	-0.010	-0.003	0
	0.1	0	0.025	-0.012	-0.026	-0.028	-0.024	-0.019	-0.013	-0.007	-0.002	0
	0.2	0	0.017	0.061	0.026	0.006	-0.004	-0.008	-0.007	-0.005	-0.001	0
	0.3	0	0.009	0.035	0.080	0.042	0.019	0.006	0.001	0.000	0.000	0
	0.4	0	0.002	0.012	0.039	0.085	0.048	0.025	0.013	0.006	0.002	0
	0.5	0	0.002	0.006	0.013	0.025	0.048	0.085	0.039	0.012	0.002	0
	0.6	0	0.000	0.00	0.001	0.006	0.019	0.042	0.080	0.035	0.009	0
	0.7	0	-0.001	-0.005	-0.007	-0.008	-0.004	0.006	0.026	0.061	0.017	0
	0.8	0	-0.002	-0.007	-0.013	-0.019	-0.024	-0.028	-0.026	-0.012	0.025	0
	0.9	0	-0.003	-0.010	-0.018	-0.029	-0.044	-0.061	-0.77	-0.084	-0.066	0
	1.0	0	-0.066	-0.084	-0.077	-0.061	-0.044	-0.029	-0.018	-0.010	-0.003	0

Table 5.7 Moment M_A. Coefficients $\bar{p}$ for obtaining soil pressure $p = \bar{p}\dfrac{M_A}{bL^2}$

α	β						ξ					
		0.0	0.1	0.2	0.3	0.4	0.5	0.6	0.7	0.8	0.9	1.0
0	0–1	7.500	4.920	3.060	1.740	0.780	0	-0.780	-1.740	-3.060	-4.920	-7.500
	0.0	8.759	5.395	3.002	1.363	0.269	-0.494	-1.139	-1.877	2.922	-4.487	-6.783
	0.1	8.751	5.371	3.001	1.365	0.273	-0.490	-1.135	-1.875	-2.923	-4.469	-6.791
	0.2	8.697	5.367	2.999	1.379	0.297	-0.463	-1.111	-1.861	-2.925	-4.515	-6.845
	0.3	8.545	5.297	2.993	1.419	0.363	-0.387	-1.045	-1.821	-2.931	-4.585	-6.997
	0.4	8.253	5.163	2.981	1.495	0.492	-0.241	-0.916	-1.745	-2.943	-4.719	-7.289
25	0.5	7.771	4.941	2.962	1.620	0.704	0	-0.704	-1.620	-2.962	-4.941	-7.771
	0.6	7.289	4.719	2.943	1.745	0.916	0.241	-0.492	-1.495	-2.981	-5.163	-8.253
	0.7	6.997	4.585	2.931	1.821	1.045	0.387	-0.363	-1.419	-2.993	-5.297	-8.545
	0.8	6.845	4.515	2.925	1.861	1.111	0.463	-0.297	-1.379	-2.999	-5.367	-8.697
	0.9	6.791	4.469	2.923	1.875	1.135	0.490	-0.273	-1.365	-3.001	-5.371	-8.751
	1.0	6.783	4.487	2.922	1.877	1.139	0.494	-0.269	-1.363	-3.002	-5.395	-8.759
	0.0	9.917	5.829	2.943	1.015	-0.197	-0.940	-1.457	-1.993	-2.793	-4.099	-6.157
	0.1	9.903	5.799	2.943	1.019	-0.191	-0.933	-1.451	-1.989	-2.793	-4.083	-6171
	0.2	9.797	5.774	2.938	1.046	-0.144	-0.880	-1.404	-1.962	-2.798	-4.154	-6.277
	0.3	9.511	5.642	2.927	1.121	-0.019	-0.737	-1.279	-1.887	-2.809	-4.286	-6.563
	0.4	8.955	5.386	2.905	1.265	0.226	-0.459	-1.034	-1.741	-2.831	-4.542	-7.119
50	0.5	8.037	4.964	2.868	1.501	0.630	0	-0.630	-1.501	-2.868	-4.964	-8.037
	0.6	7.119	4.542	2.831	1.741	1.034	0.459	-0.226	-1.265	-2.905	-5.386	-8.955
	0.7	6.563	4.286	2.809	1.887	1.279	0.737	0.019	-1.121	-2.927	-5.642	-9.511
	0.8	6.277	4.154	2.798	1.962	1.404	0.880	0.144	-1.046	-2.938	-5.774	-9.797
	0.9	6.171	4.083	2.793	1.989	1.451	0.933	0.191	-1.019	-2.943	-5.799	-9.903
	1.0	6.157	4.099	2.793	1.993	1.457	0.940	0.197	-1.015	-2.943	5.829	-9.917

Table 5.7 Con't

						ξ					
α	β	0.0	0.1	0.2	0.3	0.4	0.5	0.6	0.7	0.8	0.9	1.0
	0.0	11.976	6.581	2.820	0.389	-1.021	-1.714	-1.995	-2.171	-2.546	-3.427	-5.120
	0.1	11.948	6.568	2.819	0.396	-1.009	-1.700	-1.983	-2.164	-2.547	-3.440	-5.148
	0.2	11.756	6.480	2.811	0.446	-0.925	-1.604	-1.899	-2.114	-2.555	-3.528	-5.340
	0.3	11.234	6.240	2.790	0.582	-0.695	-1.343	-1.669	-1.978	-2.576	-3.768	-5.862
	0.4	10.222	5.773	2.750	0.845	-0.249	-0.836	-1.223	-1.715	-2.616	-4.235	-6.876
100	0.5	8.548	5.004	2.683	1.280	0.487	0	-0.487	-1.280	-2.683	-5.004	-8.548
	0.6	6.876	4.235	2.616	1.715	1.223	0.836	0.249	-0.845	-2.750	-5.773	-10.22
	0.7	5.862	3.768	2.576	1.978	1.669	1.343	0.695	-0.582	-2.790	-6.240	-11.23
	0.8	5.340	3.528	2.555	2.114	1.899	1.604	0.925	-0.446	-2.811	-6.48	-11.75
	0.9	5.148	3.440	2.547	2.164	1.983	1.700	1.009	-0.396	-2.819	-6.568	-11.94
	1.0	5.120	3.427	2.546	2.171	1.995	1.714	1.021	-0.389	-2.820	-6.581	-11.97
	0.0	13.758	7.216	2.696	-0.163	-1.728	-2.361	-2.428	-2.293	-2.318	-2.872	-4.314
	0.1	13.720	7.199	2.694	-0.153	-1.711	-2.342	-2.411	-2.283	-2.320	-2.889	-4.352
	0.2	13.456	7.077	2.684	-0.084	-1.595	-2.210	-2.295	-2.214	-2.330	-3.011	-4.616
	0.3	12.738	6.747	2.655	0.102	-1.279	-1.851	-1.979	-2.028	-2.359	3.341	-5.334
	0.4	11.340	6.104	2.599	0.466	-0.664	-1.152	-1.364	-1.664	-2.415	-3.984	-6.732
150	0.5	9.036	5.044	2.507	1.065	0.350	0	-0.350	-1.065	-2.507	-5.044	-9.036
	0.6	6.732	3.984	2.415	1.664	1.364	1.152	0.664	-0.466	-2.599	-6.104	-11.34
	0.7	5.334	3.341	2.359	2.028	1.979	1.851	1.279	-0.102	-2.655	-6.747	-12.74
	0.8	4.616	3.011	2.330	2.214	2.295	2.210	1.595	0.084	-2.684	-7.077	-13.46
	0.9	4.352	2.889	2.320	2.283	2.411	2.342	1.711	0.153	-2.694	-7.199	-13.72
	1.0	4.314	2.872	2.318	2.293	2.428	2.361	1.728	0.163	-2.696	-7.216	-13.76
	0.0	15.323	7.758	2.572	-0.655	-2.343	-2.911	-2.781	-2.373	-2.106	-2.402	-3.679
	0.1	15.275	7.736	2.570	-0.642	-2.322	-2.887	-2.760	-2.360	-2.108	-2.424	-3.727
	0.2	14.949	7.586	2.557	-0.557	-2.178	-2.724	-2.616	-2.275	-2.121	-2.574	-4.053
200	0.3	14.065	7.179	2.522	-0.328	-1.789	-2.282	-2.227	-2.046	-2.156	-2.981	-4.937
	0.4	12.341	6.386	2.453	0.121	-2.343	-1.420	-1.469	-1.597	-2.225	-3.774	-6.661
	0.5	9.501	5.080	2.339	0.859	0.219	0	-0.219	-0.859	-2.339	-5.080	-9.501
	0.6	6.661	3.774	2.225	1.597	1.469	1.420	1.031	-0.121	-2.453	-6.386	-12.341

	0.7	4.937	2.981	2.156	2.046	2.227	2.282	1.789	0.328	-2.522	-7.179	-14.065
	0.8	4.053	2.574	2.121	2.275	2.616	2.724	2.178	0.557	-2.557	-7.586	-14.949
	0.9	3.727	2.424	2.108	2.360	2.760	2.887	2.322	0.642	-2.570	7.736	-15.275
	1.0	3.679	2.402	2.106	2.373	2.781	2.911	2.343	0.655	-2.572	-7.758	-15.323
250	0.0	16.713	8.227	2.450	-1.096	-2.882	-3.383	-3.072	-2.422	-1.908	-2.003	-3.181
	0.1	16.659	8.203	2.447	-1.082	-2.858	-3.356	-3.048	-2.408	-1.911	-2.027	-3.235
	0.2	16.281	8.029	2.432	-0.984	-2.692	-3.167	-2.882	-2.310	-1.926	2.201	-3.613
	0.3	15.251	7.555	2.391	-0.716	-2.239	-2.652	-2.429	-2.042	-1.967	-2.675	-4.643
	0.4	13.249	6.634	2.311	-0.196	-1.358	-1.651	-1.548	-1.522	-2.047	-3.596	-6.645
	0.5	9.947	5.115	2.179	0.663	0.095	0	-0.095	-0.663	-2.179	-5.115	-9.947
	0.6	6.645	3.596	2.047	1.522	1.548	1.651	1.358	0.196	-2.311	-6.634	-13.249
	0.7	4.643	2.675	1.967	2.042	2.429	2.652	2.239	0.716	-2.391	-7.555	-15.251
	0.8	3.613	2.201	1.926	2.310	2.882	3.167	2.692	0.984	-2.432	-8.029	-16.281
	0.9	3.235	2.027	1.911	2.408	3.048	3.356	2.858	1.082	-2.447	-8.203	-16.659
	1.0	3.181	2.003	1.908	2.422	3.072	3.383	2.882	1.096	-2.450	-8.227	-16.713
300	0.0	17.963	8.640	2.329	-1.498	-3.364	-3.794	-3.314	-2.448	-1.721	-1.660	-2.787
	0.1	17.903	8.613	2.326	-1.482	-3.337	-3.764	-3.287	-2.432	-1.724	-1.687	-2.847
	0.2	17.477	8.417	2.309	-1.372	-3.150	-3.551	-3.100	-2.322	-1.741	-1.883	-3.273
	0.3	16.323	7.886	2.263	-1.071	-2.642	-2.974	-2.592	-2.021	-1.787	-2.414	-4.427
	0.4	14.079	6.854	2.173	-0.488	-1.655	-1.852	-1.605	-1.438	-1.877	-3.446	-6.671
	0.5	10.375	5.150	2.025	0.475	-0.025	0	0.025	-0.475	-2.025	-5.150	-10.375
	0.6	6.671	3.446	1.877	1.438	1.605	1.852	1.655	0.488	-2.173	-6.854	-14.079
	0.7	4.427	2.414	1.787	2.021	2.592	2.974	2.642	1.071	-2.263	-7.886	-16.323
	0.8	3.273	1.883	1.741	2.322	3.1	3.551	3.150	1.372	-2.309	-8.417	-17.477
	0.9	2.847	1.687	1.724	2.432	3.287	3.764	3.337	1.482	-2.326	-8.613	-17.903
	1.0	2.787	1.660	1.721	2.448	3.314	3.794	3.364	1.498	-2.329	-8.640	-17.963

Table 5.7 Con't

α	β						ξ					
		0.0	0.1	0.2	0.3	0.4	0.5	0.6	0.7	0.8	0.9	1.0
	0.0	19.092	9.005	2.209	-1.865	-3.795	-4.154	-3.517	-2.455	-1.545	-1.361	-2.476
	0.1	19.026	8.974	2.207	-1.848	-3.765	-4.121	-3.487	-2.438	-1.547	-1.392	-2.542
	0.2	18.560	8.760	2.188	-1.727	-3.560	-3.888	-3.282	-2.317	-1.566	-1.606	-3.008
	0.3	17.298	8.179	2.138	-1.399	-3.005	-3.257	-2.727	1.989	-1.616	2.187	-4.270
	0.4	14.838	7.048	2.039	-0.759	-1.923	-2.027	-1.645	-1.349	-1.715	-3.318	-6.730
350	0.5	10.784	5.183	1.877	0.295	-0.139	0	0.139	-0.295	-1.877	-5.183	-10.784
	0.6	6.730	3.318	1.715	1.349	1.645	2.027	1.923	0.759	-2.039	-7.048	-14.838
	0.7	4.270	2.187	1.616	1.989	2.727	3.257	3.005	1.399	-2.138	-8.179	-17.298
	0.8	3.008	1.606	1.566	2.317	3.282	3.888	3.560	1.727	-2.188	-8.760	-18.560
	0.9	2.542	1.392	1.547	2.438	3.487	4.121	3.765	1.848	-2.207	-8.974	-19.026
	1.0	2.476	1.361	1.545	2.455	3.517	4.154	3.795	1.865	-2.209	-9.005	-19.092
	0	20.122	9.329	2.094	-2.204	-4.185	-4.473	-3.687	-2.448	-1.378	-1.099	-2.230
	0.1	20.050	9.296	2.091	-2.185	-4.154	-4.437	-3.656	-2.429	-1.381	-1.132	-2.302
	0.2	19.550	9.066	2.071	-2.055	-3.934	-4.187	-3.436	-2.299	-1.401	-1.362	-2.802
	0.3	18.190	8.440	2.017	-1.702	-3.335	-3.507	-2.837	-1.946	-1.455	-1.988	-4.162
	0.4	15.542	7.222	1.911	-1.013	-2.170	-2.183	-1.672	-1.257	-1.561	-3.206	-6.810
400	0.5	11.176	5.214	1.736	0.122	-0.249	0	0.249	-0.122	-1.736	-5.214	-11.176
	0.6	6.810	3.206	1.561	1.257	1.672	2.183	2.170	1.013	-1.911	-7.222	-15.542
	0.7	4.162	1.988	1.455	1.946	2.837	3.507	3.335	1.702	-2.017	8.440	-18.190
	0.8	2.802	1.362	1.401	2.299	3.436	4.187	3.934	2.055	-2.071	-9.066	-19.550
	0.9	2.302	1.132	1.381	2.429	3.656	4.437	4.154	2.185	-2.091	-9.296	-20.050
	1.0	2.230	1.099	1.378	2.448	3.687	4.473	4.185	2.204	-2.094	-9.329	-20.122
	0	21.065	9.620	1.981	-2.516	-4.540	-4.756	-3.830	-2.430	-1.221	-0.868	-2.041
	0.1	20.991	9.585	1.979	-2.497	-4.508	-4.719	-3.798	-2.411	-1.223	-0.903	-2.115
	0.2	20.457	9.340	1.957	-2.358	-4.273	-4.452	-3.563	-2.272	-1.245	-1.148	-2.649
450	0.3	19.011	8.675	1.899	-1.982	-3.637	-3.729	-2.927	-1.896	-1.303	-1.813	-4.095
	0.4	16.195	7.379	1.787	-1.250	-2.397	-2.397	-1.687	-1.164	-1.415	-3.109	-6.911
	0.5	11.553	5.244	1.601	-0.043	-0.355	0	0.355	0.043	-1.601	-5.244	-11.553
	0.6	6.911	3.109	1.415	1.164	1.687	2.397	2.397	1.250	-1.787	-7.379	-16.195

	0.7	4.095	1.813	1.303	1.896	2.927	3.729	3.637	1.982	−1.899	−8.675	−19.011
	0.8	2.649	1.148	1.245	2.272	3.563	4.452	4.273	2.358	−1.957	−9.340	−20.457
	0.9	2.115	0.903	1.223	2.411	3.798	4.719	4.508	2.497	−1.979	−9.585	−20.991
	1.0	2.041	0.868	1.221	2.430	3.830	4.756	4.540	2.516	−1.981	−9.620	−21.065
	0	21.937	9.883	1.871	−2.809	−4.867	−5.011	−3.953	−2.403	−1.069	−0.663	−1.893
	0.1	21.857	9.846	1.868	−2.788	−4.831	−4.971	−3.917	−2.382	−1.072	−0.700	−1.973
	0.2	21.295	9.588	1.845	−2.642	−4.584	−4.690	−3.670	2.236	−1.095	−0.958	−2.535
	0.3	19.771	8.887	1.784	−2.246	−3.914	−3.928	−3.000	−1.840	−1.156	−1.659	−4.059
	0.4	16.805	7.525	1.666	−1.474	−2.609	−2.445	−1.695	−1.068	−1.274	−3.024	−7.025
500	0.5	11.915	5.273	1.470	−0.203	−0.457	0	0.457	0.203	−1.470	−5.273	−11.915
	0.6	7.025	3.024	1.274	1.068	1.695	2.445	2.609	1.474	−1.666	−7.525	−16.805
	0.7	4.059	1.659	1.156	1.840	3.000	3.928	3.914	2.246	−1.784	−8.887	−19.771
	0.8	2.535	0.958	1.095	2.236	3.670	4.690	4.584	2.642	−1.845	−9.588	−21.295
	0.9	1.973	0.700	1.072	2.382	3.917	4.971	4.831	2.788	−1.868	−9.846	−21.857
	1.0	1.893	0.663	1.069	2.403	3.953	5.011	4.867	2.809	−1.871	−9.883	−21.937

Table 5.8 Moment M_A. Coefficients $\overline{Q}$ for obtaining the shear forces $Q = \overline{Q}\dfrac{M_A}{L}$

α	β	0.0	0.1	0.2	0.3	0.4	0.5	0.6	0.7	0.8	0.9	1.0
0	0-1	0	0.614	1.008	1.244	1.368	1.406	1.368	1.244	1.008	0.614	0
	0.0	0	0.699	1.112	1.324	1.402	1.389	1.308	1.158	0.922	0.557	0
	0.1	0	0.699	1.111	1.323	1.402	1.389	1.308	1.159	0.923	0.557	0
	0.2	0	0.695	1.106	1.319	1.399	1.389	1.311	1.163	0.928	0.561	0
	0.3	0	0.684	1.091	1.306	1.392	1.389	1.318	1.176	0.943	0.572	0
	0.4	0	0.663	1.063	1.281	1.378	1.389	1.332	1.201	0.971	0.593	0
25	0.5	0	0.628	1.017	1.241	1.355	1.389	1.355	1.241	1.017	0.628	0
	0.6	0	0.593	0.971	1.201	1.332	1.389	1.378	1.281	1.063	0.663	0
	0.7	0	0.572	0.943	1.176	1.318	1.389	1.392	1.306	1.091	0.684	0
	0.8	0	0.561	0.928	1.163	1.311	1.389	1.399	1.319	1.106	0.695	0
	0.9	0	0.557	0.923	1.159	1.308	1.389	1.402	1.323	1.111	0.699	0
	1.0	0	0.557	0.922	1.158	1.308	1.389	1.402	1.324	1.112	0.699	0
	0.0	0	0.776	1.205	1.396	1.433	1.373	1.253	1.080	0.845	0.506	0
	0.1	0	0.775	1.204	1.395	1.433	1.373	1.253	1.081	0.846	0.507	0
	0.2	0	0.768	1.194	1.386	1.427	1.373	1.259	1.090	0.856	0.514	0
	0.3	0	0.747	1.167	1.362	1.414	1.373	1.272	1.114	0.883	0.535	0
	0.4	0	0.707	1.113	1.315	1.387	1.373	1.299	1.161	0.937	0.575	0
50	0.5	0	0.641	1.025	1.238	1.343	1.373	1.343	1.238	1.025	0.641	0
	0.6	0	0.575	0.937	1.161	1.299	1.373	1.387	1.315	1.113	0.707	0
	0.7	0	0.535	0.883	1.114	1.272	1.373	1.414	1.362	1.167	0.747	0
	0.8	0	0.514	0.856	1.090	1.259	1.373	1.427	1.386	1.194	0.768	0
	0.9	0	0.507	0.846	1.081	1.253	1.373	1.433	1.395	1.204	0.775	0
	1.0	0	0.506	0.845	1.080	1.253	1.373	1.433	1.396	1.205	0.776	0
	0.0	0	0.913	1.371	1.521	1.483	1.341	1.153	0.945	0.713	0.419	0
	0.1	0	0.909	1.368	1.519	1.481	1.341	1.155	0.947	0.716	0.423	0
100	0.2	0	0.897	1.350	1.502	1.472	1.341	1.164	0.964	0.734	0.435	0
	0.3	0	0.859	1.300	1.459	1.447	1.341	1.189	1.007	0.784	0.473	0
	0.4	0	0.786	1.203	1.373	1.398	1.341	1.238	1.093	0.881	0.546	0

Group	x										
	0.5	0	0.666	1.042	1.233	1.318	1.341	1.318	1.233	1.042	0.666	0
	0.6	0	0.786	1.203	1.373	1.398	1.341	1.238	1.093	0.881	0.546	0
	0.7	0	0.859	1.300	1.459	1.447	1.341	1.189	1.007	0.784	0.473	0
	0.8	0	0.897	1.350	1.502	1.472	1.341	1.164	0.964	0.734	0.435	0
	0.9	0	0.909	1.368	1.519	1.481	1.341	1.155	0.947	0.716	0.423	0
	1.0	0	0.913	1.371	1.521	1.483	1.341	1.153	0.945	0.713	0.419	0
150	0.0	0	0.350	0.604	0.831	1.068	1.310	1.522	1.625	1.510	1.030	0
	0.1	0	0.353	0.607	0.835	1.070	1.310	1.520	1.621	1.507	1.027	0
	0.2	0	0.372	0.633	0.857	1.083	1.310	1.507	1.599	1.481	1.008	0
	0.3	0	0.423	0.702	0.917	1.117	1.310	1.473	1.539	1.412	0.957	0
	0.4	0	0.524	0.836	1.034	1.184	1.310	1.406	1.422	1.278	0.856	0
	0.5	0	0.690	1.057	1.228	1.295	1.310	1.295	1.228	1.057	0.690	0
	0.6	0	0.856	1.278	1.422	1.406	1.310	1.184	1.034	0.836	0.524	0
	0.7	0	0.957	1.412	1.539	1.473	1.310	1.117	0.917	0.702	0.423	0
	0.8	0	1.008	1.481	1.599	1.507	1.310	1.083	0.857	0.633	0.372	0
	0.9	0	1.027	1.507	1.621	1.520	1.310	1.070	0.835	0.607	0.353	0
	1.0	0	1.030	1.510	1.625	1.522	1.310	1.068	0.831	0.604	0.350	0
200	0.0	0	0.294	0.513	0.734	0.993	1.281	1.551	1.712	1.631	1.132	0
	0.1	0	0.297	0.518	0.738	0.995	1.281	1.549	1.708	1.626	1.129	0
	0.2	0	0.321	0.549	0.765	1.010	1.281	1.534	1.681	1.595	1.105	0
	0.3	0	0.384	0.634	0.840	1.053	1.281	1.491	1.606	1.510	1.042	0
	0.4	0	0.509	0.799	0.984	1.136	1.281	1.408	1.462	1.345	0.917	0
	0.5	0	0.713	1.072	1.223	1.272	1.281	1.272	1.223	1.072	0.713	0
	0.6	0	0.917	1.345	1.462	1.408	1.281	1.136	0.984	0.799	0.509	0
	0.7	0	1.042	1.510	1.606	1.491	1.281	1.053	0.840	0.634	0.384	0
	0.8	0	1.105	1.595	1.681	1.534	1.281	1.010	0.765	0.549	0.321	0
	0.9	0	1.129	1.626	1.708	1.549	1.281	0.995	0.738	0.518	0.297	0
	1.0	0	1.132	1.631	1.712	1.551	1.281	0.993	0.734	0.513	0.294	0

Table 5.8 Con't

α	β	ξ										
		0.0	0.1	0.2	0.3	0.4	0.5	0.6	0.7	0.8	0.9	1.0
250	0.0	0	1.223	1.736	1.787	1.575	1.253	0.925	0.651	0.436	0.249	0
	0.1	0	1.219	1.730	1.783	1.572	1.253	0.928	0.655	0.442	0.253	0
	0.2	0	1.192	1.694	1.751	1.554	1.253	0.946	0.687	0.478	0.280	0
	0.3	0	1.118	1.595	1.665	1.505	1.253	0.995	0.773	0.577	0.354	0
	0.4	0	0.974	1.403	1.491	1.408	1.253	1.092	0.937	0.769	0.498	0
	0.5	0	0.736	1.086	1.219	1.250	1.253	1.250	1.219	1.086	0.736	0
	0.6	0	0.498	0.769	0.937	1.092	1.253	1.408	1.491	1.403	0.974	0
	0.7	0	0.354	0.577	0.773	0.995	1.253	1.505	1.665	1.595	1.118	0
	0.8	0	0.280	0.478	0.687	0.946	1.253	1.554	1.751	1.694	1.192	0
	0.9	0	0.253	0.442	0.655	0.928	1.253	1.572	1.783	1.730	1.219	0
	1.0	0	0.249	0.436	0.651	0.925	1.253	1.575	1.787	1.736	1.223	0
300	0.0	0	1.303	1.828	1.851	1.594	1.227	0.866	0.577	0.372	0.211	0
	0.1	0	1.299	1.823	1.846	1.591	1.227	0.869	0.582	0.377	0.215	0
	0.2	0	1.268	1.782	1.811	1.571	1.227	0.869	0.617	0.418	0.246	0
	0.3	0	1.185	1.671	1.714	1.516	1.227	0.944	0.714	0.529	0.329	0
	0.4	0	1.024	1.456	1.521	1.408	1.227	1.052	0.899	0.744	0.490	0
	0.5	0	0.757	1.100	1.214	1.230	1.227	1.230	1.214	1.100	0.757	0
	0.6	0	0.490	0.744	0.899	1.052	1.227	1.408	1.521	1.456	1.024	0
	0.7	0	0.329	0.529	0.714	0.944	1.227	1.516	1.714	1.671	1.185	0
	0.8	0	0.246	0.418	0.617	0.889	1.227	1.571	1.811	1.782	1.268	0
	0.9	0	0.215	0.377	0.582	0.869	1.227	1.591	1.846	1.823	1.299	0
	1.0	0	0.211	0.372	0.577	0.866	1.227	1.594	1.851	1.828	1.303	0
350	0.0	0	1.375	1.911	1.908	1.610	1.201	0.812	0.512	0.315	0.179	0
	0.1	0	1.370	1.904	1.902	1.607	1.201	0.815	0.518	0.322	0.184	0
	0.2	0	1.337	1.859	1.863	1.584	1.201	0.838	0.557	0.367	0.217	0
	0.3	0	1.246	1.738	1.757	1.524	1.201	0.898	0.663	0.488	0.308	0
	0.4	0	1.069	1.502	1.551	1.406	1.201	1.016	0.869	0.724	0.485	0
	0.5	0	0.777	1.113	1.210	1.211	1.201	1.211	1.210	1.113	0.777	0

Group	x											
	0.6	0	0.485	0.724	0.869	1.016	1.201	1.406	1.551	1.502	1.069	0
	0.7	0	0.308	0.488	0.663	0.898	1.201	1.524	1.757	1.738	1.246	0
	0.8	0	0.217	0.367	0.557	0.838	1.201	1.584	1.863	1.859	1.337	0
	0.9	0	0.184	0.322	0.518	0.815	1.201	1.607	1.902	1.904	1.370	0
	1.0	0	0.179	0.315	0.512	0.812	1.201	1.610	1.908	1.911	1.375	0
400	0.0	0	1.440	1.985	1.957	1.621	1.176	0.763	0.455	0.267	0.152	0
	0.1	0	1.435	1.978	1.951	1.618	1.176	0.766	0.461	0.274	0.157	0
	0.2	0	1.399	1.930	1.909	1.594	1.176	0.790	0.503	0.322	0.193	0
	0.3	0	1.301	1.799	1.795	1.529	1.176	0.855	0.617	0.453	0.291	0
	0.4	0	1.110	1.545	1.573	1.402	1.176	0.982	0.839	0.707	0.482	0
	0.5	0	0.796	1.126	1.206	1.192	1.176	1.192	1.206	1.126	0.796	0
	0.6	0	0.482	0.707	0.839	0.982	1.176	1.402	1.573	1.545	1.110	0
	0.7	0	0.291	0.453	0.617	0.855	1.176	1.529	1.795	1.799	1.301	0
	0.8	0	0.193	0.322	0.503	0.790	1.176	1.594	1.909	1.930	1.399	0
	0.9	0	0.1578	0.274	0.461	0.766	1.176	1.618	1.951	1.978	1.435	0
	1.0	0	0.152	0.267	0.455	0.763	1.176	1.621	1.957	1.985	1.440	0
450	0.0	0	1.500	2.051	2.000	1.630	1.153	0.716	0.402	0.225	0.130	0
	0.1	0	1.495	2.044	1.994	1.626	1.153	0.720	0.408	0.232	0.135	0
	0.2	0	1.456	1.993	1.949	1.600	1.153	0.746	0.453	0.283	0.174	0
	0.3	0	1.352	1.854	1.827	1.531	1.153	0.815	0.575	0.422	0.278	0
	0.4	0	1.149	1.584	1.591	1.396	1.153	0.950	0.811	0.692	0.481	0
	0.5	0	0.815	1.138	1.201	1.173	1.153	1.173	1.201	1.138	0.815	0
	0.6	0	0.481	0.692	0.811	0.950	1.153	1.396	1.591	1.584	1.149	0
	0.7	0	0.278	0.422	0.575	0.815	1.153	1.531	1.827	1.854	1.352	0
	0.8	0	0.174	0.283	0.453	0.746	1.153	1.600	1.949	1.993	1.456	0
	0.9	0	0.135	0.232	0.408	0.720	1.153	1.626	1.994	2.044	1.495	0
	1.0	0	0.130	0.225	0.402	0.716	1.153	1.630	2.000	2.051	1.500	0

Table 5.8 Con't

α	β	0.0	0.1	0.2	0.3	0.4	ξ 0.5	0.6	0.7	0.8	0.9	1.0
	0.0	0	1.555	2.112	2.040	1.637	1.130	0.675	0.356	0.188	0.111	0
	0.1	0	1.549	2.104	2.033	1.633	1.130	0.679	0.363	0.196	0.117	0
	0.2	0	1.508	2.050	1.986	1.606	1.130	0.706	0.410	0.250	0.158	0
	0.3	0	1.399	1.904	1.858	1.533	1.130	0.779	0.538	0.396	0.267	0
	0.4	0	1.185	1.619	1.609	1.393	1.130	0.921	0.787	0.681	0.481	0
500	0.5	0	0.833	1.150	1.198	1.156	1.130	1.156	1.198	1.150	0.833	0
	0.6	0	0.481	0.681	0.787	0.921	1.130	1.391	1.609	1.619	1.185	0
	0.7	0	0.267	0.396	0.538	0.779	1.130	1.533	1.858	1.904	1.399	0
	0.8	0	0.158	0.250	0.410	0.706	1.130	1.606	1.986	2.050	1.508	0
	0.9	0	0.117	0.196	0.363	0.679	1.130	1.633	2.033	2.104	1.549	0
	1.0	0	0.111	0.188	0.356	0.675	1.130	1.637	2.040	2.112	1.555	0

Table 5.9 Moment M_A. Coefficients $\overline{M}$ for obtaining moments $M = \overline{M}M_A$

α	β	\multicolumn{11}{c}{ξ}										
		0.0	0.1	0.2	0.3	0.4	0.5	0.6	0.7	0.8	0.9	1.0
0	0.0	−1	−0.967	−0.895	−0.770	−0.639	−0.500	−0.361	−0.230	−0.105	−0.033	0
	0.1	0	$\frac{0.033}{-0.967}$	−0.895	−0.770	−0.639	−0.500	−0.361	−0.230	−0.105	−0.033	0
	0.2	0	0.033	$\frac{0.115}{-0.895}$	−0.770	−0.639	−0.500	−0.361	−0.230	−0.105	−0.033	0
	0.3	0	0.033	0.105	$\frac{0.230}{-0.770}$	−0.639	−0.500	−0.361	−0.230	−0.105	−0.033	0
	0.4	0	0.033	0.105	0.230	$\frac{0.361}{-0.639}$	−0.500	−0.361	−0.230	−0.105	−0.033	0
	0.5	0	0.033	0.105	0.230	0.361	$\frac{0.500}{-0.500}$	−0.361	−0.230	−0.105	−0.033	0
	0.6	0	0.033	0.105	0.230	0.361	0.500	$\frac{0.639}{-0.361}$	−0.230	−0.105	−0.033	0
	0.7	0	0.033	0.105	0.230	0.361	0.500	0.639	$\frac{0.770}{-0.230}$	−0.105	−0.033	0
	0.8	0	0.033	0.105	0.230	0.361	0.500	0.639	0.770	$\frac{0.895}{-0.105}$	−0.033	0
	0.9	0	0.033	0.105	0.230	0.361	0.500	0.639	0.770	0.895	$\frac{0.967}{-0.033}$	0
	1.0	0	0.033	0.105	0.230	0.361	0.500	0.639	0.770	0.895	0.967	1
25	0.0	−1	−0.962	−0.880	−0.747	−0.610	−0.469	−0.334	−0.209	−0.094	−0.030	0
	0.1	0	$\frac{0.038}{-0.962}$	−0.880	−0.747	−0.610	−0.469	−0.334	−0.209	−0.094	−0.030	0
	0.2	0	0.038	$\frac{0.119}{-0.881}$	−0.749	−0.611	−0.471	−0.335	−0.211	−0.095	−0.030	0
	0.3	0	0.037	0.117	$\frac{0.248}{-0.752}$	−0.616	−0.476	−0.340	−0.214	−0.097	−0.031	0
	0.4	0	0.036	0.113	0.242	$\frac{0.376}{-0.624}$	−0.485	−0.348	−0.220	−0.101	−0.032	0
	0.5	0	0.034	0.107	0.231	0.362	$\frac{0.500}{-0.500}$	−0.362	−0.231	−0.107	−0.034	0
	0.6	0	0.032	0.101	0.22	0.348	0.485	$\frac{0.624}{-0.376}$	−0.242	−0.113	−0.036	0
	0.7	0	0.031	0.097	0.214	0.340	0.476	0.616	$\frac{0.752}{-0.248}$	−0.117	−0.037	0
	0.8	0	0.030	0.095	0.211	0.335	0.471	0.611	0.749	$\frac{0.881}{-0.119}$	−0.038	0
	0.9	0	0.030	0.094	0.209	0.334	0.469	0.610	0.747	0.880	$\frac{0.962}{-0.038}$	0
	1.0	0	0.030	0.094	0.209	0.334	0.469	0.610	0.747	0.880	0.962	1

Table 5.9 Con't

α	β	ξ										
		0.0	0.1	0.2	0.3	0.4	0.5	0.6	0.7	0.8	0.9	1.0
50	0.0	−1	−0.957	−0.863	−0.725	−0.583	−0.441	−0.309	−0.193	−0.089	−0.027	0
	0.1	0	$\frac{0.043}{-0.957}$	−0.863	−0.725	−0.583	−0.442	−0.309	−0.193	−0.089	−0.027	0
	0.2	0	0.042	$\frac{0.136}{-0.864}$	−0.727	−0.586	−0.445	−0.312	−0.195	−0.090	−0.028	0
	0.3	0	0.041	0.132	$\frac{0.267}{-0.733}$	−0.595	−0.454	−0.321	−0.201	−0.094	−0.029	0
	0.4	0	0.039	0.125	0.254	$\frac{0.389}{-0.611}$	−0.471	−0.337	−0.214	−0.101	−0.031	0
	0.5	0	0.035	0.113	0.234	0.363	$\frac{0.500}{-0.500}$	−0.363	−0.234	−0.113	−0.035	0
	0.6	0	0.031	0.101	0.214	0.337	0.471	$\frac{0.611}{-0.389}$	−0.254	−0.125	−0.039	0
	0.7	0	0.029	0.094	0.201	0.321	0.454	0.595	$\frac{0.733}{-0.267}$	−0.132	−0.041	0
	0.8	0	0.028	0.090	0.195	0.312	0.445	0.586	0.727	$\frac{0.864}{-0.136}$	−0.042	0
	0.9	0	0.027	0.089	0.193	0.309	0.442	0.583	0.725	0.863	$\frac{0.957}{-0.043}$	0
	1.0	0	0.027	0.089	0.193	0.309	0.441	0.583	0.725	0.863	0.957	1
100	0.0	−1	−0.950	−0.839	−0.686	−0.534	−0.393	−0.268	−0.162	−0.073	−0.022	0
	0.1	0	$\frac{0.050}{-0.950}$	−0.839	−0.687	−0.535	−0.394	−0.269	−0.163	−0.073	−0.022	0
	0.2	0	0.049	$\frac{0.158}{-0.842}$	−0.691	−0.541	−0.400	−0.275	−0.167	−0.076	−0.023	0
	0.3	0	0.047	0.151	$\frac{0.297}{-0.703}$	−0.556	−0.416	−0.290	−0.179	−0.083	−0.025	0
	0.4	0	0.043	0.138	0.275	$\frac{0.415}{-0.585}$	−0.448	−0.319	−0.201	−0.096	−0.029	0
	0.5	0	0.036	0.117	0.238	0.367	$\frac{0.500}{-0.500}$	−0.367	−0.238	−0.117	−0.036	0
	0.6	0	0.029	0.096	0.201	0.319	0.448	$\frac{0.585}{-0.415}$	−0.275	−0.138	−0.043	0
	0.7	0	0.025	0.083	0.179	0.290	0.416	0.556	$\frac{0.703}{-0.297}$	−0.151	−0.047	0
	0.8	0	0.023	0.076	0.167	0.275	0.400	0.541	0.691	$\frac{0.842}{-0.158}$	−0.049	0
	0.9	0	0.022	0.073	0.163	0.269	0.394	0.535	0.687	0.839	$\frac{0.950}{-0.050}$	0
	1.0	0	0.022	0.073	0.162	0.268	0.393	0.534	0.686	0.839	0.950	1

Table 5.9 Con't

α	β	ξ										
		0.0	**0.1**	**0.2**	**0.3**	**0.4**	**0.5**	**0.6**	**0.7**	**0.8**	**0.9**	**1.0**
	0.0	−1	−0.943	−0.817	−0.653	−0.495	−0.352	−0.233	−0.139	−0.063	−0.019	0
	0.1	0	$\frac{0.057}{-0.943}$	−0.817	−0.654	−0.496	−0.354	−0.234	−0.140	−0.063	−0.019	0
	0.2	0	0.056	$\frac{0.180}{-0.820}$	−0.654	−0.504	−0.362	−0.242	−0.146	−0.066	−0.020	0
	0.3	0	0.053	0.170	$\frac{0.325}{-0.675}$	−0.524	−0.384	−0.262	−0.161	−0.076	−0.023	0
	0.4	0	0.047	0.152	0.294	$\frac{0.435}{-0.565}$	−0.428	−0.303	−0.192	−0.094	−0.029	0
150	0.5	0	0.038	0.123	0.243	0.369	$\frac{0.500}{-0.500}$	−0.369	−0.243	−0.123	−0.038	0
	0.6	0	0.029	0.094	0.192	0.393	0.428	$\frac{0.565}{-0.435}$	−0.294	−0.152	−0.047	0
	0.7	0	0.023	0.076	0.161	0.262	0.384	0.524	$\frac{0.675}{-0.325}$	−0.170	−0.053	0
	0.8	0	0.020	0.066	0.146	0.242	0.362	0.504	0.660	$\frac{0.820}{-0.180}$	−0.056	0
	0.9	0	0.019	0.063	0.140	0.234	0.354	0.496	0.654	0.817	$\frac{0.943}{-0.057}$	0
	1.0	0	0.019	0.063	0.139	0.233	0.352	0.495	0.653	0.817	0.943	1
	0.0	−1	0.937	−0.796	−0.625	−0.460	−0.318	−0.204	−0.119	−0.054	−0.015	0
	0.1	0	$\frac{0.062}{-0.938}$	−0.797	−0.626	−0.462	−0.320	−0.206	−0.120	−0.055	−0.016	0
	0.2	0	0.061	$\frac{0.199}{-0.801}$	−0.633	−0.471	−0.330	−0.215	−0.127	−0.059	−0.017	0
	0.3	0	0.057	0.187	$\frac{0.348}{-0.652}$	−0.497	−0.357	−0.241	−0.146	−0.071	−0.021	0
	0.4	0	0.050	0.165	0.310	$\frac{0.454}{-0.546}$	−0.411	−0.290	−0.184	−0.093	0.028	0
200	0.5	0	0.039	0.129	0.247	0.372	$\frac{0.500}{-0.500}$	−0.372	−0.247	−0.129	−0.039	0
	0.6	0	0.028	0.093	0.184	0.290	0.411	$\frac{0.546}{-0.454}$	−0.310	−0.165	−0.050	0
	0.7	0	0.021	0.071	0.146	0.241	0.357	0.497	$\frac{0.652}{-0.348}$	−0.187	−0.057	0
	0.8	0	0.017	0.059	0.127	0.215	0.330	0.471	0.633	$\frac{0.801}{-0.199}$	−0.061	0
	0.9	0	0.016	0.055	0.120	0.206	0.320	0.462	0.626	0.797	$\frac{0.938}{-0.062}$	0
	1.0	0	0.015	0.054	0.119	0.204	0.318	0.460	0.625	0.796	0.937	1

Table 5.9 Con't

α	β	ξ 0.0	0.1	0.2	0.3	0.4	0.5	0.6	0.7	0.8	0.9	1.0
250	0.0	−1	−0.932	−0.778	−0.600	−0.430	−0.289	−0.180	−0.102	−0.048	−0.014	0
	0.1	0	$\frac{0.068}{-0.932}$	−0.779	−0.601	−0.432	−0.290	−0.182	−0.103	−0.049	−0.014	0
	0.2	0	0.067	$\frac{0.216}{-0.784}$	−0.609	−0.443	−0.302	−0.193	−0.111	0.054	−0.015	0
	0.3	0	0.062	0.203	$\frac{0.368}{-0.632}$	−0.472	−0.334	−0.222	−0.134	−0.067	−0.020	0
	0.4	0	0.054	0.177	0.324	$\frac{0.470}{-0.530}$	−0.397	−0.280	−0.178	−0.093	−0.028	0
	0.5	0	0.041	0.135	0.251	0.375	$\frac{0.500}{-0.500}$	−0.375	−0.251	−0.135	−0.041	0
	0.6	0	0.028	0.093	0.178	0.280	0.397	$\frac{0.530}{-0.470}$	−0.324	−0.177	−0.054	0
	0.7	0	0.020	0.067	0.134	0.222	0.334	0.472	$\frac{0.632}{-0.368}$	−0.203	−0.062	0
	0.8	0	0.015	0.054	0.111	0.193	0.302	0.443	0.609	$\frac{0.784}{-0.216}$	−0.067	0
	0.9	0	0.014	0.049	0.103	0.182	0.290	0.432	0.601	0.779	$\frac{0.932}{-0.068}$	0
	1.0	0	0.014	0.048	0.102	0.180	0.289	0.430	0.600	0.778	0.932	1
300	0.0	−1	−0.927	−0.763	−0.578	−0.404	−0.263	−0.158	−0.088	−0.043	−0.011	0
	0.1	0	$\frac{0.072}{-0.928}$	−0.764	−0.579	−0.406	−0.265	−0.160	−0.089	−0.043	−0.012	0
	0.2	0	0.071	$\frac{0.231}{-0.769}$	−0.588	−0.418	−0.278	−0.172	−0.098	−0.049	−0.013	0
	0.3	0	0.066	0.216	$\frac{0.386}{-0.614}$	−0.452	−0.314	−0.206	−0.124	−0.064	−0.018	0
	0.4	0	0.057	0.187	0.337	$\frac{0.484}{-0.516}$	−0.384	−0.270	−0.173	−0.093	−0.027	0
	0.5	0	0.042	0.140	0.255	0.377	$\frac{0.500}{-0.500}$	−0.377	−0.255	−0.140	−0.042	0
	0.6	0	0.027	0.093	0.173	0.270	0.384	$\frac{0.516}{-0.484}$	−0.337	−0.187	−0.057	0
	0.7	0	0.018	0.064	0.124	0.206	0.314	0.452	$\frac{0.614}{-0.386}$	−0.216	−0.066	0
	0.8	0	0.013	0.049	0.098	0.172	0.278	0.418	0.588	$\frac{0.769}{-0.231}$	−0.071	0
	0.9	0	0.012	0.043	0.089	0.160	0.265	0.406	0.579	0.764	$\frac{0.928}{-0.072}$	0
	1.0	0	0.011	0.043	0.088	0.158	0.263	0.404	0.578	0.763	0.927	1

Table 5.9 Con't

α	β	ξ										
		0.0	0.1	0.2	0.3	0.4	0.5	0.6	0.7	0.8	0.9	1.0
	0.0	−1	−0.923	−0.750	−0.559	−0.382	−0.240	−0.140	−0.075	−0.038	−0.009	0
	0.1	0	0.076 / −0.924	−0.751	−0.560	−0.384	−0.242	−0.142	−0.076	−0.039	−0.010	0
	0.2	0	0.074	0.244 / −0.756	−0.571	−0.397	−0.257	−0.155	−0.087	−0.044	−0.012	0
	0.3	0	0.069	0.227	0.402 / −0.598	−0.433	−0.296	−0.191	−0.114	−0.061	−0.017	0
	0.4	0	0.059	0.196	0.347	0.496 / −0.504	−0.373	−0.262	−0.169	−0.093	−0.027	0
350	0.5	0	0.043	0.144	0.258	0.379	0.500 / −0.500	−0.379	−0.258	−0.144	−0.043	0
	0.6	0	0.027	0.093	0.169	0.262	0.373	0.504 / −0.496	−0.347	−0.196	−0.059	0
	0.7	0	0.017	0.061	0.114	0.191	0.296	0.433	0.598 / −0.402	−0.227	−0.069	0
	0.8	0	0.012	0.044	0.087	0.155	0.257	0.397	0.571	0.244 / −0.756	−0.074	0
	0.9	0	0.010	0.039	0.076	0.142	0.242	0.384	0.560	0.751	0.924 / −0.076	0
	1.0	0	0.009	0.038	0.075	0.140	0.240	0.382	0.559	0.750	0.923	1
	0.0	−1	−0.919	−0.735	−0.541	−0.360	−0.220	−0.124	−0.065	−0.035	−0.009	0
	0.1	0	0.081 / −0.919	−0.736	−0.542	−0.362	−0.223	−0.126	−0.066	−0.036	−0.009	0
	0.2	0	0.079	0.257 / −0.743	−0.553	−0.377	−0.238	−0.141	−0.077	−0.043	−0.011	0
	0.3	0	0.073	0.240	0.417 / −0.583	−0.416	−0.281	−0.180	−0.107	−0.060	−0.017	0
	0.4	0	0.063	0.206	0.358	0.508 / −0.492	−0.364	−0.255	−0.166	−0.094	−0.027	0
400	0.5	0	0.045	0.150	0.262	0.382	0.500 / −0.500	−0.382	−0.262	−0.150	−0.045	0
	0.6	0	0.027	0.094	0.166	0.255	0.364	0.492 / −0.508	−0.358	−0.206	−0.063	0
	0.7	0	0.017	0.060	0.107	0.180	0.281	0.416	0.583 / −0.417	−0.240	−0.−73	0
	0.8	0	0.011	0.043	0.077	0.141	0.238	0.377	0.553	0.743 / −0.257	−0.079	0
	0.9	0	0.009	0.036	0.066	0.126	0.223	0.362	0.542	0.736	0.919 / −0.081	0
	1.0	0	0.009	0.035	0.065	0.124	0.220	0.360	0.541	0.735	0.919	1

Table 5.9 Con't

α	β	\multicolumn{11}{c}{ξ}									
		0.0	0.1	0.2	0.3	0.4	0.5	0.6	0.7	0.8	0.9	1.0
	0.0	-1	-0.915	-0.725	-0.525	-0.342	-0.203	-0.110	-0.055	-0.031	-0.007	0
	0.1	0	$\frac{0.084}{-0.916}$	-0.726	0.527	-0.344	-0.205	-0.112	-0.057	-0.032	-0.008	0
	0.2	0	0.082	$\frac{0.267}{-0.733}$	-0.539	-0.360	-0.222	-0.128	-0.069	-0.039	-0.010	0
	0.3	0	0.076	0.248	$\frac{0.429}{-0.571}$	-0.401	-0.267	-0.169	-0.101	-0.058	-0.016	0
	0.4	0	0.065	0.212	0.367	$\frac{0.518}{-0.482}$	-0.355	-0.250	-0.163	-0.094	-0.027	0
450	0.5	0	0.046	0.153	0.265	0.384	$\frac{0.500}{-0.500}$	-0.384	-0.265	-0.153	-0.046	0
	0.6	0	0.027	0.094	0.163	0.250	0.355	$\frac{0.482}{-0.518}$	-0.367	-0.212	-0.065	0
	0.7	0	0.016	0.058	0.101	0.169	0.267	0.401	$\frac{0.571}{-0.429}$	-0.248	-0.076	0
	0.8	0	0.010	0.039	0.069	0.128	0.222	0.360	0.539	$\frac{0.733}{-0.267}$	-0.082	0
	0.9	0	0.008	0.032	0.057	0.112	0.205	0.344	0.527	0.726	$\frac{0.916}{-0.084}$	0
	1.0	0	0.007	0.031	0.055	0.110	0.203	0.342	0.525	0.725	0.915	1
	0.0	-1	-0.912	-0.714	-0.511	-0.325	-0.187	-0.097	-0.047	-0.30	-0.006	0
	0.1	0	$\frac{0.087}{-0.013}$	-0.715	-0.513	-0.328	-0.189	-0.100	-0.049	-0.031	-0.007	0
	0.2	0	0.085	$\frac{0.278}{-0.722}$	-0.525	-0.344	-0.207	-0.116	-0.061	-0.038	-0.009	0
	0.3	0	0.079	0.259	$\frac{0.441}{-0.559}$	-0.388	-0.255	-0.160	-0.095	-0.057	-0.015	0
	0.4	0	0.067	0.221	0.376	$\frac{0.527}{-0.473}$	-0.347	-0.245	-0.160	-0.094	-0.027	0
500	0.5	0	0.047	0.158	0.267	0.386	$\frac{0.500}{-0.500}$	-0.386	-0.267	-0.158	-0.047	0
	0.6	0	0.027	0.094	0.160	0.245	0.347	$\frac{0.473}{-0.527}$	-0.376	-0.221	-0.067	0
	0.7	0	0.015	0.057	0.095	0.160	0.255	0.388	$\frac{0.559}{-0.441}$	-0.259	-0.079	0
	0.8	0	0.009	0.038	0.061	0.116	0.207	0.344	0.525	$\frac{0.722}{-0.278}$	-0.085	0
	0.9	0	0.007	0.031	0.049	0.100	0.189	0.328	0.513	0.715	$\frac{0.913}{-0.087}$	0
	1.0	0	0.006	0.030	0.047	0.097	0.187	0.325	0.511	0.714	0.912	1

Table 5.10 Example 5.1

ξ	0.0	0.1	0.2	0.3	0.4	0.5	0.6	0.7	0.8	0.9	1.0
$\bar{p}$	0.018	0.143	0.231	0.303	0.372	0.458	0.578	0.749	0.991	1.319	1.750
p	36.00	283.0	462.0	606.0	744.0	916.0	1,156	1,598	1,982	2,638	3,500
$\bar{Q}$	0	0.009	0.028	0.053	0.090	0.029	−0.020	−0.053	−0.068	−0.053	0
Q	0	1,080	3,360	6,360	10,800	3,480	−2,400	−6,360	−8,160	−6,360	0
$\bar{M}$	0	0.001	0.002	0.007	0.014	0.019	0.019	0.015	0.009	0.003	0
M	0	4,800	9,600	33,600	67,200	45,600	45,600	36,000	21,600	7,200	0

Table 5.11 Example 5.2 *q* plus fictitious load

ξ	0	0.1	0.2	0.3	0.4	0.5	0.6	0.7	0.8	0.9	1
$\bar{p}$	0.535	0.565	0.575	0.577	0.588	0.626	0.706	0.843	1.053	1.354	1.761
p	1,070	1,130	1,150	1,154	1,176	1,252	1,412	1,686	2,106	2,706	3,522
$\bar{Q}$	0	0.055	0.112	0.069	0.028	−0.012	−0.046	−0.069	−0.074	−0.055	0
Q	0	6,600	13,440	8,280	3,360	−1,140	−5,520	−8,280	−8,880	−6,600	0
$\bar{M}$	0	0.003	0.011	0.020	0.025	0.026	0.022	0.017	0.009	0.003	0
M	0	7,200	26,400	48,000	60,000	62,400	52,800	40,800	21,600	7,200	0

Table 5.12 Extended fictitious load (−*q*)

ξ	0	0.1	0.2	0.3	0.4	0.5	0.6	0.7	0.8	0.9	1
$\bar{p}$	−0.177	−0.085	−0.030	0.006	0.037	0.082	0.155	0.272	0.448	0.703	1.049
p	354	170	60	−12	−74	−164	−310	−544	−896	−1,406	−2,098
$\bar{Q}$	0	−0.013	−0.020	−0.019	−0.018	−0.012	0.000	0.021	0.056	0.013	0
Q	0	1,560	2,400	2,280	2,160	1,440	0.000	−2,520	−6,720	−7,560	0
$\bar{M}$	0	0.001	−0.002	−0.005	−0.006	−0.008	−0.009	−0.008	−0.004	−0.001	0
M	0	2,400	4,800	12,000	14,400	19,200	21,600	19,200	9,600	2,400	0

Table 5.13 Example 5.2 Final soil pressure

ξ	0	0.1	0.2	0.3	0.4	0.5	0.6	0.7	0.8	0.9	1
p_1	1,070	1,130	1,150	1,154	1,176	1,252	1,412	1,686	2,106	2,706	3,522
p_2	354	170	60	−12	−74	−164	−310	−544	−896	−1,406	−2,098
$\sum p$	1,424	1,300	1,210	1,142	1,102	1,088	1,102	1,142	1,210	1,300	1,424

Table 5.14 Example 5.2 Final moments

ξ	0	0.1	0.2	0.3	0.4	0.5	0.6	0.7	0.8	0.9	1
M_1	0	7,200	26,400	48,000	60,000	62,400	52,800	40,800	21,600	7,200	0
M_2	0	2400	4,800	12,000	14,400	19,200	21,600	19,200	9,600	2,400	0
$\sum M$	0	9,600	31,200	60,000	74,000	81,600	74,400	60,000	31,200	9,600	0

Table 5.15 Example 5.3

ξ	0	0.1	0.2	0.3	0.4	0.5	0.6	0.7	0.8	0.9	1
$\overline{p}$	2.076	1.694	1.374	1.114	0.912	0.768	0.680	0.644	0.664	0.732	0.852
p	8.141	6.643	5.388	4.369	3.576	3.011	2.667	2.525	2.604	2.871	3.341
$\overline{Q}$	0	0.188	0.342	0.465	$\dfrac{0.566}{-0.434}$	−0.351	−0.2	−0.213	−0.148	−0.080	0
Q	0	41.36	75.24	102.30	$\dfrac{124.52}{-95.48}$	−77.22	−61.60	−46.86	−32.56	−17.60	0
$\overline{M}$	0	0.011	0.037	0.078	0.130	0.090	0.059	0.034	0.016	0.005	0
M	0	41.14	138.38	291.72	486.20	336.60	220.66	127.16	59.84	18.70	0

Table 5.16 Example 5.4

ξ	0	0.1	0.2	0.3	0.4	0.5	0.6	0.7	0.8	0.9	1
$\overline{p}$	9.797	5.774	2.938	1.046	−0.144	−0.880	−1.404	−1.962	−2.798	−4.154	−6.277
p	0.435	0.257	0.131	0.047	−0.006	−0.004	−0.062	−0.087	−0.124	−0.185	−0.279
$\overline{Q}$	0	0.768	1.194	1.386	1.427	1.373	1.259	1.090	0.856	0.514	0
Q	0	5.12	7.96	9.240	9.513	9.153	8.393	7.267	5.707	2.933	0
$\overline{M}$	0	0.042	$\dfrac{0.136}{-0.864}$	−0.727	−0.586	−0.445	−0.312	−0.195	−0.090	−0.028	0
M	0	8.400	$\dfrac{27.30}{-172.80}$	−145.40	−117.20	−89.00	−62.40	−39.00	−18.00	−0.560	0

Table 5.17 Rotations of beams with uniformly distributed loads

Coefficients $\bar{\varphi}$ for beams with uniformly distributed loads q

α	β	ξ 0.0	0.1	0.2	0.3	0.4	0.5	0.6	0.7	0.8	0.9	1.0
1	0	0.0115	0.0113	0.0102	0.0078	0.0012	0.0000	-0.0012	-0.0078	-0.0102	-0.0113	-0.0115
	0.1	0.5762	0.5750	0.5740	0.5710	0.5670	0.5620	0.5570	0.5540	0.5510	0.54497	0.5495
	0.2	1.0134	1.0132	1.0112	1.0000	1.0050	0.9998	0.9950	0.9910	0.9880	0.9870	0.9860
	0.3	1.3240	1.3240	1.3230	1.3210	1.3170	1.3130	1.3080	1.3050	1.3020	1.3010	1.3010
	0.4	1.5080	1.5080	1.5075	1.5060	1.5040	1.5010	1.4970	1.4940	1.4920	1.4910	1.4900
	0.5	1.5670	1.5670	1.5670	1.5660	1.5650	1.5630	1.5610	1.5590	1.5570	1.5560	1.5560
	0.6	1.5020	1.5020	1.5020	1.5020	1.5010	1.5010	1.5000	1.4990	1.4970	1.4970	1.4970
	0.7	1.3121	1.3122	1.3122	1.3123	1.3125	1.3127	1.3128	1.3129	1.3125	1.9123	1.3122
	0.8	0.9980	0.998	0.9990	0.9990	0.9990	1.0000	1.0000	1.0010	1.0010	1.0020	1.0020
	0.9	0.5610	0.561	0.5610	0.5610	0.5620	0.5620	0.5030	0.5640	0.5640	0.5650	0.5650
2	0	0.0230	0.0226	0.0203	0.0155	0.0084	0.0000	-0.0083	-0.0155	-0.0203	-0.0226	-0.0230
	0.1	0.5900	0.5890	0.5860	0.5800	0.5710	0.5620	0.5520	0.5470	0.5300	0.5370	0.5370
	0.2	1.0270	1.0260	1.0240	1.0180	1.0090	0.9996	0.9900	0.9820	0.9770	0.9740	0.9740
	0.3	1.3350	1.3350	1.3330	1.3290	1.3220	1.3130	1.3040	1.2970	1.2920	1.2890	1.2890
	0.4	1.5160	1.5160	1.5150	1.5130	1.5080	1.5010	1.4940	1.4880	1.4830	1.4810	1.4800
	0.5	1.5720	1.5720	1.5720	1.5700	1.5680	1.5640	1.5590	1.5550	1.5510	1.5490	1.5490
	0.6	1.5040	1.5040	1.5040	1.5030	1.5030	1.5010	1.5000	1.4970	1.4950	1.4940	1.4930
	0.7	1.3120	1.3120	1.3120	1.3120	1.3130	1.3130	1.3130	1.3130	1.3130	1.3120	1.3120
	0.8	0.9970	0.9970	0.9970	0.9980	0.9980	1.0000	1.0000	1.0020	1.0030	1.0040	1.0040
	0.9	0.5700	0.5600	0.5600	0.5600	0.5610	0.5620	0.5630	0.5650	0.5660	0.5670	0.5670

Table 5.17 Con't

Coefficients $\bar{\varphi}$ for beams with uniformly distributed loads q

α	β						ξ					
		0.0	0.1	0.2	0.3	0.4	0.5	0.6	0.7	0.8	0.9	1.0
5	0	0.0569	0.0560	0.0503	0.0384	0.0208	0.0000	-0.0208	-0.0384	-0.0503	-0.0560	-0.0569
	0.1	0.6300	0.6290	0.6210	0.6060	0.5850	0.5610	0.5380	0.5180	0.5050	0.4990	0.4980
	0.2	1.0660	1.0650	1.0590	1.0440	1.0230	0.9990	0.9750	0.9560	0.9430	0.9370	0.9360
	0.3	1.3680	1.3680	1.3630	1.3530	1.3350	1.3140	1.2920	1.2730	1.2610	1.2550	1.2540
	0.4	1.5410	1.5400	1.5380	1.5310	1.5200	1.5030	1.4860	1.4700	1.4590	1.4530	1.4530
	0.5	1.5860	1.5860	1.5850	1.5810	1.5760	1.5670	1.5550	1.5430	1.5340	1.5300	1.5290
	0.6	1.5090	1.5090	1.5090	1.5080	1.5060	1.5030	1.4990	1.4930	1.4870	1.4840	1.4840
	0.7	1.3110	1.3110	1.3110	1.3120	1.3130	1.3140	1.3140	1.3140	1.3130	1.3120	1.3110
	0.8	0.9920	0.9930	0.9930	0.9940	0.9960	0.9990	1.0020	1.0050	1.0080	1.0090	1.0090
	0.9	0.5550	0.5550	0.5560	0.5570	0.5580	0.5610	0.5640	0.5670	0.5710	0.5730	0.5730
10	0	0.1120	0.1100	0.0993	0.0757	0.0410	0.0000	-0.0410	-0.0757	-0.0993	-0.1100	-0.1120
	0.1	0.6960	0.6940	0.6780	0.6480	0.6060	0.5590	0.5130	0.4750	0.4500	0.4380	0.4360
	0.2	1.1300	1.1280	1.1160	1.0870	1.0480	0.9980	0.9520	0.9130	0.8870	0.8750	0.8730
	0.3	1.4220	1.4210	1.4130	1.3920	1.3570	1.3150	1.2720	1.2350	1.2110	1.1990	1.1970
	0.4	1.5800	1.5790	1.5740	1.5610	1.5390	1.5060	1.4720	1.4400	1.4190	1.4080	1.4060
	0.5	1.6090	1.6090	1.6060	1.6000	1.5890	1.5710	1.5480	1.5240	1.5070	1.4990	1.4970
	0.6	1.5190	1.5190	1.5180	1.5160	1.5130	1.5070	1.4980	1.4860	1.4750	1.4690	1.4680
	0.7	1.3090	1.3090	1.3100	1.3110	1.3130	1.3150	1.3160	1.3160	1.3130	1.3110	1.3100
	0.8	0.9850	0.9850	0.9860	0.9890	0.9930	0.9980	1.0040	1.0120	1.0170	1.0170	1.0180
	0.9	0.5480	0.5480	0.5490	0.5510	0.5540	0.5590	0.5650	0.5720	0.5700	0.5830	0.5840
15	0	0.1660	0.1630	0.1470	0.1120	0.0607	0.0000	-0.0607	-0.1120	-0.1470	-0.163	-0.166
	0.1	0.7610	0.7570	0.7430	0.6800	0.6270	0.5580	0.4900	0.4340	0.3960	0.3780	0.3750
	0.2	1.1930	1.1900	1.1720	1.1290	1.0680	0.9970	0.9280	0.8710	0.8330	0.8150	0.8120
	0.3	1.4750	1.4730	1.4610	1.4300	1.3790	1.3160	1.2520	1.1980	1.1620	1.1450	1.1420

0.4	1.6180	1.6170	1.6000	1.5910	1.5580	1.5100	1.4580	1.4120	1.3790	1.3640	1.3610
0.5	1.6320	1.6310	1.6270	1.6180	1.6010	1.5750	1.5400	1.5060	1.4810	1.4680	1.4660
0.6	1.5270	1.5270	1.5260	1.5240	1.5190	1.5110	1.4970	1.4790	1.4620	1.4540	1.4520
0.7	1.3080	1.3080	1.3090	1.3100	1.3130	1.3160	1.3180	1.3180	1.3140	1.3090	1.3090
0.8	0.9780	0.9780	0.9800	0.9830	0.9890	0.9970	1.0070	1.0170	1.0250	1.0260	1.0270
0.9	0.5410	0.5420	0.5430	0.5460	0.5510	0.5580	0.5670	0.5770	0.5870	0.5930	0.5950

20											
0	0.2190	0.2150	0.1930	0.1470	0.0800	0.0000	-0.0800	-0.1470	-0.1930	-0.2150	-0.2190
0.1	0.8230	0.8180	0.7880	0.7290	0.6480	0.5560	0.4670	0.3930	0.34300	0.3200	0.3160
0.2	1.2540	1.2500	1.2260	1.1700	1.0890	0.9960	0.9060	0.8310	0.7800	0.7570	0.7530
0.3	1.5260	1.5230	1.5070	1.4670	1.4000	1.3170	1.2330	1.1630	1.1140	1.0920	1.0880
0.4	1.6550	1.6530	1.6440	1.6190	1.5760	1.5140	1.4450	1.3840	1.3410	1.3210	1.3170
0.5	1.6530	1.6530	1.6480	1.6350	1.6130	1.5790	1.5340	1.4880	1.4650	1.4380	1.4350
0.6	1.5360	1.5360	1.5350	1.5310	1.5250	1.5140	1.4960	1.4720	1.4510	1.4390	1.4360
0.7	1.3060	1.3070	1.3080	1.3100	1.3130	1.3170	1.3200	1.3200	1.3140	1.3090	1.3070
0.8	0.9710	0.9710	0.9730	0.9780	0.9860	0.9960	1.0000	1.0220	1.0320	1.0350	1.0350
0.9	0.5350	0.5350	0.5370	0.5400	0.5470	0.5560	0.5680	0.5810	0.5940	0.6030	0.6050

25											
0	0.2590	0.2550	0.2380	0.1820	0.0980	0.0000	-0.0980	-0.1820	-0.2380	-0.2550	-0.2590
0.1	0.8840	0.8780	0.8400	0.7670	0.0680	0.5650	0.4450	0.3540	0.2020	0.2640	0.2600
0.2	1.3120	1.3080	1.2780	1.2090	1.1100	0.9950	0.8840	0.7910	0.7290	0.7000	0.6950
0.3	1.6750	1.6720	1.6520	1.6040	1.4210	1.3180	1.2150	1.1280	1.0680	1.0400	1.0350
0.4	1.6910	1.6890	1.6770	1.6470	1.6040	1.6170	1.4320	1.3570	1.3040	1.2700	1.2750
0.5	1.6740	1.6730	1.6670	1.6520	1.6250	1.6830	1.6270	1.4710	1.4290	1.4090	1.4050
0.6	1.6440	1.6440	1.6430	1.6380	1.6300	1.6170	1.4960	1.4660	1.4390	1.4240	1.4210
0.7	1.3040	1.3050	1.3060	1.3090	1.3130	1.3180	1.3220	1.3220	1.3140	1.3080	1.3060
0.8	0.9640	0.9650	0.9670	0.9730	0.9820	0.9950	1.0110	1.0280	1.0400	1.0430	1.0430
0.9	0.5280	0.5290	0.5310	0.5350	0.5430	0.5650	0.5910	0.5860	0.6020	0.6130	0.6150

Table 5.17 Con't

Coefficients $\overline{\varphi}$ for beams with uniformly distributed loads q

α	β	0.0	0.1	0.2	0.3	0.4	ξ 0.5	0.6	0.7	0.8	0.9	1.0
50	0	0.5050	0.4970	0.4470	0.0340	0.1840	0.0000	-0.1840	-0.3400	-04470	-0.4970	-0.5050
	0.1	1.1670	1.1550	1.0840	0.9460	0.7680	0.5470	0.3410	0.1710	0.0560	0.0030	-0.0060
	0.2	1.5860	1.5780	1.5220	1.3930	1.2050	0.9900	0.7820	0.6090	0.4920	0.4380	0.4290
	0.3	1.8040	1.7990	1.7620	1.6710	1.6160	1.3240	1.1300	0.9660	0.8540	0.8020	0.7930
	0.4	1.8550	1.8520	1.8310	1.7770	1.6780	1.6340	1.3740	1.2320	1.1320	1.0850	1.0760
	0.5	1.7700	1.7690	1.7680	1.7310	1.6810	1.6030	1.4970	1.3910	1.3110	1.2720	1.2650
	0.6	1.6820	1.6810	1.6790	1.6720	1.6580	1.6340	1.4940	1.4360	1.3840	1.3560	1.3500
	0.7	1.2980	1.2990	1.3010	1.3060	1.3140	1.3240	1.3310	1.3310	1.3160	1.3020	1.2990
	0.8	0.9340	0.9350	0.9390	0.9490	0.9660	0.9960	1.0210	1.0520	1.0760	1.0810	1.0810
	0.9	0.4990	0.5000	0.5030	0.5110	0.5260	0.5470	0.5740	0.6060	06370	0.6680	0.6620
100	0	0.8980	0.8830	0.7940	0.6050	0.3280	0.0000	-0.3280	-0.6050	-0.7940	-0.0030	-0.8980
	0.1	1.6420	1.6190	1.4900	1.2420	0.9080	0.5320	1.5780	1.3320	-3.3360	-4.2080	-4.4610
	0.2	2.0430	2.0280	1.9300	1.6980	1.3630	0.9810	0.6110	0.3040	0.0980	0.0030	-0.0140
	0.3	2.1810	2.1720	2.1110	1.9530	1.6780	1.3350	0.9880	0.6960	0.4970	0.4040	0.3880
	0.4	2.1240	2.119	2.0850	1.9940	1.8230	1.5660	1.2780	1.0230	0.8430	0.7680	0.7430
	0.5	1.9250	1.9230	1.9060	1.8610	1.7770	1.6410	1.4500	1.2560	1.1120	1.0400	1.0270
	0.6	1.6410	1.6400	1.6370	1.6280	1.6060	1.5660	1.4950	1.3890	1.2910	1.2370	1.2260
	0.7	1.2860	1.2870	1.2910	1.3010	1.3160	1.3350	1.3500	1.3480	1.3170	1.2900	1.2830
	0.8	0.8840	0.8850	0.8920	0.9100	0.9390	0.9810	1.0350	1.0930	1.1360	1.1450	1.1450
	0.9	0.4520	0.4530	0.4580	0.4720	0.4960	0.5320	0.5800	0.6380	0.6960	0.7370	0.7440
150	0	1.2120	1.1910	1.0720	0.8170	0.4430	0.0000	-0.4430	-0.8170	-1.0720	-1.1910	-1.2120
	0.1	2.0250	1.9940	1.8160	1.4780	1.0240	0.5170	0.0280	-0.3760	-0.6480	-0.7740	-0.7960
	0.2	2.0250	2.3880	2.2570	1.9430	1.4890	0.9730	0.4740	0.0610	-0.2160	-0.3450	-0.3670
	0.3	2.4080	2.4670	2.3890	2.1810	1.8100	1.3450	0.8770	0.4810	0.2110	0.0850	0.0630
	0.4	2.4790	2.3260	2.2840	2.1690	1.9450	1.5970	1.2040	0.8550	0.6100	0.4930	0.4730

	0.5	2.3320	2.0410	2.0200	1.9640	1.8560	1.6750	1.4140	1.1470	0.9480	0.8500	0.8320
	0.6	2.0440	1.6840	1.6810	1.6720	1.6470	1.5970	1.5020	1.3520	1.2120	1.1350	1.1210
	0.7	1.6850	1.2760	1.2830	1.2970	1.3190	1.3450	1.3680	1.3640	1.3170	1.2760	1.2670
	0.8	1.2750	0.8470	0.8560	0.8780	0.9160	0.9730	1.0460	1.1260	1.1860	1.1970	1.1960
	0.9	0.8450	0.4170	0.4240	0.4400	0.4700	0.5170	0.5820	0.6620	0.7440	0.8030	0.8130
200	0	1.4690	1.4440	1.2990	0.9900	0.5360	0.0000	-0.5360	-0.9000	-1.2990	-1.4440	-1.4690
	0.1	2.3420	2.3040	2.0840	1.6700	1.1170	0.5030	0.0880	-0.5750	-0.9020	-1.0550	-1.0810
	0.2	2.7070	2.6840	2.5260	2.1430	1.5910	0.9650	0.3510	0.1370	-0.4720	-0.6280	-0.6550
	0.3	2.7180	2.7050	2.6150	2.3690	1.9210	1.3560	0.7860	0.3040	0.0230	-0.1770	-0.2040
	0.4	2.4970	2.4910	2.4440	2.3130	2.0490	1.6270	1.1460	0.7180	0.4170	0.2740	0.2490
	0.5	2.1380	2.1350	2.1110	2.0480	1.9230	1.7070	1.3870	1.0680	0.8120	0.6910	0.6690
	0.6	1.7170	1.7180	1.7160	1.7070	1.6830	1.6270	1.6120	1.3240	1.1460	1.0470	1.0280
	0.7	1.2650	1.2670	1.2760	1.2940	1.3220	1.3560	1.3850	1.3790	1.3170	1.2610	1.2490
	0.8	0.8140	0.8160	0.8270	0.8520	0.8970	0.9650	1.0540	1.1580	1.2270	1.2410	1.2390
	0.9	0.3880	0.3890	0.3960	0.4140	0.4480	0.5030	0.5810	0.6800	0.7850	0.8600	0.8740
250	0	1.6830	1.6540	1.4880	1.1340	0.6140	0.0000	-0.6140	-1.1340	-1.4880	-1.6540	-1.6830
	0.1	2.6100	2.5650	2.3070	1.8280	1.1930	0.4890	0.1850	-0.7410	-1.1130	-1.2870	-1.3170
	0.2	2.9570	2.9310	2.7510	2.3100	1.6750	0.9570	0.3030	-0.6850	-0.8620	-0.8930	-0.9350
	0.3	2.9130	2.9000	2.8030	2.5280	2.0160	1.3670	0.7110	0.1570	-0.2190	-0.3950	-0.4260
	0.4	2.6290	2.6230	2.5750	2.4350	2.1390	1.6550	1.1000	0.6030	0.2550	0.0890	0.0600
	0.5	2.2130	2.2090	2.1840	2.1160	1.9790	1.7350	1.3650	0.9820	0.6960	0.5650	0.5300
	0.6	1.7420	1.7430	1.7430	1.7370	1.7140	1.6550	1.5250	1.3010	1.0880	0.9690	0.9460
	0.7	1.2560	1.2580	1.2690	1.2910	1.3250	1.3670	1.4020	1.3940	1.3150	1.2460	1.2310
	0.8	0.7890	0.7910	0.8030	0.8310	0.8810	0.9570	1.0610	1.1760	1.2630	1.2770	1.2740
	0.9	0.3650	0.3670	0.3740	0.3940	0.4290	0.4890	0.5780	0.6940	0.8200	0.9110	0.9270

Table 5.17 Con't

Coefficients $\overline{\varphi}$ for beams with uniformly distributed loads q

α	β											
							ξ					
		0.0	0.1	0.2	0.3	0.4	0.5	0.6	0.7	0.8	0.9	1.0
300	0	1.8640	1.8320	1.6480	1.2560	0.6800	0.0000	-0.6800	-1.2560	-1.649	-1.832	-1.864
	0.1	2.8300	2.7880	2.4970	1.9610	1.2540	0.4760	0.2680	-0.8800	-1.291	-1.832	-1.516
	0.2	3.1680	3.1400	2.9420	2.4520	1.7460	0.9600	0.1860	-0.4430	-0.864	-1.483	-1.095
	0.3	3.0750	3.0620	2.9610	2.5650	2.0990	1.3770	0.6480	0.0330	-0.385	-1.061	-0.615
	0.4	2.7360	2.7310	2.6840	2.5390	2.2200	1.6830	1.0620	0.5050	0.1160	-0.581	-0.102
	0.5	2.2750	2.2710	2.2440	2.1730	2.0280	1.7610	1.3470	0.0170	0.5960	0.0690	0.4110
	0.6	1.7620	1.7620	1.7640	1.7610	1.7420	1.6830	1.5390	1.2830	1.0360	0.4390	0.8720
	0.7	1.2480	1.2500	1.2630	1.2880	1.3280	1.3770	1.4180	1.4090	1.3130	0.8990	1.2120
	0.8	0.7690	0.7710	0.7830	0.8130	0.8660	0.9600	1.0650	1.1060	1.2950	1.2300	1.3040
	0.9	0.3480	0.3490	0.3560	0.3750	0.4120	0.4760	0.5740	0.7050	0.8500	1.3080	0.9750
350	0	2.0190	1.9840	1.7580	1.3600	0.7370	0.0000	-0.7370	-1.3600	-1.785	-1.984	-2.019
	0.1	3.0380	2.9820	2.6610	2.0740	1.3060	0.4630	0.3400	-1.0000	-1.443	-1.649	-1.685
	0.2	3.3490	3.3190	3.1070	2.5740	1.8060	0.9430	0.1170	-0.5630	-1.018	-1.230	-1.267
	0.3	3.2110	3.1990	3.0960	2.7830	2.1720	1.3880	0.5950	0.0740	-0.528	-0.741	-0.778
	0.4	2.8240	2.8190	2.7760	2.6290	2.2920	1.7090	1.0300	0.4210	0.0040	-.2070	-0.242
	0.5	2.3260	2.3220	2.2940	2.2210	2.0700	1.7850	1.3330	0.8610	0.5090	0.3370	0.3070
	0.6	1.7760	1.7770	1.7800	1.7820	1.7670	1.7090	1.5550	1.2690	0.9910	0.3500	0.8050
	0.7	1.2410	1.2430	1.2570	1.2860	1.3320	1.3880	1.4350	1.4230	1.3110	1.2140	1.1920
	0.8	0.7520	0.7540	0.7670	0.7980	0.8540	0.9430	1.0690	1.2140	1.3220	1.3350	1.3300
	0.9	0.3330	0.3350	0.3420	0.3600	0.3970	0.4630	0.5600	0.7130	0.8760	0.9970	1.0190
400	0	2.1530	2.1160	1.9040	1.4510	0.7860	0.0000	-0.7860	-1.4510	-1.9040	-2.1160	-2.1530
	0.1	3.2120	3.1510	2.8030	2.1700	1.3490	0.4510	0.4030	-1.1040	-1.5740	-1.7930	-1.8310
	0.2	3.5050	3.4750	3.2510	2.6800	1.8580	0.9370	0.0560	0.6660	-1.1500	-1.3760	-1.4160
	0.3	3.3260	3.3140	3.2130	2.8880	2.2370	1.3980	0.5490	0.1660	-0.6520	-0.8800	-0.9190

	0.4	2.8970	2.8930	2.8530	2.7080	2.3580	1.7350	1.0040	0.3490	0.1100	-0.3270	-0.3650
	0.5	2.3690	2.3650	2.3360	2.2610	2.1060	1.8070	1.3200	0.8110	0.4320	0.2480	0.2160
	0.6	1.7880	1.7890	1.7940	1.7990	1.790	1.7350	1.5710	1.2570	0.9490	0.7770	0.7440
	0.7	1.2340	1.2370	1.2520	1.2850	1.3360	1.3980	1.4510	1.4370	1.3090	1.1980	1.1730
	0.8	0.7380	0.7400	0.7630	0.7850	0.8430	0.9370	1.0720	1.2290	1.3470	1.3590	1.3520
	0.9	0.3220	0.3230	0.3300	0.3470	0.3830	0.4510	0.5630	0.7200	0.8990	1.0350	1.0580
450	0	2.2710	2.2320	2.0080	1.5300	0.8290	0.0000	-0.8290	-1.5300	-2.0080	-2.2320	-2.2710
	0.1	3.3670	3.3020	2.9280	2.2540	1.3850	0.4390	0.4590	-1.1950	-1.6880	-1.9180	-1.9580
	0.2	3.6420	3.6110	3.3770	2.7730	1.9030	0.9310	0.0040	-0.7570	-1.2660	-1.5040	-1.5450
	0.3	3.4240	3.4130	3.3140	2.9810	2.2960	1.4090	0.5100	0.2470	-0.7610	-1.0020	-1.0440
	0.4	2.9570	2.9550	2.9190	2.7780	2.4180	1.7600	0.9820	0.2840	0.2030	-0.4340	-0.4740
	0.5	2.4060	2.4010	2.3720	2.2960	2.1380	1.8260	1.3090	0.7660	0.3640	0.1690	0.1360
	0.6	1.7970	1.7980	1.8050	1.8140	1.8110	1.760	1.5890	1.2480	0.9120	0.7230	0.6870
	0.7	1.2270	1.2300	1.2470	1.2830	1.3390	1.4090	1.4670	1.4500	1.3060	1.1810	1.1530
	0.8	0.7250	0.7280	0.7410	0.7730	0.8330	0.9310	1.0740	1.2430	1.3700	1.3790	1.3710
	0.9	0.3120	0.3130	0.320	0.3360	0.370	0.4390	0.5560	0.7240	0.9200	1.0700	1.0960
500	0	2.3750	2.3340	2.0990	1.6000	0.8670	0.0000	-0.8670	-1.6000	-2.0990	-2.3340	-2.3750
	0.1	3.5050	3.4360	3.0390	2.3280	1.4160	0.4270	-0.5080	-1.2740	-1.7880	-2.0280	-2.0700
	0.2	3.7630	3.7320	3.4890	2.8550	1.9430	0.9260	0.0430	-0.8360	-1.3680	-1.6160	-1.6600
	0.3	3.5080	3.4980	3.4030	3.0640	2.3500	1.4190	0.4760	0.3190	-0.8570	-1.1100	-1.1540
	0.4	3.0080	3.0060	2.9760	2.8410	2.4730	1.7830	0.9630	0.2270	-0.2860	-0.5290	-0.5710
	0.5	2.4390	2.4330	2.4020	2.3260	2.1660	1.8430	1.2990	0.7260	0.3030	0.0990	0.0640
	0.6	1.8040	1.8050	1.8130	1.8270	1.8300	1.7830	1.6060	1.2410	0.8770	0.6730	0.6330
	0.7	1.2210	1.2240	1.2420	1.2810	1.3430	1.4190	1.4830	1.4640	1.3030	1.1640	1.1330
	0.8	0.7150	0.7170	0.7310	0.7630	0.8240	0.9260	1.0760	1.2550	1.3900	1.3980	1.3880
	0.9	0.3040	0.3060	0.3110	0.3260	0.3590	0.4270	0.5490	0.7280	0.9390	1.1020	1.1310

Table 5.18 Relative settlements of beams with uniformly distributed loads

Coefficients $\bar{y}$ for beams with uniformly distributed loads q

α	β	0.0	0.1	0.2	0.3	0.4	0.5	0.6	0.7	0.8	0.9	1.0
	0	0	0.00115	0.00223	0.00314	0.00374	0.00395	0.00374	0.00314	0.00223	0.00115	0.000
	0.1	0	0.0580	0.115	0.172	0.229	0.286	0.342	0.397	0.453	0.508	0.563
	0.2	0	0.1010	0.203	0.304	0.404	0.505	0.604	0.704	0.803	0.901	1.000
	0.3	0	0.1330	0.265	0.397	0.592	0.660	0.791	0.922	1.052	1.182	1.312
1	0.4	0	0.1510	0.302	0.452	0.603	0.753	0.903	1.053	1.202	1.351	1.500
	0.5	0	0.1570	0.313	0.470	0.627	0.783	0.939	1.096	1.251	1.407	1.562
	0.6	0	0.1502	0.300	0.451	0.601	0.751	0.901	1.051	1.201	1.350	1.500
	0.7	0	0.1310	0.262	0.394	0.525	0.656	0.787	0.919	1.050	1.181	1.312
	0.8	0	0.0990	0.199	0.299	0.399	0.499	0.599	0.699	0.799	0.899	1.000
	0.9	0	0.0561	0.112	0.168	0.224	0.281	0.337	0.393	0.450	0.506	0.563
	0	0	0.00229	0.00445	0.00626	0.00747	0.00790	0.00747	0.00626	0.00445	0.00229	0.000
	0.1	0	0.0590	0.118	0.176	0.234	0.290	0.346	0.401	0.455	0.509	0.563
	0.2	0	0.1030	0.205	0.307	0.409	0.509	0.609	0.707	0.805	0.903	1.000
	0.3	0	0.1330	0.267	0.400	0.533	0.664	0.795	0.925	1.055	1.184	1.312
2	0.4	0	0.1520	0.303	0.465	0.606	0.756	0.906	1.055	1.204	1.352	1.500
	0.5	0	0.1570	0.214	0.471	0.628	0.785	0.941	1.097	1.252	1.407	1.562
	0.6	0	0.1500	0.301	0.451	0.601	0.752	0.902	1.051	1.201	1.350	1.500
	0.7	0	0.1330	0.262	0.394	0.525	0.656	0.787	0.919	1.050	1.181	1.312
	0.8	0	0.0997	0.199	0.299	0.399	0.499	0.599	0.699	0.799	0.899	1.000
	0.9	0	0.0560	0.112	0.168	0.223	0.280	0.336	0.393	0.449	0.506	0.563
	0	0	0.00567	0.0110	0.0155	0.0185	0.0196	0.0185	0.0155	0.0110	0.00567	0.000
5	0.1	0	0.053	0.126	0.187	0.247	0.304	0.359	0.412	0.463	0.513	0.563
	0.2	0	0.107	0.213	0.318	0.421	0.523	0.621	0.718	0.813	0.907	1.000
	0.3	0	0.137	0.273	0.400	0.544	0.676	0.806	0.935	1.061	1.187	1.312

		0										
	0.4	0	0.154	0.308	0.461	0.614	0.765	0.915	1.062	1.209	1.354	1.500
	0.5	0	0.159	0.317	0.476	0.633	0.791	0.947	1.102	1.255	1.409	1.562
	0.6	0	0.151	0.302	0.453	0.603	0.754	0.904	1.054	1.203	1.354	1.500
	0.7	0	0.131	0.262	0.393	0.525	0.656	0.787	0.919	1.050	1.181	1.312
	0.8	0	0.099	0.199	0.298	0.397	0.497	0.597	0.698	0.798	0.899	1.000
	0.9	0	0.056	0.111	1.670	0.222	0.278	0.335	0.391	0.448	0.505	0.563
10	0	0	0.0112	0.0218	0.0306	0.0365	0.0.386	0.0365	0.0306	0.0218	0.0112	0.000
	0.1	0	0.0696	0.138	0.205	0.268	0.326	0.370	0.429	0.475	0.519	0.563
	0.2	0	0.113	0.225	0.336	0.442	0.545	0.642	0.735	0.825	0.913	1.000
	0.3	0	0.142	0.284	0.424	0.562	0.596	0.825	0.950	1.072	0.193	1.312
	0.4	0	0.158	0.316	0.473	0.628	0.780	0.920	1.074	1.217	1.359	1.499
	0.5	0	0.161	0.322	0.482	0.642	0.800	0.955	1.109	1.261	1.411	1.561
	0.6	0	0.152	0.304	0.455	0.609	0.758	0.908	1.057	1.205	1.352	1.499
	0.7	0	0.131	0.262	0.393	0.524	0.556	0.787	0.919	1.050	1.181	1.312
	0.8	0	0.099	0.197	0.296	0.395	0.494	0.595	0.695	0.797	0.899	1.000
	0.9	0	0.055	0.110	0.165	0.220	0.276	0.332	0.389	0.446	0.504	0.563
15	0	0	0.017	0.032	0.045	0.054	0.057	0.054	0.045	0.032	0.017	0.000
	0.1	0	0.076	0.151	0.222	0.288	0.347	0.399	0.446	0.487	0.525	0.563
	0.2	0	0.119	0.237	0.353	0.463	0.566	0.662	0.752	0.837	0.919	1.000
	0.3	0	0.147	0.294	0.439	0.580	0.714	0.843	0.965	1.083	1.198	1.312
	0.4	0	0.162	0.323	0.483	0.641	0.794	0.943	1.086	1.226	1.363	1.499
	0.5	0	0.1.63	0.326	0.488	0.649	0.808	0.964	1.116	1.266	1.413	1.560
	0.6	0	0.153	0.305	0.458	0.610	0.762	0.912	1.060	1.208	1.354	1.499
	0.7	0	0.131	0.262	0.393	0.524	0.655	0.787	0.919	1.050	1.182	1.312
	0.8	0	0.098	0.196	0.294	0.392	0.492	0.592	0.693	0.795	0.898	1.000
	0.9	0	0.054	0.108	0.163	0.218	0.273	0.329	0.386	0.445	0.504	0.563

Table 5.18 Con't

Coefficients $\bar{y}$ for beams with uniformly distributed loads q

α	β	ξ 0.0	0.1	0.2	0.3	0.4	0.5	0.6	0.7	0.8	0.9	1.0
	0	0	0.0218	0.0423	0.0596	0.071	0.075	0.071	0.0596	0.0423	0.0218	0.000
	0.1	0	0.0822	0.163	0.239	0.308	0.368	0.419	0.462	0.498	0.531	0.563
	0.2	0	0.125	0.249	0.369	0.482	0.587	0.682	0.768	0.849	0.925	1.000
	0.3	0	0.153	0.304	0.453	0.507	0.733	0.860	0.980	1.093	1.203	1.312
20	0.4	0	0.165	0.330	0.494	0.654	0.808	0.956	1.098	1.234	1.367	1.498
	0.5	0	0.165	0.330	0.495	0.657	0.817	0.973	1.124	1.271	1.415	1.559
	0.6	0	0.134	0.307	0.460	0.613	0.865	0.916	1.064	1.210	1.355	1.498
	0.7	0	0.131	0.261	0.392	0.523	0.655	0.787	0.919	1.051	1.182	1.312
	0.8	0	0.097	0.194	0.292	0.390	0.489	0.589	0.691	0.794	0.897	1.000
	0.9	0	0.053	0.107	0.161	0.215	0.270	0.327	0.384	0.443	0.503	0.563
	0	0	0.027	0.052	0.073	0.088	0.093	0.088	0.073	0.052	0.027	0.000
	0.1	0	0.088	0.174	0.255	0.327	0.388	0.438	0.478	0.510	0.537	0.563
	0.2	0	0.131	0.274	0.385	0.502	0.607	0.701	0.784	0.860	0.931	1.001
	0.3	0	0.157	0.314	0.467	0.613	0.750	0.877	0.994	1.103	1.209	1.312
25	0.4	0	0.169	0.337	0.504	0.666	0.822	0.969	1.108	1.241	1.370	1.498
	0.5	0	0.167	0.335	0.501	0.665	0.825	0.981	1.131	1.275	1.417	1.558
	0.6	0	0.154	0.309	0.463	0.616	0.769	0.919	1.068	1.213	1.356	1.498
	0.7	0	0.131	0.261	0.392	0.523	0.655	0.787	0.919	1.050	1.182	1.312
	0.8	0	0.096	0.193	0.290	0.388	0.487	0.587	0.689	0.792	0.896	1.001
	0.9	0	0.053	0.106	0.159	0.213	0.268	0.324	0.382	0.441	0.502	0.563
	0	0	0.050	0.098	0.138	0.164	0.174	0.164	0.138	0.098	0.050	0.000
50	0.1	0	0.116	0.229	0.331	0.416	0.482	0.526	0.551	0.562	0.565	0.564
	0.2	0	0.158	0.314	0.460	0.591	0.700	0.789	0.858	0.912	0.958	1.002
	0.3	0	0.180	0.359	0.531	0.691	0.833	0.955	1.060	1.150	1.233	1.312

	0.4	0	0.185	0.370	0.551	0.724	0.884	1.030	1.160	1.278	1.388	1.496
	0.5	0	0.177	0.353	0.528	0.699	0.863	1.019	1.163	1.298	1.426	1.553
	0.6	0	0.158	0.316	0.474	0.630	0.785	0.937	1.083	1.224	1.361	1.496
	0.7	0	0.130	0.260	0.390	0.521	0.652	0.786	0.919	1.051	1.182	1.312
	0.8	0	0.093	0.187	0.281	0.377	0.475	0.575	0.679	0.786	0.893	1.002
	0.9	0	0.050	0.100	0.151	0.203	0.250	0.312	0.371	0.433	0.498	0.564
100	0	0	0.089	0.174	0.245	0.292	0.309	0.292	0.245	0.174	0.089	0.000
	0.1	0	0.164	0.320	0.458	0.566	0.638	0.672	0.673	0.649	0.610	0.566
	0.2	0	0.204	0.403	0.585	0.739	0.856	0.936	0.981	1.000	1.004	1.003
	0.3	0	0.218	0.433	0.637	0.819	0.970	1.086	1.170	1.228	1.273	1.312
	0.4	0	0.212	0.423	0.627	0.819	0.989	1.131	1.246	1.338	1.418	1.492
	0.5	0	0.192	0.384	0.573	0.755	0.926	1.081	1.216	1.334	1.441	1.544
	0.6	0	0.164	0.328	0.491	0.653	0.812	0.965	1.110	1.243	1.369	1.492
	0.7	0	0.129	0.257	0.387	0.518	0.650	0.785	0.920	1.053	1.183	1.392
	0.8	0	0.088	0.177	0.267	0.360	0.455	0.556	0.663	0.774	0.889	1.003
	0.9	0	0.045	0.091	0.137	0.185	0.237	0.292	0.353	0.420	0.492	0.566
150	0	0	0.121	0.235	0.330	0.304	0.417	0.394	0.330	0.235	0.121	0.000
	0.1	0	0.202	0.394	0.560	0.685	0.763	0.789	0.771	0.719	0.646	0.567
	0.2	0	0.240	0.474	0.685	0.858	0.981	1.053	1.079	1.070	1.041	1.005
	0.3	0	0.248	0.491	0.721	0.922	1.080	1.191	1.258	1.291	1.305	1.312
	0.4	0	0.233	0.464	0.687	0.894	1.072	1.212	1.314	1.387	1.441	1.489
	0.5	0	0.204	0.408	0.607	0.799	0.976	1.131	1.259	1.363	1.452	1.535
	0.6	0	0.168	0.337	0.505	0.571	0.833	0.989	1.132	1.259	1.376	1.489
	0.7	0	0.128	0.255	0.384	0.515	0.648	0.784	0.921	1.055	1.185	1.312
	0.8	0	0.085	0.170	0.256	0.346	0.440	0.541	0.650	0.765	0.885	1.005
	0.9	0	0.042	0.084	0.127	0.172	0.221	0.276	0.338	0.409	0.486	0.567

Table 5.18 Con't

Coefficients $\bar{y}$ for beams with uniformly distributed loads q

α	β	0.0	0.1	0.2	0.3	0.4	0.5	ξ 0.6	0.7	0.8	0.9	1.0
200	0	0	0.146	0.285	0.400	0.478	0.505	0.478	0.4000	0.285	0.146	0.000
	0.1	0	0.233	0.454	0.614	0.784	0.865	0.885	0.851	0.775	0.676	0.569
	0.2	0	0.270	0.532	0.768	0.955	1.083	1.149	1.159	1.127	1.070	1.006
	0.3	0	0.271	0.538	0.789	1.005	1.170	1.276	1.330	1.342	1.331	1.311
	0.4	0	0.250	0.497	0.736	0.955	1.140	1.278	1.371	1.426	1.460	1.485
	0.5	0	0.214	0.426	0.635	0.834	1.016	1.171	1.293	1.386	1.460	1.527
	0.6	0	0.172	0.343	0.515	0.684	0.850	1.008	1.150	1.273	1.382	1.485
	0.7	0	0.127	0.254	0.382	0.513	0.647	0.784	0.923	1.058	1.186	1.311
	0.8	0	0.081	0.154	0.247	0.335	0.428	0.528	0.639	0.758	0.882	1.006
	0.9	0	0.039	0.078	0.118	0.161	0209	0.263	0.326	0.399	0.482	0.569
250	0	0	0.168	0.326	0.459	0.547	0.578	0.547	0.459	0.326	0.168	0.000
	0.1	0	0.260	0.506	0.714	0.866	0.950	0.965	0.917	0.823	0.701	0.570
	0.2	0	0.295	0.581	0.836	1.037	1.169	1.229	1.226	1.175	1.096	1.007
	0.3	0	0.291	0.577	0.846	1.075	1.244	1.348	1.390	1.385	1.353	1.311
	0.4	0	0.263	0.523	0.775	1.005	1.196	1.334	1.418	1.459	1.475	1.482
	0.5	0	0.221	0.441	0.657	0.862	1.049	1.205	1.322	1.404	1.466	1.520
	0.6	0	0.174	0.349	0.523	0.695	0.864	1.024	1.166	1.285	1.387	1.482
	0.7	0	0126	0.252	0.380	0.511	0.645	0.784	0.924	1.060	1.188	1.311
	0.8	0	0.079	0.159	0.240	0.326	0.417	0.518	0.630	0.752	0.880	1.007
	0.9	0	0.037	0.074	0.112	0.153	0.198	0.252	0.315	0.391	0.478	0.570
300	0	0	0.186	0.361	0.508	0.606	0.641	0.606	0.508	0.361	0.186	0.000
	0.1	0	0.283	0.540	0.774	0.936	1.022	1.032	0.973	0.863	0.722	0.572
	0.2	0	0.316	0.622	0.894	1.106	1.240	1.297	1.282	1.215	1.117	1.009
	0.3	0	0.307	0.610	0.893	1.133	1.308	1.408	1.441	1.422	1.372	1.311

		0										
	0.4	0	0273	0.545	0.807	1.047	1.243	1.381	1.458	1.487	1.488	1.479
	0.5	0	0.227	0.453	0.675	0.886	1.076	1.233	1.345	1.420	1.470	1.512
	0.6	0	0.176	0.353	0.529	0.704	0.876	1.038	1.180	1.295	1.391	1.479
	0.7	0	0.125	0.250	0.380	0.509	0.644	0.784	0.926	1.063	1.180	1.311
	0.8	0	0.077	0.155	0.234	0.318	0.409	0.509	0.622	0.747	0.878	1.009
	0.9	0	0.035	0.070	0.197	0.148	0.190	0.242	0.306	0.384	0.474	0.572
350	0	0	0.201	0.391	0.550	0.657	0.694	0.657	0.550	0.301	0.201	0.000
	0.1	0	0.302	0.587	0.826	0.996	1.084	1.090	1.021	0.897	0.740	0.573
	0.2	0	0.334	0.658	0.944	1.165	1.303	1.355	1,331	1.250	1.135	1.010
	0.3	0	0.321	0.636	0.933	1.183	1.362	1.460	1.485	1.453	1.388	1.311
	0.4	0	0.282	0.563	0.834	1.082	1.284	1.421	1.492	1.511	1.499	1.476
	0.5	0	0.232	0.464	0.690	0.905	1.099	1.256	1.365	1.432	1.473	1.505
	0.6	0	0.178	0.356	0.534	0.711	0.886	1.050	1.192	1.304	1.394	1.476
	0.7	0	0.124	0.249	0.376	0.507	0.643	0.784	0.928	1.065	1.191	1.311
	0.8	0	0.075	0.151	0.229	0.312	0.401	0.502	0.616	0.743	0.877	1.010
	0.9	0	0.083	0.067	0.102	0.140	0.183	0.234	0.298	0.377	0.471	0.573
400	0	0	0.214	0.417	0.587	0.700	0.740	0.700	0.587	0.417	0.214	0.000
	0.1	0	0.320	0.620	0.871	1.048	1.138	1.140	1.068	0.935	0.756	0.574
	0.2	0	0.350	0.689	0.088	1.216	1.356	1.405	1.373	1.280	1.151	1.011
	0.3	0	0.332	0.660	0.967	1.226	1.409	1.506	1.523	1.480	1.401	1.310
	0.4	0	0.290	0.577	0.857	1.112	1.319	1.456	1.522	1.532	1.508	1.473
	0.5	0	0.237	0.472	0.703	0.922	1.119	1.276	1.382	1.443	1.475	1.498
	0.6	0	0.179	0.358	0.538	0.717	0.894	1.061	1.03	1.313	1.398	1.473
	0.7	0	0.123	0.248	0.374	0.505	0.620	0.785	0.930	1.068	1.192	1.310
	0.8	0	0.074	0.148	0.225	0.306	0.395	0.495	0.610	0.740	0.875	1.011
	0.9	0	0.032	0.065	0.099	0.135	0.176	0.226	0.290	0.371	0.468	0.574

Table 5.18 Con't

Coefficients $\bar{y}$ for beams with uniformly distributed loads q

α	β	0.0	0.1	0.2	0.3	0.4	0.5	0.6	0.7	0.8	0.9	1.0
	0	0	0.226	0.440	0.619	0.739	0.781	0.739	0.619	0.440	0.226	0.000
	0.1	0	0.335	0.649	0.911	1.094	1.185	1.183	1.099	0.952	0.770	0.575
	0.2	0	0.363	0.716	1.260	1.261	1.403	1.449	1.410	1.306	1.165	1.012
	0.3	0	0.342	0.680	0.997	1.264	1.450	1.545	1.557	1.504	1.414	1.310
450	0.4	0	0.296	0.590	0.876	1.138	1.49	1.486	1.548	1.550	1.516	1.471
	0.5	0	0.241	0.479	0.713	0.936	1.136	1.294	1.397	1.452	1.477	1.491
	0.6	0	0.180	0.360	0.541	0.722	0.901	1.070	1.213	1.320	1.400	1.470
	0.7	0	0.123	0.246	0.373	0.504	0.641	0.785	0.932	1.071	1.194	1.310
	0.8	0	0.073	0.146	0.222	0.302	0.389	0.489	0.605	0.737	0.875	1.012
	0.9	0	0.031	0.063	0.096	0.131	0.171	0.220	0.284	0.366	0.466	0.575
	0	0	0.236	0.460	0.647	0.772	0.816	0.772	0.647	0.460	0.236	0.000
	0.1	0	0.349	0.676	0.946	1.134	1.227	1.222	1.131	0.975	0.782	0.576
	0.2	0	0.376	0.739	1.060	1.301	1.445	1.488	1.442	1.329	1.178	1.013
	0.3	0	0.351	0.697	1.023	1.297	1.486	1.580	1.586	1.525	1.424	1.310
500	0.4	0	0.301	0.600	0.803	1.161	1.376	1.513	1.571	1.566	1.523	1.467
	0.5	0	0.244	0.486	0.723	0.948	1.151	1.309	1.409	1.459	1.477	1.485
	0.6	0	0.180	0.361	0.543	0.726	0.908	1.079	1.222	1.327	1.403	1.467
	0.7	0	0.122	0.245	0.371	0.502	0.640	0.786	0.934	1.073	1.196	1.310
	0.8	0	0.072	0,144	0.218	0.297	0.385	0.484	0.601	0.734	0.874	1.013
	0.9	0	0.030	0.061	0.093	0.127	0.166	0.214	0.278	0.361	0.454	0.576

ξ

Table 5.19 Rotations of beams with concentrated vertical loads

Coefficients $\overline{\varphi}$ for beams with concentrated vertical loads P

α	β	ξ										
		0.0	0.1	0.2	0.3	0.4	0.5	0.6	0.7	0.8	0.9	1.0
1	0	-6.288	-6.283	-6.274	-6.264	-6.253	-6.244	-6.237	-6.232	-6.229	-6.228	-6.228
	0.1	-5.008	-5.009	-5.009	-5.006	-5.003	-4.999	-4.996	-4.994	-4.993	-4.993	-4.993
	0.2	-3.736	-3.737	-3.741	-3.747	-3.751	-2.753	-3.754	-3.755	-3.755	-3.755	-3.756
	0.3	-2.473	-2.474	-2.477	-2.486	-2.497	-2.505	-2.511	-2.515	-2.517	-2.518	-2.517
	0.4	-1.217	-1.217	-1.220	-1.227	-1.240	-1.254	-1.264	-1.270	-1.274	-1.276	-1.276
	0.5	0.032	0.032	0.029	0.024	0.014	0.000	-0.014	-0.024	-0.029	-0.032	-0.032
	0.6	1.276	1.276	1.274	1.270	1.263	1.254	1.240	1.228	1.220	1.217	1.217
	0.7	2.517	2.518	2.517	2.515	2.511	2.505	2.497	2.486	2.477	2.474	2.473
	0.8	3.756	3.756	3.755	3.755	3.754	3.753	3.751	3.747	3.741	3.737	3.736
	0.9	4.993	4.993	4.993	4.994	4.996	4.999	5.003	5.006	5.009	5.009	5.008
	1.0	6.228	6.228	6.229	6.232	6.237	6.244	6.253	6.264	6.274	6.283	6.288
2	0	-6.326	-6.318	-6.299	-6.278	-6.257	-6.239	-6.224	-6.215	-6.209	-6.205	-6.205
	0.1	-5.015	-5.017	-5.017	-5.012	-5.005	-4.998	-4.993	-4.989	-4.986	-4.985	-4.985
	0.2	-3.722	-3.724	-3.733	-3.745	-3.752	-3.756	-3.759	-3.761	-3.762	-3.762	-3.762
	0.3	-2.446	-2.447	-2.455	-2.473	-2.494	-2.510	-2.521	-2.520	-2.535	-2.535	-2.535
	0.4	-1.184	-1.185	-1.191	-1.205	-1.230	-1.258	-1.278	-1.291	-1.299	-1.302	-1.303
	0.5	0.064	0.063	0.059	0.048	0.029	0.000	-0.029	-0.048	-0.059	-0.063	-0.064
	0.6	1.303	1.302	1.299	1.291	1.278	1.258	1.230	1.205	1.191	1.185	1.184
	0.7	2.535	2.535	2.533	2.529	2.521	2.510	2.494	2.473	2.455	2.447	2.446
	0.8	3.762	3.762	3.762	3.761	3.750	3.756	3.752	3.745	3.733	3.724	3.723
	0.9	4.985	4.985	4.986	4.989	4.993	4.998	5.005	5.012	5.017	5.017	5.723
	1.0	6.204	6.207	6.209	6.215	6.224	6.239	6.257	6.278	6.299	6.318	6.326

Table 5.19 Con't

Coefficients $\bar{\varphi}$ for beams with concentrated vertical loads P

α	β		0.0	0.1	0.2	0.3	0.4	0.5 (ξ)	0.6	0.7	0.8	0.9	1.0
5	0		-6.439	-6.418	-6.373	-6.319	-6.266	-6.222	-6.187	-6.162	-6.143	-6.142	-6.141
	0.1		-5.038	-5.042	-5.013	-5.030	-5.013	-4.996	-4.982	-4.972	-4.966	-4.964	-4.963
	0.2		-3.682	-3.686	-3.708	-3.736	-3.754	-3.765	-3.772	-3.777	-3.779	-3.780	-3.780
	0.3		-2.367	-2.370	-2.388	-2.433	-2.485	-2.525	-2.553	-2.572	-2.582	-2.587	-2.587
	0.4		-1.086	-1.088	-1.103	-1.139	-1.201	-1.260	-1.319	-1.352	-1.371	-1.379	-1.380
	0.5		0.159	0.158	0.146	0.119	0.071	0.000	-0.071	-0.119	-0.146	-0.158	-0.159
	0.6		1.380	1.379	1.371	1.352	1.319	1.269	1.201	1.139	1.103	1.088	1.086
	0.7		2.587	2.587	2.582	2.572	2.553	2.525	2.485	2.433	2.388	2.370	2.367
	0.8		3.780	3.780	3.779	3.777	3.772	3.765	3.754	3.736	3.708	3.686	3.682
	0.9		4.963	4.964	4.966	4.972	4.982	4.996	5.013	5.030	5.043	5.042	5.038
	1.0		6.141	6.142	6.143	6.162	6.187	6.222	6.266	6.319	6.373	6.418	6.439
10	0		-6.623	-6.583	-6.492	-6.385	-6.282	-6.193	-6.125	-6.077	-6.049	-6.038	-6.036
	0.1		-5.075	-5.083	-5.085	-5.059	-5.025	-4.992	-4.965	-4.945	-4.933	-4.928	-4.927
	0.2		-3.616	-3.623	-3.667	-3.724	-3.758	-3.781	-3.795	-3.803	-3.807	-3.809	-3.809
	0.3		-2.236	-2.242	-2.279	-2.367	-2.472	-2.549	-2.605	-2.642	-2.662	-2.071	-2.672
	0.4		-0.926	-0.930	-0.959	-1.030	-1.154	-1.288	-1.386	-1.451	-1.488	-1.504	-1.506
	0.5		0.314	0.310	0.288	0.235	0.140	0.000	-0.140	-0.235	-0.288	-0.310	-0.314
	0.6		1.506	1.504	1,488	1.451	1.386	1.288	1.154	1.030	0.959	0.930	0.926
	0.7		2.672	2.671	2.662	2.642	2.605	2.549	2.472	2.367	2.279	2.242	2.236
	0.8		3.809	3.809	3.807	3.803	3.795	3.781	3.758	3.724	3.667	3.623	3.616
	0.9		4.927	4.928	4.933	4.945	4.965	4.992	5.025	5.059	5.085	5.083	5.075
	1.0		6.036	6.038	6.049	6.077	6.125	6.193	6.282	6.385	6.492	6.583	6.623

15

0	-6.805	-6.744	-6.608	-6.449	-6.296	-6.165	-6.064	-5.994	-5.954	-5.936	-5.934
0.1	-5.111	-5.122	-5.126	-5.088	-5.037	-4.988	-4.947	-4.919	-4.901	-4.984	-4.893
0.2	-3.551	-3.561	-3.627	-3.712	-3.763	-3.796	-3.816	-3.828	-3.835	-3.837	-3.837
0.3	-2.109	-2.118	-2.172	-2.304	-2.459	-2.574	-2.654	-2.710	-2.740	-2.752	-2.754
0.4	-0.770	-0.777	-0.820	-0.925	-1.110	-1.306	-1.451	-1.547	-1.602	-1.625	-1.628
0.5	0.465	0.459	0.426	0.317	0.208	0.000	-0.208	-0.347	-0.426	-0.459	-0.465
0.6	1.628	1.625	1.602	1.547	1.451	1.306	1.110	0.925	0.820	0.777	0.770
0.7	2.754	2.752	2.740	2.709	2.654	2.574	2.459	2.304	2.172	2.118	2.109
0.8	3.837	3.837	3.835	3.828	3.816	3.796	3.763	3.712	3.627	3.561	3.551
0.9	4.893	4.894	4.901	4.919	4.947	4.988	5.037	5.088	5.126	5.123	5.111
1.0	5.934	5.936	5.954	5.994	6.064	6.165	6.296	6.449	6.608	6.744	6.805

20

0	-6.982	-6.901	-6.722	-6.511	-6.308	-6.137	-6.005	-5.914	-5.861	-5.838	-5.835
0.1	-5.145	-5.162	-5.166	-5.116	-5.049	-4.984	-4.931	-4.893	-4.870	-4.860	-4.859
0.2	-3.487	-3.500	-3.588	-3.700	-3.768	-3.811	-3.837	-3.853	-3.861	-3.864	-3.864
0.3	-1.985	-1.996	-2.068	-2.243	-2.447	-2.598	-2.705	-2.776	-2.815	-2.831	-2.833
0.4	-0.618	-0.627	-0.684	-0.822	-1.064	-1.325	-1.515	-1.641	-1.713	-1.742	-1.747
0.5	0.611	0.604	0.561	0.457	0.274	0.000	-0.274	-0.457	-0.561	-0.604	-0.611
0.6	1.747	1.742	1.713	1.641	1.515	1.325	1.064	0.822	0.684	0.627	0.618
0.7	2.833	2.831	2.815	2.776	2.705	2.598	2.447	2.243	2.068	1.996	1.985
0.8	3.864	3.864	3.861	3.853	3.837	3.811	3.768	3.700	3.588	3.500	3.487
0.9	4.859	4.860	4.870	4.893	4.931	4.984	5.049	5.116	5.166	5.162	5.145
1.0	8.835	5.838	5.861	5.914	6.005	6.137	6.308	6.511	6.722	6.901	6.982

Table 5.19 Con't

Coefficients $\bar{\varphi}$ for beams with concentrated vertical loads P

α	β	ξ										
		0.0	0.1	0.2	0.3	0.4	0.5	0.6	0.7	0.8	0.9	1.0
	0	-7.157	-7.056	-6.832	-6.570	-6.320	-6.109	-5.947	-5.835	-5.771	-5.743	-5.739
	0.1	-5.170	-5.200	-5.202	-5.143	-5.060	-4.980	-4.914	-4.868	-4.840	-4.828	-4.826
	0.2	-3.423	-3.441	-3.550	-3.690	-3.774	-3.826	-3.858	-3.877	-3.887	-3.890	-3.891
	0.3	-1.863	-1877	-1.966	-2.183	-2.436	-2.622	-2.754	-2.840	-2.888	-2.907	-2.910
	0.4	-0.471	-0.481	-0.551	-0.721	-1.021	-1.343	-1.577	-1.732	-1.820	-1.850	-1.862
25	0.5	0.762	0.744	0.692	0.564	0.339	0.000	-0.339	-0.564	-0.692	-0.744	-0.752
	0.6	1.892	1.856	1.820	1.732	1.577	1.343	1.021	0.721	0.551	0.481	0.471
	0.7	2.910	2.907	2.888	2.840	2.754	2.622	2.436	2.183	1.966	1.877	1.863
	0.8	3.891	3.890	3.887	3.877	3.858	3.836	3.774	3.690	3.550	3.441	3.423
	0.9	4.826	4.828	4.840	4.867	4.914	4.980	5.060	5.143	5.205	5.200	5.179
	1.0	5.739	5.743	5.771	5.835	5.947	6.109	6.320	6.570	6.832	7.056	7.157
	0	-7.984	-7.784	-7.345	-6.839	-6.364	-5.970	-5.673	-5.473	-5.358	-5.310	-5.302
	0.1	-5.332	-5.374	-5.387	-5.270	-5.112	-4.960	-4.838	-4.753	-4.702	-4.580	-4.576
	0.2	-3.122	-3.156	-3.372	-3.645	-3.806	-3.901	-3.957	-3.988	-4.003	-4.008	-4.009
	0.3	-1.292	-1.319	-1.490	-1.908	-2.393	-2.742	-2.983	-3.137	-3.221	-3.254	-3.259
	0.4	0.214	0.194	0.066	-0.252	-0.821	-1.432	-1.868	-2.153	-2.313	-2.378	-2.388
50	0.5	1.405	1.390	1.296	1.061	0.641	0.000	-0.641	-1.061	-1.296	-1.390	-1.405
	0.6	2.388	2.378	2.313	2.153	1.868	1.432	0.821	0.252	-0.066	-0.194	-0.214
	0.7	3.259	3.254	3.221	3.137	2.983	2.742	2.393	1.908	1.490	1.319	1.292
	0.8	4.009	4.008	4.003	3.988	3.957	3.901	3.806	3.645	3.372	3.156	3.122
	0.9	4.576	4.580	4.702	4.753	4.838	4.960	5.112	5.270	5.387	5.374	5.332
	1.0	5.302	5.310	5.358	5.473	5.673	5.790	6.364	6.839	7.345	7.784	7.984

100

0	-4.601	-4.613	-4.688	-4.870	-5.197	-5.699	-6.392	-7.256	-8.208	-9.057	-9.452
0.1	-4.430	-4.437	-4.474	-4.561	-4.709	-4.924	-5.196	-5.485	-5.701	-5.665	-5.580
0.2	-4.102	-4.192	-4.189	-4.173	-4.133	-4.048	-3.888	-3.593	-3.065	-2.642	-2.576
0.3	-3.819	-3.811	-3.761	-3.629	-3.381	-2.973	-2.356	-1.456	-0.671	-0.356	-0.306
0.4	-3.242	-3.226	-3.119	-2.851	-2.364	-1.598	-0.489	0.546	1.108	1.327	1.359
0.5	-2.475	-2.452	-2.296	-1.893	-1.160	0.000	1.160	1.893	2.296	2.452	2.475
0.6	-1.359	-1.327	-1.108	-0.546	0.489	1.598	2.364	2.851	3.119	3.226	3.242
0.7	0.306	0.356	0.671	1.456	2.356	2.973	3.381	3.629	3.761	3.811	3.819
0.8	2.576	2.642	3.065	3.593	3.888	4.048	4.133	4.173	4.189	4.192	4.192
0.9	5.580	5.665	5.701	5.485	5.196	4.924	4.709	4.561	4.474	4.437	4.430
1.0	9.452	9.057	8.208	7.256	6.392	5.699	5.197	4.870	4.688	4.613	4.601

150

0	-4.070	-4.084	-4.171	-4.389	-4.792	-5.436	-6.360	-7.554	-8.909	-10.145	-10.731
0.1	-4.239	-4.247	-4.295	-4.408	-4.602	-4.889	-5.261	-5.663	-5.963	-5.898	-5.768
0.2	-4.324	-4.325	-4.327	-4.320	-4.286	-4.193	-3.987	-3.577	-2.809	-2.187	-2.089
0.3	-4.240	-4.232	-4.175	-4.021	-3.718	-3.196	-2.366	-1.103	0.011	0.454	0.521
0.4	-3.903	-3.882	-3.747	-3.405	-2.771	-1.749	-0.222	1.204	1.955	2.237	2.278
0.5	-3.312	-3.284	-3.091	-2.576	-1.593	0.000	1.593	2.576	3.091	3.284	3.312
0.6	-2.278	-2.237	-1.955	-1.204	0.222	1.749	2.771	3.405	3.747	3.882	3.903
0.7	-0.521	-0.454	-0.011	1.103	2.366	3.196	3.718	4.021	4.175	4.232	4.240
0.8	2.089	2.187	2.809	3.577	3.987	4.193	4.286	4.320	4.327	4.325	4.324
0.9	5.768	5.898	5.963	5.663	5.261	4.889	4.602	4.408	4.295	4.247	4.239
1.0	10.731	10.145	8.909	7.554	6.360	5.436	4.792	4.389	4.171	4.084	4.070

Table 5.19 Con't

Coefficients $\overline{\varphi}$ for beams with concentrated vertical loads P

α	β	0.0	0.1	0.2	0.3	0.4	0.5	0.6	0.7	0.8	0.9	1.0
	0	-11.871	-11.097	-9.489	7.766	-6.287	-5.182	-4.442	-3.997	-3.766	-3.675	-3.660
	0.1	-5.911	-6.087	-6.186	-5.812	-5.312	-4.858	-4.512	-4.283	-4.152	-4.096	-4.087
	0.2	-1.645	-1.773	-2.589	-3.588	-4.099	-4.334	-4.422	-4.440	-4.432	-4.422	-4.419
	0.3	1.233	1.149	0.595	-0.821	-2.408	-3.410	-4.009	-4.340	-4.499	-4.556	-4.564
	0.4	3.028	2.984	2.660	1.761	-0.009	-1.886	-3.112	-3.854	-4.247	-4.402	-4.426
200	0.5	3.980	3.950	3.735	3.141	1.965	0.000	-1.965	-3.141	-3.735	-3.950	-3.980
	0.6	4.426	4.402	4.247	3.854	3.112	1.886	0.009	-1.761	-2.660	-2.984	-3.028
	0.7	4.564	4.556	4.499	4.340	4.009	3.410	2.408	0.821	-0.595	-1.149	-1.233
	0.8	4.419	4.422	4.432	4.440	4.422	4.334	4.099	3.588	2.589	1.773	1.645
	0.9	4.087	4.096	4.152	4.283	4.512	4.858	5.312	5.812	6.186	6.087	5.911
	1.0	3.660	3.675	3.766	3.997	4.442	5.182	6.287	7.766	9.489	11.097	11.871
	0	-12.91	-11.945	-9.979	-7.912	-6.184	-4.935	-4.132	-3.672	-3.443	-3.355	-3.341
	0.1	-6.021	-6.243	-6.381	-5.942	-5.355	-4.828	-4.436	-4.179	-4.035	-3.974	-3.963
	0.2	-1.235	-1.392	-2.398	-3.619	-4.219	-4.472	-4.545	-4.539	-4.511	-4.492	-4.488
	0.3	1.855	1.757	1.103	-0.591	-2.473	-3.615	-4.264	-4.603	-4.756	-4.808	-4.816
	0.4	3.649	3.604	3.256	2.243	0.189	-2.009	-3.400	-4.223	-4.653	-4.822	-4.849
250	0.5	4.520	4.490	4.256	3.620	2.290	0.000	-2.290	-3.620	-4.256	-4.490	-4.520
	0.6	4.849	4.822	4.653	4.223	3.400	2.009	-0.189	-1.243	-3.256	-3.604	-3.649
	0.7	4.816	4.808	4.756	4.603	4.264	3.615	2.473	0.591	-1.103	-1.757	-1.855
	0.8	4.488	4.492	4.511	4.539	4.545	4.472	4.219	3.619	2.398	1.392	1.235
	0.9	3.963	3.974	4.035	4.179	4.436	4.828	5.355	5.942	6.381	6.243	6.021
	1.0	3.341	3.355	3.443	3.672	4.132	4.935	6.184	7.912	9.979	11.945	12.905

ξ

300	0	0.1	0.2	0.3	0.4	0.5	0.6	0.7	0.8	0.9	1.0
0	-13.856	-12.713	-10.399	-8.009	-6.060	-4.695	-3.856	-3.399	-3.182	-3.101	-3.088
0.1	-6.104	-6.374	-6.553	-6.055	-5.390	-4.801	-4.370	-4.093	-3.938	-3.872	-3.861
0.2	-0.850	-1.037	-2.229	-3.664	-4.346	-4.607	-4.657	-4.622	-4.572	-4.542	-4.536
0.3	2.407	2.297	1.552	-0.402	-2.556	-3.814	-4.492	-4.823	-4.962	-5.008	-5.014
0.4	4.169	4.126	3.768	2.668	0.356	-2.120	-3.648	-4.531	-4.988	-5.168	-5.197
0.5	4.962	4.935	4.712	4.032	2.580	0.000	-2.580	-4.032	-4.712	-1.935	-4.962
0.6	5.197	5.166	4.988	4.531	3.648	2.120	-0.356	-2.068	-3.768	-4.126	-4.169
0.7	5.014	5.008	4.962	4.823	4.492	3.814	2.556	0.402	-1.552	-2.297	-2.407
0.8	4.536	4.542	4.572	4.622	4.657	4.607	4.346	3.664	2.229	1.037	0.850
0.9	3.861	3.872	3.938	4.093	4.370	4.801	5.390	6.056	6.553	6.374	6.104
1.0	3.088	3.101	3.182	3.399	3.856	4.695	6.060	8.009	10.399	12.713	13.856

350	0	0.1	0.2	0.3	0.4	0.5	0.6	0.7	0.8	0.9	1.0
0	-14.742	-13.417	-10.764	-8.068	-5.918	-4.462	-3.606	-3.167	-2.971	-2.899	-2.888
0.1	-6.167	-6.484	-6.708	-6.156	-5.420	-4.776	-4.312	-4.019	-3.857	-3.788	-3.777
0.2	-0.487	-0.701	-2.077	-3.722	-4.478	-4.740	-4.761	-4.692	-4.618	-4.576	-4.569
0.3	2.903	2.783	1.954	-0.245	-2.650	-4.005	-4.696	-5.009	-5.130	-5.166	-5.171
0.4	4.607	4.568	4.212	3.048	0.507	-2.219	-3.860	-4.791	-5.268	-5.458	-5.489
0.5	5.328	5.303	5.089	4.394	2.841	0.000	-2.841	-4.394	-5.089	-5.303	-5.328
0.6	5.489	5.458	5.268	4.791	3.860	2.219	-0.507	-3.048	-4.212	-4.569	-4.607
0.7	5.171	5.166	5.130	5.009	4.696	4.005	2.050	0.245	-1.954	-2.783	-2.903
0.8	4.569	4.576	4.618	4.692	4.761	4.740	4.478	3.722	2.077	0.701	0.487
0.9	3.777	3.788	3.857	4.019	4.312	4.776	5.420	6.156	6.708	6.484	6.167
1.0	2.888	2.899	2.971	3.167	3.606	4.462	5.918	8.068	10.764	13.417	14.742

Table 5.19 Con't

Coefficients $\overline{\varphi}$ for beams with concentrated vertical loads P

α	β	0.0	0.1	0.2	0.3	0.4	0.5	0.6	0.7	0.8	0.9	1.0
	0	-15.573	-14.067	-11.083	-8.095	-5.764	-4.235	-3.378	-2.067	-2.709	-2.738	-2.728
	0.1	-6.212	-6.578	-6.849	-6.247	-5.445	-4.753	-4.261	-3.995	-3.787	-3.716	-3.703
	0.2	-0.140	-0.383	-1.940	-3.789	-4.613	-4.871	-4.858	-4.752	-4.652	-4.599	-4.589
	0.3	3.353	3.223	2.319	-0.113	-2.754	-4.189	-4.881	-5.168	-5.266	-5.293	-5.297
	0.4	4.978	4.945	4.601	3.392	0.645	-2.307	-4.045	-5.012	-5.505	-5.703	-5.737
400	0.5	5.632	5.612	5.412	4.713	3.080	0.000	-3.080	-4.713	-5.412	-5.612	-5.632
	0.6	5.737	5.703	5.505	5.012	4.045	2.307	-0.645	-3.392	-4.601	-4.945	-4.978
	0.7	5.297	5.293	5.266	5.168	4.881	4.189	2.754	0.113	-2.319	-3.223	-3.358
	0.8	4.589	4.599	4.652	4.752	4.858	4.871	4.613	3.789	1.940	0.383	0.140
	0.9	3.703	3.716	3.787	3.955	4.261	4.753	5.445	6.247	6.849	6.578	0.212
	1.0	2.728	2.738	2.797	2.997	3.378	4.235	5.764	8.095	11.083	14.067	15.573
	0	-16.360	-14.675	-11.617	-8.098	-5.600	-4.013	-3.168	-2.794	-2.655	-2.608	-2.600
	0.1	-6.243	-6.657	-6.977	-6.330	-5.468	-4.731	-4.216	-3.900	-3.728	-3.654	-3.641
	0.2	0.192	-0.078	-1.814	-3.864	-4.751	-4.999	-4.948	-4.804	-4.677	-4.612	-4.601
	0.3	3.764	3.628	2.652	-0.009	-2.865	-4.367	-5.050	-5.205	-5.378	-5.395	-5.398
	0.4	5.292	5.268	4.945	3.707	0.775	-2.385	-4.205	-5.201	-5.707	-5.915	-5.951
450	0.5	5.886	5.871	5.691	4.999	3.301	0.000	-3.301	-4.999	-5.691	-5.871	-5.886
	0.6	5.951	5.915	5.707	5.201	4.205	2.385	-0.775	-3.707	-4.045	-5.268	-5.292
	0.7	5.398	5.395	5.378	5.305	5.050	4.367	2.865	0.009	-2.652	-3.628	-3.764
	0.8	4.601	4.612	4.677	4.804	4.948	4.999	4.751	3.864	1.814	0.078	-0.192
	0.9	3.641	3.654	3.728	3.900	4.216	4.731	5.468	6.330	6.977	6.657	6.243
	1.0	2.600	2.608	2.655	2.794	3.168	4.013	5.600	8.098	11.365	14.675	16.360

	0										
	0.1										
	0.2										
	0.3										
	0.4										
500	0.5										
	0.6										
	0.7										
	0.8										
	0.9										
	1.0										

0	−17.100	−15.246	−11.617	−8.080	−5.429	−3.798	−2.974	−2.644	−2.537	−2.504	−2.497
0.1	−6.263	−6.726	−7.096	−6.407	−5.488	−4.712	−4.176	−3.852	−3.677	−3.601	−3.587
0.2	0.512	0.216	−1.697	−3.945	−4.890	−5.125	−5.034	−4.849	−4.695	−4.618	−4.605
0.3	4.143	3.999	2.959	0.094	−2.980	−4.539	−5.205	−5.423	−5.471	−5.478	−5.479
0.4	5.559	5.545	5.251	3.999	0.899	−2.453	−4.344	−5.363	−5.882	−6.100	−6.138
0.5	6.100	6.090	5.933	5.2558	3.507	0.000	−3.507	−5.258	−5.933	−6.090	−6.100
0.6	6.138	6.100	5.882	5.363	4.344	2.453	−0.890	−3.999	−5.251	−5.545	−5.559
0.7	5.479	5.478	5.471	5.423	5.205	4.539	2.980	−0.094	−2.959	−3.999	−4.143
0.8	4.605	4.618	4.695	4.849	5.034	5.125	4.890	3.945	1.697	−0.216	−0.512
0.9	3.587	3.601	3.677	3.852	4.176	4.712	5.488	6.407	7.096	6.726	6.263
1.0	2.497	2.504	2.537	2.644	2.974	3.798	5.429	8.080	11.617	15.246	17.109

Table 5.20 Relative settlements of beams with concentrated vertical loads

α	β	Coefficients $\bar{y}$ for beams with concentrated vertical loads P										
		ξ										
		0.0	0.1	0.2	0.3	0.4	0.5	0.6	0.7	0.8	0.9	1.0
1	0	0	-0.6286	-1.2570	-1.884	-2.509	-3.134	-3.134	-4.382	-5.005	-5.628	-6.251
	0.1	0	-0.5010	-1.0020	-1.502	-2.003	-2.503	-2.503	-3.502	-4.002	-4.501	-5.000
	0.2	0	-0.3740	-0.7480	-1.122	-1.497	-1.872	-1.872	-2.623	-2.999	-3.374	-3.750
	0.3	0	-0.2470	-0.4950	-0.743	-0.992	-1.242	-1.242	-1.744	-1.996	-2.248	-2.499
	0.4	0	-0.1220	-0.2440	-0.366	-0.489	-0.614	-0.614	-0.807	-0.994	-1.121	-1.249
	0.5	0	0.0032	0.0063	0.009	0.011	0.012	0.012	0.009	0.006	0.003	0.000
	0.6	0	0.1280	0.2550	0.382	0.509	0.635	0.635	0.883	1.006	1.127	1.249
	0.7	0	0.2520	0.5030	0.755	1.006	1.257	1.257	1.756	2.005	2.252	2.499
	0.8	0	0.3760	0.7510	1.127	1.502	1.878	1.878	2.628	3.002	3.376	3.750
	0.9	0	0.4990	0.9990	1.498	1.997	2.497	2.497	3.498	3.998	4.499	5.000
	1.0	0	0.6230	1.2460	1.869	2.492	3.116	3.116	4.367	4.994	5.622	6.251
2	0	0	-0.6320	-1.2630	-1.892	-2.519	-3.144	-3.144	-4.380	-5.010	-5.630	-6.251
	0.1	0	-0.5020	-1.0030	-1.505	-2.006	-2.506	-2.506	-3.505	-4.003	-4.502	-5.000
	0.2	0	-0.3720	-0.7450	-1.119	-1.494	-1.869	-1.869	-2.621	-2.997	-3.373	-3.750
	0.3	0	-0.2450	-0.4900	-0.756	-0.984	-1.235	-1.235	-1.739	-1.992	-2.245	-2.499
	0.4	0	0.1180	-0.2370	-0.357	-0.478	-0.603	-0.603	-0.858	-0.988	-1.118	-1.248
	0.5	0	0.0064	0.0126	0.018	0.022	0.023	0.023	0.018	0.013	0.006	0.000
	0.6	0	0.1300	0.2600	0.370	0.518	0.645	0.645	0.901	1.011	1.129	1.248
	0.7	0	0.254	0.5070	0.760	1.013	1.264	1.264	1.763	2.009	2.254	2.499
	0.8	0	0.3760	0.7520	1.129	1.505	1.880	1.880	2.631	3.004	3.377	3.750
	0.9	0	0.4990	0.9970	1.496	1.995	2.494	2.494	3.495	3.997	4.499	5.000
	1.0	0	0.6210	1.2410	1.862	2.484	3.108	3.108	4.359	4.989	5.619	6.281

5

	0										
0	0	-0.6430	-1.2830	-1.917	-2.546	-3.171	-3.791	-4.409	-5.024	-5.639	-6.253
0.1	0	-0.5040	-1.0080	-1.512	-2.014	-2.515	-3.013	-3.511	-4.008	-4.504	-5.000
0.2	0	-0.3680	-0.7370	-1.110	-1.484	-1.861	-2.238	-2.615	-2.993	-3.371	-3.749
0.3	0	-0.2370	-0.4740	-0.715	-0.961	-1.212	-1.466	-1.722	-1.979	-2.238	-2.497
0.4	0	-0.1090	-0.2180	-0.330	-0.447	-0.570	-0.700	-0.834	-0.970	-1.107	-1.245
0.5	0	0.0160	0.0310	0.045	0.054	0.058	0.054	0.045	0.031	0.016	0.000
0.6	0	0.1370	0.2760	0.412	0.545	0.675	0.797	0.915	1.027	1.137	1.245
0.7	0	0.2590	0.5170	0.775	1.031	1.285	1.536	1.782	2.022	2.260	2.497
0.8	0	0.3780	0.7560	1.134	1.511	1.888	2.264	2.639	3.011	3.380	3.749
0.9	0	0.4960	0.9920	1.490	1.987	2.486	2.987	3.489	3.992	4.497	5.000
1.0	0	0.6140	1.2290	1.844	2.461	3.082	3.706	4.335	4.970	5.610	6.252

10

	0										
0	0	-0.660	-1.315	-1.959	-2.592	-3.216	-3.831	-4.441	-5.047	-5.652	-6.255
0.1	0	-0.508	-1.016	-1.524	-2.028	-2.529	-3.027	-3.522	-4.016	-4.509	-5.002
0.2	0	-0.362	-0.726	-1.096	-1.470	-1.847	-2.226	-2.606	-2.986	-3.367	-3.748
0.3	0	-0.224	-0.449	-0.681	-0.924	-1.175	-1.433	-1.695	-1.960	-2.227	-2.494
0.4	0	-0.093	-0.187	-0.286	-0.395	-0.517	-0.651	-0.793	-0.940	-1.090	-1.241
0.5	0	0.031	0.062	0.088	0.107	0.115	0.107	0.088	0.062	0.031	0.000
0.6	0	0.151	0.300	0.448	0.590	0.724	0.845	0.955	1.054	1.148	1.241
0.7	0	0.267	0.534	0.799	1.062	1.320	1.571	1.813	2.045	2.271	-2.494
0.8	0	0.381	0.762	1.142	1.522	1.901	2.278	2.652	3.022	3.386	3.748
0.9	0	0.493	0.986	1.480	1.975	2.473	2.974	3.478	3.985	4.494	5.002
1.0	0	0.604	1.208	1.814	2.424	3.0396	3.663	4.296	4.940	5.594	6.255

Table 5.20 Con't

Coefficients $\bar{y}$ for beams with concentrated vertical loads P

α	β	0.0	0.1	0.2	0.3	0.4	0.5	0.6	0.7	0.8	0.9	1.0
	0	0	-0.678	-1.346	-1.999	-2.636	-3.259	-3.870	-4.473	-5.070	-5.664	-6.258
	0.1	0	-0.511	-1.024	-1.535	-2.041	-2.543	-3.039	-3.533	-4023	-4.513	-5.000
	0.2	0	-0.355	-0.714	-1.081	-1.455	-1.833	-2.214	-2.596	-2.980	-3.363	-3.747
	0.3	0	-0.211	-0.425	-0.648	-0.887	-1.139	-1.400	-1.669	-1.941	-2.216	-2.492
	0.4	0	-0.077	-0.157	-0.243	-0.344	-0.465	-0.603	-0.753	-0.911	-1.073	-1.236
15	0.5	0	0.046	0.091	0.130	0.160	0.170	0.160	0.130	0.091	0.046	0.000
	0.6	0	0.163	0.324	0.482	0.632	0.771	0.892	0.993	1.079	1.160	1.236
	0.7	0	0.275	0.550	0.823	1.091	1.353	1.605	1.843	2.066	2.280	2.492
	0.8	0	0.384	0.767	1.150	1.533	1.914	2.292	2.666	3.033	3.392	3.747
	0.9	0	0.489	0.979	1.470	1.968	2.460	2.961	3.467	3.978	4.491	5.002
	1.0	0	0.593	1.188	1.785	2.388	2.999	3.622	4.259	4.912	5.580	6.258
	0	0	-0.695	-1.377	-2.039	-2.679	-3.301	-3.908	-4.504	-5.092	-5.677	-6.260
	0.1	0	-0.514	-1.032	-1.546	-2.055	-2.556	-3.052	-3.543	-4.031	-4.517	-5.000
	0.2	0	-0.349	-0.703	-1.007	-1.441	-1.820	-2.202	-2.587	-2.973	-3.360	-3.746
	0.3	0	-0.199	-0.401	-0.616	-0.851	-1.103	-1.369	-1.643	-1.923	-2.205	-2.489
	0.4	0	-0.062	-0.127	-0.202	-0.295	-0.415	-0.557	-0.716	-0.864	-1.057	-1.231
20	0.5	0	0.061	0.120	0.171	0.208	0.223	0.208	0.171	0.120	0.061	0.000
	0.6	0	0.175	0.348	0.516	0.674	0.817	0.937	1.030	1.104	1.169	1.231
	0.7	0	0.283	0.566	0.845	1.120	1.385	1.038	1.873	2.087	2.290	2.489
	0.8	0	0.386	0.773	1.159	1.543	1.926	2.305	2.678	3.043	3.397	3.746
	0.9	0	0.486	0.972	1.460	1.951	2.447	2.949	3.457	3.971	4.488	5.000
	1.0	0	0.584	1.168	1.757	2.352	2.959	3.581	4.222	4.883	5.565	6.260

25											
0	0	-0.712	-1.407	-2.077	-2.721	-3.343	-3.945	-3.534	-5.114	-5.689	-6.263
0.1	0	-0.518	-1.039	-1.557	-2.067	-2.569	-3.064	-3.553	-4.038	-4.521	-5.004
0.2	0	-0.343	-0.691	-1.054	-1.427	-1.808	-2.192	-2.579	-2.967	-3.356	-3.745
0.3	0	-0.187	-0.378	-0.584	-0.816	-1.069	-1.338	-1.618	-1.905	-2.195	-2.486
0.4	0	-0.047	-0.098	-0.161	-0.247	-0.366	-0.512	-0.679	-0.857	-1.041	-1.227
0.5	0	0.075	0.147	0.211	0.257	0.275	0.257	0.211	0.147	0.075	0.000
0.6	0	0.186	0.370	0.548	0.714	0.861	0.980	1.066	1.128	1.180	1.227
0.7	0	0.291	0.580	0.868	1.148	1.417	1.670	1.902	2.108	2.299	2.486
0.8	0	0.389	0.778	1.166	1.553	1.937	2.318	2.691	3.054	3.402	3.745
0.9	0	0.483	0.966	1.451	1.940	2.435	2.937	3.447	3.965	4.486	5.004
1.0	0.	0.574	1.149	1.729	2.318	2.920	3.541	4.186	4.856	5.551	6.263

50											
0	0	-0.791	-1.549	-2.258	-2.918	-3.534	-4.115	-4.672	-5.212	-5.745	-6.276
0.1	0	-0.534	-1.074	-1.607	-2.127	-2.630	-3.120	-3.599	-4.071	-4.540	-5.008
0.2	0	-0.313	-0.637	-0.989	-1.363	-1.748	-2.141	-2.539	-2.939	-3.339	-3.740
0.3	0	-0.130	-0.269	-0.436	-0.653	-0.910	-1.197	-1.504	-1.822	-2.146	-2.472
0.4	0	0.021	0.035	0.028	-0.024	-0.138	-0.304	-0.506	-0.731	-0.966	-1.204
0.5	0	0.140	0.275	0.395	0.481	0.515	0.481	0.395	0.275	0.140	0.000
0.6	0	0.239	0.474	0.698	0.900	1.066	1.181	1.232	1.240	1.225	1.204
0.7	0	0.326	0.650	0.968	1.275	1.562	1.820	2.036	2.204	2.342	2.472
0.8	0	0.401	0.801	1.201	1.599	1.992	2.378	2.751	3.103	3.427	3.740
0.9	0	0.468	0.937	1.409	1.888	2.378	2.881	3.400	3.934	4.474	5.008
1.0	0	0.530	1.063	1.604	2.161	2.742	3.358	4.018	4.727	5.484	6.276

Table 5.20 Con't

Coefficients $\bar{y}$ for beams with concentrated vertical loads P

α	β	0.0	0.1	0.2	0.3	0.4	0.5	0.6	0.7	0.8	0.9	1.0
	0	0	-0.931	-1.796	-3.251	-3.251	-3.854	-4.397	4.899	-5.376	-5.850	-6.300
	0.1	0	-0.560	-1.132	-2.220	-2.226	-2.732	-3.213	-3.676	-4.127	-4.572	-5.016
	0.2	0	-0.250	-0.540	-1.251	-1.251	-1.649	-2.058	-2.474	-2.892	-3.311	-3.730
	0.3	0	-0.032	-0.080	-0.375	-0.375	-0.644	-0.963	-1.315	-1.685	-2.064	-2446
	0.4	0	0.135	0.259	0.345	0.352	0.245	0.044	-0.219	-0.519	-0.837	-1.161
100	0.5	0	0.247	0.486	0.854	0.854	0.916	0.854	0.698	0.486	0.247	0.000
	0.6	0	0.324	0.642	1.205	1.205	1.406	1.513	1.506	1.420	1.296	1.161
	0.7	0	0.382	0.761	1.483	1.483	1.802	2.070	2.264	2.366	2.414	2.446
	0.8	0	0.419	0.838	1.672	1.672	2.082	2.479	2.855	3.190	3.471	3.730
	0.9	0	0.443	0.888	1.803	1.803	2.284	2.789	3.323	3.884	4.455	5.016
	1.0	0	0.460	0.925	1.904	1.904	2.447	3.050	3.731	4.504	5.369	6.300
	0	0	-1.052	-2.008	-2.830	-3.524	-4.111	-4.621	-5.078	-5.505	-5.917	-6.324
	0.1	0	-0.580	-1.178	-1.761	-2.307	-2.814	-3.288	-3.737	-4.172	-4.599	-5.023
	0.2	0	-0.211	-0.455	-0.778	-1.159	-1.599	-1.993	-2.424	-2.856	-3.289	-3.722
	0.3	0	0.050	0.078	0.030	-0.147	-0.429	-0.776	-1.165	-1.576	-1.997	-2.420
	0.4	0	0.227	0.439	0.602	0.658	0.554	0.325	0.013	-0.347	-0.729	-1.119
150	0.5	0	0.331	0.651	0.938	1.151	1.236	1.151	0.938	0.651	0.331	0.000
	0.6	0	0.390	0.773	1.132	1.444	1.673	1.777	1.721	1.559	1.346	1.119
	0.7	0	0.424	0.845	1.256	1.644	1.992	2.273	2.451	2.499	2.471	2.420
	0.8	0	0.432	0.865	1.298	1.728	2.153	2.563	2.943	3.267	3.510	3.722
	0.9	0	0.424	0.851	1.285	1.735	2.209	2.716	3.262	3.845	4.443	5.023
	1.0	0	0.407	0.819	1.246	1.703	2.212	2.800	3.494	4.316	5.272	6.324

200											
0	0	-1.160	-2.192	-3.054	-3.754	-4.324	-4.802	-5.222	-5.609	-5.980	-6.346
0.1	0	-0.596	-1.215	-1.817	-2.374	-2.881	-3.349	-3.788	-4.209	-4.621	-5.030
0.2	0	-0.168	-0.378	-0.692	-1.079	-1.503	-1.941	-2.385	-2.828	-3.271	-3.713
0.3	0	0.121	0.214	0.211	0.044	-0.251	-0.624	-1.044	-1.487	-1.940	-2.396
0.4	0	0.302	0.587	0.815	0.911	0.810	0.566	0.204	-0.204	-0.637	-1.079
0.5	0	0.397	0.784	1.132	1.393	1.408	1.393	1.132	0.784	0.397	0.000
0.6	0	0.442	0.876	1.283	1.635	1.890	1.990	1.894	1.667	1.381	1.079
0.7	0	0.456	0.909	1.353	1.772	2.145	2.440	2.608	2.610	2.517	2.396
0.8	0	0.442	0.885	1.328	1.772	2.210	2.634	3.021	3.335	3.545	3.713
0.9	0	0.409	0.821	1.242	1.681	2.148	2.656	3.212	3.814	4.434	5.030
1.0	0	0.366	0.738	1.124	1.544	2.022	2.593	3.292	4.154	5.187	6.346

250											
0	0	-1.256	-2.356	-3.249	-3.950	-4.502	-4.952	-5.340	-5.694	-6.033	-6.368
0.1	0	-0.608	-1.246	-1.865	-2.430	-2.938	-3.400	-3.830	-4.240	-4.640	-5.036
0.2	0	-0.127	-0.307	-0.614	-1.010	-1.447	-1.899	-2.353	-2.806	-3.256	-3.705
0.3	0	0.183	0.333	0.369	0.208	-0.101	-0.499	-0.944	-1.413	-1.892	-2.373
0.4	0	0.364	0.711	0.993	1.125	1.026	0.750	0.365	-0.082	-0.557	-1.041
0.5	0	0.451	0.892	1.290	1.593	1.717	1.593	1.290	0.892	0.451	0.000
0.6	0	0.484	0.960	1.406	1.791	2.067	2.166	2.034	1.752	1.405	1.041
0.7	0	0.481	0.960	1.429	1.875	2.272	2.581	2.742	2.706	2.556	2.373
0.8	0	0.449	0.899	1.351	1.806	2.258	2.694	3.090	3.398	3.577	3.705
0.9	0	0.397	0.796	1.206	1.636	2.098	2.606	3.171	3.790	4.429	5.036
1.0	0	0.334	0.673	1.028	1.416	1.866	2.418	3.119	4.012	5.112	6.368

Table 5.20 Con't

Coefficients $\bar{y}$ for beams with concentrated vertical loads P

α	β	0.0	0.1	0.2	0.3	0.4	0.5	0.6	0.7	0.8	0.9	1.0
	0	0	-1.345	-2.505	-3.423	-4.121	-4.654	-5.078	-5.438	-5.766	-6.079	-6.388
	0.1	0	-0.617	-1.272	-1.906	-2.478	-2.987	-3.444	-3.866	-4.266	-4.656	-5.043
	0.2	0	-0.090	-0.241	-0.544	-0.949	-1.399	-1.864	-2.328	-2.788	-3.243	-3.697
	0.3	0	0.238	0.438	0.508	0.351	0.027	-0.392	-0.860	-1.351	-1.850	-2.351
	0.4	0	0.416	0.815	1.145	1.308	1.210	0.916	0.502	0.023	-0.486	-1.005
300	0.5	0	0.496	0.980	1.423	1.761	1.901	1.761	1.423	0.980	0.496	0.000
	0.6	0	0.519	1.029	1.507	1.921	2.215	2.313	2.150	1.820	1.421	1.005
	0.7	0	0.501	1.000	1.491	1.958	2.377	2.702	2.859	2.789	2.589	2.351
	0.8	0	0.454	0.909	1.369	1.833	2.297	2.748	3.153	3.456	3.607	3.697
	0.9	0	0.386	0.776	1.177	1.599	2.056	2.564	3.137	3.770	4.425	5.043
	1.0	0	0.309	0.623	0.950	1.310	1.734	2.267	2.966	3.884	5.044	6.388
	0	0	-1.427	-2.640	-3.570	-4.272	-4.786	-5.185	-5.521	-5.826	-6.119	-6.408
	0.1	0	-0.625	-1.294	-1.941	-2.520	-3.028	-3.481	-3.896	-4.289	-4.671	-5.049
	0.2	0	-0.054	-0.179	-0.479	-0.894	-1.358	-1.834	-2.307	-2.773	-3.232	-3.689
	0.3	0	0.287	0.533	0.632	0.477	0.137	-0.302	-0.790	-1.297	-1.813	-2.330
	0.4	0	0.460	0.903	1.275	1.467	1.370	1.059	0.622	0.116	-0.422	-0.970
350	0.5	0	0.532	1.054	1.534	1.905	2.059	1.905	1.534	1.054	0.532	0.000
	0.6	0	0.548	1.086	1.592	2.029	2.341	2.437	2.246	1.874	1.430	0.970
	0.7	**0**	**0.517**	**1.032**	**1.540**	**2.028**	**2.467**	**2.807**	**2.962**	**2.863**	**2.617**	**2.330**
	0.8	0	0.457	0.916	1.382	1.855	2.331	2.795	3.210	3.510	3.635	3.689
	0.9	0	0.378	0.759	1.152	1.568	2.020	2.529	3.108	3.754	4.424	5.049
	1.0	0	0.289	0.582	0.887	1.223	1.622	2.136	2.829	3.768	4.981	6.408

400

0	0	-1.503	-2.765	-3.721	-4.407	-4.901	-5.276	-5.591	-5.878	-6.154	-6.427
0.1	0	-0.630	-1.313	-1.972	-2.556	-3.065	-3.514	-3.923	-4.309	-4.684	-5.054
0.2	0	-0.020	-0.121	-0.418	-0.845	-1.322	-1.810	-2.291	-2.761	-3.223	-3.682
0.3	0	0.332	0.619	0.745	0.589	0.234	-0.224	-0.729	-1.251	-1.780	-2.309
0.4	0	0.497	0.979	1.389	1.606	1.511	1.185	0.727	0.198	-0.364	-0.937
0.5	0	0.563	1.117	1.629	2.028	2.196	2.028	1.629	1.117	0.563	0.000
0.6	0	0.573	1.135	1.664	2.122	2.447	2.513	2.325	1.916	1.434	0.937
0.7	0	0.530	1.058	1.581	2.085	2.543	2.898	3.054	2.928	2.641	2.309
0.8	0	0.459	0.921	1.391	1.872	2.360	2.837	3.264	3.561	3.662	3.682
0.9	0	0.371	0.745	1.131	1.541	1.990	2.498	3.083	3.741	4.424	5.054
1.0	0	0.273	0.549	0.836	1.151	1.527	2.020	2.706	3.662	4.924	6.427

450

0	0	-1.575	-2.882	-3.851	-4.528	-5.002	-5.356	-5.651	-5.922	-6.185	-6.445
0.1	0	-0.635	-1.329	-1.999	-2.589	-3.097	-3.542	-3.947	-4.327	-4.696	-5.060
0.2	0	0.012	-0.065	-0.361	-0.799	-1.290	-1.789	-2.277	-2.751	-3.215	-3.675
0.3	0	0.373	0.697	0.847	0.600	0.320	-0.156	-0.676	-1.211	-1.750	-2.290
0.4	0	0.529	1.044	1.487	1.729	1.635	1.296	0.821	0.272	-0.311	-0.905
0.5	0	0.588	1.169	1.710	2.135	2.316	2.135	1.710	1.169	0.588	0.000
0.6	0	0.594	1.177	1.726	2.201	2.539	2.634	2.392	1.949	1.433	0.905
0.7	0	0.540	1.079	1.614	2.134	2.609	2.980	3.137	2.987	2.663	2.290
0.8	0	0.460	0.924	1.398	1.886	2.385	2.876	3.314	3.610	3.687	3.675
0.9	0	0.364	0.733	1.113	1.518	1.963	2.471	3.061	3.731	4.425	5.060
1.0	0	0.260	0.523	0.794	1.089	1.443	1.917	2.594	3.563	4.870	6.445

Table 5.20 Con't

Coefficients $\bar{y}$ for beams with concentrated vertical loads P

α	β	ξ										
		0.0	0.1	0.2	0.3	0.4	0.5	0.6	0.7	0.8	0.9	1.0
	0	0	-1.644	-2.992	-3.971	-4.638	-5.092	-5.425	-5.703	-5.961	-6.213	-6.463
	0.1	0	-0.638	-1.343	-2.023	-2.617	-3.125	-3.568	-3.968	-4.343	-4.706	-5.065
	0.2	0	0.044	-0.011	-0.307	-0.757	-1.262	-1.771	-2.266	-2.742	-3.207	-3.668
	0.3	0	0.411	0.770	0.942	0.782	0.396	-0.097	-0.630	-1.176	-1.723	-2.271
	0.4	0	0.556	1.100	1.574	1.838	1.745	1.396	0.905	0.340	-0.261	-0.874
500	0.5	0	0.610	1.213	1.779	2.229	2.422	2.229	1.779	1.213	0.610	0.000
	0.6	0	0.613	1.214	1.779	2.270	2.610	2.712	2.448	1.974	1.430	0.874
	0.7	0	0.548	1.095	1.641	2.174	2.667	3.053	3.213	3.041	2.682	2.271
	0.8	0	0.461	0.926	1.402	1.897	2.406	2.911	3.361	3.657	3.712	3.668
	0.9	0	0.359	0.722	1.098	1.498	1.940	2.448	3.043	3.722	4.427	5.065
	1.0	0	0.250	0.502	0.760	1.038	1.371	1.824	2.491	3.471	4.819	6.463

Table 5.21 Rotations of beams loaded with concentrated moments

α	β	Coefficients $\bar{\varphi}$ for beams loaded with a concentrated moment M										
							ξ					
		0.0	0.1	0.2	0.3	0.4	0.5	0.6	0.7	0.8	0.9	1.0
1	0	-12.851	-12.752	-12.660	-12.577	-12.506	-12.450	-12.407	-12.377	-12.361	-12.354	-12.352
	0.1	-12.756	-12.758	-12.665	-12.582	-12.511	-12.455	-12.412	-12.382	-12.367	-12.358	-12.357
	0.2	-12.671	-12.673	-12.680	-12.597	-12.526	-12.470	-12.427	-12.397	-12.380	-12.373	-12.372
	0.3	-12.597	-12.598	-12.605	-12.622	-12.551	-12.495	-12.452	-12.422	-12.405	-12.398	-12.397
	0.4	-12.532	-12.533	-12.540	-12.557	-12.586	-12.530	-12.487	-12.457	-12.440	-12.433	-12.432
	0.5	-12.477	-12.478	-12.485	-12.502	-12.532	-12.575	-12.432	-12.502	-12.485	-12.478	-12.477
	0.6	-12432	-12.433	-12.440	-12.457	-12.487	-12.530	-12.586	-12.557	-12.540	-12.533	-12.532
	0.7	-12.397	-12.398	-12.405	-12.422	-12.452	-12.495	-12.551	-12.622	-12.605	-12.598	-12.597
	0.8	-12.372	-12.373	-12.380	-12.397	-12.427	-12.470	-12.526	-12.597	-12.680	-12.673	-12.671
	0.9	-12.357	-12.358	-12.367	-12.382	-12.412	-12.455	-12.511	-12.582	-12.665	-12.758	-12.756
	1.0	-12.352	-12.354	-12.361	-12.377	-12.407	-12.450	-12.506	-12.577	-12.660	-12.752	-12.851
2	0	-12.202	-13.004	-12.818	-12.653	-12.512	-12.399	-12.314	-12.256	-12.222	-12.208	-12.206
	0.1	-13.120	-13.014	-12.828	-12.663	-12.522	-12.409	-12.324	-12.266	-12.232	-12.218	-12.216
	0.2	-12.842	-12.844	-12.850	-12.693	-12.552	-12.439	-12.354	-12.295	-12.262	-12.248	-12.245
	0.3	-12.692	-12.695	-12.709	-12.743	-12.603	-12.489	-12.404	-12.345	-12.311	-12.297	-12.295
	0.4	-12.563	-12.565	-12.580	-12.614	-12.673	-12.559	-12.473	-12.415	-12.381	-12.367	-12.364
	0.5	-12.454	-12.456	-12.470	-12.504	-12.563	-12.649	-12.563	-12.504	-12.470	-12.456	-12.454
	0.6	-12.364	-12.367	-12.381	-12.415	-12.473	-12.559	-12.673	-12.614	-12.580	-12.565	-12.563
	0.7	-12.295	-12.297	-12.311	-12.345	-12.404	-12.489	-12.603	-12.743	-12.709	-12.695	-12.692
	0.8	-12.245	-12.248	-12.262	-12.295	-12.354	-12.439	-12.552	-12.693	-12.859	-12.844	-12.842
	0.9	-12.216	-12.218	-12.232	-12.266	-12.324	-12.409	-12.522	-12.663	-12.828	-13.014	-13.120
	1.0	-12.206	-12.208	-12.222	-12.256	-12.314	-12.399	-12.512	-12.653	-12.818	-13.004	-13.202

Table 5.21 Con't

Coefficients $\bar{\varphi}$ for beams loaded with a concentrated moment M

α	β	ξ										
		0.0	0.1	0.2	0.3	0.4	0.5	0.6	0.7	0.8	0.9	1.0
5	0	-14.246	-13.752	-13.289	-12.877	-12.528	-12.248	-12.038	-11.894	-11.812	-11.777	-11.772
	0.1	-13.772	-13.778	-13.314	-12.902	-12.553	-12.273	-12.062	-11.919	-11.836	-11.802	-11.796
	0.2	-13.348	-13.354	-13.971	-12.978	-12.629	-12.348	-12.137	-11.993	-11.910	-11.875	-11.870
	0.3	-12.976	-12.982	-13.018	-13.105	-12.755	-12.473	-12.261	-12.116	-12.033	-11.998	-11.992
	0.4	-12.654	-12.660	-12.696	-12.783	-12.931	-12.648	-12.435	-12.289	-12.204	-12.169	-12.164
	0.5	-12.384	-12.390	-12.425	-12.511	-12.658	-12.873	-12.658	-12.511	-12.425	-12.390	-12.384
	0.6	-12.164	-12.169	-12.204	-12.289	-12.435	-12.648	-12.931	-12.783	-12.696	-12.660	-12.654
	0.7	-11.992	-11.998	-12.033	-12.116	-12.261	-12.473	-12.755	-13.105	-13.018	-12.982	-12.976
	0.8	-11.870	-11.875	-11.910	-11.993	-12.137	-12.348	-12.629	-12.978	-13.391	-13.354	-13.348
	0.9	-11.796	-11.802	-11.836	-11.919	-12.062	-12.273	-12.553	-12.902	-13.314	-13.778	-13.772
	1.0	-11.772	-11.777	-11.812	-11.894	-12.038	-12.248	-12.528	-12.877	-13.289	-13.752	-14.246
10	0	-15.967	-14.979	-14.056	-13.238	-12.548	-11.997	-11.585	-11.306	-11.145	-11.079	-11.068
	0.1	-15.019	-15.031	-14.107	-13.289	-12.599	-12.047	-11.634	-11.354	-11.198	-11.127	-11.116
	0.2	-14.175	-14.187	-14.263	-13.443	-12.751	-12.197	-11.782	-11.500	-11.338	-11.271	-11.260
	0.3	-13.435	-13.447	-13.522	-13.700	-13.005	-12.447	-12.028	-11.743	-11.579	-11.511	-11.500
	0.4	-12.799	-12.811	-12.884	-13.059	-13.360	-12.797	-12.373	-12.084	-11.916	-11.847	-11.836
	0.5	-12.267	-12.279	-12.350	-12.522	-12.816	-13.247	-12.816	-12.522	-12.350	-12.279	-12.267
	0.6	-11.836	-11.847	-11.916	-12.084	-12.373	-12.797	-13.360	-13.059	-12.884	-12.811	-12.799
	0.7	-11.500	-11.511	-11.579	-11.743	-12.028	-12.447	-13.005	-13.700	-13.522	-13.447	-13.435
	0.8	-11.260	-11.271	-11.338	-11.500	-11.782	-12.197	-12.751	-13.443	-14.263	-14.187	-14.175
	0.9	-11.116	-11.127	-11.193	-11.354	-11.634	-12.047	-12.509	-13.289	-14.107	-15.031	-15.019
	1.0	-11.068	-11.079	-11.145	-11.306	-11.585	-11.997	-12.548	-13.238	14.056	-14.979	-15.997

15											
0	-17.662	-16.182	-14.801	-13.582	-12.593	-11.746	-11.141	-10.733	-10.499	-10.403	-10.387
0.1	-16.242	-16.261	-14.879	-13.660	-12.636	-11.821	-11.215	-10.805	-10.570	-10.473	-10.458
0.2	-14.980	-14.999	-15.116	-13.893	-12.866	-12.046	-11.435	-11.021	-10.783	-10.685	-10.669
0.3	-13.878	-13.896	-14.011	-14.284	-13.249	-12.421	-11.802	-11.381	-11.138	-11.038	-11.022
0.4	-12.934	-12.952	-13.064	-13.330	-13.786	-12.946	-12.315	-11.885	-11.636	-11.532	-11.515
0.5	-12.150	-12.167	-12.275	-12.533	-12.975	-13.621	-12.975	-12.533	-12.275	-12.107	-12.150
0.6	-11.515	-11.532	-11.636	-11.885	-12.315	-12.946	-13.786	-13.330	-13.064	-12.952	-12.934
0.7	-11.022	-11.038	-11.138	-11.381	-11.802	-12.421	-13.249	-14.284	-14.011	-13.896	-13.878
0.8	-10.669	-10.685	-10.783	-11.021	-11.435	-12.046	-12.866	-13.893	-15.116	-14.999	-14.980
0.9	-10.458	-10.473	-10.570	-10.805	-11.215	-11.821	-12.636	-13.660	-14.879	-16.261	-16.242
1.0	-10.387	-10.403	-10.499	-10.733	-11.141	-11.746	-12.593	-13.582	-14.801	-16.182	-17.662

20											
0	-19.335	-17.361	-15.525	-13.911	-12.563	-11.496	-10.707	-10.175	-9.873	-9.748	-9.729
0.1	-17.442	-17.469	-15.632	-14.016	-12.666	-11.596	-10.804	-10.271	-9.966	-9.841	-9.821
0.2	-15.766	-15.792	-15.953	-14.332	-12.974	-11.896	-11.095	-10.555	-10.245	-10.117	-10.097
0.3	-14.306	-14.331	-14.488	-14.859	-13.489	-12.396	-11.581	-11.029	-10.711	-10.579	-10.558
0.4	-13.061	-13.085	-13.236	-13.596	-14.209	-13.096	-12.261	-11.691	-11.362	-11.225	-11.203
0.5	-12.032	-12.055	-12.200	-12.544	-13.135	-13.996	-13.135	-12.544	-12.200	-12.055	-12.032
0.6	-11.203	-11.225	-11.362	-11.691	-12.261	-13.096	-14.209	-13.596	-13.236	-13.085	-13.061
0.7	-10.558	-10.579	-10.711	-11.029	-11.581	-12.396	-13.489	-14.859	-14.488	-14.331	-14.306
0.8	-10.097	-10.117	-10.245	-10.555	-11.095	-11.896	-12.974	-14.332	-15.953	-15.792	-15.766
0.9	-9.821	-9.841	-9.966	-10.271	-10.804	-11.596	-12.666	-14.016	-15.632	-17.469	-17.442
1.0	-9.729	-9.748	-9.873	-10.175	-10.707	-11.496	-12.563	-13.911	-15.525	-17.361	-19.335

Table 5.21 Con't

Coefficients $\bar{\varphi}$ for beams loaded with a concentrated moment M

α	β	ξ										
		0.0	0.1	0.2	0.3	0.4	0.5	0.6	0.7	0.8	0.9	1.0
	0	-20.984	-18.518	-16.229	-14.225	-12.559	-11.241	-10.281	-9.634	-9.266	-9.115	-9.091
	0.1	-18.621	-18.655	-16.365	-14.358	-12.688	-11.372	-10.401	-9.751	-9.381	-9.228	-9.204
	0.2	-16.533	-16.566	-16.772	-14.758	-13.077	-11.747	-10.763	-10.102	-9.724	-9.568	-9.543
	0.3	-14.718	-14.751	-14.951	-15.424	-13.724	-12.372	-11.366	-10.686	-10.295	-10.133	-10.107
	0.4	-13.178	-13.209	-13.401	-13.856	-14.630	-13.247	-12.210	-11.503	-11.095	-10.924	-10.897
25	0.5	-11.913	-11.942	-12.123	-12.555	-13.295	-14.372	-13.295	-12.555	-12.123	-11.942	-11.913
	0.6	-10.897	-10.924	-11.095	-11.503	-12.210	-13.247	-14.630	-13.856	-13.401	-13.209	-13.178
	0.7	-10.107	-10.133	-10.295	-10.686	-11.366	-12.372	-13.724	-15.424	-14.951	-14.751	-14.718
	0.8	-9.543	-9.568	-9.724	-10.102	-10.763	-11.747	-13.077	-14.758	-16.772	-16.566	-16.533
	0.9	-9.204	-9.228	-9.381	-9.751	-10.401	-11.372	-12.688	-14.358	-16.365	-18.655	-18.621
	1.0	-9.091	-9.115	-9.266	-9.634	-10.281	-11.241	-12.559	-14.225	-16.229	-18.518	-20.984
	0	-28.919	-23.998	-19.480	-15.595	-12.436	-10.010	-8.268	-7.130	-6.494	-6.237	-6.196
	0.1	-24.214	-24.293	-19.770	-15.875	-12.703	-10.260	-8.501	-7.349	-6.704	-6.442	-6.401
	0.2	-20.101	-20.177	-20.641	-16.717	-13.503	-11.010	-9.201	-8.007	-7.333	-7.058	-7.014
	0.3	-16.579	-16.651	-17.093	-18.121	-14.836	-12.260	-10.368	-9.103	-8.381	-8.084	-8.036
	0.4	-13.647	-13.714	-14.124	-15.086	-16.702	-14.010	-12.002	-10.638	-9.850	-9.520	-9.468
50	0.5	-11.307	-11.357	-11.737	-12.612	-14.102	-16.260	-14.102	-12.612	-11.737	-11.367	-11.307
	0.6	-9.468	-9.520	-9.850	-10.638	-12.002	-14.010	-16.702	-15.086	-14.124	-13.714	-13.647
	0.7	-8.036	-8.084	-8.381	-9.103	-10.368	-12.260	-14.836	-18.121	-17.093	-16.651	-16.579
	0.8	-7.014	-7.058	-7.333	-8.007	-9.201	-11.010	-13.503	-16.717	-20.641	-20.177	-20.101
	0.9	-6.401	-6.442	-6.704	-7.349	-8.501	-10.260	-12.703	-15.875	-19.770	-24.293	-24.214
	1.0	-6.196	-6.237	-6.494	-7.130	-8.268	-10.010	-12.436	-15.595	-19.480	-23.998	-28.919

100											
0	-43.497	-33.693	-24.860	-17.506	-11.775	-7.684	-4.731	-2.962	-2.019	-1.1651	-1.594
0.1	-34.159	-34.352	-25.503	-18.115	-12.334	-8.084	-5.172	-3.353	-2.376	-1.992	-1.933
0.2	-26.145	-26.329	-27.433	-19.942	-14.011	-9.584	-6.495	-4.526	-3.447	-3.014	-2.947
0.3	-19.455	-19.625	-20.649	-22.988	-16.807	-12.084	-8.699	-6.480	-5.231	-4.719	-4.637
0.4	-14.088	-14.239	-15.151	-17.252	-20.721	-15.584	-11.785	-9.216	-7.728	-7.104	-7.003
0.5	-10.046	-10.172	-10.940	-12.734	-15.753	-20.084	-15.753	-12.734	-10.940	-10.172	-10.046
0.6	-7.003	-7.104	-7.728	-9.216	-11.785	-15.584	-20.721	-17.252	-15.151	-14.239	-14.088
0.7	-4.637	-4.719	-5.231	-6.480	-8.699	-12.084	-16.807	-22.988	-20.649	-19.625	-19.455
0.8	-2.947	3.014	-3.447	-4.526	-6.495	-9.584	-14.011	-19.942	-27.433	-26.329	-26.145
0.9	-1.933	1.992	-2.376	-3.353	-5.172	-8.084	-12.334	-18.115	-25.503	-34.352	-34.159
1.0	-1.594	-1.651	-2.019	-2.962	-4.731	-7.684	-11.775	-17.500	-24.860	-33.693	-43.497

150											
0	-56.773	-42.111	-29.113	-18.592	-10.706	-5.218	-1.692	0.362	1.392	1.776	1.834
0.1	-42.850	-43.183	-30.153	-19.563	-11.576	-5.968	-2.322	-0.167	1.392	1.348	1.411
0.2	-31.085	-31.400	-33.273	-22.476	-14.185	-8.218	-4.213	-1.755	0.932	0.065	0.145
0.3	-21.474	-21.762	-23.472	-27.330	-18.533	-11.968	-7.364	-4.400	-0.448	-2.074	-1.966
0.4	-14.019	-14.267	-15.752	-19.127	-24.621	-17.218	-11.776	-8.103	-2.749	-5.068	-4.921
0.5	-8.720	-8.918	-10.111	-12.865	-17.449	-23.968	-17.440	-12.865	-5.970	-8.918	-8.720
0.6	-4921	-5.068	-5.970	-8.103	-11.776	-17.218	-24.621	-19.127	-10.111	-14.267	-14.019
0.7	-1.966	-2.074	-2.749	-4.400	-7.364	-11.968	-18.533	-27.330	-15.752	-21.762	-21.474
0.8	0.145	0.065	-0.448	-1.755	-4.213	-8.218	-14.185	-22.476	-23.472	-31.400	-31.085
0.9	1.411	1.348	0.932	-0.167	-1.232	-5.968	-11.576	-19.563	-33.273	-43.183	-42.851
1.0	1.834	1.776	1.392	0.362	-1.692	-5.218	-10.706	-18.592	-30.153	-42.111	-56.773

Table 5.21 Con't

Coefficients $\bar{\varphi}$ for beams loaded with a concentrated moment M

α	β	ξ										
		0.0	0.1	0.2	0.3	0.4	0.5	0.6	0.7	0.8	0.9	1.0
	0	-69.088	-49.590	-32.541	-19.079	-9.351	-2.907	0.977	3.070	4.038	4.373	4.421
	0.1	-50.618	-51.111	-34.009	-20.436	-10.545	-3.907	0.170	2.427	3.506	3.804	3.951
	0.2	-35.208	-35.673	-38.415	-24.507	-14.125	-6.907	-2.249	0.498	1.911	2.456	2.541
	0.3	-22.857	-23.277	-25.757	-31.292	-20.092	-11.907	-6.282	-2.717	-0.746	-0.060	0.190
	0.4	-13.565	-13.922	-16.036	-20.791	-28.446	-18.907	-11.928	-7.218	-4.467	-3.295	-3.102
200	0.5	-7.334	-7.609	-9.252	-13.005	-19.187	-27.907	-19.187	-13.005	-9.252	-7.609	-7.334
	0.6	-3.102	-3.295	-4.467	-7.218	-11.928	-18.907	-28.446	-20.791	-16.036	-13.922	-13.565
	0.7	0.190	-0.060	-0.746	-2.717	-6.282	-11.907	-20.092	-31.292	-25.757	-23.277	-22.857
	0.8	2.541	2.456	1.911	0.498	-2.249	-6.907	-14.125	-24.507	-38.415	-35.673	-35.208
	0.9	3.951	3.894	3.506	2.427	0.170	-3.907	-10.545	-20.436	-34.009	-51.111	-50.618
	1.0	4.421	4.373	4.038	3.070	0.977	-2.907	-9.351	-19.079	-32.541	-49.500	-69.088
	0	-80.673	-56.355	-35.344	-19.120	-7.789	-0.649	3.359	5.317	6.114	6.358	6.391
	0.1	-57.682	-58.351	-37.265	-20.882	-9.316	-1.899	2.386	4.578	5.635	5.854	5.900
	0.2	-38.708	-39.338	-43.027	-26.165	-13.897	-5.649	-0.532	2.362	3.798	4.341	4.426
	0.3	-23.751	-24.317	-27.631	-34.972	-21.533	-11.899	-5.397	-1.332	0.902	1.820	1.969
	0.4	-12.812	-13.287	-16.077	-22.301	-32.222	-20.649	-12.208	-6.503	-3.152	-1.710	-1.470
250	0.5	-5.801	-6.248	-8.365	-13.152	-20.965	-31.899	-20.965	-13.152	-8.365	-6.248	-5.801
	0.6	-1.470	-1.710	-3.152	-6.503	-12.208	-20.649	-32.222	-22.301	-16.077	-13.287	-12.812
	0.7	1.960	1.820	0.902	-1.332	-5.397	-11.899	-21.533	-34.972	-27.631	-24.317	-23.751
	0.8	4.426	4.341	3.798	2.362	-0.532	-5.649	-13.897	-26.165	-43.027	-39.338	-38.708
	0.9	5.900	5.854	5.635	4.578	2.386	-1.899	-9.316	-20.882	-37.265	-58.351	-57.682
	1.0	6.391	6.358	6.114	5.317	3.359	-0.649	-7.789	-19.120	-35.344	-56.355	-80.673

300	0	0.1	0.2	0.3	0.4	0.5	0.6	0.7	0.8	0.9	1.0
0	-91.686	-62.562	-37.662	-18.819	-6.075	1.561	5.516	7.206	7.758	7.881	7.894
0.1	-64.194	-65.053	-40.054	-20.999	-7.943	0.061	4.384	6.385	7.150	7.372	7.402
0.2	-41.719	-42.526	-47.228	-27.537	-13.548	-4.439	0.988	3.924	5.325	5.846	5.928
0.3	-24.261	-24.982	-29.186	-38.435	-22.889	-11.939	-4.671	-0.179	2.283	3.301	3.470
0.4	-11.820	-12.420	-15.927	-23.691	-35.966	-22.439	-12.593	-5.922	-1.976	-0.261	0.029
0.5	-4.396	-4.840	-7.452	-13.307	-22.780	-35.939	-22.780	-13.307	-7.452	-4.840	-4.396
0.6	0.029	-0.261	-1.976	-5.922	-12.593	-22.439	-35.966	-23.691	-15.927	-12.420	-11.820
0.7	3.470	3.301	2.283	-0.179	-4.671	-11.939	-22.889	-38.485	-29.186	-24.982	-24.261
0.8	5.928	5.846	5.325	3.924	0.988	-4.439	-13.548	-27.537	-47.228	-42.526	-41.719
0.9	7.402	7.372	7.150	6.385	4.384	0.061	-7.943	-20.999	-40.054	-65.053	-64.194
1.0	7.894	7.881	7.758	7.206	5.616	1.561	-6.075	-18.819	-37.662	-62.562	-91.686

350	0	0.1	0.2	0.3	0.4	0.5	0.6	0.7	0.8	0.9	1.0
0	-102.242	-68.323	-39.595	-18.250	-4.247	3.726	7.488	8.813	9.067	9.049	9.039
0.1	-70.266	-71.326	-42.472	-20.859	-6.462	1.976	6.204	7.922	8.443	8.551	8.563
0.2	-44.330	-45.333	-51.102	-28.685	-13.108	-3.274	2.350	5.248	6.574	7.059	7.137
0.3	-24.461	-25.346	-30.486	-41.729	-24.184	-12.024	-4.074	0.792	3.457	4.572	4.759
0.4	-10.632	-11.364	-15.624	-24.990	-39.691	-24.274	-13.067	-5.447	-0.905	1.090	1.429
0.5	-2.851	-3.387	-6.514	-13.468	-24.629	-40.024	-24.629	-13.468	-6.514	-3.387	-2.851
0.6	1.429	1.090	-0.905	-5.447	-13.067	-24.274	-39.691	-24.990	-15.624	-11.364	-10.632
0.7	4.759	4.572	3.457	0.792	-4.074	-12.024	-24.184	-41.729	-30.486	-25.346	-24.461
0.8	7.137	7.059	6.574	5.248	2.350	-3.274	-13.108	-28.685	-51.102	-45.333	-44.339
0.9	8.563	8.551	8.443	7.922	6.204	1.976	-6.462	-20.859	-42.472	-71.326	-70.266
1.0	9.039	9.049	9.067	8.813	7.488	3.726	-4.247	-18.250	-39.595	-68.323	-102.242

Table 5.21 Con't

Coefficients $\overline{\varphi}$ for beams loaded with a concentrated moment M

α	β	ξ										
		0.0	0.1	0.2	0.3	0.4	0.5	0.6	0.7	0.8	0.9	1.0
400	0	-112.426	-73.722	-41.218	-17.468	-2.332	5.847	9.309	10.195	10.110	9.938	9.905
	0.1	-75.970	-77.249	-44.591	-20.515	-4.899	3.847	7.877	9.242	9.483	9.465	9.458
	0.2	-46.639	-47.829	-54.712	-29.655	-12.600	-2.153	3.578	6.382	7.603	8.045	8.118
	0.3	-24.406	-25.463	-31.579	-44.889	-25.436	-12.153	-3.586	1.616	4.471	5.679	5.885
	0.4	-9.280	-10.151	-15.193	-26.216	-43.407	-26.153	-13.616	-5.057	0.085	2.367	2.759
	0.5	-1.261	-1.892	-5.554	-13.636	-26.511	-44.153	-26.511	-13.636	-5.554	-1.892	-1.261
	0.6	2.759	2.367	0.085	-5.057	-13.616	-26.153	-43.407	-26.216	-15.193	-10.151	-9.280
	0.7	5.885	5.679	4.471	1.616	-3.586	-12.153	-25.436	-44.889	-31.579	-25.463	-24.406
	0.8	8.118	8.045	7.603	6.382	3.578	-2.153	-12.600	-29.655	-54.712	-47.829	-46.639
	0.9	9.458	9.465	9.483	9.242	7.877	3.847	-4.899	-20.515	-44.591	-77.249	-75.979
	1.0	9.905	9.938	10.110	10.195	9.309	5.847	-2.332	-17.468	-41.218	-73.722	-112.426
450	0	-122.301	-78.821	-42.584	-16.514	-0.351	7.929	11.003	11.392	10.940	10.606	10.549
	0.1	-81.394	-82.882	-46.464	-20.005	-3.274	5.679	9.426	10.384	10.320	10.168	10.142
	0.2	-48.073	-50.067	-58.102	-30.481	-12.012	-10.713	4.595	7.360	8.458	8.852	8.921
	0.3	-24.138	-25.374	-32.500	-47.941	-26.657	-12.321	-3.190	2.319	5.356	6.660	6.886
	0.4	-7.789	-8.804	-14.650	-27.384	-47.117	-28.071	-14.230	-4.738	1.012	3.590	4.037
	0.5	0.374	-0.357	-4.572	-13.811	-28.071	-48.321	-28.424	-13.811	-4.572	-0.357	0.374
	0.6	4.037	3.590	1.012	-4.738	-14.230	-28.071	-47.117	-27.384	-14.656	-8.804	-7.789
	0.7	6.886	6.660	5.356	2.319	-3.190	-12.321	-26.657	-47.941	-32.500	-25.374	-24.138
	0.8	8.921	8.852	8.458	7.360	4.595	-10.713	-12.042	-30.481	-58.102	-50.067	-48.673
	0.9	10.142	10.168	10.320	10.384	9.426	5.679	-3.274	-20.005	-46.464	-82.882	-81.394
	1.0	10.549	10.606	10.940	11.392	11.003	7.929	-0.351	-16.514	-42.584	-78.821	122.301

500	0	0.1	0.2	0.3	0.4	0.5	0.6	0.7	0.8	0.9	1.0
0	11.016	11.098	11.598	12.436	12.589	9.972	1.681	−15.418	−43.738	−83.668	−131.918
0.1	10.657	10.703	10.992	11.379	10.871	7.472	−1.601	−19.361	−48.131	−88.273	−86.559
0.2	9.581	9.517	9.171	8.208	5.716	−0.028	−11.446	−31.190	−61.311	−52.087	−50.483
0.3	7.788	7.540	6.138	2.923	−28.746	−12.528	−27.856	−50.904	−33.277	−25.110	−23.690
0.4	5.277	4.773	1.891	−4.477	−14.902	−30.028	−50.829	−28.505	−14.030	−7.343	−6.179
0.5	2.049	1.215	−3.570	−13.991	−30.365	−52.528	−30.365	−13.991	−3.570	1.215	2.049
0.6	−6.179	−7.343	−14.030	−28.505	−50.829	−30.028	−14.902	−4.477	1.891	4.773	5.277
0.7	−23.690	−25.110	−33.277	−50.904	−27.856	−12.528	−28.746	2.923	6.138	7.540	7.788
0.8	−50.483	−52.087	−61.311	−31.190	−11.446	−0.028	5.716	8.208	9.171	9.517	9.581
0.9	−86.559	−88.273	−48.131	−19.361	−1.601	7.472	10.871	11.379	10.992	10.703	10.657
1.0	−131.918	−83.668	−43.738	−15.418	1.681	9.972	12.589	12.436	11.598	11.098	11.016

Table 5.22 Relative settlements of beams with concentrated moments

Coefficients $\bar{y}$ for beams loaded with a concentrated moment M

α	β	ξ										
		0.0	0.1	0.2	0.3	0.4	0.5	0.6	0.7	0.8	0.9	1.0
1	0	0	-1.280	-2.551	-3.812	-5.066	-6.314	-7.557	-8.796	-10.003	-11.268	-12.504
	0.1	0	-1.276	-2.547	-3.809	-5.063	-6.312	-7.555	-8.794	-10.032	-11.268	-12.504
	0.2	0	-1.267	-2.535	-3.798	-5.054	-6.304	-7.549	-8.790	-10.029	-11.266	-12.504
	0.3	0	-1.260	-2.520	-3.781	-5.040	-6.292	-7.539	-8.783	-10.024	-11.264	-12.504
	0.4	0	-1.253	-2.507	-3.762	-5.019	-6.274	-7.525	-8.772	-10.017	-11.260	-12.504
	0.5	0	-1.248	-2.426	-3.745	-4.997	-6.252	-7.507	-8.759	-10.008	-11.256	-12.504
	0.6	0	-1.243	-2.487	-3.732	-4.979	-6.229	-7.485	-8.742	-9.997	-11.250	-12.504
	0.7	0	-1.240	-2.480	-3.721	-4.965	-6.212	-7.464	-8.723	-9.984	-11.244	-12.504
	0.8	0	-1.237	-2.475	-3.714	-4.955	-6.199	-7.449	-8.705	-9.969	-11.236	-12.504
	0.9	0	-1.236	-2.472	-3.709	-4.949	-6.192	-7.440	-8.695	-9.957	-11.228	-12.504
	1.0	0	-1.235	-2.471	-3.708	-4.947	-6.190	-7.437	-8.691	-9.953	-11.223	-12.504
2	0	0	-1.310	-2.601	-3.875	-5.133	-6.378	-7.613	-8.842	-10.065	-11.287	-12.507
	0.1	0	-1.301	-2.593	-3.868	-5.127	-6.372	-7.609	-8.839	-10.063	-11.286	-12.507
	0.2	0	-1.284	-2.569	-3.847	-5.109	-6.358	-7.597	-8.830	-10.057	-11.283	-12.507
	0.3	0	-1.269	-2.539	-3.812	-5.079	-6.333	-7.578	-8.815	-10.047	-11.278	-12.507
	0.4	0	-1.256	-2.513	-3.773	-5.037	-6.298	-7.550	-8.794	-10.034	-11.271	-12.507
	0.5	0	-1.245	-2.492	-3.740	-4.993	-6.254	-7.514	-8.767	-10.016	-11.262	-12.507
	0.6	0	-1.237	-2.474	-3.713	-4.958	-6.209	-7.470	-8.734	-9.994	-11.251	-12.507
	0.7	0	-1.230	-2.460	-3.692	-4.930	-6.174	-7.428	-8.696	-9.968	-11.238	-12.507
	0.8	0	-1.225	-2.450	-3.678	-4.910	-6.149	-7.399	-8.661	-9.938	-11.223	-12.507
	0.9	0	-1.222	-2.444	-3.669	-4.898	-6.134	-7.381	-8.640	-9.914	-11.266	-12.507
	1.0	0	-1.221	-2.442	-3.666	-4.894	-6.129	-7.375	-8.633	-9.906	-11.197	-12.507

0	0	-1.440	-2.751	-4.059	-5.329	-6.567	-7.781	-8.977	-10.162	-11.341	-12.518
0.1	0	-1.377	-2.732	-4.042	-5.414	-6.555	-7.771	-8.970	-10.157	-11.338	-12.518
0.2	0	-1.335	-2.672	-3.990	-5.270	-6.520	-7.742	-8.948	-10.142	-11.331	-12.518
0.3	0	-1.298	-2.597	-3.903	-5.195	-6.436	-7.692	-8.911	-10.118	-11.319	-12.518
0.4	0	-1.266	-2.533	-3.807	-5.092	-6.370	-7.624	-8.859	-10.083	-11.302	-12.518
0.5	0	-1.239	-2.479	-3.725	-4.983	-6.259	-7.535	-8.793	-10.039	-11.280	-12.518
0.6	0	-1.217	-2.435	-3.660	-4.895	-6.148	-7.427	-8.712	-9.985	-11.253	-12.518
0.7	0	-1.199	-2.401	-3.608	-4.826	-6.062	-7.323	-8.615	-9.921	-11.221	-12.518
0.8	0	-1.187	-2.376	-3.571	-4.777	-6.000	-7.249	-8.528	-9.846	-11.183	-12.518
0.9	0	-1.180	-2.351	-3.549	-4.747	-5.963	-7.204	-8.476	-9.787	-11.141	-12.518
5 1.0	0	-1.777	-2.356	-3.541	-4.737	-5.951	-7.189	-8.439	-9.767	-11.118	-12.518

0	0	-1.547	-2.998	-4.362	-5.650	-6.876	-8.054	-9.107	-10.319	-11.429	-12.536
0.1	0	-1.502	-2.958	-4.327	-5.620	-6.852	-8.034	-9.183	-10.309	-11.425	-12.536
0.2	0	-1.418	-2.840	-4.224	-5.532	-6.779	-7.976	-9.139	-10.280	-11.410	-12.536
0.3	0	-1.314	-2.691	-4.052	-5.386	-6.657	-7.880	-9.067	-10.233	-11.386	-12.536
0.4	0	-1.280	-2.564	-3.860	-5.180	-6.487	-7.744	-8.966	-10.165	-11.353	-12.536
0.5	0	-1.227	-2.458	-3.700	-4.966	-6.268	-7.570	-8.836	-10.079	-11.309	-12.536
0.6	0	-1.184	-2.371	-3.570	-4.792	-6.-49	-7.356	-8.676	-9.972	-11.256	-12.536
0.7	0	-1.150	-2.304	-3.469	-4.657	-5.879	-7.151	-8.485	-9.845	-11.193	-12.536
0.8	0	-1.126	-2.256	-3.397	-4.560	-5.758	-7.004	-8.313	-9.697	-11.119	-12.536
0.9	0	-1.112	-2.227	-3.354	-4.502	-5.685	-6.916	-8.209	-9.578	-11.034	-12.536
10 1.0	0	-1.107	-2.218	-3.339	-4.483	-5.661	-6.887	-8.175	-9.538	-10.989	-12.536

Table 5.22 Con't

Coefficients $\bar{y}$ for beams loaded with a concentrated moment M

α	β	0.0	0.1	0.2	0.3	0.4	0.5	0.6	0.7	0.8	0.9	1.0
							ξ					
15	0	0	−1.692	−3.240	−4.657	−5.963	−7.176	−8.319	−9.411	−10.471	−11.515	−12.555
	0.1	0	−1.625	−3.181	−4.606	−5.910	−7.140	−8.290	−9.390	−10.457	−11.508	−12.555
	0.2	0	−1.499	−3.003	−4.452	−5.788	−7.032	−8.205	−9.326	−10.415	−11.487	−12.555
	0.3	0	−1.388	−2.783	−4.196	−5.571	−6.853	−8.062	−9.220	−10.344	−11.452	−12.555
	0.4	0	−1.294	−2.594	−3.912	−5.266	−6.601	−7.862	−9.071	−10.245	−11.403	−12.555
	0.5	0	−1.215	−2.436	−3.675	−4.949	−6.277	−7.605	−8.879	−10.118	−11.339	−12.555
	0.6	0	−1.152	−2.309	−3.484	−4.692	−5.954	−7.289	−8.643	−9.961	−11.261	−12.555
	0.7	0	−1.103	−2.210	−3.335	−4.193	−5.702	−6.984	−8.359	−9.772	−11.166	−12.555
	0.8	0	−1.007	−2.140	−3.229	−4.350	−5.522	−7.766	−8.102	−9.551	−11.056	−12.555
	0.9	0	−1.046	−2.097	−3.165	−4.264	−5.414	−6.635	−7.949	−9.374	−11.930	−12.555
	1.0	0	−1.039	−2.083	−3.144	−4.236	−5.378	−6.592	−7.897	−9.315	−10.863	−12.555
20	0	0	−1.834	−3.477	−4.947	−6.268	−7.469	−8.576	−9.619	−10.619	−11.599	−12.573
	0.1	0	−1.744	−3.398	−4.879	−6.211	−7.421	−8.539	−9.591	−10.601	−11.590	−12.573
	0.2	0	−1.577	−3.163	−4.675	−6.038	−7.279	−8.427	−9.507	−10.545	−11.562	−12.573
	0.3	0	−1.431	−2.871	−4.336	−5.751	−7.043	−8.239	−9.368	−10.433	−11.516	−12.573
	0.4	0	−1.307	−2.621	−3.961	−5.349	−6.712	−7.977	−9.173	−10.324	−11.452	−12.573
	0.5	0	−1.204	−2.415	−3.650	−4.932	−6.286	−7.640	−8.922	−10.157	−11.369	−12.573
	0.6	0	−1.121	−2.249	−3.400	−4.595	−5.861	−7.224	−8.612	−9.951	−11.266	−12.573
	0.7	0	−1.056	−2.119	−3.205	−4.333	−5.530	−6.822	−8.237	−9.702	−11.141	−12.573
	0.8	0	−1.010	−2.027	−3.065	−4.146	−5.293	−6.534	−7.897	−9.410	−10.995	−12.573
	0.9	0	−0.983	−1.972	−2.982	−4.034	−5.151	−6.362	−7.694	−9.174	−10.828	−12.573
	1.0	0	−0.973	−1.953	2.954	−3.996	−5.104	−6.305	−7626	−9.096	−10.738	−12.573

25	0	0	-1.974	-3.710	-5.230	-6.566	-7.753	-8.827	-9.820	-10.763	-11.681	-12.590
	0.1	0	-1.863	-3.612	-5.146	-6.495	-7.695	-8.781	-9.786	-10.740	-11.669	-12.590
	0.2	0	-1.654	-3.319	-4.893	-6.282	-7.520	-8.643	-9.683	-10.672	-11.636	-12.590
	0.3	0	-1.473	-2.955	-4.472	-5.927	-7.228	-8.412	-9.512	-10.559	-11.579	-12.590
	0.4	0	-1.319	-2.647	-4.008	-5.429	-6.820	-8.090	-9.273	-10.401	-11.500	-12.590
	0.5	0	-1.192	-2.394	-3.625	-4.915	-6.295	-7.676	-8.965	10.197	-11.398	-12.590
	0.6	0	-1.090	-2.190	-3.317	-4.500	-5.770	-7.161	-8.583	-9.943	-11.271	-12.590
	0.7	0	-1.011	-2.031	-3.078	-4.178	-5.362	-6.664	-8.118	-9.635	-11.118	-12.590
	0.8	0	-0.955	-1.918	-2.907	-3.948	-5.070	-6.309	-7.697	-9.271	-10.936	-12.590
	0.9	0	-0.921	-1.850	-2.805	-3.810	-4.895	-6.096	-7.445	-8.978	-10.727	-12.590
	1.0	0	-0.910	-1.827	-2.770	-3.764	-4.837	-6.024	-7.361	-8.881	-10.616	-12.590

50	0	0	-2.644	-4.813	-6.561	-7.956	-9.073	-9.981	-10.747	-11.424	-12.056	-12.679
	0.1	0	-2.423	-4.622	-6.399	-7.821	-8.964	-9.896	-10.648	-11.383	-12.038	-12.679
	0.2	0	-2.012	-4.049	-5.911	-7.416	-8.635	-9.641	-10.500	-11.259	-11.976	-12.679
	0.3	0	-1.660	-3.343	5.098	-6.740	-8.089	-9.215	-10.183	-11.058	-11.875	-12.679
	0.4	0	-1.366	-2.754	-4.210	-5.794	-7.323	-8.618	-9.745	-10.765	-11.730	-12.679
	0.5	0	-1.132	-2.284	-3.497	-4.827	-6.340	-7.852	-9.182	-10.394	-11.546	-12.679
	0.6	0	-0.948	-1.913	-2.923	-4.060	-5.355	-6.885	-8.469	-9.924	-11.312	-12.679
	0.7	0	-0.805	-1.625	-2.495	-3.464	-4.590	-5.939	-7.581	-9.376	-11.019	-12.679
	0.8	0	-0.703	-1.419	-2.182	-3.038	-4.043	-5.263	-6.768	-8.630	-10.667	-12.679
	0.9	0	-0.641	-1.296	-1.995	-2.783	-3.715	-4.857	-6.280	-8.057	-10.255	-12.679
	1.0	0	-0.621	-1.255	-1.932	-2.697	-3.606	-4.722	-6.118	-7.866	-10.034	-12.679

Table 5.22 Con't

Coefficients $\bar{y}$ for beams loaded with a concentrated moment M

α	β	ξ										
		0.0	0.1	0.2	0.3	0.4	0.5	0.6	0.7	0.8	0.9	1.0
	0	0	-3.855	-6.772	-8.876	-10.327	-11.283	-11.889	-12.265	-12.508	-12.688	-12.849
	0.1	0	-3.421	-6.403	-8.571	-10.080	-11.089	-11.741	-12.159	-12.440	-12.654	-12.849
	0.2	0	-2.619	-5.297	-7.653	-9.338	-10.506	-11.299	-11.842	-12.234	-12.553	-12.849
	0.3	0	-1.950	-3.954	-6.124	-8.101	-9.534	-10.563	-11.313	-11.891	-12.383	-12.849
	0.4	0	-1.413	-2.874	-4.483	-6.370	-8.174	-9.531	-10.572	-11.410	-12.146	-12.849
100	0.5	0	-1.008	-2.056	-3.230	-4.644	-6.425	-8.205	-9.610	-10.793	-11.841	-12.849
	0.6	0	-0.703	-1.439	-2.278	-3.318	-4.676	-6.480	-8.366	-9.976	-11.436	-12.849
	0.7	0	-0.466	-0.958	-1.537	-2.287	-3.315	-4.748	-6.725	-8.895	-10.899	-12.849
	0.8	0	-0.296	-0.615	-1.007	-1.550	-2.344	-3.511	-5.196	-7.552	-10.230	-12.849
	0.9	0	-0.195	-0.409	-0.690	-1.108	-1.761	-2.769	-4.279	-6.446	-9.428	-12.849
	1.0	0	-0.161	-0.341	-0.584	-0.961	-1.566	-2.522	-3.973	-6.078	-8.994	-12.849
	0	0	-4.936	-8.479	-10.842	-12.285	-13.063	-13.394	-13.451	-13.356	-13.194	-13.012
	0.1	0	-4.294	-7.942	-10.406	-11.942	-12.800	-13.201	-12.315	-13.270	-13.151	-13.012
	0.2	0	-3.117	-6.323	-9.099	-10.912	-12.014	-12.621	-12.908	-13.010	-13.024	-13.012
	0.3	0	-2.155	-4.441	-6.921	-9.195	-10.703	-11.654	-12.230	-12.578	-12.813	-13.012
	0.4	0	-1.408	-2.895	-4.622	-6.792	-8.867	-10.301	-11.281	-11.973	-12.516	-13.012
150	0.5	0	-0.877	-1.817	-2.952	-4.451	-6.506	-8.560	-10.060	-11.195	-12.135	-13.012
	0.6	0	-0.496	-1.039	-1.731	-2.711	-4.453	-6.220	-8.390	-10.116	-11.604	-13.012
	0.7	0	-0.199	-0.434	-0.782	-1.358	-2.309	-3.917	-6.090	-5.611	-10.857	-13.012
	0.8	0	0.012	-0.002	-0.104	-0.391	-0.998	-2.100	-3.912	-6.679	-9.895	-13.012
	0.9	0	0.140	0.258	0.303	0.189	-0.211	-1.070	-2.605	-5.070	-8.718	-13.012
	1.0	0	0.182	0.344	0.439	0.383	0.051	-0.726	-1.270	-4.533	-8.076	-13.012

200											
0	0	-5.922	-10.001	-12.551	-13.942	-14.551	-14.609	-14.395	-14.032	-13.608	-13.167
0.1	0	5.075	-9.304	-11.995	-13.515	-14.213	-14.381	-14.239	-13.935	-13.561	-13.167
0.2	0	-3.533	-7.212	-10.328	-12.232	-13.259	-13.698	-13.773	-13.643	-13.419	-13.167
0.3	0	-2.296	-4.725	-7.551	-10.093	-11.670	-12.560	-12.995	-13.156	-13.183	-13.167
0.4	0	-1.366	-2.844	-4.662	-7.099	-9.445	-10.966	-11.906	-12.475	-12.852	-13.167
0.5	0	-0.740	-1.568	-2.662	-4.250	-6.584	-8.917	-10.506	-11.599	-12.427	-13.167
0.6	0	-0.315	-0.692	-1.261	-2.201	-3.723	-6.068	-8.506	-10.323	-11.802	-13.167
0.7	0	0.016	-0.911	-0.173	-0.607	-1.498	-3.074	-5.617	-8.442	-10.871	-13.167
0.8	0	0.252	0.476	0.605	0.531	0.092	-0.936	-2.839	-5.955	-9.635	-13.167
0.9	0	0.394	0.078	1.072	1.214	1.046	0.347	-1.172	-3.863	-8.093	-13.167
1.0	0	0.441	0.865	1.228	1.442	1.363	0.775	-0.617	-3.166	-7.245	-13.167

250											
0	0	-6.835	-11.384	-14.065	-15.372	-15.763	-15.606	-15.159	-14.581	-13.954	-13.316
0.1	0	-5.785	-10.531	-13.397	-14.808	-15.398	-15.352	-14.991	-14.478	-13.905	-13.316
0.2	0	-3.887	-7.972	-11.392	-13.359	-14.306	-14.593	-14.486	-14.169	-13.756	-13.316
0.3	0	-2.390	-4.957	-8.051	-10.843	-12.485	-13.327	-13.646	-13.654	-13.509	-13.316
0.4	0	-1.293	-2.736	-4.624	-7.320	-9.936	-11.554	-12.468	-12.933	-13.163	-13.316
0.5	0	-0.598	-1.309	-2.361	-4.041	-6.658	-9.275	-10.955	-12.007	-12.718	-13.316
0.6	0	-0.153	-0.383	-0.847	-1.762	-3.380	-5.996	-8.692	-10.580	-12.022	-13.316
0.7	0	0.193	0.338	0.330	0.011	-0.831	-2.473	-5.265	-8.350	-10.926	-13.316
0.8	0	0.440	0.853	1.170	1.276	0.990	0.043	-1.924	-5.344	-9.429	-13.316
0.9	0	0.589	1.162	1.675	2.037	2.083	1.552	0.081	-2.785	-7.531	-13.316
1.0	0	0.638	1.265	1.843	2.290	2.447	2.056	0.749	-1.932	-6.481	-13.316

Table 5.22 Con't

Coefficients $\bar{y}$ for beams loaded with a concentrated moment M

α	β	ξ 0.0	0.1	0.2	0.3	0.4	0.5	0.6	0.7	0.8	0.9	1.0
	0	0	-7.691	-12.656	-15.427	-16.624	-16.813	-16.435	-15.785	-15.031	-14.247	-13.458
	0.1	0	-6.442	-11.652	-14.652	-16.052	-16.410	-16.163	-15.610	-14.927	-14.198	-13.458
	0.2	0	-4.103	-8.638	-12.327	-14.336	-15.200	-15.347	-15.085	-14.613	-14.049	-13.458
	0.3	0	-2.445	-5.115	-8.452	-11.476	-13.183	-13.987	-14.209	-14.090	-13.801	-13.458
	0.4	0	-1.197	-2.583	-4.526	-7.472	-10.359	-12.083	-12.984	-13.358	-13.454	-13.458
300	0.5	0	-0.451	-1.042	-2.050	-3.824	-6.729	-9.634	-11.408	-12.417	-13.007	-13.458
	0.6	0	-0.005	-0.100	-0.474	-1.376	-3.099	-5.986	-8.932	-10.876	-12.261	-13.458
	0.7	0	0.343	0.032	0.751	0.528	-0.275	-1.982	-5.007	-8.343	-11.014	-13.458
	0.8	0	0.591	1.155	1.627	1.888	1.741	0.878	-1.131	-4.820	-9.266	-13.458
	0.9	0	0.739	1.468	2.152	2.705	2.952	2.594	1.194	-1.807	-7.017	-13.458
	1.0	0	0.789	1.573	2.327	2.977	3.355	3.166	1.969	-0.802	-5.767	-13.458
	0	0	-8.502	-13.841	-16.670	-17.738	-17.721	-17.133	-16.304	-15.405	-14.499	-13.595
	0.1	0	-7.054	-12.688	-15.792	-17.102	-17.284	-16.847	-16.126	-15.302	-14.451	-13.595
	0.2	0	-4.460	-9.229	-13.159	-15.195	-15.973	-15.991	-15.593	-14.993	-14.306	-13.595
	0.3	0	-2.469	-5.214	-8.772	-12.010	-13.788	-14.563	-14.706	-14.478	-14.066	-13.595
	0.4	0	-1.082	-2.393	-4.379	-7.569	-10.730	-12.564	-13.463	-13.756	-13.729	-13.595
350	0.5	0	-0.299	-0.766	-1.730	-3.600	-6.797	-9.994	-11.864	-12.829	-13.296	-13.595
	0.6	0	0.134	0.162	-0.132	-1.030	-2.865	-6.025	-9.216	-11.202	-12.513	-13.595
	0.7	0	0.471	0.883	1.111	0.068	0.194	-1.577	-4.823	-8.881	-11.126	-13.595
	0.8	0	0.712	1.398	1..999	2.396	2.378	1.601	-0.435	-4.365	-9.135	-13.595
	0.9	0	0.856	1.707	2.531	3.252	3.689	3.507	2.197	-0.906	-6.541	-13.595
	1.0	0	0.904	1.810	2.709	3.538	4.126	4.143	3.075	0.247	-5.093	-13.595

			0									
400	0	0	-9.276	-14.956	-17.814	-18.738	-18.513	-17.726	-16.737	-15.718	-14.717	-13.725
	0.1	0	-7.631	-13.657	-16.837	-18.043	-18.047	-17.431	-16.560	-15.619	-14.671	-13.725
	0.2	0	-4.605	-9.760	-13.908	-15.959	-16.649	-16.547	-16.031	-15.322	-14.535	-13.725
	0.3	0	-2.468	-5.265	-9.026	-12.485	-14.319	-15.073	-15.148	-14.827	-14.309	-13.725
	0.4	0	-0.950	-2.172	-4.190	-7.622	-11.057	-13.009	-13.912	-14.135	-13.991	-13.725
	0.5	0	-0.142	-0.481	-1.402	-3.369	-6.863	-10.357	-12.324	-13.244	-13.583	-13.725
	0.6	0	0.266	0.409	0.187	-0.716	-2.669	-6.104	-9.535	-11.553	-12.775	-13.725
	0.7	0	0.583	1.102	1.423	1.347	0.594	-1.240	-4.700	-8.461	-11.258	-13.725
	0.8	0	0.810	1.597	2.305	2.821	2.924	2.233	0.183	-3.966	-9.031	-13.725
	0.9	0	0.946	1.894	2.835	3.705	4.322	4.318	3.112	-0.069	-6.095	-13.725
	1.0	0	0.991	1.993	3.011	4.000	4.788	5.012	4.089	1.230	-4.449	-13.725
450	0	0	-10.019	-16.011	-18.870	-19.646	-19.212	-18.233	-17.100	-15.982	-14.907	-13.851
	0.1	0	-8.178	-14.568	-17.806	-18.895	-18.721	-17.933	-16.928	-15.889	-14.806	-13.851
	0.2	0	-4.903	-10.240	-14.588	-16.644	-17.246	-17.031	-16.410	-15.611	-14.741	-13.851
	0.3	0	-2.446	-5.275	-9.226	-12.891	-14.789	-15.528	-15.547	-15.147	-14.534	-13.851
	0.4	0	-0.805	-1.926	-3.968	-7.637	-11.340	-13.424	-14.339	-14.497	-14.243	-13.851
	0.5	0	0.018	-0.190	-1.065	-3.132	-6.926	-10.710	-12.786	-13.661	-13.870	-13.851
	0.6	0	0.392	0.646	0.488	-0.427	-2.502	-6.214	-9.886	-11.925	-13.046	-13.851
	0.7	0	0.683	1.295	1.696	1.677	0.938	-0.960	-4.625	-8.576	-11.405	-13.851
	0.8	0	0.890	1.760	2.559	3.180	3.395	2.793	0.737	-3.611	-8.948	-13.851
	0.9	0	1.015	2.038	3.077	4.032	4.870	5.044	3.955	0.717	-5.673	-13.851
	1.0	0	1.056	2.131	3.249	4.382	5.361	5.795	5.028	2.160	-3.832	-13.851

Table 5.22 Con't

Coefficients $\bar{y}$ for beams loaded with a concentrated moment M

α	β	ξ										
		0.0	0.1	0.2	0.3	0.4	0.5	0.6	0.7	0.8	0.9	1.0
	0	0	-10.737	-17.017	-19.876	-20.477	-19.834	-18.672	-17.407	-16.206	-15.075	-13.971
	0.1	0	-8.700	-15.432	-18.709	-19.674	-19.320	-18.368	-17.241	-16.121	-15.039	-13.971
	0.2	0	-5.090	-10.677	-15.211	-17.263	-17.778	-17.457	-16.743	-15.866	-14.928	-13.971
	0.3	0	-2.406	-5.252	-9.380	-13.245	-15.209	-15.940	-15.911	-15.441	-14.744	-13.971
	0.4	0	-0.648	-1.657	-3.716	-7.620	-11.611	-13.815	-14.748	-14.845	-14.486	-13.971
500	0.5	0	0.183	0.108	-0.720	-2.889	-6.986	-11.083	-13.252	-14.080	-14.155	-13.971
	0.6	0	0.515	0.874	0.776	-0.157	-2.361	-6.351	-10.256	-12.315	-13.323	-13.971
	0.7	0	0.772	1.469	1.940	1.968	1.237	-0.726	-4.592	-8.720	-11.566	-13.971
	0.8	0	0.956	1.894	2.771	3.486	3.806	3.291	1.239	-3.295	-8.882	-13.971
	0.9	0	1.067	2.149	3.269	4.396	5.348	5.702	4.738	1.460	-5.272	-13.971
	1.0	0	1.104	2.234	3.436	4.700	5.862	6.505	5.904	3.046	-3.234	-13.971

Table 5.23 Example 5.7

ζ	0	0.1	0.2	0.3	0.4	0.5	0.6	0.7	0.8	0.9	1.0
$\overline{y}$	0	0.517	1.032	1.540	2.028	2.467	2.807	2.962	2.863	2.617	2.330
y	0	0.549	1.096	1.635	2.154	2.620	2.981	3.146	3.041	2.779	2.474

In conclusion, it is important to note that the method of analysis of beams on elastic half-space described above is developed only for plane strain analysis. It cannot be used for free beams supported on elastic half-space.

References

1. Flamant. 1892. *Comptes Rendus* (Paris).
2. Gersevanov, N. M. 1933, 1934. Stepped functions and their application in structural mechanics. VIOS, nos. 1, 2, ONTI, Gosstroiizdat.
3. Simvulidi, I. A. 1949. General formula of the elastic line of a beam. Collection of Papers. Metallurgizdat.
4. Simvulidi, I. A. 1958. *Analysis of beams on elastic foundation*. Soviet Science.
5. Simvulidi, I. A. 1973. *Analysis of engineering structures on elastic foundation*. 3rd ed. Moscow.
6. Tsudik, E. A. 1975. *Analysis of 2D and 3D frames on elastic foundation*. Engineering School of Bridges and Underground Structures, pp 15–18.
7. Tsudik, E. A. 2006. *Analysis of beams and frames on elastic foundation*. Canada/UK: Trafford Publishing, pp. 198–248.

6

Numerical Analysis of Beams Supported on Elastic Half-Space and Elastic Layer

6.1 Introduction

Methods of analysis of beams supported on elastic half-space described in Chapters 4 and 5 are developed for hand calculations. Simple equations and tables allow analyzing beams working as 3D systems and also under plane strain conditions. As shown earlier, tables and equations for simple beam analysis are also used for analysis of some complex problems such as beams with various boundary conditions, continuous beams, and others. However, analysis of complex beams, in most cases, is very difficult. Even analysis of a simple beam with both fixed ends, shown in Figure 6.1, using the Gorbunov-Posadov technique, is very time consuming. It requires removal of all restraints and the replacement of the removed restraints with unknown reactions and analysis of the beam loaded with the given loads and one unit loads applied to both ends of the beam. It also requires solving a system of linear equations and final analyses of the beam loaded successively with all reactions and each of the given loads. The beam shown in Figure 6.2 has to be analyzed six times: four times under reactions X_1, X_2, X_3, X_4 and two times under the given loads P and M. In addition, it requires obtaining combinations of performed analyses. It is obvious that analysis of continuous beams is even more difficult because usually it contains more given loads and more restraints. It is also obvious that methods of hand analysis described in Chapters 4 and 5 can be applied only to the analysis of simple, mostly free supported beams.

Chapter 6 describes a numerical method that allows analyzing various types of simple and complex beams supported on elastic half-space. The method was developed by Zhemochkin (1937) and Zhemochkin and Sinitsin (1962). It is based on the idea that continuous contact of the beam and soil can be replaced with a series of individual contacts-supports. By equating the settlements of the beam to the settlements of the half-space, a system of linear equations is built and unknown reactions are found.

All tables can be found at the end of the chapter.

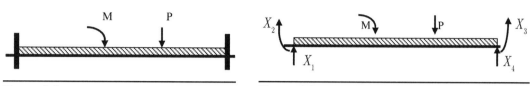

Figure 6.1　　　　　　　　　　　　　**Figure 6.2**

Zhemochkin's method is described below in detail. A similar method of analysis was developed by Ohde (1942). His method was used by Kany (1959), Sommer (1965), Barden (1962, 1963), Lousberg (1957), and Brown (1975). A good review of these publications was conducted by Selvadurai (1979). The author of this work developed Zhemochkin's method further by applying it to the analysis of complex beams such as pin-connected beams, interconnected beams, frames, and mat foundations (see Tsudik 2003, 2006).

This chapter includes a method of analysis of 2D and 3D frames with individual foundations supported on elastic half-space developed by the author of this book. It also includes the application of Zhemochkin's method to analysis of beams supported on an elastic layer.

6.2 Settlements of an Elastic Half-Plane, Plane Strain Analysis

A practicing engineer is usually dealing with two types of analyses: plane strain analysis and 3D analysis.

A beam works as a 2D system on two occasions:

1. When the beam is supported on the top of an elastic half-plane. In this case it is assumed that the half-plane is infinitely large in horizontal and vertical directions and its thickness is equal to one unit. In this case all stresses in the direction perpendicular to both sides of the beam and elastic half-space are equal to zero while deflections are not.
2. When the beam represents a band that is cut off from a long plate in the short direction along with an infinitely large half-space. In this case deflections in the direction perpendicular to the sides of the beam and half-space are equal to zero while stresses are not.

The actual settlements of the elastic half-plane cannot be found. In accordance to Flamant (1892), the actual settlement of the elastic half-plane is infinitely large because the half-plane itself is infinitely large. When performing plane strain analysis we obtain the differential settlement between the point the load P is applied and any point located far enough at distance d from that point. The settlement at point K located at distance r from the point the load is applied, as shown in Figure 6.3, is obtained as follows:

$$y = \frac{2P}{\pi E_0} \ln \frac{d}{r} \tag{6.1}$$

where E_0 is the modulus of elasticity of the elastic half-space when $r = 0$ $y = \infty$. Therefore, assuming $P = 1$, the given load is replaced with a uniformly distributed load

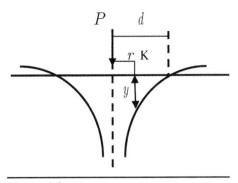

Figure 6.3

$p = \dfrac{1}{c}$, as shown in Figure 6.4, and the settlement at point K can be found from equation 6.2:

By integration of equation 6.1 from $(x - c/2)$ to $(x + c/2)$ we have:

$$y = \frac{1}{c} \int\limits_{x-\frac{c}{2}}^{x+\frac{c}{2}} \frac{2}{\pi E_0} \ln \frac{d}{\xi} d\xi \tag{6.2}$$

where d is the distance between point K and any other point located far enough from that point and not shown in Figure 6.4 After integration, equation 6.2 looks as follows:

$$y = \frac{1}{\pi E_0} (F + C) \tag{6.3}$$

where

$$F = -2\frac{x}{c} \ln \left[\frac{2\frac{x}{c} + 1}{2\frac{x}{c} - 1} \right] - \ln\left[\left(2\frac{x}{c} + 1\right)\left(2\frac{x}{c} - 1\right)\right] \tag{6.4}$$

$$C = 2 \ln \frac{d}{c} + 2 + 2 \ln 2 \tag{6.5}$$

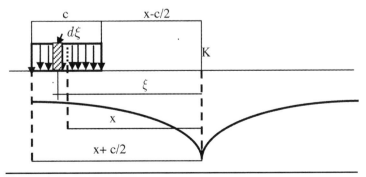

Figure 6.4

Equation 6.4 is obtained when point K is located outside of the loaded area. If point K is located inside the loaded area, the vertical deflection at that point is found as follows:

$$F = -2\frac{x}{c}\ln\left|\frac{1 + 2\frac{x}{c}}{1 - 2\frac{x}{c}}\right| - \ln\left[\left(1 + 2\frac{x}{c}\right)\left(1 - 2\frac{x}{c}\right)\right] \tag{6.6}$$

Coefficient F is obtained from equation 6.4 when point K is located outside of the loaded area and from equation 6.6 when point K is located inside of that area. Now, the settlement of the elastic half-plane at the center of the loaded length, when $x = 0$, is not equal to ∞. It is equal to:

$$y = \frac{1}{\pi E_0} \tag{6.7}$$

When the beam represents a band that is cut off from a plate, the settlement from equation 6.3 is multiplied by $\left(1 - \mu_0^2\right)$:

$$y = \frac{1}{\pi E_0}\left(F_{ki} + C\right)\left(1 - \mu_0^2\right) \tag{6.8}$$

Coefficient F can also be obtained from Table 6.1. In Table 6.1, x is the distance between the point the settlement is obtained and the center of the uniformly distributed load located along the length c. Table 6.1 is developed for a load $P = 1$ distributed over the length c.

6.3 Settlements of an Elastic Half-Space, 3D Analysis

Settlements of an elastic half-space with various types of vertical loads are found from equation 6.9 (see Chapter 1, Figures 1.4 and 1.5):

$$y = \frac{(1 - \mu_0^2)}{\pi E_0 c}F \tag{6.9}$$

where F is obtained from equation 6.10 or equation 6.11:

$$F = \frac{c}{b}\left[2\ln\frac{b}{c} - \ln\left(\frac{4x^2}{c^2} - 1\right) - 2\frac{x}{c}\ln\frac{2\frac{x}{c} + 1}{2\frac{x}{c} - 1} + \frac{b}{c} - \ln\frac{\left(2\frac{x}{b} + \frac{c}{b}\right) + \sqrt{\left(2\frac{x}{b} + \frac{c}{b}\right)^2 + 1}}{\left(2\frac{x}{b} - \frac{c}{b}\right) + \sqrt{\left(2\frac{x}{b} - \frac{c}{b}\right)^2 + 1}}\right.$$

$$+ \frac{c}{b}\left[2\frac{x}{c}\ln\frac{1 + \sqrt{\left(2\frac{x}{b} + \frac{c}{b}\right)^2 + 1}}{1 + \sqrt{\left(2\frac{x}{b} - \frac{c}{b}\right)^2 + 1}} + \ln\left\{1 + \sqrt{\left(2\frac{x}{b} + \frac{c}{b}\right)^2 + 1}\right\}\cdot\left\{1 + \sqrt{\left(2\frac{x}{b} - \frac{c}{b}\right)^2 + 1}\right\}\right.$$

$$+ \frac{c}{b}\left[2\frac{x}{c}\ln\frac{1 + \sqrt{\left(2\frac{x}{b} + \frac{c}{b}\right)^2 + 1}}{1 + \sqrt{\left(2\frac{x}{b} - \frac{c}{b}\right)^2 + 1}} + \ln\left\{1 + \sqrt{\left(2\frac{x}{b} + \frac{c}{b}\right)^2 + 1}\right\}\cdot\left\{1 + \sqrt{\left(2\frac{x}{b} - \frac{c}{b}\right)^2 + 1}\right\}\right]$$

$$\tag{6.10}$$

$$F = 2\frac{c}{b}\left\{\ln\left(\frac{b}{c}\right) + \frac{b}{c}\ln\left[\frac{c}{b} + \sqrt{\left(\frac{c}{b}\right)^2 + 1}\right] + \ln\left[1 + \sqrt{\left(\frac{c}{b}\right)^2 + 1}\right]\right\} \quad (6.11)$$

Equation 6.10 is used for obtaining the settlement at point k due to the load applied at point i, and equation 6.11 is used for obtaining the settlement at point k due to the load applied at the same point k, when $x = 0$. Coefficient F can also be obtained from Table 6.2. In Chapter 1, equations 1.26–1.30 are introduced for settlements of an elastic half-space loaded with a uniformly distributed load q of a circular shape with a radius a. If a load of a circular shape is applied to the half-space at any point 0, the settlement of the half-space at any point M located at distance r from point 0 is found from equation 6.12:

$$\Delta_{MO} = \frac{4(1 - \mu^2)}{\pi E}(E - K) \quad (6.12)$$

By assuming in equation 1.26 (from chapter 1) $r = 0$, the settlement at the center of the circular load can be found:

$$\Delta_{r=0} = \frac{2(1 - \mu^2)}{\pi E}qa \quad (6.13)$$

Assuming in equation 1.26 $r = a$, we obtain the settlement at the edge of the circular load:

$$\Delta_{r=a} = \frac{4(1 - \mu^2)}{\pi E}qa \quad (6.14)$$

Equation 6.13 is used for obtaining the settlement at the center of the load and equation 6.14 is used for obtaining the settlement at the edge of the load. Settlements of an elastic half-space can also be found the following two equations:

1. For a uniformly distributed load of a square shape, the average settlement can be found from equation 6.15:

$$y_{av} = \frac{1.9qa(1 - \mu^2)}{E} \quad (6.15)$$

2. For a uniformly distributed load of a rectangular shape, the average settlement can be obtained from equation 6.16 proposed by Schleicher (1926):

$$y_{av} = \frac{mP(1 - \mu^2)}{E\sqrt{A}} \quad (6.16)$$

where m is the coefficient taken from Table 6.3 and A is the area of the total load. Coefficient m depends on the shape of the load and the relationship between the lengths of the sides of the rectangular load.

6.4 Analysis of Beams Supported on Elastic Half-Plane

Using equations for obtaining settlements of an elastic half-space shown previously, Zhemochkin (1937, 1962) proposed a numerical method for analysis of beams on elastic half-space for plane strain analysis as well as for 3D analysis. The method was developed for analysis of beams free supported on elastic half-space.

Analysis of a beam is performed as follows: Continuous contact between the beam and soil, as shown in Figure 6.5, is replaced with a series of individual totally rigid supports, as shown in Figure 6.6. The first and last supports are usually located at distance 0.5 c from the ends of the beam, where c is the spacing of supports.

Two fictitious restraints are applied to one of the beam ends: one against vertical deflection and the second against rotation. Both restraints are usually applied to the left end, although they can also be applied to the right end as well. For analysis of symmetrical beams it is convenient to apply both restraints at the center of the beam. Analysis of symmetrical beams is discussed later in this chapter.

Now, the given beam supported on elastic half-space is replaced with a cantilevered beam loaded with the given loads P and soil reactions $X_1, X_2, \ldots, X_n$, as shown in Figure 6.7. The half-space is loaded with the same reactions acting in the opposite direction. Since concentrated loads applied to the elastic half-space produce infinitely large settlements, they are replaced with uniformly distributed loads q. The distributed load at any point i is equal to $q_i = \dfrac{X_i}{cb}$, where b is the width of the beam and X_i is the soil reaction at point i. The settlement of the half-space at any point i can be found using equations 6.9–6.11 as follows:

$$y_k = y_{k1}X_1 + y_{k2}X_2 + \ldots y_{kn}X_n \tag{6.17}$$

where y_{ki} is the settlement of the half-space at point k due to one unit load applied to the half-space at point i.

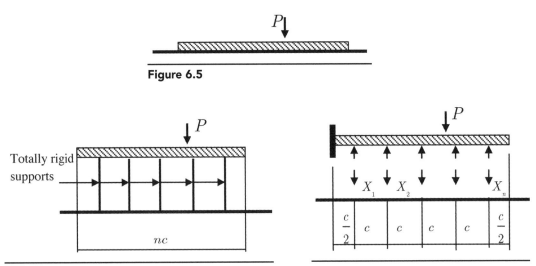

Figure 6.5

Figure 6.6

Figure 6.7

The vertical deflection of the beam due to the given loads and unknown reactions at any point k can be obtained as follows:

$$v_k = v_{k1}X_1 + v_{k2}X_2 + \ldots v_{kn}X_n + v_{kP} \tag{6.18}$$

where v_{ki} is deflection of the beam at point k due to one unit load applied to the beam at point i; v_{kP} is the deflection of the beam at point k due to all given loads applied to the beam.

However, equation 6.18 does not take into account that the beam, in reality, is not fixed at the left end. The beam will experience an additional vertical deflection due to vertical deflection and rotation of the beam at the left end. This deflection is equal to $(y_0 + a_k\varphi_0)$, as shown in Figure 6.8. Now, the total vertical deflection of the beam at any point k will look as follows:

$$v_k = v_{k1}X_1 + v_{k2}X_2 + \ldots v_{kn}X_n + v_{kP} + y_0 + a_k\varphi_0 \tag{6.19}$$

In this equation, y_0 and φ_0 are, respectively, the vertical deflection and rotation at the left end of the beam, a_k is the distance between point k and the left end of the beam.

Taking into account that vertical deflections of the half-space and the beam at points of contact are the same and assuming $\delta_{ki} = (v_{ki} + y_{ki})$, the following equation is written:

$$\delta_{k1}X_1 + \delta_{k2}X_2 + \ldots + \delta_{kn}X_n + y_0 + a_k\varphi_0 + v_{kP} = 0 \tag{6.20}$$

The same equation can be written for all points of contact between the beam shown in Figure 6.7 and the top of the elastic half-space:

$$\left.\begin{array}{l} \delta_{11}X_1 + \delta_{12}X_2 + \ldots + \delta_{1n}X_n + y_0 + a_1\varphi_0 + v_{1P} = 0 \\ \delta_{21}X_1 + \delta_{22}X_2 + \ldots + \delta_{2n}X_n + y_0 + a_2\varphi_0 + v_{2P} = 0 \\ \cdots\cdots\cdots\cdots\cdots\cdots\cdots\cdots\cdots\cdots\cdots\cdots\cdots \\ \delta_{n1}X_1 + \delta_{n2}X_2 + \ldots + \delta_{nn}X_n + y_0 + a_n\varphi_0 + v_{nP} = 0 \end{array}\right\} \tag{6.21}$$

This system of equations contains $(n + 2)$ unknowns: n reactions and two deflections, y_0 and φ_0 while the total number of equations in this system is equal only to n.

In order to solve this problem two additional equations are needed. Taking into account that two restraints applied to the left end of the beam are fictitious and, in reality, do not exist, the following two equations are written:

$$\left.\begin{array}{l} X_1 + X_2 + \ldots + X_n - \sum P = 0 \\ a_1X_1 + a_2X_2 + \ldots + a_nX_n - \sum Pa_P = 0 \end{array}\right\} \tag{6.22}$$

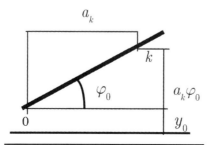

Figure 6.8

The first and the second equations reflect, respectively, the equilibrium of the shear forces and moments at the left end of the beam, a_i is the distance between the point i and the left end of the beam, and a_p is the distance between the point the given load is applied and the left end of the beam. The final system of equations contains $(n + 2)$ equations with $(n + 2)$ unknowns, and looks as shown below:

$$\left.\begin{array}{l} \delta_{11}X_1 + \delta_{12}X_2 + ... + \delta_{1n}X_n + y_0 + a_1\varphi_0 + v_{1P} = 0 \\ \delta_{21}X_1 + \delta_{22}X_2 + ... + \delta_{2n}X_n + y_0 + a_2\varphi_0 + v_{2P} = 0 \\ \cdots\cdots\cdots\cdots\cdots\cdots\cdots\cdots\cdots\cdots\cdots\cdots\cdots \\ \delta_{n1}X_1 + \delta_{n2}X_2 + ... + \delta_{nn}X_n + y_0 + a_n\varphi_0 + v_{nP} = 0 \\ X_1 + X_2 + ... + X_n - \sum P = 0 \\ a_1X_1 + a_2X_2 + ... + a_nX_n - \sum Pa_p = 0 \end{array}\right\} \qquad (6.23)$$

By solving this system of equations all soil reactions, and deflections at the left end of the beam, are found. Final analysis of the beam is performed by applying to the beam the given loads and soil reactions. The settlements of the half-space are obtained by applying the found reactions to the half-space. Reactions are applied to the half-space as uniformly distributed loads $q_i = \dfrac{X_i}{cb}$ or $q_i = \dfrac{X_i}{c}$, when $b = 1$.

It is important to mention that replacing a free supported beam with a cantilever, taking into account only the given vertical concentrated loads, simplifies calculations and allows obtaining deflections of the beam. Deflections of the beam are found from Table 6.4, which also simplifies the hand calculations.

In order to simplify the calculations, Zhemochkin (1962) developed Tables 6.1 and 6.2 for obtaining settlements of the elastic half-space F_{ik}, and Table 6.4 for obtaining vertical deflections of a cantilevered beam loaded only with a concentrated vertical load. Deflections of the beam are obtained from equation 6.24:

$$v_{ki} = \int \frac{M_k M_i}{EI}dx \qquad (6.24)$$

Deflection of the beam shown in Figure 6.9 is obtained using the moment diagrams shown in Figures 6.10 and 6.11.

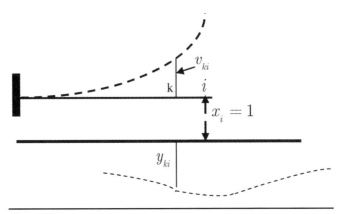

Figure 6.9

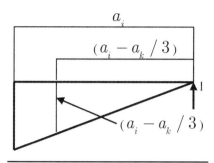

Figure 6.10 M_i Diagram

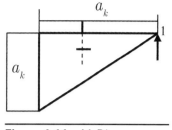

Figure 6.11 M_k Diagram

When $a_i > a_k$:

$$v_{ki} = \frac{a_k^2}{2}\left(a_i - \frac{a_k}{3}\right)\frac{1}{EI} = \frac{a_k^2(3a_i - a_k)}{6EI} \tag{6.25}$$

When $a_i < a_k$:

$$v_{ki} = \frac{a_i^2(3a_k - a_i)}{6EI} \tag{6.26}$$

Equation 6.26 can be rewritten as follows:

$$v_{ki} = \frac{c^3}{6EI}w_{ki} \tag{6.27}$$

where:

$$w_{ki} = \left(\frac{a_k}{c}\right)^3\cdot\left(3\frac{a_i}{c} - \frac{a_k}{c}\right) \tag{6.28}$$

Table 6.4 gives numerical values of w_{ki} depending on $\frac{a_i}{c}$ and $\frac{a_k}{c}$ that simplify hand calculations. Since $w_{ki} = w_{ik}$, all numerical values are given only for half of Table 6.4. The total vertical deflection at any point of the beam can be found from this equation:

$$\delta_{ki} = \frac{1}{\pi E_0}(F_{ki} + C) + \frac{c^3}{6EI}w_{ki} \tag{6.29}$$

In this equation the constant C depends only on the distance d and can be assumed the same for all points of the beam. Now we have:

$$\delta_{ki} = \frac{1}{\pi E_0}F_{ki} + \frac{c^3}{6EI}w_{ki} \tag{6.30}$$

In this case equation 6.20 can be rewritten as follows:

$$\delta_{k1}X_1 + \delta_{k2}X_2 + \delta_{k3}X_3 + \dots + \frac{1}{\pi E_0}(X_1 + X_2 + X_3 + \dots)C + y_0 + a_k\varphi_0 + \delta_{kP} = 0 \quad (6.31)$$

But since $\sum X = \sum P$, equation 6.31 can be rewritten as follows:

$$\delta_{k1}X_1 + \delta_{k2}X_2 + \delta_{k3}X_3 + ... + \left(y_0 + \frac{C\sum P}{\pi E_0}\right) + a_k\varphi_0 + \delta_{kP} = 0 \tag{6.32}$$

where $\left(y_0 + \dfrac{C\sum P}{\pi E_0}\right)$ is the same in all equations of the system 6.21. Now, by designating this term of the equation again as y_0, we obtain an equation that looks exactly like all equations of the system 6.21:

$$\delta_{k1}X_1 + \delta_{k2}X_2 + ... + \delta_{kn}X_n + y_0 + a_k\varphi_0 + v_{kP} = 0 \tag{6.33}$$

Now, coefficients of this equation can be found not from equation 6.29, but from 6.30. By multiplying equation 6.30 by πE_0 we have:

$$\delta_{ki} = F_{ki} + \alpha w_{ki} \tag{6.34}$$

where

$$\alpha = \frac{\pi E_0 c^3}{6EI} \tag{6.35}$$

Now, all equations will look exactly like the system 6.21, but all δ are not equal to the real deflections; they are πE_0 times larger. Plane strain analysis for beams that are cut off from plates is performed the same way, but in this case coefficient α is obtained from the following equation:

$$\alpha = \frac{\pi E_0 c^3}{6D(1 - \mu_0^2)} \tag{6.36}$$

where

$$D = \frac{1 \cdot h^3 E}{12(1 - \mu_0^2)}$$

6.5 3D Analysis of Beams Supported on Elastic Half-Space

The 3D analysis of beams supported on elastic half-space is performed exactly like plane strain analysis, but in this case the settlement of the elastic half-space F_{ki} is obtained from equations 6.10 and 6.11. w_{ki} is obtained from Table 6.4. The total vertical deflection of the beam at any point is obtained from the following equation:

$$\delta_{ki} = \frac{1 - \mu_0^2}{\pi E_0 c}F_{ki} + \frac{c^3}{6EI}w_{ki} \tag{6.37}$$

In this equation, $\dfrac{1 - \mu_0^2}{\pi E_0 c}F_{ki}$ is the settlement of the half-space at point k due to one unit load applied to the half-space at point i, and $\dfrac{c^3}{6EI}w_{ki}$ is the deflection of the beam at point k due to one unit load applied to the beam at point i. In order to simplify calculations, the first n equations in the system 6.23 are multiplied by $\dfrac{\pi E_0 c}{(1 - \mu_0^2)}$, and, by designating $y_0 \cdot \dfrac{\pi E_0 c^4}{6EI\left(1 - \mu_0^2\right)}$ again as y_0, equation 6.37 is changed and looks as shown below:

$$\hat{\delta}_{ki} = F_{ki} + \alpha w_{ki} \tag{6.38}$$

where:

$$\alpha = \frac{\pi E_0 c^4}{6EI\left(1 - \mu_0^2\right)} \tag{6.39}$$

Coefficient α is obtained only one time for all beams. In this case we obtain the actual settlements of the beam.

It is useful to note that analysis can be simplified when the beam is symmetrical. By taking into account the symmetry of the beam the number of unknowns can be significantly reduced. For example, let us say we have to analyze the beam shown in Figure 6.12. Analysis is performed as follows:

1. Continuous contact between the beam and the soil is replaced with a series of individual supports and these supports, in turn, are replaced with unknown soil reactions as shown in Figure 6.13.
2. Since the beam is symmetrical we can analyze only one-half of the beam, the left or the right half. The center of the beam, because of the symmetry, will experience only a settlement and will not experience any rotation. In other words, $\varphi_0 = 0$, $y_0 \neq 0$. Therefore, we apply at point 0 only one restraint against the vertical deflection y_0. The system of equations for the left half of the beam will look as follows:

$$\left.\begin{aligned}
\delta_{11}X_1 + \delta_{12}X_2 + \delta_{13}X_3 + y_0 + v_{1P} &= 0 \\
\delta_{21}X_1 + \delta_{22}X_2 + \delta_{23}X_3 + y_0 + v_{2P} &= 0 \\
\delta_{31}X_1 + \delta_{32}X_2 + \delta_{33}X_n + y_0 + v_{3P} &= 0 \\
X_1 + X_2 + X_3 - P &= 0
\end{aligned}\right\} \tag{6.40}$$

Figure 6.12

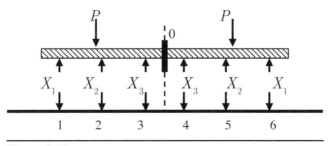

Figure 6.13

In this system all one unit deflections include not only deflections of the beam and the half-space at the left half of the beam, but also settlements of the half-space located under the right half of the beam. For example, $\delta_{11} = (y_{11} + y_{16} + v_{11})$. Analogously $\delta_{21} = (y_{21} + y_{26} + v_{21})$.

Now we can solve a numerical example.

Example 6.1

The beam shown in Figure 6.14 is cut off from a plate and works under plane strain conditions. The length of the beam is 2.8m, the width is equal to 1m, and the depth of the beam is 0.20m.

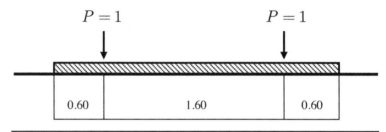

Figure 6.14

Let us divide the beam into seven elements, locating under the center of each element totally rigid columns-supports and replacing these supports with unknown reactions as shown in Figure 6.15.

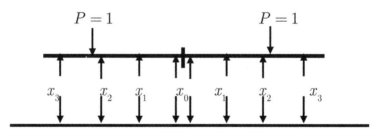

Figure 6.15

The spacing of supports is equal to $c = \frac{2.8}{7} = 0.4\text{m}$. The modulus of elasticity of the material of the half-space $E_0 = 1,000\text{t/m}^2$. The modulus of elasticity of the beam $E = 2,100,000\text{t/m}^2$. Poisson's ratio of the soil $\mu_0 = 0.35$. Poisson's ratio of the beam $\mu = 0.167$.

The moment of inertia of the beam $I = \frac{1 \cdot 0.2^3}{12} = 0.00067\text{m}^4$. The rigidity of the beam is equal to: $D = \frac{2,100,000 \cdot 0.00067}{1 - 0.167^2} = 1,428\text{tm}^2$.

1. Find coefficient α:

$$\alpha = \frac{\pi E_0 c^3}{6D(1 - \mu_0^2)} = \frac{3.14 \cdot 1000 \cdot 0.4^3}{6 \cdot 1,428 \cdot (1 - 0.35^2)} = 0.0267$$

2. From Tables 6.2 and 6.4 all deflections are found:

$\delta_{01} = -2 \cdot 3{,}296 = -6.592$, $\delta_{02} = -2 \cdot 4.751 = -9.502$, $\delta_{02} = -2 \cdot 4.751 = -9.502$, $\delta_{03} = -2 \cdot 5.574 = -11.148$, $\delta_{11} = -4{,}751 + 0.0267 \cdot 2 = -4.698$, $\delta_{12} = -3.296$ $- 5.574 + 0.0267 \cdot 5 = -8.736$, $\delta_{13} = -4.751 - 6.154 + 0.0267 \cdot 8 = -10.691$, $\delta_{22} = -6.154 + 0.0267 \cdot 16 = -5.727$, $\delta_{23} = -3.296 - 6.602 + 0.0267 \cdot 28 = -9.150$, $\delta_{33} = -6.967 + 0.0267 \cdot 54 = -5.525$, $\delta_{1P} = -0.0267 \cdot 5 = -0.134$, $\delta_{2P} = -0.0267 \cdot 16 = -0.427$, $\delta_{3P} = -0.0267 \cdot 28 = -0.748$

3. Now, the following system of equations is written:

$$\left. \begin{array}{l} -4.698X_1 - 8.736X_2 - 10.691X_3 - 6.592X_0 + 10y_0 - 0.134 = 0 \\ -8.736X_1 - 5.727X_2 - 9.150X_3 - 9.502X_0 + 10y_0 - 0.427 = 0 \\ -10.691X_1 - 9.150X_2 - 5.525X_3 - 11.148X_0 + 10y_0 - 0.748 = 0 \\ -6.592X_1 + -9.502X_2 - 11.148X_3 + 10y_0 = 0 \\ 10X_1 + 10X_2 + 10X_3 + 10X_0 - 10 = 0 \end{array} \right\} \quad (6.41)$$

From this system we find:

$$y_0 = +0.87305$$
$$X_0 = 0.09820$$
$$X_1 = 0.20647$$
$$X_2 = 0.23189$$
$$X_3 = 0.46335$$

Now the soil pressures, moments, and shear forces can be found. The soil pressures:

$$p_0 = \frac{2X_0}{0.4} = 0.0982/2 = 0.491 \text{t/m}^2$$

$$p_1 = \frac{X_1}{0.4} = 0.20647/0.4 = 0.516 \text{t/m}^2$$

$$p_2 = \frac{X_2}{0.4} = 0.23189/0.4 = 0.580 \text{t/m}^2$$

$$p_3 = \frac{X_3}{0.4} = 0.46335/0.4 = 1.158 \text{t/m}^2$$

Moments:

$$M_0 = X_3 \cdot 1.2 + X_2 \cdot 0.8 + X_1 \cdot 0.4 + X_0 \cdot 0.4/4 - 1 \cdot 0.80 = 0.033 \text{tm}$$

$$M_1 = X_3 \cdot 0.8 + X_2 \cdot 0.4 + \frac{X_1 \cdot 0.4}{2 \cdot 4} - 1 \cdot 0.80 = 0.074 \text{tm}$$

$$M_2 = X_3 \cdot 0.4 + \frac{X_2 \cdot 0.4}{2 \cdot 4} = 0.197 \text{tm}$$

$$M_3 = \frac{X_3 \cdot 0.4}{2 \cdot 4} = 0.0232 \text{tm}$$

$$M_0 = X_3 \cdot 1.2 + X_2 \cdot 0.8 + X_1 \cdot 0.4 + X_0 \cdot 0.4/4 - 1 \cdot 0.80 = 0.033\text{tm}$$

$$M_1 = X_3 \cdot 0.8 + X_2 \cdot 0.4 + \frac{X_1 \cdot 0.4}{2 \cdot 4} - 1 \cdot 0.80 = 0.074\text{tm}$$

$$M_2 = X_3 \cdot 0.4 + \frac{X_2 \cdot 0.4}{2 \cdot 4} = 0.197\text{tm}$$

$$M_3 = \frac{X_3 \cdot 0.4}{2 \cdot 4} = 0.0232\text{tm}$$

Shear forces:

$$Q_0 = 0$$
$$Q_1 = -X_3 - X_2 - 0.5X_1 + 1 = 0.201\text{t}$$
$$Q_2 = -X_3 - 0.5X_2 + 1 = +0.421\text{t}$$
$$Q_2' = -0.579\text{t}$$
$$Q_3 = -0.5X_3 = -0.232\text{t}$$

The soil pressures, moments, and shear forces diagrams are shown in Figures 6.16, 6.17, and 6.18, respectively.

Three-dimensional analysis of a beam supported on elastic half-space is performed analogously, but the total one unit deflection at each point of the beam is obtained from equation 6.42. In that equation the settlement of the half-space F_{ki} is obtained from equations 6.9 and 6.10 or Table 6.2. A numerical example shown on the next page illustrates the application of these formulae to 3D analysis of beams supported on elastic half-space.

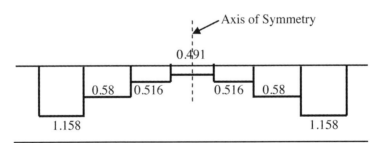

Figure 6.16   Soil pressure diagram

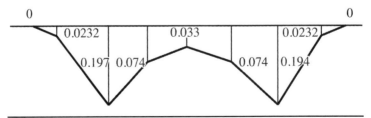

Figure 6.17   Moment diagram

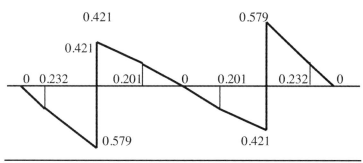

Figure 6.18   Shear forces diagram

Example 6.2

The given concrete beam is shown in Figure 6.19. The length of the beam is equal to L = 90m, the width of the beam b = 30m, the moment of inertia I = 3,200m⁴, the modulus of elasticity of the beam material is equal to E = 2,100,000t/m², the modulus of elasticity of the soil E_0 = 20,000t/m², and Poisson's ratio of the soil μ_0 = 0.3. The middle area has five concentrated loads P = 100t. Obtain the reactions of the half-space.

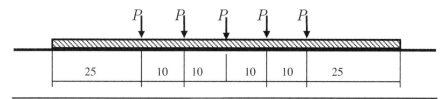

Figure 6.19

Solution:

The spacing of supports is 10m; the first and last supports are located at 5m from the beam ends. Since the beam is symmetrical the number of unknowns is significantly reduced. The soil reaction at the center of the beam is divided into two reactions X_0. Both reactions are acting at the same point, but shown at right and left from the center (see Figure 6.20). The center of the beam does not experience any rotation and, therefore, only one fictitious restraint against vertical deflection is applied at the center. Deflections are obtained from equation 6.38.

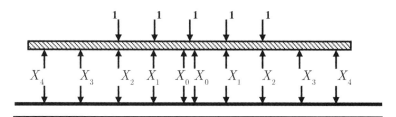

Figure 6.20

The numerical values of F_{ki} are taken from Table 6.2 and w_{ki} are taken from Table 6.4. Coefficient α is obtained from equation 6.39 and is equal to:

$$\alpha = \frac{\pi E_0 c^4}{6EI\left(1 - \mu_0^2\right)} = \frac{3.14 \cdot 20,000 \cdot 10^4}{6 \cdot 2,100,000 \cdot 3,200\left(1 - 0.3^2\right)} = 0.0171$$

Now, taking into account that $\dfrac{b}{c} = \dfrac{30}{10} = 3$ and using Table 6.4, all deflections are found:

$\delta_{00} = 2 \cdot 1.867 = 3.734$, $\delta_{01} = 2 \cdot 0.829 = 1.658$, $\delta_{02} = 2 \cdot 0.469 = 0.938$

$\delta_{03} = 2 \cdot 0.323 = 0.646$, $\delta_{04} = 2 \cdot 0.246 = 0.492$, $\delta_{11} = 1.867 + 0.469 + 0.0171 \cdot 2 = 2.370$

$\delta_{12} = 0.829 + 0.323 + 0.0171 \cdot 5 = 1.238$, $\delta_{13} = 0.469 + 0.246 + 0.0171 \cdot 8 = 0.852$

$\delta_{14} = 0.323 + 0.197 + 0.0171 \cdot 11 = 0.708$, $\delta_{22} = 1.867 + 0.246 + 0.0171 \cdot 16 = 2.387$

$\delta_{23} = 0.829 + 0.197 + 0.0171 \cdot 28 = 1.505$, $\delta_{24} = 0.469 + 0.165 + 0.0171 \cdot 40 = 1.318$

$\delta_{33} = 1.867 + 0.165 + 0.0171 \cdot 54 = 2.955$, $\delta_{34} = 0.829 + 0.142 + 0.0171 \cdot 81 = 2.356$

$\delta_{44} = 1.867 + 0.124 + 0.0171 \cdot 128 = 4.180$, $\delta_{0P} = 0$, $\delta_{1P} = -0.0171 \cdot (2 + 5) \cdot 100 = -12$, $\delta_{2P} = -0.0171 \cdot (5 + 16) \cdot 100 = -35.9$, $\delta_{3P} = -0.0171 \cdot (8 + 28) \cdot 100 = -61.6$, $\delta_{4P} = -0.0171 \cdot (11 + 40) \cdot 100 = -87.2$

Now, the following system of equations can be written:

$$\left.\begin{aligned}
3.734X_0 + 1.658X_1 + 0.938X_2 + 0.646X_3 + 0.492X_4 + 1.000y_0 &= 0\\
1.658X_0 + 2.370X_1 + 1.238X_2 + 0.852X_3 + 0.708X_4 + 1.000y_0 &= -0.120\\
0.938X_0 + 1.238X_1 + 2.387X_2 + 1.505X_3 + 1.318X_4 + 1.000y_0 &= -0.359\\
0.646X_0 + 0.852X_1 + 1.505X_2 + 2.955X_3 + 2.356X_4 + 1.000y_0 &= -0.616\\
0.492X_0 + 0.708X_1 + 1.318X_2 + 2.356X_3 + 4.180X_4 + 1.000y_0 &= -0.872\\
1.000X_0 + 1.000X_1 + 1.000X_2 + 1.000X_3 + 1.000X_4 + 0.000y_0 &= -2.500
\end{aligned}\right\} \quad (6.42)$$

From this system of equations we find:

$$y_0 = -346.0$$
$$2X_0 = 2 \cdot 34.6 = +69.2t$$
$$X_1 = 66.6t$$
$$X_2 = 58.4t$$
$$X_3 = 46.9t$$
$$X_4 = 43.5t$$

The settlement y_0 found above is not the real settlement of the beam at point 0. The actual settlement at point 0 is found as follows:

$$y_{0act} = \frac{1 - \mu_0^2}{\pi E_0 c} y_0 = \frac{1 - 0.3^2}{3.14 \cdot 20,000 \cdot 10} \cdot (-346) = -0.502\text{mm}$$

Settlements of the half-space at any point are found as the sum of settlements produced at that point by all loads applied to the half-space. For example, the total settlement at point 3 is equal to:

$y_3 = 2X_0 \cdot 0.646 + 0.715 \cdot X_1 + 1.026 \cdot X_2 + 2.032 \cdot X_3 + 0.972X_4 = 0.646 \cdot 69.2 + 0.715 \cdot 66.6 + 1.026 \cdot 58.4 + 2.032 \cdot 46.9 + 0.972 \cdot 43.5 = 289.823$

The actual settlement at point 3 is equal to:

$$y_{3,act} = \frac{(1 - \mu_0^2)}{\pi E_0 c} y_3 = \frac{1 - 0.3^2}{3.14 \cdot 20,000 \cdot 10} \cdot 289.823 = 0.42\text{mm}$$

Settlements of the beam at any other point are obtained analogously. Reactions of the soil found above allow building the soil pressure, shear, and moment diagrams. Equilibrium of all forces applied to the beam is checked below:

$$\sum X + \sum P = 69.2 + 66.6 + 58.4 + 43.5 - 250 = 0$$

It is important to mention that the Zhemochkin method was developed for hand analysis at a time when the practicing engineer did not have the modern tools to perform fast and accurate analysis and, therefore, the method has some serious limitations such as:

1. The method was developed only for free supported beams; analysis of beams with various boundary conditions was not discussed and was not developed.
2. The method was developed only for analysis of beams with concentrated vertical loads. It was not developed for other types of loads such as moments, distributed loads, and other complex loads.
3. Trying to simplify hand calculations, Zhemochkin developed some tables that are helpful although, in some cases, they require the use of interpolation that may reduce the accuracy of the final results.
4. Zhemochkin's method, like any numerical method, requires the building and solution of large systems of linear equations that make hand analysis, in most cases, practically impossible. Even analysis of simple beams, as demonstrated above, is time consuming.

However, Zhemochkin's method remains one of the best, if not the only method, for practical analysis of beams supported on elastic half-space, when used for computer analysis in which all limitations mentioned above do not exist. Moreover, the method, as it was shown by the author of this work (2003, 2006, 2008), can be applied to analysis of not only beams, but also to analysis of mat foundations as well as analysis of frames supported on elastic half-space. The application of the method to analysis of various complex beams on elastic half-space is presented below.

6.6 Analysis of Beams with Various Boundary Conditions

As mentioned above, Zhemochkin's method was developed only for free supported beams. By applying two fictitious restraints to one of the ends of the beam, the given beam is replaced with a cantilever that allows obtaining vertical deflections of the beam, building and solving a system of linear equations, and performing final analysis. However, boundary conditions are not limited to beams with free supported ends. Beams supported on the elastic foundation may have various boundary conditions, as shown in Table 6.5. Any of these beams can also be replaced with a cantilever beam by adding fictitious restraints and replacing some of the existing restraints with unknown reactions. Table 6.6 shows all beams taken from Table 6.5 and replaced with cantilever beams.

However, replacement of any given beam with a cantilever beam is not always necessary. Computer analysis, in many cases, does not require application of fictitious restraints at all. Analysis of some beams may require application of fictitious restraints, but without replacing the given beam with a cantilever. For example, analysis of a beam with both simple supported ends, as shown in Figure 6.21, can be performed without applying any fictitious additional restraints. Analysis, in principle, is not different from analysis described above for a cantilever beam and is performed as follows:

1. Continuous contact between the beam and the half-space is replaced with a series of totally rigid individual column supports, and all supports, in turn, are replaced with unknown soil reactions, as shown in Figure 6.22.
2. Settlements of the half-space are obtained using the same equations 6.9–6.11.
3. Vertical deflections of the beam are found using well-known solutions for a beam with two simple non-yielding supports.
4. The total vertical deflection at any point i of the beam is equal to $\delta_{ki} = (v_{ki} + y_{ki})$, where y_{ki} is the settlement of the half-space and v_{ki} is the vertical deflection of the beam. But in this case deflection of the beam v_{ki} is obtained for a beam with both non-yielding supports. The number of unknowns is equal only to the number of reactions n. The system of equations is shown below:

$$\left.\begin{aligned}
\delta_{11}X_1 + \delta_{12}X_2 + \ldots + \delta_{1n}X_n + v_{1P} = 0\\
\delta_{21}X_1 + \delta_{22}X_2 + \ldots + \delta_{2n}X_n + v_{2P} = 0\\
\cdots\cdots\cdots\cdots\cdots\cdots\cdots\cdots\cdots\cdots\cdots\cdots\cdots\cdots\\
\delta_{n1}X_1 + \delta_{n2}X_2 + \ldots + \delta_{nn}X_n + v_{nP} = 0
\end{aligned}\right\} \qquad (6.43)$$

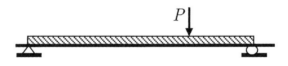

Figure 6.21

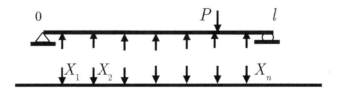

Figure 6.22

where v_{kP} is the settlement of the beam at point k due to the given loads applied to the beam.

However, in some cases, obtaining vertical deflections of the beam may still require application of fictitious restraints in addition to the existing ones. For example, a beam with one simple supported end and one free end, shown in Figure 6.23a, can be analyzed as a cantilever beam by adding an additional restraint against rotation at the left end of the beam using Tables 6.5 and 6.6. The same beam can also be analyzed as a beam with two non-yielding vertical supports by applying an additional fictitious restraint against vertical deflections at the right end of the beam. By replacing the soil pressure with unknown reactions, the given beam supported on elastic half-space will look as shown in Figure 6.23b. Taking into account that the support at the right end of the beam, in reality, does not exist, the beam will experience a vertical deflection v_l at the right end as shown in Figure 6.24. Vertical deflection of the beam at any point due to this deflection is equal to $\dfrac{v_l \cdot a_k}{l}$, where a_k is the distance between any point k of the beam and the left end of the beam, and l is the length of the beam. The system of equations looks as follows:

$$\left.\begin{array}{l} \delta_{11}X_1 + \delta_{12}X_2 + \ldots + \delta_{1n}X_n + v_l\dfrac{a_1}{l} + v_{1P} = 0 \\[2mm] \delta_{21}X_1 + \delta_{22}X_2 + \ldots + \delta_{2n}X_n + v_l\dfrac{a_2}{l} + v_{2P} = 0 \\[2mm] \cdots\cdots\cdots\cdots\cdots\cdots\cdots\cdots\cdots\cdots\cdots\cdots\cdots\cdots \\[2mm] \delta_{n1}X_1 + \delta_{n2}X_2 + \ldots + \delta_{nn}X_n + v_l\dfrac{a_n}{l} + v_{nP} = 0 \\[2mm] a_1X_1 + a_2X_2 + \ldots + a_nX_n - \sum Pa_p = 0 \end{array}\right\} \qquad (6.44)$$

a)

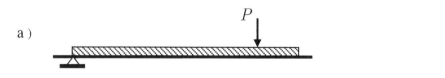

Figure 6.23a   The beam with a fictitious restraint at the left end

b)

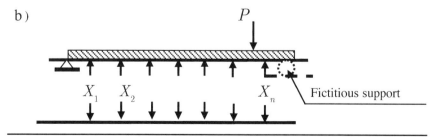

Figure 6.23b   The beam with a fictitious restraint at the right end

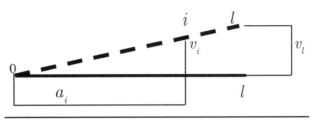

Figure 6.24

The last equation shows the equilibrium of the moments at the left end of the beam. The total number of equations is equal to $(n + 1)$, n soil reactions, and one vertical deflection v_l at the right end of the beam.

6.7 Analysis of Complex Continuous Beams

Zhemochkin's method can also be applied to analysis of complex continuous beams. Let us illustrate application of the method to analysis of the beam shown in Figure 6.25. The beam is supported on soil and two intermediate supports and loaded with various loads. Analysis is performed as follows:

1. Continuous contact between the beam and the half-space is replaced with a series of individual totally rigid supports, as shown in Figure 6.26, and all supports are replaced with unknown reactions $X_1, X_2, \ldots, X_n$.
2. Intermediate supports at points j and m are replaced with unknown reactions R_j and R_m, respectively.

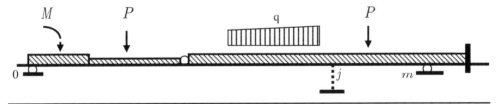

Figure 6.25

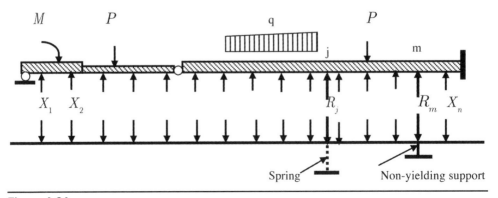

Figure 6.26

3. Now, the matrix of deflections of the beam can be built:

$$
\left.\begin{array}{l}
v_{11}, v_{12}, \ldots, v_{1n}, v_{1R_j}, v_{1R_m}, v_{1P} \\
v_{21}, v_{22}, \ldots, v_{2n}, v_{2R_j}, v_{2R_m}, v_{2P} \\
\cdots\cdots\cdots\cdots\cdots\cdots\cdots \\
v_{n1}, v_{n2}, \ldots, v_{nn}, v_{nR_j}, v_{nR_m}, v_{nP}
\end{array}\right\}
\tag{6.45}
$$

In this matrix of deflections 6.45, v_{ki} is the vertical deflection of the beam at point k due to one unit load applied at point i, v_{kR_j} and v_{kR_m} are deflections of the beam at point k due to support reactions R_j and R_m, respectively, and v_{kP} is the deflection of the beam at point k due to all given loads applied to the beam. The matrix of the settlements of the half-space can be obtained as follows:

$$
\left.\begin{array}{l}
y_{11}, y_{12}, \ldots, y_{1n}, y_{1R_j}, y_{1R_m} \\
y_{21}, y_{22}, \ldots, y_{2n}, y_{2R_j}, y_{2R_m} \\
\cdots\cdots\cdots\cdots\cdots\cdots\cdots \\
y_{n1}, y_{n2}, \ldots, y_{nn}, y_{nR_j}, y_{nR_m}
\end{array}\right\}
\tag{6.46}
$$

In this matrix, y_{ki} is the settlement of the half-space at point k due to one unit load applied at point i.

Using the matrices 6.45 and 6.46, the following system of equations is built:

$$
\left.\begin{array}{l}
\delta_{11}X_1 + \delta_{12}X_2 + \ldots + \delta_{1n}X_n + \delta_{1j}R_j + \delta_{1m}R_m + \delta_{1P} = 0 \\
\delta_{21}X_1 + \delta_{22}X_2 + \ldots + \delta_{2n}X_n + \delta_{2j}R_j + \delta_{2m}R_m + \delta_{2P} = 0 \\
\cdots\cdots\cdots\cdots\cdots\cdots\cdots\cdots\cdots\cdots\cdots \\
\delta_{n1}X_1 + \delta_{n2}X_2 + \ldots + \delta_{nn}X_n + \delta_{nj}R_j + \delta_{nm}R_m + \delta_{nP} = 0
\end{array}\right\}
\tag{6.47}
$$

The total number of unknowns in this system is equal to the number of unknown reactions including reactions at points j and m or $(n + 2)$ while the total number of equations is equal only to n. The additional two equations show that the actual settlement of the beam at point j is equal to the settlement of the spring at that point and the actual settlement at point m is equal to zero.

$$
\begin{array}{l}
v_{j1}X_1 + v_{j2}X_2 + \ldots + v_{jn}X_n + v_{jj}R_j + v_{jm}R_m + v_{nP} = R_j/C_j \\
v_{m1}X_1 + v_{m2}X_2 + \ldots + v_{mn}X_n + v_{mj}R_j + v_{mm}R_m + v_{mP} = 0
\end{array}
\tag{6.48}
$$

In this system, R_j/C_j is the settlement of the beam at point j, and C_j is the rigidity of spring.

The final system of equations looks as follows:

$$
\left.\begin{array}{l}
\delta_{11}X_1 + \delta_{12}X_2 + \ldots + \delta_{1n}X_n + \delta_{1j}R_j + \delta_{1m}R_m + \delta_{1P} = 0 \\
\delta_{21}X_1 + \delta_{22}X_2 + \ldots + \delta_{2n}X_n + \delta_{2j}R_j + \delta_{2m}R_m + \delta_{2P} = 0 \\
\cdots\cdots\cdots\cdots\cdots\cdots\cdots\cdots\cdots\cdots\cdots \\
\delta_{n1}X_1 + \delta_{n2}X_2 + \ldots + \delta_{nn}X_n + \delta_{nj}R_j + \delta_{nm}Rm + \delta_{nP} = 0 \\
v_{j1}X_1 + v_{j2}X_2 + \ldots + v_{jn}X_n + v_{jj}R_j + v_{jm}R_m + v_{nP} = R_j/C_j \\
v_{m1}X_1 + v_{m2}X_2 + \ldots + v_{mn}X_n + v_{mj}R_j + v_{mm}R_m + v_{mP} = 0
\end{array}\right\}
\tag{6.49}
$$

This system contains $(n + 2)$ equations with $(n + 2)$ unknowns. By solving this system of equations, all soil reactions $X_1, X_2, \ldots, X_n$ and support reactions R_j and R_m at points j and m are found. Final analysis of the beam is performed by applying to the beam the given loads and found reactions.

6.8 Analysis of 2D Frames with Continuous Foundations

Zhemochkin's method can also be applied to analysis of frames supported on elastic half-space. Figure 6.27 shows a frame with a continuous foundation supported on elastic half-space. Analysis of the system frame-foundation is performed in the following order:

1. Continuous contact of the foundation and the soil is replaced with a series of individual totally rigid supports and all supports, in turn, are replaced with unknown reactions $X_1, X_2, \ldots, X_n$, as shown in Figure 6.28.
2. Two fictitious restraints are applied to the left end of the foundation at point 0: one against vertical deflection y_0 and the second one against rotation φ_0, as shown in Figure 6.29.
3. Now, we have a statically indeterminate system frame-foundation loaded with the given loads, and soil reactions and a half-space, loaded with the same reactions acting in the opposite direction.
4. The vertical deflection of the foundation at any point i will look as follows:

$$v_k = v_{k1}X_1 + v_{k2}X_2 + \ldots v_{kn}X_n + v_{kP} + y_0 + a_k\varphi_0 \tag{6.50}$$

In this equation, v_{ki} is the vertical deflection of the system frame-foundation at point k due to one unit vertical load applied at point i, and v_{kP} is the vertical deflection at the same point due to all given loads applied to the system frame-foundation. Settlements v_{ki} and v_{kP} are found using any available computer software for 2D frame analysis.

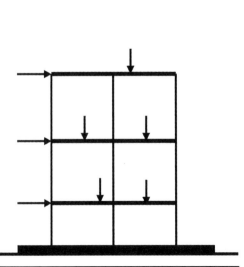

Figure 6.27   **Figure 6.28**

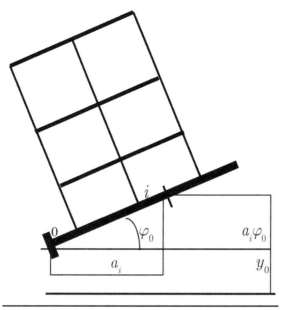

Figure 6.29

Since both restraints applied to the left end of foundation are fictitious, the foundation will experience a vertical deflection y_0 and rotation φ_0. In equation 6.50, the last two terms show these two settlements, y_0 and $a_i\varphi_0$, where a_k is the distance between point 0 and point k. Settlements of the half-space are obtained the same way they are found for a beam. The system of equations looks as shown below:

$$
\left.
\begin{aligned}
&\delta_{11}X_1 + \delta_{12}X_2 + \ldots + \delta_{1n}X_n + y_0 + a_1\varphi_0 + v_{1P} = 0 \\
&\delta_{21}X_1 + \delta_{22}X_2 + \ldots + \delta_{2n}X_n + y_0 + a_2\varphi_0 + v_{2P} = 0 \\
&\qquad\qquad\qquad\qquad\ldots\ldots\ldots\ldots\ldots\ldots\ldots \\
&\delta_{n1}X_1 + \delta_{n2}X_2 + \ldots + \delta_{nn}X_n + y_0 + a_n\varphi_0 + v_{nP} = 0 \\
&X_1 + X_2 + \ldots + X_n - \sum P = 0 \\
&a_1X_1 + a_2X_2 + \ldots + a_nX_n - \sum M_{0P}
\end{aligned}
\right\}
\qquad (6.51)
$$

The last two equations reflect the actual boundary conditions at the left end of the foundation. They show that the total shear force and the total moment at point 0 are equal to zero. In this system, $\sum P$ is the sum of all given vertical loads applied to the frame and foundation, and $\sum M_{0P}$ is the total moment at point 0 due to all given vertical and horizontal loads applied to the frame and foundation. As shown above, analysis, in principle, is not different from analysis of a beam. However, all frame deflections as well as final analysis of the frame are found using computer analysis.

6.9 Analysis of a Group of Foundations

When several continuous foundations supported on elastic half-space are located close to each other, they affect each other producing additional settlements of all foundations. Figure 6.30 shows a group of three foundations: two continuous foundations

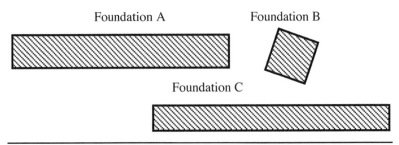

Figure 6.30   Plan of three foundations A, B, and C

(A and C) and one individual foundation (B). Additional settlements of foundations A and C produced by the individual foundation B can be easily obtained from equations 6.9–6.13 shown earlier, or by using Table 6.7, which is much easier. Let's say we have to perform analysis of three foundations: beam A, beam C, and B. Since all three foundations are located very close to each other, the soil pressure produced by one of the foundations will cause additional settlements of other foundations. We will analyze foundations A and B taking into account loads produced by all three foundations. Analysis is performed as follows:

1. Continuous contact of the beams A and C and elastic half-space are replaced with a series of individual totally rigid supports. Beams A and C are divided into a system of rectangular elements as shown in Figure 6.31. Supports are located at the centers of all these elements. As shown in Figure 6.31, the beam A is supported on supports and beam C is supported on $(n - m)$ supports.
2. All supports of both beams, A and C, are replaced with unknown reactions x_1, $x_2, \ldots, x_n$. The same reactions acting in the opposite direction are applied to the half-space as uniformly distributed loads.
3. Two fictitious restraints are applied to the left ends of both beams: one against vertical deflection and the second against rotation. The left end of beam A is restrained against vertical deflection y_{0A} and rotation φ_{0A}; the left end of beam C is restrained against vertical deflection y_{0C} and rotation φ_{0C}.
4. Now, both given beams supported on half-space are replaced with two cantilevered beams loaded with the given loads and soil reactions. Beam A is loaded with reactions $x_1, x_2, \ldots, x_m$; beam C is loaded with the given loads and reactions $x_{m+1}, x_{m+2}, \ldots, x_n$, as shown in Figure 6.32.

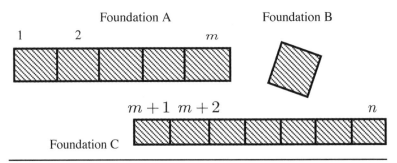

Figure 6.31

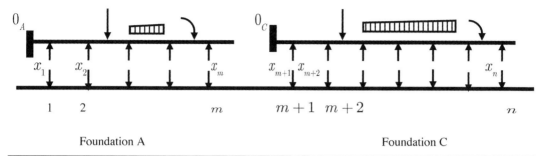

Figure 6.32

The vertical deflection of beam A at any point i due to the given loads and soil reactions is found as follows:

$$y_i = y_{i1}x_1 + y_{i2}x_2 + \dots + y_{im}x_m + y_{iP} \tag{6.52}$$

Taking into account that the left end of the beam, in reality, is not restrained against any deflections, beam A will experience a settlement y_{0A} and rotation φ_{0A} at that point 0_A, as shown in Figure 6.33.

The total vertical deflection of beam A at any point i is equal to:

$$y_i = y_{i1}x_1 + y_{i2}x_2 + \dots + y_{im}x_m + y_{iP} + y_{0A} + a_i\varphi_{0A} \tag{6.53}$$

where a_i is the distance between the left end of the beam and point i.

The total vertical deflection of beam C at point k can be obtained analogously:

$$y_k = y_{k(m+1)}x_{(m+1)} + y_{k(m+2)}x_{(m+2)} + \dots + y_{kn}x_n + y_{0C} + a_k\varphi_{0C} + y_{kP} \tag{6.54}$$

where a_k is the distance between the left end of beam C and point k.

In both equations, 6.53 and 6.54, y_{ij} and y_{kj} are vertical deflections of beams A and C, respectively, due to one unit loads applied at point j.

The total settlement of the elastic half-space at the same point i is obtained from the following equation:

$$w_i = w_{i1}x_1 + w_{i2}x_2 + \dots + w_{im}x_m + w_{i(m+1)}x_{m+1} + w_{i(m+2)}x_2 + \dots + w_{in}x_n + w_{iB} \tag{6.55}$$

As we can see from this equation, the total settlement of the half-space at any point i is affected by loads applied to the soil by both beams, A and C, and individual foundation B. In this equation, w_{ij} is the settlement of the half-space at point i due to one unit

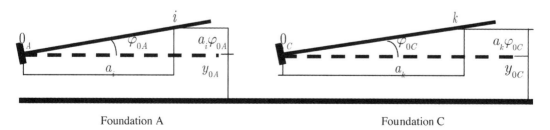

Figure 6.33

vertical load applied to the half-space at point j; w_{iB} is the settlement of the half-space at point i produced by the individual foundation B. Taking into account that the settlement of the half-space and beam A at the same point are the same, and by designating $\delta_{ik} = (y_{ik} + w_{ik})$ for any point of beam A, the following equation can be written:

$$\delta_{i1}x_1 + \delta_{i2}x_2 + \ldots + \delta_{im}x_m + w_{i(m+1)}x_{m+1} + w_{i(m+2)}x_{m+2}$$
$$+ \ldots + w_{in}x_n + y_{0A} + a_i\varphi_{0A} + y_{iP} + w_{iB} = 0 \qquad (6.56)$$

A similar equation can be written for any point of beam C. For beam A, the following group of equations is written below:

$$\delta_{11}x_1 + \delta_{12}x_2 + \ldots + \delta_{1m}x_m + w_{1(m+1)}x_{m+1} + w_{1(m+2)}x_{m+2} + \ldots + w_{1n}x_n$$
$$+ y_{0A} + a_1\varphi_{0A} + y_{1P} + w_{1B} = 0$$
$$\delta_{21}x_1 + \delta_{22}x_2 + \ldots + \delta_{2m}x_m + w_{2(m+1)}x_{m+1} + w_{2(m+2)}x_{m+2} + \ldots + w_{2n}x_n$$
$$+ y_{0A} + a_2\varphi_{0A} + y_{2P} + w_{2B} = 0 \qquad (6.57)$$
$$\ldots\ldots\ldots\ldots\ldots\ldots\ldots\ldots\ldots\ldots\ldots\ldots\ldots\ldots\ldots\ldots\ldots\ldots$$
$$\delta_{m1}x_1 + \delta_{m2}x_2 + \ldots + \delta_{mm}x_m + w_{m(m+1)}x_{m+1} + w_{m(m+2)}x_{m+2} + \ldots + w_{mn}x_n$$
$$+ y_{0A} + a_m\varphi_{0A} + y_{mP} + w_{mB} = 0$$

Analogously, for each point of beam C, from point $(m + 1)$ to point n a similar group of equations can also be written. This system of equations is written below (see equations 6.58). The total number of equations in both groups, 6.57 and 6.58, is equal to n or to the total number of unknown reactions, while the total number of unknowns is equal to $(n + 4)$. The four additional unknowns are deflections at the left ends of beams A and C: y_{0A}, φ_{0A}, y_{0C} and φ_{0C}.

The additional four equations 6.59 reflect the actual boundary conditions of beams A and C at their left ends: the shear forces and moments are equal to zero. In these equations, $\sum P_A$ and $\sum P_C$ are the shear forces at the left ends of the beams produced by all given loads applied to beams A and C, respectively. $\sum M_A$ and $\sum M_C$ are the moments at the left ends of the beams produced by all given loads applied to beams A and C, respectively; a_i is the distance between the left end of the beam and the point where the reaction x_i is applied:

$$w_{(m+1)1}x_1 + w_{(m+1)2}x_2 + \ldots + w_{(m+1)m}x_m + \delta_{(m+1)(m+1)}x_{m+1} + \delta_{(m+1)(m+2)}x_{m+2}$$
$$+ \ldots + \delta_{(m+1)n}x_n + y_{0C} + a_{(m+1)}\varphi_{0C} + y_{(m+1)P} + w_{(m+1)B} = 0$$
$$w_{(m+2)1}x_1 + w_{(m+2)2}x_2 + \ldots + w_{(m+2)m}x_m + \delta_{(m+2)(m+1)}x_{m+1} + \delta_{(m+2)(m+2)}x_{m+2}$$
$$+ \ldots + \delta_{(m+2)n}x_n + y_{0C} + a_{(m+2)}\varphi_{0C} + y_{(m+2)P} + w_{(m+2)B} = 0 \qquad (6.58)$$
$$\ldots\ldots\ldots\ldots\ldots\ldots\ldots\ldots\ldots\ldots\ldots\ldots\ldots\ldots\ldots\ldots\ldots\ldots$$
$$w_{n1}x_1 + w_{n2}x_2 + \ldots + w_{nm}x_m + \delta_{n(m+1)}x_{m+1} + \delta_{n(m+2)}x_{m+2} + \ldots + \delta_{nn}x_n$$
$$+ y_{0C} + a_n\varphi_{0C} + y_{nP} + w_{nB} = 0$$

$$\left.\begin{array}{l} x_1 + x_2 + \ldots + x_m - \sum P_A = 0 \\ a_1x_1 + a_2x_2 + \ldots + a_mx_m + \sum M_A = 0 \\ x_{m+1} + x_{m+2} + \ldots + x_n - \sum P_C = 0 \\ a_{m+1}x_{m+1} + a_{m+2}x_{m+2} + \ldots + a_nx_n + \sum M_C = 0 \end{array}\right\} \qquad (6.59)$$

Now, the total number of equations is equal to the total number of unknowns and equal to $(n + 4)$. All three groups of equations, 6.57, 6.58, and 6.59, are shown in Table 6.8. By solving this system of equations soil reactions and deflections of both beams are found. Settlement of the individual foundation B is obtained from equation 6.60 shown:

$$w_B = w_{B1}x_1 + w_{B2}x_2 + \ldots + w_{Bm}x_m + w_{B(m+1)}x_{m+1} + w_{B(m+2)}x_{m+2} + \ldots + w_{Bm}x_n + w_{BN} \quad (6.60)$$

In this equation, w_{Bi} is the settlement of the individual foundation B due to the settlement of half-space produced by one unit load applied to the half-space at any point i under foundations A and C, and w_{BN} is the settlement of the foundation due to the given load applied to the foundation B.

   It is important to mention that the proposed method described in this chapter allows performing analysis of a group of beams and individual totally rigid foundations supported on elastic half-space as one structural system. A group of foundations, working as one system, produces not only additional settlements of the neighboring foundations, but also affects the forces of interaction between all foundations and the elastic half-space. It is obvious that combined analysis also produces additional settlements of the individual totally rigid foundations, but cannot change the total soil reaction under the foundation. However, it changes the shape of the soil pressure under the foundation.

6.10 Analysis of Interconnected Beams

Foundations of commercial and industrial tall buildings, in some cases, are designed as a system of interconnected beams. This type of foundation is used not only for tall buildings, but also for lower story, heavy buildings when individual foundations cannot be used. It is also recommended for framed buildings designed in earthquake areas, since it increases the stability of the first floor columns as well as the stability of the total structure. It also reduces differential settlements of the columns and stresses in the superstructure compared to individual foundations. Analysis of such type of foundation can be performed easily using Winkler's soil model by replacing the soil with a series of independently working elastic supports. Analysis, in this case, is performed by using any well-known computer program for analysis of 3D statically indeterminate systems. More accurate analytical methods of analysis of interconnected beams on the Winkler foundation were developed by M. I. Gorbunov-Posadov (1953) and later by the author of this book (Tsudik 1975). Gorbunov-Posadov proposed to divide the given system of interconnected beams into a series of simple beams loaded with the given loads and unknown forces of interaction between the beams. Using the method of forces and solutions for simple free supported beams, they write a system of linear equations and obtain all forces and moments of interaction between the beams. Once these forces and moments are found, each beam can be analyzed using any available method for analysis of simple beams on Winkler foundation. Taking into account that beams running in one direction resist the bending of the beams running in a perpendicular direction, all areas of beam intersection are assumed totally rigid. The number of unknowns at each node is usually equal to 6 and the total number of unknowns is equal to $6n$, where n is the total number of points of intersection. The author of this book solved the same problem using the stiffness method taking into account the totally rigid elements. The number of unknowns, in this case, is always equal to $3n$. Final analysis of each beam is performed using the method of initial parameters or method of superposition. In our method the number of unknowns is two times less than in the method proposed by Gorbunov-Posadov. The stiffness method is described in Chapter 2 of this book and earlier in our works (Tsudik 1970, 1975, 1983). Both methods use Winkler's soil model. Behavior of a system of interconnected beams on elastic foundation was also investigated by Glassman (1972). Here, a method of analysis of a system of interconnected beams supported on elastic half-space is introduced. Figure 6.34 shows a simple system of interconnected beams. Areas of intersection of the beams are assumed to be totally rigid. Analysis of the foundation is performed as follows:

1. All beams of the given system are divided into a series of elements as shown in Figure 6.34.
2. Continuous contact of the foundation and the half-space is replaced with a series of individual totally rigid column supports located at the centers of all elements as shown in Figures 6.35 and 6.36.
3. The origin 0 is connected with the beams A and 1 with two fictitious totally rigid beams 0-A and 0-1.
4. Three fictitious restraints are applied to the node 0: one restraint against vertical deflection y_0, a second restraint against rotation φ_{0X} along axis X, and a third restraint against rotation φ_{0Y} along axis Y.
5. All introduced totally rigid supports are removed and replaced with unknown vertical reactions x_i applied to the system of interconnected beams. The same reactions acting in the opposite direction are applied to the elastic half-space. They are applied as uniformly distributed loads that at any point i are equal to $q_i = x_i/cb$ where c is the spacing of supports and b is the width of the beam.

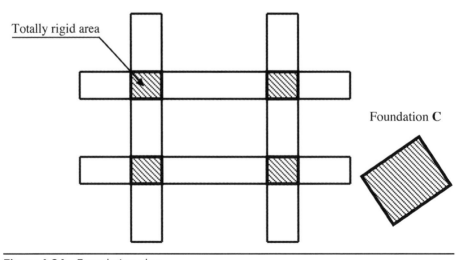

Figure 6.34 Foundation plan

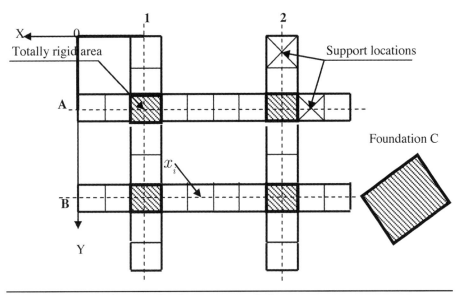

Figure 6.35   Foundation with tributary areas of all supports

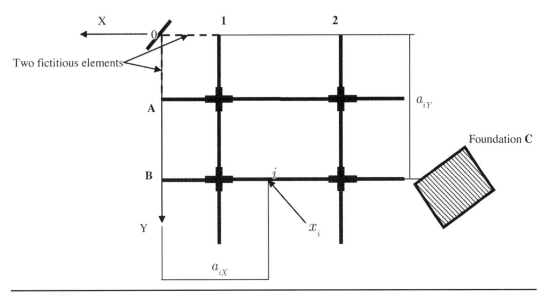

Figure 6.36

6. Now, the given system of interconnected beams supported on elastic half-space is divided into two systems: a system of interconnected beams restrained against vertical deflection y_0, rotations φ_{0X} and φ_{0Y} at point 0 with the given loads and unknown soil reactions; and elastic half-space loaded with the same reactions acting in the opposite direction and given loads applied to the half-space and located outside of the foundation.

By applying to the system of interconnected beams successively the given loads acting downward and one unit vertical reactions acting upward, the following matrix of vertical deflections is obtained:

$$
\begin{bmatrix}
y_{11} y_{12}, \ldots, y_{1n} y_{1P} \\
y_{21} y_{22}, \ldots, y_{2n} y_{2P} \\
\cdots\cdots\cdots\cdots \\
y_{n1} y_{n2}, \ldots, y_{nn} y_{nP}
\end{bmatrix}
\tag{6.61}
$$

In this matrix, y_{ik} is the vertical deflection of the foundation at point k due to one unit vertical load applied to the foundation at point k; y_{iP} is the vertical deflection at the same point i due to all given loads applied to the foundation. All deflections of this matrix can be found using any available software for analysis of 3D statically indeterminate systems. The matrix of settlements of the elastic half-space is obtained by applying to the soil one unit uniformly distributed loads acting downward.

The matrix of settlements shown below also includes settlements w_{iC} at any point i produced by the totally rigid individual foundation C located outside of the system of interconnected beams.

$$
\begin{bmatrix}
w_{11} w_{12} \ldots w_{1n} w_{1C} \\
w_{21} w_{22} \ldots w_{2n} w_{1C} \\
\cdots\cdots\cdots\cdots \\
w_{n1} w_{n2} \ldots w_{nn} w_{nC}
\end{bmatrix}
\tag{6.62}
$$

In this matrix, w_{ik} is the settlement of the elastic half-space at point i due to one unit vertical load applied to the half-space at point k; w_{iC} is the settlement of the half-space at the same point i due to the soil pressure produced by foundation C located outside of the system of the interconnected beams. The vertical deflection of the foundation at any point will look as follows:

$$
y_i = y_{i1} x_1 + y_{i2} x_2 + \ldots + y_{in} x_n + y_{iP}
\tag{6.63}
$$

where y_{ik} is the vertical deflection of the foundation at point i due to one unit load applied to the foundation at point k, and y_{iP} is the vertical deflection of the foundation at point i due to the given loads applied to the foundation. Equation 6.63 shows the vertical deflection of the foundation restrained against the three deflections: y_0, φ_{0X}, and φ_{0Y}. Since all these restraints are fictitious deflections, y_0, φ_{0X}, and φ_{0Y} have to be taken into account. Assuming that both rotations are acting in clockwise direction when looking in directions X and Y, they will produce at point i two additional vertical deflections, $a_{iX}\varphi_{0X}$ and $a_{iY}\varphi_{0Y}$, where a_{iX} and a_{iY} are coordinates of point i. Now the total vertical deflection of the foundation at point i will look as follows:

$$
y_i = y_{i1} x_1 + y_{i2} x_2 + \ldots + y_{in} x_n + y_0 + a_{iX} \varphi_X + a_{iY} \varphi_Y + y_{iP}
\tag{6.64}
$$

Vertical deflections of the foundation at all points, from 1 to n, are shown on the next page:

$$\left.\begin{array}{l} y_1 = y_{11}x_1 + y_{12}x_2 + \ldots + y_{1n}x_n + y_0 + a_{1X}\varphi_{0X} + a_{1Y}\varphi_{0Y} + y_{1P} \\ y_2 = y_{21}x_1 + y_{22}x_2 + \ldots + y_{2n}x_n + y_0 + a_{2X}\varphi_{0X} + a_{2Y}\varphi_{0Y} + y_{2P} \\ \ldots\ldots\ldots\ldots\ldots\ldots\ldots\ldots\ldots\ldots\ldots\ldots\ldots\ldots\ldots\ldots\ldots\ldots\ldots\ldots \\ y_n = y_{n1}x_1 + y_{n2}x_2 + \ldots + y_{nn}x_n + y_0 + a_{nX}\varphi_{0X} + a_{nY}\varphi_{0Y} + y_{nP} \end{array}\right\} \quad (6.65)$$

Settlements of the elastic half-space at the same points are written below:

$$\left.\begin{array}{l} w_1 = w_{11}x_1 + w_{12}x_2 + \ldots + w_{1n}x_n + w_{1C} \\ w_2 = w_{21}x_1 + w_{22}x_2 + \ldots + w_{2n}x_n + w_{2C} \\ \ldots\ldots\ldots\ldots\ldots\ldots\ldots\ldots\ldots\ldots\ldots\ldots\ldots\ldots\ldots\ldots\ldots \\ w_n = w_{n1}x_1 + w_{n2}x_2 + \ldots + w_{nn}x_n + w_{nC} \end{array}\right\} \quad (6.66)$$

Since the settlements of the half-space and the settlements of the foundation are the same, assuming $\delta_{ik} = (y_{ik} + w_{ik})$ the following system of equations is written:

$$\left.\begin{array}{l} \delta_1 = \delta_{11}x_1 + \delta_{12}x_2 + \ldots + \delta_{1n}x_n + y_0 + a_{1X}\varphi_{0X} + a_{1Y}\varphi_{0Y} + y_{1P} + w_{1P} \\ \delta_2 = \delta_{21}x_1 + \delta_{22}x_2 + \ldots + \delta_{2n}x_n + y_0 + a_{2X}\varphi_{0X} + a_{2Y}\varphi_{0Y} + y_{2P} + w_{2P} \\ \ldots\ldots\ldots\ldots\ldots\ldots\ldots\ldots\ldots\ldots\ldots\ldots\ldots\ldots\ldots\ldots\ldots\ldots\ldots\ldots \\ \delta_n = \delta_{n1}x_1 + \delta_{n2}x_2 + \ldots + \delta_{nn}x_n + y_0 + a_{nX}\varphi_{0X} + a_{nY}\varphi_{0Y} + y_{nP} + w_{nP} \end{array}\right\} \quad (6.67)$$

The total number of equations in this system is equal to the number of reactions n, but the number of unknowns is equal to $(n + 3)$. The system of equations 6.67 contains three additional unknowns: y_0, φ_{0X}, and φ_{0Y}. So, three additional equations are needed. These equations can be written by taking into account the actual boundary conditions at point 0:

$$\left.\begin{array}{l} a_{1X}x_1 + a_{2X}x_2 + \ldots + a_{nX}x_n + \sum M_X = 0 \\ a_{1Y}x_1 + a_{2Y}x_2 + \ldots + a_{nY}x_n + \sum M_Y = 0 \\ x_1 + x_1 + \ldots + x_n - \sum P = 0 \end{array}\right\} \quad (6.68)$$

$\sum M_X$ and $\sum M_Y$ are the total moments at point 0 in directions X and Y, respectively, due to the given loads applied to the foundation; $\sum P$ is the total given vertical load applied to the foundation. The first two equations reflect the equilibrium of all moments at point 0 in X and Y directions respectively. The third equation shows the equilibrium of all shear forces at the same point. By solving two systems of equations, 6.67 and 6.68, as one system, all soil reactions $x_1, x_2, \ldots, x_n$ and deflections y_0, φ_{0X}, and φ_{0Y}, are found.

Applying to the elastic half-space all reactions acting in the opposite direction, all settlements are obtained. Final analysis of the foundation is performed by applying to the foundation the given loads and obtained settlements. Analysis of the foundation is performed using any available computer software for analysis of 3D statically indeterminate systems.

6.11 Other Methods of Analysis

As shown above, analysis of various beams, including interconnected beams, supported on elastic half-space, is performed in several steps: applying to the given beam fictitious restraints, writing two groups of linear equations, solving a system of linear equations, finding all soil reactions, obtaining the soil settlements, and final analysis of the foundation. However, the method described above is not the only method that is recommended for analysis of beams, frames, and interconnected beams supported on elastic half-space. There are some other methods that can be successfully used in practical applications. Two methods developed by the author of this book are presented below. We shall explain the method using a simple beam supported on elastic half-space shown in Figure 6.37. Analysis is performed in the following order:

1. The given beam is divided into a group of equally spaced small line elements and all centers of these elements are supported on non-yielding supports, as shown in Figure 6.38. The spacing of supports is equal to c. The first and last supports are located at distance $0.5\ c$ from the ends of the beam.
2. By applying successively the given loads and one unit settlements to all points of support, from 1 to n the following matrix of reactions is obtained:

$$\left. \begin{array}{c} r_{11}r_{12}\cdots r_{1n}r_{1P} \\ r_{11}r_{12}\cdots r_{1n}r_{1P} \\ \cdots\cdots\cdots\cdots \\ r_{n1}r_{n2}\cdots r_{nn}r_{nP} \end{array} \right\} \tag{6.69}$$

where r_{ik} is the reaction at point i due to a vertical one unit settlement applied to the beam at point k and r_{iP} is the reaction of the beam at point i due to the given loads applied to the beam. Reactions at all points of support can be expressed as follows:

$$\left. \begin{array}{l} r_1 = r_{11}y_1 + r_{12}y_2 + \ldots + r_{1n}y_n + r_{1P} \\ r_2 = r_{21}y_1 + r_{22}y_2 + \ldots + r_{2n}y_n + r_{2P} \\ \cdots\cdots\cdots\cdots\cdots\cdots\cdots\cdots\cdots \\ r_n = r_{n1}y_1 + r_{n2}y_2 + \ldots + r_{nn}y_n + r_{nP} \end{array} \right\} \tag{6.70}$$

The same reactions acting in the opposite direction and applied to the elastic half-space as uniformly distributed loads (Figure 6.39) will produce settlements of the half-space that are equal to the settlements of the beam and equal to:

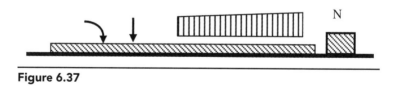

Figure 6.37

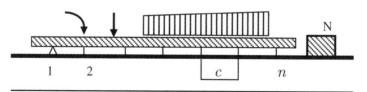

Figure 6.38

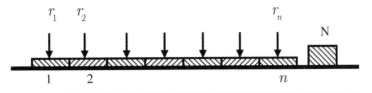

Figure 6.39

$$
\left.\begin{aligned}
y_1 &= w_{11}r_1 + w_{12}r_2 + \ldots + w_{1n}r_n + w_{1N} \\
y_2 &= w_{21}r_1 + w_{22}r_2 + \ldots + w_{2n}r_n + w_{2N} \\
&\cdots\cdots\cdots\cdots\cdots\cdots\cdots\cdots\cdots\cdots\cdots \\
y_n &= w_{n1}r_1 + w_{n2}r_2 + \ldots + w_{nn}r_n + w_{nN}
\end{aligned}\right\} \tag{6.71}
$$

In this system w_{ik} is the settlement of the elastic half-space at point i due to one unit vertical load applied at point k to the half-space; w_{iN} is the settlement of the half-space at point i due to the foundation N located outside of the beam. Both systems, 6.70 and 6.71, contain $2n$ equations and $2n$ unknowns (n settlements and n reactions). By solving both systems as one system of equations, all settlements and soil reactions are found. Final analysis of the beam is performed by applying to the beam the given loads and obtained settlements.

The method described above can be simplified by reducing the number of unknowns by half. By introducing reactions r_i from system 6.70 into system 6.71, we obtain a system of equations with only n unknown settlements. By applying the obtained settlements along with the given loads to the foundation, final analysis is performed. The same goal can be achieved by introducing settlements y_i from the system of equations 6.71 into system 6.70. By doing that, we again obtain a system of equations with only n unknown reactions. By applying found reactions to the half-space, settlements are obtained. Final analysis of the foundation is performed by applying to the foundation found settlements and given loads. Both methods can also be applied to the analysis of any complex beams or group of beams, including interconnected beams supported on elastic half-space. The same analysis can be performed by using the iterative method described next.

First iteration

Analysis of the foundation is performed assuming that all supports do not experience any settlements. From this analysis all reactions are found. Obtained reactions acting in the opposite direction are applied to the half-space and from this analysis settlements of the half-space are found.

Second iteration

The foundation loaded with the given loads and obtained settlements is analyzed and new soil reactions are obtained. Found reactions acting in the opposite direction are applied to the elastic half-space and new settlements are found.

Third iteration

The third iteration repeats the same procedures mentioned in the second iteration and so on. Results of analysis are considered satisfactory when settlements and reactions obtained in iteration n are close enough to the results obtained in iteration $(n - 1)$. It takes several iterations for obtaining good practical results.

6.12 Analysis of 2D and 3D Frames with Individual Foundations Supported on Elastic Half-Space

Analysis of 2D and 3D frames with individual foundations supported on the Winkler foundation was discussed in Chapter 3. It was shown that one of the simplest methods of analysis is modeling the system soil-foundation with equivalent line elements that behave similarly to the behavior of the given soil and foundation. This way of analysis replaces the given frame with individual foundations with a 2D or 3D frame that is analyzed using well-known software. In Chapter 3, in order to replace the given foundation with line elements, we used equations for obtaining deflections of footings supported on the Winkler foundation. When the foundations are supported on an elastic half-space the same method is used, but deflections of the foundations are obtained for an elastic half-space. All figures illustrating the method of obtaining equations are the same as those given in Chapter 3 and, therefore, they are not shown here. The settlement of an individual foundation supported on elastic half-space is found from the following equation:

$$y_z = \frac{\left(1 - \mu_0^2\right)}{\pi E_0 c} FN_z \tag{6.72}$$

Rotations of a rectangular foundation are obtained from equations:

$$\varphi_X = \frac{\left(1 - \mu_0^2\right)}{E_0} \cdot \frac{K_1 M_X}{a^3}, \quad \varphi_Y = \frac{\left(1 - \mu_0^2\right)}{E_0} \cdot \frac{K_2 M_Y}{b^3} \tag{6.73}$$

In equations 6.72 and 6.73, $c = 2a$ is the length of the foundation, $2b$ is the width of the foundation, M_X is the moment applied to the foundation in direction X, and M_Y is the moment applied to the foundation in direction Y. E_0 is the modulus of elasticity of the soil. Coefficients K_1 and K_2 are found from Table 1.5 from Chapter 1.

Now, by equating deflections of the foundation to the deflections of the beam we obtain the following three equations:

$$\left.\begin{array}{c} \dfrac{\left(1 - \mu_0^2\right)}{\pi E_0 c} FN_z = \dfrac{N_z l^3}{192\,(EI)} \\[3mm] \dfrac{N_X}{A_F K_X} = \dfrac{N_X l}{4EA} \\[3mm] \dfrac{\left(1 - \mu_0^2\right)}{E_0} \cdot \dfrac{K_1 M_X l}{16\,(EI)} = \dfrac{M_X l}{16\,(EI)} \end{array}\right\} \tag{6.74}$$

From these equations the length and rigidities of the equivalent beam are found:

$$\left. \begin{array}{l} l = a\sqrt{\dfrac{6F}{\pi K_1}} \\[4mm] (EI) = \dfrac{lE_0 a^3}{16\left(1 - \mu_0^2\right)K_1} \\[4mm] (EA) = \dfrac{lA_F K_X}{4} \end{array} \right\} \qquad (6.75)$$

In these equations, A_F is the area of the foundation, and EA is the rigidity of the equivalent beam under tension compression. Obtained equations allow analyzing 2D frames supported on elastic half-space by replacing the given system soil-foundation with a simple beam fixed at both ends. When the given foundation and soil are replaced with an equivalent column, the following three equations will look as follows:

$$\left. \begin{array}{l} \dfrac{(1 - \mu_0^2)}{\pi E_0 c} FN_z = \dfrac{N_z l}{EA} \\[4mm] \dfrac{N_X}{A_F K_X} = \dfrac{N_X l^3}{3EI} \\[4mm] \dfrac{(1 - \mu_0^2)}{E_0} \cdot \dfrac{K_1 M_X}{a^3} = \dfrac{M_X l}{EI} \end{array} \right\} \qquad (6.76)$$

In these equations, EI is the flexural rigidity of the equivalent column, EA is the rigidity of the column under tension compression, and l is the height of the column.

From these equations we find the length and rigidities of the equivalent column:

$$l = \sqrt{\dfrac{3E_0 a^3}{A_F K_X (1 - \mu_0^2)K_1}}, EI = \dfrac{lE_0 a^3}{(1 - \mu_0^2)K_1}, EA = \dfrac{2al\pi E_0}{(1 - \mu_0^2)F} \qquad (6.77)$$

Obtained equations 6.77 allow analyzing 2D frames supported on elastic half-space by replacing the given system soil-foundation with a simple column.

Equations for 3D frames are the same as equations written in Chapter 3 for frames supported on Winkler foundation. In Chapter 3, we used equations for foundations supported on Winkler foundation. Equations written below are equations for foundations supported on elastic half-space. They are written for 3D frames supported on elastic half-space by replacing the system soil-foundation with a beam fixed at both ends.

$$\left. \begin{array}{l} \dfrac{(1 - \mu_0^2)N_z}{\pi E_0 c} F = \dfrac{N_z l^3}{192 (EI)_X} \\[4mm] \dfrac{N_X}{A_F K_X} = \dfrac{N_X l}{4EA} \\[4mm] \dfrac{N_Y}{A_F K_Y} = \dfrac{N_z l^3}{192 (EI)_Y} \\[4mm] \dfrac{(1 - \mu_0^2)}{E_0} \cdot \dfrac{K_1 M_X}{a^3} = \dfrac{M_X l}{16 (EI)_X} \\[4mm] \dfrac{(1 - \mu_0^2)}{E_0} \cdot \dfrac{K_2 M_Y}{b^3} = \dfrac{M_Y l}{4GI_T} \end{array} \right\} \qquad (6.78)$$

From these equations the length and rigidities of the beam are obtained:

$$l = a\sqrt{\frac{6F}{\pi K_1}}, \quad EI_X = \frac{l^3\pi E_0 a}{96(1-\mu_0^2)}, \quad EA = \frac{lA_F K_X}{4}, \quad EI_Y = \frac{l^3 A_F K_Y}{192},$$

$$GI_T = \frac{lE_0 b^3}{4(1-\mu_0^2)K_2}$$

(6.79)

When the system soil-foundation is replaced with an equivalent column, the following system of equations is written:

$$\left.\begin{array}{l}\dfrac{(1-\mu_0^2)N_Z}{\pi E_0 c}F = \dfrac{N_Z l}{EA} \\[3mm] \dfrac{N_X l}{A_F K_X} = \dfrac{N_X l^3}{3EI_X} \\[3mm] \dfrac{N_Y}{A_F K_Y} = \dfrac{N_Y l^3}{EI_Y} \\[3mm] \dfrac{(1-\mu_0^2)K_1 M_X}{E_0 a^3} = \dfrac{M_X l}{EI_X} \\[3mm] \dfrac{(1-\mu_0^2)K_2 M_Y}{E_0 b^3} = \dfrac{M_Y l}{EI_Y}\end{array}\right\}$$

(6.80)

From these equations the length of the equivalent column and the column rigidities are found:

$$\left.\begin{array}{l}l = \sqrt{\dfrac{3E_0 a^3}{A_F K_X (1-\mu_0^2)K_1}} \\[4mm] EI_X = \dfrac{E_0 a^3 l}{(1-\mu_0^2)K_1} \\[4mm] EI_Y = \dfrac{E_0 b^3 l}{(1-\mu_0^2)K_2} \\[4mm] EA = \dfrac{2\pi E_0 al}{(1-\mu_0^2)F}\end{array}\right\}$$

(6.81)

As we can see, modeling the system soil-foundation with a beam with both ends fixed or with a column allows replacing a 2D or 3D frame with individual foundations with a regular frame, and performing analysis of that frame as statically indeterminate systems using available computer software. Equations obtained above are different from similar equations obtained in Chapter 3 for frames supported on Winkler foundation because they are obtained for frames with foundations supported on elastic half-space. It can be useful, in some cases, to perform analysis of frames using two soil models, Winkler foundation and elastic half-space, compare the results of the analyses and choose the safest results for final design.

6.13 Analysis of Beams Supported on Elastic Layer

Zhemochkin's method for analysis of beams supported on elastic half-space can also be applied to analysis of beams supported on an elastic layer. It is well known that settlements of foundations on an elastic layer produce much smaller settlements compared to analysis on elastic half-space. This affects the soil pressure under the foundation, moments, and shear forces in the foundation and usually leads to a more economical design compared to a design on elastic half-space. Analysis is performed exactly as analysis on elastic half-space, but settlements of the elastic layer are found using equations obtained by Shechter (1937). The equation for obtaining settlement of the foundation on the elastic layer for plane strain analysis looks as follows:

$$y_i = \frac{q(1-\mu_0^2)}{E_0} \cdot \frac{4H}{\pi} \int_0^\infty \frac{sh^2\alpha \cos\frac{x}{H}\alpha \sin\frac{a}{H}\alpha}{(sh\alpha ch\alpha + \alpha)\alpha^2} \cdot d\alpha \qquad (6.82)$$

This equation can be simplified by taking into account that:

$$\cos\frac{x}{H}\alpha \sin\frac{a}{H}\alpha = \frac{1}{2}\left[\sin\frac{a+x}{H}\alpha - \sin\frac{x-a}{H}\alpha\right]; \quad \frac{\pi H}{l} = \Delta\alpha$$

and by designating:

$$\Omega\left(\frac{x\pm a}{H}\right) = \frac{y}{\pi} \int_0^\infty \frac{sh^2\alpha \sin\frac{x\pm a}{H}\alpha}{(sh\alpha ch\alpha + \alpha)\alpha^2} d\alpha \qquad (6.83)$$

equation 6.83 will look as follows:

$$y_i = -\frac{qH}{E_0} \cdot \frac{(1-\mu_0^2)}{2}\left[\Omega\left(\frac{x+a}{H}\right) - \Omega\left(\frac{x-a}{H}\right)\right]$$

where integral $\Omega\left(\frac{x\pm a}{H}\right)$ can be found using Simpson's formula.

Now, by designating $\Omega\left(\frac{x+a}{H}\right) - \Omega\left(\frac{x-a}{H}\right) = w'$ equation 6.82 will look as shown below:

$$y_i = -\frac{qH}{\pi E_0} \cdot \frac{c}{H}\left(w'_i \cdot \frac{\pi H}{c}\right)(1-\mu_0^2) \qquad (6.84)$$

or

$$y_i = \frac{qc}{\pi E_0}(1-\mu_0^2)\left(\overline{F}_i - \overline{y}_0\right) \qquad (6.85)$$

where:

$$\overline{F}_i = \overline{y}_0 - \overline{y}_i = \frac{\pi H}{c}\left(\overline{w}_0 - \overline{w}_i\right) \qquad (6.86)$$

In this equation, $\overline{y}_0$ is the settlement of the elastic layer when $\frac{x}{c} = 0$, as shown in Figure 6.40.

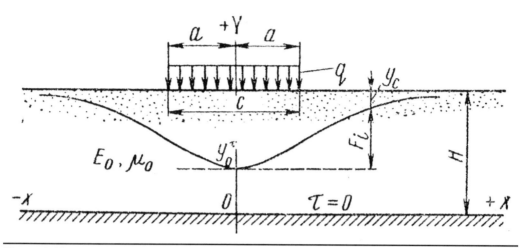

Figure 6.40

Analysis is performed using Zhemochkin's method and Tables 6.9 and 6.10 developed by Krasheninnikova (1967). Settlement of the top of the elastic layer at any point i located at distance $\dfrac{x}{c}$ from the center of the loaded area is obtained from the following equation:

$$y_x = \frac{qc(1 - \mu_0^2)}{\pi E_0} \cdot \overline{y}_i \tag{6.87}$$

When $qc = 1$ we have:

$$y_i = \frac{(1 - \mu_0^2)}{\pi E_0} \cdot \overline{y}_i \tag{6.88}$$

Analogously, the difference of settlements is equal to $(\Delta y)_x = y_0 - y_x$ or

$$(\Delta y)_x = \frac{qc(1 - \mu_0^2)}{\pi E_0} \cdot F_x \tag{6.89}$$

When $qc = 1$ we have:

$$(\Delta y)_{ki} = F_{ki} \cdot \frac{(1 - \mu_0^2)}{\pi E_0} \tag{6.90}$$

where F_{ki} is obtained from Table 6.10.

All equations presented above are obtained without taking into account the friction stresses at the bottom of the elastic layer, when $\tau_{y=H} \neq 0$, which practically does not affect the final results. Friction stresses may significantly affect the results of analysis only when the height of the elastic layer is small, when $H < 0.25l$.

The system of equations 6.91 looks exactly like the system for analysis of a beam on elastic half-space:

$$\left.\begin{array}{l} \delta_{11}x_1 + \delta_{12}x_2 + \ldots + \delta_{1n}x_n + y_0 + a_1\varphi_0 + y_{1P} = 0 \\ \delta_{21}x_1 + \delta_{22}x_2 + \ldots + \delta_{2n}x_n + y_0 + a_2\varphi_0 + y_{2P} = 0 \\ \cdots\cdots\cdots\cdots\cdots\cdots\cdots\cdots\cdots\cdots\cdots\cdots\cdots\cdots\cdots\cdots\cdots\cdots\cdots\cdots \\ \delta_{n1}x_1 + \delta_{n2}x_2 + \ldots + \delta_{nn}x_n + y_0 + a_n\varphi_0 + y_{nP} = 0 \end{array}\right\}$$ (6.91)

The additional two equations that reflect the equilibrium of the shear forces and moments at the left end of the beam are the same as analysis of beams on elastic half-space. However, it is important to remember that settlements in this case are obtained using equations for an elastic layer shown above. The vertical deflection at any point k in direction of the reaction x_k is found from equation 6.92:

$$\delta_{ki} = (y_{ki})_{\frac{c}{H}} + f_{ki}$$ (6.92)

where $(y_{ki})_{\frac{c}{H}}$ is the deflection of the beam due to the vertical settlement of the elastic layer, obtained from equation 6.89, in which $\bar{y}_i$ is taken from Table 6.9, f_{ki} is the vertical deflection of the beam taken from Table 6.4, and $(y_{ki}) = \dfrac{1 \cdot (1 - \mu_0^2)}{\pi E_0} \cdot \bar{y}_{ki}$ where $\bar{y}_{ki}$ is taken from Table 6. 9.

When performing practical analysis, reliable results are obtained by dividing the beam into 16 sections and $c = \dfrac{1}{16}2l = \dfrac{l}{8}$, where l is the half length of the beam. When $H = 4l$, results of analysis are the same as results obtained from analysis on elastic half-space.

When the height of the elastic layer is small and $0 < H < \dfrac{1}{8}l$, results of analysis are close to results obtained from analysis on the Winkler foundation. So, it can be concluded that the equations given above and Table 6.9 allow analyzing beams supported not only on the elastic layer, but also supported on elastic half-space and on Winkler foundation. It is useful to note that analysis of beams supported on elastic layer cannot take into account variable soil properties because the elastic layer, like elastic half-space, is isotropic. The only realistic way to take into account variable soil properties is by employing the finite element method, modeling the soil with plane finite elements for plane strain analysis and 3D elements for 3D analysis.

Table 6.1 Numerical values of F for plane strain analysis

$\dfrac{x}{c}$	F	$\dfrac{x}{c}$	F	$\dfrac{x}{c}$	F	$\dfrac{x}{c}$	F
0	0	6	−6.967	11	−8.181	16	−8.931
1	−3.296	7	−7.276	12	−8.356	17	−9.052
2	−4.751	8	−7.544	13	−8.516	18	−9.167
3	−5.574	9	−7.780	14	−8.664	19	−9.275
4	−6.154	10	−7.991	15	−8.802	20	−9.378
5	−6.602						

Table 6.2 Numerical values of F for 3D analysis of beams supported on elastic half–space

$\dfrac{x}{c}$	$\dfrac{c}{x}$	$\dfrac{b}{c}=\dfrac{2}{3}$	$\dfrac{b}{c}=1$	$\dfrac{b}{c}=2$	$\dfrac{b}{c}=3$	$\dfrac{b}{c}=4$	$\dfrac{b}{c}=5$
0	∞	4.265	3.525	2.406	1.867	1.542	1.322
1	1.000	1.069	1.038	0.929	0.829	0.746	0.678
2	0.500	0.508	0.505	0.490	0.469	0.446	0.424
3	0.333	0.336	0.335	0.330	0.323	0.315	0.305
4	0.250	0.251	0.251	0.249	0.246	0.242	0.237
5	0.200	0.200	0.200	0.199	0.197	0.196	0.193
6	0.167	0.167	0.167	0.166	0.165	0.164	0.163
7	0.143	0.143	0.143	0.143	0.142	0.141	0.140
8	0.125	0.125	0.124	0.125	0.124	0.124	0.123
9	0.111	0.111	0.111	0.111	0.111	0.111	0.110
10	0.100	0.100	0.100	0.100	0.100	0.100	0.099
11	0.091			0.091			
12	0.083			0.083			
13	0.077			0.077			
14	0.071			0.071			
15	0.067			0.067			
16	0.063			0.063			
17	0.059			0.059			
18	0.056			0.056			
19	0.053			0.053			
20	0.050			0.050			

Table 6.3 Coefficients m

	Circle	Square	Rectangular with various a/b					
			1.5	2	3	5	10	100
m	0.96	0.95	0.94	0.92	0.88	0.82	0.71	0.37

Table 6.4 Vertical deflections w of a cantilever beam with a one-unit vertical load

n ＼ m	0.5	1	1.5	2	2.5	3	3.5	4	4.5	5	5.5	6	6.5	7	7.5	8	8.5	9	9.5	10
0.5	0.25	0.625	1	1.375	1.75	2.125	2.5	2.875	3.25	3.625	4	4.375	4.75	5.125	5.5	5.875	6.25	6.625	7	7.375
1.0		2	3.5	5	6.5	8	9.5	11	12.5	14	15.5	17	18.5	20	21.5	23	24.5	26	27.5	29
1.5			6.75	10.13	13.5	16.87	20.25	23.62	27	30.37	33.75	37.12	40.5	43.87	47.25	50.62	54	57.38	60.75	64.13
2.0				16	22	28	34	40	46	52	58	64	70	76	82	88	94	100	106	112
2.5					31.25	40.62	50	59.37	68.75	78.12	87.5	96.87	106.2	115.6	125	134.4	143.7	153.1	162.5	171.9
3.0						54	67.5	81	94.5	108	121.5	135	148.5	162	175.5	189	202.5	216	229.5	243
3.5							85.75	104.1	122.5	140.9	159.2	177.6	196	214.4	232.7	251.1	269.5	287.9	306.25	324.6
4.0								128	152	176	200	224	248	272	296	320	344	368	392	416
4.5									182.2	212.6	243	273.4	303.7	334.1	364.5	394.9	425.25	455.6	486	516.4
5.0										250	287.5	325	362.5	400	437.5	475	512.5	550	587.5	625
5.5											332.7	373.1	423.5	468.9	514.2	559.6	605	650.4	695.8	741.1
6.0												432	486	540	594	648	702	756	810	864
6.5													549.2	612.6	676	739	802.7	866.1	929.5	992.9
7.0														686	759.5	833	906.5	980	1,053	1,127
7.5															843.7	928.1	1,012	1,096	1,181	1,265
8.0																1024	1120	1216	1312	1408
8.5																	1228.2	1,336	1445	1,553
9.0																		1458	1,579	1,701
9.5																			1,715	1,850
10.0																				2,000

Notes

1. In table 6.4 $m = \dfrac{a_i}{c}$ and $n = \dfrac{a_k}{c}$ where a_i is the distance between the applied load and the left fixed end of the beam, a_k is the distance between the point the deflection is obtained and the left end of the beam, c is the distance between the centers of neighboring elements of the beam.

2. Deflections w_{ki} are not the real deflections of the beam. The real deflections are obtained from the following equation $v_{ki} = \dfrac{c^3}{6EI}\, w_{ki}$. If the beam represents a band cut off from a plate (plane-strain analysis) deflection of the beam is found from $v_{ki} = \dfrac{c^3}{6D}\, w_{ki}$, where $D = \dfrac{h^3 E}{12(1 - \mu^2)}$ and h is the thickness of the plate, and μ is the Poison's Ratio of the plate material.

3. The load is equal to one unit.

Table 6.5 Beams with various boundary conditions

Right end / Left end	Free	Simple supported	Fixed	Guided	Vertical spring	Rotational spring
Free						
Simple supported						
Fixed						
Guided						
Vertical spring						
Rotational spring						

Table 6.6 Cantilever beams with fictitious unknowns

Left end \ Right end	Free end	Simple supported	Fixed	Guided	Vertical spring	Rotational spring
Free	y_0 0	l		y_l	l y_l	y_l l
Simple supported	0	x_0 l	x_0	x_0 y_l	0 x_l	0 m_l
Fixed		x_l	x_l m_l	m_l	x_l	m_l
Guided	y_0	y_0 x_l	m_0	y_0 x_l	y_0 x_l	y_0 m_l
Vertical spring	y_0 0	x_0 l	x_0	x_0 y_l	x_0 l y_l	y_0 0 m_l
Rotational spring	0 y_0	m_0 l	m_0	m_0 y_l	m_0 y_l l	m_0 y_l l

Table 6.7 One unit settlements of an elastic half-space with a circular load

$\dfrac{X}{c}$	F	$\dfrac{X}{c}$	F	$\dfrac{X}{c}$	F	$\dfrac{X}{c}$	F
0	3.545	1.6	0.635	2.8	0.359	4.0	0.251
0.5	2.682	1.7	0.597	2.9	0.346	4.1	0.244
0.6	1.984	1.8	0.563	3.0	0.335	4.2	0.239
0.7	1.589	1.9	0.532	3.1	0.324	4.3	0.233
0.8	1.348	2.0	0.505	3.2	0.314	4.4	0.228
0.9	1.176	2.1	0.481	3.3	0.304	4.5	0.223
1.0	1.045	2.2	0.458	3.4	0.295	4.6	0.218
1.1	0.942	2.3	0.438	3.5	0.287	4.7	0.213
1.2	0.859	2.4	0.420	3.6	0.279	4.8	0.209
1.3	0.789	2.5	0.403	3.7	0.271	4.9	0.204
1.4	0.730	2.6	0.387	3.8	0.264	5.0	0.200
1.5	0.679	2.7	0.372	3.9	0.257		X/c

1. X—distance between the center of the load and point the settlement is obtained
   c—the length of the square equivalent to the area of the circle
   If the area of load application is a rectangle, this area is replaced with an equivalent square area and the side of this square c is used.
2. The load used in this table is equal to one unit uniformly distributed over the circular area.
3. The actual settlement is obtained from the following Equation

$$y_{kl} = \frac{(1 - \mu_0^2)}{\pi E_0 c} F_{kl}$$

Where E_0 is the modulus of elasticity of the half-space and μ_0 is the poison's ratio of the half-space

Table 6.8 Analysis of a group of foundations

x_1	x_2	$\cdots$	x_m	x_{m+1}	x_{m+2}	$\cdots$	x_n	y_{0A}	φ_{0A}	y_{0C}	φ_{0C}			
δ_{11}	δ_{12}	$\cdots$	δ_{1m}	$w_{1(m+1)}$	$w_{1(m+2)}$	$\cdots$	w_{1n}	1	a_1	0	0	$y_{1P}+w_{1B}$	$=$	0
δ_{21}	δ_{22}	$\cdots$	δ_{2m}	$w_{2(m+1)}$	$w_{2(m+2)}$	$\cdots$	w_{2n}	1	a_2	0	0	$y_{2P}+w_{2B}$	$=$	0
δ_{m1}	δ_{m2}	$\cdots$	δ_{mm}	$w_{m(m+1)}$	$w_{m(m+2)}$	$\cdots$	w_{mn}	1	a_m	0	0	$y_{mP}+w_{mB}$	$=$	0
$w_{(m+1)1}$	$w_{(m+1)2}$	$\cdots$	$w_{(m+1)m}$	$\delta_{(m+1)(m+1)}$	$\delta_{(m+1)(m+2)}$	$\cdots$	$\delta_{(m+1)n}$	0	0	1	$a_{(m+1)}$	$y_{(m+1)P}+w_{(m+1)B}$	$=$	0
$w_{(m+2)1}$	$w_{(m+2)2}$	$\cdots$	$w_{(m+2)m}$	$\delta_{(m+2)(m+1)}$	$\delta_{(m+2)(m+2)}$	$\cdots$	$\delta_{(m+2)n}$	0	0	1	$a_{(m+2)}$	$y_{(m+2)P}+w_{(m+2)B}$	$=$	0
w_{n1}	w_{n2}	$\cdots$	w_{nm}	$\delta_{n(m+1)}$	$\delta_{n(m+2)}$	$\cdots$	δ_{nn}	0	0	1	a_n	$y_{nP}+w_{nB}$	$=$	0
1	1	$\cdots$	1	0	0	$\cdots$	0	0	0	0	0	$-\sum P_A$	$=$	0
a_1	a_2	$\cdots$	a_m	0	0	$\cdots$	0	0	0	0	0	$\sum M_A$	$=$	0
0	0	$\cdots$	0	1	1	$\cdots$	1	0	0	0	0	$-\sum P_C$	$=$	0
0	0	$\cdots$	0	a_{m+1}	a_{m+2}	$\cdots$	a_n	0	0	0	0	$\sum M_C$	$=$	0

Table 6.9 Coefficients $\bar{y}_i$ for analysis of beams supported on elastic layer

c/H x/c	1/32	1/16	1/8	1/4	1/2	1	2	4
0.00	9.61447	8.22843	6.84315	5.46092	4.09071	2.76654	1.59831	0.79497
0.25	9.19978	7.96705	6.84315	5.20334	4.09071	2.56161	1.49611	0.79268
0.50	8.22844	6.84315	6.58254	4.09073	3.84481	1.58832	0.79498	0.39268
0.75	6.98781	5.59341	5.46091	2.85965	2.76655	0.59029	0.11280	0.00000
1.00	6.31965	4.93666	4.21497	2.22868	1.59058	0.22567	0.0000	
2.00	4.86718	3.49329	3.56351	0.95247	1.03250	0.00000		
3.00	4.04948	2.69054	2.15578	0.39496	0.14791			
4.00	3.47704	2.13885	1.40509	0.13000	0.00000			
5.00	3.03793	1.72583	0.93878	0.01200				
6.00	2.68329	1.40228	0.61036	0.00000				
7.00	2.38732	1.14199	0.38642					
8.00	2.13469	0.92915	0.23143					
9.00	1.91554	0.75335	0.12561					
10.0	1.72310	0.60732	0.05514					
11.0	1.55256	0.48567	0.00998					
12.0	1.40034	0.38430	0.00000					
13.0	1.22350	0.29993						
14.0	1.18074	0.22992						
15.0	1.02909	0.17206						
16.0	0.92800	0.12452						
17.0	0.83610	0.08573						
18.0	0.75243	0.05438						
19.0	0.67615	0.02025						
20.0	0.60656	0.00943						
21.0	0.54304	0.00000						
25.0	0.33970							
30.0	0.17170							
35.0	0.06900							
40.0	0.00930							
42.0	0.00000							

Table 6.10 One unit settlements F_{ki} of the top of the elastic layer

c/H x/c	0 Elastic Half-Space	1/32	1/16	1/8	1/4	1/2	1	2
0	0	0	0	0	0	0	0	0
1	−3.296	−3.29482	−3.29177	−3.27964	−3.23244	−3.05821	−2.54087	−1.59831
2	−4.751	−4.74729	−4.73514	−4.68737	−4.50845	−.3.94280	−2.76654	−1.59831
3	−5.574	−5.56504	−5.53789	−5.43808	−5.06596	−4.09071	−2.76654	
4	−6.154	−6.13743	−6.08958	−5.90438	−5.33092	−4.09071		
5	−6.602	−6.57650	−6.50260	−6.23279	−5.44872			
6	−6.967	−6.93118	−6.82615	−6.45672	−5.46092			
7	−7.276	−7.22715	−7.08644	−6.61172	−5.46092			
8	−7.544	−7.47978	−7.29928	−6.71734				
9	−7.780	−7.69893	−7.47508	−6.78801				
10	−7.991	−7.89137	−7.62111	−6.83317				
11	−8.181	−8.06191	−7.74276	−6.84315				
12	−8.356	−8.21413	−7.84413	−6.84315				
13	−8.516	−8.39097	−7.92850					
14	−8.664	−8.43373	−7.99851					
15	−8.802	−8.58538	−8.05637					
16	−8.931	−8.68747	−8.10391					
17	−9.052	−8.77837	−8.14270					
18	−9.167	−8.86204	−8.17405					
19	−9.275	−8.93832	−8.19918					
20	−9.378	−9.00791	−8.21900					
21	−9.475	−9.07143	−8.22843					
25	−9.824	−9.27479	−8.22843					
30	−10.186	−9.44273						
35	Elastic	−9.54549						
40	Half-Space	−9.60518						

References

1. Barden, L. 1962. Distribution of contact pressure under foundations. *Géotechnique* 12:181–198.
2. Barden, L. 1963. Stresses and displacements in a cross-anisotropic soil. *Géotechnique* 13:198–210.
3. Brown, R. T. 1975. Strip footing with concentrated loads on deep elastic foundations. *Geotechnical Engineering* 6:1–13.
4. Flamant. 1892. *Comptes Rendus* (Paris).
5. Glassman, A. 1972. Behavior of crossed beams on elastic foundations. *Journal of the Soil Mechanics and Foundations Division: Proceedings of ASCE* 98 (SM1): 1–11.
6. Gorbunov-Posadov, M. I. 1953. *Analysis of structures on elastic foundation.* Moscow: Gosstroiizdat.
7. Kany, M. 1959. *Berechnung von Flachengrundungen.* Berlin: W. Ernst und Sohn.
8. Krasheninnikova, G. V. 1967. Investigation of the influence of the elastic layer on behavior of beams supported on elastic layer. Thesis, Moscow Civil Engineering Institute (MISI).
9. Lousberg, E. 1957. Calculation of the distribution of soil reaction underneath eccentrically loaded footing. In *Proceedings of the 4th international conference on soil mechanics foundation engineering.* Vol. 2, 355–359. London: Butterworths.
10. Ohde, J. 1942. Die Berechnung der Sohdrucverteilung unter Grundungskorper Bauingenieur, 23:99–107.
11. Schleicher, F. 1926. *Kreisplatten auf Elastischer Unterlage.* Berlin: Julius Springer.
12. Selvadurai, A.P.S. 1979. *Elastic analysis of soil-foundation interaction.* Amsterdam/Oxford/ New York: Elsevier Scientific.
13. Shechter, O. Y. 1937. On determination of ground settlement with bedding layers below the foundation. *Journal Gidrotechnicheskoe Stroitelstvo.* 10.
14. Sommer, H. 1965. A method for the calculation of settlements, contact pressures and bending moments in a foundation including the influence of the rigidity of the superstructure. In *Proceedings of the 6th international conference on soil mechanics and foundation engineering.* Vol. 2, 197.
15. Tsudik, E. A. 1983. Soil, foundation , structure interaction, ASCE, Houston, Texas.
16. Tsudik, E. A. 1970. A general method of combined analysis frames and foundations, Stroitelstvo, 2, pp. 3–7.
17. Tsudik, E. A. 1975. *Analysis of 2D and 3D frames supported on elastic foundation.* Moscow: MADI.
18. Tsudik, E. A. 2003. Analysis of mat foundations for various soil models. In *Proceedings of first international conference on foundations,* Dundee, Scotland.
19. Tsudik, E. A. 2006. *Analysis of beams and frames on elastic foundation.* Canada/UK: Trafford.
20. Zhemochkin, B. N. 1937. *Plane strain analysis of an infinite beam on elastic foundation.* Moscow: VIA.
21. Zhemochkin, B. N., and A. P. Sinitsin. 1962. *Practical methods of analysis of beams and plates on elastic foundation.* Moscow: Gosstroiizdat.

7

Simplified Analysis of Mat Foundations

7.1. Introduction

Foundations of tall buildings are designed, in many cases, as mat or raft foundations. Mat foundations are designed not only for tall buildings. They are also used for smaller but heavy buildings, when the allowable soil bearing pressure is relatively small and loads applied to the foundation are large; when individual foundations cannot support the superstructure. As mentioned by Teng (1975), by combining all individual footings into one mat, not only is the unit pressure on the supporting soil reduced, but the bearing capacity is often increased. In the case of individual footings, the depth of the foundation is the dimension from the top of the base slab to the bottom of the footing, whereas the depth of the foundation of the mat is measured from the exterior ground surface to the bottom of the mat, as shown in Figure 7.1. The advantage of using a mat foundation are twofold: the bearing capacity increases with the depth and the width of the foundation and the settlement decreases with the increasing depth of the foundation. It is also important to add that individual foundations experience much larger settlements and rotations compared to mats, producing additional stresses in the elements of the superstructure (differential settlements of the mat foundation at points of support of the columns are much smaller) and rotations at the same points are negligibly small and practically equal to zero. Mat foundations are also used for buildings with equipment sensitive to differential settlements. In most cases the mat is directly supported on soil and, in some cases, on piles. Usually the mat foundation is flat with a uniform thickness throughout the entire area as shown in Figure 7.2. However, it may be increased under the columns loaded with large loads in order to reduce the shear stresses and stresses due to negative moments in the mat. Two ways of thickening the mat are usually used by practicing engineers: by pedestals or by increasing the thickness of the mat around the columns. However, increasing the thickness of the mat under individual columns may not be enough; large negative moments may require increasing the thickness of the mat along the column lines. More complex types of mat foundations such as cellular mats or boxed foundations are usually designed for tall

---

All tables can be found at the end of the chapter.

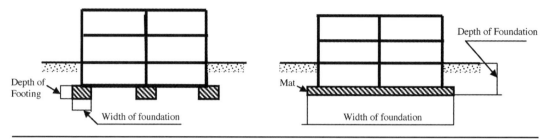

Figure 7.1   Depth and width of individual foundations and mat foundations

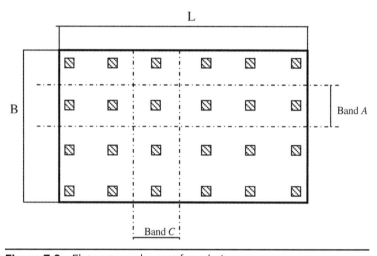

Figure 7.2   Flat rectangular mat foundation

and heavy buildings. Accurate exact analysis of mat foundations requires taking into account the soil properties as well as the rigidity of the foundation. Detailed computer analyses based on finite element or finite difference methods are costly, time consuming, and not always available.

Therefore, the use of simplified methods in many cases, while not the only methods available, are justified. Two simplified methods of analysis usually used by practicing engineers are the rigid method and the elastic method.

7.2 Rigid Analysis

This method can be used when the mat is totally rigid and all ordinates of the 3D soil pressure diagram are located in one plane; in other words, when the soil pressure has a planar distribution and when the rigidity of the mat does not affect the soil pressure distribution. Analysis is performed exactly like analysis of a totally rigid individual foundation. In accordance with the American Concrete Institute (ACI), Committee 466 Report (1966), the mat is considered totally rigid when the spacing of the columns is relatively uniform and does not exceed 1.75 λ, or when the mat is supporting a rigid superstructure. Parameter λ is obtained from equation 7.1:

$$\lambda = \sqrt[4]{\frac{k_b b}{4EI}} \tag{7.1}$$

Where k_b is the modulus of subgrade reaction, b is the width of the bands or the distance between adjacent bays of the mat, as shown in Figure 7.2, and EI is the flexural rigidity of the band. The soil pressure is obtained from the following equation:

$$p = \frac{N}{A} \pm N\frac{e_y}{x}y \pm N\frac{e_x}{y}x \tag{7.2}$$

In this equation N is the total vertical load applied to the mat, A is the total area of the mat, x and y are coordinates of any given point of the mat with respect to the X and Y axes passing through the centroid of the mat e_x, and e_y are the coordinates of the resultant vertical load applied to the mat. Coordinates e_x and e_y are obtained taking into account moments applied to the mat from the columns.

I_X and I_Y are moments of inertia of the foundation with respect to the axes X and Y, respectively, obtained from the following equations:

$$I_X = \frac{LB^3}{12}, \ I_Y = \frac{BL^3}{12} \tag{7.3}$$

Equations 7.3 are used only when the soil pressure is positive, which is usually the case in the design of buildings. Once the soil pressure is obtained, analysis of the mat can be performed as one whole system. The total vertical load N includes not only the loads applied by the columns, but also the weight of the walls supported on the mat if any. Moments from the columns should be algebraically added to the moments Ne_x and Ne_y. Moments and shear forces at any cross section of the mat, in X and Y directions, are found by applying to one part of the mat the soil pressure and given loads. Satisfactory results can be expected when the loads applied to the columns and the spacing of the columns are close enough. As can be seen, analysis is the same as analysis of individual totally rigid foundations. However, moments and shear forces along the sections of the mat will vary, especially when the given column loads and spacing of the columns vary significantly. Nevertheless, the method is used by practicing engineers, mostly for preliminary design and, in some cases, for final analyses as well.

Sometimes, analysis is performed by employing the method used for two-way slabs, by cutting off from the plate two bands, in both directions, as shown previously in Figure 7.2, and analyzing each band as a simple beam loaded with given loads and soil pressure. But almost always the total given load from the columns applied to the band is not equal to the total soil pressure on the band. In other words, the sum of all active loads applied to the band is not equal to the total soil reaction applied to the same band. The explanation is simple: loads applied to the mat at any point will produce soil reaction not only under that point, but will also affect the soil pressure at other points. Therefore, while the sum of the given loads applied to the mat is equal to the total soil pressure on the mat, active and reactive forces applied to individual bands are not equal.

Both ways of analysis have a significant advantage; they are recommended for hand calculations and easy to use, producing fast practical results. However, their application is limited: they do not take into account the real soil properties as well as the rigidity of the mat. It is also important to mention that by cutting off bands from the mat and analyzing each band as a beam, we take into account two times the load located at the area of intersection of both bands, A and C, which is incorrect. A simplified elastic method of analysis that takes into account the rigidity of the mat as well as the soil properties is presented below.

7.3 Elastic Analysis

If the mat shown in Figure 7.2 is supported on Winkler foundation, both bands, A and C, can be analyzed using any of the well-known methods described earlier in Chapter 2, such as the method of superposition, or the method of initial parameters. It is also easy to use Tables 3.1–3.14 and the equations from Chapter 3. By using any of the methods mentioned above for analysis of beams on Winkler foundation, the soil properties and flexural rigidity of the foundation are taken into account.

When the mat is supported on hard soil, analysis of the same bands can also be performed as beams supported on elastic half-space using Tables 5.1–5.9 from Chapter 5, which produces more accurate results, especially for bands cut off from plates, because these tables are developed for plane strain analysis, reflecting much better the actual conditions the band is working. This analysis also takes into account the soil properties and the rigidity of the foundation as well.

Another simplified elastic method of analysis was recommended by ACI Committee 436 (1966) for analysis of mat foundations. Developed by Hetenyi (1946), this method analyzes infinite plates supported on Winkler foundation loaded with a series of vertical concentrated loads.

Heyenyi's method is based on the assumption that a concentrated vertical load applied to the mat affects only a small area around the applied load; in other words, each load has its area or zone of influence. Analysis is performed in the following steps:

1. Obtain the flexural rigidity of the mat:

$$D = \frac{Eh^3}{12\left(1 - \mu^2\right)} \tag{7.4}$$

   where E is the modulus of elasticity of the concrete, h is the thickness of the mat, and μ is the Poisson's ratio of the concrete.

2. Find the radius of effective stiffness L:

$$L = \sqrt[4]{\frac{D}{k_b}} \tag{7.5}$$

   where k_b is the modulus of subgrade reaction and the radius of influence of each load is approximately $4L$.

3. Find the radial M_r and the tangential M_t moments, shear forces Q, and settlements w of the mat at any point from the following equations:

$$M_r = \frac{P}{4}\left[Z_4\left(\frac{r}{L}\right) - \left(1 - \mu\right)\frac{Z_3'\left(\frac{r}{L}\right)}{\left(\frac{r}{L}\right)}\right] \tag{7.6}$$

$$M_t = \frac{P}{4}\left[\mu Z_4\left(\frac{r}{L}\right) + \left(1 - \mu\right)\frac{Z_3'\left(\frac{r}{L}\right)}{\left(\frac{r}{L}\right)}\right] \tag{7.7}$$

$$Q = \frac{P}{4L}Z_4'\left(\frac{r}{L}\right) \tag{7.8}$$

$$w = \frac{PL^2}{4D}Z_3\left(\frac{r}{L}\right) \tag{7.9}$$

In these equations, P is the column load, r is the distance between the point the load is applied and the point where M_r, M_t, Q, and w are obtained. Functions Z_3 and Z_4 and their derivatives Z_3' and Z_4' are obtained from tables or curves developed by Hetenyi. The numerical values of functions Z_3, Z_4, Z_3', and Z_4' are shown in Table 7.1.

4. Convert the radial and tangential moments to rectangular coordinates using equations 7.10:

$$\left.\begin{aligned} M_x &= M_r\cos^2\varphi + M_t\sin^2\varphi \\ M_y &= M_r\sin^2\varphi + M_t\cos^2\varphi \end{aligned}\right\} \tag{7.10}$$

where φ is the angle as shown in Figure 7.3. See also Figures 7.4–7.5.

5. Moments, shear forces, and settlements are obtained at all points of investigation due to the concentrated vertical loads. When the edge of the mat is located within the radius of influence ($4L$) it is assumed that the mat is infinitely large, the moments and shear forces perpendicular to the edge are found, and in order to satisfy the boundary conditions, the same moments and shear forces, acting in the opposite direction, are applied to the same points at the edge of the mat. These loads will produce additional moments, shear forces, and settlements at all points located in the zone of influence.

6. Final moments, shear forces, and settlements at any point of the mat are obtained by superimposing all moments, shear forces, and settlements produced at those points by all loads applied to the mat.

This method of analysis is more accurate compared to elastic analyses described above, because it takes into account not only the soil properties and the rigidity of the foundation, but also analyzes the mat as a 3D system. However, the method, although simple, is still labor intensive. It is interesting to note that a similar method of analysis of plates on Winkler foundation was developed by Korenev (1962).

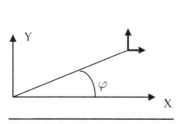

Figure 7.3

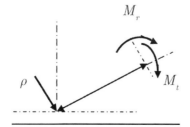

Figure 7.4

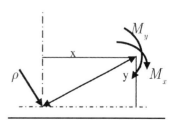

Figure 7.5

7.4 Analysis of Infinite Plates Supported on Elastic Half-Space with a Vertical Concentrated Load

A general method for analysis of infinite plates supported on elastic half-space belongs to Shechter (1939). Practical application of the method is explained below. The method was developed for analysis of infinite plates supported on elastic half-space loaded with a vertical concentrated load. In order to determine if the method can be applied, parameter δ has to be found from equation 7.11:

$$\delta = \frac{d}{L} \tag{7.11}$$

where d is the real distance between the load and the closest edge of the plate and L is the elastic characteristic of the plate that is obtained from the following equation:

$$L = \sqrt[3]{2D(1 - v_0^2)/E_0} \tag{7.12}$$

where

$$D = (E_1 h^3)/12(1 - v_1^2) \tag{7.13}$$

In equations 7.12 and 7.13, E_0 is the modulus of elasticity of the soil, v_0 is Poisson's ratio of the soil, D is the rigidity of the plate, E_1 is the modulus of elasticity of the plate material, h is the thickness of the plate, and v_1 is Poisson's ratio of the plate material. Analysis is performed using Table 7.2, but it can only be used when $\delta \geq 1.5$. In Table 7.2,

$$\rho = r/L \tag{7.14}$$

where r is the real distance between the point the load is applied and the point the soil pressure, settlements, moments, and shear forces are obtained. In Table 7.2, the following coefficients are given: $\bar{p}$ for obtaining the soil pressure, $\bar{w}$ for obtaining the settlements, $\overline{Q}_r$ for obtaining the radial shear forces, $\overline{M}_r$ for obtaining the radial moments, and $\overline{M}_t$ for obtaining the tangential moments. The tangential shear force $\overline{Q}_t = 0$. All numerical values given in Table 7.2 are dimensionless. Radial moments and shear forces are acting in the radial direction from the point the load is applied to the point of investigation. Tangential moments act perpendicular to the radial direction. The real settlements, soil pressure, moments, and shear forces are found using equations 7.15–7.19:

$$p = \bar{p}(P/L^2) \tag{7.15}$$

$$w = \bar{w}(1 - v_0^2)P/E_0 L \tag{7.16}$$

$$M_r = \overline{M}_r P \tag{7.17}$$

$$M_t = \overline{M}_t P \tag{7.18}$$

$$Q_r = -\overline{Q}_r P/L \tag{7.19}$$

All moments and shear forces given in Table 7.2 are attributed to one unit of the plate length (1m). Moments M_r and M_t are obtained for reinforced concrete assuming $v_1 = 1/6$. M_1 and M_2 shown in the last two columns of the Table 7.2 are used for material other than reinforced concrete. The real moments in such types of plates are obtained from the following equations:

$$M_r = (\overline{M}_2 + v_1\overline{M}_1)P \tag{7.20}$$

$$M_t = (\overline{M}_1 + v_1\overline{M}_2)P \tag{7.21}$$

In these equations, v_1 is the Poisson's ratio of the plate material. All other parameters such as $\overline{p}$, $\overline{w}$, and $\overline{Q}_r$ do not depend on Poisson's ratio and therefore are used for analysis of plates made from any material. It is noteworthy that when $\rho = 4$, the soil pressure, shear forces, and moments are very close to zero, while $\overline{w}$ can be large enough. When $\rho > 4$, the settlement of the plate is not affected significantly by the plate's rigidity. In this case the settlement is found using Boussinesq's equation 7.22:

$$w = P(1 - v_0^2)/\pi E_0 r \tag{7.22}$$

where r is the distance between the point where the load P is applied and the point the settlement is obtained. In order to illustrate the method described above, an example is offered.

Example 7.1

Given: $E_0 = 1{,}000 t/m^2$, $v_0 = 0.35$, $E_1 = 2{,}500{,}000 t/m^2$, $r = 6m$, $h = 1.5m$, and $P = 100t$. Find the soil pressure, settlement, moments, and shear forces.

Solution:

$$D = 2{,}500{,}000 \cdot 1.5^3/12\,(1 - 0.1666^2) = 724{,}871 tm$$
$$L = \sqrt[3]{2 \cdot 724{,}871\,(1 - 0.35^2)/1{,}000} = 10.83m$$
$$\rho = 6/10.83 \approx 0.554$$

Using Table 7.2 we find:

$$\overline{p} = 0.1195, \ \overline{w} = 0.358 \ \overline{Q}_r = 0.26 \ \overline{M}_r = 0.045 \ \overline{M}_t = 0.112$$

From equations 7.15–7.19 we find:

$$p = 0.1195 \cdot (100/10.83^2) = 0.102 t/m^2$$
$$w = 0.358(1 - 0.35^2) \cdot 100/1000 \cdot 10.83 = 0.0029m = 0.29cm$$
$$M_r = 0.045 \cdot 100 = 4.5 tm/m \ \ M_t = 0.112 \cdot 100 = 11.2 tm/m$$
$$Q_r = 0.26\frac{100}{10.83} = 2.40 t/m$$

As shown in Table 7.2, the radial shear force as well as the radial and tangential moments at the point the vertical concentrated load is applied to the plate ($\rho = 0$) are infinitely large. In reality, vertical loads are never applied as concentrated loads. They are always applied as loads distributed over a certain area.

If the load is applied as a distributed load over a circular area and located at a distance equal to $\delta \geq 1.5$ from the edge of the plate, the bending moments can be found from the following equation:

$$M_r = M_t = P\left(0.0592 - 0.09284 \ln \alpha\right) \tag{7.23}$$

where $\alpha = a/L$ and a is the actual radius of the circular area. If the load is applied to a square area, the moments at the center of that area acting parallel to the square edges are found as follows:

$$M_X = M_Y = P\left(0.1123 - 0.09284 \ln \gamma\right) \tag{7.24}$$

where $\gamma = c/L$ and c is the length of the side of the square. If the area of load application is of a different shape, this area should be replaced with an equivalent circular area and moments obtained from equation 7.23.

When $\alpha \leq 0.50$ or $\gamma \leq 0.50$, numerical values obtained from equations 7.23 and 7.24 do not differ from the exact value by more than 1%. The exact solution for a circular uniformly distributed load applied to an infinite plate can be found using equation 7.33 shown below.

In order to avoid infinitely large shear forces, the shear force is not obtained at the center of the loaded area. It is obtained at the edge of the loaded area, per one linear unit of the column perimeter, from equations 7.25 and 7.26. For a circular area:

$$Q_r = -\left(P - p_0 \pi a^2\right)/2\pi a \tag{7.25}$$

For a square area:

$$Q_X = Q_Y = -\left(P - p_0 c^2\right)/4c \tag{7.26}$$

In these equations, p_0 is the soil pressure under the applied load and is obtained from Table 7.2 for $\rho = 0$. In equations 7.25 and 7.26, $p_0\pi a^2$ and $p_0 c^2$ are the total soil pressures under the base of the column. The numerical values of these pressures are usually small compared to the applied load P and may not be taken into account. At all other points of the plate, where $\rho > 0$, equations 7.15–7.19 and coefficients from Table 7.2 are used. Described analysis allows obtaining p, w, M_r, M_t, and Q_r at any point of an infinite plate loaded with any number of concentrated vertical loads.

7.5 Analysis of Infinite Plates Supported on Elastic Layer with a Vertical Concentrated Load

The settlement and soil pressure of an infinite plate with a vertical concentrated load P supported on the elastic layer with the thickness equal to H, in accordance with Shechter (1948), is obtained from the following equations:

$$w = \frac{\left(1 - v_0^2\right)}{\pi E_0} \cdot \frac{P}{L} \int_0^\infty \frac{J_0(pt)\,dt}{\Phi(\gamma t)} = \bar{w} \frac{\left(1 - v_0^2\right)}{E_0} \cdot \frac{P}{L} \tag{7.27}$$

$$p = \frac{1}{2\pi} \cdot \frac{P}{L^2} \int_0^\infty \frac{t J_0(pt)}{\Phi(\gamma t) + t^3} = \bar{p} \frac{P}{L^2} \tag{7.28}$$

In these equations, L is the elastic characteristic of the plate obtained from equation 7.12, $\gamma = H/L$, t is the parameter of integration, and $\rho = r/L$ where r is the actual distance between the point the load is applied and the point the settlement and soil pressure are found. Parameter Φ depends on the height H of the elastic layer and is obtained from the following equation:

$$\Phi(\gamma t) = (sh2\gamma t + 2\gamma t)/(ch2\gamma t - 1) \qquad (7.29)$$

As can be seen, when $\gamma t \to \infty$ parameter $\Phi(\gamma t) \to 1$. In this case we have a solution for an elastic half-space ($H \to \infty$). Equations for obtaining moments and shear forces are shown below:

$$M_r = \left[\overline{V}(\rho) + v_1 \overline{U}(\rho)\right]P \qquad (7.30)$$

$$M_t = \left[\overline{U}(\rho) + v_1 \overline{V}(\rho)\right]P \qquad (7.31)$$

$$Q_r = -\overline{Q}_r(\rho) \cdot P/L \qquad (7.32)$$

where $\overline{w}$, $\overline{p}$, $\overline{U}$, $\overline{V}$, and $\overline{Q}_r$ are taken from Tables 7.3–7.7. Tables are also used for obtaining moments under the center of a circular column supported on a plate. They look as follows:

$$M_0 = M_r = M_t = P(1 + v_1)\overline{U}(\alpha) \qquad (7.33)$$

where $\alpha = a/L$ and a is the radius of the column. The same equation is used for a rectangular column by replacing the rectangular area with an equivalent circular area. Table 7.3 does not include numerical values of $\overline{w}$ when $\gamma < 0.3$ because when $\gamma < 0.3$, results of analysis are close to results obtained from analysis of plates supported on Winkler foundation. Table 7.3 also does not include numerical values of $\overline{w}$ when $\gamma > 3$ because in this case results of analysis are close to results of analysis of plates supported on elastic half-space. Tables 7.4–7.7 also do not include $\gamma < 0.3$ for the same reason. They do not include values of $\overline{p}$, $\overline{V}$, $\overline{U}$, and $\overline{N}_t$ for $\gamma > 1$ because results of analysis are close to results obtained from analysis on elastic half-space. When $\gamma < 0.3$ and Winkler foundation is used, the modulus of subgrade reaction is found using one of two equations:

$$k = \frac{E_0}{(1 - v_0)H} \qquad (7.34)$$

$$k = \frac{(1 - v)E_0}{(1 + v_0)(1 - 2v_0)H} \qquad (7.35)$$

Analysis in this case is performed using Table 7.8 and equations 7.36–7.41. Analysis is performed in several steps. First, parameter ρ is obtained from the following equation:

$$\rho = r/C \qquad (7.36)$$

where r is the real distance between the load and the point the settlement, soil pressures, moments, and shear forces are found, and C is the radius of rigidity obtained from equation 7.37:

$$C = \sqrt[4]{D/k} \qquad (7.37)$$

The actual settlements, moment, and shear forces are found from the equations shown below:

$$w = \overline{w}\frac{PC^2}{D} \tag{7.38}$$

$$M_r = \overline{M}_r P \tag{7.39}$$

$$M_t = \overline{M}_t P \tag{7.40}$$

$$Q_r = -\overline{Q}_r \frac{P}{C} \tag{7.41}$$

The soil pressure is obtained from the well-known equation:

$$p = kw \tag{7.42}$$

Moments under the center of a circular column can be found from equation 7.43:

$$M_r = M_t = P(0.0578 - 0.09288 \ln \alpha) \tag{7.43}$$

where $\alpha = a/C$.

Moments under the center of a square column, acting in a direction parallel to the sides of the column, are found as follows:

$$M_X = M_Y = P(0.1109 - 0.09288 \ln \gamma) \tag{7.44}$$

where $\gamma = c/C$ and c is the side of the square column.

Some numerical examples illustrate application of the method to practical analysis of infinitely large plates on elastic half-space, the elastic layer and the Winkler foundation.

Example 7.2

Given: A circular column supported on an infinitely large plate on elastic half-space is loaded with a vertical load $P = 300$ton. The thickness of the plate $h = 1.0$m, the modulus of elasticity $E_1 = 2,650,000$t/m², and the Poisson's ratio of the plate material is equal to $\nu_1 = 0.17$. The modulus of elasticity of the soil $E_0 = 1,000$t/m² and $\nu_0 = 0.30$. The radius of the column is equal to $a = 0.30$m. Find the settlement, the soil pressure, moments, and shear forces in the plate at point A located at a distance $r = 6$m from the applied load. Find the moments and shear forces under the column.

Solution:

1. Find the rigidity of the mat:

$$D = E_1 h^3/12(1 - \nu_1^2) = 2,650,000 \cdot 1.0^3/12(1 - 0.17^2) = 227,405.35 \text{tm}$$

2. Find the elastic characteristic of the mat:

$$L = \sqrt[3]{2D(1 - \nu_0^2)/E_0} = \sqrt[3]{2 \cdot 227,405.35 \cdot (1 - 0.3^2)/1,000} = 7.45 \text{m}$$

3. Find parameter $\rho = r/L = 6/7.45 = 0.81$.

4. Find from Table 7.2 for $\rho = 0.81$ coefficients $\overline{p} = 0.094$, $\overline{w} = 0.314$, $\overline{Q}_r = 0.15$, $\overline{M}_r = 0.016$, $\overline{M}_t = 0.074$.

5. Using equations 7.15–7.19 find:

$p = \overline{p}(P/L^2) = 0.094 \cdot 300/7.45^2 = 0.508\text{t/m}^2 = 0.0508\text{kg/cm}^2$

$w = \overline{w}(1 - v_0^2)P/E_0 L = 0.314 \cdot (1 - 0.3^2) \cdot 300/1,000 \cdot 7.45 = 0.0115\text{m} = 1.15\text{cm}$

$M_r = \overline{M}_r P = 0.016 \cdot 300 = 4.8\text{tm/m}$ $M_t = \overline{M}_t P = 0.074 \cdot 300 = 22.2\text{tm/m}$

$Q_r = -\overline{Q}_r P/L = -0.15 \cdot 300/11.18 = -4.03\text{t/m}$

6. Find the moments and shear force under the column using equations 7.23 and 7.25, respectively:

$M_r = M_t = P(0.0592 - 0.09284 \ln \alpha) = 300(0.0592 + 0.09284 \cdot 3.219)$
$= 107.42\text{tm/m}$

where $\alpha = a/L = 0.3/7.45 = 0.0403$, $\ln(4.03 \cdot 10^{-2}) = 1.3863 - 4.60517 = -3.219$

$Q_r = -(P - p_0 \pi a^2)/2\pi a = -(300 - 1.04 \cdot 3.141 \cdot 0.3^2)/6.282 \cdot 0.3 = -159.03\text{t/m}.$

The soil pressure under the column p_0 is obtained from equation 7.15 and Table 7.2 when $\rho = 0$, $p_0 = \overline{p}P/L^2 = 0.192 \cdot 300/(7.45)^2 = 1.04\text{t/m}^2$.
It is easy to notice that the shear force Q_r is practically not affected by reaction $p_0\pi a^2$. By ignoring this reaction the shear force can be obtained as follows: $Q_r = -P/2\pi a = -300/6.282 \cdot 0.3 = -159.18\text{t/m}$. Results of both calculations are nearly the same.

If the shape of the given column is square, analysis is performed analogously, but moments are found using equation 7.24 and the shear force is obtained from equation 7.26. Assuming that the given column is square and equal to $c^2 = 0.6 \cdot 0.6 = 0.36\text{m}^2$, the moments and the shear force can be found as shown below:

$M_X = M_Y = P(0.1123 - 0.09284 \ln \gamma) = 300 \cdot (0.1123 + 0.09284 \cdot 2.513)$
$= 103.68\text{tm/m}$

$\gamma = c/L = 0.60/7.45 = 0.081$ $\ln 0.081 = \ln 8.1 \cdot 10^{-2} = 2.0919 - 4.6052 = -2.513$

$Q_X = Q_Y = -(P - p_0 c^2)/4c = -(300 - 1.04 \cdot 0.6^2)/4 \cdot 0.6 = -124.84\text{t/m}$

Example 7.3

Let us perform the same analysis when the plate is supported on an elastic layer assuming that the height of the soil layer is $H = 5\text{m}$ and all other data remain the same as in Example 7.2. Analysis is performed in the following steps:

1. Find the elastic characteristic of the plate $L = 7.45$.
2. Find parameter $\gamma = H/L = 5/7.45 = 0.67$.
3. Find parameter $\rho = r/L = 6/7.45 = 0.81$.

4. Find from Table 7.3 coefficient $\overline{w} = 0.086$ and the settlement of the plate:

$$w = \overline{w}\frac{\left(1 - v_0^2\right)}{E_0} \cdot \frac{P}{L} = 0.086 \cdot \frac{\left(1 - 0.3^2\right) \cdot 300}{1,000 \cdot 7.45} = 0.0032\text{m} = 0.32\text{cm}.$$

5. Find the soil pressure using Table 7.4:

$$p = \overline{p}\frac{P}{L^2} = 0.131 \cdot \frac{300}{7.45^2} = 0.71\text{t/m}^2 = 0.071\text{kg/cm}^2$$

6. Find parameter $\overline{U}$ using Table 7.5, $\overline{U} = 0.053$.
7. Find parameter $\overline{V}$ using Table 7.6, $\overline{V} = -0.013$.
8. Find $M_r = \left[\overline{V}(\rho) + v_1\overline{U}(\rho)\right]P = \left[-0.013 + 0.17 \cdot 0.053\right]300 = -1.12\text{tm/m}$.
9. Find $M_t = \left[\overline{U}(\rho) + v_1\overline{V}(\rho)\right]P = \left[0.053 - 0.17 \cdot 0.013\right]300 = 15.24\text{tm/m}$.
10. Find the shear force using Table 7.7:

$$Q_r = -\overline{Q}_r(\rho) \cdot P/L = -0.131 \cdot 300/7.45 = -5.28\text{t/m}.$$

11. Find the moments under the column (the radius of the column $a = 0.3$m):

$$\alpha = a/L = 0.3/7.45 = 0.04$$
$$M_0 = M_r = M_t = P\left(1 + v_1\right)\overline{U}(\alpha) = 300 \cdot \left(1 + 0.17\right) \cdot 0.264 = 92.66\text{tm/m}$$

Example 7.4

Let us perform analysis of the same plate assuming that the plate is supported on Winkler foundation. All data are taken from Example 7.3 except the height of the elastic layer $H = 2$m. Analysis is performed as follows:

1. Find the modulus of subgrade reaction using formula 7.34:

$$k = \frac{E_0}{\left(1 - v_0^2\right)H} = \frac{1,000}{\left(1 - 0.3^2\right)2} = 549.45\frac{\text{t}}{\text{m}^3}$$

2. Find the rigidity of the plate:

$$D = E_1 h^3/12\left(1 - v_1^2\right) = 2,650,000 \cdot 1.0^3/12\left(1 - 0.17^2\right) = 227,409.25\text{tm}.$$

3. Find the radius of the rigidity of the plate $C = \sqrt[4]{\dfrac{D}{K}} = \sqrt[4]{\dfrac{227,409.25}{549.45}} = 4.51$.
4. Find parameter $\rho = r/C = 6/4.51 = 1.33$.
5. Using Table 7.8 for $\rho = 1.33$ find the settlement of the plate:

$$w = \overline{w}\left(PC^2/D\right) = 0.063\left(300 \cdot 4.51^2/227,405.35\right) = 0.00169\text{m} = 0.17\text{cm}.$$

The soil pressure $p = kw = 549.45 \cdot 0.00169 = 0.93\text{t/m}^2 = 0.093\text{kg/cm}^2$.
   The moments and shear force are equal to:

$$M_r = \overline{M}_r P = -0.013 \cdot 300 = -3.9\text{tm/m}$$
$$M_t = \overline{M}_t P = 0.036 \cdot 300 = 10.8\text{tm/m}$$
$$Q_r = \overline{Q}_r P/C = -0.067 \cdot 300/4.51 = -4.46\text{t/m}$$

Now, we can find the moments under the center of the circular column with the radius $a = 0.3$m:

$$M_r = M_t = P(0.0578 - 0.09288\ln\alpha) = 300(0.0578 + 0.09288 \cdot 2.735) = 93.55\text{tm/m}$$

where

$$\alpha = \frac{a}{C} = \frac{0.3}{4.51} = 0.0665$$

$$\ln\alpha = \ln 0.0665 = \ln(6.65 \cdot 10^{-2}) = 1.87 - 4.60517 = -2.735$$

If the column has a square shape, with the length of the side, for example, $c = 0.6$m moments under the center of the column acting along the column sides are found as follows:

$$M_X = M_Y = P(0.1109 - 0.09288\ln\gamma) = 300(0.1109 + 0.09288 \cdot 2.0056) = 89.15\text{tm/m}$$

where

$$\gamma = c/C = 0.6/4.51 = 0.133$$
$$\ln\gamma = \ln 0.133 = \ln(13.3 \cdot 10^{-2}) = 2.6 - 4.6056 = -2.0056$$

As shown above, analysis of infinite plates can be performed using all three soil models: Winkler foundation, elastic half-space, and elastic layer. However, these methods of analyses cannot be recommended for final design, but allow obtaining the moment and shear forces in the central area of the foundation where the largest loads are applied and maximum stresses usually occur.

As shown above, analysis of infinite plates on Winkler foundation can also be used for simplified analysis of finite plates. Another approximate method for analysis of finite rectangular plates on elastic half-space is introduced using the analysis of infinite plates described above. The method was developed by Gorbunov-Posadov (1984).

7.6 Approximate Analysis of Rectangular Mat Foundations Supported on Elastic Half-Space

This method is developed for analysis of rectangular plates supported on elastic half-space with a series of concentrated vertical loads or uniformly distributed circular loads, as shown in Figure 7.6. If the shape of the load is different and looks, for example, as an ellipse, square, or rectangle, this load has to be replaced with an equivalent circular load.

This method was developed for plates which satisfy the following condition $b > 4L$, where b is the length of the smaller side of the plate and L is the elastic characteristic of the plate obtained from equation 7.12.

The loads applied to the plate should also satisfy the following condition $a \leq 0.6L$, where a is the radius of the circular load. The number of loads is unlimited. Analyses of the plate under each load are performed independently and the final result is obtained by superposition. Analysis can be performed for any point of the plate, for points located at the centers of the column loads as well as between the columns.

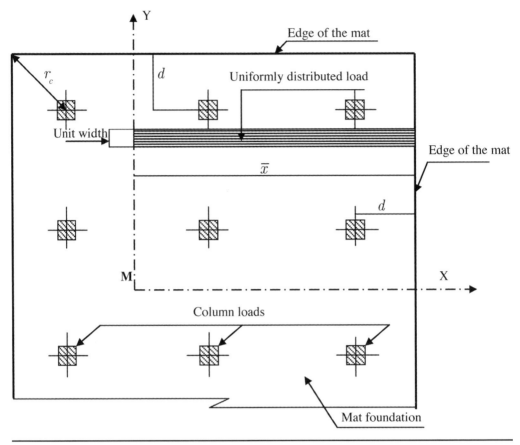

Figure 7.6

When the load is applied at a distance $r_C < L$ from the corner of the plate, analysis produces approximate but, nevertheless, acceptable practical results. It is important to also take into account the following:

1. If the load is applied at a distance equal to $d \geq 1.5L$ from the edge of the plate, Tables 7.9–7.13 are used for the analysis of infinite plates.
2. If the load is applied at distance $d < 1.5L$ from the edge of the plate, analysis is performed using Tables 7.14–7.20 for analysis of semi-infinite plates. Tables 7.9–7.13 and Table 7.2 have been developed for analysis of infinite plates supported on elastic half-space. However, Table 7.2 uses polar coordinates while Tables 7.9–7.13 use rectangular coordinates.

Tables are developed for infinite as well as semi-infinite plates and used for approximate analysis of finite rectangular plates with a series of vertical concentrated or circular loads. Analysis of infinite and semi-infinite rectangular plates supported on elastic half-space is explained on the next page.

Tables 7.9–7.13 are used for analysis of infinite plates with vertical concentrated loads. They allow obtaining the soil pressure, settlements, flexural moments, torsional moments, and shear forces at any point of the plate located at distance d from the edge of the plate when $d > 1.5L$. Tables contain coefficients $\overline{p}$, $\overline{w}$, $\overline{M}_x$, $\overline{M}_{xy}$, and $\overline{Q}$ for obtaining the soil pressure, settlements, flexural moments, torsional moments, and shear forces from equations 7.45–7.49.

Numerical values of these coefficients are functions of dimensionless coordinates $\xi = \dfrac{x}{L}$ and $\eta = \dfrac{y}{L}$, where x and y are the real coordinates. The origin of the coordinates ($\xi = 0$, $\eta = 0$) is located at the point where all numerical values are found.

$$p(\xi, \eta) = \overline{p}\frac{P_0}{L^2} \tag{7.45}$$

$$w(\xi, \eta) = \frac{\overline{w}\left(1 - v_0^2\right)P_0}{E_0 L} \tag{7.46}$$

$$M_X(\xi, \eta) = \overline{M}_x P_0 \tag{7.47}$$

$$M_{xy}(\xi, \eta) = \overline{M}_{xy}(1 - v)P_0 \tag{7.48}$$

$$Q_x(\xi, \eta) = \overline{Q}_x \frac{P_0}{L} \tag{7.49}$$

In equations 7.45–7.49, P_0 is the given concentrated load applied to the plate. If the distance between the loads applied to the plate and the origin of the coordinates exceeds those given in Tables 7.9–7.13, the influence of these loads is not taken into account. When values of η and ξ are negative the following rules should be observed:

1. For values of w, p, M_x, and M_y the algebraic sign remains unchanged.
2. For the value of Q_x the algebraic sign of the value taken from the tables is reversed when ξ is negative and η is positive or negative.
3. For the value of Q_y the algebraic sign of the value taken from the tables is reversed when η is negative and ξ is positive or negative.
4. For the value of M_{xy} the algebraic sign of the value taken from the tables remains unchanged if ξ and η have the same sign, but is reversed when they have different signs. As can be seen, tables do not contain coefficients for obtaining M_y, Q_y, and M_{yx} but they can be found using tables for M_x, Q_x, and M_{xy}. M_y is obtained by interchanging ξ and η in M_x. Q_y is obtained by interchanging ξ and η in Q_x, and $M_{yx} = -M_{xy}$.

If analysis is performed for a point of the plate located at the center of a column, in order to avoid infinitely large moments and shear forces produced by the same column, it has to be taken into account that the vertical load is distributed over the area of the column (or over the area of the column pedestal). The flexural moments and shear forces under a circular column with radius a can be found from the following equations:

$$M_x = M_Y = P_0(0.0592 - 0.09284\ln\alpha) \tag{7.50}$$

$$M_{xy} = -M_{yx} = 0.0332P_0 \tag{7.51}$$

$$Q_x = Q_y = \frac{P_0 - p_0 \pi a^2}{2\pi a} \approx \frac{P_0}{2\pi a} \qquad (7.52)$$

where $\alpha = a/L$.

For a square column with a side length equal to c, the same equations look as follows:

$$M_x = M_y = P_0 \left(0.1123 - 0.09284 \ln \gamma \right) \qquad (7.53)$$

$$M_{xy} = -M_{yx} = 0.0332 P_0 \qquad (7.54)$$

$$Q_x = Q_y = \frac{P_0 - p_0 c^2}{4c} \approx \frac{P_0}{4c} \qquad (7.55)$$

where $\gamma = \frac{c}{L}$ and p_0 is the soil pressure under the center of the column. In formulae 7.52 and 7.55, $p_0 \pi a^2$ and $p_0 c^2$ are usually small compared to P_0 and therefore can be ignored. If a plate is loaded with a series of vertical loads from the columns, all loads will affect the settlements, the soil pressure, moments, and shear forces at other points of the plate. Analysis can be performed for any point of the plate. If, for example, we want to find the soil pressure, settlement, moments, and shear forces at point M, as shown in Figure 7.8, the origin of the system of coordinates is located at point M and both axes, X and Y, are located parallel to the centerlines of the columns. For each column are found coordinates $\xi = \frac{x}{L}$ and $\eta = \frac{y}{L}$, where x and y are the actual coordinates in linear units. Using the coordinates ξ and η, coefficients $\overline{w}$, $\overline{p}$, $\overline{M}_x$, $\overline{M}_{xy}$, and $\overline{Q}_x$ are found. Using equations 7.45–7.49, actual settlements, soil pressures, moments, and shear forces due to each vertical load are found. Affect of the loads located at the distance $\delta < 1.5L$ from the edge of the plate also have to be taken into account.

Tables 7.9–7.13 are similar to Table 7.2, but Table 7.2 was developed for the analysis of infinite plates, while Tables 7.9–7.13 were developed only for the analysis of rectangular plates. From Table 7.2 we obtain settlements, radial and tangential moments, and shear forces using the distance-radius $\rho = r/L$, while analysis of rectangular plates is performed using rectangular coordinates parallel to the plate edges $\xi = \frac{x}{L}$ and $\eta = \frac{y}{L}$. Final results are obtained by simple superposition. If coordinates of some loads are not included in Tables 7.9–7.13, these loads are not taken into account. Coefficients for obtaining settlements, moments, and shear forces for points located at distance $\delta < 1.5L$ from the edge of the plate are obtained from Tables 7.14–7.20 for analysis of semi-infinite plates. However, equations 7.45–7.49 are used for infinite and semi-infinite plates as well.

It is also important to mention that vertical loads applied to the plates are not limited to loads distributed over circular and square areas. Any load of a different shape can be replaced with an equivalent distributed circular or square area load.

The analysis described above does not include results produced by the weight of the plate. Equations 7.56 and coefficients shown in Table 7.21 allow obtaining the soil pressure, flexural moments, and shear forces produced by the weight of the plate:

$$q_w(\xi) = \overline{q}_w p_w, \quad M_{xw}(\xi) = \overline{M}_{xw} L^2 p_w, \quad Q_{xw}(\xi) = \overline{Q}_{xw} L p_w \qquad (7.56)$$

where $p_w = \gamma_c h$, γ_c is the unit weight of the plate material, and h is the height of the plate.

7.7 Approximate Analysis of Semi-Infinite Rectangular Plates Supported on Elastic Half-Space

When the load or a row of loads is applied to the mat at a distance $\delta < 1.5L$ from the edge of the mat, analysis of the mat under each column is performed separately. The affect of other column loads is also taken into account and final results are obtained by simple superposition.

The origin of the coordinates where $X = Y = 0$ is located at the point of intersection of two lines: one perpendicular to the edge of the plate (X) and the second (Y) that coincides with the edge of the plate, as shown in Figures 7.7 and 7.8.

Tables 7.14–7.20 give numerical coefficients for obtaining the soil pressure, settlements, moments, and shear forces at points located close to the edge of semi-infinite plates. All coefficients depend on coordinates $\xi = x/L$, $\eta = y/L$, and parameter $\delta = d/L$. When choosing the direction of the axes X and Y we have to observe the following rules:

1. If the load is applied to the right from the free edge, the axis X is directed from left to right while the axis Y is directed downward in the plane of the plate, as shown in Figure 7.7.

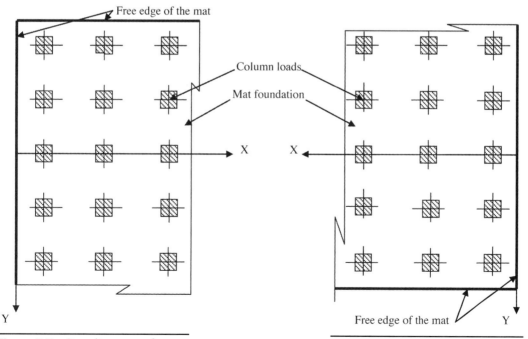

Figure 7.7 Coordinates configuration for left-hand edge region of a semi-infinite plate

Figure 7.8 Coordinates configuration for right-hand region of a semi-infinite plate

2. If the load is applied close to the right-hand edge of the plate, the origin of co-ordinates is located at the intersection of the axis X directed from right to left toward the interior of the plate, and axis Y is located along the edge of the plate and directed downward, as shown in Figure 7.8.

We have to remember that the tables are developed only for the left-hand edge, but they can be used for the right-hand end as well. In this case, the algebraic signs for Q_x and M_{xy} have to be reversed.

To determine the signs of the required quantities in the upper part of the plate, above axis X, the axis Y is directed upward and the signs of Q_y and M_{xy} are reversed. The values of w, p, M_x, and M_y do not depend on the directions of axes X and Y. Tables 7.14–7.20 were developed for various values of $\delta = d/L$ from 0 to 2.0. When $\delta < 0.9$, the values taken from the tables have to be rounded off to the closest value given in these tables. When $0.9 < \delta < 1.6$, two cases have to be distinguished:

1. When the plate is loaded with one single load. In this case the value of $\delta = d/L$ may be rounded off to the closest value given in the tables.
2. If the plate is loaded with a series of loads, the rounding-off value should not exceed the value of 0.10. If, after that, the obtained value cannot be used, linear interpolation between two values of $\delta = d/L$ is recommended.

Bending moments at the center of the loaded column is found from the following equations:

$$\left. \begin{aligned} M_x &= P_0\left(A - 0.0928\ln\alpha\right) \\ M_y &= P_0\left(B - 0.0928\ln\alpha\right) \end{aligned} \right\} \tag{7.57}$$

where P_0 is the concentrated vertical load applied to the column, $\alpha = a/L$, a is the radius of the column, and A and B are coefficients obtained from the curves shown in Figure 7.9.

Shear forces exactly under the center of the column are replaced with the shear forces per one linear unit of the column perimeter. They are obtained from equations 7.58 and used when $\delta = d/L > 0.4$. The first formula is used for circular columns; the second is used for square columns.

$$\left. \begin{aligned} Q_x &= Q_y = \frac{P_0 - p_0\pi a^2}{2\pi a} \approx \frac{P_0}{2\pi a} \\ Q_x &= Q_y = \frac{P_0 - p_0 c^2}{4c} \approx \frac{P_0}{4c} \end{aligned} \right\} \tag{7.58}$$

When $\delta = d/L < 0.4$, the same equations are used, but the shear forces in the direction perpendicular to the edge of the plate are multiplied by coefficients shown in Table 7.22. Coefficients are given for a load located in the vicinity of the left edge. For a column located in the vicinity of the right edge these coefficients should be reversed.

Example 7.5

Figure 7.10 shows a mat foundation supported on elastic half-space loaded with a system of regularly located columns. The mat has 35 column loads equal to 195,000kg.

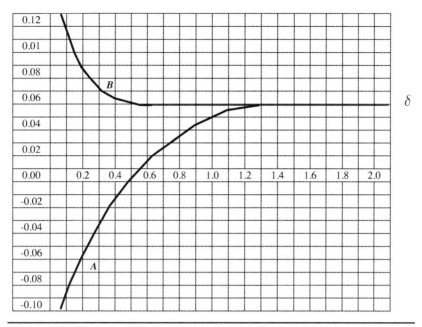

Figure 7.9   Analysis of infinite plates on elastic half-space

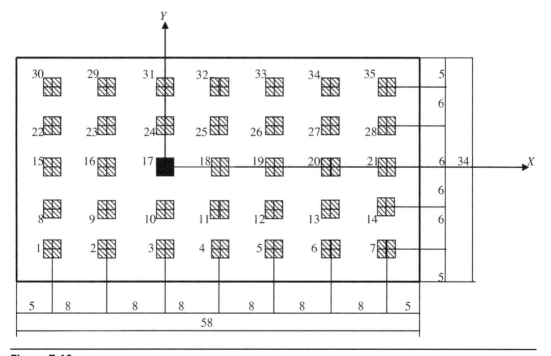

Figure 7.10

The height of the mat is $h = 1.2$m. All dimensions are given in meters. The properties of the mat are $E_1 = 180,000$kg/cm^2 and $v_1 = 0.167$.

The unit weight of the mat is equal to $\gamma = 2,400$kg/m^3. The soil parameters are $E_0 = 1,400$kg/cm^2 and $v_0 = 0.25$. Find the settlement, soil pressure, bending moments, torsional moments, and shear forces at point 17.

Solution:

Find parameter L:

$$L = \sqrt[3]{2D(1 - v_0^2)/E_0} = \sqrt[3]{\frac{2E_1 h^3 (1 - v_0^2)}{12(1 - v_1^2)E_0}} = \sqrt[3]{\frac{2 \cdot 180,000 \cdot 120^3 \cdot 0.9375}{12 \cdot 0.9722 \cdot 1,400}} = 3.293 \text{m}$$

Now, using equation 7.11 we can check parameter $\delta = \dfrac{d}{L}$ taking into account that the location of the column closest to the edge of the mat is equal to 5m. $\delta = \dfrac{d}{L} = \dfrac{5}{3.293} = 1.5184 > 1.5$, which means that the mat can be analyzed as an infinite one. As shown earlier, analysis of an infinite mat is performed using formulae 7.45–7.49. Let us obtain the settlement at point 17. The origin of the coordinates is located at the point shown in Figure 7.10. All values obtained at point 17 (settlements, soil pressure, moments, and shear forces) depend not only on the load applied to the mat at point 17, but also on loads applied at points located at the zone of influence of point 17. As seen from Figure 7.10, these loads are applied at points 9, 10, 11, 16, 18, 23, 24, and 25. In our analysis we take into account the load applied at point 17 and loads applied at points mentioned above. Loads applied at other points will not significantly affect the values found at point 17. Analysis is performed in Table 7.23. Coordinates of all loads as well as coordinates ζ and η are shown in the left side of the Table 7.23. Settlements, soil pressure, moments, and shear forces are estimated and shown in the right part of the table. Settlements are given in cm, soil pressure in kg/m^2, moments in kgm/m, and shear forces in kg/m. The first line of the table shows the estimated settlements, soil pressures, moments, and shear forces produced by the load applied at point 17. The second line shows the settlements, soil pressures, moments, and shear forces at point 17 produced by the load applied at point 9, and so on. Calculations for obtaining all numerical values in the second line are shown below. All numerical values are found using equations 7.45–7.49 and Tables 7.9–7.13.

$$w(\xi, \eta) = \frac{\overline{w}(1 - v_0^2)P_0}{E_0 L} = w_{17,9}(-2.42, -2.42) = \frac{0.106 \cdot (1 - 0.25^2) \cdot 190,000}{1,400 \cdot 329.3}$$

$$= 0.0414 \text{cm}$$

$$p(\xi, \eta) = \overline{p}\frac{P_0}{L^2} = p_{17,9}(-2.42, -2.42) = 0.003 \cdot \frac{190,000}{329.3^2} = 0.00526 \text{kg/cm}^2$$

$$= 52.6 \text{kg/m}^2$$

$$M_x(\xi,\eta) = \overline{M}_x P_0 = M_{x(17,9)}(-2.42, -2.42) = -0.004 \cdot 190,000 = -760 \text{kgm/m}$$
$$(\xi,\eta) = \overline{M}_{xy}(1-v)P_0 = M_{xy}(-2.42, -2.42) = -0.01 \cdot (1-0.1666) \cdot 190,000$$
$$= -1,583 \text{kgm/m}$$
$$Q_X(\xi,\eta) = \overline{Q}_x \frac{P_0}{L} = \overline{Q}_{x(17,9)}(-2.42, -2.42) = 0 \cdot \frac{190,000}{329.3} = 0.$$

The remainder of all numerical values is obtained analogously. By summing all results in Table 7.23, the total settlement, soil pressure, moments, and shear force are found. The same method is used for obtaining the final results at all other points of the mat. As mentioned earlier, the weight of the mat also produces additional stresses in the mat and soil pressure. Using equations 7.57 and Figure 7.9 we find:

$$M_{xy}q_w(\overline{\xi}) = \overline{q}_w p_w = \gamma h = 2,400 \cdot 1.2 \cdot 0.7 = 2,016 \text{kg/m}^2$$
$$M_{xw}(\overline{\xi}) = \overline{M}_{xw}L^2 p_w = 0.21 \cdot 3.3^2 \cdot 2,800 = 6,586 \text{kgm/m}$$
$$Q_{xw}(\overline{\xi}) = \overline{Q}_{xw}L p_w = 0 \cdot 3.3 \cdot 2800 = 0$$

Example 7.6

Let us perform analysis of a mat foundation $(20 \times 27)\text{m}^2$ supporting 12 columns, as shown in Figure 7.11. Loads applied to the foundation at points 6 and 7 are equal to

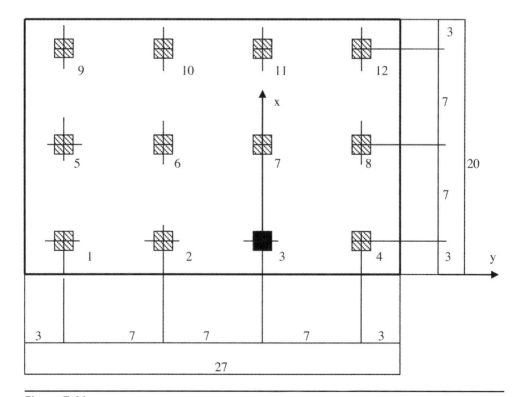

Figure 7.11

$P_6 = P_7 = 70t$. All other column loads acting along the perimeter of the foundation are equal to 45t each. The square column bases are equal to $(1.2 \times 1.2)\text{m}^2$; the height of the mat is equal to $h = 1.25\text{m}$.

The modulus of elasticity of the soil is equal to $E_0 = 600\text{kg/cm}^2$; Poisson's ratio of the soil is equal to $v_0 = 0.45$. The modulus of elasticity of the concrete is equal to $200,000\text{kg/cm}^2$; Poisson's ratio $v = 1/6$. Find the soil pressure under the center of column 3 and the flexural moment perpendicular to the nearer edge of the mat at the same point 3.

Solution:

1. The length parameter of the plate $L = 4.47\text{m}$.
2. For the first row of columns 1, 2, 3, and 4, we have $\delta = 3/4.47 = 0.67 < 1.5$, which means that analysis due to the loads from these columns is performed as a semi-infinite mat.
3. Check if $a \leq 0.6L$, where a is the radius of the base of the column. Since the column is square the equivalent loaded circular area is equal to $a = \sqrt{1.2^2/3.141}$ $= 0.68\text{m} < 0.6 \cdot 4.47 = 2.68\text{m}$, which means that this region of the mat should be treated as a semi-finite region. For columns 5, 6, 7, and 8 located at the distance 10m from the same edge of the mat, $\delta = d/L = 10/4.47 = 2.24 > 1.5$. That means all of these columns are located in the area that should be treated as an infinite region of the mat. Column loads at points 9, 10, 11, and 12 are located at distance 17m from the same edge, and $\delta = d/L = 3.8 > 1.5$. They will influence the soil pressure and moments at point 3 as loads applied to the infinite zone of the mat.
4. The origin of the coordinates is located as shown in Figure 7.11. Using Tables 7.9–7.13 for infinite plates and Tables 7.14–7.20 for semi-infinite plates, coefficients $\overline{p}$ and $\overline{M}_x$ are found with exception of the moment $\overline{M}_y$ at point 3. This moment in accordance with Table 7.17 is equal to ∞. In order to obtain a realistic result, the moment is found from equation 7.57:

$$\left[\overline{M}_x\right]_3 = (A - 0.09284\ln\alpha') = 0.02 - 0.09284 \cdot \ln 0.15$$
$$= 0.02 - 0.09284 \cdot (-1.897) = 0.196$$

$A = 0.0118$ is obtained from Figure 7.9 and $\alpha' = \dfrac{a}{L} = \dfrac{0.68}{4.47} = 0.15$. From Table 7.23 the total contact stress at point 3 is found:

$$p_3 = \frac{45,000}{L^2}(0.10 + 0.21 + 0.10 + 0.002 + 0.018) + \frac{70,000}{L^2}(0.018 + 0.041)$$
$$= 1,175\text{kg/m}^2$$

Notice that only columns located close to point 3 produce significant stresses. The total moment $(M_X)_3$ acting in the X direction at point 3 shown below is obtained from Table 7.24.

$$\left(M_X\right)_3 = 45,000\big(0.025 + 0.196 + 0.025 + 0.001 - 0.002 - 0.008 - 0.012 - 0.008\big) +$$
$$- 70,000\big(0.002 + 0.014\big) = 8,645\text{kgm/m}$$

In conclusion, we have to mention that some methods of simplified analysis of mats on an elastic foundation are not included in this chapter. For example, Baker's method (1948) is recommended by Scott (1981). A detailed analysis of various simplified, analytical, and numerical methods of analysis of mats on elastic foundation is given by Selvadurai (1979). Some methods of simplified analysis are also discussed by Bowles (1977).

Table 7.1 Functions Z_3 Z_4 Z_3' Z_4'

$\dfrac{r}{L}$	Z_3	Z_4	Z_3'	Z_4'	$\dfrac{r}{L}$	Z_3	Z_4	Z_3'	Z_4'
0.00	0.5000	$-\infty$	0.0000	$+\infty$	0.30	0.4667	−0.8513	−0.1746	2.0498
0.01	0.4999	−3.0056	−0.0166	63.6595	0.31	0.4650	−0.8312	−0.1773	1.9791
0.02	0.4997	−2.5643	−0.0288	31.8260	0.32	0.4632	−0.8117	−0.1798	1.9127
0.03	0.4993	−2.3063	−0.0394	21.2132	0.33	0.4614	−0.7929	−0.1823	1.8502
0.04	0.4989	−2.1232	−0.0488	15.9055	0.34	0.4595	−0.7747	−0.1846	1.7912
0.05	0.4984	−1.9813	−0.0575	12.7199	0.35	0.4577	−0.7571	−0.1869	1.7355
0.06	0.4978	−1.8653	−0.0655	10.5954	0.36	0.4558	−0.7400	−0.1891	1.6828
0.07	0.4971	−1.7674	−0.0730	9.0771	0.37	0.4539	−0.7234	−0.1912	1.6329
0.08	0.4963	−1.6825	−0.0800	7.9378	0.38	0.4520	−0.7073	−0.1932	1.5854
0.09	0.4955	−1.6078	−0.0866	7.0512	0.39	0.4500	−0.6917	−0.1952	1.5403
0.10	0.4946	−1.5409	−0.0929	6.3413	0.40	0.4480	−0.6765	−0.1970	1.4974
0.11	0.4936	−1.4805	−0.0989	5.7601	0.41	0.4461	−0.6617	−0.1988	1.4564
0.12	0.4926	−1.4254	−0.1046	5.2754	0.42	0.4441	−0.6474	−0.2006	1.4174
0.13	0.4915	−1.3748	−0.1100	4.8648	0.43	0.4421	−0.6334	−0.2022	1.3800
0.14	0.4904	−1.3279	−0.1152	4.5126	0.44	0.4400	−0.6198	−0.2038	1.3443
0.15	0.4892	−1.2843	−0.1201	4.2071	0.45	0.4380	−0.6065	−0.2054	1.3101
0.16	0.4880	−1.2436	−0.1248	3.9394	0.46	0.4359	−0.5935	−0.2068	1.2773
0.17	04867	−1.2054	−0.1294	3.7029	0.47	0.4338	−0.5809	−0.2082	1.2458
0.18	0.4854	−1.1695	−0.1337	3.4925	0.48	0.4318	−0.5686	−0.2096	1.2156
0.19	0.4840	−1.1355	−0.1379	3.3040	0.49	0.4297	−0.5566	−0.2100	1.1865
0.20	0.4826	−1.1034	−0.1419	3.1340	0.50	0.4275	−0.5449	−0.2121	1.1585
0.21	0.4812	−1.0728	−0.1458	2.9801	0.51	0.4254	−0.5334	−0.2133	1.1316
0.22	0.4797	−1.0437	−0.1495	2.8400	0.52	0.4233	−0.5222	−0.2144	1.1056
0.23	0.4782	−1.0160	−0.1531	2.7118	0.53	0.4211	−0.5113	−0.2155	1.0806
0.24	0.4767	−0.9894	−0.1565	2.5941	0.54	0.4190	−0.5006	−0.2165	1.0564
0.25	0.4751	−0.9640	−0.1598	2.4857	0.55	0.4168	−0.4902	−0.2175	1.0330
0.26	0.4735	−0.9397	−0.1630	2.3854	0.56	0.4146	−0.4800	−0.2184	1.0105
0.27	0.4718	−0.9163	−0.1661	2.2924	0.57	0.4124	−0.4700	−0.2193	0.9887
0.28	0.4701	−0.8938	−0.1690	2.2060	0.58	0.4102	−0.4602	−0.2201	0.9675
0.29	0.4684	−0.8722	−0.1719	2.1253	0.59	0.4080	−0.4506	−0.2209	0.9471

Table 7.1 Con't

$\dfrac{r}{L}$	z_3	z_4	z_3'	z_4'	$\dfrac{r}{L}$	z_3	z_4	z_3'	z_4'
0.60	0.4058	−0.4413	−0.2216	0.9273	1.00	0.3151	−0.1825	−0.2243	0.4422
0.61	0.4036	−0.4321	−0.2224	0.9080	1.02	0.3106	−0.1738	−0.2234	0.4273
0.62	0.4014	−0.4231	−0.2230	0.8894	1.04	0.3062	−0.1654	−0.2225	0.4130
0.63	0.3991	−0.4143	−0.2236	0.8713	1.06	0.3018	−0.1573	−0.2215	0.3992
0.64	0.3969	−0.4057	−0.2242	0.8538	1.08	0.2973	−0.1495	−0.2204	0.3859
0.65	0.3946	−0.3972	−0.2247	0.8367	1.10	0.2929	−0.1419	−0.2193	0.3730
0.66	0.3924	−0.3889	−0.2252	0.8201	1.12	0.2886	−0.1345	−0.2182	0.3606
0.67	0.3902	−0.3808	−0.2257	0.8040	1.14	0.2842	−0.1274	−0.2169	0.3486
0.68	0.3879	−0.3728	−0.2261	0.7883	1.16	0.2799	−0.1206	−0.2156	0.3370
0.69	0.3856	−0.3650	−0.2265	0.7730	1.18	0.2756	−0.1140	−0.2143	0.3258
0.70	0.3834	−0.3574	−0.2268	0.7582	1.20	0.2713	−0.1076	−0.2129	0.3149
0.71	0.3811	−0.3499	−0.2272	0.7437	1.22	0.2671	−0.1014	−0.2115	0.3044
0.72	0.3788	−0.3425	−0.2274	0.7296	1.24	0.2628	−0.0954	−0.2100	0.2942
0.73	0.3765	−0.3353	−0.2277	0.7159	1.26	0.2587	−0.0896	−0.2086	0.2844
0.74	0.3743	−0.3282	−0.2279	0.7021	1.28	0.2545	−0.0840	−0.2070	0.2748
0.75	0.3720	−0.3212	−0.2281	0.6894	1.30	0.2504	−0.0786	−0.2054	0.2656
0.76	0.3697	−0.3144	−0.2282	0.6766	1.32	0.2463	−0.0734	−0.2038	0.2567
0.77	0.3674	−0.3077	−0.2284	0.6642	1.34	0.2422	−0.0683	−0.2022	0.2480
0.78	0.3651	−0.3011	−0.2285	0.6520	1.36	0.2382	−0.0634	−0.2005	0.2396
0.79	0.3628	−0.2947	−0.2285	0.6402	1.38	0.2342	−0.0587	−0.1988	0.2314
0.80	0.3606	−2883	−0.2286	0.6286	1.40	0.2302	−0.0542	−0.1971	0.2235
0.81	0.3583	−0.2821	−0.2286	0.6172	1.42	0.2263	−0.0498	−0.1954	0.2158
0.82	0.3560	−0.2760	−0.2286	0.6062	1.44	0.2224	−0.0456	−0.1936	0.2084
0.83	0.3537	−0.2700	−0.2285	0.5953	1.46	0.2186	−0.0416	−0.1918	0.2011
0.84	0.3514	−0.2641	−0.2285	0.5847	1.48	0.2148	−0.0375	−0.1900	0.1941
0.85	0.3491	−0.2583	−0.2284	0.5744	1.50	0.2110	−0.0337	−0.1882	0.1873
0.86	0.3468	−0.2526	−0.2283	0.5642	1.52	0.2072	−0.0300	−0.1864	0.1807
0.87	0.3446	−0.2470	−0.2281	0.5543	1.54	0.2035	−0.0265	−0.1845	0.1742
0.88	0.3423	−0.2415	−0.2280	0.5446	1.56	0.1999	−0.0230	−0.1826	0.1680
0.89	0.3400	−0.2361	−0.2278	0.5351	1.58	0.1962	−0.0198	−0.1807	0.1619
0.90	0.3377	−0.2308	−0.2276	0.5258	1.60	0.1926	−0.0166	−0.1788	0.1560
0.91	0.3355	−0.2256	−0.2273	0.5166	1.62	0.1891	−0.0135	−0.1769	0.1503
0.92	0.3332	−0.2204	−0.2271	0.5077	1.64	0.1855	−0.0106	−0.1750	0.1448
0.93	0.3309	−0.2154	−0.2268	0.4989	1.66	0.1821	−0.0077	−0.1731	0.1394
0.94	0.3286	−0.2105	−0.2265	0.4904	1.68	0.1786	−0.0050	−0.1711	0.1341
0.95	0.3264	−0.2056	−0.2262	0.4819	1.70	0.1752	−0.0024	−0.1692	0.1290
0.96	0.3241	−0.2008	−0.2258	0.4737	1.72	0.1718	0.0002	−0.1672	0.1241
0.97	0.3219	−0.1961	−0.2255	0.4656	1.74	0.1685	0.0026	−0.1653	0.1193
0.98	0.3196	−0.1915	−0.2251	0.4576	1.76	0.1652	0.0050	−0.1634	0.1146
0.99	0.3174	−0.1870	−0.2247	0.4498	1.78	0.1620	0.0072	−0.1614	0.1100

Table 7.1 Con't

$\frac{r}{L}$	Z_3	Z_4	Z_3'	Z_4'	$\frac{r}{L}$	Z_3	Z_4	Z_3'	Z_4'
1.80	0.1588	0.0094	−0.1594	0.1056	3.30	0.0176	0.0376	−0.0418	−0.0195
1.82	0.1556	0.0114	−0.1575	0.1014	3.40	0.0137	0.0356	−0.0369	−0.0204
1.84	0.1525	0.0134	−0.1555	0.0972	3.50	0.0102	0.0335	−0.0325	−0.0210
1.86	0.1494	0.0153	−0.1536	0.0931	3.60	0.0072	0.0314	−0.0284	−0.0213
1.88	0.1463	0.0171	−0.1516	0.0892	3.70	0.0045	0.0293	−0.0246	−0.0213
1.90	0.1433	0.0189	−0.1496	0.0854	3.80	0.0022	0.0272	−0.0212	−0.0210
1.92	0.1404	0.0206	−0.1477	0.0817	3.90	0.0003	0.0251	−0.0180	−0.0206
1.94	0.1374	0.0222	−0.1458	0.0781	4.00	−0.0014	0.0230	−0.0152	−0.0200
1.96	0.1345	0.0237	−0.1438	0.0746	4.10	−0.0028	0.0211	−0.0127	−0.0193
1.98	0.1317	0.0251	−0.1419	0.0712	4.20	−0.0039	0.0192	−0.0104	−0.0185
2.00	0.1289	0.0265	−0.1399	0.0679	4.30	−0.0049	0.0174	−0.0083	−0.0177
1.92	0.1404	0.0206	−0.1477	0.0817	4.40	−0.0056	0.0156	−0.0065	−0.0168
1.94	0.1374	0.0222	−0.1458	0.0781	4.50	−0.0062	0.0140	−0.0049	−0.0158
1.96	0.1345	0.0237	−0.1438	0.0746	4.60	−0.0066	0.0125	−0.0035	−0.0148
1.98	0.1317	0.0251	−0.1419	0.0712	4.70	−0.0069	0.0110	−0.0023	−0.0138
2.00	0.1289	0.0265	−0.1399	0.0679	4.80	−0.0071	0.0097	−0.0012	−0.0129
2.10	0.1153	0.0325	−0.1304	0.0527	4.90	−0.0071	0.0085	−0.0003	−0.0119
2.20	0.1028	0.0371	−0.1210	0.0397	5.00	−0.0071	0.0073	0.0005	−0.0109
2.30	0.0911	0.0405	−0.1120	0.0285	5.10	−0.0070	0.0063	0.0012	−0.0100
2.40	0.0804	0.0420	−0.1032	0.0189	5.20	−0.0069	0.0053	0.0017	−0.0091
2.50	0.0705	0.0444	−0.0948	0.0108	5.30	−0.0067	0.0045	0.0022	−0.0083
2.60	0.0614	0.0451	−0.0868	0.0039	5.40	−0.0065	0.0037	0.0025	−0.0075
2.70	0.0531	0.0452	−0.0791	−0.0018	5.50	−0.0062	0.0029	0.0028	−0.0067
2.80	0.0455	0.0447	−0.0719	−0.0066	5.60	−0.0059	0.0023	0.0030	−0.0060
2.90	0.0387	0.0439	−0.0650	−0.0105	5.70	−0.0056	0.0017	0.0032	−0.0053
3.00	0.0326	0.0427	−0.0586	−0.0137	5.80	−0.0053	0.0012	0.0033	−0.0047
3.10	0.0270	0.0412	−0.0526	−0.0162	5.90	−0.0049	0.0008	0.0033	−0.0041
3.20	0.0220	0.0394	−0.0470	−0.0180	6.00	−0.0046	0.0004	0.0033	−0.0036

Table 7.2 Analysis of plates supported on elastic half-space loaded with a vertical load equal to one unit

ρ	$\overline{\rho}$	$\overline{w}$	$\overline{Q}_r$	$\overline{M}_r$	$\overline{M}_t$	$\overline{M}_1 = -\dfrac{1}{\rho}\dfrac{d\overline{w}}{d\rho}$	$\overline{M}_2 = -\dfrac{d^2\overline{w}}{d\rho^2}$
0.00	0.192	0.385	∞	∞	∞	∞	∞
0.05	0.185	0.384	3.178	0.256	0.332	0.287	0.208
0.10	0.177	0.382	1.582	0.191	0.258	0.232	0.153
0.20	0.162	0.377	0.779	0.129	0.195	0.178	0.099
0.30	0.149	0.369	0.506	0.093	0.158	0.147	0.068
0.40	0.136	0.359	0.367	0.068	0.132	0.124	0.047
0.60	0.114	0.338	0.224	0.037	0.097	0.093	0.021
0.80	0.094	0.314	0.150	0.016	0.074	0.073	0.004
1.00	0.077	0.291	0.105	0.004	0.057	0.058	−0.006
1.20	0.063	0.268	0.074	−0.005	0.045	0.047	−0.013
1.40	0.051	0.247	0.053	−0.011	0.035	0.038	−0.017
1.60	0.041	0.226	0.038	−0.014	0.028	0.031	−0.019
1.80	0.033	0.207	0.027	−0.015	0.022	0.025	−0.019
2.00	0.026	0.189	0.019	−0.016	0.018	0.021	−0.020
2.20	0.020	0.173	0.013	−0.016	0.014	0.017	−0.019
2.40	0.015	0.159	0.008	−0.016	0.011	0.014	−0.018
2.60	0.012	0.146	0.005	−0.015	0.009	0.012	−0.017
2.80	0.009	0.135	0.002	−0.014	0.007	0.010	−0.016
3.00	0.007	0.124	0.000	−0.013	0.006	0.008	−0.014
3.20	0.005	0.115	0.000	−0.012	0.005	0.007	−0.013
3.40	0.003	0.107	−0.001	−0.011	0.004	0.006	−0.012
3.60	0.002	0.099	−0.002	−0.010	0.003	0.005	−0.011
3.80	0.002	0.093	−0.002	−0.008	0.002	0.004	−0.009
4.00	0.001	0.087	−0.002	−0.007	0.002	0.003	−0.008

Table 7.3 Analysis of plates supported on elastic layer, coefficients $\overline{\omega}$

ρ	γ								
	0.3	**0.4**	**0.6**	**0.8**	**1.0**	**1.5**	**2.0**	**3.0**	**∞**
0	0.097	0.112	0.136	0.157	0.175	0.210	0.237	0.272	0.385
0.05	0.096	0.111	0.135	0.156	0.174	0.209	0.236	0.271	0.384
0.10	0.094	0.109	0.134	0.154	0.172	0.207	0.234	0.269	0.382
0.20	0.090	0.105	0.129	0.150	0.167	0.202	0.229	0.264	0.377
0.30	0.084	0.099	0.123	0.143	0.160	0.195	0.221	0.256	0.369
0.40	0.077	0.091	0.114	0.134	0.151	0.186	0.212	0.246	0.359
0.60	0.063	0.076	0.098	0.117	0.133	0.166	0.192	0.225	0.338
0.80	0.048	0.060	0.080	0.098	0.113	0.145	0.170	0.202	0.314
1.00	0.037	0.047	0.065	0.081	0.095	0.125	0.149	0.180	0.291
1.20	0.026	0.035	0.051	0.065	0.078	0.106	0.128	0.159	0.268
1.40	0.018	0.026	0.040	0.052	0.064	0.088	0.109	0.138	0.247
1.60	0.012	0.018	0.029	0.040	0.050	0.073	0.092	0.120	0.226
1.80	0.007	0.012	0.021	0.030	0.039	0.061	0.077	0.103	0.207
2.00	0.003	0.007	0.014	0.021	0.028	0.049	0.062	0.087	0.189
2.20	0.001	0.003	0.008	0.015	0.021	0.037	0.050	0.073	0.173
2.40	0.000	0.001	0.004	0.010	0.014	0.028	0.039	0.060	0.159
2.60	−0.001	0.000	0.001	0.006	0.010	0.020	0.031	0.050	0.146
2.80	−0.001	−0.001	0.000	0.004	0.006	0.014	0.024	0.041	0.135
3.00	−0.001	−0.001	0.000	0.001	0.003	0.010	0.018	0.033	0.124
3,20	−0.001	−0.001	−0.001	0.000	0.001	0.007	0.013	0.026	0.115
3.40	−0.001	−0.001	−0.001	−0.001	0.000	0.005	0.009	0.019	0.107
3.60	−0.000	−0.001	−0.002	−0.002	−0.001	0.003	0.005	0.014	0.099
3.80	−0.000	−0.001	−0.002	−0.002	−0.001	0.001	0.003	0.011	0.093
4.00	−0.000	−0.001	−0.002	−0.002	−0.002	−0.001	0.001	0.009	0.087

Table 7.4 Soil pressure $\bar{p}$

ρ	γ				
	0.3	0.4	0.6	0.8	1.0
0	0.338	0.299	0.257	0.235	0.221
0.05	0.330	0.292	0.250	0.228	0.214
0.1	0.321	0.283	0.241	0.220	0.206
0.2	0.300	0.265	0.225	0.204	0.191
0.3	0.279	0.247	0.210	0.190	0.177
0.4	0.256	0.227	0.194	0.175	0.163
0.6	0.209	0.189	0.165	0.149	0.139
0.8	0.163	0.152	0.135	0.124	0.116
1.0	0.123	0.118	0.108	0.101	0.095
1.2	0.090	0.089	0.085	0.081	0.078
1.4	0.063	0.065	0.065	0.064	0.062
1.6	0.041	0.045	0.048	0.049	0.049
1.8	0.025	0.030	0.035	0.037	0.038
2.0	0.014	0.018	0.024	0.027	0.028
2.2	0.006	0.010	0.015	0.018	0.020
2.4	0.000	0.003	0.009	0.011	0.013
2.6	−0.003	0.000	0.004	0.007	0.009
2.8	−0.004	−0.002	0.001	0.003	0.006
3.0	−0.004	−0.003	−0.001	0.001	0.003
3.2	−0.004	−0.003	−0.002	−0.001	0.001
3.4	−0.004	−0.004	0.003	−0.002	−0.001
3.6	−0.003	−0.003	−0.004	−0.003	−0.002
3.8	−0.002	−0.002	−0.003	−0.003	−0.002
4.0	−0.002	−0.002	−0.003	−0.003	−0.002

Table 7.5 Coefficients $\overline{U}$

ρ	γ				
	0.3	0.4	0.6	0.8	1.0
0	∞	∞	∞	∞	∞
0.05	0.249	0.255	0.262	0.268	0.171
0.1	0.194	0.200	0.208	0.213	0.216
0.2	0.141	0.146	0.154	0.159	0.163
0.3	0.110	0.115	0.123	0.128	0.132
0.4	0.088	0.093	0.100	0.105	0.109
0.6	0.058	0.063	0.070	0.075	0.078
0.8	0.040	0.045	0.051	0.055	0.059
1.0	0.028	0.032	0.037	0.041	0.044
1.2	0.019	0.023	0.028	0.031	0.034
1.4	0.013	0.016	0.020	0.023	0.026
1.6	0.009	0.011	0.014	0.017	0.019
1.8	0.006	0.007	0.010	0.012	0.014
2.0	0.004	0.005	0.007	0.009	0.011
2.2	0.002	0.003	0.004	0.006	0.008
2.4	0.001	0.002	0.003	0.004	0.005
2.6	0.001	0.001	0.003	0.003	0.004
2.8	0.000	0.000	0.002	0.002	0.003
3.0	0.000	0.000	0.001	0.001	0.002
3.2	0.000	0.000	0.001	0.001	0.002
3.4	0.000	0.000	0.001	0.001	0.002
3.6	0.000	0.000	0.000	0.001	0.001
3.8	−0.001	−0.001	0.000	0.000	0.001
4.0	−0.001	−0.001	−0.001	0.000	0.000

Table 7.6 Functions $\overline{V}$

ρ	γ				
	0.3	0.4	0.6	0.8	1.0
0	∞	∞	∞	∞	∞
0.05	0.170	0.176	0.183	0.189	0.192
0.10	0.115	0.121	0.129	0.134	0.137
0.20	0.062	0.068	0.075	0.080	0.084
0.30	0.032	0.038	0.045	0.049	0.053
0.40	0.013	0.018	0.025	0.029	0.032
0.60	−0.008	−0.005	0.000	0.004	0.007
0.80	−0.020	−0.017	−0.014	−0.011	−0.010
1.00	−0.024	−0.023	−0.021	−0.018	−0.017
1.20	−0.025	−0.025	−0.025	−0.023	−0.022
1.40	−0.023	−0.024	−0.025	−0.025	−0.024
1.60	−0.020	−0.022	−0.023	−0.024	−0.024
1.80	−0.016	−0.018	−0.021	−0.022	−0.022
2.00	−0.014	−0.016	−0.019	−0.021	−0.022
2.20	−0.010	−0.013	−0.015	−0.018	−0.019
2.40	−0.008	−0.010	−0.013	−0.016	−0.017
2.60	−0.006	−0.008	−0.011	−0.014	−0.015
2.80	−0.004	−0.006	−0.009	−0.011	−0.013
3.00	−0.002	−0.004	−0.006	−0.008	−0.010
3.20	−0.002	−0.003	−0.005	−0.007	−0.009
3.40	−0.001	−0.002	−0.003	−0.006	−0.008
3.60	−0.001	−0.001	−0.002	−0.005	−0.006
3.80	0.000	0.000	−0.001	−0.003	−0.004
4.00	0.000	0.000	0.000	−0.002	−0.003

Table 7.7 Coefficients $\overline{Q}_r$

ρ	γ				
	0.3	0.4	0.6	0.8	1.0
0.00	∞	∞	∞	∞	∞
0.05	3.174	3.175	3.176	3.177	3.177
0.10	1.575	1.577	1.579	1.580	1.581
0.20	0.785	0.789	0.793	0.795	0.796
0.30	0.486	0.491	0.496	0.500	0.502
0.40	0.341	0.347	0.355	0.359	0.361
0.60	0.189	0.197	0.207	0.212	0.216
0.80	0.109	0.118	0.129	0.136	0.140
1.00	0.062	0.071	0.082	0.089	0.093
1.20	0.032	0.039	0.050	0.057	0.061
1.40	0.013	0.019	0.029	0.035	0.039
1.60	0.002	0.007	0.015	0.020	0.025
1.80	−0.005	−0.001	0.006	0.011	0.014
2.00	−0.007	−0.004	0.000	0.004	0.007
2.20	−0.009	−0.006	−0.004	−0.001	0.001
2.40	−0.009	−0.008	−0.006	−0.004	−0.003
2.60	−0.007	−0.008	−0.007	−0.005	−0.005
2.80	−0.007	−0.008	−0.007	−0.007	−0.006
3.00	−0.006	−0.007	−0.007	−0.007	−0.007
3.20	−0.004	−0.005	−0.005	−0.005	−0.006
3.40	−0.003	−0.004	−0.005	−0.005	−0.006
3.60	−0.002	−0.003	−0.004	−0.005	−0.006
3.80	−0.001	−0.002	−0.003	−0.004	−0.005
4.00	0.000	−0.001	−0.002	−0.003	−0.004

Table 7.8 Infinite plates on Winkler foundation

ρ	$\overline{w}$	$\overline{M}_r$	$\overline{M}_t$	Q	ρ	$\overline{w}$	$\overline{M}_r$	$\overline{M}_t$	Q
0.01	0.125	0.405	0.472	15.915	1.80	0.040	−0.021	0.018	0.026
0.02	0.125	0.341	0.407	7.957	2.00	0.032	−0.021	0.013	0.017
0.03	0.125	0.276	0.343	3.976	2.20	0.026	−0.021	0.010	0.010
0.06	0.124	0.239	0.305	2.649	2.40	0.020	−0.020	0.007	0.005
0.08	0.124	0.212	0.278	1.984	2.60	0.015	−0.018	0.005	0.001
0.10	0.124	0.192	0.258	1.585	2.80	0.011	−0.017	0.003	−0.002
0.20	0.121	0.128	0.194	0.784	3.00	0.008	−0.015	0.002	−0.003
0.40	0.112	0.066	0.131	0.374	3.20	0.006	−0.013	0.001	−0.005
0.60	0.101	0.033	0.095	0.231	3.40	0.003	−0.011	0.001	−0.005
0.80	0.090	0.013	0.072	0.157	3.60	0.002	−0.009	0.000	−0.005
1.00	0.079	−0.001	0.054	0.110	3.80	0.001	−0.008	0.000	−0.005
1.20	0.068	−0.010	0.041	0.079	4.00	0.000	−0.007	0.000	−0.005
1.40	0.058	−0.016	0.032	0.056	5.00	−0.002	−0.002	0.000	−0.003
1.60	0.048	−0.019	0.024	0.039					

Table 7.9 Analysis of infinite plates on elastic half-space, coefficients $\bar{p}$

η	ξ 0.0	0.2	0.4	0.6	0.8	1.0	1.2	1.4	1.6	1.8	
0.0	0.192	0.162	0.136	0.114	0.094	0.077	0.063	0.051	**0.041**	0.033	
0.2	0.162	0.152	0.131	0.111	0.092	0.076	0.062	0.050	0.041	0.033	
0.4	0.136	0.131	0.117	0.102	0.086	0.072	0.059	0.048	0.039	0,031	
0.6	0.114	0.111	0.102	0.089	0.077	0.065	0.055	0.045	0.037	0.029	
0.8	0.094	0.092	0.086	0.077	0.069	0.058	0.049	0.041	0.033	0.027	
1.0	0.077	0.076	0.072	0.065	0.058	0.050	0.043	0.036	0.029	0.024	
1.2	0.063	0.062	0.059	0.055	0.049	0.043	0.037	0.031	0.026	0.021	
1.4	0.061	0.050	0.048	0.045	0.041	0.036	0.031	0.027	0.022	0.017	
1.6	0.041	0.041	0.039	0.037	0.033	0.029	0.026	0.022	**0.018**	0.015	
1.8	0.033	0.033	0.031	0.029	0.027	0.024	0.021	0.017	0.015	0.013	
2.0	0.026	0.026	0.025	0.023	0.021	0.019	0.016	0.014	0.012	0.010	
2.2	0.020	0.020	0.019	0.018	0.016	0.015	0.013	0.012	0.010	0.009	
2.4	0.015	0.015	0.015	0.014	0.013	0.012	0.010	0.009	0.008	0.007	
2.6	0.012	0.012	0.012	0.010	0.010	0.009	0.008	0.008	0.005	0.005	
2.8	0.009	0.009	0.009	0.008	0.008	0.007	0.007	0.006	0.005	0.004	
3.0	0.007	0.007	0.007	0.006	0.006	0.005	0.005	0.004	0.003	0.002	
3.2	0.005	0.005	0.005	0.004	0.004	0.004	0.003	0.002	**0.002**	0.002	
3.4	0.003	0.003	0.003	0.002	0.002	0.002	0.002	0.002	0.002	0.002	
3.6	0.002	0.002	0.002	0.002	0.002	0.002	0.002	0.002	0.002	0.001	
3.8	0.002	0.002	0.002	0.002	0.002	0.002	0.001	0.001			
4.0	0.001	0.001	0.001	0.001							

η	ξ 2.0	2.2	2.4	2.6	2.8	3.0	3.2	3.4	3.6	3.8	4.0
0.0	0.026	0.020	0.015	0.012	0.009	0.007	0.005	0.003	0.002	0.002	0.001
0.2	0.026	0.02	0.015	0.012	0.009	0.007	0.005	0.003	0.002	0.002	0.001
0.4	0.025	0.019	0.015	0.012	0.009	0.007	0.005	0.003	0.002	0.002	0.001
0.6	0.023	0.018	0.014	0.010	0.008	0.006	0.004	0.002	0.002	0.002	0.001
0.8	0.021	0.016	0.013	0.010	0.008	0.006	0.004	0.002	0.002	0.002	0.001
1.0	0.019	0.015	0.012	0.009	0.007	0.006	0.003	0.002	0.002	0.002	
1.2	0.016	0.013	0.010	0.008	0.007	0.005	0.003	0.002	0.002	0.001	
1.4	0.014	0.012	0.009	0.008	0.006	0.004	0.002	0.02	0.002	0.001	
1.6	0.012	0.010	0.008	0.007	0.005	0.003	0.002	0.002	0.002		
1.8	0.010	0.009	0.007	0.005	0.004	0.002	0.002	0.002	0.001		
2.0	0.009	0.007	0.006	0.004	0.003	0.002	0.002	0.002			
2.2	0.007	0.006	0.004	0.003	0.002	0.002	0.002	0.001			
2.4	0.006	0.004	0.003	0.002	0.002	0.002	0.001				
2.6	0.004	0.003	0.002	0.002	0.002	0.001					
2.8	0.003	0.002	0.002	0.002	0.001						
3.0	0.002	0.002	0.002	0.001							
3.2	0.002	0.002	0.001								
3.4	0.002	0.001									
3.6											
3.8											
4.0											

Table 7.10 Coefficients $\overline{\omega}$

η	ξ										
	0.0	**0.2**	**0.4**	**0.6**	**0.8**	**1.0**	**1.2**	**1.4**	**1.6**	**1.8**	
0	0.385	0.377	0.359	0.338	0.314	0.291	0.268	0.247	0.226	0.207	
0.2	0.377	0.371	0.354	0.334	0.312	0.289	0.266	0.246	0.225	0.206	
0.4	0.359	0.354	0.341	0.324	0.304	0.281	0.261	0.241	0.221	0.203	
0.6	0.338	0.334	0.324	0.308	0.291	0.271	0.253	0.234	0.215	0.198	
0.8	0.314	0.312	0.304	0.291	0.276	0.259	0.243	0.225	0.208	0.192	
1.0	0.291	0.289	0.281	0.271	0.259	0.246	0.232	0.214	0.199	0.184	
1.2	0.268	0.266	0.261	0.258	0.243	0.232	0.216	0.203	0.189	0.176	
1.4	0.247	0.246	0.241	0.234	0.225	0.214	0.203	0.191	0.179	0.167	
1.6	0.226	0.225	0.221	0.215	0.208	0.199	0.189	0.179	0.169	0.158	
1.8	0.207	0.206	0.203	0.198	0.192	0.184	0.176	0.167	0.158	0.150	
2.0	0.189	0.188	0.186	0.182	0.177	0.170	0.164	0.157	0.149	0.141	
2.2	0.173	0.172	0.170	0.167	0.163	0.158	0.152	0.145	0.139	0.133	
2.4	0.159	0.158	0.157	0.155	0.151	0.146	0.141	0.136	0.130	0.125	
2.6	0.146	0.145	0.144	0.142	0.139	0.136	0.132	0.127	0.122	0.117	
2.8	0.135	0.134	0.133	0.131	0.129	0.126	0.122	0.118	0.114	0.110	
3.0	0.124	0.124	0.123	0.121	0.119	0.117	0.114	0.105	0.107	0.103	
3.2	0.115	0.115	0.114	0.113	0.111	0.109	0.106	0.103	0.100	0.097	
3.4	0.107	0.107	0.106	0.105	0.103	0.101	0.099	0.097	0.094	0.091	
3.6	0.099	0.099	0.098	0.097	0.096	0.095	0.093	0.091	0.089	0.086	
3.8	0.093	0.093	0.092	0.091	0.090	0.089	0.087	0.086	0.084	0.082	
4.0	0.087	0.087	0.086	0.085	0.085	0.084	0.082	0.081	0.079	0.077	

η	ξ										
	2.0	**2.2**	**2.4**	**2.6**	**2.8**	**3.0**	**3.2**	**3.4**	**3.6**	**3.8**	**4.0**
0	0.189	0.173	0.159	0.146	0.135	0.124	0.115	0.107	0.099	0.093	0.087
0.2	0.188	0.172	0.158	0.145	0.134	0.124	0.115	0.107	0.099	0.093	0.087
0.4	0.186	0.170	0.157	0.144	0.133	0.123	0.114	0.106	0.098	0.092	0.086
0.6	0.182	0.167	0.155	0.142	0.131	0.121	0.113	0.105	0.097	0.091	0.085
0.8	0.177	0.163	0.151	0.139	0.129	0.119	0.111	0.103	0.096	0.090	0.085
1.0	0.170	0.158	0.146	0.136	0.126	0.117	0.109	0.101	0.095	0.089	0.084
1.2	0.164	0.152	0.141	0.132	0.122	0.114	0.106	0.099	0.093	0.087	0.082
1.4	0.157	0.145	0.136	0.127	0.118	0.110	0.103	0.097	0.091	0.086	0.081
1.6	0.149	0.139	0.130	0.122	0.114	0.107	0.100	0.094	0.089	0.084	0.079
1.8	0.141	0.133	0.125	0.117	0.110	0.103	0.097	0.091	0.086	0.082	0.077
2.0	0.133	0.126	0.119	0.112	0.106	0.099	0.094	0.089	0.084	0.079	0.075
2.2	0.126	0.119	0.113	0.107	0.101	0.096	0.091	0.086	0.081	0.077	0.074
2.4	0.118	0.112	0.106	0.101	0.096	0.092	0.087	0.083	0.079	0.075	0.072
2.6	0.113	0.107	0.102	0.097	0.092	0.088	0.084	0.080	0.076	0.073	0.070
2.8	0.106	0.101	0.096	0.092	0.088	0.084	0.080	0.077	0.074	0.071	0.068
3.0	0.099	0.096	0.092	0.088	0.084	0.081	0.077	0.075	0.071	0.068	0.066
3.2	0.094	0.091	0.087	0.084	0.080	0.077	0.074	0.071	0.069	0.066	0.062
3.4	0.089	0.086	0.083	0.080	0.077	0.075	0.071	0.070	0.067	0.062	0.060
3.6	0.084	0.081	0.079	0.076	0.074	0.071	0.069	0.067	0.062	0.060	0.059
3.8	0.079	0.077	0.075	0.073	0.071	0.068	0.066	0.062	0.060	0.059	0.057
4.0	0.075	0.074	0.072	0.070	0.068	0.066	0.062	0.060	0.059	0.57	0.056

Table 7.11 Coefficients $\overline{M}_x$

η	ξ									
	0.0	0.2	0.4	0.6	0.8	1.0	1.2	1.4	1.6	1.8
0	∞	0.129	0.068	0.036	0.016	0.004	−0.005	−0.011	−0.014	−0.015
0.2	0.194	0.133	0.072	0.039	0.018	0.005	−0.003	−0.010	−0.013	−0.014
0.4	0.132	0.110	0.071	0.042	0.021	0.006	−0.002	−0.008	−0.011	−0.013
0.6	0.096	0.087	0.063	0.042	0.023	0.009	0.001	−0.006	−0.009	−0.012
0.8	0.074	0.068	0.054	0.039	0.023	0.011	0.002	−0.004	−0.007	−0.010
1.0	0.057	0.054	0.044	0.034	0.022	0.012	0.004	−0.001	−0.005	−0.008
1.2	0.045	0.042	0.037	0.028	0.020	0.011	0.006	0.000	−0.004	−0.007
1.4	0.035	0.034	0.030	0.023	0.017	0.011	0.006	0.001	−0.003	−0.006
1.6	0.028	0.027	0.030	0.020	0.015	0.010	0.005	0.002	−0.002	−0.004
1.8	0.022	0.021	0.019	0.017	0.013	0.008	0.005	0.002	−0.001	−0.002
2.0	0.018	0.018	0.015	0.014	0.011	0.007	0.004	0.003	0.000	−0.002
2.2	0.014	0.014	0.012	0.011	0.009	0.006	0.004	0.003	0.000	−0.001
2.4	0.011	0.11	0.010	0.009	0.008	0.005	0.003	0.002	0.000	−0.001
2.6	0.000	0.009	0.009	0.008	0.007	0.005	0.003	0.002	0.000	0.000
2.8	0.007	0.007	0.006	0.006	0.005	0.003	0.003	0.002	0.001	0.000
3.0	0.006	0.006	0.006	0.005	0.004	0.003	0.002	0.002	0.001	0.000
3.2	0.005	0.005	0.005	0.004	0.003	0.002	0.002	0.002	0.001	0.000
3.4	0.004	0.004	0.004	0.004	0.003	0.002	0.002	0.002	0.001	0.000
3.6	0.003	0.003	0.003	0.003	0.002	0.002	0.002	0.001	0.001	0.000
3.8	0.003	0.003	0.003	0.003	0.002	0.002	0.002	0.001		
4.0	0.002	0.002	0.002	0.002						

η	ξ										
	2.0	2.2	2.4	2.6	2.8	3.0	3.2	3.4	3.6	3.8	4.0
0	−0.017	−0.016	−0.016	−0.015	−0.014	−0.013	−0.012	−0.011	−0.010	−0.008	−0.008
0.2	−0.017	−0.016	−0.016	−0.015	−0.014	−0.013	−0.012	−0.011	−0.010	−0.008	−0.008
0.4	−0.015	−0.015	−0.015	−0.014	−0.014	−0.013	−0.012	−0.011	−0.010	−0.008	−0.007
0.6	−0.013	−0.015	−0.014	−0.014	−0.013	−0.012	−0.011	−0.011	−0.010	−0.008	−0.007
0.8	−0.012	−0.013	−0.013	−0.013	−0.012	−0.011	−0.011	−0.010	−0.009	−0.007	
1.0	−0.011	−0.012	−0.012	−0.012	−0.011	−0.010	−0.009	−0.008	−0.008	−0.007	
1.2	−0.009	−0.010	−0.010	−0.010	−0.010	−0.009	−0.009	−0.008	−0.008	−0.007	
1.4	−0.007	−0.008	−0.009	−0.009	−0.010	−0.009	−0.009	−0.008	−0.008	−0.007	
1.6	−0.006	−0.007	−0.008	−0.008	−0.008	−0.008	−0.008	−0.007	−0.007		
1.8	−0.004	−0.006	−0.006	−0.007	−0.007	−0.007	−0.006	−0.006	−0.006		
2.0	−0.003	−0.004	−0.005	−0.006	−0.006	−0.006	−0.006	−0.006			
2.2	−0.003	−0.003	−0.004	−0.004	−0.005	−0.005	−0.005	−0.005			
2.4	−0.002	−0.003	−0.004	−0.004	−0.004	−0.004	−0.004				
2.6	−0.001	−0.002	−0.003	−0.003	−0.003						
2.8	0.001	−0.001	−0.003	−0.003							
3.0	0.000	−0.001	−0.003								
3.2	0.000	−0.001									
3.4	0.000										
3.6											
3.8											
4.0											

Table 7.12 Coefficients $\overline{M}_{xy}$

η	ξ									
	0.0	0.2	0.4	0.6	0.8	1.0	1.2	1.4	1.6	1.8
0.0	0	0	0	0	0	0	0	0	0	0
0.2	0	−0.040	−0.030	−0.022	−0.016	−0.012	−0.010	−0.008	−0.006	−0.005
0.4	0	−0.030	−0.037	−0.032	−0.027	−0.021	−0.017	−0.014	−0.011	−0.009
0.6	0	−0.022	−0.032	−0.034	−0.031	−0.027	−0.022	−0.018	−0.015	−0.013
0.8	0	−0.016	−0.027	−0.031	−0.030	−0.028	−0.024	−0.021	−0.018	−0.015
1.0	0	−0.012	−0.021	−0.027	−0.028	−0.027	−0.025	−0.022	−0.019	−0.016
1.2	0	−0.010	−0.017	−0.022	−0.024	−0.025	−0.024	−0.021	−0.019	−0.017
1.4	0	−0.008	−0.014	−0.018	−0.021	−0.022	−0.021	−0.021	−0.018	−0.017
1.6	0	−0.006	−0.011	−0.015	−0.018	−0.019	−0.019	−0.018	−0.018	−0.016
1.8	0	−0.005	−0.009	−0.013	−0.015	−0.016	−0.017	−0.017	−0.016	−0.015
2.0	0	−0.004	−0.008	−0.010	−0.013	−0.014	−0.015	−0.014	−0.014	−0.014
2.2	0	−0.003	−0.006	−0.009	−0.011	−0.012	−0.013	−0.013	−0.013	−0.013
2.4	0	−0.003	−0.005	−0.007	−0.009	−0.010	−0.011	−0.011	−0.011	−0.011
2.6	0	−0.002	−0.005	−0.006	−0.008	−0.009	−0.009	−0.009	−0.009	−0.009
2.8	0	−0.002	−0.004	−0.005	−0.006	−0.007	−0.008	−0.008	−0.009	−0.009
3.0	0	−0.001	−0.004	−0.004	−0.005	−0.005	−0.007	−0.008	−0.008	−0.008
3.2	0	−0.001	−0.003	−0.004	−0.005	−0.005	−0.006	−0.007	−0.007	−0.006
3.4	0	−0.001	−0.002	−0.003	−0.004	−0.005	−0.005	−0.005	−0.005	−0.005
3.6	0	−0.001	−0.002	−0.002	−0.003	−0.004	−0.004	−0.004	−0.005	−0.005
3.8	0	−0.001	−0.002	−0.002	−0.002	−0.003	−0.003	−0.004		
4.0	0	−0.001	−0.001	−0.002						

η	ξ										
	2.0	2.2	2.4	2.6	2.8	3.0	3.2	3.4	3.6	3.8	4.0
0.0	0	0	0	0	0	0	0	0	0	0	0
0.2	−0.004	−0.003	−0.003	−0.002	−0.002	−0.001	−0.001	−0.001	−0.001	−0.001	−0.001
0.4	−0.008	−0.006	−0.005	−0.004	−0.004	−0.003	−0.002	−0.002	−0.002	−0.001	−0.001
0.6	−0.010	−0.009	−0.007	−0.006	−0.005	−0.004	−0.004	−0.003	−0.002	−0.002	−0.002
0.8	−0.013	−0.011	−0.009	−0.008	−0.006	−0.005	−0.005	−0.004	−0.003	−0.002	
1.0	−0.014	−0.012	−0.010	−0.009	−0.007	−0.006	−0.005	−0.005	−0.004	−0.003	
1.2	−0.015	−0.013	−0.011	−0.009	−0.008	−0.007	−0.006	−0.005	−0.004	−0.003	
1.4	−0.014	−0.013	−0.011	−0.009	−0.009	−0.008	−0.007	−0.005	−0.004	−0.004	
1.6	−0.014	−0.013	−0.011	−0.009	−0.009	−0.008	−0.007	−0.005	−0.005		
1.8	−0.014	−0.013	−0.011	−0.009	−0.009	−0.008	−0.006	−0.005	−0.004		
2.0	−0.013	−0.011	−0.010	−0.010	−0.008	−0.007	−0.006	−0.005			
2.2	−0.013	−0.011	−0.010	−0.010	−0.008	−0.007	−0.006	−0.005			
2.4	−0.011	−0.011	−0.010	−0.009	−0.008	−0.007	−0.006	−0.005			
2.6	−0.010	−0.010	−0.009	−0.009	−0.007	−0.006	−0.005				
2.8	−0.010	−0.009	−0.009	−0.008	−0.007	−0.005					
3.0	−0.008	−0.008	−0.007	−0.007	−0.006						
3.2	−0.007	−0.007	−0.006	−0.005							
3.4	−0.006	−0.006	−0.005								
3.6	−0.005	−0.005									
3.8											
4.0											

Table 7.13 Coefficients $\overline{Q}_x$

η	0.0	0.2	0.4	0.6	0.8	1.0	1.2	1.4	1.6	1.8	2.0	2.2	2.4	2.6	2.8
								ξ							
0.0	∞	0.779	0.0367	0.224	0.150	0.105	0.074	0.053	0.038	0.027	0.019	0.013	0.008	0.005	0.002
0.2	0	0.385	0.294	0.203	0.141	0.100	0.071	0.052	0.037	0.027	0.019	0.013	0.008	0.005	0.002
0.4	0	0.132	0.172	0.150	0.117	0.086	0.065	0.047	0.034	0.024	0.018	0.012	0.008	0.005	0.002
0.6	0	0.064	0.100	0.098	0.085	0.068	0.053	0.041	0.030	0.022	0.015	0.011	0.007	0.004	0.001
0.8	0	0.035	0.058	0.063	0.060	0.052	0.040	0.032	0.025	0.018	0.013	0.008	0.006	0.003	0.001
1.0	0	0.020	0.034	0.041	0.041	0.037	0.032	0.025	0.019	0.015	0.011	0.007	0.005	0.002	0.000
1.2	0	0.012	0.022	0.026	0.027	0.026	0.023	0.019	0.015	0.012	0.009	0.005	0.004	0.001	
1.4	0	0.007	0.013	0.017	0.018	0.018	0.016	0.014	0.011	0.009	0.006	0.004	0.002	0.000	
1.6	0	0.005	0.008	0.011	0.013	0.012	0.011	0.010	0.008	0.006	0.005	0.002	0.001		
1.8	0	0.003	0.005	0.007	0.008	0.008	0.008	0.007	0.005	0.004	0.003	0.001	0.000		
2.0	0	0.002	0.004	0.005	0.005	0.006	0.005	0.005	0.004	0.003	0.002	0.000			
2.2	0	0.001	0.002	0.003	0.003	0.003	0.003	0.003	0.002	0.001	0.000				
2.4	0	0.001	0.001	0.002	0.002	0.002	0.002	0.001	0.001	0.000					
2.6	0	0.000	0.001	0.001	0.001	0.001	0.000	0.000							
2.8	0	0.000	0.000	0.000	0.000	0.000									

Table 7.14 Analysis of semi-infinite plates on elastic half-space, coefficients $\bar{p}$

δ	η	0.0	0.1	0.2	0.4	0.6	0.8	1.0	1.2	1.6	2.0	2.4	2.8	3.2	3.6	4.0
																ξ
0	0.0	1.06	0.81	0.63	0.39	0.25	0.17	0.11	0.06	0.00	-0.03					
	0.2	1.05	0.80	0.62	0.38	0.25	0.16	0.11	0.06	0.00	-0.03					
	0.4	1.01	0.77	0.59	0.36	0.23	0.15	0.09	0.05	-0.01	-0.03					
	0.6	0.96	0.71	0.54	0.32	0.20	0.12	0.07	0.03	-0.02	-0.03					
	0.8	0.88	0.65	0.48	0.27	0.16	0.09	0.05	0.01	-0.03	-0.03					
	1.0	0.79	0.56	0.42	0.22	0.12	0.06	0.02	-0.01	-0.04	-0.03					
	1.2	0.70	0.50	0.35	0.16	0.07	0.02	-0.01	-0.03	-0.05	-0.03					
	1.6	0.50	0.32	0.19	0.06	0.00	0.00	-0.01	-0.02	-0.02	-0.02					
	2.0	0.30	0.15	0.05	-0.03	-0.04	-0.02	-0.01	-0.01	-0.01	-0.01					
0.1	0.0	0.98	0.75	0.58	0.37	0.25	0.18	0.12	0.06	0.02	0.01					
	0.2	0.97	0.74	0.57	0.36	0.25	0.18	0.12	0.06	0.02	0.01					
	0.4	0.95	0.72	0.55	0.33	0.23	0.15	0.10	0.05	0.01	0.01					
	0.6	0.89	0.66	0.50	0.30	0.20	0.12	0.06	0.04	0.01	0.01					
	0.8	0.82	0.60	0.45	0.26	0.16	0.09	0.05	0.04	0.01	0.01					
	1.0	0.74	0.52	0.39	0.22	0.12	0.07	0.05	0.04	0.01	0.01					
	1.2	0.65	0.47	0.32	0.18	0.09	0.06	0.04	0.04	0.00	0.00					
	1.6	0.47	0.30	0.18	0.11	0.05	0.05	0.03	0.03	0.00	-0.01					
	2.0	0.28	0.14	0.07	0.04	0.03	0.03	0.03	0.02	0.00	-0.02					
0.2	0.0	0.91	0.70	0.53	0.34	0.25	0.18	0.13	0.07	0.03	0.01	0.01				
	0.2	0.90	0.69	0.52	0.34	0.25	0.18	0.13	0.07	0.03	0.01	0.01				
	0.4	0.88	0.66	0.51	0.31	0.23	0.15	0.11	0.06	0.03	0.01	0.01				
	0.6	0.82	0.61	0.47	0.29	0.20	0.13	0.08	0.06	0.03	0.01	0.01				
	0.8	0.76	0.56	0.41	0.25	0.16	0.09	0.07	0.05	0.02	0.01	0.01				
	1.0	0.69	0.48	0.36	0.21	0.12	0.08	0.07	0.05	0.02	0.01	0.01				
	1.2	0.61	0.44	0.30	0.17	0.10	0.08	0.06	0.05	0.02	0.01	0.01				
	1.6	0.44	0.28	0.17	0.12	0.07	0.07	0.05	0.04	0.02	0.01	0.01				
	2.0	0.27	0.13	0.08	0.06	0.05	0.05	0.05	0.04	0.02	0.01	0.00				

Table 7.14 Con't

δ	η	0.0	0.1	0.2	0.4	0.6	0.8	1.0	1.2	1.6	2.0	2.4	2.8	3.2	3.6	4.0
	0.0	0.76	0.58	0.45	0.30	0.23	0.18	0.14	0.11	0.06	0.03	0.02				
	0.2	0.76	0.58	0.45	0.30	0.23	0.18	0.13	0.11	0.06	0.03	0.02				
	0.4	0.74	0.55	0.43	0.27	0.22	0.17	0.12	0.10	0.05	0.02	0.02				
	0.6	0.70	0.52	0.39	0.25	0.20	0.14	0.11	0.09	0.05	0.02	0.02				
0.4	0.8	0.64	0.47	0.34	0.22	0.15	0.11	0.10	0.08	0.04	0.02	0.02				
	1.0	0.58	0.41	0.30	0.19	0.12	0.10	0.09	0.07	0.04	0.02	0.01				
	1.2	0.51	0.37	0.25	0.15	0.11	0.10	0.08	0.07	0.04	0.02	0.01				
	1.6	0.37	0.25	0.15	0.13	0.09	0.09	0.08	0.06	0.03	0.01	0.01				
	2.0	0.23	0.12	0.10	0.09	0.08	0.08	0.07	0.06	0.03	0.01	0.00				
	0.0	0.62	0.47	0.36	0.25	0.21	0.17	0.15	0.13	0.09	0.06	0.04	0.03			
	0.2	0.62	0.47	0.36	0.25	0.21	0.16	0.14	0.12	0.08	0.06	0.04	0.03			
	0.4	0.60	0.44	0.34	0.23	0.20	0.15	0.14	0.12	0.08	0.05	0.03	0.03			
	0.6	0.57	0.42	0.32	0.21	0.18	0.14	0.12	0.11	0.08	0.05	0.03	0.03			
0.6	0.8	0.51	0.39	0.28	0.19	0.14	0.12	0.11	0.10	0.07	0.04	0.03	0.03			
	1.0	0.47	0.34	0.24	0.17	0.12	0.11	0.10	0.09	0.06	0.03	0.02	0.02			
	1.2	0.42	0.31	0.21	0.14	0.11	0.10	0.09	0.08	0.05	0.03	0.02	0.01			
	1.6	0.31	0.21	0.15	0.13	0.11	0.10	0.08	0.07	0.05	0.02	0.01	0.00			
	2.0	0.20	0.10	0.10	0.10	0.10	0.09	0.08	0.07	0.04	0.02	0.00	-0.01			
	0.0	0.48	0.36	0.28	0.21	0.18	0.17	0.16	0.14	0.11	0.07	0.05	0.03			
	0.2	0.48	0.36	0.28	0.20	0.18	0.17	0.16	0.14	0.11	0.07	0.05	0.03			
	0.4	0.47	0.34	0.27	0.19	0.17	0.16	0.15	0.13	0.10	0.07	0.04	0.03			
	0.6	0.44	0.33	0.26	0.18	0.16	0.15	0.13	0.12	0.10	0.06	0.03	0.03			
0.8	0.8	0.40	0.31	0.22	0.15	0.14	0.13	0.12	0.11	0.08	0.06	0.03	0.02			
	1.0	0.37	0.26	0.19	0.14	0.12	0.11	0.10	0.10	0.07	0.05	0.02	0.02			
	1.2	0.33	0.25	0.16	0.12	0.11	0.10	0.09	0.08	0.06	0.04	0.02	0.02			
	1.6	0.25	0.18	0.11	0.11	0.10	0.10	0.09	0.08	0.05	0.03	0.02	0.01			
	2.0	0.17	0.09	0.10	0.10	0.09	0.09	0.08	0.07	0.05	0.03	0.01	0.00			

ξ

1.2	0.0	0.24	0.19	0.15	0.12	0.13	0.14	0.16	0.17	0.16	0.12	0.07	0.03	0.03		
	0.2	0.24	0.19	0.15	0.12	0.12	0.14	0.15	0.16	0.15	0.11	0.07	0.03	0.03		
	0.4	0.23	0.18	0.14	0.11	0.11	0.13	0.14	0.15	0.14	0.10	0.06	0.03	0.03		
	0.6	0.22	0.17	0.13	0.09	0.10	0.11	0.12	0.13	0.13	0.09	0.05	0.03	0.03		
	0.8	0.21	0.16	0.11	0.08	0.08	0.09	0.10	0.11	0.10	0.08	0.05	0.03	0.03		
	1.0	0.18	0.14	0.10	0.06	0.06	0.07	0.08	0.09	0.08	0.06	0.04	0.03	0.03		
	1.2	0.17	0.13	0.09	0.05	0.04	0.05	0.06	0.07	0.07	0.05	0.03	0.02	0.01		
	1.6	0.15	0.11	0.07	0.05	0.04	0.05	0.05	0.06	0.05	0.04	0.03	0.01	0.00		
	2.0	0.11	0.06	0.06	0.05	0.04	0.05	0.05	0.05	0.04	0.03	0.02	0.01	0.00		
1.6	0.0	0.10	0.09	0.09	0.09	0.10	0.13	0.14	0.16	0.18	0.16	0.12	0.07	0.04	0.04	0.03
	0.2	0.10	0.09	0.09	0.09	0.10	0.12	0.14	0.16	0.18	0.16	0.12	0.07	0.04	0.04	0.03
	0.4	0.10	0.09	0.08	0.08	0.10	0.11	0.13	0.15	0.16	0.15	0.11	0.07	0.04	0.04	0.03
	0.6	0.09	0.08	0.07	0.07	0.08	0.10	0.12	0.13	0.14	0.13	0.09	0.06	0.04	0.04	0.03
	0.8	0.08	0.07	0.07	0.06	0.07	0.08	0.10	0.11	0.12	0.11	0.08	0.05	0.04	0.04	0.03
	1.0	0.08	0.06	0.06	0.05	0.06	0.07	0.08	0.09	0.10	0.08	0.07	0.04	0.03	0.04	0.02
	1.2	0.07	0.06	0.05	0.05	0.05	0.05	0.06	0.07	0.07	0.06	0.05	0.03	0.03	0.03	0.02
	1.6	0.06	0.06	0.05	0.04	0.04	0.04	0.05	0.05	0.05	0.04	0.03	0.02	0.02	0.02	0.02
	2.0	0.05	0.05	0.05	0.04	0.04	0.04	0.04	0.04	0.05	0.04	0.03	0.01	0.01	0.01	0.01
2.0	0.0	0.07	0.07	0.07	0.06	0.07	0.08	0.09	0.11	0.15	0.17	0.15	0.11	0.07	0.05	0.05
	0.2	0.07	0.07	0.07	0.06	0.06	0.07	0.09	0.11	0.15	0.17	0.15	0.11	0.07	0.05	0.05
	0.4	0.07	0.07	0.06	0.06	0.06	0.07	0.09	0.10	0.14	0.15	0.14	0.10	0.07	0.05	0.05
	0.6	0.06	0.06	0.06	0.06	0.06	0.07	0.08	0.09	0.12	0.13	0.12	0.09	0.06	0.04	0.04
	0.8	0.05	0.06	0.06	0.06	0.06	0.06	0.07	0.08	0.10	0.11	0.10	0.08	0.06	0.04	0.04
	1.0	0.05	0.05	0.05	0.05	0.05	0.05	0.06	0.07	0.08	0.09	0.08	0.07	0.05	0.03	0.03
	1.2	0.04	0.04	0.04	0.05	0.05	0.05	0.05	0.06	0.07	0.07	0.07	0.06	0.04	0.03	0.02
	1.6	0.04	0.04	0.04	0.04	0.05	0.04	0.04	0.05	0.05	0.05	0.05	0.05	0.03	0.02	0.01
	2.0	0.04	0.04	0.03	0.03	0.04	0.04	0.04	0.04	0.04	0.05	0.04	0.03	0.02	0.01	0.01

Table 7.15 Coefficients $\bar{\omega}$

δ	η	0.0	0.2	0.4	0.6	0.8	1.0	1.2	1.6	2.0	2.4	2.8	3.2	3.6	4.0
								ξ							
0.0	0.0	0.66	0.57	0.49	0.42	0.36	0.30	0.27	0.20	0.16	0.15	0.13	0.12	0.10	0.09
	0.2	0.64	0.55	0.48	0.41	0.35	0.30	0.26	0.20	0.16	0.14	0.13	0.12	0.10	0.09
	0.4	0.59	0.52	0.45	0.39	0.34	0.29	0.25	0.19	0.16	0.14	0.12	0.12	0.10	0.09
	0.6	0.53	0.47	0.41	0.36	0.31	0.27	0.24	0.18	0.15	0.13	0.12	0.12	0.10	0.09
	0.8	0.47	0.41	0.37	0.32	0.28	0.25	0.22	0.17	0.14	0.12	0.12	0.11	0.09	0.09
	1.0	0.40	0.36	0.32	0.29	0.25	0.22	0.20	0.16	0.13	0.12	0.11	0.11	0.09	0.08
	1.2	0.33	0.31	0.28	0.25	0.22	0.20	0.17	0.14	0.11	0.11	0.11	0.10	0.09	0.08
	1.6	0.22	0.20	0.19	0.18	0.16	0.14	0.13	0.11	0.10	0.10	0.10	0.09	0.08	0.07
	2.0	0.14	0.13	0.13	0.12	0.11	0.10	0.10	0.10	0.09	0.09	0.08	0.09	0.08	0.07
0.1	0.0	0.61	0.54	0.48	0.42	0.36	0.31	0.28	0.22	0.18	0.16	0.14	0.12	0.10	0.09
	0.2	0.59	0.53	0.47	0.41	0.35	0.30	0.28	0.22	0.18	0.16	0.14	0.12	0.10	0.09
	0.4	0.56	0.51	0.46	0.40	0.35	0.30	0.27	0.21	0.18	0.15	0.13	0.12	0.10	0.09
	0.6	0.50	0.48	0.43	0.39	0.34	0.30	0.26	0.20	0.17	0.15	0.12	0.12	0.10	0.09
	0.8	0.44	0.41	0.39	0.34	0.30	0.27	0.24	0.19	0.16	0.14	0.12	0.11	0.09	0.09
	1.0	0.38	0.37	0.34	0.32	0.28	0.25	0.22	0.18	0.15	0.13	0.11	0.11	0.09	0.08
	1.2	0.32	0.31	0.30	0.28	0.25	0.22	0.19	0.16	0.13	0.12	0.11	0.10	0.08	0.08
	1.6	0.21	0.22	0.21	0.20	0.19	0.17	0.15	0.14	0.12	0.11	0.10	0.10	0.08	0.07
	2.0	0.14	0.15	0.15	0.14	0.13	0.13	0.12	0.12	0.11	0.10	0.09	0.09	0.08	0.07
0.2	0.0	0.57	0.52	0.48	0.43	0.37	0.32	0.30	0.24	0.20	0.17	0.15	0.12	0.11	0.09
	0.2	0.55	0.51	0.47	0.42	0.36	0.32	0.30	0.24	0.20	0.17	0.15	0.12	0.11	0.09
	0.4	0.52	0.50	0.46	0.41	0.36	0.31	0.28	0.22	0.19	0.16	0.13	0.12	0.11	0.09
	0.6	0.47	0.48	0.44	0.40	0.35	0.30	0.27	0.21	0.18	0.16	0.13	0.11	0.10	0.08
	0.8	0.41	0.41	0.40	0.36	0.32	0.29	0.25	0.20	0.17	0.15	0.12	0.11	0.10	0.08
	1.0	0.36	0.37	0.35	0.33	0.30	0.27	0.23	0.19	0.16	0.14	0.12	0.11	0.10	0.08
	1.2	0.30	0.32	0.31	0.30	0.27	0.24	0.21	0.18	0.15	0.14	0.12	0.10	0.09	0.08
	1.6	0.21	0.23	0.22	0.22	0.21	0.19	0.17	0.16	0.14	0.13	0.12	0.10	0.09	0.08
	2.0	0.14	0.15	0.15	0.15	0.14	0.14	0.14	0.14	0.13	012	0.11	0.10	0.09	0.08

0.4	0.0	0.49	0.48	0.48	0.46	0.42	0.38	0.35	0.29	0.24	0.20	0.17	0.14	0.12	0.10
	0.2	0.47	0.47	0.47	0.44	0.41	0.38	0.35	0.29	0.24	0.20	0.17	0.14	0.12	0.10
	0.4	0.45	0.46	0.45	0.43	0.40	0.37	0.34	0.29	0.24	0.19	0.16	0.14	0.12	0.10
	0.6	0.41	0.44	0.42	0.40	0.39	0.36	0.34	0.28	0.23	0.18	0.15	0.13	0.11	0.10
	0.8	0.36	0.40	0.39	0.37	0.36	0.35	0.33	0.26	0.22	0.18	0.15	0.13	0.11	0.10
	1.0	0.32	0.35	0.36	0.33	0.32	0.30	0.30	0.24	0.21	0.17	0.14	0.12	0.11	0.10
	1.2	0.28	0.31	0.32	0.30	0.28	0.27	0.27	0.22	0.20	0.16	0.14	0.12	0.11	0.09
	1.6	0.19	0.23	0.23	0.22	0.22	0.22	0.21	0.19	0.18	0.15	0.13	0.11	0.10	0.09
	2.0	0.13	0.17	0.18	0.18	0.18	0.18	0.18	0.17	0.16	0.14	0.12	0.11	0.09	0.09
0.6	0.0	0.42	0.43	0.46	0.48	0.46	0.43	0.39	0.33	0.27	0.23	0.19	0.16	0.13	0.11
	0.2	0.41	0.42	0.44	0.47	0.45	0.42	0.39	0.33	0.27	0.23	0.19	0.16	0.13	0.11
	0.4	0.39	0.41	0.43	0.44	0.42	0.39	0.37	0.32	0.26	0.22	0.18	0.15	0.13	0.11
	0.6	0.36	0.41	0.40	0.40	0.38	0.36	0.35	0.31	0.26	0.21	0.17	0.14	0.12	0.11
	0.8	0.32	0.36	0.37	0.35	0.33	0.32	0.32	0.30	0.25	0.20	0.16	0.13	0.11	0.10
	1.0	0.29	0.33	0.33	0.30	0.29	0.29	0.28	0.27	0.24	0.19	0.15	0.12	0.11	0.10
	1.2	0.25	0.30	0.30	0.28	0.27	0.27	0.24	0.25	0.22	0.17	0.13	0.12	0.11	0.10
	1.6	0.18	0.21	0.22	0.23	0.23	0.23	0.22	0.21	0.19	0.16	0.12	0.12	0.11	0.10
	2.0	0.12	0.15	0.18	0.20	0.20	0.19	0.19	0.18	0.17	0.16	0.12	0.11	0.10	0.09
0.8	0.0	0.36	0.37	0.42	0.46	0.47	0.45	0.41	0.35	0.29	0.25	0.20	0.16	0.13	0.11
	0.2	0.35	0.36	0.41	0.45	0.46	0.44	0.41	0.35	0.29	0.25	0.20	0.16	0.13	0.11
	0.4	0.34	0.36	0.40	0.42	0.44	0.42	0.40	0.34	0.29	0.24	0.20	0.16	0.13	0.11
	0.6	0.31	0.35	0.39	0.38	0.39	0.38	0.37	0.32	0.28	0.23	0.19	0.15	0.13	0.11
	0.8	0.28	0.32	0.36	0.33	0.35	0.34	0.33	0.31	0.27	0.22	0.18	0.15	0.13	0.11
	1.0	0.25	0.30	0.32	0.29	0.31	0.30	0.30	0.28	0.25	0.21	0.17	0.15	0.12	0.11
	1.2	0.22	0.27	0.28	0.27	0.29	0.28	0.28	0.26	0.24	0.20	0.16	0.14	0.12	0.10
	1.6	0.16	0.21	0.22	0.23	0.24	0.23	0.23	0.23	0.21	0.17	0.15	0.13	0.11	0.10
	2.0	0.11	0.14	0.18	0.19	0.20	0.20	0.20	0.19	0.17	0.15	0.13	0.12	0.11	0.09

Table 7.15 Con't

1.2	0.0	0.28	0.30	0.35	0.39	0.41	0.42	0.43	0.40	0.34	0.29	0.23	0.19	0.16	0.13
	0.2	0.28	0.30	0.35	0.39	0.41	0.42	0.42	0.39	0.34	0.29	0.23	0.19	0.16	0.13
	0.4	0.27	0.28	0.34	0.37	0.40	0.40	0.40	0.38	0.33	0.28	0.23	0.18	0.16	0.13
	0.6	0.26	0.27	0.34	0.35	0.37	0.38	0.38	0.36	0.32	0.27	0.22	0.18	0.16	0.13
	0.8	0.24	0.25	0.33	0.32	0.33	0.36	0.36	0.34	0.30	0.26	0.21	0.17	0.16	0.13
	1.0	0.22	0.23	0.30	0.28	0.30	0.32	0.33	0.31	0.28	0.24	0.20	0.17	0.15	0.12
	1.2	0.19	0.21	0.27	0.26	0.28	0.29	0.30	0.29	0.27	0.22	0.19	0.17	0.14	0.12
	1.6	0.15	0.17	0.21	0.22	0.23	0.24	0.24	0.24	0.21	0.19	0.17	0.15	0.13	0.12
	2.0	0.12	0.14	0.18	0.18	0.20	0.20	0.20	0.20	0.18	0.17	0.15	0.13	0.12	0.11
1.6	0.0	0.20	0.24	0.29	0.33	0.35	0.37	0.40	0.42	0.38	0.35	0.31	0.27	0.23	0.18
	0.2	0.19	0.23	0.29	0.32	0.35	0.37	0.39	0.42	0.38	0.35	0.31	0.27	0.23	0.18
	0.4	0.19	0.22	0.29	0.31	0.34	0.36	0.38	0.40	0.37	0.35	0.31	0.27	0.23	0.18
	0.6	0.18	0.21	0.28	0.30	0.32	0.34	0.36	0.38	0.35	0.33	0.30	0.26	0.22	0.17
	0.8	0.17	0.20	0.24	0.29	0.31	0.32	0.34	0.35	0.33	0.32	0.28	0.25	0.22	0.17
	1.0	0.16	0.19	0.24	0.27	0.28	0.30	0.31	0.32	0.30	0.30	0.27	0.23	0.20	0.16
	1.2	0.14	0.18	0.22	0.25	0.26	0.28	0.29	0.30	0.28	0.28	0.25	0.22	0.19	0.15
	1.6	0.11	0.16	0.19	0.21	0.23	0.23	0.24	0.25	0.23	0.23	0.21	0.18	0.16	0.13
	2.0	0.10	0.14	0.16	0.17	0.19	0.19	0.20	0.20	0.19	0.19	0.16	0.14	0.13	0.12
2.0	0.0	0.16	0.20	0.24	0.27	0.29	0.32	0.34	0.38	0.41	0.38	0.34	0.29	0.24	0.20
	0.2	0.16	0.20	0.24	0.27	0.29	0.31	0.34	0.38	0.41	0.38	0.34	0.29	0.24	0.20
	0.4	0.16	0.20	0.24	0.26	0.29	0.31	0.32	0.37	0.38	0.37	0.33	0.29	0.24	0.20
	0.6	0.15	0.18	0.24	0.26	0.28	0.30	0.31	0.35	0.36	0.34	0.31	0.28	0.24	0.19
	0.8	0.14	0.17	0.23	0.25	0.27	0.28	0.30	0.33	0.34	0.33	0.30	0.26	0.23	0.19
	1.0	0.13	0.16	0.22	0.24	0.25	0.26	0.28	0.31	0.32	0.31	0.28	0.25	0.23	0.18
	1.2	0.11	0.15	0.21	0.22	0.24	0.24	0.26	0.29	0.29	0.28	0.26	0.23	0.20	0.17
	1.6	0.10	0.14	0.18	0.19	0.21	0.21	0.22	0.24	0.24	0.23	0.22	0.19	0.17	0.14
	2.0	0.09	0.13	0.16	0.16	0.17	0.17	0.17	0.19	0.19	0.18	0.18	0.16	0.14	0.13

Table 7.16 Coefficients $\overline{M}_x$

									ξ							
δ	η	0.0	0.1	0.2	0.4	0.6	0.8	1.0	1.2	1.6	2.0	2.4	2.8	3.2	3.6	4.0
0.0	0.0	0	-0.045	-0.076	-0.103	-0.110	-0.103	-0.092	-0.078	-0.056	-0.044	-0.037	-0.028	-0.018	-0.012	-0.008
	0.2	0	-0.043	-0.074	-0.100	-0.104	-0.097	-0.087	-0.074	-0.055	-0.043	-0.037	-0.028	-0.018	-0.012	
	0.4	0	-0.039	-0.068	-0.084	-0.087	-0.084	-0.076	-0.069	-0.052	-0.042	-0.036	-0.027	-0.018		
	0.6	0	-0.034	-0.060	-0.070	-0.072	-0.069	-0.064	-0.059	-0.049	-0.041	-0.035	-0.026	-0.017		
	0.8	0	-0.027	-0.048	-0.054	-0.055	-0.054	-0.053	-0.051	-0.046	-0.039	-0.034	-0.025	-0.015		
	1.0	0	-0.019	-0.031	-0.037	-0.041	-0.044	-0.044	-0.043	-0.039	-0.037	-0.033	-0.024	-0.013		
	1.2	0	-0.011	-0.018	-0.023	-0.026	-0.029	-0.031	-0.033	-0.035	-0.035	-0.032	-0.023	-0.012		
	1.6	0	-0.003	-0.006	-0.010	-0.014	-0.018	-0.021	-0.023	-0.028	-0.031	-0.029	-0.022			
	2.0	0	-0.002	-0.003	-0.006	-0.009	-0.012	-0.015	-0.018	-0.023	-0.027	-0.025	-0.020			
0.1	0.0	0	∞	0.056	-0.019	-0.039	-0.040	-0.039	-0.035	-0.029	-0.025	-0.024	-0.021	-0.014	-0.011	-0.008
	0.2	0	0.052	0.038	-0.009	-0.029	-0.034	-0.034	-0.032	-0.027	-0.024	-0.024	-0.021	-0.014	-0.011	
	0.4	0	0.032	0.027	-0.001	-0.014	-0.021	-0.023	-0.023	-0.023	-0.024	-0.023	-0.021	-0.013	-0.011	
	0.6	0	0.018	0.020	0.008	0.004	-0.002	-0.006	-0.010	-0.018	-0.021	-0.020	-0.019	-0.012	-0.011	
	0.8	0	0.009	0.015	0.018	0.017	0.016	-0.012	0.006	-0.009	-0.019	-0.018	-0.017	-0.011	-0.010	
	1.0	0	0.006	0.012	0.021	0.022	0.019	0.017	0.012	-0.004	-0.017	-0.016	-0.015	-0.010		
	1.2	0	0.005	0.010	0.020	0.022	0.019	0.017	0.015	-0.002	-0.015	-0.014	-0.013	-0.009		
	1.6	0	0.004	0.08	0.013	0.013	0.013	0.011	0.008	-0.001	-0.011	-0.013	-0.012			
	2.0	0	0.003	0.07	0.009	0.009	0.009	0.007	0.005	0.000	-0.010	-0.012	-0.011			
0.2	0.0	0	0.074	∞	0.033	-0.008	-0.024	-0.029	-0.030	-0.028	-0.025	-0.023	-0.021	-0.014	-0.012	-0.008
	0.2	0	0.057	0.087	0.039	-0.001	-0.018	-0.026	-0.028	-0.026	-0.024	-0.023	-0.021	-0.015	-0.012	-0.008
	0.4	0	0.035	0.040	0.030	0.009	-0.007	-0.015	-0.019	-0.022	-0.023	-0.023	-0.021	-0.016	-0.012	-0.008
	0.6	0	0.018	0.026	0.026	0.019	0.008	-0.002	-0.008	-0.017	-0.020	-0.023	-0.023	-0.018	-0.011	-0.008
	0.8	0	0.015	0.021	0.023	0.023	0.021	0.013	0.005	-0.008	-0.017	-0.023	-0.025	-0.021	-0.011	-0.008
	1.0	0	0.012	0.015	0.019	0.021	0.022	0.018	0.014	0.001	-0.015	-0.023	-0.025	-0.022	-0.009	-0.007
	1.2	0	0.009	0.013	0.017	0.019	0.019	0.018	0.016	0.006	-0.013	-0.021	-0.022	-0.019	-0.010	-0.007
	1.6	0	0.006	0.010	0.013	0.014	0.016	0.014	0.012	0.006	-0.011	-0.010	-0.010	-0.009	-0.008	-0.007
	2.0	0	0.005	0.008	0.012	0.012	0.011	0.010	0.008	0.002	-0.005	-0.008	-0.007	-0.006	-0.006	-0.006

Table 7.16 Con't

δ	η	0.0	0.1	0.2	0.4	0.6	0.8	1.0	1.2	1.6	2.0	2.4	2.8	3.2	3.6	4.0
								ξ								
	0.0	0	0.018	0.052	∞	0.070	0.020	-0.002	-0.014	-0.022	-0.021	-0.020	-0.018	-0.014	-0.012	-0.010
	0.2	0	0.022	0.056	0.128	0.074	0.026	0.002	-0.010	-0.18	-0.020	-0.020	-0.018	-0.015	-0.012	-0.010
	0.4	0	0.018	0.042	0.073	0.059	0.031	0.010	-0.003	-0.016	-0.019	-0.019	-0.019	-0.015	-0.012	-0.010
	0.6	0	0.014	0.029	0.049	0.048	0.034	0.017	0.005	-0.011	-0.016	-0.017	-0.018	-0.014	-0.011	-0.010
0.4	0.8	0	0.012	0.023	0.040	0.040	0.037	0.024	0.013	-0.004	-0.013	-0.015	-0.016	-0.013	-0.011	-0.009
	1.0	0	0.009	0.018	0.031	0.034	0.034	0.030	0.021	0.000	-0.010	-0.013	-0.014	-0.012	-0.009	-0.008
	1.2	0	0.007	0.015	0.027	0.028	0.029	0.028	0.022	0.004	-0.007	-0.011	-0.011	-0.010	-0.009	-0.008
	1.6	0	0.004	0.011	0.019	0.019	0.019	0.018	0.016	0.004	-0.006	-0.008	-0.008	-0.008	-0.08	-0.007
	2.0	0	0.003	0.009	0.014	0.013	0.013	0.010	0.008	0.002	-0.003	-0.004	-0.005	-0.006	-0.006	-0.006
	0.0	0	0.005	0.019	0.079	∞	0.090	0.036	0.009	-0.015	-0.022	-0.022	-0.019	-0.015	-0.013	-0.011
	0.2	0	0.007	0.023	0.083	0.152	0.093	0.039	0.012	-0.013	-0.021	-0.022	-0.019	-0.015	-0.013	-0.011
	0.4	0	0.011	0.026	0.066	0.093	0.076	0.042	0.017	-0.010	-0.019	-0.020	-0.018	-0.015	-0.013	-0.011
	0.6	0	0.010	0.024	0.049	0.064	0.059	0.039	0.021	-0.007	-0.017	-0.020	-0.018	-0.015	-0.012	-0.011
0.6	0.8	0	0.009	0.022	0.040	0.050	0.048	0.036	0.023	-0.004	-0.012	-0.018	-0.018	-0.014	-0.011	-0.010
	1.0	0	0.009	0.020	0.035	0.042	0.040	0.033	0.024	-0.001	-0.008	-0.015	-0.015	-0.012	-0.010	-0.008
	1.2	0	0.009	0.019	0.033	0.035	0.035	0.030	0.024	0.001	-0.006	-0.012	-0.012	-0.010	-0.009	-0.008
	1.6	0	0.008	0.015	0.021	0.025	0.025	0.025	0.019	0.003	-0.004	-0.007	-0.008	-0.008	-0.008	-0.007
	2.0	0	0.007	0.013	0.014	0.016	0.015	0.014	0.013	0.005	-0.002	-0.004	-0.005	-0.006	-0.006	-0.006
	0.0	0	-0.001	0.007	0.037	0.100	∞	0.108	0.050	0.005	-0.011	-0.015	-0.017	-0.016	-0.014	-0.012
	0.2	0	0.001	0.010	0.042	0.102	0.170	0.109	0.054	0.007	-0.010	-0.014	-0.016	-0.016	-0.014	-0.012
	0.4	0	0.006	0.014	0.043	0.084	0.110	0.101	0.056	0.010	-0.008	-0.013	-0.015	-0.015	-0.014	-0.012
	0.6	0	0.009	0.017	0.040	0.065	0.078	0.071	0.051	0.013	-0.005	-0.011	-0.012	-0.013	-0.013	-0.011
0.8	0.8	0	0.011	0.018	0.036	0.052	0.060	0.057	0.044	0.016	-0.001	-0.009	-0.011	-0.011	-0.012	-0.011
	1.0	0	0.011	0.018	0.034	0.045	0.050	0.048	0.039	0.018	0.001	-0.007	-0.010	-0.010	-0.011	-0.009
	1.2	0	0.011	0.018	0.032	0.041	0.046	0.043	0.032	0.019	0.003	-0.006	-0.009	-0.010	-0.010	-0.009
	1.6	0	0.009	0.015	0.022	0.025	0.026	0.026	0.026	0.018	0.005	-0.003	-0.006	-0.007	-0.008	-0.008
	2.0	0	0.007	0.011	0.014	0.015	0.016	0.016	0.015	0.011	0.014	0.001	-0.003	-0.005	-0.006	-0.006

1.2	0.0	0	-0.002	0.000	0.008	0.025	0.057	0.118	∞	0.063	0.016	-0.003	-0.013	-0.016	-0.016	-0.014
	0.2	0	-0.001	0.001	0.010	0.028	0.062	0.120	0.186	0.067	0.018	-0.002	-0.012	-0.016	-0.015	-0.014
	0.4	0	0.000	0.004	0.014	0.032	0.063	0.100	0.125	0.068	0.021	-0.001	-0.011	-0.015	-0.014	-0.014
	0.6	0	0.003	0.007	0.019	0.033	0.058	0.083	0.092	0.061	0.023	0.001	-0.009	-0.013	-0.014	-0.013
	0.8	0	0.004	0.009	0.019	0.032	0.050	0.065	0.070	0.053	0.023	0.002	-0.007	-0.012	-0.013	-0.012
	1.0	0	0.004	0.009	0.018	0.030	0.044	0.053	0.056	0.044	0.022	0.004	-0.005	-0.011	-0.012	-0.011
	1.2	0	0.004	0.009	0.017	0.028	0.039	0.045	0.046	0.037	0.020	0.006	-0.004	-0.009	-0.010	-0.010
	1.6	0	0.003	0.008	0.016	0.024	0.030	0.030	0.030	0.024	0.015	0.005	-0.002	-0.006	-0.008	-0.008
	2.0	0	0.003	0.006	0.012	0.016	0.018	0.018	0.018	0.015	0.011	0.004	-0.000	-0.003	-0.006	-0.006
1.6	0.0	0	-0.003	-0.004	-0.001	0.005	0.016	0.034	0.067	∞	0.068	0.019	0.000	-0.014	-0.017	-0.016
	0.2	0	-0.003	-0.002	0.000	0.006	0.018	0.038	0.070	0.193	0.073	0.021	0.001	-0.013	-0.017	-0.016
	0.4	0	-0.001	-0.001	0.001	0.008	0.022	0.041	0.071	0.132	0.072	0.023	0.003	0.011	-0.015	-0.015
	0.6	0	0.000	0.000	0.004	0.012	0.025	0.042	0.065	0.097	0.063	0.025	0.005	-0.009	-0.013	-0.014
	0.8	0	0.001	0.003	0.007	0.015	0.026	0.041	0.057	0.076	0.057	0.027	0.006	-0.007	-0.012	-0.013
	1.0	0	0.002	0.004	0.009	0.017	0.027	0.038	0.048	0.059	0.047	0.026	0.007	-0.005	-0.011	-0.012
	1.2	0	0.002	0.005	0.010	0.019	0.025	0.033	0.039	0.046	0.039	0.024	0.008	-0.004	-0.009	-0.010
	1.6	0	0.002	0.004	0.009	0.013	0.019	0.025	0.027	0.028	0.026	0.017	0.008	-0.002	-0.006	-0.008
	2.0	0	0.001	0.003	0.005	0.009	0.011	0.013	0.015	0.018	0.017	0.014	0.008	0.000	-0.003	-0.005
2.0	0.0	0	-0.003	-0.004	-0.006	-0.005	-0.001	0.006	0.018	0.069	∞	0.069	0.018	-0.005	-0.014	-0.017
	0.2	0	-0.002	-0.004	-0.005	-0.004	0.000	0.007	0.020	0.073	0.195	0.073	0.020	-0.003	-0.013	-0.017
	0.4	0	-0.002	-0.004	-0.004	-0.003	0.002	0.010	0.023	0.072	0.133	0.072	0.023	-0.002	-0.011	-0.015
	0.6	0	-0.001	-0.003	-0.003	-0.001	0.004	0.013	0.025	0.063	0.098	0.063	0.025	0.001	-0.009	-0.013
	0.8	0	-0.001	-0.002	-0.002	0.001	0.006	0.015	0.026	0.055	0.076	0.055	0.025	0.002	-0.007	-0.012
	1.0	0	-0.001	-0.002	-0.001	0.003	0.007	0.016	0.025	0.045	0.059	0.045	0.025	0.004	-0.005	-0.011
	1.2	0	-0.001	-0.002	0.000	0.003	0.007	0.015	0.024	0.038	0.045	0.038	0.022	0.006	-0.004	-0.009
	1.6	0	0.000	0.000	0.000	0.002	0.006	0.010	0.015	0.025	0.028	0.024	0.015	0.005	-0.002	-0.006
	2.0	0	0.000	0.000	0.000	0.001	0.004	0.007	0.011	0.015	0.018	0.015	0.011	0.004	0.000	-0.003

Table 7.17 Coefficients $\overline{M}_y$

δ	η	ξ 0.0	0.1	0.2	0.4	0.6	0.8	1.0	1.2	1.6	2.0	2.4	2.8	3.2	3.6	4.0
0	0.0	∞	0.519	0.379	0.237	0.164	0.117	0.087	0.066	0.044	0.030	0.020	0.014	0.009	0.007	0.005
	0.2	0.273	0.265	0.249	0.195	0.146	0.107	0.083	0.062	0.041	0.029	0.020	0.014	0.009	0.007	0.005
	0.4	0.133	0.132	0.128	0.119	0.099	0.080	0.065	0.053	0.037	0.027	0.019	0.012	0.009		
	0.6	0.059	0.059	0.059	0.058	0.054	0.050	0.046	0.040	0.031	0.024	0.017	0.011	0.008		
	0.8	0.008	0.010	0.012	0.017	0.022	0.026	0.027	0.026	0.024	0.020	0.014	0.009	0.007		
	1.0	-0.020	-0.020	-0.019	-0.011	-0.002	0.007	0.012	0.014	0.016	0.015	0.010	0.008	0.006		
	1.2	-0.038	-0.038	-0.037	-0.030	-0.021	-0.010	-0.001	0.002	0.009	0.009	0.007	0.006	0.005		
	1.6	-0.056	-0.054	-0.053	-0.048	-0.041	-0.031	-0.020	-0.012	-0.001	0.001	0.002	0.002			
	2.0	-0.060	-0.059	-0.057	-0.054	-0.048	-0.039	-0.027	-0.017	-0.005	-0.003	-0.002	-0.001			
0.1	0.0	0.360	∞	0.350	0.214	0.164	0.118	0.083	0.063	0.038	0.024	0.015	0.011	0.008	0.006	0.005
	0.2	0.216	0.237	0.222	0.187	0.144	0.109	0.080	0.059	0.036	0.023	0.015	0.011	0.008	0.006	
	0.4	0.118	0.124	0.122	0.114	0.097	0.079	0.063	0.050	0.032	0.021	0.014	0.011	0.007	0.006	
	0.6	0.057	0.057	0.057	0.056	0.053	0.038	0.042	0.035	0.026	0.019	0.013	0.010	0.007	0.005	
	0.8	0.003	0.003	0.004	0.007	0.011	0.015	0.018	0.021	0.020	0.016	0.012	0.009	0.006	0.005	
	1.0	-0.023	-0.021	-0.020	-0.015	-0.010	-0.004	0.002	0.008	0.013	0.013	0.009	0.007	0.005		
	1.2	-0.040	-0.038	-0.036	-0.030	-0.026	-0.015	-0.009	0.002	0.005	0.008	0.006	0.005	0.004		
	1.6	-0.056	-0.035	-0.053	-0.047	-0.041	-0.028	-0.019	-0.013	-0.004	0.000	0.001	0.002			
	2.0	-0.060	-0.059	-0.057	-0.054	-0.047	-0.038	-0.027	-0.018	-0.007	-0.004	-0.002	-0.001			
0.2	0.0	0.272	0.311	∞	0.235	0.164	0.119	0.081	0.060	0.035	0.022	0.014	0.009	0.007	0.006	0.005
	0.2	0.197	0.220	0.212	0.178	0.144	0.103	0.073	0.054	0.031	0.021	0.014	0.009	0.007	0.005	0.005
	0.4	0.110	0.113	0.113	0.110	0.096	0.071	0.052	0.041	0.025	0.019	0.013	0.008	0.006	0.004	0.004
	0.6	0.053	0.054	0.054	0.050	0.041	0.032	0.026	0.023	0.019	0.016	0.012	0.007	0.005	0.004	0.004
	0.8	0.001	0.001	0.002	0.004	0.007	0.009	0.012	0.013	0.014	0.015	0.010	0.006	0.001	0.003	0.003
	1.0	-0.025	-0.024	-0.023	-0.019	-0.014	-0.008	-0.003	0.002	0.009	0.011	0.009	0.005	0.003	0.003	0.003
	1.2	-0.041	-0.040	-0.037	-0.033	-0.026	-0.021	-0.010	-0.007	0.004	0.008	0.007	0.004	0.002	0.002	0.002
	1.6	-0.055	-0.054	-0.051	-0.046	-0.040	-0.031	-0.023	-0.016	-0.004	0.000	0.001	0.001	0.001	0.001	0.001
	2.0	-0.059	-0.058	-0.056	-0.055	-0.047	-0.036	-0.026	-0.018	-0.008	-0.004	-0.003	-0.001	0.000	0.000	0.000

0.4	0.0	0.194	0.204	0.225	∞	0.205	0.126	0.084	0.059	0.033	0.025	0.018	0.011	0.007	0.005	0.003
	0.2	0.161	0.160	0.160	0.158	0.138	0.107	0.068	0.048	0.030	0.022	0.016	0.011	0.007	0.005	0.003
	0.4	0.102	0.102	0.101	0.097	0.084	0.060	0.045	0.034	0.022	0.019	0.015	0.010	0.006	0.005	0.003
	0.6	0.048	0.048	0.048	0.047	0.040	0.026	0.016	0.015	0.015	0.016	0.014	0.009	0.006	0.004	0.003
	0.8	-0.007	-0.007	-0.005	-0.004	-0.004	-0.002	0.000	0.002	0.005	0.011	0.011	0.008	0.005	0.003	0.002
	1.0	-0.025	-0.024	-0.023	-0.020	-0.016	-0.013	-0.005	-0.004	0.001	0.005	0.007	0.005	0.003	0.002	0.002
	1.2	-0.040	-0.037	-0.034	-0.030	-0.024	-0.020	-0.011	-0.008	-0.003	0.001	0.004	0.003	0.003	0.002	0.002
	1.6	-0.051	-0.049	-0.046	-0.043	-0.037	-0.026	-0.018	-0.014	-0.009	-0.004	0.000	0.000	0.001	0.001	0.001
	2.0	-0.057	-0.056	-0.053	-0.050	-0.045	-0.033	-0.024	-0.018	-0.012	-0.006	-0.004	-0.002	0.000	0.000	0.000
0.6	0.0	0.147	0.152	0.162	0.209	∞	0.191	0.126	0.089	0.053	0.035	0.022	0.014	0.009	0.006	0.005
	0.2	0.131	0.132	0.135	0.141	0.130	0.125	0.101	0.078	0.050	0.034	0.021	0.014	0.009	0.006	0.005
	0.4	0.091	0.088	0.088	0.086	0.084	0.069	0.057	0.051	0.038	0.029	0.019	0.012	0.009	0.006	0.005
	0.6	0.042	0.044	0.044	0.025	0.019	0.024	0.029	0.028	0.027	0.023	0.017	0.011	0.008	0.005	0.004
	0.8	0.004	0.001	0.001	-0.001	-0.001	0.008	0.012	0.012	0.018	0.017	0.013	0.009	0.007	0.004	0.003
	1.0	-0.022	-0.021	-0.021	-0.017	-0.011	-0.004	0.000	0.002	0.009	0.011	0.008	0.006	0.005	0.003	0.002
	1.2	-0.036	-0.030	-0.030	-0.027	-0.024	-0.011	-0.008	-0.005	0.001	0.006	0.005	0.004	0.003	0.002	0.002
	1.6	-0.046	-0.041	-0.041	-0.038	-0.033	-0.022	-0.017	-0.013	-0.008	-0.003	-0.001	0.000	0.000	0.001	0.001
	2.0	-0.053	-0.049	-0.049	-0.046	-0.041	-0.030	-0.022	-0.018	-0.013	-0.008	-0.005	-0.003	-0.001	0.000	0.000
0.8	0.0	0.115	0.118	0.124	0.148	0.203	∞	0.194	0.130	0.073	0.045	0.028	0.018	0.011	0.007	0.005
	0.2	0.106	0.106	0.111	0.123	0.137	0.131	0.129	0.107	0.068	0.042	0.027	0.018	0.011	0.007	0.005
	0.4	0.080	0.079	0.078	0.077	0.072	0.065	0.067	0.066	0.052	0.037	0.024	0.015	0.010	0.006	0.005
	0.6	0.047	0.044	0.042	0.036	0.030	0.026	0.028	0.032	0.035	0.030	0.020	0.014	0.009	0.006	0.004
	0.8	0.010	0.009	0.008	0.002	-0.001	-0.002	0.001	0.007	0.018	0.023	0.015	0.011	0.008	0.005	0.003
	1.0	-0.016	-0.015	-0.015	-0.015	-0.015	-0.012	-0.008	-0.005	0.006	0.017	0.011	0.007	0.005	0.003	0.002
	1.2	-0.030	-0.028	-0.026	-0.026	-0.023	-0.018	-0.014	-0.011	-0.003	0.006	0.006	0.004	0.003	0.003	0.002
	1.6	-0.041	-0.039	-0.037	-0.035	-0.032	-0.024	-0.020	-0.017	-0.011	-0.002	0.002	0.000	0.000	0.001	0.001
	2.0	-0.046	-0.045	-0.042	-0.040	-0.035	-0.027	-0.020	-0.017	-0.012	-0.009	-0.006	-0.003	-0.002	-0.001	0.001

Table 7.17 Con't

δ	η								ξ							
		0.0	0.1	0.2	0.4	0.6	0.8	1.0	1.2	1.6	2.0	2.4	2.8	3.2	3.6	4.0
1.2	0.0	0.071	0.073	0.076	0.088	0.111	0.141	0.194	∞	0.132	0.077	0.048	0.030	0.018	0.011	0.007
	0.2	0.068	0.070	0.073	0.083	0.100	0.114	0.129	0.131	0.110	0.071	0.046	0.029	0.018	0.011	0.007
	0.4	0.060	0.061	0.063	0.065	0.069	0.069	0.069	0.068	0.070	0.057	0.041	0.025	0.016	0.010	0.006
	0.6	0.043	0.042	0.042	0.042	0.042	0.042	0.039	0.036	0.040	0.042	0.032	0.021	0.015	0.009	0.006
	0.8	0.018	0.018	0.017	0.017	0.017	0.016	0.014	0.013	0.017	0.018	0.018	0.017	0.012	0.008	0.005
	1.0	-0.001	0.000	0.000	0.000	0.000	0.002	0.001	0.003	0.004	0.007	0.012	0.014	0.008	0.005	0.003
	1.2	-0.012	-0.012	-0.010	-0.009	-0.007	-0.005	-0.005	-0.004	-0.005	0.006	0.007	0.009	0.004	0.003	0.003
	1.6	-0.023	-0.022	-0.021	-0.019	-0.017	-0.014	-0.013	-0.012	-0.011	-0.007	-0.002	0.001	0.000	0.000	0.001
	2.0	-0.027	-0.026	-0.025	-0.023	-0.021	-0.019	-0.018	-0.017	-0.016	-0.013	-0.009	-0.006	-0.003	-0.002	-0.001
1.6	0.0	0.049	0.050	0.052	0.058	0.067	0.081	0.102	0.136	∞	0.134	0.077	0.048	0.030	0.018	0.011
	0.2	0.047	0.048	0.050	0.055	0.063	0.075	0.092	0.114	0.131	0.111	0.071	0.046	0.029	0.018	0.011
	0.4	0.042	0.043	0.045	0.049	0.054	0.061	0.069	0.075	0.071	0.073	0.057	0.041	0.025	0.016	0.010
	0.6	0.035	0.036	0.036	0.037	0.041	0.044	0.046	0.045	0.040	0.044	0.042	0.032	0.021	0.015	0.009
	0.8	0.025	0.027	0.027	0.027	0.027	0.027	0.027	0.024	0.020	0.026	0.028	0.023	0.017	0.012	0.008
	1.0	0.016	0.016	0.016	0.016	0.016	0.014	0.012	0.010	0.009	0.012	0.016	0.016	0.014	0.008	0.005
	1.2	0.007	0.007	0.007	0.006	0.005	0.004	0.002	0.001	-0.001	0.003	0.006	0.007	0.009	0.004	0.003
	1.6	-0.003	-0.003	-0.004	-0.005	-0.007	-0.008	0.009	-0.010	-0.011	-0.009	-0.005	-0.002	0.001	0.000	0.000
	2.0	-0.005	-0.006	-0.007	-0.009	-0.011	-0.012	-0.013	-0.015	-0.016	-0.015	-0.011	-0.009	-0.006	-0.003	-0.001
2.0	0.0	0.031	0.032	0.034	0.039	0.045	0.053	0.065	0.081	0.137	∞	0.135	0.078	0.049	0.030	0.018
	0.2	0.030	0.032	0.034	0.038	0.044	0.051	0.061	0.075	0.115	0.133	0.112	0.072	0.047	0.029	0.018
	0.4	0.028	0.029	0.031	0.035	0.040	0.045	0.053	0.061	0.077	0.073	0.074	0.057	0.040	0.024	0.016
	0.6	0.026	0.027	0.028	0.031	0.034	0.038	0.042	0.045	0.048	0.042	0.045	0.042	0.031	0.020	0.014
	0.8	0.023	0.023	0.024	0.026	0.028	0.030	0.031	0.031	0.028	0.023	0.025	0.027	0.021	0.015	0.011
	1.0	0.020	0.020	0.021	0.022	0.022	0.022	0.022	0.020	0.015	0.012	0.014	0.016	0.014	0.010	0.007
	1.2	0.018	0.018	0.018	0.017	0.016	0.015	0.014	0.012	0.007	0.002	0.004	0.006	0.006	0.005	0.004
	1.6	0.013	0.013	0.013	0.011	0.006	0.004	0.002	-0.002	-0.005	-0.009	-0.009	-0.006	-0.004	-0.002	0.000
	2.0	0.010	0.010	0.009	0.007	0.000	-0.002	-0.005	-0.009	-0.013	-0.016	-0.015	-0.010	-0.009	-0.006	-0.003

Table 7.18 Coefficients $\overline{M}_{xy}$

δ	η										ξ					
		0.0	0.1	0.2	0.4	0.6	0.8	1.0	1.2	1.6	2.0	2.4	2.8	3.2	3.6	4.0
0.0	0.0	0	0	0	0	0	0	0	0	0						
	0.2	-0.060	-0.106	-0.112	-0.092	-0.054	-0.038	-0.024	-0.017	-0.011						
	0.4	-0.120	-0.123	-0.125	-0.112	-0.086	-0.063	-0.045	-0.034	-0.019						
	0.6	-0.155	-0.142	-0.131	-0.116	-0.095	-0.073	-0.056	-0.043	-0.025						
	0.8	-0.155	-0.143	-0.131	-0.113	-0.095	-0.075	-0.059	-0.045	-0.028						
	1.0	-0.141	-0.131	-0.121	-0.104	-0.087	-0.071	-0.055	-0.044	-0.028						
	1.2	-0.107	-0.104	-0.100	-0.080	-0.076	-0.063	-0.050	-0.042	-0.027						
	1.6	-0.066	-0.066	-0.066	-0.062	-0.054	-0.042	-0.032	-0.027	-0.024						
0.1	0.0	0	0	0	0	0	0	0	0	0						
	0.2	-0.059	-0.059	-0.058	-0.055	-0.050	-0.042	-0.030	-0.020	-0.010						
	0.4	-0.098	-0.097	-0.094	-0.087	-0.077	-0.060	-0.045	-0.032	-0.018						
	0.6	-0.123	-0.115	-0.108	-0.093	-0.082	-0.070	-0.052	-0.040	-0.023						
	0.8	-0.123	-0.118	-0.111	-0.098	-0.081	-0.068	-0.054	-0.043	-0.027						
	1.0	-0.116	-0.109	-0.101	-0.087	-0.076	-0.064	-0.052	-0.042	-0.026						
	1.2	-0.088	-0.086	-0.083	-0.075	-0.066	-0.055	-0.044	-0.037	-0.024						
	1.6	-0.064	-0.061	-0.057	-0.045	-0.033	-0.029	-0.025	-0.022	-0.019						
0.2	0.0	0	0	0	0	0	0	0	0	0						
	0.2	-0.035	-0.029	-0.031	-0.049	-0.051	-0.044	-0.031	-0.020	-0.008						
	0.4	-0.065	-0.059	-0.060	-0.067	-0.068	-0.055	-0.046	-0.031	-0.017						
	0.6	-0.086	-0.084	-0.084	-0.079	-0.072	-0.063	-0.051	-0.039	-0.023						
	0.8	-0.094	-0.094	-0.093	-0.083	-0.072	-0.061	-0.051	-0.042	-0.027						
	1.0	-0.095	-0.093	-0.090	-0.078	-0.065	-0.057	-0.048	-0.039	-0.025						
	1.2	-0.075	-0.073	-0.070	-0.063	-0.056	-0.048	-0.040	-0.033	-0.021						
	1.6	-0.069	-0.060	-0.052	-0.032	-0.029	-0.023	-0.019	-0.017	-0.015						

Table 7.18 Con't

δ	η	ξ 0.0	0.1	0.2	0.4	0.6	0.8	1.0	1.2	1.6	2.0	2.4	2.8	3.2	3.6	4.0
0.4	0.0	0	0	0	0	0	0	0	0	0	0	0	0			
	0.2	-0.018	0.004	0.011	-0.021	-0.051	-0.035	-0.027	-0.016	-0.097	-0.002					
	0.4	-0.032	-0.019	-0.018	-0.038	-0.051	-0.046	-0.039	-0.031	-0.014	-0.005					
	0.6	-0.047	-0.041	-0.040	-0.046	-0.051	-0.048	-0.044	-0.033	-0.021	-0.015					
	0.8	-0.056	-0.056	-0.056	-0.054	-0.050	-0.046	-0.042	-0.035	-0.025	-0.020					
	1.0	-0.063	-0.063	-0.062	-0.052	-0.045	-0.040	-0.036	-0.031	-0.023	-0.018					
	1.2	-0.064	-0.062	-0.056	-0.044	-0.037	-0.032	-0.029	-0.024	-0.015	-0.011					
	1.6	-0.051	-0.041	-0.037	-0.022	-0.016	-0.013	-0.011	-0.009	-0.007	-0.006					
0.6	0.0	0	0	0	0	0	0	0	0	0	0	0	0			
	0.2	0.001	0.006	0.013	0.026	-0.009	-0.045	-0.033	-0.023	-0.013	-0.008	-0.005	-0.003			
	0.4	-0.006	-0.001	0.004	0.005	-0.017	-0.042	-0.042	-0.033	-0.020	-0.013	-0.008	-0.005			
	0.6	-0.020	-0.016	-0.013	-0.012	-0.023	-0.035	-0.037	-0.032	-0.025	-0.016	-0.011	-0.008			
	0.8	-0.034	-0.030	-0.026	-0.022	-0.024	-0.028	-0.029	-0.029	-0.024	-0.018	-0.013	-0.010			
	1.0	-0.044	-0.039	-0.035	-0.027	-0.021	-0.025	-0.026	-0.025	-0.023	-0.019	-0.014	-0.010			
	1.2	-0.048	-0.041	-0.035	-0.026	-0.020	-0.018	-0.014	-0.019	-0.021	-0.017	-0.015	-0.010			
	1.6	-0.042	-0.032	-0.025	-0.017	-0.011	-0.007	-0.009	-0.011	-0.016	-0.015	-0.014	-0.010			
0.8	0.0	0	0	0	0	0	0	0	0	0	0	0	0			
	0.2	0.003	0.006	0.010	0.021	0.032	-0.005	-0.042	-0.031	-0.016	-0.008	-0.003	-0.002			
	0.4	0.001	0.005	0.010	0.020	0.017	-0.009	-0.036	-0.039	-0.029	-0.015	-0.008	-0.006			
	0.6	-0.005	-0.001	0.003	0.008	0.005	-0.011	-0.028	-0.031	-0.029	-0.019	-0.012	-0.008			
	0.8	-0.013	-0.009	-0.006	-0.002	-0.003	-0.011	-0.021	-0.026	-0.028	-0.021	-0.015	-0.010			
	1.0	-0.021	-0.016	-0.013	-0.008	-0.006	-0.009	-0.014	-0.019	-0.026	-0.022	-0.015	-0.011			
	1.2	-0.030	-0.025	-0.020	-0.010	-0.005	-0.006	-0.010	-0.015	-0.021	-0.021	-0.015	-0.012			
	1.6	-0.033	-0.028	-0.023	-0.009	-0.001	0.000	-0.005	-0.009	-0.015	-0.016	-0.014	-0.011			

		C1	C2	C3	C4	C5	C6	C7	C8	C9	C10	C11	C12	C13	C14	C15
1.2	0.0	0	0	0	0	0	0	0	0	0	0	0	0	0	0	0
	0.2	0.004	0.006	0.007	0.010	0.018	0.027	0.038	0.000	-0.024	-0.015	-0.007	-0.005	-0.004		
	0.4	0.007	0.009	0.011	0.018	0.026	0.032	0.028	0.000	-0.034	-0.022	-0.012	-0.008	-0.007		
	0.6	0.008	0.010	0.012	0.019	0.025	0.027	0.020	0.000	-0.028	-0.023	-0.014	-0.010	-0.009		
	0.8	0.007	0.008	0.010	0.016	0.021	0.021	0.014	0.001	-0.019	-0.020	-0.014	-0.010	-0.010		
	1.0	0.004	0.006	0.008	0.013	0.017	0.017	0.013	0.005	-0.010	-0.014	-0.012	-0.009	-0.009		
	1.2	0.000	0.002	0.005	0.010	0.014	0.014	0.011	0.006	-0.003	-0.006	-0.007	-0.008	-0.009		
	1.6	-0.002	-0.002	0.001	0.005	0.009	0.011	0.10	0.007	0.003	-0.002	-0.005	-0.007	-0.009	0	0
1.6	0.0	0	0	0	0	0	0	0	0	0	0	0	0	0	0	0
	0.2	0.003	0.004	0.005	0.006	0.010	0.014	0.020	0.030	0.000	-0.030	-0.015	-0.008	-0.005	-0.004	
	0.4	0.005	0.007	0.008	0.012	0.016	0.022	0.029	0.035	0.000	-0.036	-0.024	-0.015	-0.010	-0.007	
	0.6	0.007	0.008	0.010	0.014	0.020	0.025	0.030	0.031	0.000	-0.030	-0.027	-0.018	-0.013	-0.010	
	0.8	0.008	0.009	0.010	0.015	0.020	0.024	0.026	0.024	0.001	-0.028	-0.025	-0.019	-0.014	-0.012	
	1.0	0.007	0.009	0.010	0.014	0.017	0.021	0.022	0.019	0.002	-0.015	-0.022	-0.018	-0.014	-0.013	
	1.2	0.006	0.007	0.008	0.012	0.016	0.018	0.018	0.016	0.003	-0.011	-0.018	-0.014	-0.012	-0.010	
	1.6	0.005	0.006	0.007	0.010	0.014	0.015	0.014	0.013	0.004	-0.006	-0.009	-0.009	-0.010	-0.009	0
2.0	0.0	0	0	0	0	0	0	0	0	0	0	0	0	0	0	0
	0.2	0.004	0.004	0.004	0.005	0.007	0.009	0.011	0.015	0.029	0.000	-0.030	-0.016	-0.010	-0.006	-0.004
	0.4	0.008	0.008	0.009	0.010	0.012	0.016	0.020	0.025	0.036	0.000	-0.037	-0.027	-0.017	-0.011	-0.008
	0.6	0.011	0.011	0.012	0.014	0.016	0.020	0.024	0.029	0.031	0.000	-0.032	-0.031	-0.022	-0.015	-0.010
	0.8	0.014	0.014	0.014	0.016	0.018	0.021	0.025	0.028	0.025	0.000	-0.027	-0.030	-0.024	-0.018	-0.013
	1.0	0.016	0.016	0.016	0.017	0.019	0.021	0.023	0.024	0.019	0.000	-0.021	-0.028	-0.025	-0.019	-0.014
	1.2	0.017	0.016	0.016	0.017	0.018	0.019	0.020	0.020	0.015	0.000	-0.017	-0.024	-0.024	-0.019	-0.015
	1.6	0.017	0.016	0.014	0.013	0.013	0.12	0.012	0.012	0.010	0.000	-0.011	-0.018	-0.019	-0.018	-0.014

Table 7.19 Coefficients $\overline{Q}_x$

δ	η	0.0	0.1	0.2	0.4	0.6	0.8	1.0	1.2	1.6	2.0	2.4	2.8	3.2	3.6	4.0
									ξ							
0.0	0.0	0.24/-∞	-2.91	-1.30	-0.50	-0.24	-0.12	-0.06	-0.02	0.00						
	0.2	0.21	-0.38	-0.52	-0.35	-0.19	-0.10	-0.05	-0.02	0.00						
	0.4	0.18	0.06	-0.07	-0.13	-0.10	-0.06	-0.03	-0.01	0.00						
	0.6	0.10	0.09	0.05	-0.01	-0.02	-0.02	-0.01	0.00	0.00						
	0.8	0.06	0.06	0.06	0.03	-0.02	0.01	-0.01	0.00	0.00						
	1.0	-0.01	0.02	0.03	0.03	0.03	0.02	0.01	0.01	0.00						
	1.2	-0.08	-0.04	-0.01	0.01	0.02	0.02	0.01	0.01	0.00						
	1.6	-0.12	-0.11	-0.08	-0.03	-0.01	0.00	0.00	0.00	0.00						
0.1	0.0	0.22	∞/-∞	-1.69	-0.54	-0.27	-0.14	-0.07	-0.14	0.03						
	0.2	0.19	-0.14	-0.40	-0.37	-0.22	-0.12	-0.06	-0.14	0.03						
	0.4	0.15	-0.05	-0.13	-0.16	-0.13	-0.07	-0.03	0.00	0.04						
	0.6	0.08	0.01	-0.02	-0.04	-0.04	-0.02	0.00	0.03	0.04						
	0.8	0.01	0.04	0.06	0.03	0.03	0.03	0.03	0.05	0.06						
	1.0	-0.04	0.08	0.11	0.08	0.07	0.07	0.07	0.07	0.07						
	1.2	-0.08	0.09	0.13	0.11	0.10	0.10	0.10	0.10	0.10						
	1.6	-0.08	0.10	0.14	0.11	0.11	0.11	0.11	0.11	0.11						
0.2	0.0	0.19	1.42	∞/-∞	-0.82	-0.36	-0.19	-0.10	-0.05	0.01						
	0.2	0.17	0.17	-0.10	-0.41	-0.28	-0.16	-0.09	-0.04	0.01						
	0.4	0.11	0.02	-0.05	-0.15	-0.15	-0.10	-0.06	-0.02	0.02						
	0.6	0.03	0.01	0.00	-0.04	-0.05	-0.04	-0.02	0.01	0.03						
	0.8	-0.04	0.00	0.04	0.03	0.01	0.01	0.02	0.02	0.04						
	1.0	-0.07	0.00	0.06	0.07	0.06	0.05	0.05	0.04	0.04						
	1.2	-0.08	0.00	0.05	0.09	0.08	0.07	0.06	0.05	0.05						
	1.6	-0.05	0.00	0.03	0.06	0.06	0.05	0.05	0.05	0.05						

0.4	0.0	0.0	0.12	0.36	0.68	∞/−∞	−0.079	−0.35	−0.20	−0.11	−0.03	0.01		
	0.2	0.10	0.22	0.29	−0.04	−0.038	−0.27	−0.17	−0.10	−0.02	0.01			
	0.4	0.04	0.06	0.08	−0.02	−0.14	−0.15	−0.11	−0.07	−0.01	0.01			
	0.6	0.00	0.02	0.04	0.01	−0.04	−0.06	−0.05	−0.03	0.00	0.01			
	0.8	−0.03	0.01	0.03	0.02	0.01	−0.01	−0.01	−0.03	0.00	0.02			
	1.0	−0.04	0.00	0.03	0.04	0.04	0.03	0.02	0.00	0.02	0.02			
	1.2	−0.03	0.00	0.02	0.04	0.04	0.04	0.05	0.02	0.02	0.02			
	1.6	−0.01	0.00	0.02	0.03	0.04	0.05	0.05	0.05	0.02	0.02			
0.6	0.0	0.05	0.16	0.28	0.74	∞/−∞	−0.01	−0.77	−0.35	−0.20	−0.13	−0.05	−0.02	0.00
	0.2	0.04	0.12	0.21	0.35	−0.01	−0.38	−0.27	−0.18	−0.11	−0.05	−0.02	0.00	
	0.4	0.02	0.06	0.11	0.12	0.00	−0.13	−0.15	−0.12	−0.09	−0.04	−0.01	0.01	
	0.6	0.01	0.03	0.06	0.06	0.01	−0.05	−0.05	−0.07	−0.06	−0.03	−0.01	0.01	
	0.8	0.00	0.02	0.05	0.05	0.02	0.00	0.00	−0.03	−0.03	−0.01	0.00	0.02	
	1.0	−0.01	0.02	0.05	0.05	0.04	0.02	0.02	0.00	−0.01	0.00	0.00	0.03	
	1.2	−0.01	0.01	0.04	0.04	0.04	0.04	0.04	0.02	0.01	0.01	0.00	0.03	
	1.6	−0.01	0.01	0.04	0.04	0.04	0.04	0.05	0.04	0.02	0.01	0.01	0.03	
0.8	0.0	0.01	0.08	0.15	0.33	0.77	∞/−∞	−0.77	−0.36	−0.20	−0.14	−0.06	−0.03	−0.01
	0.2	0.01	0.07	0.13	0.26	0.37	0.39	−0.04	−0.28	−0.18	−0.13	−0.06	−0.03	−0.02
	0.4	0.01	0.04	0.09	0.15	0.14	0.15	0.00	−0.13	−0.16	−0.10	−0.05	−0.02	−0.02
	0.6	0.00	0.03	0.06	0.08	0.07	0.08	0.01	−0.05	−0.08	−0.07	−0.04	−0.02	−0.01
	0.8	0.00	0.02	0.05	0.06	0.05	0.05	0.01	−0.02	−0.04	−0.04	−0.03	−0.02	−0.01
	1.0	−0.01	0.02	0.05	0.05	0.04	0.05	0.02	0.00	−0.01	−0.03	−0.02	−0.01	0.00
	1.2	−0.01	0.01	0.04	0.04	0.04	0.05	0.02	0.01	0.00	−0.02	−0.01	−0.01	0.00
	1.6	−0.01	0.01	0.04	0.04	0.03	0.05	0.02	0.01	0.00	−0.01	0.01	0.00	0.00
1.2	0.0	−0.01	0.03	0.07	0.13	0.22	0.27	−0.36	−0.14	−0.06	−0.03	−0.01		
	0.2	−0.01	0.03	0.06	0.12	0.19	0.29	−0.28	−0.13	−0.06	−0.03	−0.02		
	0.4	0.00	0.03	0.06	0.10	0.14	0.17	−0.16	−0.10	−0.05	−0.02	−0.02		
	0.6	0.01	0.03	0.05	0.08	0.10	0.10	−0.09	−0.07	−0.04	−0.02	−0.01		
	0.8	0.02	0.04	0.05	0.07	0.08	0.07	−0.04	−0.04	−0.03	−0.01	−0.01		
	1.0	0.03	0.03	0.05	0.05	0.06	0.05	−0.02	−0.02	−0.02	−0.01	0.00		
	1.2	0.03	0.03	0.04	0.04	0.05	0.05	0.00	0.00	−0.01	0.00	0.00		
	1.6	0.02	0.03	0.04	0.05	0.04	0.05	0.00	0.00	0.00	0.00	0.00		

Table 7.19 Con't

δ	η	ξ														
		0.0	0.1	0.2	0.4	0.6	0.8	1.0	1.2	1.6	2.0	2.4	2.8	3.2	3.6	4.0
	0.0	-0.01	0.00	0.02	0.05	0.08	0.14	0.22	0.36	∞/-∞	-0.36	-0.14	-0.06	-0.03	-0.01	
	0.2	-0.01	0.00	0.02	0.05	0.08	0.13	0.19	0.28	0.00	-0.29	-0.13	-0.06	-0.03	-0.01	
	0.4	-0.01	0.00	0.02	0.04	0.07	0.10	0.14	0.17	0.00	-0.17	-0.10	-0.05	-0.02	-0.01	
1.6	0.6	0.00	0.00	0.02	0.03	0.05	0.07	0.09	0.10	0.00	-0.09	-0.06	-0.04	-0.02	-0.01	
	0.8	0.00	0.00	0.02	0.03	0.04	0.05	0.06	0.06	0.00	-0.05	-0.05	-0.03	-0.02	0.00	
	1.0	0.00	0.00	0.02	0.03	0.03	0.04	0.04	0.04	0.00	-0.03	-0.03	-0.02	-0.01	0.00	
	1.2	0.00	0.00	0.01	0.02	0.03	0.03	0.03	0.02	0.00	-0.02	-0.02	-0.02	-0.01	0.00	
	1.6	0.00	0.00	0.00	0.01	0.01	0.02	0.02	0.02	0.00	-0.01	-0.01	-0.01	0.00	0.00	
	0.0	-0.01	0.00	0.01	0.02	0.04	0.06	0.10	0.14	0.37	∞/-∞	-0.37	-0.15	-0.07	-0.04	-0.02
	0.2	-0.01	0.00	0.01	0.02	0.04	0.06	0.09	0.13	0.29	0.00	-0.29	-0.14	-0.07	-0.04	-0.02
	0.4	-0.01	0.00	0.00	0.02	0.04	0.05	0.08	0.11	0.17	0.00	-0.17	-0.12	-0.06	-0.03	-0.02
2.0	0.6	-0.01	0.00	0.00	0.02	0.03	0.04	0.06	0.08	0.16	0.00	-0.10	-0.08	-0.05	-0.03	-0.02
	0.8	0.00	0.00	0.00	0.01	0.02	0.03	0.05	0.05	0.05	0.00	-0.05	-0.06	-0.04	-0.03	-0.02
	1.0	0.00	0.00	0.00	0.01	0.02	0.03	0.03	0.04	0.03	0.00	-0.03	-0.04	-0.03	-0.02	-0.01
	1.2	0.00	0.00	0.00	0.01	0.01	0.02	0.02	0.03	0.02	0.00	-0.02	-0.03	-0.02	-0.02	-0.01
	1.6	0.00	0.00	0.01	0.01	0.01	0.01	0.01	0.01	0.01	0.00	-0.01	-0.01	-0.01	-0.01	0.00

Table 7.20 Coefficients $\overline{Q}_y$

δ	η	0.0	0.1	0.2	0.4	0.6	0.8	1.0	1.2	1.6	2.0	2.4	2.8	3.2	3.6	4.0
																ξ
0.0	0.0	$-\infty/0$	0	0	0	0	0	0	0	0						
	0.2	-1.48	-1.17	-0.71	-0.24	-0.10	-0.04	-0.02	-0.01	-0.01						
	0.4	-0.58	-0.54	-0.46	-0.26	-0.13	-0.06	-0.03	-0.02	-0.01						
	0.6	-0.24	-0.24	-0.24	-0.17	-0.11	-0.06	-0.03	-0.02	-0.01						
	0.8	-0.04	-0.08	-0.09	-0.08	-0.06	-0.04	-0.02	-0.01	-0.01						
	1.0	0.08	0.04	0.02	-0.01	-0.02	-0.02	-0.01	-0.01	-0.02						
	1.2	0.11	0.08	0.05	0.02	0.00	-0.01	-0.01	-0.02	-0.03						
	1.6	0.07	0.06	0.04	0.02	0.00	-0.01	-0.02	-0.02	-0.03						
0.1	0.0	0	$-\infty$	$-\infty$	0	0	0	0	0	0						
	0.2	-0.74	-0.87	-0.69	-0.28	-0.13	-0.07	-0.04	-0.03	-0.01						
	0.4	-0.55	-0.52	-0.48	-0.32	-0.19	-0.12	-0.07	-0.05	-0.02						
	0.6	-0.48	-0.43	-0.38	-0.29	-0.20	-0.14	-0.10	-0.07	-0.02						
	0.8	-0.40	-0.35	-0.32	-0.25	-0.19	-0.14	-0.10	-0.07	-0.03						
	1.0	-0.30	-0.27	-0.25	-0.21	-0.17	-0.14	-0.10	-0.07	-0.03						
	1.2	-0.22	-0.20	-0.18	-0.17	-0.15	-0.12	-0.09	-0.06	-0.03						
	1.6	-0.14	-0.14	-0.14	-0.14	-0.12	-0.08	-0.05	0.00	0.02						
0.2	0.0	0	0	0	0	0	0	0	0	0						
	0.2	-0.48	-0.70	-0.84	-0.42	-0.17	-0.09	-0.05	-0.03	-0.01						
	0.4	-0.47	-0.47	-0.48	-0.37	-0.22	-0.13	-0.08	-0.05	-0.02						
	0.6	-0.42	-0.40	-0.37	-0.30	-0.22	-0.15	-0.10	-0.07	-0.02						
	0.8	-0.36	-0.33	-0.30	-0.24	-0.19	-0.14	-0.10	-0.07	-0.03						
	1.0	-0.28	-0.25	-0.23	-0.20	-0.16	-0.13	-0.10	-0.07	-0.03						
	1.2	-0.17	-0.17	-0.17	-0.16	-0.14	-0.11	-0.08	-0.06	-0.02						
	1.6	-0.10	-0.10	-0.10	-0.10	-0.09	-0.07	-0.04	-0.01	0.01						

Table 7.20 Con't

ξ

δ	η	0.0	0.1	0.2	0.4	0.6	0.8	1.0	1.2	1.6	2.0	2.4	2.8	3.2	3.6	4.0
	0.0	0	0	0	−∞	0	0	0	0	0	0					
	0.2	−0.22	−0.29	−0.43	−0.81	−0.40	−0.16	−0.08	−0.04	−0.02	−0.01					
	0.4	−0.30	−0.33	−0.37	−0.42	−0.33	−0.20	−0.11	−0.07	−0.02	−0.01					
	0.6	−0.31	−0.31	−0.30	−0.29	−0.25	−0.18	−0.12	−0.08	−0.03	−0.01					
0.4	0.8	−0.28	−0.26	−0.24	−0.22	−0.19	−0.15	−0.11	−0.08	−0.03	−0.01					
	1.0	−0.22	−0.21	−0.9	−0.17	−0.15	−0.12	−0.09	−0.07	−0.02	−0.01					
	1.2	−0.14	−0.14	−0.14	−0.13	−0.11	−0.10	−0.07	−0.05	−0.02	−0.01					
	1.6	−0.08	−0.08	−0.07	−0.07	−0.06	−0.05	−0.03	−0.01	0.00	0.00					
	0.0	0	0	0	0	−∞	0	0	0	0	0	0	0			
	0.2	−0.11	−0.13	−0.18	−0.40	−0.80	−0.39	−0.15	−0.07	−0.02	−0.01	0.00	0.00			
	0.4	−0.19	−0.21	−0.23	−0.33	−0.40	−0.31	−0.18	−0.10	−0.03	−0.01	0.00	0.01			
	0.6	−0.21	−0.21	−0.22	−0.25	−0.26	−0.22	−0.16	−0.11	−0.04	−0.01	0.00	0.01			
0.6	0.8	−0.20	−0.19	−0.19	−0.19	−0.19	−0.19	−0.13	−0.04	−0.04	−0.01	0.00	0.01			
	1.0	−0.17	−0.16	−0.16	−0.15	−0.14	−0.12	−0.10	−0.07	−0.04	−0.01	0.00	0.02			
	1.2	−0.12	−0.12	−0.12	−0.11	−0.10	−0.09	−0.07	−0.06	−0.03	−0.01	0.00	0.02			
	1.6	−0.05	−0.05	−0.05	−0.05	−0.04	−0.04	−0.03	−0.02	−0.03	0.00	0.01	0.00			
	0.0	0	0	0	0	0	−∞	0	0	0	0	0	0			
	0.2	−0.07	−0.08	−0.09	−0.16	−0.39	−0.79	−0.39	−0.15	−0.03	−0.01	0.00	0.00			
	0.4	−0.12	−0.13	−0.14	−0.20	−0.31	−0.38	−0.30	−0.18	−0.06	−0.02	0.00	0.00			
	0.6	−0.14	−0.14	−0.15	−0.18	−0.22	−0.24	−0.21	−0.15	−0.06	−0.02	−0.01	0.00			
0.8	0.8	−0.14	−0.14	−0.14	−0.15	−0.16	−0.16	−0.15	−0.12	−0.06	−0.02	−0.01	0.01			
	1.0	−0.11	−0.11	−0.11	−0.11	0.12	−0.11	−0.10	−0.09	−0.04	−0.02	0.00	0.01			
	1.2	−0.08	−0.08	−0.08	−0.08	−0.08	−0.08	−0.07	−0.06	−0.03	−0.02	0.00	0.01			
	1.6	−0.03	−0.03	−0.03	−0.03	−0.03	−0.03	−0.02	−0.02	−0.02	−0.01	0.00	0.00			

Group 1.2

0.0	0	0	0	0	0	0	0	$-\infty$	0	0	0	0	0
0.2	−0.03	−0.03	−0.03	−0.04	−0.07	−0.15	−0.38	−0.78	−0.13	−0.03	−0.01	0.00	0.00
0.4	−0.06	−0.06	−0.06	−0.07	−0.11	−0.18	−0.29	−0.37	−0.17	−0.05	−0.02	−0.01	−0.00
0.6	−0.08	−0.08	−0.08	−0.09	−0.11	−0.15	−0.20	−0.22	−0.15	−0.06	−0.02	−0.01	−0.00
0.8	−0.08	0.08	−0.08	−0.09	−0.10	−0.12	−0.14	−0.15	−0.11	−0.05	−0.02	0.00	0.01
1.0	−0.09	0.08	−0.08	−0.08	−0.09	−0.09	−0.10	−0.10	−0.09	−0.04	−0.02	0.00	0.01
1.2	−0.08	−0.08	−0.07	−0.07	−0.07	−0.07	−0.07	−0.07	−0.05	−0.03	−0.02	0.00	0.01
1.6	−0.04	−0.04	−0.04	−0.05	−0.06	−0.04	−0.05	−0.02	−0.02	−0.02	−0.01	0.00	0.01

Group 1.6

0.0	0	0	0	0	0	0	0	0	$-\infty$	0	0	0	0	0
0.2	−0.01	−0.01	−0.01	−0.01	−0.02	−0.03	−0.07	−0.14	−0.78	−0.13	−0.03	−0.01	0.00	0.00
0.4	−0.02	−0.02	−0.02	−0.03	−0.04	−0.06	−0.10	−0.17	−0.36	−0.17	−0.05	−0.02	0.00	0.00
0.6	−0.03	−0.03	−0.03	−0.03	−0.04	−0.06	−0.09	−0.14	−0.22	−0.15	−0.06	−0.02	0.00	0.00
0.8	−0.04	−0.04	−0.04	−0.04	−0.04	−0.06	−0.08	−0.11	−0.14	−0.10	−0.05	−0.02	0.00	0.00
1.0	−0.04	−0.04	−0.04	−0.04	−0.04	−0.05	−0.06	−0.08	−0.10	−0.09	−0.04	−0.02	0.00	0.00
1.2	−0.03	−0.03	−0.03	−0.03	−0.03	−0.04	−0.04	−0.05	−0.06	−0.05	−0.03	−0.02	0.00	0.00
1.6	−0.02	−0.02	−0.02	−0.02	−0.02	−0.02	−0.02	−0.02	−0.02	−0.02	−0.02	−0.01	0.00	0.00

Group 2.0

0.0	0	0	0	0	0	0	0	0	$-\infty$	0	0	0	0	0
0.2	0.00	0.00	−0.01	−0.01	−0.01	−0.02	−0.03	−0.13	−0.78	−0.13	−0.03	−0.01	0.00	0.00
0.4	−0.01	−0.01	−0.01	−0.01	−0.02	−0.03	−0.05	−0.17	−0.37	−0.17	−0.05	−0.02	−0.01	0.00
0.6	−0.01	−0.01	−0.01	−0.02	−0.03	−0.04	−0.06	−0.15	−0.32	−0.15	−0.06	−0.03	−0.01	−0.01
0.8	−0.01	−0.01	−0.01	−0.02	−0.03	−0.04	−0.05	−0.11	−0.14	−0.11	−0.05	−0.03	−0.01	−0.01
1.0	−0.01	−0.01	−0.01	−0.02	−0.03	−0.03	−0.04	−0.09	−0.10	−0.09	−0.04	−0.03	−0.01	−0.01
1.2	−0.01	−0.01	−0.01	−0.01	−0.02	−0.03	−0.03	−0.05	−0.06	−0.05	−0.04	−0.02	−0.01	−0.01
1.6	−0.01	−0.01	−0.01	−0.01	−0.01	−0.02	−0.02	−0.03	−0.04	−0.03	−0.03	−0.01	0.00	0.00

Table 7.21 Coefficients for obtaining the soil pressure, moments and shear forces due to the weight of the infinite plate

$\bar{\xi}$	0.0	0.2	0.4	0.6	0.8	1.0	1.2	1.4	1.6	1.8	2.0	2.2	2.4	2.6	2.8	3.0
$\bar{q}_w$	0	13.2	1.1	1.0	0.94	0.92	0.91	0.91	0.92	0.93	0.95	0.95	0.96	0.96	0.97	0.97
$\bar{M}_{xw}$	0	0.01	0.03	0.06	0.09	0.12	0.14	0.16	0.17	0.18	0.19	0.20	0.21	0.21	0.21	0.21
$\bar{Q}_{xw}$	0	0.09	0.13	0.14	0.13	0.12	0.10	0.09	0.07	0.06	0.04	0.03	0.02	0.02	0.01	0.0

Table 7.22 Shear forces along the perimeter of the column

$\delta = d/L$	From the right	From the left
0.4	1.07	0.93
0.3	1.19	0.81
0.2	1.35	0.65
0.1	1.61	0.39

Table 7.23 Analysis of an infinite mat (Example 7.5)

Location of the column	x	y	ζ	η	w	p	M_x	M_y	Q_x
17	0	0	0	0	0.1490	3,370.20	51,341.0	6,308.0	79,166.0
9	−8	−8	−2.42	−2.42	0.0414	52.60	−760.0	−1,583.0	0
10	0	−8	0	−2.42	0.0615	263.29	2,090.0	0	0
11	8	−8	2.42	−2.42	0.0414	52.60	−760.0	1,583.0	0
16	−8	0	−2.42	0	0.0615	263.29	−3,040.0	0	−462.0
18	8	0	2.42	0	0.0615	263.29	−3,040.0	0	−462.0
23	−8	8	2.42	2.42	0.0414	52.60	−760.0	0	0
24	0	8	0	2.42	0.0615	263.29	2,090.0	1,583.0	0
25	8	8	2.42	2.42	0.0414	52.60	−760.0	−1,583.0	0
				Σ	0.5606	4,633.76	46,401.0	6,308.0	79,166.0

Table 7.24 Analysis of a semi-infinite mat (Example 7.6)

Load	x	y	ξ	η	$\overline{p}$	$\overline{M}_x$
P_1	3	−14	0.6	3.2	0.000	0.000
P_2	3	−7	0.6	1.6	0.100	0.025
P_3	3	0	0.6	0	0.210	0.196
P_4	3	7	0.6	1.6	0.100	0.025
P_5	7	−14	1.6	3.2	0.002	0.001
P_6	7	−7	1.6	1.6	0.018	−0.002
P_7	7	0	1.6	0	0.041	−0.014
P_8	7	7	1.6	1.6	0.018	−0.002
P_9	14	−14	3.2	−3.2	0.000	0.000
P_{10}	14	−7	3.2	1.6	0.000	−0.008
P_{11}	14	0	3.2	0	0.000	−0.012
P_{12}	14	−7	3.2	1.6	0.000	−0.008

References

1. American Concrete Institute (ACI), ACI Committee 436. 1966. Suggested design procedures for combined footings and mats. *ACI Journal Proceedings* 63 (October): 1041–1058.
2. Baker, A. L. L. 1948. *Raft foundations*. 2nd ed. London: Concrete Publications.
3. Bowles, J. E. 1977. *Foundation analysis and design*. New York: McGraw-Hill.
4. Gorbunov-Posadov, M. I. 1959. *I. Tables for analysis of thin plates on elastic foundation*. Moscow: Gosstroiizdat.
5. Gorbunov-Posadov, M. I. 1984. *Analysis of structures on elastic foundation*. Moscow: Stroiizdat, p. 246.
6. Hetenyi, M. 1946. *Beams on elastic foundation*. Ann Arbor: University of Michigan Press.
7. Korenev, B. G. 1962. *Analysis of plates on elastic foundation*. Moscow: Gosstroiizdat.
8. Scott, R. F. 1981. *Foundation analysis*. Civil Engineering and Engineering Mechanics Series. Englewood Cliffs, NJ: Prentice Hall.
9. Selvadurai, A.P.S. 1979. *Elastic analysis of soil-foundation interaction*. Amsterdam/Oxford/New York: Elsevier Scientific.
10. Shechter. O. Y. 1939. *Analysis of an infinite plate supported on elastic half-space and elastic layer loaded with a concentrated vertical load*. Publications Research Co. Moscow: Fundamentstroi, Gosstroiizdat, nu. 10.
11. Shechter, O. Y. 1948. *Regarding analysis of mat foundations supported on elastic layer*, no. 11. Stroivoenmorizdat, Moscow.
12. Teng, W. C. 1975. Mat foundations. In *Foundation engineering handbook*. 1st ed. New York: Van Nostrand Reinhold, p. 528.

8

Numerical Analysis of Mat Foundations

8.1 Introduction

Although simplified methods of analysis described in Chapter 7 are still used by practicing engineers, they cannot always produce accurate and reliable results. The design of large and complex structures requires application of more accurate analytical or numerical methods of analysis. As shown by Selvadurai (1979), various methods can be employed for analysis of plates supported on elastic foundation to accomplish these goals.

However, exact analytical methods have their limitations. They were developed mostly for rectangular or circular foundations and do not take into account variable soil properties and variable foundation rigidities. Therefore, in most of the cases practicing engineers use numerical methods: finite element and finite difference methods that do not have these limitations. However, it is important to mention that, in accordance with Bowles (1977), the finite difference method has some disadvantages compared to the finite element method, such as difficulties allowing for holes, notches, reentrant corners, and others, and also applying moments to the nodes. It is also noteworthy that the accuracy of the finite difference method depends on the size and number of squares by which the mat is divided. As mentioned by Teng (1975), when the squares are significantly larger than the size of the columns, the results adjacent to the column areas are unreliable. Nevertheless, the finite difference method is still used by some authors and engineers in practical applications.

Chapter 8 introduces some methods of analysis of mats supported on elastic foundation based only on the finite element method. However, it does not include a detailed description of the finite element method. The method is widely used and already well known. The author assumes that the reader has the basic knowledge and experience in using the method for analysis of regular plates.

This chapter also includes analysis of mat foundations supported on piles, and methods of combined analysis of mats and superstructure such as frames and walls. The

---

All tables can be found at the end of the chapter.

methods described below are based on three soil models: Winkler foundation, elastic half-space, and elastic layer.

8.2 Analysis of Mats Supported on Winkler Foundation

Analysis of mat foundations is usually performed using software based on Winkler's soil model. Analysis is relatively simple:

1. The mat is divided into a system of small plates—finite elements supported on soil. So, the given mat is replaced with a series of finite elements, as shown in Figure 8.1.
2. Continuous contact between the mat and the soil is replaced with a series of elastic supports located at all nodes of the mat. Rigidities of all supports are obtained as $C_i = k_i A_i$, where A_i is the tributary area of node i and k_i is the modulus of subgrade reaction of the soil at point i.
3. Using available software for regular plate analysis, the stiffness matrix of the plate $\vec{A}$ is built.
4. Obtaining the stiffness matrix of the soil is simple. Since all support rigidities are independent and do not affect each other, the matrix $\vec{B}$ looks like a diagonal matrix that shows support rigidities at each node of the mat.
5. The final system of linear equations looks as follows:

$$(\vec{A} + \vec{B})X = \vec{C} \tag{8.1}$$

where X are unknown deflections of the mat at all nodes and $\vec{C}$ is the matrix of the given loads applied to the mat.

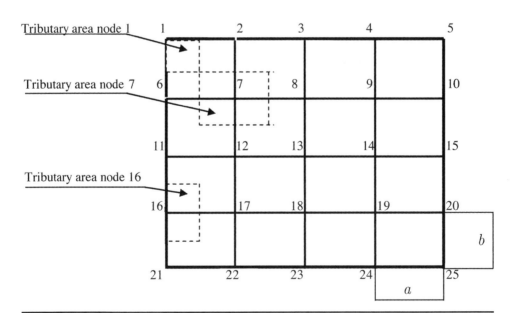

Figure 8.1 Plan of the mat with tributary areas

The total number of unknowns in this system is equal to $3n$, where n is the total number of nodes. Analysis takes into account only three deflections at each node: the settlement and two rotations about axes X and Y located in the plane of the plate. For example, the number of unknowns in the plate shown in Figure 8.1 is equal to $3n = 3 \cdot 25 = 75$. In some cases the number of unknown deflections can be less than $3n$, when some nodes are restrained against deflections. If, let's say, the right edge of the plate, shown in Figure 8.1, is simply supported, all nodes located at this edge are restrained against settlements at points 5, 10, 15, 20, and 25, and the total number of unknowns is equal to $75 - 5 = 70$. As we can see, analysis of plates on Winkler foundation is simple, and therefore attracts practicing engineers as well as developers of structural software.

Analysis of plates supported on elastic half-space or elastic layer is more complex. The stiffness matrix of the plate is no different from the matrix for analysis of plates on Winkler foundation, but obtaining the stiffness matrix of the soil is much more difficult, because the rigidity of the soil at each point of support depends not only on the soil rigidity at that point, but also on deflections produced at this point by loads applied to the soil at other points. The section below introduces a numerical method of analysis of plates supported on elastic half-space that can also be used for an elastic layer and other soil models which follow the rules of the theory of elasticity when loads produce settlements not only at the areas they are applied, but also at the areas outside of the loaded areas.

8.3 Analysis of Mat Foundations Supported on Elastic Half-Space: The First Method

Let us assume that the plate shown in Figure 8.1 is supported on elastic half-space. The plate is divided into finite elements in such a way that the plate consists of four groups of finite elements. Each group contains four elements shown in Figure 8.2. Analysis is performed in several steps:

1. Continuous contact between the plate and the soil is replaced with a series of individual non-yielding supports located not at all nodes, but only at the centers of each group of nodes (in our case, at points 7, 9, 17, and 19). By dividing the plate into groups and locating supports at the centers of each group, the total number of supports as well as the total number of equations is reduced four times.
2. Reactions in all of these supports depend on the loads applied to the plate as well as on the settlements of all supports.
3. The total reaction at any support i can be obtained from the following equation:

$$X_i = r_{i1}\Delta_1 + r_{i2}\Delta_2 + \ldots + r_{in}\Delta_n + r_{iP} \tag{8.2}$$

In this equation, r_{ik} is the reaction at point i due to one unit settlement applied to the plate at point k, Δ_i is the unknown settlement of the plate at point i, and r_{iP} is the reaction at point i due to the given loads applied to the plate. The system of equations for all points of the plate will look as follows:

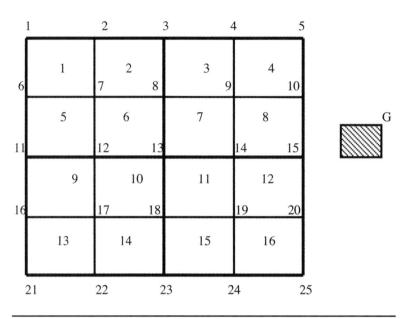

Figure 8.2 The mat supported at points 7, 9, 17, and 19

$$
\left.
\begin{aligned}
X_7 &= r_{77}\Delta_7 + r_{79}\Delta_9 + r_{7,17}\Delta_{17} + r_{7,19}\Delta_{19} + r_{7P} \\
X_9 &= r_{97}\Delta_7 + r_{99}\Delta_9 + r_{9,17}\Delta_{17} + r_{9,19}\Delta_{19} + r_{9P} \\
X_{17} &= r_{17,7}\Delta_7 + r_{17,9}\Delta_9 + r_{17,17}\Delta_{17} + r_{17,19}\Delta_{19} + r_{17P} \\
X_{19} &= r_{19,7}\Delta_7 + r_{19,9}\Delta_9 + r_{19,17}\Delta_{17} + r_{19,19}\Delta_{19} + r_{19P}
\end{aligned}
\right\}
\tag{8.3}
$$

The same reactions acting in the opposite direction and applied to the soil as active vertical loads will produce settlements of the soil. However, the loads applied to the soil cannot be applied as concentrated loads. In reality, they are applied as uniformly distributed loads. The uniformly distributed load applied to any soil area is obtained as follows:

$$q_i = X_i/A_i \tag{8.4}$$

where A_i is the area of the soil with the uniformly distributed load q_i. The total settlement of the soil is found from this equation:

$$\Delta_i = \Delta_{i1}X_1 + \Delta_{i2}X_2 + \ldots + \Delta_{in}X_n + \Delta_{iG} \tag{8.5}$$

In this equation, Δ_{ik} is the soil settlement of the area i due to one unit load uniformly distributed over area k, and Δ_{iG} is the settlement of the soil at area i due to the given load G applied to the soil outside of the plate. The number of such loads is unlimited. Settlements of the soil at all four areas supporting the plate are shown in Figure 8.3. Settlements of the half-space at all points of support are found from the following system of equations:

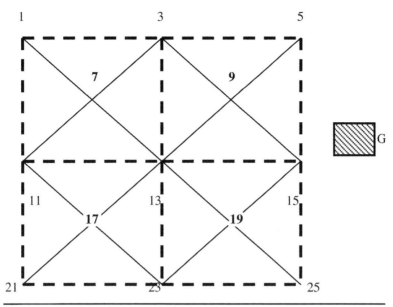

Figure 8.3 The half-space divided into bearing areas with centers at points 7, 9, 17, and 19

$$\left.\begin{aligned}
\Delta_7 &= \Delta_{77}X_7 + \Delta_{79}X_9 + \Delta_{7,17}X_{17} + \Delta_{7,19}X_{19}\Delta_{7G} \\
\Delta_9 &= \Delta_{97}X_7 + \Delta_{99}X_9 + \Delta_{9,17}X_{17} + \Delta_{9,19}X_{19}\Delta_{9G} \\
\Delta_{17} &= \Delta_{17,7}X_7 + \Delta_{17,9}X_9 + \Delta_{17,17}X_{17} + \Delta_{17,19}X_{19}\Delta_{17G} \\
\Delta_{19} &= \Delta_{19,7}X_7 + \Delta_{19,9}X_9 + \Delta_{19,17}X_{17} + \Delta_{19,19}X_{19}\Delta_{19G}
\end{aligned}\right\} \tag{8.6}$$

Both systems of equations 8.3 and 8.6 contain eight unknowns: four reactions and four settlements. The total number of equations is equal to eight. By solving both systems as one, all reactions and all settlements are found.

However, obtained concentrated soil reactions X_7, X_9, X_{17}, and X_{19} cannot be applied to the plate. These reactions, in reality, are applied to the plate also as uniformly distributed loads. Since distributed loads cannot be applied to the plate, they are applied to the plate at all nodes as shown in Table 8.1.

From Table 8.1 we can see that four reactions (X_7, X_9, X_{17}, X_{19}) are distributed between 25 nodes of the plate, and the sum of these reactions is equal to all vertical loads applied to the mat or:

$$\sum_{n=1}^{n=25} r_n = (X_7 + X_9 + X_{17} + X_{19}) = \sum P_j \tag{8.7}$$

where $\sum P_j$ is the sum of all given vertical loads applied to the plate and $\sum_{n=1}^{n=25} r_n$ is the sum of all reactions at all 25 supports. Now, by applying all found reactions acting in the opposite direction to the soil, all settlements at all nodes are found. Final analysis of the plate is performed by applying to the plate the given vertical loads and obtained settlements.

However, settlements at points 7, 9, 17, and 19 will differ from settlements obtained previously at these points. By redistributing the soil reactions between all nodes of the plate and by applying the same reactions in the opposite direction to the half-space, we obtain more accurate final results.

The described method has an obvious convenience. By supporting the plate only in some selected points and applying the uniformly distributed soil pressure to large areas of soil equal to the area of four finite elements, we significantly reduce the number of unknown reactions. But distribution of the soil pressure over a large area also has a disadvantage: it may reduce the accuracy of the analysis. Below is an introduction of the second method of analysis in which the soil pressure is distributed between all nodes of the plate and the soil pressure is applied to much smaller areas of the soil.

8.4 The Second Method

Analysis is performed as follows:

1. Continuous contact between the plate and the soil is replaced with a series of totally rigid non-yielding supports located at all 25 nodes.
2. Let us assume that the soil surface under each finite element is loaded with a uniformly distributed load equal to $q_i = 4X_i/A_i$, where A_i is the area of the finite element i and X_i is the reaction applied to each node of the finite element.
3. The total soil reactions applied to the plate at each node are shown below in Table 8.2.

All reactions X_i shown in Table 8.2 represent full reactions or parts of full reactions applied to the plate nodes. In order to simplify calculations, it is convenient to replace these reactions with new reactions Y_i applied to each node of the plate.

$$X_1 = Y_1, X_1 + X_2 = Y_2, X_2 + X_3 = Y_3, X_3 + X_4 = Y_4, X_5 = Y_5, X_1 + X_5 = Y_6,$$
$$X_1 + X_2 + X_5 + X_6 = Y_7, X_2 + X_3 + X_6 + X_7 = Y_8, X_3 + X_4 + X_7 + X_8 = Y_9,$$
$$X_4 + X_8 = Y_{10}, X_5 + X_9 = Y_{11}, X_5 + X_6 + X_9 + X_{10} = Y_{12}, X_6 + X_7 + X_{10} + X_{11} = Y_{13},$$
$$X_7 + X_8 + X_{11} + X_{12} = Y_{14}, X_8 + X_{12} = Y_{15}, X_9 + X_{13} = Y_{16}, X_9 + X_{10} + X_{13} + X_{14} = Y_{17},$$
$$X_{10} + X_{11} + X_{14} + X_{15} = Y_{18}, X_{11} + X_{12} + X_{15} + X_{16} = Y_{19}, X_{12} + X_{16} = Y_{20},$$
$$X_{13} = Y_{21}, X_{13} + X_{14} = Y_{22}, X_{14} + X_{15} = Y_{23}, X_{15} + X_{16} = Y_{24}, X_{16} = Y_{25}.$$

All reactions found above are acting in the opposite direction and applied to the soil as active uniformly distributed loads. The distributed loads can be found from the following equation:

$$q_i = Y_i/A_i \tag{8.8}$$

Settlements of the soil at any point i can be found from the following equation:

$$\Delta_i = \Delta_{i1} Y_1 + \Delta_{i2} Y_2 + \ldots + \Delta_{in} Y_n + \Delta_{iG} \tag{8.9}$$

where Δ_{ik} is the settlement of the soil at point i due to one unit load applied to the soil at point k, and Δ_{iG} is the settlement of the soil at point i due to the given loads applied to the soil outside of the area of the plate. Settlements of the soil supporting the plate, shown in Figures 8.4 and 8.5, can be obtained from the following system of equations:

$$\left.\begin{array}{l} \Delta_1 = \Delta_{11} Y_1 + \Delta_{12} Y_2 + \ldots + \Delta_{1,25} Y_{25} + \Delta_{1G} \\ \Delta_2 = \Delta_{21} Y_1 + \Delta_{22} Y_2 + \ldots + \Delta_{2,25} Y_{25} + \Delta_{2G} \\ \ldots\ldots\ldots\ldots\ldots\ldots\ldots\ldots\ldots\ldots\ldots\ldots\ldots\ldots\ldots\ldots\ldots\ldots \\ \Delta_{25} = \Delta_{25,1} Y_1 + \Delta_{25,2} Y_2 + \ldots + \Delta_{25,25} Y_{25} + \Delta_{25G} \end{array}\right\} \qquad (8.10)$$

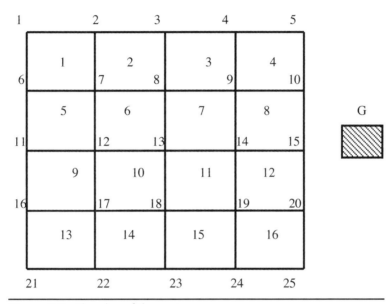

Figure 8.4 The plan of the mat

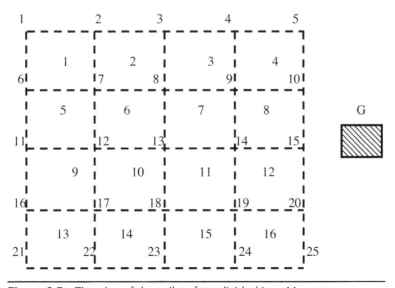

Figure 8.5 The plan of the soil surface divided into 16 areas

Reaction at any point i supporting the plate can be obtained from the following equation:

$$Y_i = r_{i1}\Delta_1 + r_{i2}\Delta_2 + \ldots + r_{in}\Delta_n + r_{iP} \qquad (8.11)$$

where r_{ik} is the reaction at point i due to one unit settlement of the plate at point k, and r_{iP} is the reaction at point i due to the given loads applied to the plate.

$$\left.\begin{array}{l} Y_1 = r_{11}\Delta_1 + r_{12}\Delta_2 + \ldots + r_{1,25}\Delta_{25} + r_{1P} \\ Y_2 = r_{21}\Delta_1 + r_{22}\Delta_2 + \ldots + r_{2,25}\Delta_{25} + r_{2P} \\ \cdots\cdots\cdots\cdots\cdots\cdots\cdots\cdots\cdots\cdots\cdots\cdots\cdots\cdots \\ Y_{25} = r_{25,1}\Delta_1 + r_{25,2}\Delta_2 + \ldots + r_{25,25}\Delta_{25} + r_{25P} \end{array}\right\} \qquad (8.12)$$

Each of the two systems of equations 8.10 and 8.12 contain 25 unknown settlements and 25 unknown reactions, totally 50 unknowns. The total number of equations is equal to 50. By solving these two systems of equations as one system, all unknown reactions and settlements are found. By applying to the plate the given loads and obtained settlements, final analysis is performed.

It is interesting to note that by solving both systems as one system with 50 unknowns, we use the so-called *combined method of structural mechanics* finding all settlements and reactions as well. If all reactions from system 8.12 are introduced into system 8.10, we have a system with only 25 unknown settlements. In other words, we are using the stiffness method of structural mechanics.

If we introduce the settlements from system 8.10 into system 8.12, we again have a system with 25 unknown reactions. In this case the method of forces is used. Any of these three ways of analysis are available.

As we can see, the analysis introduced here requires the use of two computer programs: one for regular plate analysis and the second for obtaining settlements of the elastic half-space. Both ways of analysis also require building and solving a system of linear equations.

8.5 The Iterative Method

Analysis of plates on an elastic foundation can be performed using the iterative method or method of iterations. Analysis is performed in several steps/iterations:

Step 1: Assuming that the plate is supported at all nodes, analysis of the plate with the given loads is performed. Obtained reactions at all nodes are applied to the soil and soil settlements are found.

Step 2: Analysis of the plate with the given loads and settlements found in Step 1 is performed and new support reactions are found. Obtained new reactions are applied to the soil and new specified settlements are found.

Step 3: Repeats the procedures performed in Step 2, and so on. After several iterations, results with acceptable accuracy are found.

It is easy to notice that the method of iterations explained above is nothing else but the solution of two systems of equations 8.10 and 8.12 using the same iterative method.

In fact, assuming in system 8.12 that all settlements are equal to zero, reactions at all points can be found as follows:

$$\left.\begin{array}{l} Y'_1 = r_{1P} \\ Y'_2 = r_{2P} \\ \dots\dots\dots \\ Y'_{25} = r_{25P} \end{array}\right\} \tag{8.13}$$

Now, by introducing obtained reactions from equations 8.13 into system 8.10, all settlements are found:

$$\left.\begin{array}{l} \Delta'_1 = \Delta_{11} Y'_1 + \Delta_{12} Y'_2 + \dots + \Delta_{1,25} Y'_{25} + \Delta_{1G} \\ \Delta'_2 = \Delta_{21} Y'_1 + \Delta_{22} Y'_2 + \dots + \Delta_{2,25} Y'_{25} + \Delta_{2G} \\ \dots\dots\dots\dots\dots\dots\dots\dots\dots\dots\dots\dots\dots \\ \Delta'_{25} = \Delta_{25,1} Y'_1 + \Delta_{25,2} Y'_2 + \dots + \Delta_{25,25} Y'_{25} + \Delta_{25G} \end{array}\right\} \tag{8.14}$$

Settlements from equation 8.14 introduced into system 8.12 allow obtaining new reactions Y''_{ik}, and introducing these reactions into system 8.10, new settlements Δ''_{ik} are found. After several iterations all soil settlements and all soil reactions are obtained.

   In conclusion, it is important to note that the iterative method is simple. It requires the use of available computer programs for analysis of regular plates and some hand calculations for obtaining settlements of the soil using equation 8.9. When the mat is supported on Winkler foundation, settlements are found from the well-known relationship $\Delta_i = P_i/k_iA_i$, where P_i is the load applied to the soil, k_i is the modulus of subgrade reaction, and A_i is the tributary area of loads at node i.

8.6 Zhemochkin's Method

In Chapter 6, the analysis of free supported beams on elastic half-space developed by Zhemochkin (1962) was explained. As shown by the author of this work (Tsudik 2003), the same method can be applied to analysis of plates supported on elastic half-space, elastic layer, and any other soil model.

   We shall describe the method using the plate shown in Figure 8.6. The given plate has various vertical loads and is supported on elastic half-space. Analysis is performed in the following order:

1. The given plate is divided into a series of rectangular finite elements.
2. Continuous contact between the soil and the plate is replaced with a series of individual non-yielding supports located at all nodes of the plate.
3. Three fictitious restraints are applied to one of the plate corners, for example, at node 1, with one restraint against vertical deflection Δ_1 and two others against rotations φ_{1X} and φ_{1Y} along axes X and Y.
4. All individual supports are replaced with unknown reactions X_2, X_3, . . ., X_{25} applied upward to the plate as concentrated vertical loads. Reaction X_1 is not taken into account because the plate at node 1 is restrained against any deflections. The same reactions are applied to the soil acting in the opposite direction

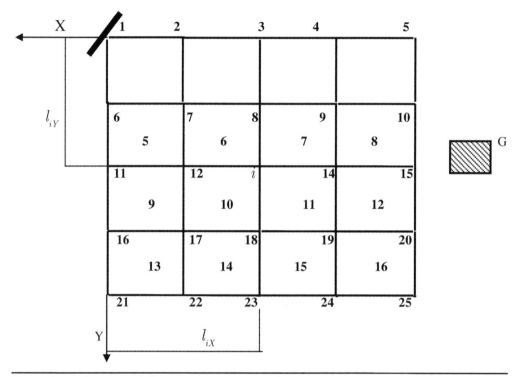

Figure 8.6   The mat with fictitious restraints at node 1

as uniformly distributed loads. All tributary areas of these loads have a rectangular shape. For example, the tributary area of the load at point 8 is equal to $\frac{1}{4}(A_2 + A_3 + A_6 + A_7)$, the tributary area of the load at point 3 is equal to $\frac{1}{4}(A_2 + A_3)$, and so on.

5. Now, the given system soil-plate is divided into two parts: the plate with the given loads and unknown soil reactions, and the soil loaded with a series of unknown uniformly distributed loads.

The soil has loads located outside of the plate area (loads G). Now, each part of the system can be investigated separately. The vertical linear deflection of the plate at any node i can be found from the following equation:

$$\Delta_i' = \Delta_{i2}X_2 + \ldots + \Delta_{i24}X_{24} + \Delta_{i25}X_{25} + \Delta_{iP} \tag{8.15}$$

$\Delta_{i1}X_1$ is not included in this equation because the plate at node 1 is restrained against any deflections including settlements. In this equation, Δ_{ik} is the vertical deflection of the plate at point i due to one unit load applied to the plate at point k, and Δ_{iP} is the vertical deflection of the plate at the same point i due to the given loads applied to the plate. However, formula 8.15 does not take into account that restraints applied at point 1 are fictitious and, in reality, do not exist. The plate at node 1 will experience one vertical deflection Δ_1 and two rotations, φ_{1X} and φ_{1Y}, that will produce additional vertical deflections of the plate. Assuming that the rotation of the plate in direction X about axes Y and rotation in direction Y about axes X are positive, the plate at any point i will experience three additional vertical deflections equal to Δ_1, $\varphi_{1X}l_{iX}$ and $\varphi_{1Y}l_{iY}$, respectively. All of these vertical deflections are positive because it is assumed that Δ_1 is directed upward and rotations are directed clockwise when looking in directions X or Y from the origin of the coordinates at point 1. Now, the vertical deflection of the plate at any point i will look as follows:

$$\Delta_i = \Delta_{i2}X_2 + \ldots + \Delta_{i25}X_{25} + \Delta_{iP} + \Delta_1 + \varphi_{1X}l_{iX} + \varphi_{1Y}l_{iY} \tag{8.16}$$

where the last three members are vertical deflections of the plate due to the vertical deflection and rotations of the plate at point 1. The settlements of the soil at the same point i are found as shown below:

$$\omega_i = \omega_{i2}X_2 + \ldots + \omega_{i25}X_{25} + \omega_{iG} \tag{8.17}$$

where ω_{ik} is the settlement of the soil at point i due to one unit load applied to the soil at point k, and ω_{iG} is the settlement of the plate at the same point due to the load applied to the soil outside of the plate.

When writing equation 8.17, the soil reaction at point 1 is not taken into account because the plate is restrained against any deflections at that point. Taking into account that vertical linear deflections of the soil and the plate at any point are the same and acting in opposite directions, the following equation can be written:

$$\delta_{i2}X_2 + \ldots + \delta_{i25}X_{25} + \Delta_1 + \varphi_{1X}l_{iX} + \varphi_{1Y}l_{iY} + \Delta_{iP} + w_{iG} = 0 \tag{8.18}$$

where $(\Delta_{ik} + \omega_{ik}) = \delta_{ik}$. Analogously for all nodes of the plate shown above in Figure 8.6, the following system of equations can be written:

$$\left.\begin{array}{l} \delta_{12}X_2 + \ldots + \delta_{1,25}X_{25} + \Delta_1 + \varphi_{1X}l_{1X} + \varphi_{1Y}l_{1Y} + \Delta_{1P} + w_{1G} = 0 \\ \delta_{22}X_2 + \ldots + \delta_{2,25}X_{25} + \Delta_1 + \varphi_{1X}l_{2X} + \varphi_{1Y}l_{2Y} + \Delta_{2P} + w_{2G} = 0 \\ \cdots\cdots\cdots\cdots\cdots\cdots\cdots\cdots\cdots\cdots\cdots\cdots\cdots\cdots\cdots\cdots\cdots\cdots\cdots\cdots\cdots\cdots \\ \delta_{25,2}X_2 + \ldots + \delta_{25,25}X_{25} + \Delta_1 + \varphi_{1X}l_{25X} + \varphi_{1Y}l_{25Y} + \Delta_{25P} + w_{25G} = 0 \end{array}\right\} \tag{8.19}$$

The total number of unknowns in this system of equations is equal to $(n + 3)$, where n is the number of unknown reactions and 3 is the number of unknown deflections at node 1. In our case $n = 24$ and the total number of unknowns is equal to $24 + 3 = 27$; 24 reactions plus 3 deflections at node 1. The additional three equations will look as shown below:

$$\left.\begin{array}{l} X_2 + \ldots + X_{25} + \sum P = 0 \\ X_2 l_{2X} + \ldots + X_{25}l_{25X} + \sum M_{PX} = 0 \\ X_2 l_{2Y} + \ldots + X_{25}l_{25Y} + \sum M_{PY} = 0 \end{array}\right\} \tag{8.20}$$

In this system, the first equation shows the equilibrium of all vertical loads and reactions applied to the plate, the second equation shows the equilibrium of all moments in the X direction at point 1, and the third equation shows the equilibrium of all moments in the Y direction at point 1. Now, we have a system of 27 equations with 27 unknowns. By solving this system all soil reactions and deflections at point 1 are found.

By applying to the plate the given loads, obtained reactions, and three deflections at point 1, final analysis of the plate is performed. This method is developed for analysis of plates free supported on elastic half-space and elastic layer and can be used for any other soil model.

Equations for obtaining settlements of the elastic half-space and elastic layer are given in Chapter 1. It is also important to mention that the method leads to the solution of a system of $(n + 3)$ equations, while the second method described earlier leads to a much larger system of equations with $2n$ unknowns.

8.7 Mats with Various Boundary Conditions

All methods described above are developed for analysis of plates free supported on soil when all nodes located along the perimeter of the plate have no restraints against rotations and vertical linear deflections. However, in some cases we deal with various boundary conditions; when some edges are supported on non-yielding vertical supports, other edges may be fixed (restrained against linear vertical deflections and rotations in two directions) and other edges are free. In most cases analysis of such types of plates is simpler because the number of unknowns is smaller. In Figure 8.7 some types of such plates are shown:

Type 1 is a plate with all edges simply supported on non-yielding supports.

Type 2 is a plate with three edges simply supported on non-yielding supports and the remainder of the area including the fourth edge supported on soil.

Type 3 is a plate with all fixed edges restrained against any deflections and the remainder of the plate supported on soil.

Type 4 is a plate with three fixed edges restrained against any deflections and the remainder of the plate supported on soil.

Type 5 is a plate with two parallel edges supported on non-yielding supports and the remainder of the plate supported on soil.

Type 6 is a plate with two parallel fixed edges and the remainder of the plate supported on soil.

Type 7 is a plate with one fixed edge restrained against any deflections, the second edge parallel to the first one simply supported on non-yielding supports, and the remainder of the plate supported on soil.

Type 8 is a plate with two fixed edges perpendicular to each other restrained against any deflections and the remainder of the plate supported on soil.

Type 9 is a plate with one fixed edge restrained against deflections and the remainder of the plate supported on soil.

Analysis of any plate shown above is simple. The number of unknowns is much smaller than the number of unknowns for analysis of plates free supported on soil. The number of unknowns is equal only to the number of nodes supported on soil. For example,

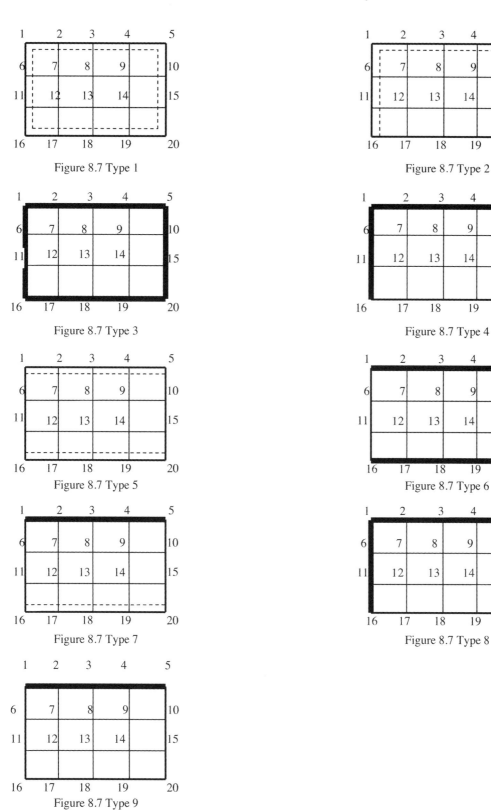

Figure 8.7 Type 1

Figure 8.7 Type 2

Figure 8.7 Type 3

Figure 8.7 Type 4

Figure 8.7 Type 5

Figure 8.7 Type 6

Figure 8.7 Type 7

Figure 8.7 Type 8

Figure 8.7 Type 9

Figure 8.7

analysis of the plate shown in Figure 8.7 (Type 3) requires solving a system of six equations. Analysis of the same plate with all four free supported on soil edges requires solving a system with 22 equations: 19 soil reactions and 3 deflections at point 1. Analysis is performed as follows:

1. Continuous contact between the plate and the soil is replaced with a series of individual totally rigid supports at all nodes supported on soil.
2. All individual supports are replaced with a series of unknown reactions X_{ik} acting upward and applied to the nodes of the plate.
3. The same reactions acting in the opposite direction as uniformly distributed loads $q_{ik} = X_{ik}/A_i$ are applied to the soil.

Now, the given system plate-soil is divided into two parts: the plate with the given loads and soil reactions, and the soil loaded with distributed soil reactions acting downward and loads located outside of the plate area. Settlement of the plate at any point i is obtained from the following equation:

$$\Delta_i = \Delta_{i1}X_1 + \Delta_{i2}X_2 + \dots + \Delta_{in}X_n + \Delta_{iP} \tag{8.21}$$

Settlement of the soil at the same point is obtained from equation 8.22:

$$\omega_i = \omega_{i1}X_1 + \omega_{i2}X_2 + \dots + \omega_{in}X_n + \omega_{iG} \tag{8.22}$$

By taking into account that the settlements of the soil and the plate are the same and by designating $\delta_{ik} = (\Delta_{ik} + \omega_{ik})$ for any plate shown above, the following system of equations can be written:

$$\left.\begin{array}{l} \delta_{11}X_1 + \delta_{12}X_2 + \dots + \delta_{1n}X_n + \Delta_{1P} + \omega_{1G} = 0 \\ \delta_{21}X_1 + \delta_{22}X_2 + \dots + \delta_{2n}X_n + \Delta_{2P} + \omega_{2G} = 0 \\ \dots\dots\dots\dots\dots\dots\dots\dots\dots\dots\dots\dots\dots\dots\dots\dots\dots\dots\dots \\ \delta_{n1}X_1 + \delta_{n2}X_2 + \dots + \delta_nX_n + \Delta_{nP} + \omega_{nG} = 0 \end{array}\right\} \tag{8.23}$$

As we can see, analysis always leads to a simple system of linear equations of the method of forces: all unknowns are the soil reactions $X_1, X_2, \dots, X_n$. The same method can be used when the plate is supported on other soil models. When the plate is supported on Winkler foundation, where $\omega_{ii} \neq 0$ and $\omega_{ik} = 0$, the system of equations looks as follows:

$$\left.\begin{array}{l} \delta_{11}X_1 + \Delta_{12}X_2 + \dots + \Delta_{1n}X_n + \Delta_{1P} = 0 \\ \Delta_{21}X_1 + \delta_{22}X_2 + \dots + \Delta_{2n}X_n + \Delta_{2P} = 0 \\ \dots\dots\dots\dots\dots\dots\dots\dots\dots\dots\dots\dots\dots\dots\dots\dots\dots\dots\dots \\ \Delta_{n1}X_1 + \Delta_{n2}X_2 + \dots + \delta_{nn}X_n + \Delta_{nP} = 0 \end{array}\right\} \tag{8.24}$$

System 8.24 reflects the essence of Winkler foundation: loads applied to the soil at any point produce settlements only at points they are applied. Therefore, load G located outside of the area of the plate do not produce any additional settlements of the plate. It is easy to understand that the same system of equations can be applied to analysis of any plate shown above, supported on a series of elastic columns or any other elastic or totally rigid supports. In conclusion, let us write a system of linear equations for the plate named Type 1. The system will include only six unknown reactions X_7, X_8, X_9, X_{12}, X_{13}, X_{14} and will look as follows:

$$\left.\begin{aligned}
\delta_{77}X_7 + \delta_{78}X_8 + \delta_{79}X_9 + \delta_{7,12}X_{12} + \delta_{7,13}X_{13} + \delta_{7,14}X_{14} + \Delta_{7P} + \omega_{7G} &= 0 \\
\delta_{87}X_7 + \delta_{88}X_8 + \delta_{89}X_9 + \delta_{8,12}X_{12} + \delta_{8,13}X_{13} + \delta_{8,14}X_{14} + \Delta_{8P} + \omega_{8G} &= 0 \\
\delta_{97}X_7 + \delta_{98}X_8 + \delta_{99}X_9 + \delta_{9,12}X_{12} + \delta_{9,13}X_{13} + \delta_{9,14}X_{14} + \Delta_{9P} + \omega_{9G} &= 0 \\
\delta_{12,7}X_7 + \delta_{12,8}X_8 + \delta_{12,9}X_9 + \delta_{12,12}X_{12} + \delta_{12,13}X_{13} + \delta_{12,14}X_{14} + \Delta_{12P} + \omega_{12G} &= 0 \\
\delta_{13,7}X_7 + \delta_{13,8}X_8 + \delta_{13,9}X_9 + \delta_{13,12}X_{12} + \delta_{13,13}X_{13} + \delta_{13,14}X_{14} + \Delta_{13P} + \omega_{13G} &= 0 \\
\delta_{14,7}X_7 + \delta_{14,8}X_8 + \delta_{14,9}X_9 + \delta_{14,12}X_{12} + \delta_{14,13}X_{13} + \delta_{14,14}X_{14} + \Delta_{14P} + \omega_{14G} &= 0
\end{aligned}\right\} \quad (8.25)$$

By solving this system of equations all soil reactions are found. Now, by applying obtained reactions to the soil, settlements at all nodes are obtained, and by applying the settlements to the plate along with the given loads, final analysis of the plate is performed.

It is important to remind the reader that the system of equations, in general, is not symmetrical because settlements of the soil depend on the tributary areas of the applied loads, and these areas are usually different. For example, in plate Type 9 the tributary area node 14 is equal to ab while the tributary area at node 20 is equal only to $ab/4$, where a and b are dimensions of the finite element.

8.8 Combined Analysis of Mat Foundations and Superstructures

Analyses of mat foundations, as well as analyses of other types of foundations, and superstructures are usually performed separately. Analysis of the foundation does not take into account the rigidity of the superstructure, and analysis of the superstructure does not take into account differential settlements and the rigidity of the foundation.

Deflections of individual foundations may significantly affect the superstructure, producing additional stresses in its elements, while the rigidity of the superstructure may reduce significantly differential settlements of the foundations.

Since mat foundations are very rigid, they usually do not produce large differential settlements, and therefore do not produce significant additional stresses in the superstructure. However, in many cases, when loads applied to the central area of the mat are much larger than loads applied along the perimeter, the differential settlements may be large enough. Therefore, analysis of mat foundations that takes into account the rigidity of the superstructure can reduce the differential settlements and specify the stresses in the mat as well as in the superstructure. The superstructure, in most cases, represents a 3D frame, or a combination of a 3D frame and a system of walls, or just a system of walls running in two directions. A method of combined analysis of mat foundations and 3D frames is introduced below. Figure 8.8 shows a 3D frame supported on a mat foundation.

Combined analysis of the mat and 3D frame is performed in the following steps:

1. Assuming that the columns of the given frame are restrained against any deflections, analysis of the frame with the given loads is performed and vertical reactions r_{9P}, r_{11P}, . . . r_{41P} are found at all n points of support.

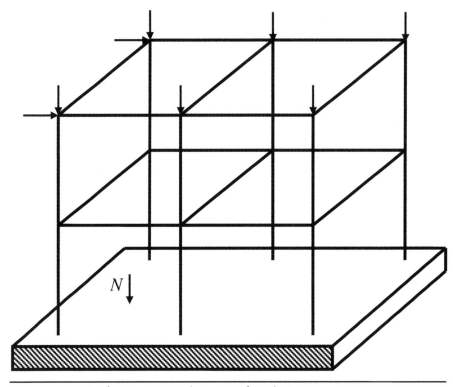

Figure 8.8   A 3D frame supported on a mat foundation

2. The given mat is divided into a series of rectangular finite elements and, as shown in Figure 8.9, the points of column supports are located at nodes 9, 11, 13, 37, 39, and 41.

3. By removing successively all restraints against vertical deflections at all points and by applying at all of these points vertical deflections equal to one unit, reactions at all frame columns are found as follows:

$$\left.\begin{aligned}
X_9 &= r_{9,9}\Delta_9 + r_{9,11}\Delta_{11} + r_{9,13}\Delta_{13} + r_{9,13}\Delta_{13} + r_{9,37}\Delta_{13} + r_{9,39}\Delta_{39} + r_{9,41}\Delta_{41} + r_{9P}\\
X_{11} &= r_{11,9}\Delta_9 + r_{11,11}\Delta_{11} + r_{11,13}\Delta_{13} + r_{11,13}\Delta_{13} + r_{11,37}\Delta_{13} + r_{11,39}\Delta_{39} + r_{11,41}\Delta_{41} + r_{11P}\\
X_{13} &= r_{13,9}\Delta_9 + r_{13,11}\Delta_{11} + r_{13,13}\Delta_{13} + r_{13,13}\Delta_{13} + r_{13,37}\Delta_{13} + r_{13,39}\Delta_{39} + r_{13,41}\Delta_{41} + r_{13P}\\
X_{37} &= r_{37,9}\Delta_9 + r_{37,11}\Delta_{11} + r_{37,13}\Delta_{13} + r_{37,13}\Delta_{13} + r_{37,37}\Delta_{13} + r_{37,39}\Delta_{39} + r_{37,41}\Delta_{41} + r_{37P}\\
X_{39} &= r_{39,9}\Delta_9 + r_{39,11}\Delta_{11} + r_{39,13}\Delta_{13} + r_{39,13}\Delta_{13} + r_{39,37}\Delta_{13} + r_{39,39}\Delta_{39} + r_{39,41}\Delta_{41} + r_{39P}\\
X_{41} &= r_{41,9}\Delta_9 + r_{41,11}\Delta_{11} + r_{41,13}\Delta_{13} + r_{41,13}\Delta_{13} + r_{41,37}\Delta_{13} + r_{41,39}\Delta_{39} + r_{41,41}\Delta_{41} + r_{41P}
\end{aligned}\right\} \quad (8.26)$$

4. Assuming that the given plate is supported at all nodes on non-yielding supports, the total vertical reaction at any of the supports can be expressed as follows:

$$Y_i = r_{i1}\Delta_1 + r_{i2}\Delta_2 + \ldots + r_{in}\Delta_n + r_{iN} + r_{i9}X_9 + r_{i,11}X_{11} + r_{i,13}X_{13} + r_{i,37}X_{37} + r_{i,39}X_{39} + r_{i,41}X_{41} \quad (8.27)$$

where r_{ij} is the vertical reaction at point i due to one unit settlement applied at point j, r_{iN} is the reaction at point i due to the loads N applied to the plate and located

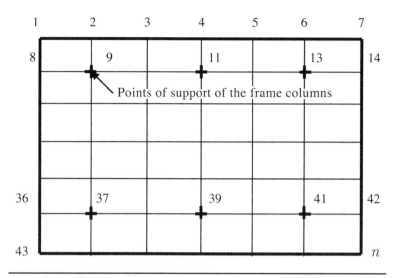

Figure 8.9 The plan of the foundation

between the columns, and r_{i9}, r_{i11}, r_{i13}, r_{i37}, r_{i39}, r_{i41} are reactions at point i due to one unit loads applied at points 9, 11, 13, 37, 39, and 41. X_9, X_{11}, X_{13}, X_{37}, X_{39}, X_{41} are unknown reactions acting downward applied to the foundation.

Reactions Y_i at all points of support can be obtained from the following system of equations:

$$Y_1 = r_{11}\Delta_1 + r_{12}\Delta_2 + \ldots + r_{1n}\Delta_n + r_{1N} + r_{1,9}X_9 + r_{1,11}X_{11} + r_{1,13}X_{13} + r_{1,37}X_{37} + r_{1,39}X_{39} + r_{1,41}X_{41}$$

$$Y_2 = r_{21}\Delta_1 + r_{22}\Delta_2 + \ldots + r_{2n}\Delta_n + r_{2N} + r_{2,9}X_9 + r_{2,11}X_{11} + r_{2,13}X_{13} + r_{2,37}X_{37} + r_{2,39}X_{39} + r_{2,41}X_{41}$$

$$\ldots\ldots\ldots\ldots\ldots\ldots\ldots\ldots\ldots\ldots\ldots\ldots\ldots\ldots\ldots\ldots\ldots\ldots\ldots\ldots\ldots\ldots\ldots$$

$$Y_n = r_{n1}\Delta_1 + r_{n2}\Delta_2 + \ldots + r_{nn}\Delta_n + r_{nN} + r_{n9}X_9 + r_{n11}X_{11} + r_{n13}X_{13} + r_{n37}X_{37} + r_{n39}X_{39} + r_{n41}X_{41} \quad (8.28)$$

Reactions Y_i acting in opposite directions applied to the soil as active uniformly distributed loads will produce settlements equal to the settlements of the plate at all n nodes. These settlements are found from the following equations:

$$\left.\begin{aligned}
\Delta_1 &= \Delta_{11}Y_1 + \Delta_{12}Y_2 + \ldots + \Delta_{1n}Y_n + \Delta_{1G} \\
\Delta_2 &= \Delta_{21}Y_1 + \Delta_{22}Y_2 + \ldots + \Delta_{2n}Y_n + \Delta_{2G} \\
&\ldots\ldots\ldots\ldots\ldots\ldots\ldots\ldots\ldots\ldots\ldots\ldots\ldots \\
\Delta_n &= \Delta_{n1}Y_1 + \Delta_{n2}Y_2 + \ldots + \Delta_{nn}Y_n + \Delta_{nG}
\end{aligned}\right\} \quad (8.29)$$

In this system, Δ_{iG} is the settlement of the soil at any point i due to the given load applied to the soil outside of the plate area.

Now, we have three systems of equations: 8.26, 8.28, and 8.29. These systems contain the following unknowns: n settlements (Δ_i), n soil reactions (Y_i), and m reactions (X_i) at the frame supports, where m is the number of frame columns. In our case $m = 6$. The total number of unknowns is equal to $2n + m$. The total number of equations is also equal to $2n + m$. By solving all three systems of equations we find reactions in the columns, settlements of the plate, and soil reactions at all nodes. Final analysis of the frame is performed by applying to the frame the given loads and settlements of the columns. Final analysis of the plate is performed by applying to the plate all given loads located between the columns (N), and reactions of the columns (X_i).

This method used can be applied to various soil models that have the ability to distribute stresses and deflections in the soil such as elastic half-space, elastic layer, and others. When the plate is supported on Winkler foundation analysis becomes much easier. System 8.29 is replaced with a series of simple equations as shown below:

$$Y_1 = \frac{\Delta_1}{\Delta_{11}} \quad Y_2 = \frac{\Delta_2}{\Delta_{22}} \cdots Y_n = \frac{\Delta_n}{\Delta_{nn}} \tag{8.30}$$

By introducing equations 8.30 into the system of equations 8.28, and by considering that reactions $Y_1, Y_2, \ldots Y_n$ are applied to the soil as active uniformly distributed loads, the following system of equations is obtained:

$$\left.\begin{array}{l} (r_{11} + k_1 A_1)\Delta_1 + r_{12}\Delta_2 + \ldots + r_{1n}\Delta_n + r_{1N} + r_{1,9}X_9 + r_{1,11}X_{11} + r_{1,13}X_{13} + r_{1,37}X_{37} + r_{1,39}X_{39} + r_{1,41}X_{41} = 0 \\ r_{21}\Delta_1 + (r_{22} + k_2 A_2)\Delta_2 + \ldots + r_{2n}\Delta_n + r_{2N} + r_{2,9}X_9 + r_{2,11}X_{11} + r_{2,13}X_{13} + r_{2,37}X_{37} + r_{2,39}X_{39} + r_{2,41}X_{41} = 0 \\ \cdots\cdots\cdots\cdots\cdots\cdots\cdots\cdots\cdots\cdots\cdots\cdots\cdots\cdots\cdots\cdots\cdots\cdots\cdots\cdots\cdots\cdots\cdots\cdots\cdots\cdots\cdots\cdots\cdots \\ r_{n1}\Delta_1 + r_{n2}\Delta_2 + \ldots + (r_{nn} + k_n A_n)\Delta_n + r_{nN} + r_{n9}X_9 + r_{n11}X_{11} + r_{n13}X_{13} + r_{n37}X_{37} + r_{n39}X_{39} + r_{n41}X_{41} = 0 \end{array}\right\} \tag{8.31}$$

In this system of equations k_i is the modulus of subgrade reaction of the soil at point i and A_i is the tributary area at point i. Now, we have two systems of equations 8.26 and 8.31. The total number of equations in both systems is equal to $(n + m)$, the total number of unknowns is also equal to $(n + m)$: n settlements and m reactions at points of column supports. In our case the total number of unknowns is equal to $(n + 6)$. As can be seen, the number of unknowns is significantly less than in the analysis of frames supported on elastic half-space and elastic layer.

8.9 Combined Analysis of Walls, Frames, and Mat Foundations

In many cases the superstructure represents a series of shear walls running in two orthogonal directions bearing the vertical loads and resisting the horizontal loads. Figure 8.10 shows a mat supported on elastic half-space and loaded with two walls A and B and a 3D frame. Analysis is similar to analysis of frames and mat foundations described above and performed using the finite element method:

1. It is assumed that the given mat is divided into a series of rectangular finite elements and supported on all n non-yielding supports replacing continuous contact between the mat and the soil.
2. Continuous contact between walls A and B and the mat is replaced with a series of non-yielding supports located at all nodes of the foundation, as shown in Figures 8.11 and 8.12. The 3D frame is shown in Figure 8.13.

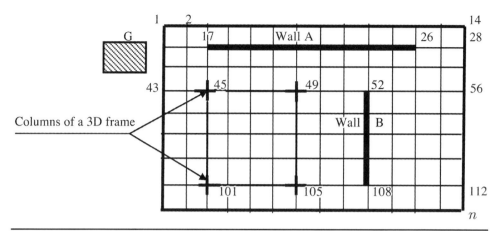

Figure 8.10 The plan of the mat with walls and frame

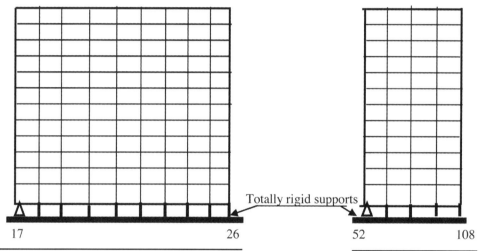

Figure 8.11 Wall A

Figure 8.12 Wall B

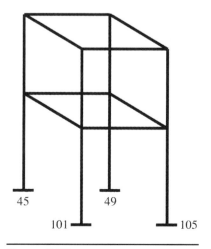

Figure 8.13

3. All columns of the frame at points of support are restrained against any deflections.
4. Using any available computer program for wall analysis, the method of finite element or method of finite difference, both walls A and B are analyzed successively under the given loads and one unit settlements applied to their points of support.
5. Using any available computer program for analysis of 3D statically indeterminate systems, analysis of the frame is performed successively under the given loads and one unit settlements applied to all points of support.
6. Using the results of performed analysis of wall A, the following system of equations is written:

$$
\left.\begin{aligned}
X_{17} &= r_{17,17}\Delta_{17} + r_{17,18}\Delta_{18} + \ldots + r_{17,26}\Delta_{26} + r_{17P_A} \\
X_{18} &= r_{18,17}\Delta_{17} + r_{18,18}\Delta_{18} + \ldots + r_{18,26}\Delta_{26} + r_{18P_A} \\
X_{19} &= r_{19,17}\Delta_{17} + r_{19,18}\Delta_{18} + \ldots + r_{19,26}\Delta_{26} + r_{19P_A} \\
X_{20} &= r_{20,17}\Delta_{17} + r_{20,18}\Delta_{18} + \ldots + r_{20,26}\Delta_{26} + r_{20P_A} \\
X_{21} &= r_{21,17}\Delta_{17} + r_{21,18}\Delta_{18} + \ldots + r_{21,26}\Delta_{26} + r_{21P_A} \\
X_{22} &= r_{22,17}\Delta_{17} + r_{22,18}\Delta_{18} + \ldots + r_{22,26}\Delta_{26} + r_{22P_A} \\
X_{23} &= r_{23,17}\Delta_{17} + r_{23,18}\Delta_{18} + \ldots + r_{23,26}\Delta_{26} + r_{23P_A} \\
X_{24} &= r_{24,17}\Delta_{17} + r_{24,18}\Delta_{18} + \ldots + r_{24,26}\Delta_{26} + r_{24P_A} \\
X_{25} &= r_{25,17}\Delta_{17} + r_{25,18}\Delta_{18} + \ldots + r_{25,26}\Delta_{26} + r_{25P_A} \\
X_{26} &= r_{26,17}\Delta_{17} + r_{26,18}\Delta_{18} + \ldots + r_{26,26}\Delta_{26} + r_{26P_A}
\end{aligned}\right\}
\tag{8.32}
$$

In these equations, X_i are reactions at points 17, 18, ... 26 due to the settlements and loads applied to wall A, r_{ik} is the reaction at any point i due to one unit settlement of the wall at point k, and r_{iP_A} is the reaction at the same point i due to the given loads applied to wall A.

A similar system of equations is written for wall B:

$$
\left.\begin{aligned}
X_{52} &= r_{52,52}\Delta_{52} + r_{52,66}\Delta_{66} + r_{52,70}\Delta_{70} + r_{52,94}\Delta_{94} + r_{52,108}\Delta_{108} + r_{52P_B} \\
X_{66} &= r_{66,52}\Delta_{52} + r_{66,66}\Delta_{66} + r_{66,70}\Delta_{70} + r_{66,94}\Delta_{94} + r_{66,108}\Delta_{108} + r_{66P_B} \\
X_{80} &= r_{80,52}\Delta_{52} + r_{80,66}\Delta_{66} + r_{80,70}\Delta_{70} + r_{80,94}\Delta_{94} + r_{80,108}\Delta_{108} + r_{80P_B} \\
X_{94} &= r_{94,52}\Delta_{52} + r_{94,66}\Delta_{66} + r_{94,70}\Delta_{70} + r_{94,94}\Delta_{94} + r_{94,108}\Delta_{108} + r_{94P_B} \\
X_{108} &= r_{108,52}\Delta_{52} + r_{108,66}\Delta_{66} + r_{108,70}\Delta_{70} + r_{108,94}\Delta_{94} + r_{108,108}\Delta_{108} + r_{108P_B}
\end{aligned}\right\}
\tag{8.33}
$$

In this system, r_{iP_B} is the reaction at point i due to the given loads applied to wall B.

7. Using the results of frame analysis, all reactions at points of support can be found from the following equations:

$$
\left.\begin{aligned}
X_{45} &= r_{45,45}\Delta_{45} + r_{45,49}\Delta_{49} + r_{45,101}\Delta_{101} + r_{45,105}\Delta_{105} + r_{45P_F} \\
X_{49} &= r_{49,45}\Delta_{45} + r_{49,49}\Delta_{49} + r_{49,101}\Delta_{101} + r_{49,105}\Delta_{105} + r_{49P_F} \\
X_{101} &= r_{101,45}\Delta_{45} + r_{101,49}\Delta_{49} + r_{101,101}\Delta_{101} + r_{101,105}\Delta_{105} + r_{101P_F} \\
X_{105} &= r_{105,45}\Delta_{45} + r_{105,49}\Delta_{49} + r_{105,101}\Delta_{101} + r_{105,105}\Delta_{105} + r_{105P_F}
\end{aligned}\right\}
\tag{8.34}
$$

In this system of equations, r_{iP_F} is the reaction at support i due to the given loads applied to the frame.

8. Now, the given mat can be investigated. Continuous contact of the mat and soil is replaced with a series of individual totally rigid supports located at all n nodes.

9. The mat is analyzed under the given loads located between the two walls and frame columns. Reactions r_{iP} due to these loads are found at all n nodes.

10. By releasing successively all supports from restraints against vertical linear deflections, the mat is analyzed under one unit settlements applied to each support of the mat. Reactions r_{ik} at all supports of the mat are found.

11. Reactions at all supports produced by walls A and B and the frame columns obtained from equations 8.32–8.34, acting in the opposite direction and applied to the mat as active vertical loads $X_{17}, X_{18}, \ldots, X_{26}, X_{52}, X_{66}, X_{80}, X_{94}, X_{108}, X_{45}, X_{49}, X_{101}, X_{105}$, will produce reactions only at all points they are applied.

12. The total reaction at any node of the mat can be found from the following equation:

$$Y_i = r_{i1}\Delta_1 + r_{i2}\Delta_2 + \ldots + r_{in}\Delta_n + X_i + r_{iP} \tag{8.35}$$

where r_{ik} is the reaction at any node i of the mat due to one unit settlement at node k, X_i is the reaction at node i produced by the walls or frame, and r_{iP} is the reaction at the same node due to the given loads applied to the mat. Reactions at all supports of the mat can be found from the following system of equations:

$$\left.\begin{array}{l} Y_1 = r_{11}\Delta_1 + r_{12}\Delta_2 + \ldots + r_{1n}\Delta_n + X_1 + r_{1P} \\ Y_2 = r_{21}\Delta_1 + r_{22}\Delta_2 + \ldots + r_{2n}\Delta_n + X_2 + r_{2P} \\ \cdots\cdots\cdots\cdots\cdots\cdots\cdots\cdots\cdots\cdots\cdots\cdots \\ Y_n = r_{n1}\Delta_1 + r_{n2}\Delta_2 + \ldots + r_{nn}\Delta_n + X_n + r_{nP} \end{array}\right\} \tag{8.36}$$

13. These reactions acting in the opposite direction and applied to the soil as uniformly distributed loads will produce settlements at all nodes of the mat. The settlements are obtained from the following equations:

$$\left.\begin{array}{l} \Delta_1 = \Delta_{11}Y_1 + \Delta_{12}Y_2 + \ldots + \Delta_{1n}Y_n + \Delta_{1G} \\ \Delta_2 = \Delta_{21}Y_1 + \Delta_{22}Y_2 + \ldots + \Delta_{2n}Y_n + \Delta_{2G} \\ \cdots\cdots\cdots\cdots\cdots\cdots\cdots\cdots\cdots\cdots\cdots\cdots \\ \Delta_n = \Delta_{n1}Y_1 + \Delta_{n2}Y_2 + \ldots + \Delta_{nn}Y_n + \Delta_{nG} \end{array}\right\} \tag{8.37}$$

Now, we have five groups of equations: 8.32, 8.33, 8.34, 8.36, and 8.37. They include the following unknowns: n settlements, n soil reactions, and m reactions at the frame and wall supports. The total number of unknowns is equal to $(2n + m)$ The number of equations is also equal to $(2n + m)$. By solving all five systems as one system of equations we find the settlements, soil reactions, and reactions under the walls and frame supports. Final analyses of the mat foundation, walls, and frame are performed as follows: By applying to the mat the given loads, and loads X_i, final analysis of the mat is performed. Analyses of the walls and frame are performed by applying to them the given loads and supports settlements.

Analysis is much simpler when the mat is supported on Winkler foundation. In this case, settlements of the soil at any point depend only on the load applied at that point. Equations 8.37 are replaced with simple equations:

$$\left.\begin{array}{l} \Delta_1 = \Delta_{11} Y_1 \\ \Delta_2 = \Delta_{22} Y_2 \\ \cdots\cdots\cdots\cdots \\ \Delta_n = \Delta_{nn} Y_n \end{array}\right\} \tag{8.38}$$

From the equations in 8.38 we find:

$$Y_1 = \frac{\Delta_1}{\Delta_{11}} = A_1 k_1 \Delta_1, \; Y_2 = \frac{\Delta_2}{\Delta_{22}} = k_2 A_2 \Delta_2, \ldots Y_n = \frac{\Delta_n}{\Delta_{nn}} = k_n A_n \Delta_n \tag{8.39}$$

where k_i is the modulus of subgrade reaction at point i, A_i is the tributary area of the soil at node i, and $k_i A_i$ is the rigidity of the soil at point i. By introducing reactions Y_i from equation 8.39 into the system of equations 8.36, and taking into account that these reactions act in the opposite direction we have:

$$\left.\begin{array}{l} (r_{11} + k_1 A_1)\Delta_1 + r_{12}\Delta_2 + \ldots + r_{1n}\Delta_n + X_1 + r_{1P} = 0 \\ r_{21}\Delta_1 + (r_{22} + k_2 A_2)\Delta_2 + \ldots + r_{2n}\Delta_n + X_2 + r_{2P} = 0 \\ \cdots\cdots\cdots\cdots\cdots\cdots\cdots\cdots\cdots\cdots\cdots\cdots\cdots\cdots\cdots\cdots\cdots\cdots \\ r_{n1}\Delta_1 + r_{n2}\Delta_2 + \ldots + (r_{nn} + k_n A_n)\Delta_n + X_n + r_{nP} = 0 \end{array}\right\} \tag{8.40}$$

This system contains $(n + m)$ unknowns: n settlements and m reactions, while the total number of equations is equal to n.

Additional m equations are given above: 8.32, 8.33, and 8.34. By solving four systems of equations, 8.32, 8.33, 8.34, and 8.40 as one system, settlements at all nodes and all reactions at points of support of the walls and frame columns are found. Now, analysis of the frame and walls can be performed by applying to the walls and frame the given loads and support settlements. Final analysis of the foundation is performed by applying to the mat the given loads and obtained reactions acting downward.

As shown above, combined analysis of a structural system *soil-mat foundation-frame-walls* can be performed using available computer programs for separate analyses of walls, frames, and mat foundations. Analysis also includes building and solving a system of linear equations for obtaining the forces of interaction between the elements of the structural system and its settlements. The methods of analysis developed above allow using soil models such as elastic half-space and elastic layer and any other soil model.

Combined analysis of mat foundations supporting 3D frames and walls can be performed by using the *iterative method*. This method is probably the easiest and the simplest. If a superstructure contains one or several 3D frames and several walls, analysis is performed as follows:

1. Continuous contact of the wall and the mat foundation is replaced with a series of individual totally rigid supports. Analysis of the wall with the given loads is performed and reactions at all points of support are found. If several walls are supported on the mat, all walls are analyzed and reactions at all supports are found.

2. Assuming that the frame columns at their points of support are restrained against any deflections, analysis of the frame is performed and all reactions are found.
3. All wall and frame reactions acting in the opposite direction are applied to the mat foundation along with other given loads, and analysis of the mat is performed.
4. Obtained settlements of the mat are applied to the walls and frames along with the given loads. New reactions are found.
5. These new reactions and the given loads are applied to the mat foundation and the mat is analyzed again, and so on. After several iterations, when reactions or settlements obtained in iteration n are close enough to the reactions and settlements obtained in iteration $(n - 1)$, the iterative process is stopped and the analysis is finished.

8.10 Analysis of Mat Foundations Supported on Piles

Mat foundations cannot always be supported directly on soil. In some cases, when the soil bearing capacity is not high enough, the mat foundation is supported on piles that transfer all loads or part of the loads to the better soils. Mat foundations supported on piles are usually used when a soil layer with a reliable bearing capacity is located at a greater depth or the superstructure transmits to the foundation large vertical loads.

Analysis of mat foundations supported on piles is no different from the analysis of mat foundations supported on soil discussed above. Analysis is similar to the analysis of plates on Winkler foundation or the analysis of plates supported on a series of elastic independently working supports. When analyzing a regular mat foundation we usually need the modulus of subgrade reaction of the soil. Analogously, when analyzing a mat foundation supported on piles we need the rigidity of the pile. One of the practical methods for obtaining the rigidity of the piles is field testing of single piles. The rigidity of a single pile is obtained from the following equation:

$$K_P = \frac{N}{\Delta}$$
(8.41)

where N is the load applied to the pile and Δ is the settlement of the pile due to the load N. Testing a single pile is performed analogously to testing soil with stamps. Using results of field testing, two curves can be built: one that shows the settlement of the pit of the pile and the second that shows the settlement of the head of the pile. Figure 8.14 shows the results of field testing.

Curves 1 and 2 are used for obtaining settlements as well as for obtaining the rigidities of the piles using equation 8.41. However, results of field testing of single piles are not always reliable: other piles supporting the same mat foundation may work under different conditions and behave differently compared to the behavior of the tested piles. Detailed recommendations for pile analysis can be found in Klepikov (1967) and Kezdi (1975).

Settlements of piles can be obtained analytically. The settlement of a pile consists of two parts: settlement due to the pile compression and settlement due to soil

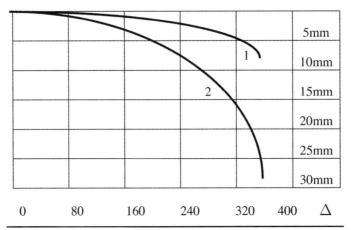

Figure 8.14   Settlement of the pit of the pile (1) and settlement of the head of the pile (2)

compression under the pit of the pile. Both curves reflect the results of testing of a
30cm diameter, 20.6m long pile. As can be seen from Figure 8.14, the head of the pile
settled about 3cm, mostly due to the compression of the pile material. When a pile
is supported on hard soil and the pit does not experience any settlements, the actual
settlement of the top of the pile can be found from the following equation:

$$\Delta = \frac{NL}{EA} \tag{8.42}$$

where N is the vertical load applied to the pile, L is the length of the pile, E is the
modulus of elasticity of the pile material, and A is the area of the cross section of the
pile. By introducing the settlement from equation 8.42 into equation 8.41 we have:

$$K_p = \frac{EA}{L} \tag{8.43}$$

When the friction stresses around the shaft of the pile are taken into account, the
settlement of the pile looks as follows:

$$\Delta = \frac{L}{EA}\left(N - \frac{TL}{4}\right) \tag{8.44}$$

In this equation, T is the resistance of the friction stresses per one meter of the pile's
length at the depth equal to L obtained from formula 8.45:

$$T = Vf \tag{8.45}$$

where V is the perimeter of the cross section of the shaft of the pile, and f is the allowable friction stress between the shaft of the pile and soil in t/m^2 at the depth equal to
L and obtained from field testing. Equation 8.44 is obtained assuming that the value
of T changes between the top of the pile and the point located at depth L, as shown
in Figure 8.15.

Let us investigate some practical ways for analysis of mat foundations supported on piles. Let us assume a mat shown in Figure 8.16 is supported on a system of piles. The given piles are supported on hard soil and the rigidities of the piles are equal to K_1, $K_2, \ldots K_n$. The mat is divided into a series of rectangular finite elements and piles are located at all nodes of the mat, as shown in Figure 8.17.

If the mat is supported on weak lower-bearing capacity soil, it is assumed that the mat is supported only on piles in all n nodes. Analysis of the mat, in this case, is performed as the analysis of a regular plate supported on a series of elastic supports with rigidities equal to $K_1, K_2, \ldots K_n$.

Analysis is the same as the analysis of mats supported on Winkler foundation discussed earlier. The only difference is the way the rigidity of supports is found. When the analysis of a mat on Winkler foundation is performed, rigidities of all supports are

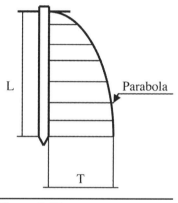

Figure 8.15

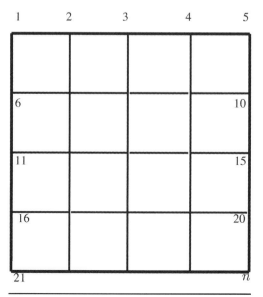

Figure 8.16

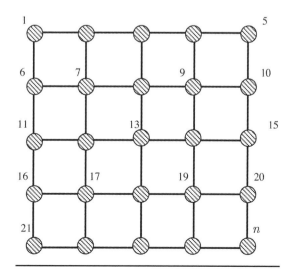

Figure 8.17   Location of the piles at all nodes

obtained as $C_i = k_i A_i$. When the mat is supported on piles, rigidities of all pile supports are obtained from equation 8.41. The analysis of mats supported on piles can be performed using the same software for the analysis of mat foundations supported on Winkler foundation.

When a part of the mat foundation is supported on piles and the remainder of the foundation is supported on elastic half-space, analysis is more complex. If the foundation shown in Figure 8.18 is supported on elastic half-space, analysis is performed as follows:

1. Assuming that the mat is supported at all nodes including piles, analysis of the mat is performed under the given loads and reactions at all supports are found $r_{1p}, r_{2p}, \ldots, r_{np}$.

2. Analysis of the mat is performed by applying successively one unit settlements to all supports, including pile supports, and the following matrix of reactions is found:

$$\left.\begin{array}{l} r_{11}, r_{12}, \ldots, r_{1n} \\ r_{21}, r_{12}, \ldots, r_{2n} \\ \ldots\ldots\ldots\ldots \\ r_{n1}, r_{n2}, \ldots, r_{nn} \end{array}\right\} \quad (8.46)$$

3. Total support reactions can be found from the following equations:

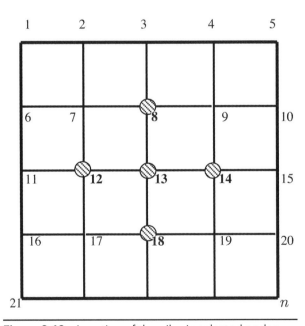

Figure 8.18   Location of the piles in selected nodes

$$
\left.
\begin{array}{l}
R_1 = r_{11}\Delta_1 + r_{12}\Delta_2 + \ldots + r_{1n}\Delta_n + r_{1p} \\
R_2 = r_{21}\Delta_1 + r_{22}\Delta_2 + \ldots + r_{2n}\Delta_n + r_{2p} \\
\cdots\cdots\cdots\cdots\cdots\cdots\cdots\cdots\cdots \\
R_n = r_{n1}\Delta_1 + r_{n2}\Delta_2 + \ldots + r_{nn}\Delta_n + r_{np}
\end{array}
\right\}
\qquad (8.47)
$$

4. Settlements of the half-space, including settlements of piles embedded in the half-space, are found from the following group of equations:

$$
\left.
\begin{array}{l}
\Delta_1 = \Delta_{11}R_1 + \Delta_{12}R_2 + \ldots + \Delta_{1n}R_n + \Delta_{1G} \\
\Delta_2 = \Delta_{21}R_1 + \Delta_{22}R_2 + \ldots + \Delta_{2n}R_n + \Delta_{2G} \\
\cdots\cdots\cdots\cdots\cdots\cdots\cdots\cdots\cdots \\
\Delta_n = \Delta_{n1}R_1 + \Delta_{n2}R_2 + \ldots + \Delta_{nn}R_n + \Delta_{nG}
\end{array}
\right\}
\qquad (8.48)
$$

The total number of unknowns in two systems of equations 8.47 and 8.48 is equal to $2n$. The total number of equations is equal to $2n$: n soil and pile reactions and n settlements. By solving both systems of equations as one, all settlements and all soil reactions are found. The final analysis of the mat foundation is performed by applying to the mat obtained settlements and the given loads.

Table 8.1 Concentrated vertical reactions applied to all nodes of the foundation. First method.

Node #	Load	Node #	Load	Node#	Load
1	$X_7/16$	10	$X_9/8$	19	$X_{19}/4$
2	$X_7/8$	11	$X_7 + X_{17}/16$	20	$X_{19}/8$
3	$X_7 + X_9/16$	12	$X_7 + X_{17}/8$	21	$X_{17}/16$
4	$X_7 + X_9/16$	13	$X_7 + X_9 + X_{17} + X_{19}/16$	22	$X_{17}/8$
5	$X_9/16$	14	$X_9 + X_{19}/8$	23	$X_{17} + X_{19}/16$
6	$X_7/8$	15	$X_8 + X_{19}/16$	24	$X_{19}/8$
7	$X_7/4$	16	$X_{17}/8$	25	$X_{19}/16$
8	$X_7 + X_9/8$	17	$X_{17}/4$		
9	$X_9/4$	18	$X_{17} + X_{19}/8$		

Table 8.2 Concentrated vertical reactions applied to all nodes of the foundation. Second method.

Node #	Load	Node #	Load	Node#	Load
1	X_1	10	$X_4 + X_8$	19	$X_{11} + X_{12} + X_{15} + X_{16}$
2	$X_1 + X_2$	11	$X_5 + X_9$	20	$X_{12} + X_{16}$
3	$X_2 + X_3$	12	$X_5 + X_6 + X_9 + X_{10}$	21	X_{13}
4	$X_3 + X_4$	13	$X_6 + X_7 + X_{10} + X_{11}$	22	$X_{13} + X_{14}$
5	X_5	14	$X_7 + X_8 + X_{11} + X_{12}$	23	$X_{14} + X_{15}$
6	$X_1 + X_5$	15	$X_8 + X_{12}$	24	$X_{15} + X_{16}$
7	$X_1 + X_2 + X_5 + X_6$	16	$X_9 + X_{13}$	25	X_{16}
8	$X_2 + X_3 + X_6 + X_7$	17	$X_9 + X_{10} + X_{13} + X_{14}$		
9	$X_3 + X_4 + X_7 + X_8$	18	$X_{10} + X_{11} + X_{14} + X_{15}$		

References

1. Bowles, J. E. 1977. *Foundation analysis and design.* New York: McGraw-Hill.
2. Kezdi, A. 1975. Pile foundations; from the *Foundation engineering handbook.* 1st ed. New York: van Nostrand Reinhold, pp. 556–562.
3. Klepikov, S. N. 1967. *Analysis of structures on elastic foundation.* Kiev: Budivelnik, pp. 26–32.
4. Selvadurai, A.P.S. 1979. *Elastic analysis of soil-foundation interaction.* Amsterdam/Oxford/ New York: Elsevier Scientific.
5. Teng, W. C. 1975. Mat foundations. In *Foundation engineering handbook.* 1st ed. New York: Van Nostrand Reinhold, pp. 533–534.
6. Tsudik, E. A. 2003. Analysis of mat foundations using various soil models. In *Proceedings of first international conference on foundations,* Dundee, Scotland.
7. Zhemochkin, B. N., and A. P. Sinitsin. 1962. *Practical methods of analysis of beams and plates on elastic foundation.* Moscow: Gosstroiizdat.

9

Analysis of Circular Plates on Elastic Foundation

9.1 Introduction

Foundations of circular and cylindrical structures such as chimneys, tanks, and other similar structures are usually designed as circular plates or rings. Analytical methods of analysis of circular plates on elastic foundation are typically developed for symmetrically loaded foundations and mostly for one soil model: Winkler foundation. Analysis of non-symmetrical plates on elastic foundation is usually performed by numerical methods such as finite element or finite difference. Simplified equations and tabulated solutions are also used by practicing engineers and produce accurate enough results.

Chapter 9 includes some simplified practical methods of analysis of circular and ring plates supported on elastic foundation, as well as some numerical methods of analysis for two soil models: Winkler foundation and elastic half-space.

9.2 Simplified Analysis of Circular Plates on Elastic Foundation

Equations are given below for the analysis of circular and ring foundations developed by Beyer (1956). In all equations it is assumed that the soil pressure is uniformly distributed. The types of plates are shown in Figures 9.1–9.6. Moments and shear forces for the plate shown in Figure 9.2 are found from the following equations:

$$M_r = \frac{qa^2}{16}(3 + \mu)\phi_1 \tag{9.1}$$

$$M_r = \frac{qa^2}{16}\left[2(1 - \mu) + (1 + 3\mu)\phi_1\right] \tag{9.2}$$

$$Q_r = qa/2 \tag{9.3}$$

All tables can be found at the end of the chapter.

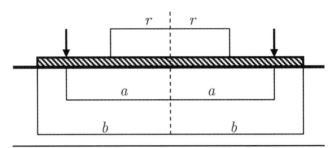

Figure 9.1

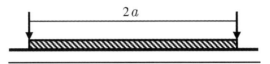

Figure 9.2   Circular plate case 1

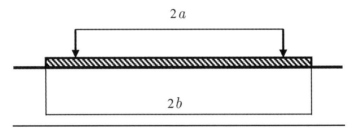

Figure 9.3   Circular plate case 2

In these equations, M_r is the radial moment, M_t is the tangential moment, q is the uniformly distributed soil pressure, Q_r is the shear per linear foot of circumference, μ is Poisson's ratio, $\phi_1 = 1 - \rho^2$, and $\rho = r/a$, and $\beta = b/a$.

Moments and shear forces for the plate shown in Figure 9.3 are found from the following equations:

a) When $\rho \le 1$

$$M_r = \frac{qa^2}{16}\left[k_1 - (3 + \mu) + (3 + \mu)\phi_1\right] \tag{9.4}$$

$$M_r = \frac{qa^2}{16}\left[k_1 - (1 + 3\mu) + (1 + 3\mu)\phi_1\right] \tag{9.5}$$

$$Q_r = \frac{qa}{2}\rho \tag{9.6}$$

b) When $\rho \ge 1$

$$M_r = \frac{qa^2}{16}\left[k_1 - (3 + \mu)\phi_0 + (3 + \mu)\phi_1 - 2(1 - \mu)\beta^2\phi_4 + 4(1 + \mu)\beta^2\phi_3\right] \tag{9.7}$$

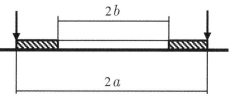

Figure 9.4 Ring plate, case 1   **Figure 9.5** Ring plate, case 2

$$M_t = \frac{qa^2}{16}\left[k_1 - (1 + 3\mu)\,\phi_0 + (1 + 3\mu)\,\phi_1 + 2(1 - \mu)\,\beta^2\phi_4 + 4(1 + \mu)\,\beta^2\phi_3\right] \quad (9.8)$$

$$Q_r = \frac{qa}{2}\left(\frac{\beta^2}{\rho} - \rho\right) \quad (9.9)$$

In these equations:

$$k_1 = 2(1 - \mu) + (1 + 3\mu)\beta^2 - 4(1 + \mu)\beta^2\ln\beta \quad (9.10)$$

$$k_2 = 2(1 - \mu) + (3 + \mu)\beta^2 - 4(1 + \mu)\beta^2\ln\beta \quad (9.11)$$

Note: ϕ_0 and ϕ_1 are negative when $\rho > 1$.

Moments and shear forces for the plate shown in Figures 9.4 and 9.5 are found as follows:

$$M_r = \frac{qa^2}{16}\left[(3 + \mu)\,\phi_1 - \beta^2 k_1\,\phi_4 + 4(1 + \mu)\,\beta^2\phi_3\right] \quad (9.12)$$

$$M_t = \frac{qa^2}{16}\left\{(1 + 3\mu)\,\phi_1 + \beta^2 k_1\,\phi_4 + 4(1 + \mu)\,\beta^2\phi_3 + 2(1 - \mu) - 2\beta^2\left[2(1 - \mu) - k_1\right]\right\} \quad (9.13)$$

$$Q_r = \frac{qa}{2}\left(\rho - \frac{\beta^2}{\rho}\right) \quad (9.14)$$

where

$$k_1 = (3 + \mu) + 4(1 + \mu)\frac{\beta^2}{1 - \beta^2}\ln\beta \quad (9.15)$$

$$k_2 = (3 + \mu) - 4(1 + \mu)\frac{\beta^2}{1 - \beta^2}\ln\beta \quad (9.16)$$

Note: Equations 9.12–9.16 are used for both when the load is applied to the outer as well as to the inner edges of the ring.

Foundation with a Pedestal

Figure 9.6 shows a circular foundation with a pedestal. The given foundation is loaded with a moment. Equations for moments and shear forces are given below. Coefficients k_1, k_2, k_3, and q_0 in these equations are obtained as follows:

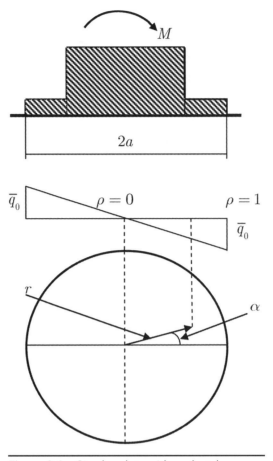

Figure 9.6   Circular plate with pedestal

$$k_1 = 3 + \mu + (1 - \mu)\beta^4 \tag{9.17}$$

$$k_2 = 4(2 + \mu) + (1 - \mu)(3 + \beta^4)\beta^2 \tag{9.18}$$

$$k_3 = 4(2 + \mu)\beta^4 - (3 + \mu)(3 + \beta^4)\beta^2 \tag{9.19}$$

$$\overline{q}_0 = \frac{4\mu}{\pi a^3} \tag{9.20}$$

Moments and shear forces are found from the following equations:

$$M_r = \frac{\overline{q}_0 a^2}{48 k_1}\left\{(5 + \mu)k_1\rho^3 - (3 + \mu)k_2\rho + 3(1 + \mu)k_1\rho^{-1} - (1 - \mu)k_3\rho^{-3}\right\}\cos\alpha \tag{9.21}$$

$$M_a = \frac{\overline{q}_0 a^2}{48 k_1}\left\{(1 + 5\mu)k_1\rho^3 - (1 + 3\mu)k_2\rho + 3(1 + \mu)k_1\rho^{-1} + (1 - \mu)k_3\rho^{-3}\right\}\cos\alpha \tag{9.22}$$

$$M_{ra} = \frac{\overline{q}_0 a^2}{48 k_1}(1 - \mu)(k_1\rho^3 - k_3\rho + 3k_1\rho^{-1} + k_3\rho^{-3})\sin\alpha \tag{9.23}$$

$$Q_r = \frac{\overline{q}_0 a}{24}\left(9\rho^2 - 2\frac{k_2}{k_1} - 3\rho^{-2}\right)\cos\alpha \tag{9.24}$$

$$Q_{ra} = \frac{\overline{q}_0 a}{24}\left(3\rho^3 - 2\frac{k_2}{k_1} + 3\rho^{-1}\right)\sin\alpha \tag{9.25}$$

Figure 9.7 shows a circular foundation of variable thickness with a vertical concentrated load and uniformly distributed soil reaction. A simplified analysis of such types of foundations can be performed using Table 9.1 recommended by Timoshenko and Woinowsky-Krieger (1959). The maximum moment at the center of the plate is obtained from the following equation:

$$M_{\max(r=0)} = M_t = M_r = M_0 + \gamma_2 P \tag{9.26}$$

where

$$M_0 = \frac{P}{4\pi}\left[(1+v)\ln\frac{a}{c} + 1 - \frac{(1=v)c^2}{4a^2}\right] \tag{9.27}$$

v is the Poisson's ratio of the plate material and c is the radius of the given load P replaced by a uniformly distributed load equal to $\dfrac{P}{\pi c^2}$.

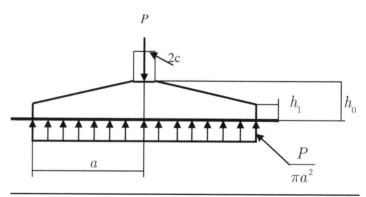

Figure 9.7   Foundations of variable thickness

9.3 Circular Plates on Elastic Half-Space: Equations and Tables

The equations and tables presented on the following pages were developed by Gorbunov-Posadov (1984) for analysis of symmetrically loaded circular plates supported on elastic half-space. The theoretical basis of this method used by the authors is not included in this work. The reader can find a detailed explanation of it in the authors' publication previously mentioned and also in Selvadurai's book (1979). This method was developed only for free supported plates without any restraints along their perimeters. The types of plates and loads discussed here are shown in Figures 9.8–9.14. Tables have been developed for totally rigid and elastic plates as well.

1. Figure 9.8 shows a circular plate with an axisymmetrical uniformly distributed load q located at the central area of the plate occupying the part of the plate area $(a < R)$ or the total area of the plate $(a = R)$.

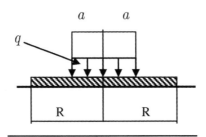

Figure 9.8

2. Figure 9.9 shows a circular plate with an axisymmetrical uniformly distributed ring shape load q.

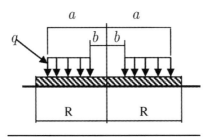

Figure 9.9

3. Figure 9.10 shows a circular plate with a uniformly distributed load P' along the circumference.

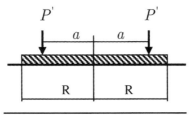

Figure 9.10

4. Figure 9.11 shows a circular plate with uniformly distributed moments along the circumference.

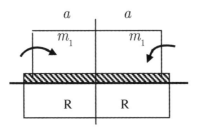

Figure 9.11

5. Figure 9.12 shows a circular plate with a vertical concentrated load P applied to the center of the plate.

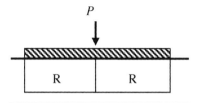

Figure 9.12

6. Figure 9.13 shows some additional information for circular flat plates.

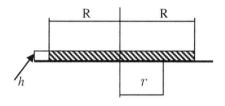

Figure 9.13

7. Figure 9.14 shows a circular plate of variable thickness.

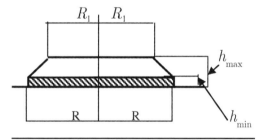

Figure 9.14

8. In order to use the following tables and equations, the following data are needed:
   a) The radius of the plate R in m
   b) The thickness of the plate h in m
   c) The modulus of elasticity of the plate material E_1
   d) Poisson's ratio of the plate material v ($v = 1/6$ for reinforced concrete)
   e) Modulus of elasticity of the soil E_0 and Poisson's ratio of the soil v_0
   f) The radius a of the load for a circular plate
   g) The radiuses a and b for ring plates
   h) The distance r between the center of the plate and the point of investigation

9. Now, using the data mentioned above, the following coefficients are found: $\rho = r/R$, $\alpha = a/R$, $\beta = b/R$, including the parameter of flexibility of the plate s from equation 9.28 shown below.

$$s = 3\frac{\left(1 - v_1^2\right)}{\left(1 - v_0^2\right)} \cdot \frac{E_0}{E_1} \cdot \frac{R^3}{h^3} \tag{9.28}$$

10. Depending on the parameter of flexibility s, the plate belongs to *totally rigid plates* or *elastic plates*. In accordance with Gorbunov-Posadov (1984), when $s \leq 1/2$ the plate is totally rigid, when $1/2 < s \leq 10$ the plate is elastic, and when $s > 10$ it can be assumed that the plate is infinite or totally flexible.

11. When the thickness of the circular plate is variable, as shown above in Figure 9.14, two parameters of flexibility are obtained: s_1 assuming that the total thickness of the plate is equal to h_{max} and s_2 assuming that the total thickness of the plate is equal to h_{min}. The coefficient $\gamma = R_1/R$ is also found.

12. If $s_1 < 0.5$ and at least one of the three conditions listed below is satisfied, the plate is analyzed as totally rigid. These three conditions are:
    a. $\gamma \leq 0.25$ and $s_1 \leq 1$
    b. $\gamma \geq 0.50$ and $s_2 \leq 1.5$
    c. $\gamma \geq 0.75$ and $s_2 \leq 2.0$
    Analysis of totally rigid circular plates is explained below.

13. If $0.50 \leq s_1 \leq 10$ and $h_{max}/h_{min} \leq 1.50$, the plate can be analyzed as a plate of uniform thickness. Analysis is performed taking into account the actual rigidity of the plate. The parameter of flexibility in this case is obtained using the thickness of the plate equal to h_{min} when $\gamma = 0.25$, or $0.5(h_{max} + h_{min})$ when $0.50 \leq \gamma \leq 0.75$, and h_{max} when $\gamma \geq 0.75$. We can use $h = h_{max}$ also when $h_{max}/h_{min} \leq 2$.

9.4 Totally Rigid Circular Plates Supported on Elastic Half-Space

If the resultant of all loads P applied to a circular totally rigid plate coincides with the center of that plate, the soil pressure is obtained by Boussinesq (1885) from equation 9.29.

$$p = P/\left(2\pi R^2 \sqrt{1 - \rho^2}\right) \tag{9.29}$$

where $\rho = r/R$ and r is the distance between the center of the plate and the point where the pressure is found. To simplify hand analysis, the soil pressure is obtained using Table 9.2 and equation 9.30:

$$p = \bar{p}\frac{P}{R^2} \tag{9.30}$$

Coefficients $\bar{p}$ are given in Table 9.2. Parameter $\bar{p}$ is given for 11 points. Point 0.0 is located at the center of the plate. If the resultant of all loads applied to the plate does not coincide with the center of the plate, the soil pressure is obtained from the following equation:

$$p = \frac{P}{2\pi R^2} \cdot \frac{3\lambda x + 1}{\sqrt{1 - \rho^2}} \tag{9.31}$$

where $\lambda = L_{cP}/R$, and L_{cP} is the distance between the center of the plate and the point the resultant P is applied, and $x = l_{cPi}/R$, where l_{cPi} is the coordinate of the point i located on the line connecting the origin of this coordinate at the center of the plate and point where the load P is applied. The settlement of the totally rigid plate with an axisymmetrical load is found as follows:

$$w = P(1 - v_0^2)/2RE_0 \qquad (9.32)$$

In the case of a non-symmetrical load the settlement is found from the following equation:

$$w = P(1 - v_0^2)(2 + 3\lambda x)/4RE_0 \qquad (9.33)$$

Rotation of the plate is obtained as follows:

$$tg\varphi = 3P\lambda(1 - v_0^2)/4E_0R^2 \qquad (9.34)$$

Equations and tables for obtaining moments and shear forces in totally rigid loaded circular plates are shown in Figures 9.15–9.17. If the plate is loaded with one vertical load applied to its center, as shown in Figure 9.15, the moments and shear forces are found using the simple equations given below and Table 9.3. The center of the plate is located at point 0.0 and the edge of the plate is located at point 1.0. The moments and shear forces are found using the following equations:

$$M_r = \overline{M}_r P_i, \ M_t = \overline{M}_t P_i, \ Q_r = -\overline{Q}_r \frac{P}{R}$$

Coefficients $\overline{M}_r$, $\overline{M}_t$, and $\overline{Q}_r$ are taken from Table 9.3. As shown in Table 9.3, moments and shear forces at the center of the plate are equal to ∞. In reality, the vertical load is always applied to a small circular area and, therefore, the moments at that point are not equal to ∞. When the vertical load is applied to the plate as a circular uniformly distributed load over a very small area, as shown in Figure 9.16, moments at that point are found from Table 9.4 and the following equations $M_r = M_t = \overline{M}P$. Analysis is performed as follows:

1. Find parameter $\alpha = a/R$ (see Figure 9.16).
2. Using obtained α find $M_r = M_t = \overline{M}P$ from Table 9.4.

To avoid large shear forces it is recommended to check the shear force applied to one linear unit of the perimeter of the applied circular load. This shear force is equal to $Q = P/2\pi a$. Tangential shear forces in symmetrically loaded circular plates are always equal to zero.

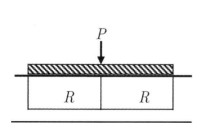

Figure 9.15

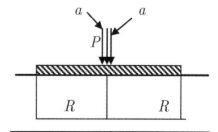

Figure 9.16

If the plate is loaded with an axisymmetrical uniformly distributed load as shown in Figure 9.17, moments and shear forces are found using coefficients from Table 9.5 and the following equations:

$$M_r = \overline{M}_r R^2 q, \; M_t = \overline{M}_t R^2 q, \; Q_r = - \overline{Q}_r R q$$

Coefficients $(\overline{M}_r, \overline{M}_t, \overline{Q}_r)$ are taken from Table 9.5 depending on $\alpha = a/R$. Note: The origin of coordinates is located at the center of the plate.

If a circular plate has uniformly distributed loads applied along the circumference, as shown in Figure 9.18, moments and shear forces are found from the following equations:

$$M_r = \overline{M}_r R P', \; M_t = \overline{M}_t R P', \; Q_r = - \overline{Q}_r P$$

Moments are given in *tm/m*, shear in *t/m*. $\overline{M}_r$, $\overline{M}_t$, and $\overline{Q}_r$ are given in Table 9.6.

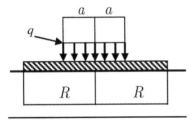

Figure 9.17

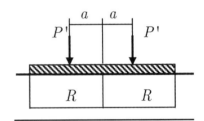

Figure 9.18

9.5 Elastic Circular Plates Supported on Elastic Half-Space

As mentioned above, when the parameter of flexibility *s* satisfies the following condition, $0.5 \le s \le 10$, analysis of the plate is performed by taking into account the actual flexural rigidity of the plate.

Simple equations and tables presented on the following pages allow performing such type of analysis. Analysis is developed for various values of *s* and for the following four types of plate loadings typical in practical applications:

1. Uniformly distributed loads P' along the circumference, as shown in Figure 9.19.
2. Uniformly distributed moments m' along a circumference, as shown in Figure 9.20.
3. Uniformly distributed loads q applied to the total area of the circular plate, as shown in Figure 9.21.
4. Concentrated vertical load P applied at the center of the plate, as shown in Figure 9.22.

The analysis for each type of loading is explained below.

Using the equations shown below and Table 9.7, we can find radial moments M_r, tangential moments M_t, shear forces Q_r, settlements w, and rotations $tg\varphi_e$ at the edge of the plate with uniformly distributed loads along the circumference. All equations are shown next:

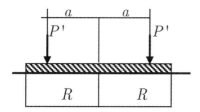

Figure 9.19   Uniformly distributed load located along the circumference of the plate

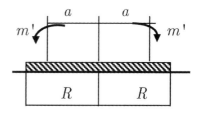

Figure 9.20   Uniformly distributed moments located along a circumference with radius *a*

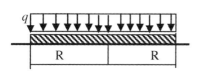

Figure 9.21   Uniformly distributed load over the total area of the plate

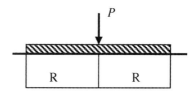

Figure 9.22   A concentrated vertical load applied to the center of the plate

$$p = \overline{p}\frac{P'}{R}, \; w = \overline{w}\frac{R^3}{D}P', \; M_r = -\overline{M}RP',$$

$$M_t = \overline{M}_t RP', \; Q_r = \overline{Q}_r P', \; tg\varphi_R = \overline{tg}\varphi_R \frac{R^2}{D}P' \tag{9.35}$$

In these equations, $\overline{p}$, $\overline{w}$, $\overline{M}_r$, $\overline{M}_t$, $\overline{Q}_r$, and $\overline{tg}\varphi_e$ are dimensionless coefficients taken from Table 9.7. D is the rigidity of the plate obtained from formula 9.36:

$$D = \frac{Eh^3}{12(1-v_1^2)} \tag{9.36}$$

In Table 9.7, $\alpha = a/R$, the load P', as well as obtained moments and shear forces, are applied to one linear unit of the circumference (see Figure 9.19). Rotation of the plate at any point can be found from the following simplified equation:

$$tg\varphi_i = (w_{i+1} - w_{i-1})/0.2R \tag{9.37}$$

All numerical values given in Table 9.7 are multiplied by 1,000 for convenience. When using formulae 9.35 for obtaining final results, these values should be divided by 1,000.

   Table 9.8 is used for analysis of circular plates loaded with uniformly distributed moments m' applied to the plate and located axisymmetrically along a circumference with radius a. The plate is shown in Figure 9.20. Analysis of a plate with uniformly distributed moments m' is performed, in principle, analogously to analysis of a plate with uniformly distributed vertical loads. When all moments are applied in a direction away from the center of the plate as shown in Figure 9.20, numerical values taken from the table are used without changing their signs. When moments act in an opposite direction all numerical values are taken with an opposite sign. Numerical values of $\overline{M}_r$ at all points where the moments m' are applied and shown with a sign (*) are used for obtaining moments inside the circle with radius a. Outside of this circle, the moment is found by adding one unit to the value M_r. Coefficients $\overline{p}$, $\overline{M}_r$, $\overline{M}_t$, $\overline{Q}_r$, $\overline{w}$, and $\overline{tg}\varphi_e$ are taken from Table 9.8. They are multiplied by 1,000 for convenience. When obtaining the final results all numerical values have to be divided by 1,000.

The actual values of p, M_r, M_t, Q_r, w, and $tg\varphi_e$ are obtained from the following formulae:

$$p = \overline{p}\frac{m'}{R^2}, \; M_r = \overline{M}_r m', \; M_t = \overline{M}_t m', \; Q_r = \overline{Q}_r \frac{m'}{R^2},$$

$$w = \overline{w}\frac{R^2}{D}, \; tg\varphi_e = \overline{tg}\varphi \frac{R^2}{D}m' \tag{9.38}$$

All tables presented are developed for reinforced concrete with the Poisson's ratio $\nu_1 = 1/6$.

When the plate has a uniformly distributed load q over the total area, as shown in Figure 9.21, analysis is performed using Table 9.8 and the following equations:

$$p = \overline{p}q, \; w = \overline{w}\frac{R^4 q}{D}, \; M_r = \overline{M}_r R^2 q, \; M = \overline{M}_t R^2 q,$$

$$Q_r = -\overline{Q}_r Rq, \; tg\varphi_e = \overline{tg}\varphi_e \frac{R^3 q}{D} \tag{9.39}$$

Coefficients $\overline{p}$, $\overline{w}$, $\overline{M}_r$, $\overline{M}_t$, $\overline{Q}_r$ in these equations are found from Tables 9.9 and $\overline{tg}\varphi_e$ from Table 9.10.

Analysis of circular elastic plates loaded with one vertical concentrated load applied to the center of the plate, shown in Figure 9.22, is performed using Tables 9.11 and 9.12 and equations 9.40:

$$p = \overline{p}\frac{P}{R^2}, \; w = \overline{w}\frac{R^2}{D}P, \; M_r = \overline{M}_r P, \; M_t = \overline{M}_t P,$$

$$Q_r = -\overline{Q}_r \frac{P}{R}, \; tg\varphi = \overline{tg}\varphi \frac{R}{D}P \tag{9.40}$$

where coefficients $\overline{p}$, $\overline{w}$, $\overline{M}_r$, $\overline{M}_t$, and $\overline{Q}_r$ are taken from Table 9.11 and $\overline{tg}\varphi$ is taken from Table 9.12. As can be seen from Table 9.11, all coefficients $\overline{M}_r$, $\overline{M}_t$, and $\overline{Q}_r$ are infinitely large at the center of the plate where the point load P is applied. By taking into account that, in reality, load P is not a concentrated point load, but a uniformly distributed load over a small circular area with a radius equal to a (Figure 9.23), moments and the shear force at the center of the plate can be found.

Moments ($M_r = M_t$) at the center of the plate are found using formula 9.41 and Tables 9.13 and 9.14. Analysis is performed in the following three steps:

1. Using the parameter of flexibility s, find $\overline{M}_A$ from Table 9.13. Using parameter $\alpha = a/R$, find $\overline{M}_B$ from Table 9.14.
3. Using formula 9.41, find the moment at the center of the plate:

$$M_0 = P(\overline{M}_A + \overline{M}_B) \tag{9.41}$$

where $P = q\pi a^2$.

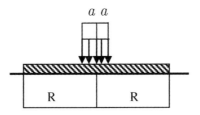

Figure 9.23 A concentrated vertical load uniformly distributed over a small circular area

Since load $P = q\pi a^2$ is actually applied to the plate as a uniformly distributed load, the shear force at the center of the plate is obtained as a shear per one linear unit of the load perimeter. This shear force is equal to:

$$Q_0 = \frac{P}{2\pi a} = \frac{q\pi a^2}{2\pi a} = q/2a \tag{9.42}$$

If the plate is loaded with two or more types of loads shown above, analyses of the plate under each load is performed. Final results are found by summing results of all analyses. Some numerical examples are given below to illustrate application of the formulae and tables to practical analysis.

Example 9.1

A circular plate shown in Figure 9.24 has a uniformly distributed load $P' = 100t/m$ along the circumference of the plate with a radius a. Find the radial and tangential moments and shear force at the point where $r = 2m$, assuming that the soil pressure applied to the plate is uniformly distributed. Use equations 9.4–9.6 and data given below:

$P = 100t/m$, $a = 3m$ $b = 4m$, $\nu = 1/6$ $\rho = r/a = 2/3 = 0.67$ $\beta = b/a = 4/3 = 1.33$,

$\phi_1 = 1 - \rho^2 = 1 - (0.67)^2 = 0.551$, $q = \dfrac{100 \cdot 2 \cdot 3.141 \cdot 4}{3.141 \cdot 4^2} = 50t/m^2$

Solution:

$$M_r = \frac{qa^2}{16}\left[k_1 - (3 + \mu) + (3 + \mu)\phi_1\right]$$

$$M_t = \frac{qa^2}{16}\left[k_1 - (1 + 3\mu) + (1 + 3\mu)\phi_1\right]$$

$$Q_r = \frac{qa}{2}\rho$$

$k_1 = 2(1 - \mu) + (1 + 3\mu)\beta^2 - 4(1 + \mu)\beta^2\ln\beta$

$k_2 = 2(1 - \mu) + (3 + \mu)\beta^2 - 4(1 + \mu)\beta^2\ln\beta$

$k_1 = 1.66 + 1.51 \cdot 1.33^2 - 4.68 \cdot 1.33^2 \cdot 0.2852 = 1.97$

$k_2 = 2(1 - 0.17) + 3.17 \cdot 1.33^2 - 4.68 \cdot 1.33^2 \cdot 0.2852 = 4.91$

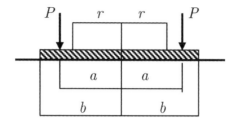

Figure 9.24 A uniformly distributed load along the circumference (example 9.1)

$$M_r = \frac{50 \cdot 3^2}{16}\left[1.97 - 3.17 + 3.17 \cdot 0.551\right] = 15.8 tm/m$$

$$M_t = \frac{50 \cdot 3^2}{16}\left[1.97 - 1.51 + 1.51 \cdot 0.551\right] = 36.34 tm/m$$

$$Q_r = 50 \cdot 3 \cdot 0.67 \cdot 0.5 = 50.25 t/m$$

Example 9.2

The example shown below demonstrates analysis of a totally rigid circular plate supported on elastic half-space.

Given: A circular plate shown in Figure 9.25. The plate is loaded with a concentrated vertical load $P = 500t$ applied to the center of the plate. The radius of the plate is equal to $R = 3m$; the thickness of the plate is equal to $h = 1m$, $E_0 = 6,000t/m^2$ and $E_1 = 3,000,000t/m^2$. Find the soil pressure and the radial moments in the plate.

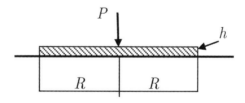

Figure 9.25 A plate loaded with a vertical concentrated load (example 9.2)

Solution:

1. Find the parameter of flexibility of the plate:

$$S = 3\frac{1 - v_1^2}{1 - v_0^2} \cdot \frac{E_0}{E_1} \cdot \frac{R^3}{h^3} = 3\frac{1 - 0.17^2}{1 - 0.4^2} \cdot \frac{6000}{3,000,000} \cdot \frac{3^3}{1^3} = 0.062 < 0.5$$

Since $S < 0.5$ the plate is considered totally rigid.

2. Using equation 9.30 and Table 9.2, find the soil pressure p.

Note: The soil pressure shown above is obtained in 11 points starting from the center of the plate at point 0 up to the edge of the plate.

3. Using formula $M_r = \overline{M}_r P$ and Table 9.2, find the radial moments in the plate.

In order to obtain the radial moment at the center of the plate, the given concentrated load is replaced with a uniformly distributed load applied to a small circular area with a radius $a = 0.3m$. The equivalent uniformly distributed load is equal to $q = 500/3.141 \cdot 3^2 = 17.687 t/m^2$ and $\alpha = a/R = 0.3/3 = 0.1$. From Table 9.4, $\overline{M}_r = 0.254$ and the moment at the center of the plate is equal to $M_r = M_t = \overline{M}_r R^2 q = 0.254 \cdot 3^2 \cdot 17.687 = 40.432 tm/m$.

Example 9.3

This example illustrates application of the method for analysis of plates on elastic half-space taking into account the actual rigidity of the plate, when $s > 0.50$. Let us perform analysis of the same plate shown in Figure 9.25 using the same data except the thickness of the plate, which is now $h = 0.58m$. Find the settlements of and radial moments in the plate. First we check the parameter of flexibility of the plate:

$$s = 3 \frac{1 - v_1^2}{1 - v_0^2} \cdot \frac{E_0}{E_1} \cdot \frac{R^3}{h^3} = 3 \frac{1 - 0.17^2}{1 - 0.4^2} \cdot \frac{6000}{3,000,000} \cdot \frac{3^3}{0.58^3} = 096 > 0.5$$

Since $s = 0.96 > 0.5$, the plate is analyzed taking into account its actual flexural rigidity that is obtained as follows:

$$D = Eh^3/12(1 - v_1^2) = 3,000,000 \cdot 0.58^3/12(1 - 0.17^2) = 50,230.5 tm$$

Now, using the second equation from 9.40 $w = \overline{w} R^2 P/D$ and Table 9.9, settlements of the plate are found.

As can be seen, the radial moment at the center of the plate is equal to ∞. In order to obtain its realistic value, the concentrated point load P is replaced with a uniformly distributed load q over a small circular area with a radius, let's say, equal to $R = 0.3m$, so $q = P/\pi a^2$. Now, using Tables 9.13, 9.14, and formula 9.41, the actual moment at the center of the plate can be found:

$$M_r(0) = M_t(0) = P(\overline{M}_A + \overline{M}_B) = 500(-0.056 + 0.293) = 118.5 tm.$$

Coefficients $\overline{M}_A = -0.056$ and $\overline{M}_B = 0.293$ are taken from Tables 9.13 and 9.14 successively.

9.6 Numerical Analysis of Circular and Ring Plates Supported on Elastic Foundation

Analysis of circular plates supported on elastic foundation, as shown above, is developed only for symmetrically loaded plates free supported on soil. Analysis of non-symmetrically loaded plates is usually performed by numerical methods such as the finite difference or finite element methods using Winkler's soil model. Analyses based on other soil models such as elastic half-space or elastic layer have not been developed. A method of analysis of circular plates loaded with symmetrical and non-symmetrical loads supported on elastic half-space and elastic layer is introduced on the following pages. The method, in principle, is not different from analysis of beams developed by Zhemochkin (1962) and analysis of plates proposed by by the author of this work (2006). It allows analyzing circular and ring foundations with various boundary conditions. Figure 9.26 shows a circular plate replaced with an octagon of equivalent area

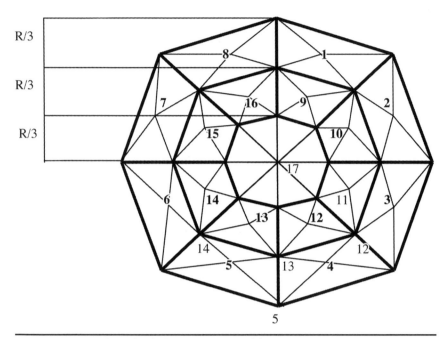

Figure 9.26   A plate divided into a system of finite triangular elements

divided into a system of finite elements. The total area of the octagon is equal to the area of the circular plate or $A_{oct} = A_{cir} = \pi R^2$ where R is the radius of the given foundation. Analysis is performed as follows:

1. The area of the octagon is divided into a system of triangular finite elements, and continuous contact between the plate and soil is replaced with a series of non-yielding individual supports located only at some selected nodes: at the centers of all 16 trapezoids (points 1, 2, . . . 16) and at the center of the foundation at point 17.

2. Using a computer program based on the finite element method, analysis of the plate loaded with the given loads is performed, and reactions $r_{1P}, r_{2P}, \ldots r_{nP}$ at all 17 points of support are found.

3. Using the same program the plate is analyzed by applying to each support one unit settlements, and a matrix of reactions at all supports is obtained. The matrix of reactions looks as follows:

$$\left.\begin{array}{l} r_{11} r_{12} \ldots r_{1n} \\ r_{21} r_{22} \ldots r_{2n} \\ \cdots\cdots\cdots \\ r_{n1} r_{n2} \ldots r_{nn} \end{array}\right\} \tag{9.43}$$

The total reaction at any point of support i can be obtained from the following equation:

$$X_i = r_{i1}\delta_1 + r_{i2}\delta_2 + \ldots + r_{in}\delta_n + r_{iP} \tag{9.44}$$

where δ_i is the unknown settlement at point i. Reactions at all n points of support look as follows:

$$\left.\begin{array}{l} X_1 = r_{11}\delta_1 + r_{12}\delta_2 + \ldots + r_{1n}\delta_n + r_{1P} \\ X_2 = r_{21}\delta_1 + r_{22}\delta_2 + \ldots + r_{2n}\delta_n + r_{2P} \\ \ldots\ldots\ldots\ldots\ldots\ldots\ldots\ldots\ldots\ldots\ldots \\ X_n = r_{n1}\delta_1 + r_{n2}\delta_2 + \ldots + r_{nn}\delta_n + r_{nP} \end{array}\right\} \qquad (9.45)$$

The same reactions acting in the opposite direction are applied to the half-space as active vertical loads. However, these loads cannot be applied to the soil as concentrated loads: they will produce infinitely large settlements. Therefore, they are applied as uniformly distributed loads. Each load is applied to a certain tributary area of the soil surface. In order to simplify the analysis, all tributary areas (trapezoids) are replaced with areas of circular shape:

1. The area of trapezoid 1. $A_1 = \left[\dfrac{2\pi R}{8} + \dfrac{2\pi\left(\frac{2}{3}R\right)}{8}\right]\dfrac{1}{2}\cdot\dfrac{R}{3} = \dfrac{2\pi R^2}{8}\left(1 + \dfrac{2}{3}\right)\dfrac{1}{6}$

   $= 0.218R^2$

2. The area of trapezoid 9. $A_9 = \left(\dfrac{2\pi\frac{2}{3}R}{8} + \dfrac{2\pi\frac{1}{3}R}{8}\right)\dfrac{1}{2}\cdot\dfrac{R}{3} = \dfrac{2\pi R^2}{3\cdot 8}(2+1)\cdot\dfrac{1}{6}$

   $= 0.131R^2$

3. The area of octagon 17. $A_{17} = \pi\left(\dfrac{R}{3}\right)^2 = 0.349R^2$

As we can see, the total area of the octagon is equal to $(0.218 + 0.131) \cdot R^2 \cdot 8 + 0.349R^2 = 3.141R^2$ that is equal to the area of the circular plate πR^2.

Now, we can obtain the radiuses of the equivalent tributary circular areas of the loads applied to the soil:

1. Supports 1, 2, ..., 8. $r_1 = \sqrt{0.218R^2/\pi} = 0.2635R$.
2. Supports 9, 10, ..., 16. $r_2 = 0.2043R$.
3. Support 17. $r_3 = 0.333R$.

The sum of all circular tributary areas is equal to:

$$(0.333R)^2 + (0.2635R)^2 8 + (0.2043R)^2 8 = (0.111 + 0.555 + 0.334)\pi R^2 = \pi R^2$$

Settlements of the soil (half-space) at any point i can be found as shown below:

$$\delta_i = \delta_{i1}X_1 + \delta_{i2}X_2 + \ldots + \delta_{in}X_n \qquad (9.46)$$

where δ_{ik} is the settlement of the half-space at point i due to one unit load applied to the soil at point k.

If the half-space is loaded with a load G located close to the plate, this load may produce an additional settlement of plate equal to δ_{iG}. In this case equation 9.46 will look as follows:

$$\delta_i = \delta_{i1}X_1 + \delta_{i2}X_2 + \ldots + \delta_{in}X_n + \delta_{iG} \qquad (9.47)$$

where δ_{iG} is the settlement of the half-space at point i due to the load located outside of the area of the plate. Using equation 9.47, settlements of the soil can be found at all points of support as follows:

$$\left.\begin{array}{l} \delta_1 = \delta_{11}X_1 + \delta_{12}X_2 + \dots + \delta_{1n}X_n + \delta_{1P} \\ \delta_2 = \delta_{21}X_1 + \delta_{22}X_2 + \dots + \delta_{2n}X_n + \delta_{2P} \\ \dots\dots\dots\dots\dots\dots\dots\dots\dots\dots\dots\dots \\ \delta_n = \delta_{n1}X_1 + \delta_{n2}X_2 + \dots + \delta_{nn}X_n + \delta_{nP} \end{array}\right\} \tag{9.48}$$

All settlements of the half-space δ_{ik} due to one unit loads are found from equations 1.24–1.27 given in Chapter 1. Now, by solving two systems of equations, 9.45 and 9.48, as one system, all settlements and soil reactions are found. Final analysis of the plate is performed by applying to the plate all given loads and obtained settlements. The total number of unknowns is equal to $2n$.

Analysis is much simpler when the plate is supported on Winkler foundation. Since settlements of the soil in this case do not affect each other, the system of equations 9.48 is replaced with the following equations:

$$\delta_1 = \frac{-X_1}{k_1 A_1}, \quad \delta_2 = \frac{-X_2}{k_2 A_2}, \quad \dots, \quad \delta_n = \frac{-X_n}{k_n A_n} \tag{9.49}$$

where k_i is the modulus of subgrade reaction of the soil at point i. By introducing equations 9.49 into the system of equations 9.45, and taking into account that loads applied to the soil act in the opposite direction to the direction of reactions, we have:

$$\left.\begin{array}{l} (r_{11} + k_1 A_1)\delta_1 + r_{12}\delta_2 + \dots + r_{1n}\delta_n + r_{1P} = 0 \\ r_{21}\delta_1 + (r_{22} + k_2 A_2)\delta_2 + \dots + r_{2n}\delta_n + r_{2P} = 0 \\ \dots\dots\dots\dots\dots\dots\dots\dots\dots\dots\dots\dots \\ r_{n1}\delta_1 + r_{n2}\delta_2 + \dots + (r_{nn} + k_n A_n)\delta_n + r_{nP} = 0 \end{array}\right\} \tag{9.50}$$

By solving the system of equations 9.50, all settlements are found and final analysis of the plate is performed by applying to the plate the given loads and obtained settlements. As we can see, the system 9.50 contains only n unknown settlements or two times less than the analysis of the plate on elastic half-space.

This analysis can also be performed by introducing settlements $\delta_1, \delta_2, \dots, \delta_n$ from equation 9.49 into the system 9.45. Now we have:

$$\left.\begin{array}{l} \left(\dfrac{r_{11}}{k_1 A_1} + 1\right)X_1 + \dfrac{r_{12}}{k_2 A_2}X_2 + \dots + \dfrac{r_{1n}}{k_n A_n}X_n + r_{1P} = 0 \\[2mm] \dfrac{r_{21}}{k_1 A_1}X_1 + \left(\dfrac{r_{22}}{k_2 A_2} + 1\right)X_2 + \dots + \dfrac{r_{2n}}{k_n A_n}X_1 + r_{2P} = 0 \\[2mm] \dots\dots\dots\dots\dots\dots\dots\dots\dots\dots\dots\dots \\[2mm] \dfrac{r_{1n}}{k_1 A_1}X_1 + \dfrac{r_{n2}}{k_2 A_2}X_2 + \dots + \left(\dfrac{r_{nn}}{k_n A_n} + 1\right)X_n + r_{nP} = 0 \end{array}\right\} \tag{9.51}$$

This system of equations also contains only n unknowns. In this case unknowns are the soil reactions. Obtained reactions acting in the opposite direction are applied to the soil as uniformly distributed loads, and settlements of the soil are found. Final analysis of the plate is performed by applying to the foundation the given loads and found settlements.

9.7 Analysis of Ring Foundations

Analysis of ring foundations, in principle, is no different from analysis of circular plates. Let us explain how to perform analysis of a ring plate using the finite element method. Analysis is performed as shown below:

1. The given ring foundation of a circular shape is replaced with a polygon type foundation, as shown in Figure 9.27, and the total area of the foundation is divided into a system of trapezoids.
2. By connecting the centers of gravity of all trapezoids with the four nodes, the total area of the foundation is divided into a system of triangular finite elements.
3. Continuous contact between the plate and the half-space is replaced with a series of individual totally rigid supports located at points 1, 2, ..., 12.
4. The plate supported on a system of non-yielding vertical supports is analyzed successively under the given vertical loads and one unit settlements applied to each support.
   The total reaction at any support can be found as follows:

$$X_i = r_{i1}\delta_1 + r_{i2}\delta_2 + \ldots + r_{in}\delta_n + r_{iP} \tag{9.52}$$

where r_{ik} is the reaction at support i due to one unit settlement applied to support k, and r_{iP} is the reaction at support i due to the given loads applied to the

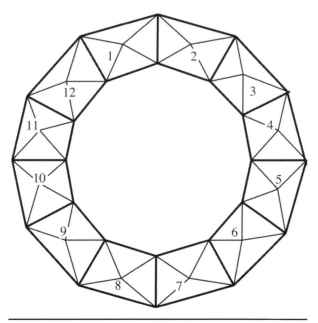

Figure 9.27 A ring plate divided into a system of triangular finite elements

plate. The following system of equations shows all n reactions at all points of support:

$$\left.\begin{array}{l} X_1 = r_{11}\delta_1 + r_{12}\delta_2 + \ldots + r_{1n}\delta_n + r_{1P} \\ X_2 = r_{21}\delta_1 + r_{22}\delta_2 + \ldots + r_{2n}\delta_n + r_{2P} \\ \cdots\cdots\cdots\cdots\cdots\cdots\cdots\cdots\cdots\cdots \\ X_n = r_{n1}\delta_1 + r_{n2}\delta_2 + \ldots + r_{nn}\delta_n + r_{nP} \end{array}\right\} \quad (9.53)$$

5. All reactions acting in the opposite direction and applied to the top of the half-space will produce settlements of the half-space. However, these loads, as mentioned earlier, cannot be applied as concentrated point loads: such loads will produce infinitely large settlements. They have to be applied as uniformly distributed circular loads, as shown in Figure 9.28.

6. Assuming that the outer radius of the ring foundation is equal to R_o and the inner radius is equal to R_i, the actual tributary area at each support is equal to

$$A = \left(R_o^2 - R_i^2\right)\frac{\pi}{12}.$$

The radius of the equivalent circular area at each support is equal to $r = \sqrt{\left(R_o^2 - R_i^2\right)/12}$. The settlement of the half-space at any point i loaded with a series of uniformly distributed loads $\left(q_i = \dfrac{X_i}{A}\right)$, where X_i is the reaction at point i, is obtained from 9.54:

Tributary loaded circular areas of the soil $r = \sqrt{\left(R_o^2 - R_i^2\right)/12}$

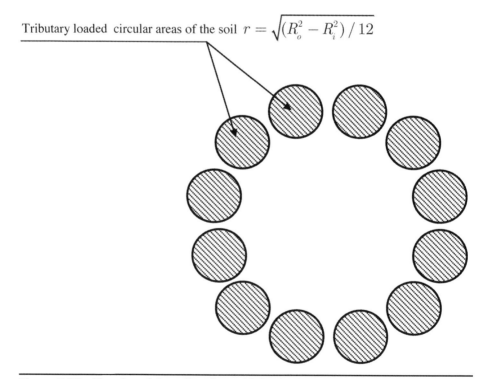

Figure 9.28   The plan of the soil surface with load tributary areas

$$\delta_i = \delta_{i1} X_1 + \delta_{i2} X_2 + \ldots + \delta_{in} X_n + \delta_{iP} \tag{9.54}$$

Settlements at all supports are found from the following equations:

$$\left.\begin{aligned}
\delta_1 &= \delta_{11} X_1 + \delta_{12} X_2 + \ldots + \delta_{1n} X_n + \delta_{1G} \\
\delta_2 &= \delta_{21} X_1 + \delta_{22} X_2 + \ldots + \delta_{2n} X_n + \delta_{2G} \\
&\cdots\cdots\cdots\cdots\cdots\cdots\cdots\cdots\cdots\cdots\cdots\cdots \\
\delta_n &= \delta_{n1} X_1 + \delta_{n2} X_2 + \ldots + \delta_{nn} X_n + \delta_{nG}
\end{aligned}\right\} \tag{9.55}$$

In equation 9.55, δ_{iG} is the soil settlement at point i due to the load G located outside of the loaded area. Both systems of equations 9.53 and 9.55 contain $2n$ equations with $2n$ unknowns. Final analysis of the plate is performed by applying to the foundation the given loads and found settlements.

9.8 Analysis of Narrow Ring Foundations

Analysis of narrow ring foundations can be performed as a circular beam supported on elastic foundation. The foundation represents a 3D statically indeterminate system with vertical and horizontal loads. Analysis is performed as follows:

1. The given ring foundation is replaced with a polygon with 12 nodes, as shown in Figure 9.29.
2. Continuous contact of the foundation and soil is replaced with a series of totally rigid non-yielding supports located at all 12 nodes.
3. The foundation is analyzed successively under the given loads and one unit settlements applied to each of the nodes.
4. As a result the following reactions are found:

$$\left.\begin{aligned}
X_1 &= r_{11}\Delta_1 + r_{12}\Delta_2 + \cdots + r_{1,12}\Delta_{12} + r_{1P} \\
X_2 &= r_{21}\Delta_1 + r_{22}\Delta_2 + \cdots + r_{2,12}\Delta_{12} + r_{2P} \\
&\cdots\cdots\cdots\cdots\cdots\cdots\cdots\cdots\cdots\cdots\cdots\cdots \\
X_{12} &= r_{12,1}\Delta_1 + r_{12,2}\Delta_2 + \cdots + r_{12,12}\Delta_{12} + r_{12P}
\end{aligned}\right\} \tag{9.56}$$

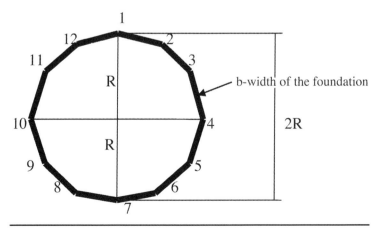

Figure 9.29   Plan of the narrow ring foundation

where r_{ik} is the reaction at point i due to one unit settlement of the foundation at point k, and r_{ip} is the reaction at the same point i due to the given loads applied to the foundation. The same reactions acting in the opposite direction and applied to the soil as uniformly distributed circular loads, shown in Figure 9.30, will produce the following settlements:

$$\left.\begin{array}{l} \Delta_1 = \Delta_{11}X_1 + \Delta_{12}X_2 + \cdots + \Delta_{1,12}X_{12} + \Delta_{1G} \\ \Delta_2 = \Delta_{21}X_1 + \Delta_{22}X_2 + \cdots + \Delta_{2,12}X_{12} + \Delta_{2G} \\ \cdots\cdots\cdots\cdots\cdots\cdots\cdots\cdots\cdots\cdots\cdots\cdots\cdots\cdots \\ \Delta_{12} = \Delta_{12,1}X_1 + \Delta_{12,2}X_2 + \cdots + \Delta_{12,12}X_{12} + \Delta_{12G} \end{array}\right\} \qquad (9.57)$$

where Δ_{ik} is the settlement of the soil support at point i due to one unit load applied to the soil at point k, and Δ_{iG} is the settlement at the same point i due to the load G located outside of the foundation.

For convenience, loads applied to the soil are replaced with circular distributed loads. The actual tributary area of the load applied to the soil at any node i looks as shown in Figure 9.31. The equivalent tributary circular area is equal to: $A_i = 2\pi Rb/12 = \pi Rb/6$ where R is the radius of the given foundation and b is the width of the foundation. The radius of the equivalent tributary area of each support is equal to $r = \sqrt{Rb/6}$.

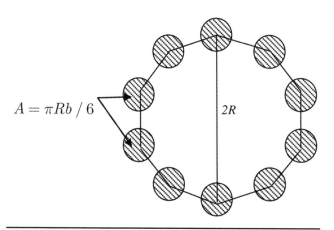

Figure 9.30   Plan of tributary areas of ring foundation supports

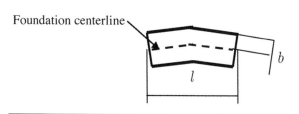

Figure 9.31

By solving the two systems of equations 9.56 and 9.57, all reactions and all settlements are found. Final analysis of the foundation is performed by applying to the foundation all settlements and the given loads. All analyses of the foundation (loaded with one unit settlements as well as the final analysis) are performed using any available computer program for analysis of 3D statically indeterminate systems. The method of iterations can also be employed. Analysis is performed in the following order:

1. The foundation supported at all 12 non-yielding supports and loaded with the given loads is analyzed, and all reactions $X_1, X_2, \ldots, X_{12}$ are found.
2. Obtained reactions acting downward as active uniformly distributed loads are applied to the soil, and settlements of the soil $\Delta_1, \Delta_2, \ldots, \Delta_{12}$ are found.
3. Obtained settlements and the given loads are applied again to the foundation and new reactions are found.

The third iteration is no different from the second one. After several iterations, when reactions or settlements obtained in iteration n are close enough to the reactions or settlements in iteration $n - 1$, analysis is finished. This method of analysis is probably the simplest and easiest.

When the foundation is supported on Winkler foundation, analysis is much simpler. In this case, continuous contact between the soil and foundation is replaced with a series of individual spring supports. The rigidity of the spring is obtained as $C = kA$, where k is the modulus of subgrade reaction and A is the tributary area of the spring. Analysis does not require any special software. Any program for the analysis of regular 3D statically indeterminate systems can be used.

9.9 Analytic-Numerical Analysis of Circular Plates on Elastic Foundation

As shown above, the analysis of circular plates on elastic foundation, based on the finite element method, requires building and solving large systems of linear equations. Although the analysis produces reasonable results, it cannot compete with more accurate analytical methods when such methods are available. A method is introduced below that uses closed analytical formulae for analysis of circular plates on Winkler foundation as well as on elastic half-space. It was developed to analyze two types of circular plates: plates with edges continuously supported on non-yielding supports, and plates with edges restrained against any linear and angular deflections. The method was developed for uniformly distributed loads applied to small circular areas with radius r_0, as shown in Figure 9.32. This figure shows a circular plate supported on a non-yielding circular foundation. At point f the plate is given a load P distributed over the area equal to πr_0^2. Analysis of the plate is performed using the following formulae:

$$\sigma_{\max r} = \sigma_{\max \theta} = -\frac{3P}{2\pi mh^2}\left[m + (m + 1)\ln\frac{a - \rho}{r_0} - (m - 1)\left(\frac{r_0^2}{4(a - \rho)^2}\right)\right] \quad (9.58)$$

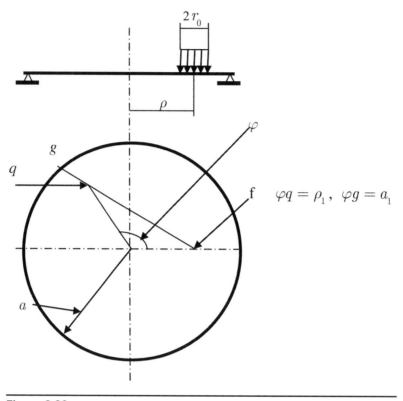

Figure 9.32

At point q stresses are found from the following formulae:

$$\sigma_r = (\sigma_{r\,max})\left|\frac{(m+1)\ln\dfrac{a_1}{\rho_1}}{1+(m+1)\ln\dfrac{a_1}{r_0}}\right| \quad \sigma_\theta = (\sigma_{max\,\theta})\left|\frac{(m+1)\ln\dfrac{a_1}{\rho_1}+(m+1)}{m+(m+1)\ln\dfrac{a_1}{r_0}}\right| \quad (9.59)$$

$$y = K_0(\rho^3 - \beta_0 a\rho^2 + c_0 a^3) + K_1(\rho^4 - b_1 a\rho^3 + c_1 a^3\rho)\cos\phi + \quad (9.60)$$
$$K_2(\rho^4 - b_2 a\rho^3 + c_2 a^2\rho^2)\cos 2\phi$$

where

$$K_0 = \frac{2(m+1)P(\rho^3 - b_0 a\rho^2 + c_0 a^4)}{9(5m+1)K\pi a^4}, \quad K_1 = \frac{2(3m+1)P(\rho^4 - b_1 a\rho^3 + c_1 a^3\rho)}{3(9m+1)K\pi a^6},$$

$$K_2 = \frac{(4m+1)^2 P(\rho^4 - b_2 a\rho^3 + c_2 a^2\rho^2)}{(9m+1)(5m+1)K\pi a^6}, \quad K = \frac{m^2 Eh^3}{12(m^2+1)}, \quad b_0 = \frac{3(2m+1)}{2(m+1)},$$

$$b_0 = \frac{3(2m+1)}{2(m+1)} \quad b_1 = \frac{3(4m+1)}{2(3m+1)} \quad b_2 = \frac{2(5m+1)}{4m+1} \quad c_0 = \frac{4m+1}{2(m+1)}$$

$$c_1 = \frac{6m+1}{2(3m+1)} \quad c_2 = \frac{6m+1}{4m+1} \quad m = \frac{1}{\mu}$$

E is the modulus of elasticity and μ is Poisson's ratio of the plate material. Formulae 9.58–9.60 allow performing analysis of regular circular plates, loaded with concentrated vertical loads distributed over small circular areas. Analysis of the same plate supported on an elastic foundation is more complex and is performed as follows.

Continuous contact of the plate with the soil is replaced with a series of individual supports with the accuracy of analysis depending on the number of supports. The more supports replace the area of contact, the higher is the accuracy of the results. The plate shown in Figure 9.33 is supported on a non-yielding ring foundation along the perimeter. The remainder of the plate is supported on soil. The location of all supports is chosen by the engineer and depends on his/her judgment. In this case the total area of the plate supported on the soil is equal to πr^2, where r is the radius of the total circular area. The total area of the plate is divided into 16 smaller areas equal to $\pi r^2/16$, where $r = 7a/8$. Individual soil supports are located at points 1, 2, ..., n.

Tributary areas of each support are shown in Figure 9.34 (A_1, A_2, A_3, A_4).

$$A_1 = \frac{1}{16}\left[\pi\left(\frac{7a}{8}\right)^2 - \pi\left(\frac{5a}{8}\right)^2\right] = \frac{3\pi a^2}{16\cdot 8}, \quad A_2 = \left[\pi\left(\frac{5a}{8}\right)^2 - \pi\left(\frac{3a}{8}\right)^2\right] = \frac{\pi a^3}{16\cdot 4}$$

$$A_3 = \left[\pi\left(\frac{3a}{8}\right)^2 - \pi\left(\frac{a}{8}\right)^2\right] = \frac{\pi a^2}{16\cdot 8}, \quad A_4 = \pi\frac{a^2}{16\cdot 64}$$

The total area of the plate supported on soil is equal to $A_{plate} = \pi\left(\frac{7a}{8}\right)^2$. The same result is obtained by summing the tributary areas of all individual supports:
$$\frac{3\pi a^2}{8} + \frac{\pi a^2}{4} + \frac{\pi a^2}{8} + \pi\frac{a^2}{64} = \pi\left(\frac{49a^2}{64}\right) = \pi\left(\frac{7a}{8}\right)^2$$ that is equal to A_{plate} found above.
As we can see, the given plate is supported on 16 supports type A_1, 16 supports type A_2, 16 supports type A_3, and 16 supports type A_4. All 16 supports type A_4 can be replaced with one circular area equal to $A_4^0 = \pi\frac{a^2}{64}$.

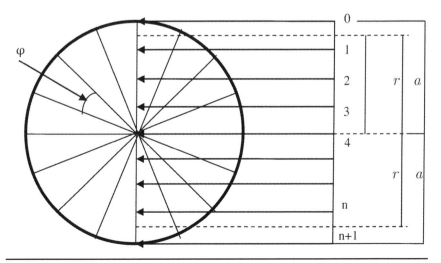

Figure 9.33

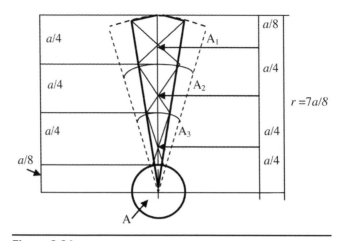

Figure 9.34

If the plate is supported on Winkler foundation and the modulus of subgrade reaction is equal to K_Z, rigidities of all supports are found as follows:

$$C_1 = K_Z A_1, \; C_2 = K_Z A_2, \; C_3 = K_Z A_3 \text{ and } C_4^0 = K_Z A_4^0$$

Settlements of the soil at any point under the given plate can be found using the following simple formula:

$$w_i = \frac{X_i}{C_i} \tag{9.61}$$

where X_i is the load applied to the soil at point i and C_i is the rigidity of the soil at point i. The settlement of the plate at any point i can obtained using formula 9.60. The total settlement of the plate will look as follows:

$$y_i = y_{i1} X_1 + y_{i2} X_2 + \dots + y_{im} X_m + y_{iP} \tag{9.62}$$

where y_{ik} is the settlement of the plate at point i due to one unit load applied to the plate at point k and y_{iP} is the settlement of the plate due to the given load applied to the plate. Taking into account that the forces of interaction between the plate and soil act in opposite directions and vertical deflections of the soil and the plate at points of contact are the same, the following equation is written:

$$(y_{ii} + 1/C_i) X_i + y_{i1} X_1 + y_{i2} X_2 + \dots + y_{im} X_m + y_{iP} = 0 \tag{9.63}$$

For all points of contact between the soil and the plate, the system of equations 9.64 is written below.

$$\left. \begin{array}{l} \left(y_{11} + \dfrac{1}{C_1}\right) X_1 + y_{22} X_2 + \dots + y_{2m} X_m + y_{1P} = 0 \\[2mm] y_{21} X_1 + \left(y_{22} + \dfrac{1}{C_2}\right) X_2 + \dots + y_{2m} X_m + y_{2P} = 0 \\[2mm] \dots\dots\dots\dots\dots\dots\dots\dots\dots\dots\dots\dots\dots \\[2mm] y_{m1} X_1 + y_{m2} X_2 + \dots + \left(y_{mm} + \dfrac{1}{C_2}\right) X_m + y_{mP} = 0 \end{array} \right\} \tag{9.64}$$

By solving the system of equations 9.64, all forces of interaction between the plate and the soil are found.

Final analysis of the plate is performed by applying to the plate the given loads and soil reactions X_i found above. Soil reactions are applied to the plate as uniformly distributed loads over small circular areas using formula 9.60. When the plate is supported on an elastic half-space, analysis is more complex but still available. Analysis of the same plate shown in Figure 9.32 is performed as shown.

1. Continuous contact of the plate and the soil is replaced with a series of individual supports. Individual supports are totally rigid and located at the same points as shown above for analysis on Winkler foundation.
2. All supports are replaced with m unknown reactions applied to the plate (see Figure 9.33).
3. The same reactions acting in the opposite direction as active loads are applied to the half-space. These loads are applied to the soil not as concentrated loads, but as uniformly distributed loads over tributary circular areas. Radiuses of these circular areas are found as shown below. Taking into account that $A_1 = \dfrac{3\pi a^2}{16 \cdot 8}$,

$A_2 = \dfrac{\pi a^2}{16 \cdot 4}$, $A_3 = \dfrac{\pi a^2}{16 \cdot 8}$, $A_4^0 = \pi \dfrac{a^2}{64}$ radiuses of the equivalent circular areas

R_1, R_2, R_3, and R_4 can be found. Radiuses are found from these formulae:

$$\pi R_1^2 = \frac{\pi 3 a^2}{16 \cdot 8}, \ \pi R_2^2 = \frac{\pi a^2}{16 \cdot 4}, \ \pi R_3^2 = \frac{\pi a^2}{16 \cdot 8}, \ \pi R_4^2 = \frac{\pi a^2}{64} \tag{9.65}$$

From equations 9.65, we find $R_1 = a\dfrac{\sqrt{1.5}}{8}$, $R_2 = \dfrac{a}{8}$, $R_3 = \dfrac{a}{8\sqrt{2}}$ and $R_4 = \dfrac{a}{8}$.

By summing tributary areas of all supports we have:

$$16\left(\frac{3\pi a^2}{16 \cdot 8} + \frac{\pi a^2}{16 \cdot 4} + \frac{\pi a^2}{16 \cdot 8}\right) + \frac{\pi a^2}{64} = \pi\left(\frac{7a}{8}\right)^2$$

that is equal to the total area of the plate supported on soil. Deflections of the plate are found using the same formula 9.62 and deflections of the half-space are found using the following formula:

$$w_i = w_{i1} X_1 + w_{i2} X_2 + \ldots + w_{im} X_m \tag{9.66}$$

Settlements of the half-space at all points of support can be expressed by the system of equations 9.67.

$$\left.\begin{array}{l} w_1 = w_{11} X_1 + w_{12} X_2 + \ldots + w_{1m} X_m \\ w_2 = w_{21} X_1 + w_{22} X_2 + \ldots + w_{2m} X_m \\ \ldots\ldots\ldots\ldots\ldots\ldots\ldots\ldots\ldots\ldots\ldots \\ w_m = w_{m1} X_1 + w_{m2} X_2 + \ldots + w_{mm} X_m \end{array}\right\} \tag{9.67}$$

The system of equations 9.68 shows settlements of the plate at the same points.

$$y_1 = y_{11}X_1 + y_{12}X_2 + \ldots + y_{1m}X_m + y_{1P}$$
$$y_2 = y_{21}X_1 + y_{22}X_2 + \ldots + y_{2m}X_m + y_{2P}$$
$$\ldots\ldots\ldots\ldots\ldots\ldots\ldots\ldots\ldots\ldots\ldots\ldots\ldots\ldots\ldots$$
$$y_m = y_{m1}X_1 + y_{m2}X_2 + \ldots + y_{mm}X_m + y_{mP}$$

(9.68)

Now, taking into account that deflections of the plate and deflections of the half-space at the same points are the same, and applied in opposite directions, the following system of equations is written:

$$(y_{11} + w_{11})X_1 + (y_{12} + w_{12})X_2 + \ldots + (y_{1m} + w_{1m})X_m + y_{1P} = 0$$
$$(y_{21} + w_{21})X_1 + (y_{22} + w_{22})X_2 + \ldots + (y_{2m} + w_{2m})X_m + y_{2P} = 0$$
$$\ldots\ldots\ldots\ldots\ldots\ldots\ldots\ldots\ldots\ldots\ldots\ldots\ldots\ldots\ldots$$
$$(y_{m1} + w_{m1})X_1 + (y_{m2} + w_{m2})X_2 + \ldots + (y_{mm} + w_{mm})X_m + y_{mP} = 0$$

(9.69)

By solving the system of equations 9.69, all forces of interaction between the plate and the half-space are obtained. Final analysis of the plate is performed by applying to the plate the given loads and found reactions. All reactions are applied as uniformly distributed loads over small circular areas.

The method described above can also be applied to a circular plate with fixed edges (totally restrained against any deflections), as shown in Figure 9.35. The analysis is the same as the analysis of a free supported plate, but the formulae for obtaining stresses

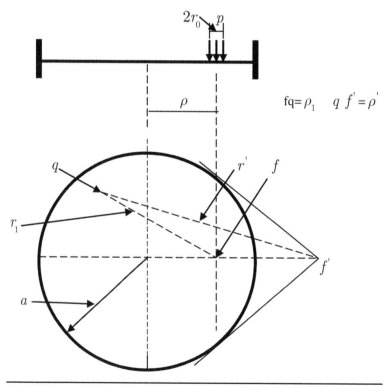

Figure 9.35

and settlements are different. The maximum stress on the plate at the point the load is applied is obtained from formula 9.70:

$$\sigma_r = \sigma_{max} = \frac{-3P}{2\pi mh^2}\left[(m+1)\ln\frac{a-\rho}{r_0} + (m+1)\frac{r_0^2}{4(a-\rho)^2}\right] \quad (9.70)$$

Formula 9.70 is used when $r_0 < 0.6(a - \rho)$. For points located along the perimeter of the plate when $r_0 \geq 0.6(a - \rho)$ formula 9.71 is used.

$$\sigma_r = \sigma_{max} = \frac{3P}{2\pi h^2}\left[1 - \frac{r_0^2}{2(a-\rho)^2}\right] \quad (9.71)$$

Settlement of the plate at the point where the load is applied is obtained from formula 9.72:

$$y = \frac{3P(m^2-1)(a^2-\rho^2)^2}{4\pi Em^2h^3a^2} \quad (9.72)$$

Settlement of the plate at any point q is obtained from formula 9.73:

$$y = -\frac{3P(m^2-1)}{2\pi Em^2h^3}\left[\frac{1}{2}\left(\frac{\rho^2 r'^2}{a^3} - r_1^2\right) - r_1^2\ln\frac{\rho r'}{ar_1}\right] \quad (9.73)$$

Using formulae 9.72 and 9.73 and the method described above, a circular plate with a fixed perimeter supported on Winkler foundation or elastic half-space can be analyzed.

If the given load applied to the plate is uniformly distributed over the plate or over a part of the plate, that load can be replaced with a series of loads distributed over small circular areas and the same method of analysis can be used.

Table 9.1 Moments in a circular plate loaded symmetrically with a vertical load

$\dfrac{h_0}{h}$	$M_r = M_t$	$M_r = \beta P$	$M_t = \beta_1 P$	
	$r = 0$	$r = a/2$	$r = a/2$	$r = a$
	γ_2	β	β_1	β_1
1.00	−0.065	0.021	0.073	0.030
1.50	−0.053	0.032	0.068	0.016
2.33	−0.038	0.040	0.063	0.007

Table 9.2 Soil pressure under a circular totally rigid plate loaded with any symmetrical load

ρ	0.0	0.1	0.2	0.3	0.4	0.5	0.6	0.7	0.8	0.9	1.0
$\bar{p}$	0.159	0.160	0.162	0.167	0.174	0.184	0.199	0.223	0.265	0.365	∞

Table 9.3 Coefficients for obtaining radial and tangential moments and shear forces (Figure 9.15)

ρ	0.0	0.1	0.2	0.3	0.4	0.5	0.6	0.7	0.8	0.9	1.0
$\overline{M}_r$	∞	0.175	0.111	0.075	0.051	0.033	0.020	0.010	0.004	0.000	0.000
$\overline{M}_t$	∞	0.245	0.177	0.140	0.114	0.095	0.080	0.068	0.058	0.050	0.044
$\overline{Q}_r$	∞	1.583	0.780	0.506	0.365	0.276	0.212	0.162	0.119	0.075	0.000

Table 9.4 Radial and tangential moments $\overline{M} = \overline{M}_r = \overline{M}_t$

α	0.005	0.010	0.020	0.030	0.040	0.050	0.075	0.100	0.150
$\overline{M}$	0.532	0.468	0.403	0.366	0.339	0.318	0.280	0.254	0.215

Table 9.5 Radial, tangential moments and shear forces for totally rigid plates with symmetrical uniformly distributed loads (See Figure 9.17)

α	0	0.1	0.2	0.3	0.4	0.5	0.6	0.7	0.8	0.9	1.0
						ρ					
						$\overline{M}_r$					
0.1	0.0080	0.0060	0.0036	0.0024	0.0016	0.0010	0.0006	0.0003	0.0001	0.0000	0
0.2	0.0237	0.0218	0.0159	0.0102	0.0068	0.0045	0.0026	0.0014	0.0005	0.0000	0
0.3	0.0426	0.0406	0.0349	0.0253	0.0166	0.0106	0.0064	0.0033	0.0012	0.0001	0
0.4	0.0616	0.0598	0.0543	0.0449	0.0324	0.0205	0.0124	0.0066	0.0025	0.0002	0
0.5	0.0788	0.0771	0.0718	0.0630	0.0511	0.0357	0.0215	0.0114	0.0047	0.0007	0
0.6	0.0922	0.0906	0.0857	0.0775	0.0664	0.0519	0.0345	0.0186	0.0080	0.0016	0
0.7	0.1001	0.0988	0.0942	0.0866	0.0764	0.0632	0.0475	0.0289	0.0126	0.0029	0
0.8	0.1008	0.0993	0.0995	0.0886	0.0794	0.0677	0.0537	0.0374	0.0193	0.0049	0
0.9	0.0926	0.0914	0.0879	0.0819	0.0740	0.0639	0.0519	0.0381	0.0231	0.0080	0
1.0	0.0737	0.0727	0.0697	0.0647	0.0582	0.0499	0.0401	0.0291	0.0177	0.0068	0
						$\overline{M}_t$					
0.1	0.0080	0.0070	0.0054	0.0043	0.0035	0.0031	0.0025	0.0021	0.0018	0.0016	0.0014
0.2	0.0237	0.0228	0.0200	0.0165	0.0138	0.0116	0.0097	0.0083	0.0070	0.0062	0.0054
0.3	0.0426	0.0116	0.0390	0.0343	0.0292	0.0247	0.0210	0.0179	0.0152	0.0132	0.0117
0.4	0.0616	0.0608	0.0583	0.0537	0.0479	0.0410	0.0352	0.0299	0.0255	0.0221	0.0195
0.5	0.0788	0.0780	0.0756	0.0714	0.0658	0.0585	0.0503	0.0432	0.0370	0.0320	0.0288
0.6	0.0922	0.0914	0.0891	0.0851	0.0799	0.0731	0.0648	0.0558	0.0481	0.0415	0.0365
0.7	0.1001	0.0994	0.0974	0.0936	0.0889	0.0826	0.0750	0.0662	0.0570	0.0491	0.0431
0.8	0.1008	0.0999	0.0983	0.0950	0.0907	0.0851	0.0784	0.0705	0.0616	0.0529	0.0412
0.9	0.0926	0.0920	0.0905	0.0875	0.0839	0.0789	0.0732	0.0664	0.0590	0.0510	0.0441
1.0	0.0737	0.0731	0.0714	0.0693	0.0663	0.0623	0.0576	0.0522	0.0462	0.0402	0.0348
						$\overline{Q}_r$					
0.1	0	0.050	0.025	0.016	0.012	0.009	0.007	0.006	0.004	0.003	0
0.2	0	0.050	0.100	0.066	0.048	0.037	0.029	0.023	0.017	0.011	0
0.3	0	0.050	0.099	0.148	0.109	0.084	0.066	0.051	0.038	0.025	0
0.4	0	0.050	0.098	0.146	0.193	0.149	0.117	0.091	0.068	0.044	0
0.5	0	0.049	0.097	0.144	0.190	0.233	0.183	0.143	0.106	0.068	0
0.6	0	0.049	0.096	0.142	0.185	0.226	0.264	0.206	0.153	0.098	0
0.7	0	0.049	0.095	0.139	0.180	0.217	0.251	0.280	0.208	0.134	0
0.8	0	0.048	0.094	0.135	0.173	0.207	0.236	0.258	0.272	0.175	0
0.9	0	0.048	0.092	0.131	0.166	0.196	0.219	0.234	0.238	0.221	0
1.0	0	0.047	0.090	0.127	0.158	0.183	0.200	0.207	0.200	0.168	0

Table 9.6 Radial, tangential moments and shear forces for rigid plates with uniformly distributed loads along the circumference (See Figure 9.18)

α	ρ										
	0	0.1	0.2	0.3	0.4	0.5	0.6	0.7	0.8	0.9	1.0
						$\overline{M}_r$					
0.1	0.130	0.130	0.075	0.049	0.033	0.021	0.013	0.007	0.002	0.000	0
0.2	0.178	0.178	0.180	0.111	0.072	0.046	0.028	0.015	0.006	0.000	0
0.3	0.133	0.194	0.195	0.198	0.125	0.079	0.047	0.025	0.010	0.001	0
0.4	0.184	0.185	0.188	0.192	0.197	0.123	0.074	0.040	0.017	0.003	0
0.5	0.156	0.157	0.160	0.165	0.172	0.182	0.109	0.059	0.026	0.006	0
0.6	0.110	0.111	0.114	0.121	0.129	0.140	0.115	0.086	0.039	0.011	0
0.7	0.043	0.048	0.052	0.059	0.069	0.082	0.099	0.120	0.056	0.017	0
0.8	-0.035	-0.033	-0.028	-0.020	-0.009	0.006	0.026	0.049	0.078	0.025	0
0.9	-0.132	-0.131	-0.125	-0.116	-0.104	-0.086	-0.065	-0.038	-0.005	0.036	0
1.0	-0.248	-0.246	-0.240	-0.239	-0.216	-0.197	-0.173	-0.144	-0.107	-0.062	0
						$\overline{M}_t$					
0.1	0.130	0.130	0.106	0.085	0.070	0.058	0.049	0.042	0.035	0.031	0.027
0.2	0.178	0.178	0.179	0.155	0.131	0.110	0.094	0.080	0.060	0.059	0.052
0.3	0.133	0.194	0.194	0.195	0.175	0.151	0.129	0.130	0.094	0.081	0.072
0.4	0.184	0.185	0.186	0.188	0.190	0.172	0.150	0.110	0.111	0.096	0.084
0.5	0.156	0.157	0.158	0.160	0.164	0.169	0.153	0.133	0.115	0.099	0.087
0.6	0.110	0.110	0.112	0.115	0.119	0.124	0.131	0.119	0.102	0.088	0.077
0.7	0.043	0.047	0.049	0.052	0.057	0.063	0.071	0.081	0.071	0.061	0.051
0.8	-0.035	-0.034	-0.031	-0.028	-0.023	-0.016	-0.006	0.005	0.017	0.013	0.009
0.9	-0.132	-0.132	-0.120	-0.125	-0.119	-0.110	-0.101	-0.088	-0.074	-0.056	-0.054
1.0	-0.248	-0.247	-0.244	-0.239	-0.233	-0.224	-0.213	-0.200	-0.183	-0.164	-0.139

						$\bar{Q}_r$					
0.1	0	0.001*	-0.498	-0.329	-0.242	-0.187	-0.147	-0.114	-0.085	-0.055	0
0.2	0	0.001	0.004*	-0.658	-0.483	-0.373	-0.293	-0.228	-0.170	-0.109	0
0.3	0	0.002	0.006	0.014*	-0.725	-0.500	-0.440	-0.343	-0.255	-0.164	0
0.4	0	0.002	0.008	0.018	0.033*	-0.746	-0.587	-0.457	-0.340	-0.219	0
0.5	0	0.003	0.010	0.023	0.042	0.067*	-0.733	-0.571	-0.425	-0.273	0
0.6	0	0.003	0.012	0.028	0.050	0.080	0.120*	-0.686	-0.510	-0.328	0
0.7	0	0.004	0.014	0.032	0.058	0.094	0.140	0.200*	-0.595	-0.383	0
0.8	0	0.004	0.016	0.037	0.067	0.107	0.160	0.229	0.320*	-0.438	0
0.9	0	0.005	0.018	0.042	0.075	0.121	0.180	0.257	0.360	0.508*	0
1.0	0	0.005	0.020	0.046	0.084	0.134	0.200	0.286	0.400	0.564	1*

*Note: When $\rho = \alpha$ the shear force is inside the edge of the plate. Outside of the edge $\bar{Q}_r = 0$.

Table 9.7 Uniformly distributed load located along the circumference of an elastic plate (See Figure 9.19)

$S = 1$

Coef.	α	0	0.1	0.2	0.3	0.4	0.5	0.6	0.7	0.8	0.9	1	$\bar{tg}\varphi_e$
$\bar{p}$		209	204	194	180	168	161	160	165	182	235	383	
$\bar{M}_r$		120	120	065	041	025	015	008	003	000	-001	000	-023
$\bar{M}_t$	0.1	120	120	096	076	061	050	042	035	030	026	023	
$\bar{Q}_r$		000	010*	-480	-304	-213	-156	-115	-084	-057	-032	000	
$\bar{w}$		0.96	096	094	0.92	090	088	085	083	080	078	075	
$\bar{p}$		396	381	370	353	337	324	321	334	377	480	709	
$\bar{M}_r$		158	158	161	094	057	034	018	007	001	-002	000	-044
$\bar{M}_t$	0.2	158	158	159	137	114	094	079	066	056	048	043	
$\bar{Q}_r$		000	019	0.38*	-611	-428	-313	-231	-168	-114	-061	000	
$\bar{w}$		190	189	187	184	180	175	171	166	161	157	152	
$\bar{p}$		526	524	519	510	496	484	483	507	581	736	1031	
$\bar{M}_r$		164	165	169	174	103	060	0.32	014	003	-002	000	-060
$\bar{M}_t$	0.3	164	165	166	169	149	127	107	091	077	066	059	
$\bar{Q}_r$		000	026	0.52	078*	-648	-474	-351	-255	-173	-092	000	
$\bar{w}$		279	279	276	273	268	262	256	250	243	237	231	
$\bar{p}$		634	635	637	636	633	631	638	677	781	1003	1440	
$\bar{M}_r$		151	152	156	162	171	100	055	026	007	-001	000	-070
$\bar{M}_t$	0.4	151	151	153	156	160	144	124	106	090	077	068	
$\bar{Q}_r$		000	0.32	064	095	127*	-642	-477	-348	-237	-127	000	
$\bar{w}$		363	362	360	357	352	346	340	332	325	318	311	

		715	717	723	731	742	757	784	843	979	1286	1949	
0.5	$\bar{p}$	715	717	723	731	742	757	784	843	979	1286	1949	−071
	$\bar{M}_r$	121	122	127	134	144	157	088	044	015	001	000	
	$\bar{M}_t$	121	122	124	127	132	138	124	108	092	079	069	
	$\bar{Q}_r$	000	036	072	109	146	184*	−609	−447	−307	−167	000	
	$\bar{w}$	441	440	439	436	432	427	421	414	407	399	392	
0.6	$\bar{p}$	772	775	783	800	825	862	916	1009	1193	1595	2475	−061
	$\bar{M}_r$	076	078	082	090	101	115	133	096	027	004	000	
	$\bar{M}_t$	076	077	079	083	088	095	103	093	079	068	059	
	$\bar{Q}_r$	000	039	078	118	159	203	251*	−553	−382	−210	000	
	$\bar{w}$	514	513	512	511	508	505	500	495	489	483	477	
0.7	$\bar{p}$	805	810	824	849	888	945	1032	1172	1427	1934	2978	−0.36
	$\bar{M}_r$	017	018	023	041	043	058	078	102	043	010	000	
	$\bar{M}_t$	0.17	0.17	020	023	029	036	045	056	049	041	035	
	$\bar{Q}_r$	000	040	081	124	169	218	272	335*	−461	−254	000	
	$\bar{w}$	583	582	582	582	581	580	578	576	573	569	566	
0.8	$\bar{p}$	815	822	842	876	930	1008	1127	1320	1659	2299	3563	006
	$\bar{M}_r$	−057	−055	−050	−042	−030	−014	006	032	065	017	000	
	$\bar{M}_t$	−057	−056	−054	−050	−044	−037	−028	−016	−002	−004	−005	
	$\bar{Q}_r$	000	041	083	127	174	226	286	358	452*	−303	000	
	$\bar{w}$	646	647	647	648	650	652	653	655	656	657	658	
0.9	$\bar{p}$	803	811	837	882	952	1053	1207	1452	1880	2683	4258	066
	$\bar{M}_r$	−145	−143	−138	−130	−118	−102	−081	−054	−019	027	000	
	$\bar{M}_t$	−145	−144	−142	−138	−132	−125	−115	−103	−088	−069	−064	
	$\bar{Q}_r$	000	040	082	126	176	230	295	376	483	641*	000	
	$\bar{w}$	705	706	708	711	715	720	726	732	739	746	753	

Table 9.7 Con't

		ρ											$\overline{tg}\varphi_e$
Coef.	α	0	0.1	0.2	0.3	0.4	0.5	0.6	0.7	0.8	0.9	1	
							$S=1$						
$\overline{p}$	1.0	769	780	812	869	956	1084	1275	1579	2105	3084	4994	149
$\overline{M}_r$		−247	−246	−241	−233	−222	−206	−184	−157	−120	−071	000	
$\overline{M}_t$		−247	−247	−245	−241	−235	−228	−219	−207	−191	−171	−145	
$\overline{Q}_r$		000	039	079	123	172	229	199	388	510	694	1000*	
$\overline{w}$		760	761	765	770	777	786	797	809	823	837	852	
							$S=2$						
$\overline{p}$	0.1	295	286	263	234	206	186	173	164	162	195	342	−021
$\overline{M}_r$		114	114	060	036	022	012	006	002	000	−001	000	
$\overline{M}_t$		114	114	090	071	056	0.46	038	032	027	023	021	
$\overline{Q}_r$		000	015*	−472	−294	−201	−144	−103	−073	−049	−027	000	
$\overline{w}$		054	054	052	050	048	046	044	042	039	037	035	
$\overline{p}$	0.2	527	518	492	455	415	378	351	339	350	408	568	−040
$\overline{M}_r$		146	147	150	085	050	028	014	005	000	−002	000	
$\overline{M}_t$		146	147	148	126	104	086	071	060	051	044	039	
$\overline{Q}_r$		000	026	051*	−593	−407	−290	−208	−147	−096	−050	000	
$\overline{w}$		106	105	103	100*	097	092	088	084	080	076	072	
$\overline{p}$	0.3	689	684	670	645	607	564	528	519	555	643	781	−054
$\overline{M}_r$		149	150	155	161	093	052	027	010	001	−002	000	
$\overline{M}_t$		149	150	152	155	136	115	097	081	069	059	053	
$\overline{Q}_r$		000	034	068	100*	−620	−444	−320	−226	−148	−075	000	
$\overline{w}$		153	153	151	148	143	138	132	127	121	115	110	

0.4	$\bar{p}$	792	793	791	781	758	726	696	695	752	897	1161
	$\bar{M}_r$	134	136	140	148	159	091	048	022	005	-002	000
	$\bar{M}_t$	134	135	137	141	146	131	112	095	081	070	062
	$\bar{Q}_r$	000	040	079	118	156*	-608	-442	-315	-208	-108	000
	$\bar{w}$	196	195	194	191	186	181	175	168	162	155	149
												-064
0.5	$\bar{p}$	850	851	855	858	857	852	848	863	943	1176	1732
	$\bar{M}_r$	106	108	113	121	133	148	082	040	014	001	000
	$\bar{M}_t$	106	107	109	113	119	126	114	098	084	072	063
	$\bar{Q}_r$	000	043	085	128	171	214*	-577	-416	-280	-150	000
	$\bar{w}$	233	233	231	229	226	221	215	209	202	195	189
												-065
0.6	$\bar{p}$	863	865	873	887	909	937	973	1031	1163	1503	2329
	$\bar{M}_r$	065	067	072	081	093	109	129	066	026	004	000
	$\bar{M}_t$	065	066	068	073	078	086	095	085	073	062	055
	$\bar{Q}_r$	000	043	087	131	177	225	275*	-529	-361	-197	000
	$\bar{w}$	266	266	265	263	261	258	254	249	244	238	232
												-056
0.7	$\bar{p}$	834	839	855	882	924	983	1066	1195	1425	1889	2872
	$\bar{M}_r$	011	013	018	027	039	055	075	100	042	009	000
	$\bar{M}_t$	011	012	015	019	024	032	041	052	046	038	033
	$\bar{Q}_r$	000	042	084	129	175	226	282	337*	-450	-246	000
	$\bar{w}$	295	295	295	294	294	293	291	289	286	283	280
												-033
0.8	$\bar{p}$	766	774	798	839	901	991	1123	1330	1670	2320	3569
	$\bar{M}_r$	-053	-052	-047	-040	-028	-013	007	032	065	017	000
	$\bar{M}_t$	-053	-053	-051	-047	-042	-035	-026	-014	000	-003	-004
	$\bar{Q}_r$	000	038	078	120	166	218	278	352	448*	-305	000
	$\bar{w}$	319	320	320	321	323	324	326	328	320	330	330
												005

Table 9.7 Con't

Coef.	α	0	0.1	0.2	0.3	0.4	0.5	0.6	0.7	0.8	0.9	1	$tg\varphi_e$
							ρ **S = 2**						
$\bar{p}$	0.9	658	669	702	758	844	968	1149	1432	1911	2787	4477	060
$\bar{M}_r$		-130	-129	-125	-118	-108	-094	-075	-050	-017	027	000	
$\bar{M}_t$		-130	-129	-127	-124	-119	-113	-105	-094	-080	-063	-058	
$\bar{Q}_r$		000	033	068	106	150	201	264	346	457	625	000	
$\bar{w}$		340	340	342	345	348	353	358	364	371	377	383	
$\bar{p}$	1.0	513	527	569	643	755	917	1156	1526	2149	3284	5462	137
$\bar{M}_r$		-220	-219	-216	-210	-201	-189	-173	-149	-117	-070	000	
$\bar{M}_t$		-220	-219	-218	-215	-211	-206	-199	-189	-176	-158	-133	
$\bar{Q}_r$		000	026	054	087	126	176	241	330	459	600	1000*	
$\bar{w}$		357	358	360	365	371	380	389	401	413	426	440	
							S = 5						
$\bar{p}$	0.1	501	481	425	354	288	237	198	157	111	105	288	-016
$\bar{M}_r$		101	102	049	026	014	007	002	000	-001	-001	000	
$\bar{M}_t$		101	101	078	059	046	037	030	025	021	018	016	
$\bar{Q}_r$		000	025*	-454	-270	-175	-116	-077	-050	-031	-018	000	
$\bar{w}$		027	027	026	024	022	020	019	017	015	013	012	
$\bar{p}$	0.2	851	829	768	680	581	487	407	341	289	257	279	-030
$\bar{M}_r$		122	123	128	066	034	016	005	000	-002	-002	000	
$\bar{M}_t$		122	123	125	105	084	068	055	046	038	033	030	
$\bar{Q}_r$		000	042	081*	-553	-360	-240	-150	-102	-060	-027	000	
$\bar{w}$		052	051	050	047	044	041	038	034	031	028	025	

			218	454	505	541	618	730	845	938	999	1031	1040
0.3	$\bar{p}$		218	454	505	541	618	730	845	938	999	1031	1040
	$\bar{M_r}$	−041	000	−003	−003	003	014	035	070	134	125	118	116
	$\bar{M_t}$		040	045	052	062	074	090	109	125	120	117	116
	$\bar{Q_r}$		000	−038	−095	−164	−253	−378	−561	149*	102	052	000
	$\bar{w}$		039	043	048	052	056	061	065	069	071	073	073
0.4	$\bar{p}$		531	675	695	732	919	930	1027	1089	1115	1119	1118
	$\bar{M_r}$	−049	000	−003	001	012	034	071	134	119	108	101	099
	$\bar{M_t}$		047	053	062	073	087	103	116	108	103	100	099
	$\bar{Q_r}$		000	−066	−147	−244	−368	−538	218*	166	112	056	000
	$\bar{w}$		054	059	064	069	074	079	083	087	089	090	091
0.5	$\bar{p}$		1274	936	961	908	990	1061	1105	1124	1125	1120	1118
	$\bar{M_r}$	−052	000	000	010	031	068	130	110	094	083	076	074
	$\bar{M_t}$		050	057	066	078	091	101	091	084	078	0.75	074
	$\bar{Q_r}$		000	−114	−222	−348	−509	277*	224	169	112	056	000
	$\bar{w}$		069	075	080	085	090	095	098	101	103	104	104
0.6	$\bar{p}$		2040	1295	1094	1085	1106	1108	1090	1066	1048	1039	1036
	$\bar{M_r}$	−046	000	004	023	059	118	094	075	061	050	044	042
	$\bar{M_t}$		045	051	060	070	078	067	058	051	046	043	042
	$\bar{Q_r}$		000	−170	−315	−476	326*	269	213	158	104	052	000
	$\bar{w}$		087	092	097	101	105	108	110	112	113	113	114
0.7	$\bar{p}$		2629	1786	1426	1257	1151	1068	997	942	904	882	875
	$\bar{M_r}$	−028	000	009	040	096	069	047	030	017	008	003	001
	$\bar{M_t}$		028	032	039	045	033	023	015	009	004	002	001
	$\bar{Q_r}$		000	−228	−425	372*	304	242	187	136	089	044	000
	$\bar{w}$		109	112	114	117	119	120	120	120	121	121	121

Table 9.7 Con't

Coef.	α	0	0.1	0.2	0.3	0.4	0.5	0.6	0.7	0.8	0.9	1	$\bar{tg}\varphi_o$
							$S = 5$						
							ρ						
$\bar{p}$	0.8	648	659	694	752	836	954	1119	1359	1733	2368	3550	
$\bar{M}_r$		-046	-044	-041	-034	-024	-010	009	033	065	017	000	
$\bar{M}_t$		-046	-045	-043	-040	-035	-029	-021	-010	003	000	-002	002
$\bar{Q}_r$		000	033	067	105	148	199	261	338	440*	-307	000	
$\bar{w}$		124	125	125	126	127	129	130	131	132	133	133	
$\bar{p}$	0.9	357	372	418	496	614	782	1024	1387	1976	3011	4948	
$\bar{M}_r$		-099	-098	-096	-092	-085	-076	-062	-042	-014	028	000	
$\bar{M}_t$		-099	-098	-097	-095	-093	-088	-083	-075	-064	-048	-046	047
$\bar{Q}_r$		000	018	039	064	096	140	199	282	402	589*	000	
$\bar{w}$		125	126	127	129	132	135	139	144	149	154	159	
$\bar{p}$	1.0	-004	015	073	175	330	556	887	1395	2231	3716	6500	
$\bar{M}_r$		-161	-161	-161	-160	-158	-154	-146	-133	-109	-069	000	
$\bar{M}_t$		-161	-161	-161	-161	-160	-159	-156	-151	-143	-130	-109	112
$\bar{Q}_r$		000	000	003	013	031	065	120	207	349	585	1000*	
$\bar{w}$		123	124	126	129	134	140	148	156	166	177	188	
							$S = 10$						
$\bar{p}$	0.1	752	713	611	482	363	277	213	139	043	009	323	
$\bar{M}_r$		088	089	037	017	007	002	-001	-001	-001	000	000	
$\bar{M}_t$		088	088	066	048	036	028	022	018	016	014	012	-011
$\bar{Q}_r$		000	037*	-432	-243	-146	-088	-051	-027	-015	-012	000	
$\bar{w}$		017	016	015	014	012	011	010	008	007	006	005	

0.2	$\bar{p}$	1212	1173	1064	909	739	579	444	331	224	116	036	−021
	$\bar{M}_r$	098	100	107	047	019	005	−002	−004	−004	−002	000	
	$\bar{M}_t$	098	090	102	083	064	050	040	032	027	023	021	
	$\bar{Q}_r$	000	060	114*	−509	−310	0189	−111	−059	−026	−007	000	
	$\bar{w}$	031	031	029	027	025	022	020	017	015	013	011	
0.3	$\bar{p}$	1386	1373	1325	1228	1074	880	690	555	474	295	−401	−029
	$\bar{M}_r$	085	087	096	109	049	018	002	−005	−006	−004	000	
	$\bar{M}_t$	085	086	090	096	082	066	053	043	035	030	027	
	$\bar{Q}_r$	000	069	136	197	−502	−313	−190	−106	−044	−002	000	
	$\bar{w}$	042	042	041	039	036	033	030	027	024	021	018	
0.4	$\bar{p}$	1401	1408	1414	1385	1291	1128	934	768	656	480	−161	−035
	$\bar{M}_r$	065	068	076	0.90	109	051	020	004	−003	−004	000	
	$\bar{M}_t$	065	066	070	077	086	076	063	052	043	037	033	
	$\bar{Q}_r$	000	070	141	211	276*	−471	−298	−177	−089	−024	000	
	$\bar{w}$	051	051	050	048	046	043	039	035	032	028	025	
0.5	$\bar{p}$	1331	1339	1358	1370	1352	1279	1143	957	774	694	853	−039
	$\bar{M}_r$	044	047	055	068	087	111	055	023	007	000	000	
	$\bar{M}_t$	044	046	049	056	065	076	069	059	050	043	038	
	$\bar{Q}_r$	000	067	135	204	272	336*	−442	−281	−166	−079	000	
	$\bar{w}$	057	056	056	055	053	051	047	043	039	035	032	
0.6	$\bar{p}$	1154	1161	1182	1219	1265	1293	1264	1153	1012	1067	1824	−036
	$\bar{M}_r$	021	023	030	042	058	080	107	053	021	004	000	
	$\bar{M}_t$	021	022	025	031	038	049	061	056	048	041	036	
	$\bar{Q}_r$	000	058	117	178	242	309	375*	−423	−269	−144	000	
	$\bar{w}$	060	060	060	059	058	057	054	051	048	044	040	

Table 9.7 Con't

S = 10

Coef.	α	0	0.1	0.2	0.3	0.4	0.5	0.6	0.7	0.8	0.9	1	$tg\varphi_e$
$\bar{p}$	0.7	867	878	912	973	1058	1159	1258	1345	1444	1675	2357	-023
$\bar{M}_r$		-008	-006	-001	008	021	039	062	091	038	000	000	
$\bar{M}_t$		-008	-007	-005	000	006	014	024	037	033	027	023	
$\bar{Q}_r$		000	044	089	138	192	254	322	397*	-397	-207	000	
$\bar{w}$		061	061	061	061	061	061	060	059	057	055	053	
$\bar{p}$	0.8	508	524	571	650	763	918	1126	1411	1810	2416	3452	000
$\bar{M}_r$		-038	-037	-034	-028	-019	-007	010	033	064	016	000	
$\bar{M}_t$		-038	-037	-036	-033	-029	-024	-017	-007	006	003	000	
$\bar{Q}_r$		000	026	054	087	127	177	241	324	434*	-306	000	
$\bar{w}$		060	060	061	062	063	064	065	066	067	067	067	
$\bar{p}$	0.9	085	103	157	251	393	599	895	1337	2037	3231	5402	035
$\bar{M}_r$		-069	-069	-068	-067	-064	-058	-049	-034	-010	028	000	
$\bar{M}_t$		-069	-069	-069	-068	-067	-065	-061	-056	-047	-034	-033	
$\bar{Q}_r$		000	005	012	025	047	082	136	219	348	554*	000	
$\bar{w}$		058	058	059	060	062	065	068	071	075	079	082	
$\bar{p}$	1.0	-433	-414	-352	-241	-066	199	600	1230	2279	4138	7591	087
$\bar{M}_r$		-106	-106	-109	-112	-116	-120	-120	-115	-100	-067	000	
$\bar{M}_t$		-106	-106	-107	-109	-111	-113	-115	-115	-111	-103	-085	
$\bar{Q}_r$		000	-021	-039	-051	-052	-036	006	088	238	508	1000*	
$\bar{w}$		053	053	055	057	060	064	070	076	083	092	100	

Table 9.8 Elastic plate loaded with uniformly distributed moments along the circumference (See Figure 9.20)

Coef.	α	ρ											$\bar{tg}\varphi_e$
		0	0.1	0.2	0.3	0.4	0.5	0.6	0.7	0.8	0.9	1	
							S = 1						
$\bar{p}$	0.1	−135	−115	−068	−016	015	019	012	027	066	040	−315	
$\bar{M}_r$		−587	−587*	100	041	021	011	006	003	001	000	000	
$\bar{M}_t$		−7	−587	−108	−050	−031	−021	−017	−014	−012	−010	−009	009
$\bar{Q}_r$		000	−006	−010	−010	−007	−004	−002	000	004	010	000	
$\bar{w}$		−009	−007	−003	−001	001	002	003	004	005	006	007	
$\bar{p}$	0.2	−345	−307	−209	−097	−019	009	014	049	126	105	−449	
$\bar{M}_r$		−591	−592	−593*	173	090	090	029	016	007	002	000	
$\bar{M}_t$		−591	−591	−592	−195	−195	−079	−060	−048	−041	−036	−033	032
$\bar{Q}_r$		000	−016	−027	−031	−027	−022	−017	−012	−003	−010	000	
$\bar{w}$		−031	−029	−021	−013	−007	−002	002	006	010	013	016	
$\bar{p}$	03	−486	−465	−401	−305	−190	−080	001	037	047	116	436	
$\bar{M}_r$		−591	−592	−594	−598*	215	125	075	044	023	009	000	
$\bar{M}_t$		−591	−591	−592	−594	−248	−166	−123	−098	−082	−072	−064	066
$\bar{Q}_r$		000	−024	−044	−059	−066	−064	−057	−047	−037	−026	000	
$\bar{w}$		−066	−063	−055	−043	−030	−019	−010	−002	005	01'2	019	
$\bar{p}$	04	−640	−626	−580	−491	−360	−208	−081	−017	019	229	1096	
$\bar{M}_r$		−597	−598	−602	−608	−615	227	138	082	044	017	000	
$\bar{M}_t$		−597	−598	−599	−602	−606	−293	−216	−172	−144	−125	−112	117
$\bar{Q}_r$		000	−032	−061	−085	−101	−106	−101	−091	−080	−062	000	
$\bar{w}$		−101	−098	−091	−078	−060	−041	−025	−011	002	014	026	

Table 9.8 Con't

Coef.	α	0	0.1	0.2	0.3	0.4	0.5	0.6	0.7	0.8	0.9	1	$\overline{tg}\varphi_e$
							$S = 1$						
$\bar{p}$	05	-744	-742	-723	-661	-535	-356	-172	037	067	365	1447	186
$\bar{M}_r$		-615	-617	-621	-628	-638	-649*	213	126	067	027	000	
$\bar{M}_t$		-615	-616	-618	-621	-626	-632	-343	-273	-229	-200	-179	
$\bar{Q}_r$		000	-037	-074	-107	-133	-146	-146	-134	-116	-085	000	
$\bar{w}$		-130	-128	-120	-106	-088	-064	-039	-016	004	023	042	
$\bar{p}$	06	-837	-829	-800	-743	-0646	-496	-286	-021	278	584	909	276
$\bar{M}_r$		-651	-653	-658	-666	-676	-689	-702*	174	092	036	000	
$\bar{M}_t$		-651	-652	-654	-658	-663	-669	-676	-406	-342	-298	-268	
$\bar{Q}_r$		000	-042	-082	-119	-150	-172	-179	-168	-135	-079	000	
$\bar{w}$		-148	-145	-146	-122	-103	-077	-046	-013	017	046	074	
$\bar{p}$	0.7	-932	-918	-880	-823	-714	-621	-415	-092	339	770	974	380
$\bar{M}_r$		-691	-693	-698	-707	-719	-733	-749	-765	124	048	000	
$\bar{M}_t$		-691	-692	-695	-699	-705	-712	-720	-728	-472	-412	-370	
$\bar{Q}_r$		000	-046	-091	-131	-167	-195	-211	-205	-168	-096	000	
$\bar{w}$		-168	-165	-156	-141	-120	-093	-059	-020	022	062	100	
$\bar{p}$	0.8	-1034	-1023	-990	-933	-849	-727	-547	-273	162	877	2075	497
$\bar{M}_r$		-732	-734	-740	-750	-763	-779	-797	-816	-834*	066	000	
$\bar{M}_t$		-732	-733	-736	-740	-747	-755	-764	-774	-784	-539	-483	
$\bar{Q}_r$		000	-051	-101	-148	-189	-222	-244	-248	-223	-152	000	
$\bar{w}$		-196	-192	-183	-167	-145	-116	-081	-038	011	063	113	

		1	2	3	4	5	6	7	8	9	10	11
0.9 (638)	$\bar{p}$	−1120	−1109	−1074	−1014	−922	−788	−595	−305	158	942	2341
	$\bar{M}_r$	−792	−794	−001	−812	−826	−844	−863	−884	−903	−917*	000
	$\bar{M}_t$	−792	−793	−796	−802	−808	−817	−827	−838	−849	−859	−620
	$\bar{Q}_r$	000	−056	−110	−100	−205	−241	−264	−269	−244	−167	000
	$\bar{w}$	−212	−209	−199	−182	−158	−127	−088	−042	011	073	137
1.0 (796)	$\bar{p}$	−1227	−1205	−1167	−1100	−1000	−855	−646	−333	165	1020	2570
	$\bar{M}_r$	−860	−863	−870	−881	−897	−916	−937	−960	−981	−996	−1,000*
	$\bar{M}_t$	−860	−861	−865	−870	−878	−887	−898	−910	−922	−933	−941
	$\bar{Q}_r$	000	−061	−119	−174	−222	−262	−287	−292	−265	−183	000
	$\bar{w}$	−231	−227	−216	−197	−171	−138	−096	−046	012	079	154

S = 2

		1	2	3	4	5	6	7	8	9	10	11
0.1 (009)	$\bar{p}$	−269	−230	−135	−033	029	036	044	053	132	081	−632
	$\bar{M}_r$	−585	−586*	100	041	010	009	004	001	−001	−001	000
	$\bar{M}_t$	−585	−586	−107	−050	−031	−022	−017	−015	−013	−011	−011
	$\bar{Q}_r$	000	−012	−020	−020	−014	−008	−004	−001	008	019	000
	$\bar{w}$	−009	−007	−003	−001	000	002	003	004	005	006	007
0.2 (029)	$\bar{p}$	−677	−600	−406	−185	−031	023	031	098	251	205	−917
	$\bar{M}_r$	−583	−584	−588*	176	091	050	028	014	005	001	000
	$\bar{M}_t$	−583	−584	−585	−190	−111	−076	−057	−047	−040	−035	−032
	$\bar{Q}_r$	−030	−032	−053	−059	−053	−042	−033	−023	004	−022	000
	$\bar{w}$	−030	−028	−020	−012	−006	−001	003	006	010	013	016
0.3 (057)	$\bar{p}$	−918	−875	−753	−566	−344	−133	018	078	082	199	803
	$\bar{M}_r$	−565	−567	−572	−580*	229	135	082	048	025	009	000
	$\bar{M}_t$	−565	−566	−569	−572	−228	−149	−108	−085	−071	−062	−056
	$\bar{Q}_r$	000	−045	−083	−110	−122	−119	−103	−083	−065	−047	000
	$\bar{w}$	−061	−059	−051	−039	−027	−017	−009	−002	005	011	017

Table 9.8 Con't

Coef.	α	ρ											$tg\bar{\varphi}_e$
		0	0.1	0.2	0.3	0.4	0.5	0.6	0.7	0.8	0.9	1	
							S = 2						
$\bar{p}$	0.4	-1181	-1156	-1069	-902	-652	-365	-131	025	017	394	2063	100
$\bar{M}_r$		-554	-556	-563	-573	-587*	249	153	091	049	019	000	
$\bar{M}_t$		-554	-555	-558	-563	-570	-261	-187	-146	-122	-106	-095	
$\bar{Q}_r$		000	-058	-113	-157	-186	-194	-183	-163	-143	-113	000	
$\bar{w}$		-092	-090	-083	-071	-054	-037	-023	-110	001	011	021	
$\bar{p}$	0.5	-1351	-1349	-1320	-1208	-974	-639	-298	-061	104	640	2712	162
$\bar{M}_r$		-556	-559	-567	-580	-597	-617*	236	140	075	029	000	
$\bar{M}_t$		-556	-557	-561	-567	-576	-586	-303	-237	-198	-173	-155	
$\bar{Q}_r$		000	-068	-134	-195	-242	-266	-264	-241	-209	-156	000	
$\bar{w}$		-118	-116	-109	-097	-080	-058	-035	-015	003	020	036	
$\bar{p}$	0.6	-1518	-1504	-1455	-1357	-1183	-909	-519	026	524	1067	1600	249
$\bar{M}_r$		-584	-587	-596	-610	-629	-652	-677*	188	097	036	000	
$\bar{M}_t$		-584	-585	-589	-596	-606	-617	-630	-366	-307	-269	-242	
$\bar{Q}_r$		000	-076	-149	-217	-274	-314	-327	-306	-244	-141	000	
$\bar{w}$		-134	-132	-124	-112	-094	-070	-041	-011	016	042	067	
$\bar{p}$	0.7	-1676	-1655	-1589	-1493	-1359	-1143	-767	-164	641	1419	1682	348
$\bar{M}_r$		-613	-616	-626	-642	-663	-680	-718	-747*	131	048	000	
$\bar{M}_t$		-613	-614	-619	-627	-637	-650	-664	-680	-431	-377	-339	
$\bar{Q}_r$		000	-083	-163	-227	-303	-355	-385	-374	-305	-172	000	
$\bar{w}$		-152	-149	-141	-128	-109	-085	-054	-017	021	057	092	

0.8	$\bar{p}$	-1839	-1822	-1768	-1675	-1534	-1326	-1012	-520	277	1599	3818	421
	$\bar{M}_r$	-636	-640	-651	-668	-692	-721	-754	-788	-820*	068	000	
	$\bar{M}_t$	-636	-638	-643	-652	-663	-677	-694	-712	-730	-492	443	
	$\bar{Q}_r$	000	-092	-180	-264	-338	-400	-441	-450	-408	-278	000	
	$\bar{w}$	-176	-173	-165	-161	-131	-106	-074	-036	010	057	103	
0.9	$\bar{p}$	-1991	-1973	-1917	-1818	-1665	-1438	-1100	-580	265	1716	4320	593
	$\bar{M}_r$	-688	-692	-704	-723	-748	-780	-816	-853	-888	-915*	000	
	$\bar{M}_t$	-688	-690	-695	-705	-717	-732	-750	-770	-790	-808	-576	
	$\bar{Q}_r$	000	-099	-195	-286	-367	-433	-478	-489	-445	-307	000	
	$\bar{w}$	-191	-188	-179	-164	-143	-115	-081	-039	010	066	126	
1.0	$\bar{p}$	-2163	-2143	-2082	-1973	-1806	-1559	-1194	-635	275	1856	4746	747
	$\bar{M}_r$	-747	-751	-764	-785	-813	-847	-885	-926	-964	-993	-1000*	
	$\bar{M}_t$	-747	-749	-755	-765	-778	-795	-815	-836	-858	-878	-893	
	$\bar{Q}_r$	000	-108	-212	-311	-398	-470	-519	-531	-484	-336	000	
	$\bar{w}$	-207	-204	-194	-178	-155	-125	-088	-043	010	072	142	

S = 5

0.1	$\bar{p}$	-667	-572	-338	-086	066	082	052	130	336	215	-1593	009
	$\bar{M}_r$	-582	-583*	101	039	016	005	-001	-004	-006	-004	000	
	$\bar{M}_t$	-582	-583	-105	-049	-031	-023	-019	-017	-016	-014	-014	
	$\bar{Q}_r$	000	-031	-049	-049	-037	-022	-012	-004	018	048	000	
	$\bar{w}$	-009	-007	-003	-001	000	002	003	004	005	006	007	
0.2	$\bar{p}$	-1611	-1422	-947	-411	-042	078	086	247	622	487	-2408	024
	$\bar{M}_r$	-563	-567	-575	184	093	048	023	007	-002	-005	000	
	$\bar{M}_t$	-563	-565	-509	-176	-101	-069	-053	-044	-039	-035	-033	
	$\bar{Q}_r$	000	-076	-126	-139	-121	-094	-071	-047	-001	060	000	
	$\bar{w}$	-028	-025	-018	-010	-005	-001	003	007	010	013	015	

Table 9.8 Con't

							ρ S = 5						
Coef.	α	0	0.1	0.2	0.3	0.4	0.5	0.6	0.7	0.8	0.9	1	$\overline{tg}\varphi_e$
$\overline{p}$	0.3	-1992	-1894	-1611	-1179	-669	-192	129	214	143	319	1649	
$\overline{M}_r$		-509	-513	-524	-540*	259	155	094	054	028	010	000	039
$\overline{M}_t$		-509	-511	-516	-524	-185	-111	-076	-057	-047	-041	-037	
$\overline{Q}_r$		000	-097	-180	-236	-257	-243	-203	-156	-120	-089	000	
$\overline{w}$		-051	-049	-042	-031	-020	-012	-006	-001	004	008	012	
$\overline{p}$	0.4	-2417	-2365	-2184	-1820	-1269	-641	-157	-017	075	642	4483	
$\overline{M}_r$		-459	-464	-478	-499	-527*	294	185	113	062	023	000	064
$\overline{M}_t$		-459	-461	-468	-478	-492	-191	-126	-092	-074	-064	-057	
$\overline{Q}_r$		000	-120	-221	-321	-376	-385	-255	-309	-275	-230	000	
$\overline{w}$		074	-072	-067	-056	-042	-028	-017	-008	-001	006	012	
$\overline{p}$	0.5	-2640	-2651	-2621	-2413	-1926	-1209	-501	087	100	1116	5823	
$\overline{M}_r$		-429	-434	-450	-476	-510	-549	284	172	092	035	000	113
$\overline{M}_t$		-429	-431	-439	-451	-468	-488	-215	-160	-130	-114	-102	
$\overline{Q}_r$		000	-132	-264	-387	-481	-526	-514	-464	-406	-316	000	
$\overline{w}$		-093	-092	-086	-076	-063	-045	-027	-012	001	012	024	
$\overline{p}$	0.6	-2954	-2935	-2864	-2698	-2374	-1827	-1018	016	1137	2124	2836	
$\overline{M}_r$		-439	-445	-463	-491	-529	-575	-625*	218	107	036	000	192
$\overline{M}_t$		-439	-442	-150	-464	-483	-505	-531	-280	-234	-206	-186	
$\overline{Q}_r$		000	-147	-291	-426	-542	-624	-651	-605	-474	-264	000	
$\overline{w}$		-106	-104	-098	-088	-074	-056	-032	-008	014	034	054	

$S = 10$

0.7	$\bar{p}$	-3212	-3171	-3063	-2914	-2704	-2323	-1577	-303	1405	2900	2775	279
	$\bar{M}_r$	-446	-452	-471	-502	-543	-594	-651	-709*	144	047	000	
	$\bar{M}_t$	-446	-449	-458	-473	-493	-517	-546	-577	-343	-303	-274	
	$\bar{Q}_r$	000	-160	-313	-458	-589	-699	-764	-745	-001	-323	000	
	$\bar{w}$	-118	-116	-110	-100	-086	-067	-043	-013	018	048	076	
0.8	$\bar{p}$	-3386	-3363	-3294	-3168	-2965	-2635	-2087	-1155	443	3170	7764	366
	$\bar{M}_r$	-435	-442	-462	-495	-539	-594	-658	-725	-790*	074	000	
	$\bar{M}_t$	-435	-439	-448	-464	-485	-512	-543	-578	-614	-393	-355	
	$\bar{Q}_r$	000	-169	-334	-492	-638	-763	-583	-885	-813	-561	000	
	$\bar{w}$	-134	-132	-126	-116	-102	-084	-060	-030	006	015	082	
0.9	$\bar{p}$	-3656	-3634	-3565	-3433	-3212	-2852	-2266	-1290	391	3386	8854	495
	$\bar{M}_r$	-469	-477	-498	-533	-582	-642	-710	-783	-853	-907*	000	
	$\bar{M}_t$	-469	-473	-483	-500	-523	-552	-586	-624	-663	-699	-480	
	$\bar{Q}_r$	000	-182	-361	-532	-690	-826	-924	-960	-888	-622	000	
	$\bar{w}$	-145	-143	-137	-126	-111	-091	-065	-033	006	053	103	
1.0	$\bar{p}$	-3973	-3948	-3871	-3725	-3482	-3090	-2459	-1413	394	3658	9750	640
	$\bar{M}_r$	-509	-517	-541	-579	-631	-696	-771	-850	-926	-985	-1000*	
	$\bar{M}_t$	-509	-513	-524	-543	-568	-599	-636	-677	-719	-759	-789	
	$\bar{Q}_r$	000	-198	-392	-578	-749	-896	-1002	-1042	-966	-680	000	
	$\bar{w}$	-157	-155	-148	-137	-121	-099	-071	-036	006	057	117	
0.1	$\bar{p}$	-1311	-1126	-668	-176	115	142	085	253	688	461	-3236	008
	$\bar{M}_r$	-576	-578*	103	038	011	-002	-009	-013	-014	-009	000	
	$\bar{M}_t$	-576	-577	-101	-047	-031	-025	-022	-021	-020	-018	-018	
	$\bar{Q}_r$	000	-061	-097	-098	-073	-045	-028	-011	034	096	000	
	$\bar{w}$	-009	-007	-003	-001	000	001	003	004	005	006	007	

Table 9.8 Con't

Coef.	α		0	0.1	0.2	0.3	0.4	0.5	0.6	0.7	0.8	0.9	1	$\bar{tg\varphi_e}$
								$S = 10$						
$\bar{p}$	0.2		-3035	-2669	-1749	-718	-021	185	177	491	1244	945	-5078	
$\bar{M_r}$			-538	-544	-559	190	091	041	012	-006	-016	-013	000	019
$\bar{M_t}$			-538	-541	-549	-161	-090	-062	-050	-044	-041	-038	-037	
$\bar{Q_r}$			000	-142	-235	-256	-219	-165	-121	-077	-012	133	000	
$\bar{w}$			-025	-023	-016	-009	-004	000	004	007	010	012	015	
$\bar{p}$	0.3		-3371	-3193	-2677	-1892	-970	-122	406	453	161	298	2638	
$\bar{M_r}$			-449	-455	-474	-501*	285	171	102	058	030	011	000	021
$\bar{M_t}$			-449	-452	-461	-475	-142	-075	-045	-032	-025	-022	-019	
$\bar{Q_r}$			000	-164	-302	-391	-417	-380	-299	-213	-157	-127	000	
$\bar{w}$			-041	-039	-034	-024	-014	-008	-003	000	003	005	008	
$\bar{p}$	0.4		-3779	-3700	-3408	-2793	-1840	-759	005	045	348	618	7726	
$\bar{M_r}$			-364	-372	-393	-427	-470*	336	215	134	076	030	000	030
$\bar{M_t}$			-364	-368	-378	-395	-416	-123	-066	-039	-027	-023	-020	
$\bar{Q_r}$			000	-187	-361	-500	-577	-577	-509	-428	-391	-360	000	
$\bar{w}$			-057	-055	-051	-042	-030	-019	-011	-006	-002	001	004	
$\bar{p}$	0.5		-3848	-3898	-3924	-3649	-2866	-1670	-539	-052	-125	1293	9869	
$\bar{M_r}$			-305	-312	-336	-374	-426	-484*	-331	203	112	043	000	067
$\bar{M_t}$			-305	-308	-319	-338	-363	-392	-130	-085	-064	-056	-049	
$\bar{Q_r}$			000	-194	-391	-579	-721	-781	-747	-661	-585	-490	000	
$\bar{w}$			-069	-068	-064	-057	-047	-033	-019	-009	-001	005	012	

0.6	139	$\bar{p}$	3434	3207	1938	181	-1505	-2799	-3625	-4058	-4232	-4279	-4284
		$\bar{M}_r$	000	034	114	242	-576*	-502	-433	-376	-334	-308	-300
		$\bar{M}_t$	-134	-147	-164	-198	-436	-397	-363	-336	-316	-304	-300
		$\bar{Q}_r$	000	-361	-688	-903	-982	-939	-810	-631	-426	-214	000
		$\bar{w}$	040	026	012	-004	-023	-042	-056	-066	-073	-077	-078
0.7	213	$\bar{p}$	2761	4565	2436	-400	-2497	-3617	-4075	-4256	-4383	-4494	-4542
		$\bar{M}_r$	000	044	152	-676*	-587	-501	-427	-367	-323	-297	-288
		$\bar{M}_t$	-214	-233	-260	-479	-433	-390	-354	-326	-305	-292	-288
		$\bar{Q}_r$	000	-442	-888	-1124	-1148	-1035	-857	-657	-446	-226	000
		$\bar{w}$	060	038	016	-009	-032	-050	-064	-074	-080	-084	-086
0.8	278	$\bar{p}$	12073	4740	416	-2012	-3303	-3950	-4256	-4394	-4455	-4480	-4487
		$\bar{M}_r$	000	082	-755*	-658	-558	-468	-391	-329	-285	-259	-250
		$\bar{M}_t$	-270	-295	-500	-448	-398	-354	-317	-288	-267	-254	-250
		$\bar{Q}_r$	000	-862	-1227	-1306	-1230	-1074	-879	-667	-447	-224	000
		$\bar{w}$	061	033	003	-024	-046	-062	-074	-083	-089	-092	-063
0.9	399	$\bar{p}$	13931	5026	273	-2250	-3578	-4263	-4600	-4750	-4807	-4823	-4826
		$\bar{M}_r$	000	-898*	-814	-708	-601	-503	-420	-354	-306	-277	-268
		$\bar{M}_t$	-387	-593	-538	-481	-427	-380	-340	-308	-286	-272	-268
		$\bar{Q}_r$	000	-960	-1340	-1414	-1328	-1160	-949	-720	-482	-241	000
		$\bar{w}$	080	040	003	-027	-050	-068	-081	-090	-096	-100	-101
1.0	536	$\bar{p}$	15390	5416	236	-2466	-3879	-4613	-4980	-5150	-5218	-5240	-5244
		$\bar{M}_r$	15390	5416	236	-2466	-3879	-4613	-4980	-5150	-5218	-301	-290
		$\bar{M}_t$	-687	-643	-584	-522	-464	-112	-369	-335	-310	-295	-290
		$\bar{Q}_r$	000	-1051	-1457	-1534	-1439	-1257	-1030	-781	-523	-202	000
		$\bar{w}$	092	043	003	-029	-054	-074	-088	-098	-105	-108	-110

Table 9.9 Elastic plate with uniformly distributed load (See Figure 9.21)

α	ρ										
	0	0.1	0.2	0.3	0.4	0.5	0.6	0.7	0.8	0.9	1
	$\overline{p}$										
0.5	0.570	0.580	0.590	0.600	0.620	0.660	0.720	0.820	1.020	1.400	–
1	0.610	0.610	0.620	0.630	0.650	0.680	0.730	0.830	1.010	1.380	–
2	0.660	0.660	0.660	0.670	0.680	0.710	0.760	0.830	1.000	1.340	–
3	0.700	0.700	0.700	0.700	0.710	0.730	0.770	0.840	0.990	1.310	–
5	0.760	0.760	0.760	0.760	0.760	0.770	0.790	0.850	0.980	1.270	–
10	0.860	0.860	0.850	0.840	0.830	0.830	0.830	0.860	0.960	1.200	–
	$\overline{w}$										
0.5	0.830	0.830	0.829	0.828	0.826	0.825	0.822	0.820	0.817	0.815	0.813
1	0.420	0.420	0.419	0.418	0.416	0.415	0.413	0.411	0.408	0.406	0.404
2	0.214	0.214	0.214	0.213	0.211	0.210	0.208	0.206	0.204	0.202	0.200
3	0.145	0.145	0.145	0.144	0.142	0.141	0.140	0.138	0.135	0.134	0.132
5	0.090	0.089	0.089	0.086	0.087	0.086	0.084	0.083	0.081	0.079	0.078
10	0.047	0.046	0.046	0.045	0.044	0.044	0.043	0.042	0.040	0.039	0.038
	$\overline{M}_r$										
0.5	0.054	0.053	0.051	0.046	0.041	0.034	0.026	0.017	0.009	0.002	0
1	0.051	0.050	0.048	0.044	0.039	0.032	0.025	0.017	0.008	0.002	0
2	0.046	0.045	0.043	0.040	0.035	0.030	0.023	0.015	0.008	0.002	0
3	0.042	0.041	0.040	0.037	0.032	0.027	0.021	0.014	0.007	0.002	0
5	0.035	0.035	0.034	0.031	0.028	0.024	0.018	0.013	0.007	0.002	0
10	0.025	0.025	0.024	0.023	0.021	0.018	0.014	0.010	0.006	0.001	0
	$\overline{M}_t$										
0.5	0.054	0.053	0.052	0.050	0.048	0.044	0.040	0.035	0.031	0.027	0.024
1	0.051	0.051	0.050	0.048	0.045	0.042	0.038	0.034	0.030	0.026	0.022
2	0.046	0.046	0.045	0.043	0.041	0.038	0.035	0.031	0.027	0.023	0.020
3	0.042	0.042	0.041	0.039	0.037	0.035	0.032	0.029	0.025	0.021	0.019
5	0.035	0.035	0.034	0.033	0.032	0.030	0.027	0.025	0.022	0.018	0.016
10	0.025	0.025	0.025	0.024	0.023	0.022	0.020	0.018	0.016	0.014	0.012
	$\overline{Q}_r$										
0.5	0	0.021	0.042	0.062	0.080	0.096	0.109	0.115	0.109	0.080	0
1	0	0.020	0.039	0.057	0.075	0.090	0.103	0.109	0.104	0.075	0
2	0	0.017	0.034	0.051	0.066	0.081	0.092	0.099	0.095	0.070	0
3	0	0.015	0.030	0.045	0.059	0.073	0.084	0.091	0.088	0.065	0
5	0	0.012	0.024	0.036	0.048	0.060	0.070	0.077	−0.077	0.058	0
10	0	0.007	0.014	0.022	0.031	0.040	0.049	0.057	0.059	0.046	0

Table 9.10 Rotations of a circular elastic plate with a uniformly distributed load

$tg\varphi_o$						
S	0.5	1	2	3	5	10
$tg\varphi_o$	−0.0241	−0.0230	−0.0210	−0.0192	−0.0165	−0.0123

Table 9.11 Elastic circular plate with a vertical concentrated load P (See Figure 9.22)

s	0	0.1	0.2	0.3	0.4	0.5	0.6	0.7	0.8	0.9	1
						$\bar{p}$					
0.5	0.26	0.26	0.25	0.24	0.23	0.23	0.24	0.26	0.30	0.41	—
1	0.35	0.34	0.31	0.29	0.26	0.25	0.25	0.26	0.28	0.37	—
2	0.50	0.48	0.43	0.38	0.32	0.29	0.27	0.26	0.24	0.30	—
3	0.63	0.60	0.53	0.45	0.37	0.32	0.29	0.25	0.21	0.24	—
5	0.87	0.82	0.71	0.57	0.45	0.37	0.31	0.24	0.14	0.14	—
10	1.33	1.25	1.04	0.78	0.56	0.42	0.33	0.20	0.00	−0.03	—
						w					
0.5	0.285	0.284	0.281	0.278	0.274	0.270	0.266	0.261	0.258	0.253	0.249
1	0.154	0.153	0.150	0.147	0.143	0.139	0.136	0.131	0.127	0.124	0.121
2	0.087	0.086	0.083	0.081	0.077	0.073	0.070	0.066	0.062	0.058	0.121
3	0.064	0.063	0.060	0.058	0.054	0.051	0.048	0.044	0.041	0.038	0.054
5	0.044	0.043	0.040	0.038	0.035	0.032	0.029	0.026	0.023	0.020	0.035
10	0.027	0.026	0.024	0.022	0.019	0.017	0.015	0.012	0.010	0.008	0.005
						$\bar{M}_r$					
0.5	∞	0.163	0.101	0.065	0.042	0.026	0.014	0.006	0.001	0.000	0
1	∞	0158	0.096	0.062	0.039	0.023	0.012	0.005	0.001	0.000	0
2	∞	0.149	0.088	0.055	0.033	0.019	0.009	0.003	0.000	0.000	0
3	∞	0.141	0.081	0.049	0.029	0.015	0.007	0.002	0.000	0.000	0
5	∞	0.129	0.070	0.040	0.022	0.010	0.004	0.001	0.000	0.000	0
10	∞	0.109	0.052	0.025	0.011	0.003	0.000	0.001	0.000	0.000	0
						$\bar{M}_t$					
0.5	∞	0.230	0.166	0.129	0.104	0.086	0.071	0.060	0.051	0.044	0.030
1	∞	0224	0.161	0.124	0.100	0.082	0.067	0.057	0.049	0.042	0.037
2	∞	0.215	0.152	0.116	0.093	0.075	0.062	0.052	0.044	0.038	0.034
3	∞	0.207	0.145	0.109	0.085	0.069	0.057	0.048	0.041	0.035	0.031
5	∞	0195	0.133	0.098	0.077	0.061	0.049	0.041	0.035	0.030	0.027
10	∞	0.174	0113	0.080	0.061	0.047	0.037	0.031	0.027	0.023	0.021
						$\bar{Q}_r$					
0.5	∞	1.58	0.77	0.49	0.35	0.26	0.19	0.14	0.10	0.06	0
1	∞	1.58	0.76	0.48	0.34	0.25	0.18	0.13	0.09	0.05	0
2	∞	1.57	0.75	0.47	0.32	0.23	0.16	0.12	0.08	0.04	0
3	∞	1.56	0.74	0.45	0.30	0.21	0.15	0.10	0.07	0.04	0
5	∞	1.55	0.72	0.43	0.27	0.18	0.12	0.08	0.06	0.03	0
10	∞	1.53	0.68	0.38	0.23	0.14	0.08	0.04	0.03	0.03	0

Table 9.12 Rotations at the edge of an elastic circular plate with a concentrated vertical load applied to the center of the plate

$\overline{tg}\varphi_e$						
s	0.5	1	2	3	5	10
$\overline{tg}\varphi_e$	0.039	0.038	0.035	0.032	0.028	0.021

Table 9.13 Numerical values of $\overline{M}_A$

S	0.5	1	2	3	5	10
$\overline{M}_A$	−0.052	−0.056	−0.056	−0.074	−0.086	−0.108

Table 9.14 Numerical values of $\overline{M}_B$

α	0.005	0.01	0.02	0.03	0.04	0.05	0.075	0.10	0.15
$\overline{M}_B$	0.571	0.507	0.443	0.405	0.378	0.358	0.370	0.293	0.255

References

1. Beyer, K. 1956. *Die Statik im Stahlbetonbau*. Berlin: Springler-Verlag.
2. Boussinesq, J. 1885. *Application des potentiels a l'etude de l'equilibre et du movement des solides elastique*. Paris: Gauthier-Villars.
3. Gorbunov-Posadov, M. I. 1984. *Analysis of structures on elastic foundation*. Moscow: Stroiizdat, pp. 180–210.
4. Selvadurai, A.S.P. 1979. Elastic *Analysis of soil-foundation interaction*. Amsterdam/Oxford/New York: Elsevier Scientific.
5. Timoshenko, S. P., and S. Woinowsky-Krieger. 1959. *Theory of plates and shells*. 2nd ed. New York: McGraw-Hill.
6. Tsudik, E. A. 2006. *Analysis of beams and frames on elastic foundation*. Oxford UK/Canada: Trafford.
7. Varvak, P. M. 1971. *Theory of elasticity handbook*. Kiev: Budivelnik, pp. 370–371.
8. Zhemochkin B. N., and A. P. Sinitsin. 1962. *Practical methods of analysis of beams and plates on elastic foundation*. Moscow: Gosstroiizdat.

10

Special Structures on Elastic Foundation

10.1 Introduction

Chapter 10 is devoted to the solution of some special problems related to analysis of structures on elastic foundation. It includes analysis of composite beams and plates, analysis of boxed foundations, and analysis of walls on elastic foundation. Methods of analysis are developed for three soil models: Winkler foundation, elastic half-space, and elastic layer. As shown below, the analysis of composite beams is performed using direct methods of structural mechanics as well as the method of iterations. The analysis of boxed foundations that represents a series of interconnected walls—deep beams, supported on a mat foundation—is developed without taking into account their rigidity or the rigidity of the superstructure. Analysis of walls on elastic foundation is developed for walls supported on continuous foundations as well as those supported on frames, including frames with individual footings and continuous foundations. Some of the material included in this chapter is taken from the author's other publications (Tsudik 2008, 2010).

10.2 Analysis of Composite Beams on Elastic Foundation

A composite beam that consists of two simple beams, A and B, is shown in Figure 10.1. The beam, with various vertical loads is supported on elastic half-space. Both beams are restrained against horizontal deflections. Analysis of the beam is performed in the following order:

1. Continuous contact between the given composite beam and the half-space is replaced with a series of individual totally rigid supports. Continuous contact between both beams analogously is replaced with a series of totally rigid column supports as shown in Figure 10.2.
2. Now, assuming that all first level supports, from 1 to n, are restrained from any deflections, the total system composite beam-half-space can be divided into two parts: the composite beam that represents a two-story frame with the given loads and unknown settlements $\Delta_1, \Delta_2, ..., \Delta_n$ at points of support, as

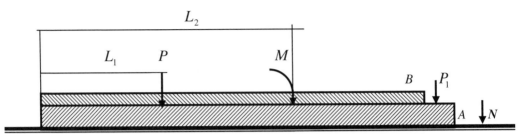

Figure 10.1   The given composite beam supported on elastic half-space

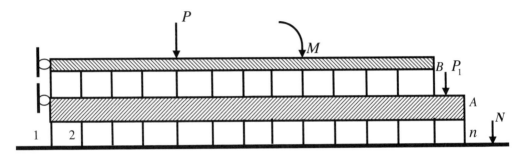

Figure 10.2   The composite beam replaced with two beams with individual supports

shown in Figure 10.3, and the half-space loaded with soil reactions, acting in the opposite direction, and with other given loads located outside of the area of the beam (see Figure 10.4). Reactions at any point of the frame support i can be found from the following equation:

$$X_i = r_{i1}\Delta_1 + r_{i2}\Delta_2 + \dots + r_{in}\Delta_n + r_{iP} \qquad (10.1)$$

In this equation, r_{ik} is the reaction at support i due to one unit settlement at support k and r_{iP} is the vertical reaction at the same support i due to the given loads applied to the frame. Analogously, reactions at all supports, from 1 to n, can be obtained as shown below:

$$\left.\begin{aligned}
X_1 &= r_{11}\Delta_1 + r_{12}\Delta_2 + \dots + r_{1n}\Delta_n + r_{1P} \\
X_2 &= r_{21}\Delta_1 + r_{22}\Delta_2 + \dots + r_{2n}\Delta_n + r_{2P} \\
&\qquad\qquad\cdots\cdots\cdots\cdots\cdots \\
X_n &= r_{n1}\Delta_1 + r_{n2}\Delta_2 + \dots + r_{nn}\Delta_n + r_{nP}
\end{aligned}\right\} \qquad (10.2)$$

Reactions X_1, X_2, ..., X_n acting in the opposite direction, downward, along with the given load N applied to the half-space will produce settlements Δ_1, Δ_2, ..., Δ_n at all points of support. However, these reactions cannot be applied to the half-space as vertical concentrated loads. They are replaced with equivalent uniformly distributed loads. For example, reaction X_i is replaced with a uniformly distributed load equal to $q_i = X_i/ab$, where a is the spacing of supports and b is the width of the foundation. Loads X_1 and X_n are replaced, respectively, with uniformly distributed loads $q_1 = X_1/2ab$ and $q_n = X_n/2ab$. Settlements are found from the following equations:

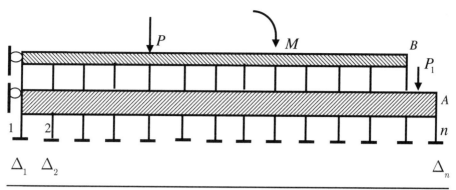

Figure 10.3 The composite beam with the given loads and soil settlements

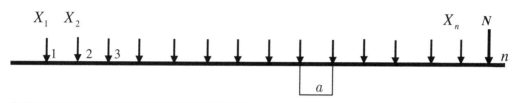

Figure 10.4 The half-space with given loads and soil reactions acting downward

$$
\left.\begin{array}{l}
\Delta_1 = \Delta_{11}X_1 + \Delta_{12}X_2 + \ldots + \Delta_{1n}X_n + \Delta_{1N} \\
\Delta_2 = \Delta_{21}X_1 + \Delta_{22}X_2 + \ldots + \Delta_{2n}X_n + \Delta_{2N} \\
\ldots\ldots\ldots\ldots\ldots\ldots\ldots\ldots\ldots\ldots\ldots\ldots\ldots \\
\Delta_n = \Delta_{n1}X_1 + \Delta_{n2}X_2 + \ldots + \Delta_{nn}X_n + \Delta_{nN}
\end{array}\right\} \tag{10.3}
$$

In equations 10.3, Δ_{ik} is the settlement of the half-space at point i due to one unit load applied to the half-space at point k, and Δ_{iN} is the settlement of the half-space at point i due to the load N.

By solving both systems of equations, 10.2 and 10.3, all settlements and soil reactions are found. Final analysis of the composite beam is performed by applying to the beam the given loads and obtained settlements. Equations for obtaining settlements of the half-space are given in Chapter 1 (see equation 1.23). The same method of analysis can be applied when the composite beam is supported on an elastic layer. In this case, settlements of the soil are obtained using equations 1.55 and 1.58 from Chapter 1. When the beam is supported on Winkler foundation, analysis is much simpler. The system of equations 10.3 in this case is replaced with a group of formulae that looks as follows:

$$
\Delta_1 = \Delta_{11}X_1, \Delta_1 = \Delta_{11}X_1, \ldots, \Delta_n = \Delta_{nn}/X_n \tag{10.4}
$$

Or

$$
X_1 = \Delta_1/\Delta_{11}, X_2 = \Delta_2/\Delta_{22}, \ldots, X_n = \Delta_n/\Delta_{nn} \tag{10.5}
$$

The settlements of the soil in this case at any point i are found from equation 10.6:

$$
\Delta_i = X_i/A_i \cdot k_i \tag{10.6}
$$

where k_i is the modulus of subgrade reaction, A_i is the soil area where the load is applied, and X_i is the actual load applied to the soil at point i. By introducing reactions X_i from equations 10.5 into system 10.2, we obtain a simple system of n equations with n unknown settlements. Final analysis of the composite beam is performed by applying to the beam the given loads and found settlements.

10.3 Analysis of Composite Plates on Elastic Foundation

Let us assume that we have a composite plate that consists of two plates supported on elastic half-space. Dimensions of both plates, the lower one and the upper one, are the same except for rigidities. Analysis of the plate is performed as follows:

1. The given composite plate is divided into a system of finite elements, as shown in Figures 10.5–10.8.
2. Continuous contact between the plates, as well as continuous contact between the lower plate and soil, are replaced with a system of individual totally rigid column supports located at all n nodes.
3. Now, the upper plate has the given loads and unknown settlements at all n nodes, the lower plate is loaded with unknown reactions from the upper plate (X_i) acting in the opposite direction, and the same unknown settlements. The soil supporting the lower plate is loaded with soil reactions (Y_i) acting in the opposite direction. The settlements of the soil are also affected by the loads G located outside of the plate area, as shown in Figure 10.5. Reaction X_i at any point i can be found from equation 10.7:

$$X_i = r_{i1}\Delta_1 + r_{i2}\Delta_2 + \ldots + r_{in}\Delta_n + r_{iP} \tag{10.7}$$

Analogously, reactions at all node supports of the upper plate are obtained as shown below:

$$\left.\begin{aligned}
X_1 &= r_{11}\Delta_1 + r_{12}\Delta_2 + \ldots + r_{1n}\Delta_n + r_{1P}\\
X_2 &= r_{21}\Delta_1 + r_{22}\Delta_2 + \ldots + r_{2n}\Delta_n + r_{2P}\\
&\cdots\cdots\cdots\cdots\cdots\cdots\cdots\cdots\cdots\cdots\cdots\\
X_n &= r_{n1}\Delta_1 + r_{n2}\Delta_2 + \ldots + r_{nn}\Delta_n + r_{nP}
\end{aligned}\right\} \tag{10.8}$$

where r_{ik} is the reaction at any support i of the upper plate due to one unit settlement applied at any support k of the same plate, Δ_i is the unknown settlement at point i, and r_{iP} is the reaction at the same point i due to the given loads applied to the upper plate. The same reactions acting in the opposite direction are applied to the lower plate. The lower plate loaded with reactions X_1, X_2, ..., X_n and settlements at each point of support will produce the following reactions Y_1, Y_2, ..., Y_n that are found from the equations 10.9:

$$\left.\begin{aligned}
Y_1 &= y_{11}\Delta_1 + y_{12}\Delta_2 + \ldots + y_{1n}\Delta_n + x_{1,1}X_1 + x_{1,2}X_2 + \ldots + x_{1,n}X_n\\
Y_2 &= y_{21}\Delta_1 + y_{22}\Delta_2 + \ldots + y_{2n}\Delta_n + x_{2,1}X_8 + x_{2,2}X_2 + \ldots + x_{2,n}X_n\\
&\cdots\cdots\cdots\cdots\cdots\cdots\cdots\cdots\cdots\cdots\cdots\cdots\cdots\cdots\\
Y_n &= y_{n1}\Delta_1 + y_{n2}\Delta_2 + \ldots + y_{nn}\Delta_n + x_{n,1}X_1 + x_{n,2}X_2 + \ldots + x_{n,n}X_n
\end{aligned}\right\} \tag{10.9}$$

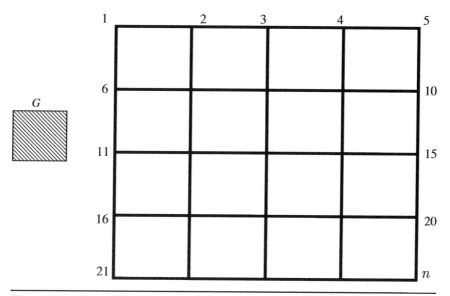

Figure10.5   Plan of the plate divided into finite elements

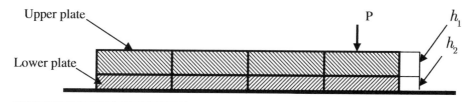

Figure10.6   The cross section of the composite plate

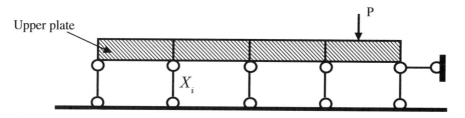

Figure 10.7   The upper plate supported on individual supports

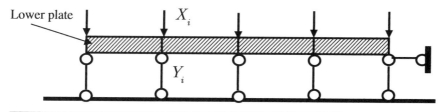

Figure 10.8   The lower plate supported on individual supports

In this system of equations y_{ik} is the reaction at support i due to one unit settlement at support k. It is easy to see that all reactions Y_i depend only on all settlements of the half-space Δ_1, Δ_2, ..., Δ_n. Reactions Y_i acting in the opposite direction and applied to the half-space will produce settlements Δ_1, Δ_2, ..., Δ_n. These settlements depend not only on reactions Y_i, but also on the load G applied to the half-space outside of the area of the plate. Settlements of the half-space shown below are found from equations 10.10.

$$\left.\begin{array}{l} \Delta_1 = \Delta_{11}Y_1 + \Delta_{12}Y_2 + \ldots + \Delta_{1n}Y_n + \Delta_{1G} \\ \Delta_2 = \Delta_{21}Y_1 + \Delta_{22}Y_2 + \ldots + \Delta_{2n}Y_n + \Delta_{2G} \\ \cdots\cdots\cdots\cdots\cdots\cdots\cdots\cdots\cdots\cdots \\ \Delta_n = \Delta_{n1}Y_1 + \Delta_{n2}Y_2 + \ldots + \Delta_{nn}Y_n + \Delta_{nG} \end{array}\right\} \qquad (10.10)$$

In equation 10.10, Δ_{ik} is the settlement of the plate at point i due to one unit load applied to the half-space at point k. The one unit load is applied to the half-space as a uniformly distributed load. The distributed load is equal to $\omega_i = 1/A_i$, where A_i is the tributary area at point i and Δ_{iG} is the settlement of the half-space at point i due to the load G applied to the half-space outside of the plate area. By solving three systems of equations 10.8–10.10, n reactions X_i, n reactions Y_i, and n settlements Δ_i are found.

When the plate is supported on Winkler foundation, loads applied to the soil do not affect each other, and the system of equations 10.10 is replaced with a series of simple equations 10.11:

$$Y_1 = \Delta_1/\Delta_{11}, Y_2 = \Delta_2/\Delta_{22}, \ldots, Y_n = \Delta_n/\Delta_{nn} \qquad (10.11)$$

By introducing equations 10.11 into the system of equations 10.9 and taking into account that reactions Y_i are applied to the soil in the opposite direction, the following system of equations is written:

$$\left.\begin{array}{l} \left(\dfrac{1}{\Delta_{11}} + y_{11}\right)\Delta_1 + y_{12}\Delta_2 + \ldots + y_{1n}\Delta_n + x_{1,1}X_1 + x_{1,2}X_2 + \ldots + x_{1,n}X_n = 0 \\[2mm] y_{21}\Delta_1 + \left(\dfrac{1}{\Delta_{22}} + y_{22}\right)\Delta_2 + \ldots + y_{2n}\Delta_n + x_{2,1}X_8 + x_{2,2}X_2 + \ldots + x_{2,n}X_n = 0 \\[2mm] \cdots\cdots\cdots\cdots\cdots\cdots\cdots\cdots\cdots\cdots\cdots\cdots\cdots\cdots\cdots\cdots \\[2mm] y_{n1}\Delta_1 + y_{n2}\Delta_2 + \ldots + \left(\dfrac{1}{\Delta_{nn}} + y_{nn}\right)\Delta_n + x_{n,1}X_1 + x_{n,2}X_2 + \ldots + x_{n,n}X_n = 0 \end{array}\right\} \qquad (10.12)$$

Now, by solving two systems of equations 10.8 and 10.12, we obtain n settlements and n reactions X_1, X_2, ..., X_n. Final analysis of the upper plate is performed by applying to the plate the given loads and obtained settlements. Final analysis of the lower plate is performed by applying to the lower plate reactions X_1, X_2, ..., X_n acting in opposite directions and the found settlements Δ_1, Δ_2, ..., Δ_n.

The composite plate does not always consist of two plates with the same plan dimensions. The plate may look as shown in Figure 10.9. In this case the given loads may be applied not only to the upper plate, but to the lower plate as well. The plate shown in Figure 10.9 consists of two plates. Analysis of the composite plate is performed as follows:

1. The upper and the lower plates are divided into a system of rectangular finite elements as shown in Figures 10.10 and 10.11.

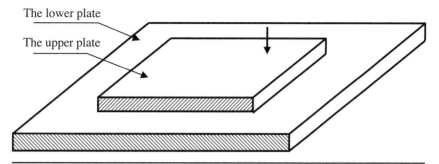

Figure 10.9   The composite plate

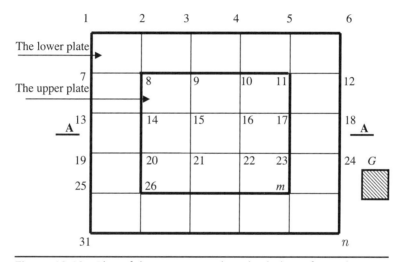

Figure 10.10   Plan of the composite plate divided into finite elements

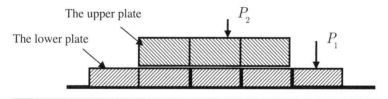

Figure 10.11   The cross section of the plate A-A

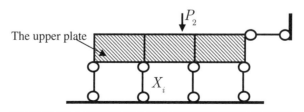

Figure 10.12   The upper plate supported on individual supports

2. Assuming that the upper plate is supported at all nodes on a system of totally rigid supports, the plate is analyzed successively under the given loads and one unit settlements applied to each node. As shown in Figure 10.10, the upper plate is supported at points: 8, 9, 10, 11, 14, 15, 16, 17, 20, 21, 22, 23, 26, ... , m.

3. From these analyses all m support reactions of the upper plate are found as:

$$
\left.
\begin{aligned}
X_8 &= r_{88}\Delta_8 + \ldots + r_{8,11}\Delta_{11} + r_{8,14}\Delta_{14} + \ldots + r_{8,17}\Delta_{17} + \\
&\quad r_{8,20}\Delta_{20} + \ldots + r_{8,23}\Delta_{23} + r_{8,26}\Delta_{26} + \ldots + r_{8,m}\Delta_m + r_{8P} \\[4pt]
&\cdots\cdots\cdots\cdots\cdots\cdots\cdots\cdots\cdots\cdots \\[4pt]
X_{11} &= r_{11,8}\Delta_8 + \ldots + r_{11,11}\Delta_{11} + r_{11,14}\Delta_{14} + \ldots + r_{11,17}\Delta_{17} + \\
&\quad r_{11,20}\Delta_{20} + \ldots + r_{11,23}\Delta_{23} + r_{11,26}\Delta_{26} + \ldots + r_{11,m}\Delta_m + r_{11P} \\
X_{14} &= r_{14,8}\Delta_8 + \ldots + r_{14,11}\Delta_{11} + r_{14,14}\Delta_{14} + \ldots + r_{14,17}\Delta_{17} + \\
&\quad r_{14,20}\Delta_{20} + \ldots + r_{14,23}\Delta_{23} + r_{14,26}\Delta_{26} + \ldots + r_{14,m}\Delta_m + r_{14,P} \\[4pt]
&\cdots\cdots\cdots\cdots\cdots\cdots\cdots\cdots\cdots\cdots \\[4pt]
X_{17} &= r_{17,8}\Delta_8 + \ldots + r_{17,11}\Delta_{11} + r_{17,14}\Delta_{14} + \ldots + r_{17,17}\Delta_{17} + \\
&\quad r_{17,20}\Delta_{20} + \ldots + r_{17,23}\Delta_{23} + r_{17,26}\Delta_{26} + \ldots + r_{17,m}\Delta_m + r_{17,P} \\
X_{20} &= r_{20,8}\Delta_8 + \ldots + r_{20,11}\Delta_{11} + r_{20,14}\Delta_{14} + \ldots + r_{20,17}\Delta_{17} + \\
&\quad r_{20,20}\Delta_{20} + \ldots + r_{20,23}\Delta_{23} + r_{20,26}\Delta_{26} + \ldots + r_{20,m}\Delta_m + r_{20P} \\[4pt]
&\cdots\cdots\cdots\cdots\cdots\cdots\cdots\cdots\cdots\cdots \\[4pt]
X_{23} &= r_{23,8}\Delta_8 + \ldots + r_{23,11}\Delta_{11} + r_{23,14}\Delta_{14} + \ldots + r_{23,17}\Delta_{17} + \\
&\quad r_{23,20}\Delta_{20} + \ldots + r_{23,23}\Delta_{23} + r_{23,26}\Delta_{26} + \ldots + r_{23,m}\Delta_m + r_{23P} \\
X_{26} &= r_{26,8}\Delta_8 + \ldots + r_{26,11}\Delta_{11} + r_{26,14}\Delta_{14} + \ldots + r_{26,17}\Delta_{17} + \\
&\quad r_{26,20}\Delta_{20} + \ldots + r_{26,23}\Delta_{23} + r_{26,26}\Delta_{26} + \ldots + r_{26,m}\Delta_m + r_{26P} \\[4pt]
&\cdots\cdots\cdots\cdots\cdots\cdots\cdots\cdots\cdots\cdots \\[4pt]
X_m &= r_{m,8}\Delta_8 + \ldots + r_{m,11}\Delta_{11} + r_{m,14}\Delta_{14} + \ldots + r_{m,17}\Delta_{17} + \\
&\quad r_{m,20}\Delta_{20} + \ldots + r_{m,23}\Delta_{23} + r_{m,26}\Delta_{26} + \ldots + r_{m,m}\Delta_m + r_{mP}
\end{aligned}
\right\} \tag{10.13}
$$

In these equations r_{ik} is the reaction at support i due to a settlement equal to one unit applied at support k, and r_{iP} is the reaction at the same support i due to the given loads applied to the upper plate.

4. Now, continuous contact between the lower plate and soil is replaced with a series of individual totally rigid supports, as shown in Figure 10.13. Supports

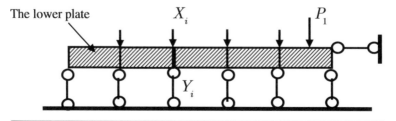

Figure 10.13   The lower plate supported on individual supports

are located at all n nodes. The plate is loaded with n unknown settlements and given loads applied to the lower plate.

5. Unknown reactions at all of these supports can be obtained by applying to the lower plate unknown settlements $\Delta_1, \Delta_2, ..., \Delta_n$ and unknown reactions $X_8, X_9, X_{10}, X_{11}, X_{14}, X_{15}, X_{16}, X_{17}, X_{20}, X_{21}, X_{22}, X_{23}, X_{26}, ..., X_m$ acting in opposite directions with the given loads applied to the lower plate. These reactions will look as shown below in system 10.14:

$$
\left.
\begin{aligned}
Y_1 &= y_{11}\Delta_1 + y_{12}\Delta_2 + ... + y_{1n}\Delta_n + x_{1,8}X_8 + x_{1,9}X_9 + x_{1,10}X_{10} + ... + x_{1,m}X_m + y_{1P} \\
Y_2 &= y_{21}\Delta_1 + y_{22}\Delta_2 + ... + y_{2n}\Delta_n + x_{2,8}X_8 + x_{2,9}X_9 + x_{2,10}X_{10} + ... + x_{2,m}X_m + y_{2P} \\
&\cdots\cdots\cdots\cdots\cdots\cdots\cdots\cdots\cdots\cdots\cdots\cdots\cdots\cdots\cdots\cdots\cdots\cdots\cdots\cdots\cdots\cdots\cdots\cdots\cdots\cdots\cdots\cdots \\
Y_n &= y_{n1}\Delta_1 + y_{n2}\Delta_2 + ... + y_{nn}\Delta_n + x_{n,8}X_8 + x_{n,9}X_9 + x_{n,10}X_{10} + ... + x_{n,m}X_m + y_{nP}
\end{aligned}
\right\} \quad (10.14)
$$

As can be seen, both systems 10.13 and 10.14 contain $(m + n)$ equations, but the number of unknowns is equal to $(m + 2n)$. Additional n equations can be obtained by expressing the settlements of the half-space as functions of the unknown reactions Y_i and given loads G applied to the half-space. These equations are shown below:

$$
\left.
\begin{aligned}
\Delta_1 &= \Delta_{11}Y_1 + \Delta_{12}Y_2 + ... + \Delta_{1n}Y_n + \Delta_{1G} \\
\Delta_2 &= \Delta_{21}Y_1 + \Delta_{22}Y_2 + ... + \Delta_{2n}Y_n + \Delta_{2G} \\
&\cdots\cdots\cdots\cdots\cdots\cdots\cdots\cdots\cdots\cdots\cdots\cdots\cdots\cdots\cdots \\
\Delta_n &= \Delta_{n1}Y_1 + \Delta_{n2}Y_2 + ... + \Delta_{nn}Y_n + \Delta_{nG}
\end{aligned}
\right\} \quad (10.15)
$$

Now, we have three systems of equations, 10.13, 10.14, and 10.15, that contain $(m + 2n)$ equations with $(m + 2n)$ unknowns: $(m + n)$ reactions and n settlements. By solving these three systems of equations, all settlements and all reactions are found. Final analysis of the lower plate is performed by applying to the plate the given loads, and all settlements. Final analysis of the upper plate is performed by applying to the plate the given loads and m settlements at points where the upper plate is supported. When the plate is supported on Winkler foundation, analysis is much simpler. As mentioned earlier, the system of equations 10.15 is replaced with a series of equations 10.16:

$$
Y_1 = \Delta_1/\Delta_{11}, \; Y_2 = \Delta_2/\Delta_{22}, ..., Y_n = \Delta_n/\Delta_{nn} \quad (10.16)
$$

By introducing reactions Y_i from equations 10.16 into the system of equations 10.14, we have a system of equations 10.17 shown below:

$$
\left.
\begin{aligned}
\left(\frac{1}{\Delta_{11}} + y_{11}\right)\Delta_1 + y_{12}\Delta_2 + ... + y_{1n}\Delta_n + x_{1,8}X_8 + x_{1,9}X_9 + x_{1,10}X_{10} + ... + x_{1,m}X_m + y_{1P} &= 0 \\
y_{21}\Delta_1 + \left(\frac{1}{\Delta_{22}} + y_{22}\right)\Delta_2 + ... + y_{2n}\Delta_n + x_{2,8}X_8 + x_{2,9}X_9 + x_{2,10}X_{10} + ... + x_{2,m}X_m + y_{2P} &= 0 \\
\cdots\cdots\cdots\cdots\cdots\cdots\cdots\cdots\cdots\cdots\cdots\cdots\cdots\cdots\cdots\cdots\cdots\cdots\cdots\cdots\cdots\cdots\cdots\cdots\cdots \\
y_{n1}\Delta_1 + y_{n2}\Delta_2 + ... + \left(\frac{1}{\Delta_{nn}} + y_{nn}\right)\Delta_n + x_{n,8}X_8 + x_{n,9}X_9 + x_{n,10}X_{10} + ... + x_{n,m}X_m + y_{nP} &= 0
\end{aligned}
\right\}
$$

$$
(10.17)
$$

Now, we have two systems of equations 10.13 and 10.17, with $(m + n)$ unknowns: m unknown reactions and n settlements. By solving both systems of equations, reactions and settlements are found. Final analysis of the upper plate is performed by applying to the plate the given loads and m settlements. Analysis of the lower plate is performed by applying to the plate the given loads and settlements $\Delta_1, \Delta_2, ..., \Delta_n$.

10.4 Analysis of Boxed Foundations

Foundations of tall and heavy buildings on many occasions are designed as a system of interconnected walls supported on a flat mat foundation, as shown below in Figures 10.14 and 10.15. As noted by Teng (1975), boxed foundations that consist of slabs and basement walls are capable of resisting large flexural stresses, but analysis presents a complex problem.

A practical method of analysis of such types of foundations is introduced below. The given boxed foundation consists of two parts: a mat foundation and a system of interconnected walls-beams. Analysis of the foundation is performed as follows:

The system of interconnected beams and the flat mat are separated. It is assumed that the system of interconnected walls-beams is not continuously supported on the mat, but supported on a system of totally rigid individual supports, as shown in Figure 10.16.

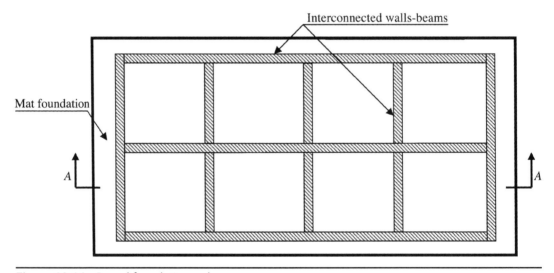

Figure 10.14   Boxed foundations, plan

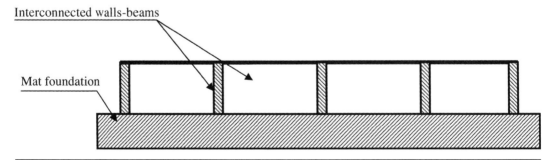

Figure 10.15   Cross section of the boxed foundation A-A

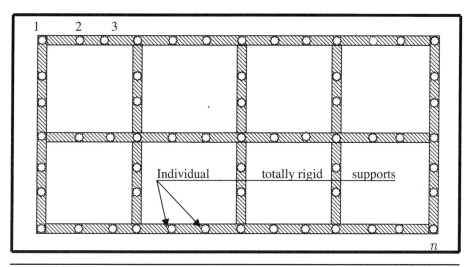

Figure 10.16   Interconnected beams supported on individual totally rigid supports

Now, the system of interconnected walls-beams is loaded with the given loads and unknown settlements at each point of support from 1 to n. Reactions X_1, X_2, ..., X_n at all points of support are found from the following system of equations:

$$\left.\begin{aligned}
X_1 &= r_{11}\Delta_1 + r_{12}\Delta_2 + \ldots + r_{1n}\Delta_n + r_{1P} \\
X_2 &= r_{21}\Delta_1 + r_{22}\Delta_2 + \ldots + r_{2n}\Delta_n + r_{2P} \\
X_n &= r_{n1}\Delta_1 + r_{n2}\Delta_2 + \ldots + r_{nn}\Delta_n + r_{nP}
\end{aligned}\right\} \qquad (10.18)$$

In this system, Δ_1, Δ_2, ..., Δ_n are unknown settlements of the foundation from point 1 to point n, r_{ik} is the reaction at point i due to one unit settlement of the system of interconnected beams at point k, and r_{iP} is the reaction at point i due to the given loads applied to the system of interconnected walls-beams. The total number of unknowns in this system of equations is equal to $2n$ while the number of equations is equal only to n. The mat separated from the system of beams and loaded with unknown reactions acting in the opposite direction will produce settlements of the mat equal to the settlements of supports of the interconnected beams Δ_1, Δ_2, ..., Δ_n. When the mat is supported on elastic half-space, these settlements can be found from the following equations (Zhemochkin and Sinitsin, 1962).

$$\left.\begin{aligned}
\Delta_1 &= \Delta_{11}X_1 + \Delta_{12}X_2 + \ldots + \Delta_{1n}X_n + \Delta_{1P} \\
\Delta_2 &= \Delta_{21}X_1 + \Delta_{22}X_2 + \ldots + \Delta_{2n}X_n + \Delta_{2P} \\
&\qquad\ldots\ldots\ldots\ldots\ldots\ldots\ldots\ldots\ldots \\
\Delta_n &= \Delta_{n1}X_1 + \Delta_{n2}X_2 + \ldots + \Delta_{nn}X_n + \Delta_{nP}
\end{aligned}\right\} \qquad (10.19)$$

In this system of equations, Δ_{ik} is the settlement of the mat at point i due to the one unit load applied to the mat at point k, and Δ_{iP} is the settlement of the mat at point i due to the given loads applied to the mat.

Now, both systems of equations 10.18 and 10.19 contain $2n$ equations with $2n$ unknowns. By solving these two systems of equations, all settlements and all reactions are found. Now, final analysis of the mat is performed by applying to the mat the found reactions and the given loads.

Final analysis of the system of interconnected beams is performed by applying to the system of interconnected beams the given loads and settlements at all n points of support. Analysis requires the use of two programs: one for analysis of mat foundations and the second one for analysis of 3D statically indeterminate systems. Any available software for analysis of mat foundations can be used for obtaining the settlements and for performing the final analysis of the mat.

The same analysis can be performed using the iterative method in several steps:

- **First iteration:** The system of interconnected beams is separated from the mat and supported on a system of individual totally rigid supports. Using any available computer software for analysis of 3D statically indeterminate systems, analysis of the system of beams with the given loads is performed. Reactions at all points of support acting in the opposite direction are applied to the mat, analysis of the mat is performed, and settlements of the mat at all points of support are found.
- **Second iteration:** The given loads and obtained settlements are applied to the system of interconnected beams and new reactions from this analysis are found. New reactions applied to the mat in the opposite direction will produce new settlements of the mat.
- **Third iteration:** Repeat the procedures from the second iteration, and so on. Analysis is stopped when settlements and reactions obtained in iteration m are close enough to the settlements and reactions found in the previous iteration $(m - 1)$. This method of analysis is more attractive since it does not require building and solving systems of equations and allows using available software for analysis of interconnected beams and mat foundations.

10.5 Combined Analysis of Boxed Foundation and a 3D Frame-Superstructure

If the foundation is supporting a 3D frame as shown in Figures 10.17 and 10.18, analysis, in principle, is no different from analysis described above. However, now the given system consists of a mat foundation and a 3D frame that also includes a system of interconnected walls-beams. Reactions X_1, X_2, ..., X_n are obtained from the same system of equations 10.18. In this system, r_{ik} is the reaction at point of support i of the frame due to one unit settlement applied at point k and r_{iP} is the reaction at point of support i due to all given loads applied to the frame including the system of interconnected beams. By solving both systems of equations, 10.18 and 10.19, the settlements at all n supports and support reactions are found.

Final analysis of the mat is performed by applying to the mat the given loads and found reactions acting in the opposite direction. The final analysis of the 3D frame

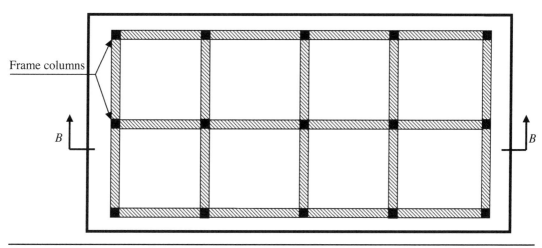

Figure 10.17   Boxed foundations supporting a 3D frame

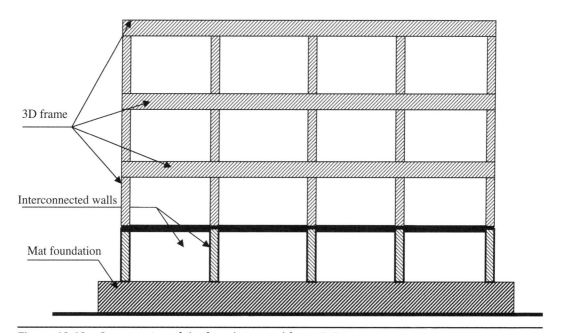

Figure 10.18   Cross section of the foundation and frame B-B

is performed by applying to the frame, which includes the system of interconnected walls, all given loads and settlements.

As we can see, the described method requires the use of two programs: one for analysis of 3D frames and the second for analysis of mat foundations. It also requires the building and solving of two systems of equations that make analysis more complex. The easiest method of analysis is the method of iterations described earlier. The method allows the use of widely available software for frames and mat foundations analysis and does not requires building and solving large systems of linear equations.

10.6 Analysis of Walls on Elastic Foundation

Walls of commercial and industrial buildings are usually supported on continuous foundations. The wall and foundation under applied loads work as one whole structural system affecting each other. When the wall does not contain openings or contains few openings, the system wall-foundation is assumed totally rigid and works as one rigid system. But, when the wall is weakened with many large openings, the wall becomes sensitive to the differential settlements of the foundation that produce additional stresses in the wall. At the same time, the rigidity of the wall affects the behavior of the foundation reducing the differential settlements. It is obvious that combined analysis of walls and foundations produces more realistic, reliable results compared to separate analyses. A method of analysis of walls with continuous foundations is proposed below.

The method is developed for two soil models: elastic half-space and Winkler foundation, but can also be used for an elastic layer and other soil models.

The method of analysis, in principle, is similar to the method used for analysis of composite beams described earlier in this chapter. We shall illustrate application of the method using the wall shown in Figure 10.19. It is assumed that the wall is supported on elastic half-space. Analysis is performed in the following steps:

- **Step 1:** The total system is divided into three parts: soil, foundation, and wall. The wall is divided into a system of rectangular or triangular finite elements.
- **Step 2:** Continuous contacts between the soil and foundation and between the foundation and the wall are replaced with a series of individual totally rigid vertical supports, as shown in Figure 10.20.

All supports, in turn, are replaced with unknown forces of interaction X_i, as shown in Figure 10.21. The total number of unknown reactions and supports is equal to $(m + n)$. The soil is loaded with forces X_i and given loads G, located outside of the foundation.

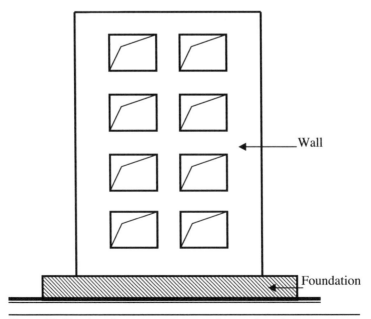

Figure 10.19 A wall supported on continuous foundation

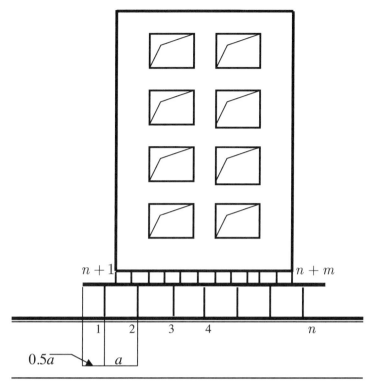

Figure 10.20

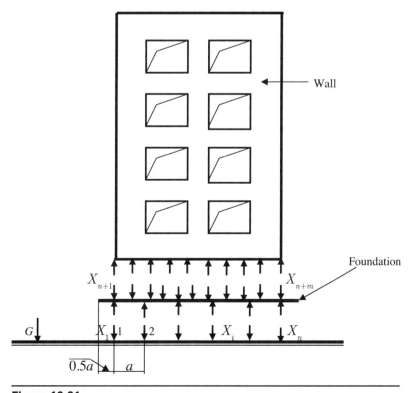

Figure 10.21

All loads are applied to the soil as uniformly distributed loads. These loads can be found from the following equation:

$$q_i = X_i/a \cdot b' \qquad (10.20)$$

In this equation, a is the spacing of supports, b' is the width of the foundation. Now, all three parts of the system wall-foundation-soil can be investigated separately. Assuming that the wall is supported on m totally rigid supports, we perform successively the analysis of the wall with the given loads and one unit settlements applied to all wall supports. Reactions at all supports can be expressed as follows:

$$\left.\begin{aligned}
X_{n+1} &= r_{(n+1)(n+1)}\Delta_{(n+1)} + r_{(n+1)(n+2)}\Delta_{(n+2)} + \ldots + r_{(n+1)(n+m)}\Delta_{n+m} + r_{(n+1)P} \\
X_{n+2} &= r_{(n+2)(n+1)}\Delta_{(n+1)} + r_{(n+2)(n+2)}\Delta_{(n+2)} + \ldots + r_{(n+2)(n+m)}\Delta_{n+m} + r_{(n+2)P} \\
&\quad\ldots\ldots\ldots\ldots\ldots\ldots\ldots\ldots\ldots\ldots\ldots\ldots\ldots\ldots\ldots\ldots\ldots\ldots\ldots\ldots\ldots \\
X_{n+m} &= r_{(n+m)(n+1)}\Delta_{(n+1)} + r_{(n+m)(n+2)}\Delta_{(n+2)} + \ldots + r_{(n+m)(n+m)}\Delta_{n+m} + r_{(n+m)P}
\end{aligned}\right\} \qquad (10.21)$$

In this system of equations, r_{ik} is the reaction at point i due to one unit settlement applied to the wall at point k and r_{iP} is the reaction at the same point i due to the given loads applied to the wall. The total number of equations in this system is equal to m.

Reactions X_{n+1}, X_{n+2}, ..., X_{n+m} acting in the opposite direction and applied to the foundation will produce vertical deflections of the foundation at points $(n + 1)$, $(n + 2)$, ..., $(n + m)$ that look as follows:

$$\left.\begin{aligned}
\Delta_{(n+1)} &= \delta_{(n+1)(n+1)}X_{n+1} + \delta_{(n+1)(n+2)}X_{n+2} + \ldots + \delta_{(n+1)(n+m)}X_{n+m} + \delta_{(n+1)P} \\
\Delta_{(n+2)} &= \delta_{(n+2)(n+1)}X_{n+1} + \delta_{(n+2)(n+2)}X_{n+2} + \ldots + \delta_{(n+2)(n+m)}X_{n+m} + \delta_{(n+2)P} \\
&\quad\ldots\ldots\ldots\ldots\ldots\ldots\ldots\ldots\ldots\ldots\ldots\ldots\ldots\ldots\ldots\ldots\ldots\ldots\ldots\ldots \\
\Delta_{(n+m)} &= \delta_{(n+m)(n+1)}X_{n+1} + \delta_{(n+m)(n+2)}X_{n+2} + \ldots + \delta_{(n+m)(n+m)}X_{n+m} + \delta_{(n+m)P}
\end{aligned}\right\} \qquad (10.22)$$

In this system of equations, δ_{ik} is the vertical deflection of the foundation at point i due to one unit load applied to the foundation at point k and δ_{iP} is the vertical deflection of the foundation at point i due to the given loads applied to the wall.

The same loads X_{n+1}, X_{n+2}, ..., X_{n+m} applied to the foundation will produce the following reactions at points of support 1, 2, ..., n:

$$\left.\begin{aligned}
X_1 &= r_{1(n+1)}X_{(n+1)} + r_{1(n+2)}X_{(n+2)} + \ldots + r_{1(n+m)}X_{n+m} + r_{1P} \\
X_2 &= r_{2(n+1)}X_{(n+1)} + r_{2(n+2)}X_{(n+2)} + \ldots + r_{2(n+m)}X_{n+m} + r_{2P} \\
&\quad\ldots\ldots\ldots\ldots\ldots\ldots\ldots\ldots\ldots\ldots\ldots\ldots\ldots\ldots\ldots\ldots\ldots\ldots\ldots\ldots \\
X_n &= r_{n(n+1)}X_{(n+1)} + r_{n(n+2)}X_{(n+2)} + \ldots + r_{(n+m)(n+m)}X_{n+m} + r_{nP}
\end{aligned}\right\} \qquad (10.23)$$

In the system of equations r_{ik} is the reaction at support i due to one unit load applied to the foundation at point k and r_{iP} is the reaction at the same support due to the given loads applied to the wall.

Reactions X_1, X_2, ..., X_n acting in the opposite direction, downward, will produce settlements of the half-space at points 1, 2, ..., n equal to:

$$\left.\begin{aligned}
\Delta_1 &= \Delta_{11}X_1 + \Delta_{12}X_2 + \ldots + \Delta_{1n}X_n + \Delta_{1P} + \Delta_{1G} \\
\Delta_2 &= \Delta_{21}X_1 + \Delta_{22}X_2 + \ldots + \Delta_{2n}X_n + \Delta_{2P} + \Delta_{2G} \\
&\quad\ldots\ldots\ldots\ldots\ldots\ldots\ldots\ldots\ldots\ldots\ldots\ldots\ldots\ldots\ldots\ldots \\
\Delta_n &= \Delta_{n1}X_1 + \Delta_{n2}X_2 + \ldots + \Delta_{nn}X_n + \Delta_{nP} + \Delta_{nG}
\end{aligned}\right\} \qquad (10.24)$$

where Δ_{ik} is the settlement of the half-space at point i due to one unit load applied to the half-space at point k and Δ_{iG} is the settlement of the half-space at point i due to the given load G applied to the half-space outside of the foundation. All four systems of equations 10.21–10.24 contain $(n + m)$ reactions and $(n + m)$ vertical deflections. The total number of equations is equal to $2(n + m)$. By solving all of these systems of equations, all reactions and all settlements are found. Final analysis of the foundation is performed by applying to the foundation obtained reactions and settlements $\Delta_1, \Delta_2, ..., \Delta_n, \Delta_{n+1}, \Delta_{n+2}, ..., \Delta_{n+m}$. Final analysis of the wall is performed by applying to the wall the given loads and all settlements.

When the foundation is supported on Winkler foundation, the system of equations 10.24 is simplified and looks as follows:

$$X_1 = \Delta_1/\Delta_{11}, X_2 = \Delta_2/\Delta_{22}, ..., X_n = \Delta_n/\Delta_{nn} \tag{10.25}$$

10.7 Analysis of Walls with a Soft Story Supported on Elastic Foundation

Walls of commercial and industrial buildings are not always supported directly on foundations. In many cases, walls are supported on frames. Analysis of such types of structural systems can be performed by applying the method used for combined analysis of walls with continuous foundations. Analysis, in principle, is the same as analysis discussed above, but it is more complex because it includes the analysis of a frame. The frame shown in Figure 10.22 is a one-story frame, but in some cases the wall is supported on two- or even three-story frames. If the foundation of the frame is designed

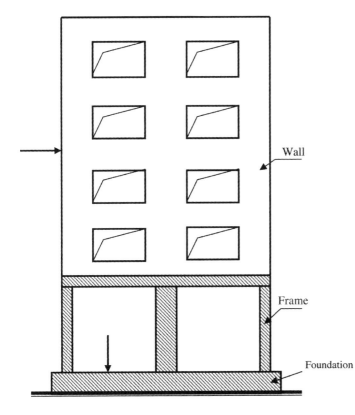

Wall

Frame

Foundation

Figure 10.22 A wall supported on a frame with a continuous foundation

as a continuous footing, the foundation becomes a part of the frame. Analysis of such a type of system is performed in the following order:

1. The given system soil-frame-wall is divided into three parts: soil, frame, and wall.
2. Continuous contact between the foundation and soil is replaced with a series of individual totally rigid supports. Supports are replaced with unknown soil reactions X_1, X_2, ..., X_n, as shown in Figure 10.23.

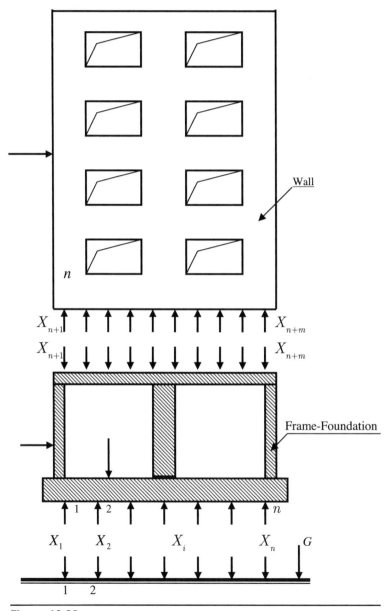

Figure 10.23

3. Continuous contact between the bottom of the wall and the top of the frame is replaced with a series of individual totally rigid supports, and these supports, in turn, are replaced with unknown reactions X_{n+1}, X_{n+2}, ..., X_{n+m}.

4. Now, the wall is successively analyzed under the given loads and one unit settlements applied to all points of support. From these analyses reactions at all m supports are found:

$$\left.\begin{array}{l}X_{n+1} = r_{(n+1)(n+1)}\Delta_{(n+1)} + r_{(n+1)(n+2)}\Delta_{(n+2)} + ... + r_{(n+1)(n+m)}\Delta_{n+m} + r_{(n+1)P} \\[4pt] X_{n+2} = r_{(n+2)(n+1)}\Delta_{(n+1)} + r_{(n+2)(n+2)}\Delta_{(n+2)} + ... + r_{(n+2)(n+m)}\Delta_{n+m} + r_{(n+2)P} \\[4pt] \hspace{2em}\cdots\cdots\cdots\cdots\cdots\cdots\cdots\cdots\cdots\cdots\cdots\cdots\cdots\cdots\cdots\cdots\cdots \\[4pt] X_{n+m} = r_{(n+m)(n+1)}\Delta_{(n+1)} + r_{(n+m)(n+2)}\Delta_{(n+2)} + ... + r_{(n+m)(n+m)}\Delta_{n+m} + r_{(n+m)P}\end{array}\right\} \quad (10.26)$$

The same reactions acting downward and applied to the frame will produce vertical deflections of the top of the frame Δ_{n+1}, Δ_{n+2}, ..., Δ_{n+m} and reactions X_1, X_2, ..., X_n at the points of supports of the frame foundation.

Vertical deflections at the top of the frame are found from the following equations:

$$\left.\begin{array}{l}\Delta_{n+1} = \Delta_{(n+1)(n+1)}X_{n+1} + \Delta_{(n+1)(n+2)}X_{n+2} + ... + \Delta_{(n+1)(n+m)}X_{n+m} + \Delta_{(n+1)P} \\[4pt] \Delta_{n+2} = \Delta_{(n+2)(n+1)}X_{n+1} + \Delta_{(n+2)(n+2)}X_{n+2} + ... + \Delta_{(n+2)(n+m)}X_{n+m} + \Delta_{(n+2)P} \\[4pt] \hspace{2em}\cdots\cdots\cdots\cdots\cdots\cdots\cdots\cdots\cdots\cdots\cdots\cdots\cdots\cdots\cdots\cdots\cdots \\[4pt] \Delta_{n+m} = \Delta_{(n+m)(n+1)}X_{n+1} + \Delta_{(n+m)(n+2)}X_{n+2} + ... + \Delta_{(n+m)(n+m)}X_{n+m} + \Delta_{(n+m)P}\end{array}\right\} \quad (10.27)$$

Reactions at points of support of the foundation are found from the equations shown below:

$$\left.\begin{array}{l}X_1 = r_{11}\Delta_1 + r_{12}\Delta_2 + ... + r_{1n}\Delta_n + r_{1(n+1)}X_{n+1} + r_{1(n+2)}X_{n+2} + ... + r_{1(n+m)}X_{n+m}\, r_{1P} \\[4pt] X_2 = r_{21}\Delta_1 + r_{22}\Delta_2 + ... + r_{2n}\Delta_n + r_{2(n+1)}X_{n+1} + r_{2(n+2)}X_{n+2} + ... + r_{2(n+m)}X_{n+m}\, r_{2P} \\[4pt] \hspace{2em}\cdots\cdots\cdots\cdots\cdots\cdots\cdots\cdots\cdots\cdots\cdots\cdots\cdots\cdots\cdots\cdots\cdots \\[4pt] X_n = r_{n1}\Delta_1 + r_{n2}\Delta_2 + ... + r_{nn}\Delta_n + r_{n(n+1)}X_{n+1} + r_{n(n+2)}X_{n+2} + ... + r_{n(n+m)}X_{n+m}\, r_{nP}\end{array}\right\} \quad (10.28)$$

where r_{ik} is the reaction at point i at the foundation due to one unit settlement applied to the foundation at point k, or reaction at point i at the foundation due to one unit load applied to the top of the frame at point k. In all of these equations the first n members reflect reactions due to settlements of the frame supports, while the remainder of the members reflect reactions due to the unknown loads and given loads applied to the frame. Settlements of the half-space are found from the system of equations 10.29. This system of equations is no different from the system of equations used earlier for analysis of walls directly supported on continuous foundations:

$$\left.\begin{array}{l}\Delta_1 = \Delta_{11}X_1 + \Delta_{12}X_2 + ... + \Delta_{1n}X_n + \Delta_{1P} + \Delta_{1G} \\[4pt] \Delta_2 = \Delta_{21}X_1 + \Delta_{22}X_2 + ... + \Delta_{2n}X_n + \Delta_{2P} + \Delta_{2G} \\[4pt] \hspace{2em}\cdots\cdots\cdots\cdots\cdots\cdots\cdots\cdots\cdots\cdots\cdots\cdots\cdots\cdots\cdots \\[4pt] \Delta_n = \Delta_{n1}X_1 + \Delta_{n2}X_2 + ... + \Delta_{nn}X_n + \Delta_{nP} + \Delta_{nG}\end{array}\right\} \quad (10.29)$$

By solving all four systems of equations, 10.26–10.29, all settlements and reactions are found. However, vertical deflections of the frame Δ_{n+1}, Δ_{n+2}, ..., Δ_{n+m} from this analysis are not the real settlements of the frame at points $(n+1)$, $(n+2)$, ..., $(n+m)$. The actual settlements at these points are found from the final analysis of the frame. Final analysis of the frame is performed by applying to the frame settlements $\Delta_1, \Delta_2, ..., \Delta_n$ $\Delta_{n+1}, \Delta_{n+2}, ..., \Delta_{n+m}$. Final analysis of the wall is performed by applying to the wall the given loads and actual settlements of the frame.

10.8 Analysis of Walls Supported on Frames with Individual Foundations

When the wall is supported on a frame with individual foundations, as shown in Figure 10.24, analysis is performed as shown below:

1. Continuous contact between the wall and the frame is replaced with a series of non-yielding supports as shown in Figure 10.25. The total number of supports is equal to n.

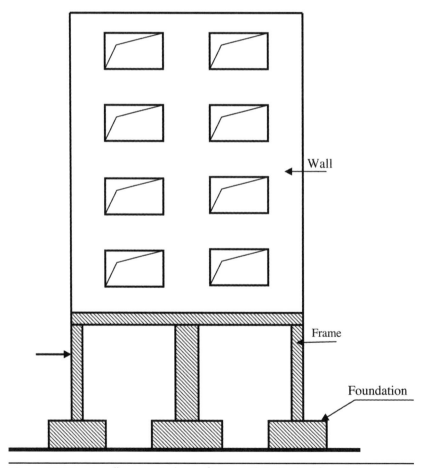

Figure 10.24   A wall supported on a frame with individual foundations

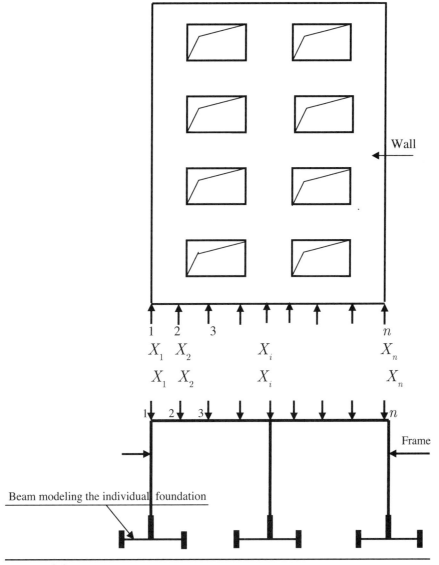

Figure 10.25

2. Analysis of the wall is performed successively under the given loads and one unit settlements applied to all n nodes and reactions at all supports are found and shown below:

$$
\left.
\begin{aligned}
X_1 &= r_{11}\Delta_1 + r_{12}\Delta_2 + \ldots + r_{1n}\Delta_n + r_{1P} \\
X_2 &= r_{21}\Delta_1 + r_{22}\Delta_2 + \ldots + r_{2n}\Delta_n + r_{2P} \\
&\quad\ldots\ldots\ldots\ldots\ldots\ldots\ldots\ldots\ldots\ldots\ldots\ldots \\
X_n &= r_{n1}\Delta_{m+1} + r_{n2}\Delta_2 + \ldots + r_{nn}\Delta_n + r_{nP}
\end{aligned}
\right\}
$$

(10.30)

The same reactions acting in the opposite direction, downward, and applied to the top of the frame, as shown in Figure 10.25, produce settlements of the frame and individual foundations. The given individual foundations are replaced with equivalent simple beams with both ends fixed. Modeling of the individual foundations with simple beams was proposed and explained in Chapter 3.

Properties of an equivalent beam modeling an individual foundation supported on Winkler foundation are obtained from the following equations:

$$l = a\sqrt{\frac{K_\varphi}{K_Z}}, \quad EI = \frac{I_F K_\varphi l}{16}, \quad EA_B = \frac{AK_X l}{4} \tag{10.31}$$

In these equations, l is the length of the beam, EI is the flexural rigidity of the beam, EA_B is the rigidity of the beam under tension-compression, a is the length of the given foundation, K_Z, K_φ, K_X are the moduli of subgrade reaction of the soil experiencing compression, rotation and sliding, respectively, I_F is the moment of inertia of the given foundation, and A is the area of the given foundation.

When the foundation is supported on elastic half-space, properties of the equivalent beam are found by equating deflections of an individual foundation supported on elastic half-space to deflections of a simple beam with both ends fixed. In this case the length and the rigidities of the equivalent beam, as shown in Chapter 6, are found from the following equations:

$$l = a\sqrt{\frac{6F}{\pi K_1}}, \quad (EI) = \frac{lE_0 a^3}{16(1-\mu_0^2)K_1}, \quad (EA) = \frac{lA_F K_X}{4} \tag{10.32}$$

Reactions X_i acting in the opposite direction, downward, and applied to the top of the frame (see Figure 10.25) will produce vertical deflections of the frame at all n points that can be found from equations 10.33:

$$\left.\begin{aligned}
\Delta_1 &= \Delta_{11}X_1 + \Delta_{12}X_2 + \ldots + \Delta_{1n}X_n + \Delta_{1P} \\
\Delta_2 &= \Delta_{21}X_1 + \Delta_{22}X_2 + \ldots + \Delta_{2n}X_n + \Delta_{2P} \\
&\cdots\cdots\cdots\cdots\cdots\cdots\cdots\cdots\cdots\cdots\cdots\cdots\cdots\cdots\cdots \\
\Delta_n &= \Delta_{n1}X_1 + \Delta_{n2}X_2 + \ldots + \Delta_{nn}X_n + \Delta_{nP}
\end{aligned}\right\} \tag{10.33}$$

From two systems of equations, 10.30 and 10.33, all deflections and reactions are found. Final analysis of the wall is performed by applying to the wall the given loads and wall settlements. Final analysis of the frame is performed by applying to the frame the given loads and found reactions. Loads obtained from the frame analysis are applied to the actual foundations and final analysis of the foundations is performed.

It is easy to notice that replacing the individual foundations with equivalent beams significantly reduces the number of unknowns and simplifies all calculations. This analysis does not take into account the affect of closely located foundations. These foundations do not produce additional settlements when Winkler's soil model is used. However, when the frame is supported on elastic half-space, in some cases, they have to be taken into account.

References

1. Teng, W. C. 1975. Mat foundations. In *Foundation engineering handbook*. New York: van Nostrand Reinhold, pp. 528–529.

2. Tsudik, E. A. 2008. Combined analysis of building walls and foundations. In *Proceedings of the second British Geotechnical Association international conference on foundations, ICOF 2008*, Dundee, Scotland, UK, pp. 1111–1123.
3. Tsudik, E. A. 2010. Combined analysis of 3D frames and foundations. Presented at the SEAOC Convention, Indian Wells, CA.
4. Zhemochkin, B. N., and A. P. Sinitsin. 1962. *Practical methods of analysis of beams and plates on elastic foundation*. Moscow: Gosstroiizdat, pp. 26–28.

Index